Structural Analysis of
POLYMERIC
Composite Materials

Second Edition

Structural Analysis of
POLYMERIC
Composite Materials

Second Edition

Mark E. Tuttle

CRC Press
Taylor & Francis Group
Boca Raton London New York

CRC Press is an imprint of the
Taylor & Francis Group, an **informa** business

CRC Press
Taylor & Francis Group
6000 Broken Sound Parkway NW, Suite 300
Boca Raton, FL 33487-2742

First issued in paperback 2019

© 2013 by Taylor & Francis Group, LLC
CRC Press is an imprint of Taylor & Francis Group, an Informa business

No claim to original U.S. Government works

ISBN-13: 978-1-4398-7512-4 (hbk)
ISBN-13: 978-0-367-38058-8 (pbk)

Library of Congress Cataloging-in-Publication Data

Tuttle, M. E.
  Structural analysis of polymeric composite materials / Mark E. Tuttle. -- Second edition.
    pages cm
    Includes bibliographical references and index.
    ISBN 978-1-4398-7512-4 (hardback)
    1. Polymeric composites. 2. Structural analysis (Engineering) I. Title.

TA418.9.C6T88 2012
620.1'18--dc23   2012032579

Visit the Taylor & Francis Web site at
http://www.taylorandfrancis.com

and the CRC Press Web site at
http://www.crcpress.com

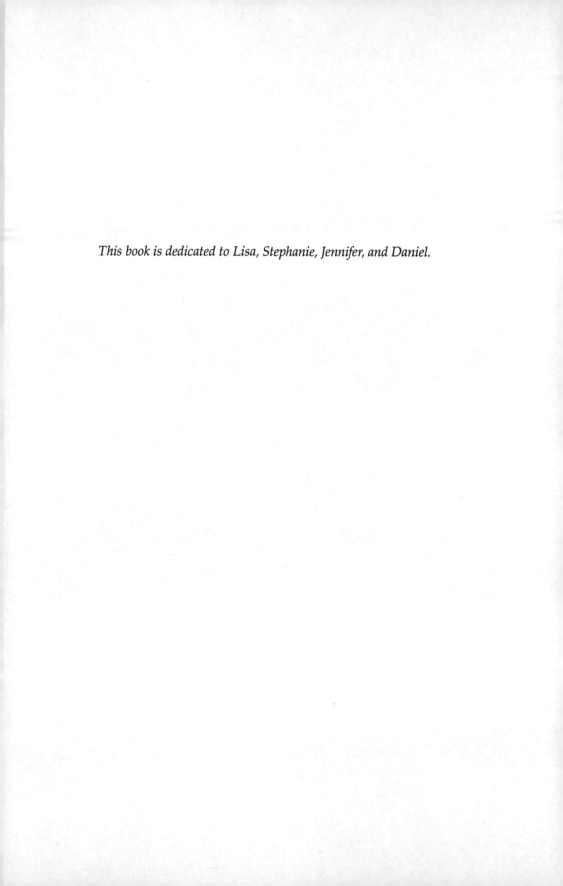

This book is dedicated to Lisa, Stephanie, Jennifer, and Daniel.

Contents

Preface

The primary objective of this book is to provide an introduction to the mechanics of laminated composite materials and structures. It is intended for use in a senior- or graduate-level course devoted to polymeric composites. It is assumed that the student has already completed one or more courses devoted to the mechanics of isotropic materials and structures. Concepts learned during these earlier studies are extended to anisotropic composites in a natural and easily understandable way. Many suggested homework problems are included, and a solutions manual is available for those who wish to adopt the book for their courses.

Although new material has been added throughout, this second edition is similar in scope to the first. Chapter 1 provides a broad overview of polymer science, advanced fibrous materials, commercially important fiber/polymer combinations, and composite manufacturing processes. Discussion in all remaining chapters is framed at the macromechanical (structural) level. Chapter 2 is devoted to a review of force, stress, and strain tensors and tensor coordinate transformations. Although the student has certainly encountered these concepts in earlier studies, it has been the author's experience that a thorough review is important before embarking on a study of anisotropic composites.

Materials properties required to predict the structural performance of anisotropic composites are introduced in Chapter 3, and the 3- and 2-dimensional forms of Hooke's law are developed in Chapters 4 and 5, respectively. The effects of environment, particularly changes in temperature, are introduced at this point and included in all subsequent chapters. Spatially varying stresses predicted to occur in unidirectional composite structural elements using the finite-element method are presented and shown to be entirely consistent with predicted strains (and vice versa) using Hooke's law.

Elements of plate theory are then developed and combined with Hooke's law in Chapter 6. This results in the composite analysis method commonly known as classical lamination theory (CLT). The spatially varying effective stresses, as well as ply stresses, predicted by a typical finite-element analysis of a structural element produced using a multiangle composite, are presented and shown to be entirely consistent with strains predicted by CLT (and vice versa).

Chapter 7 describes composite failure modes and mechanisms, including a qualitative description of composite fatigue behaviors and free-edge effects. This chapter also describes how macroscopic failure criteria may be combined with CLT to predict first-ply and last-ply failure loads and failure envelopes.

Chapter 8 is devoted to statically determinate and indeterminate composite beams. It is first shown that CLT reduces to fundamental beam theory if isotropic material properties and appropriate dimensions are assumed. CLT is then used to predict the effective axial and bending rigidities of laminated composite beams with several cross-sections (rectangular, I-, T-, hat-, and box-beams).

Chapter 9 is an entirely new chapter added to this second edition and is devoted to effective stress concentrations in multiangle composite laminates. The well-known solutions for stresses in anisotropic panels developed by Lekhnitskii and Savin are used to calculate the stresses near elliptical holes in multiangle composite laminates induced by an arbitrary combination of in-plane loads. Several example analyses for circular holes in unidirectional and multiangle composites are presented to illustrate typical results.

Chapters 10 through 12 are devoted to flat composite plates subjected to varying loads. Discussion is limited to rectangular plates with symmetric stacking sequences. The equations that govern the behavior of symmetric composite plates are first developed in Chapter 10. Several closed-form solutions that are valid for a limited range of stacking sequences (i.e., specially orthotropic plates) are then presented in Chapter 11. Approximate numerical solutions for more common generally orthotropic plates (e.g., quasi-isotropic plates) are presented in Chapter 12. Both Chapters 11 and 12 include solutions for the deflections of transversely loaded plates as well as for mechanical and/or thermal buckling due to in-plane loads.

The author has created a suite of computer programs that implement most of the analysis methods described in the book. Both English and Spanish versions are available (Spanish translations are courtesy of Professor Carlos Navarro of the Universidad Carlos III de Madrid). Executable versions of these programs suitable for use with the Windows™ operating system may be downloaded free of charge from the following website:

http://depts.washington.edu/amtas/computer.html

These programs are meant to enhance the text and are referenced at appropriate points throughout the book. Of course, composites analyses are also commercially available, so the student may opt to use resources other than those downloaded from the website listed.

Acknowledgments

I have been fortunate to work with talented and inspirational colleagues and students throughout my career. This book is a reflection of what I have learned from them.

Special thanks are extended to Professor Carlos Navarro Ugena of the Universidad Carlos III de Madrid (Spain) for his Spanish translations of the computer programs referenced throughout the book. There are many others that I would like to acknowledge by name, but space is a constraint. Sincere thanks are extended to my teachers, friends, and colleagues at the University of Washington, Virginia Tech, Michigan Tech, Battelle Memorial Institute, the Boeing Company, and the Society for Experimental Mechanics.

Author

Mark E. Tuttle is professor of mechanical engineering at the University of Washington. He is also the director of the UW Center of Excellence for Advanced Materials in Transport Aircraft Structures (AMTAS). He holds a BS in mechanical engineering and an MS in engineering mechanics, both from Michigan Technological University, and a PhD in engineering mechanics from Virginia Polytechnic Institute and State University.

Professor Tuttle is a member of several professional engineering societies, and in particular is fellow and past president of the Society for Experimental Mechanics (SEM). He has authored or coauthored about 130 journal and conference proceedings articles, and he has been a keynote lecturer at several international engineering conferences. His research has been funded by the National Science Foundation, the Office of Naval Research, the Federal Aviation Administration, NASA-Langley Research Center, the Boeing Company, and the Ford Motor Company.

1

Introduction

1.1 Basic Definitions

Just what is a "composite material?" A casual definition might be: "a composite material is one in which two (or more) materials are bonded together to form a third material." Although not incorrect, upon further reflection it becomes clear that this definition is far too broad, since it implies that essentially *all* materials can be considered to be a "composite." For example, the (nominal) composition of the 2024 aluminum alloy is 93.5% Al, 4.4% Cu, 0.6% Mn, and 1.5% Mg [1]. Hence, according to the broad definition stated above, this common aluminum alloy could be considered to be a "composite," since it consists of four materials (aluminum, copper, manganese, and magnesium) bonded together at the atomic level to form the 2024 alloy. In a similar sense, virtually all metal alloys, polymers, and ceramics satisfy this broad definition of a composite, since all of these materials contain more than one type of elemental atom.

An important characteristic that is missing in the initial broad definition is a consideration of physical scale. Another definition of a "composite material," which includes a reference to a physical scale and is appropriate for present purposes, is as follows.

A composite material is a material *system* consisting of two (or more) materials which are distinct at a physical scale greater than about 1×10^{-6} m (1 μm), and which are bonded together at the atomic and/or molecular levels.

As a point of reference, the diameter of a human hair ranges from about 30 to 60 μm. Objects of this size are easily seen with the aid of an optical microscope. Hence, when composite materials are viewed under an optical microscope the distinct constituent materials (or distinct material phases) that form the composite are easily distinguished.

Structural composites typically consist of a high strength, high stiffness *reinforcing material*, embedded within a relatively low strength, low stiffness *matrix material*. Ideally, the reinforcing and matrix materials interact to produce a composite whose properties are superior to either of the two constituent materials alone. Many naturally occurring materials can be viewed as composites. A good example is wood and laminated wood products.

Wood is a natural composite, with a readily apparent grain structure. Grains are formed by long parallel strands of reinforcing cellulose fibers, bonded together by a glue-like matrix material called lignin. Since cellulose fibers have a substantially higher stiffness and strength than lignin, wood exhibits higher stiffness and strength parallel to the grain than transverse to the grains. In *laminated* wood products (which range from the large laminated beams used in a church cathedral to a common sheet of plywood) relatively thin layers of wood are adhesively bonded together. The individual layers are called "plies," and are arranged such that the grain direction varies from one layer to the next. Thus, *laminated* wood products have high stiffness/ strength in more than one direction.

Although composites have been used in a variety of structural applications for centuries, modern (or advanced) composites are a relatively recent development, having been in existence for about 70 years. Modern composites may be classified according to the size or shape of the reinforcing material used. Four common classifications of reinforcements are

- *Particulates*, which are roughly spherical particles with diameters typically ranging from about 1 to 100 μm
- *Whiskers*, with lengths less than about 10 mm
- *Short* (or *chopped*) *fibers*, with a length ranging from about 10 to 200 mm
- *Continuous fibers*, whose length are, in effect, infinite

Whiskers, short fibers, and continuous fibers all have very small diameters relative to their length; the diameter of these products range from about 5 to 200 μm.

Distinctly different types of advanced composites can be produced using any of the above reinforcements. For example, three types of composites based on continuous fibers are shown in Figure 1.1: *unidirectional* composites, *woven* composites, and *braided* composites. In a unidirectional composite all fibers are aligned in the same direction and embedded within a matrix material. In contrast, woven composites are formed by first weaving continuous fibers into a fabric and then embedding the fabric in a matrix. Hence, a single layer of a woven composite contains fibers in two orthogonal directions. In contrast, a single layer in a braided composite typically contains two or three nonorthogonal fiber directions. Braided composites are then formed by embedding the fabric in a matrix. Additional discussion of these types of composites is provided in Section 1.4.

As implied in Figure 1.1, composite products based on continuous fibers are usually produced in the form of thin layers. A single layer of these products is called a *lamina* or *ply*, following the nomenclature used with laminated wood products. The thickness of a single ply formed using unidirectional fibers ranges from about 0.12 to 0.20 mm (0.005–0.008 in.), whereas the thickness of a single ply of a woven or braided fabric ranges from about 0.25 to 0.40 mm

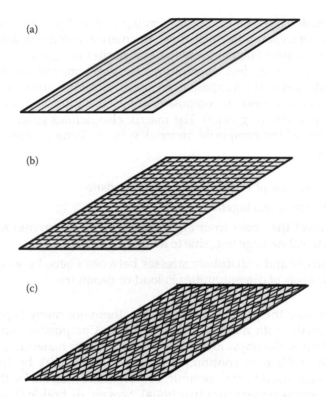

FIGURE 1.1
Different types of composites based on continuous fibers. (a) A single layer (or "ply") of a uni-directional composite; (b) a single layer (or "ply") of a woven fabric composite; (c) a single layer (or "ply") of a braided fabric composite.

(0.010–0.16 in.). Obviously, a single composite ply is quite thin. To produce a composite structure with a significant thickness many plies are stacked together to form a composite *laminate*. Conceptually any number of plies may be used in the laminate, but in practice the number of plies usually ranges from about 10 plies to (in unusual cases) perhaps as many as 200 plies. The fiber represents the reinforcing material in these composites. Hence, the orientation of the fibers is (in general) varied from one ply to the next, so as to provide high stiffness and strength in more than one direction (as is the case in plywood). It is also possible to use a combination of unidirectional, woven, and/or braided plies within the same laminate. For example, it is common to use a woven or braided fabric as the two outermost *facesheets* of a laminate, and to use unidirectional plies at interior positions.

In all composites the reinforcement is embedded within a matrix material. The matrix may be polymeric, metallic, or ceramic. In fact, composite materials are often classified on the basis of the matrix material used, rather than the reinforcing material. That is, modern composites can be categorized into three main types: polymer–matrix, metal–matrix, or ceramic–matrix composite materials. Usually, the reinforcing material governs the stiffness and strength of a composite. In contrast, the matrix material usually governs the thermal stability. Polymeric–matrix composites are used in applications involving relatively modest temperatures (service temperatures of 200°C or less, say). Metal–matrix composites are used at temperatures up to about 700°C, while ceramic–matrix composites are used at ultra-high temperatures (up to about 1200°C or greater). The matrix also defines several additional characteristics of the composite material system. Some additional roles of the matrix are

- To provide the physical form of the composite
- To bind the fibers together
- To protect the fibers from aggressive (chemical) environments, or mechanical damage (e.g., due to abrasion)
- To transfer and redistribute stresses between fibers, between plies, and in areas of discontinuities in load or geometry

To summarize the preceding discussion, there are many types of composite materials, both natural and man-made. Composites can be classified according to the physical form of the reinforcing material (particulate, whisker, short fiber, or continuous fiber reinforcement), by the type of matrix material used (metal, ceramic, or polymeric matrix), by the orientation of the reinforcement (unidirectional, woven, or braided), or by some combination thereof. The temperature the composite material/structure will experience in service often dictates the type of composite used in a given application.

The primary focus of this textbook is the structural analysis of *polymeric* composite materials and structures. Metal– and ceramic–matrix composites will not be further discussed, although many of the analysis methods developed herein may be applied to these types of composites as well. Since our focus is the *structural* analysis of polymeric composites, we will not be greatly considered with the behavior of the individual *constituent* materials. That is, we will not be greatly concerned with the behavior of an unreinforced polymer, nor with the behavior of an individual reinforcing fiber. Instead, we will be concerned with the behavior of the composite formed by combining these two constituents. Nevertheless, a structural engineer who wishes to use polymeric composites effectively in practice must understand at least the rudiments of polymer and fiber science, in much the same way as an engineer working with metal alloys must understand at least the rudiments of metallurgy. Towards that end a brief introduction to polymeric and fibrous materials is provided in the following two sections, respectively. At minimum, the reader should become acquainted with the terminology used to describe polymeric and fibrous materials, since such terms have naturally been carried over to the polymeric composites technical community.

1.2 Polymeric Materials

A brief introduction to polymeric materials is provided in this section. This introduction is necessarily incomplete. The reader interested in a more detailed discussion is referred to the many available texts and/or web-based resources devoted to modern polymers (see, e.g., [2–5]).

1.2.1 Basic Concepts

The term "polymer" comes from the Greek words *poly* (meaning many) and *mers* (meaning units). Quite literally, a polymer consists of "many units." Polymer molecules are made up of thousands of repeating chemical units, and have molecular weights ranging from about 10^3 to 10^7.

As an illustrative example, consider the single chemical mer shown in Figure 1.2. This mer is called *ethylene* (or *ethene*), and consists of two carbon atoms and four hydrogen atoms. The two lines between the carbon (C) atoms indicate a double covalent bond,* whereas the single line between the hydrogen (H) and carbon atoms represents a single covalent bond. The chemical composition of the ethylene mer is sometimes written C_2H_4 or $CH_2=CH_2$.

Under the proper conditions the double covalent bond between the two carbon atoms can be converted to a single covalent bond, which allows each of the two carbon atoms to form a new covalent bond with a suitable neighboring atom. A suitable neighboring atom would be a carbon atom in a neighboring ethylene mer, for example. If "*n*" ethylene mers join together in this way, the chemical composition of the resulting molecule can be represented as $C_{2n}H_{4n}$, where n is any positive integer. Hence, a "chain" of ethylene mers join together to form the well-known polymer *polyethylene* (literally, many ethylenes), as shown in Figure 1.3. The process of causing a monomer to chemically react and form a long molecule in this fashion is called *polymerization*.

The single ethylene unit is an example of a *monomer* (one mer). At room temperature a bulk sample of the ethylene monomer is a low-viscosity fluid.

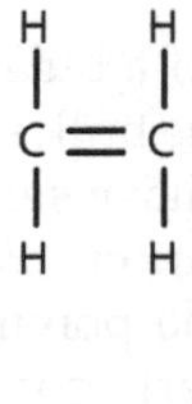

$$\begin{array}{ccc} \text{H} & & \text{H} \\ | & & | \\ \text{C} & \!\!=\!\! & \text{C} \\ | & & | \\ \text{H} & & \text{H} \end{array}$$

FIGURE 1.2
The monomer "ethylene."

---

* As fully described in any introductory chemistry text, a "covalent bond" is formed when two atoms share an electron pair, so as to fill an incompletely filled valence level.

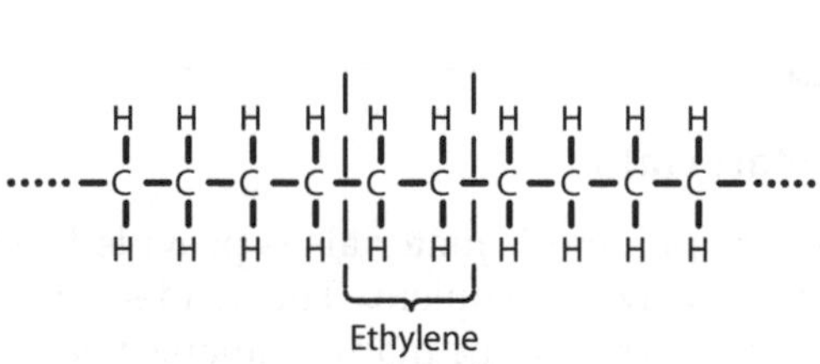

Ethylene
repeat unit

FIGURE 1.3
The polymer "polyethylene."

If two ethylene monomers bond together the resulting chemical entity has two repeating units and is called a *dimer*. Similarly, the chemical entity formed by three repeating units is called a *trimer*. The molecular weight of a dimer is twice that of the monomer, the molecular weight of a trimer is three times that of the monomer, etc. Prior to polymerization, most polymeric materials exist as relatively low-viscosity fluids known as *oligomers* (a few mers). The individual molecules within an oligomer possess a range of molecular weights, typically containing perhaps 2–20 mers.

It should be clear from the above discussion that a specific molecular weight cannot be assigned to a polymer. Rather, the molecules within a bulk sample of a polymer are of differing lengths and hence exhibit a range in molecular weight. The *average molecular weight* of a bulk sample of a polymer is increased as the polymerization process is initiated and progresses. Another measure of the "size" of the polymeric molecule is the *degree of polymerization*, defined as the ratio of the average molecular weight of the polymer molecule divided by the molecular weight of the repeating chemical unit within the molecular chain.

The average molecular weight of a polymeric sample (or, equivalently, the degree of polymerization) depends on the conditions under which it was polymerized. Now, virtually *all* physical properties exhibited by a polymer (e.g., strength, stiffness, density, thermal expansion coefficient, etc.) are dictated by the average molecular weight. Therefore, a fundamental point that must be appreciated by the structural engineer is that the properties exhibited by any polymer (or any polymeric composite) *depend on the circumstances under which it was polymerized.*

As a general rule, the volume of a bulk sample of a monomer decreases during polymerization. That is, the bulk sample shrinks as the polymerization process proceeds. This may have serious ramifications if the polymer is to be used in structural applications. For example, if a fiber(s) is embedded within the sample during the polymerization process (as is the case for most fiber-reinforced polymeric composite systems), then shrinkage of the matrix causes residual stresses to develop during polymerization. This effect contributes to the so-called *cure stresses*, which are present in most polymeric composites. As will be seen in later chapters, cure stresses arise from two primary sources. The first is shrinkage of the matrix during polymerization, as just described. The second is stresses that arise due to temperature effects. In many cases the composite is polymerized at an elevated temperature (175°C, say), and then cooled to room temperature (20°C). The thermal expansion coefficient of the matrix is typically much higher than the fibers, so during cooldown the matrix is placed in tension while the fiber is placed in compression. Cure stresses due to shrinkage during cure and/or temperature effects can be quite high relative to the strength of the polymer itself, and ultimately contribute towards failure of the composite.

1.2.2 Addition versus Condensation Polymers

Although in the case of polyethylene the repeat unit is equivalent to the original ethylene monomer, this is not always the case. In fact, in many instances the repeat unit is derived from two (or more) monomers. A typical example is Nylon 6,6. The polymerization process for this polymer is shown schematically in Figure 1.4. Two monomers are used to produce Nylon 6,6: hexamethylene diamine (chemical composition: $C_6H_{16}N_2$) and adipic acid (chemical composition: $CO_2H(CH_2)_4CO_2H$). Note that the repeat unit of Nylon 6,6 (hexamethylene adipamide) is not equivalent to either of the two original monomers.

A low-molecular weight byproduct (i.e., H_2O) is produced during the polymerization of Nylon 6,6. This is characteristic of *condensation polymers*.

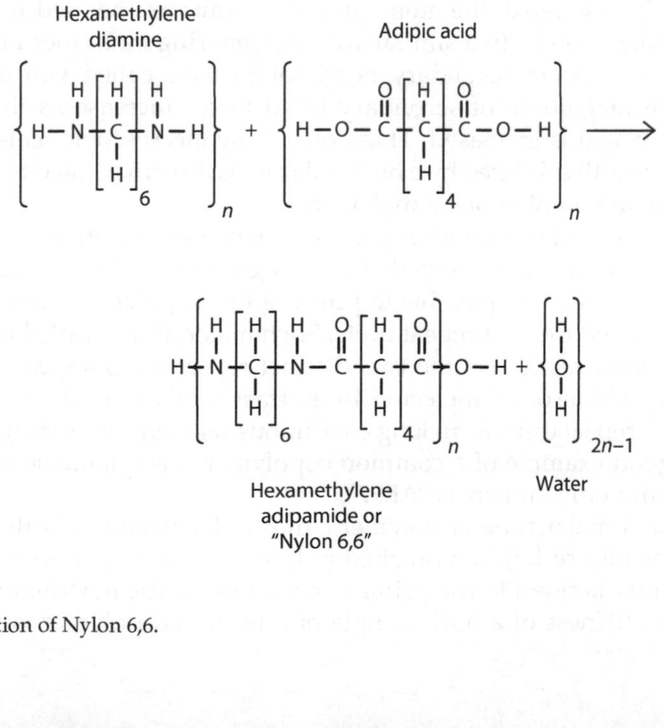

FIGURE 1.4
Polymerization of Nylon 6,6.

That is, if both a high-molecular weight polymer as well as a low-molecular weight byproduct is formed during the polymerization process, the polymer is classified as a condensation polymer. Conversely, *addition polymers* are those for which no byproduct is formed, which implies that all atoms present in the original monomer(s) occur somewhere within the repeat unit. Generally speaking, condensation polymers shrink to a greater extent during the polymerization process than do addition polymers. Residual stresses caused by shrinkage during polymerization are often a concern in structural composites, and hence difficulties with residual stresses can often be minimized if an addition polymer is used in these applications.

1.2.3 Molecular Structure

The molecular structure of a fully polymerized polymer can be roughly grouped into one of three major types: *linear, branched,* or *crosslinked* polymers. The three types of molecular structure are shown schematically in Figure 1.5.

A single molecule of a *linear polymer* can be visualized as beads on a string, where each bead represents a repeat unit. It should be emphasized that the length of these "strings" is enormous; if a typical linear molecule were scaled up so as to be 10 mm in diameter, it would be roughly 4 km long. In a bulk sample these long *macromolecules* become entangled and twisted together, much like a bowl of cooked spaghetti. Obviously, as the average molecular weight (i.e., the average length) of the polymer molecule is increased, the number of entanglements is increased.

As already discussed, the atoms in repeat units are bonded together by strong covalent bonds. In a similar way, neighboring molecules are bonded together by so-called secondary bond forces (also called van der Waals forces). The magnitude of secondary bond forces increase as the average molecular weight is increased. Therefore, at the macroscopic scale the stiffness and strength exhibited by a bulk polymer is directly related to its molecular weight and number of entanglements.

If all of the repeat units within a linear polymer are identical, the polymer is called as a *homopolymer.* Polyethylene is a good example of a linear homopolymer. However, it is possible to produce linear polymers that consist of two separate and distinct repeat units. Such materials are called *copolymers.* In *linear random copolymers* the two distinct repeat units appear randomly along the backbone of the molecule. In contrast, for *linear block copolymers,* the two distinct repeat units form long continuous segments within the polymer chain. A good example of a common copolymer is acrylonitrile-butadiene-styrene, commonly known as "ABS."

The second major type of polymeric molecular structure is the *branched polymer* (see Figure 1.5). In branched polymers relatively short side chains are covalently bonded to the primary backbone of the macromolecule. As before, the stiffness of a bulk sample of a branched polymer is related to

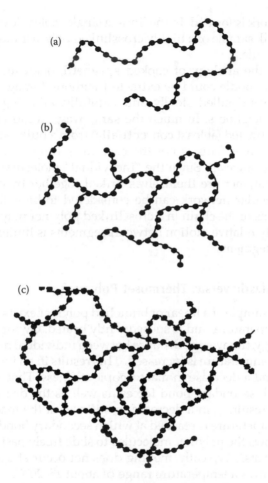

FIGURE 1.5
Types of polymer molecular structure. (a) Linear; (b) branched; (c) crosslinked.

secondary bond forces and the number of entanglements between molecules. Since the branches greatly increase the number of entanglements, the macroscopic stiffness of a branched polymer will, in general, be greater than the macroscopic stiffness of a linear polymer of similar average molecular weight. In many branched polymers the branches consist of the same chemical repeat unit as the backbone of the molecular chain. However, in some cases the branch may have a distinctly different chemical repeat unit than the main backbone of the molecule. Such materials are called *graft copolymers.*

Finally, the third major type of molecular structure is the *crosslinked or network polymer* (see Figure 1.5). During polymerization of such polymers a *crosslink* (i.e., a covalent bond) is formed between individual molecular chains. Hence, once polymerization (i.e., crosslinking) is complete a vast molecular network is formed. In the limit, a single molecule can no longer be identified. A bulk sample of a highly crosslinked polymer may be thought of as a single molecule.

Returning to the analogy of cooked spaghetti, one can imagine that a single spaghetti noodle could be extracted without damage from the bowl if the noodle were pulled slowly and carefully, allowing the noodle to "slide" past its neighbors. In much the same way, an individual molecule could also be extracted (at least conceptually) from a bulk sample of a linear or branched polymer. This is not the case for a fully polymerized cross-linked polymer, however. Since the "individual" molecular chains within a crosslinked polymer are themselves linked together by covalent bonds, the entire molecular network can be considered to be a single molecule. Although regions of the chain in a crosslinked polymer may slide past each other, eventually relative motion between segments is limited by the cross-links between segments.

1.2.4 Thermoplastic versus Thermoset Polymers

Suppose a bulk sample of a linear or branched polymer exists as a solid material at room temperature, and is subsequently heated. Owing to the increase in thermal energy, the average distance between individual molecular chains is increased as temperature is increased. This results in an increase in molecular mobility and a decrease in macroscopic stiffness. That is, as molecules move apart both secondary bond forces as well as the degree of entanglement decrease, resulting in a decrease in stiffness at the macroscopic level. Eventually a temperature is reached at which secondary bond forces are negligible. This allows the polymer molecules to slide freely past each other and the polymer "melts." Typically, melting does not occur at a single temperature, but rather over a temperature range of about 15–20°C.

A polymer that can be melted (i.e., a linear or branched polymer) is called a *thermoplastic polymer*. In contrast, a crosslinked polymer cannot be melted and is called a *thermoset polymer*. If a crosslinked polymer is heated it will exhibit a decrease in stiffness at the macroscopic structural level, since the average distance between individual segments of the molecular network is increased as temperature is increased. The crosslinks do not allow indefinite relative motion between segments, however, and eventually limit molecular motion. Therefore, a crosslinked polymer will not melt.

Of course, if the temperature increase is excessive then the covalent bonds which form both the backbone of the molecule as well as any crosslinks are broken and chemical degradation occurs. That is, both thermoplastic and thermoset polymers are destroyed at excessively high temperatures.

A point of confusion, especially for the nonspecialist, is that polymers are often classified according to some characteristic chemical linkage within the molecular chain, and in many cases polymers within a given classification may be produced as either a thermoplastic or thermoset. Thus, for example, polyester can be produced as either a thermoplastic polymer or a thermoset polymer. A thermoplastic polyester can be melted, whereas a thermoset polyester cannot.

1.2.5 Amorphous versus Semicrystalline Thermoplastics

The molecular structure of a thermoplastic polymer may be *amorphous* or *semicrystalline.** The molecular structure of an amorphous thermoplastic is completely random; that is, the molecular chains are randomly oriented and entangled, with no discernible pattern. In contrast, in a semicrystalline thermoplastic there exist regions of highly ordered molecular arrays. An idealized representation of a crystalline region is shown in Figure 1.6. As indicated, in the crystalline region the main backbone of the molecular chain undulates back and forth such that the thickness of the crystalline region is usually (about) 10 nm. The crystalline region may extend over an area with a length dimension ranging from (about) 100 to 1000 nm. Hence, the crystalline regions are typically plate-like. The high degree of order within the crystalline array allows for close molecular spacing, and hence exceptionally high bonding forces between molecules in the crystalline region. Therefore, at the macroscale a semicrystalline thermoplastic typically has a higher strength, stiffness, and density than an otherwise comparable amorphous thermoplastic. No thermoplastic is completely crystalline, however. Instead, regions of crystallinity are surrounded by amorphous regions, as shown schematically in Figure 1.7. Most semicrystalline thermoplastic are 10–50% amorphous (by volume).

Since secondary bond forces are higher in crystalline regions, amorphous regions melt at a lower temperature than crystalline regions. Therefore, semicrystalline thermoplastics may exhibit crystalline regions in both solid and liquid forms, and are known as *liquid-crystal polymers*.

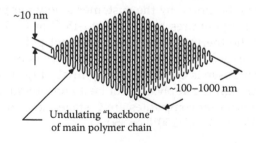

FIGURE 1.6
An idealized representation of a crystalline region in a thermoplastic polymer.

---

* In practice, a semicrystalline thermoplastic is often called a *crystalline* polymer.

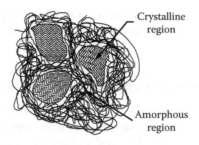

FIGURE 1.7
Molecular structure of a semicrystalline thermoplastic, showing crystalline and amorphous regions.

1.2.6 A-, B-, and C-Staged Thermosets

Three more-or-less distinct conditions are recognized during polymerization of a thermoset polymer. The original resin or oligomer is typically a low-viscosity, low-molecular weight fluid, consisting of molecules with perhaps 2–10 repeat units. A thermoset resin is said to be *A-Staged* when in this form. As the polymerization process is initiated (usually by the introduction of a catalyst, by an increase in temperature, or by some combination thereof), the molecular weight and viscosity of the oligomer increase rapidly. If the temperature of the partially polymerized thermoset is suddenly reduced the polymerization process will stop, or at least be dramatically slowed, and the polymer will exist in an intermediate stage. The thermoset resin is said to be *B-Staged* when in this form. If the B-staged thermoset is subsequently reheated the polymerization process resumes and continues until the maximum possible molecular weight has been reached. The thermoset is then said to be *C-Staged*, that is, the polymer is *fully polymerized*.

Suppliers of composites based on thermoset polymers often B-stage their product and sell it to their customers in this form. This requires that the B-staged composite be stored by the customer at low temperatures prior to use (typically at temperatures below about –15°C or 0°F). Refrigeration is required so that the thermosetting resin does not polymerize beyond the B-stage during storage. The polymerization process is reinitiated and completed (i.e., the composite is C-staged) during final fabrication of a composite part, typically through the application of heat and pressure. Most commercially available thermoset composites are C-staged (or cured) at a temperature of either 120°C or 175°C (250°F or 350°F).

1.2.7 The Glass Transition Temperature

Stiffness and strength are physical properties of obvious importance to the structural engineer. Both of these properties are temperature dependent. The effect of temperature on the stiffness of a polymer is summarized in Figure 1.8. Thermoset and thermoplastic polymers exhibit the same general

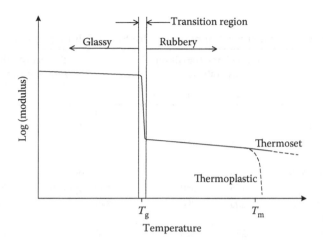

FIGURE 1.8
Effects of temperature on polymer stiffness.

temperature dependence, except that at high temperatures thermoplastics melt whereas thermosets do not. All polymers exhibit a sudden decrease in stiffness as temperature is increased to a value called the *glass-transition temperature*, T_g. At temperatures well below the T_g polymer stiffness decreases gradually with an increase in temperature. At these low (relative to the T_g) temperatures polymers are said to be in a "glassy" state and are relatively brittle. In contrast, at temperatures well above the T_g all polymers are "rubbery" and ductile. Thus, the T_g denotes the transition between glassy and rubbery regimes. This transition is associated with a sudden increase in mobility of segments within the molecular chain, and typically occurs over a range of 10–15°C. At temperatures well below the T_g the polymer molecules are firmly bonded together and cannot easily slide past each other. Consequently the polymer is "glassy," and exhibits a high stiffness and strength but relatively low ductility. Conversely, at temperatures well above the T_g the molecular spacing is increased such that the molecular chains (or segments of those chains) are mobile and can readily slide past each other if force is applied. Consequently the polymer is "rubbery" and exhibits a relatively lower stiffness and strength but higher ductility. As implied in Figure 1.8, for amorphous thermoplastics the change in stiffness (and other physical properties) that occurs as the T_g is approached may be 1–2 *orders of magnitude*. This astonishing decrease in stiffness occurs over a temperature range of only 10–20°. A marked decrease in stiffness also occurs for semi-crystalline thermoplastics and crosslinked thermosets, although in general the magnitude of the decrease is less than for amorphous thermoplastics. If temperature is raised high enough then a thermoplastic polymer will melt and stiffness tends towards zero. The temperature region at which melting occurs is denoted by T_m in Figure 1.8, although as previously discussed

TABLE 1.1

Approximate Glass Transition Temperatures for Some Polymers Used in PMCs

Polymer	Typical Glass Transition Temperature, T_g		Typical Melting Temperature, T_m	
	°C	°F	°C	°F
Polyester (thermoplastic)	80	175	250	480
Polyphenylene-sulfide (thermoplastic)	85	185	280	540
Polyester (thermoset)	150	300	Not applicable	Not applicable
Epoxy (thermoset)	175	350	Not applicable	Not applicable
Polyetheretherketone (thermoplastic)	200	390	340	650
Polyetherimide (thermoplastic)	215	420	370	700
Bismaleimide (thermoset)	230	450	Not applicable	Not applicable
Polyimide (thermoset)	260	500	Not applicable	Not applicable
Liquid-crystal polyester (thermoplastic)	360	650	370	700

Note: All temperatures approximate and depend on specific polymerization conditions.

a thermoplastic does not exhibit a unique melting temperature but rather melts over a temperature range that depends on average molecular weight. Thermoset polymers cannot be melted.

The T_g has been illustrated in Figure 1.8 by demonstrating the change in stiffness as temperature is increased. Most other macroscopic physical characteristics (density, strength, thermal expansion coefficient, heat capacity, etc.) also change sharply at this transition. Hence, the T_g can be measured by monitoring any of these physical properties as a function of temperature. The T_g exhibited by a few polymers used in PMCs is listed in Table 1.1.

1.3 Fibrous Materials

The fiber types most commonly used in PMCs are

- Glass
- Aramid
- Graphite or carbon
- Ultra high-density polyethylene

In all cases the fiber diameters are quite small, ranging from about 5 to 12 μm for glass, aramid, or graphite fibers, and from about 25 to 40 μm for polyethylene fibers.

Some of the terminology used to describe fibers will be defined here. The terms *fiber* and *filament* are used interchangeably. An *end* (also called a *strand*) is a collection of a given number of fibers gathered together. If the fibers are twisted, the collection of fibers is called a *yarn*. The ends are themselves gathered together to form a *tow* (also called *roving*). The fibers are usually coated with a *size* (also called a *finish*). The size is applied for several reasons, including

- To bind the fibers in the strand
- To lubricate the fibers during fabrication
- To serve as a coupling and wetting agent to insure a satisfactory adhesive bond between the fiber and matrix materials

Tow sizes are specified in terms of thousands of fibers/filaments per tow. For example, a "6 k tow" implies that the tow consists of about 6000 individual fibers.

1.3.1 Glass Fibers

Eight different types of glass fibers are defined in an ASTM standard [6]. These are A-glass, AR-glass, C-glass, D-glass, E-glass, E-CR-glass, R-glass, and S-glass fibers. In all cases the glass fibers consist primarily of silica (SiO_2), and additional oxides are added to enhance properties particularly suited for the intended application. For example, AR-glass fibers exhibit an increased resistance to corrosion by akali, whereas C-glass fibers have increased resistance to corrosion by acids.

The fibers are produced by blending all constituents in a large vat, followed by melting in a furnace at a temperature of about 1260°C (2300°F). The molten glass is then drawn through platinum bushings with hundreds of small diameter holes, followed by rapid cooldown. A sizing is applied to the fibers, which are then combined into a strand and wound onto a spool.

The glass fibers most commonly used in PMCs are either E-glass or S-glass. E-glass ("e"lectrical glass) is so named because of its high electrical resistivity, whereas S-glass ("s"tructural glass) is so named because it exhibits higher high stiffness and strength than other types of glass fibers. E-glass fibers have very good properties and are relatively inexpensive, however, so E-glass fibers are widely used in structural composite applications despite the improved mechanical properties provided by S-glass fibers. The compositions of E- and S-glass fibers are listed in Table 1.2, and typical mechanical properties are listed in Table 1.3.

TABLE 1.2

Constituents Used to Produce E- and S-Glass Fibers

Constituent	E-Glass Weight%	S-Glass Weight%
SiO_2 (Silica)	52–56	64–66
Al_2O_3 (aluminum oxide or alumina)	12–16	24–25
B_2O_3 (boron trioxide)	5–10	—
CaO (calcium oxide or quicklime)	16–25	0–0.2
MgO (magnesium oxide)	0–5	9–10
Na_2O (sodium oxide)	0–2	0–0.2
K_2O (potassium oxide)	0–2	0–0.2
Fe_2O_3 (ferric oxide)	0–0.8	0–0.1
CaF_2 (fluorite)	0–1	—

TABLE 1.3

Typical Properties of Glass Fibers

Property	E-Glass	S-Glass
Density, g/cm³	2.60	2.50
(lb/in.³)	(0.094)	(0.090)
Young's modulus, GPa	72	87
(Msi)	(10.5)	(12.6)
Tensile strength, MPa	3450	4310
(ksi)	(500)	(625)
Tensile elongation, %	4.8	5.0
Coefficient of thermal expansion, μm/m/°C	5.0	5.6
(μin./in./°F)	(2.8)	(3.1)

1.3.2 Aramid Fibers

Aramid is a generic name for a class of synthetic organic *polymeric* fibers (also called aromatic polyamide fibers). The aramid fiber produced by the DuPont Corp and marketed under the trade name Kevlar™ is probably the best-known aramid fiber. This fiber is based on poly [*p*-phenylene tere-phthalamide], which is a member of the aramid family of polymers. The basic repeat unit is shown in Figure 1.9. Kevlar fibers are produced using a mechanical drawing process that aligns the backbone of the polymeric molecular chain with the fiber axis. Although covalent bonding (i.e., cross-linking) does not occur between the elongated polymer chains, a strong secondary bond forms between adjacent hydrogen (H) and oxygen (O) atoms. The resulting fibers are, therefore, highly anisotropic since strong covalent bonds are formed in the fiber axial direction, whereas relatively weaker hydrogen bonds are formed in directions transverse to the fiber axis. That

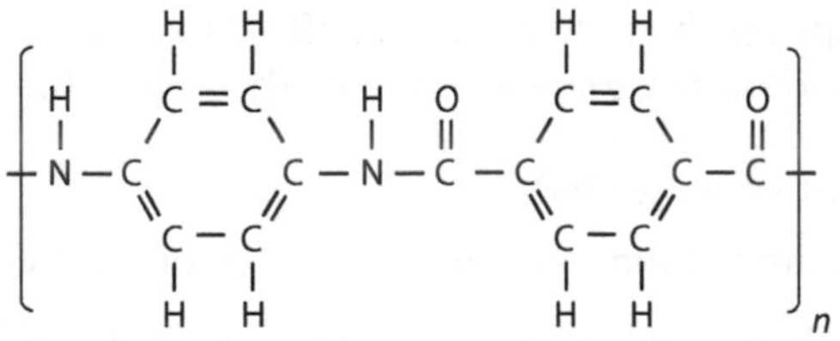

FIGURE 1.9
Repeat unit of poly[*p*-phenylene terephthalamide], or Kevlar™.

TABLE 1.4

Typical Properties of Kevlar 29 and Kevlar 49 Fibers

Property	Kevlar 29	Kevlar 49
Density, g/cm³	1.44	1.44
(lb/in.³)	(0.052)	(0.052)
Young's modulus, GPa	70	112
(Msi)	(10)	(16)
Tensile strength, MPa	2920	3000
(ksi)	(424)	(435)
Tensile elongation, %	3.6	2.4
Coefficient of thermal expansion, μm/m/°C	−3.9	−4.9
(μin./in./°F)	(−2.2)	(−2.7)

Note: All properties measured in axial direction of fiber.

is, Kevlar fibers have very high tensile strength and stiffness in the axial direction of the fiber, but relatively low tensile strength and stiffness in the transverse direction.

New forms of Kevlar and other aramid fibers are introduced almost continuously to the marketplace. At present there are at least eight different grads of Kevlar fiber, each with a different combination of properties and cost. Kevlar 49 is most commonly used in PMCs, although occasionally the less-expensive Kevlar 29 fiber is used as well. Nominal mechanical and physical properties for Kevlar 29 and 49 fibers are listed in Table 1.4. Of particular interest is the *negative* coefficient of thermal expansion these fibers exhibit in the axial direction.

1.3.3 Graphite and Carbon Fibers

The terms "graphite" and "carbon" are often used interchangeably within the composites community. The elemental carbon content of either type of fiber is above 90%, and the stiffest and strongest fibers have elemental carbon contents approaching 100%. Some effort has been made to standard-ize these terms by defining *graphite fibers* as those that have

- A carbon content above 95%
- Been heat treated at temperatures in excess of 1700°C (3100°F)
- Been stretched during heat treatment to produce a high degree of preferred crystalline orientation
- A Young's modulus on the order of 345 GPa (50 Msi)

Fibers that do not satisfy all of the above conditions are called *carbon fibers* under this standard. However, as stated above in practice this definition is not widely followed, and the terms "graphite" and "carbon" are often used interchangeably.

Both graphite and carbon fibers are produced by thermal decomposition of an organic (i.e., polymeric) fiber or "precursor" at high pressures and tem-peratures. The three most common precursors are

- Polyacrylonitrile (PAN), a synthetic polymer with a repeat unit of $[C_3H_3N]_n$
- Rayon, a semisynthetic fiber produced using the natural polymer cellulose with a repeat unit of $[C_6H_{10}O_5]_n$
- Pitch, a generic name for a range of highly viscous solid polymers

Pitch can be derived either from petroleum products, in which case it is sometimes called "petroleum pitch," or from plants.

Details of the specific steps followed during fabrication of a specific car-bon or graphite fiber are proprietary and can only be described in a general manner. During the fabrication process the precursor is drawn into a thread and then oxidized at temperatures ranging from about 230°C to 260°C (450–500°F) to form an extended carbon network. The precursor is then subjected to a carbonization treatment, during which noncarbon atoms are driven off. This step typically involves temperatures ranging from 700°C to 1000°C (1290–1800°F), and is conducted in an inert atmosphere. Finally, during the graphitization process the fibers are subjected to a combination of high tem-perature and tensile elongation. Maximum temperatures reached during this step range from 1500°C to 2000°C (2700–3600°F), and determine in large part the strength and/or stiffness exhibited by the fiber.

These processing steps results in a carbon-based crystalline structure called "graphite." The graphite crystalline structure is shown in Figure 1.10. Carbon atoms are arranged in parallel planes or sheets called graphene, as shown in Figure 1.10a. Each carbon atom within a graphene plane is covalently bonded to three neighboring carbon atoms, resulting in a pat-tern of hexagonal rings of carbon atoms within the graphene plane, as seen most clearly in Figure 1.10b,c. High fiber strength and stiffness is achieved

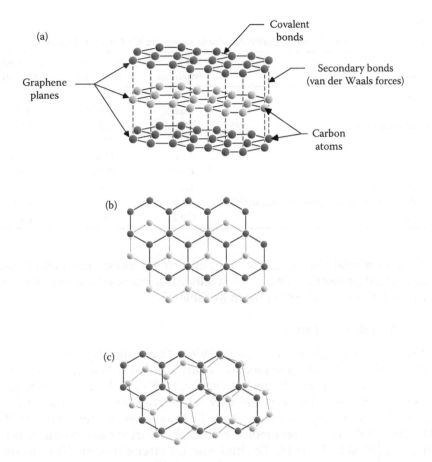

FIGURE 1.10
Graphite crystalline structures present in carbon or graphite fibers. (a) Graphite crystalline structure (side view); (b) graphitic graphite crystalline structure (orthogonal view); (c) turbostratic graphite crystalline structure (orthogonal view).

by causing the graphene planes to be aligned with the fiber axial direction. Although the carbon atoms within graphene planes are covalently bonded, adjacent graphene sheets are bonded by relatively weaker van der Waals forces. Consequently graphite/carbon fibers are highly anisotropic (as are Kevlar fibers), since the weaker secondary forces act transverse to the fiber axis.

As implied in Figure 1.10b,c, the graphene sheets may be tightly packed with a high degree of regularity (called *graphitic* graphite), or may be relatively loosely packed with less regularity (called *turbostratic* graphite). Fiber produced using the PAN precursor tends to exhibit a high-volume percentage of turbostratic graphite, leading to high tensile strength. In contrast, fiber produced using the pitch precursor tends to exhibit a high-volume percentage of graphitic graphite, leading to high tensile stiffness. Fibers are available with varying levels of both type of crystalline graphite, and

TABLE 1.5

Typical Properties of Commercially Available Graphite (or Carbon) Fibers

Property	Low Modulus	Intermediate Modulus	Ultra-High Modulus
Density, g/cm^3	1.8	1.9	2.2
(lb/in.^{3})	(0.065)	(0.069)	(0.079)
Young's modulus, GPa	230	370	900
(Msi)	(34)	(53)	(130)
Tensile strength, MPa	3450	2480	3800
(ksi)	(500)	(360)	(550)
Elongation, %	1.1	0.5	0.4
Coefficient of thermal expansion, μm/m/°C	−0.4	−0.5	−0.5
(μin./in./°F)	(−0.2)	(−0.3)	(−0.3)

hence commercial fibers exhibit a range of stiffness and strength. Mechanical and physical properties typical of low-modulus, intermediate modulus, and ultra-high modulus fibers are listed in Table 1.5.

1.3.4 Polyethylene Fibers

A high strength, high modulus polyethylene fiber called Spectra was developed at Allied Signal Technologies during the 1980s. Spectra is based on ultra high molecular weight polyethylene (UHMWPE). It has a specific gravity of 0.97, meaning that it is the only reinforcing fiber available which is lighter than water. Spectra is available in three classifications, Spectra 900, 1000, and 2000, and several grades are available in each class. Nominal properties are listed in Table 1.6. The high specific strength of the fiber makes it particularly attractive for tensile applications. The glass transition temperature of UHMWPE is in the range from −20°C to 0°C, and hence the fiber is in the rubbery state at room temperatures, and exhibits time-dependent (viscoelastic) behavior. This feature imparts outstanding impact resistance and

TABLE 1.6

Approximate Properties of Spectra Fibers

Property	Spectra 900	Spectra 1000	Spectra 2000
Specific gravity	0.97	0.97	0.97
Young's modulus, GPa	70	105	115
(Msi)	(10)	(15)	(17)
Tensile strength, MPa	2600	3200	3400
(ksi)	(380)	(465)	(490)
Elongation, %	3.8	3.0	3.0
Coefficient of thermal expansion, μm/m/°C	>70	>70	>70
(μin./in./°F)	(>38)	(>38)	(>38)

toughness, but may lead to undesirable creep effects under long-term sustained loading. The melting temperature of the fiber is about 150°C (300°F), and hence the use of polyethylene fibers is limited to relatively modest temperatures. The thermal expansion coefficients of Spectra fibers have apparently not been measured; values listed in Table 1.6 are estimates based on the properties of bulk high-molecular-weight polyethylene.

1.4 Commercially Available Forms

1.4.1 Discontinuous Fibers

Virtually all of the continuous fibers described in Section 1.3 are also available in the form of discontinuous fibers. Discontinuous fibers are embedded within a matrix, and may be randomly oriented (in which case the composite is isotropic at the macroscale), or may be oriented to some extent (in which case the composite is anisotropic at the macroscale). Orientation of discontinuous fibers, if it occurs, is usually induced during the fabrication process used to create the composite material/structure; fiber alignment often mirrors the flow direction during injection molding, for example. Discontinuous fibers are roughly classified according to length, as follows:

- *Milled fibers* are produced by grinding the continuous fiber into very short lengths. For example, milled graphite fibers are available with lengths ranging from about 0.3 to 3 mm (0.0012–0.12 in.), and milled glass fibers are available with lengths ranging from about 0.4 to 6 mm (0.0016–0.24 in).

- *Chopped fibers (or strands)* have a longer length than milled fibers, and composites produced using chopped fibers usually have higher strengths and stiffnesses than those produced using milled fibers. Chopped graphite fibers are available with lengths ranging from about 3 to 50 mm (0.12–2.0 in.), while chopped glass fibers are available with lengths ranging from about 6 to 50 mm (0.24–2.0 in.).

In general, the mechanical properties of the composite produced using discontinuous fibers (the strength or stiffness, say) are not as good as those which can be obtained using continuous fibers. However, discontinuous fibers allow the use of relatively inexpensive, high-speed manufacturing processes such as injection molding or compression molding, and have, therefore, been widely used in applications in which extremely high strength or stiffness is not required.

One of the most widely used composites systems based on the use of discontinuous fibers is known generically as "sheet-molding compound" (SMC). In its most common form SMC consists of chopped glass fibers embedded within a thermosetting polyester resin. However, other resins (e.g., vinyl esters or epoxies), and/or other fibers (e.g., chopped graphite or aramid fibers) are also used in SMC material systems. Components produced using SMC are normally manufactured using compression molding.

1.4.2 Roving Spools

Most continuous fiber types are available in the form of spools of roving, that is, roving wound onto a cylindrical tube, and ultimately resembling a large spool of thread. As mentioned in Section 1.3, roving is also known as tow. The size of tow (or roving) is usually expressed in terms of the number of fibers contained in a single tow. For example, a specific glass fiber might be available in the form of 2k, 3k, 6k, or 12k tow. In this case the product is available in tows containing from 2000 to 12,000 fibers. Fibers purchased in this form are usually "dry," and are combined with a polymer, metal, or ceramic matrices during a subsequent manufacturing operation such as filament winding or pultrusion.

1.4.3   Woven Fabrics

Most types of high-performance continuous fibers can be woven to form a fabric. Weaving is accomplished using looms specially modified for use with high-performance fibers, which are stiffer than those customarily used in the textile industry. The weaving process is illustrated in Figure 1.11.

FIGURE 1.11
A loom producing a glass fiber fabric using the plain weave pattern. (Adapted from www. nauticexpo.com/prod/nida-core-corporation/glass-fiber-fabric-multiaxial-27841-192139.html.)

Woven fabrics are produced in various widths up to about 120 cm (48 in.), and are available in (essentially) infinite lengths. Two terms associated with woven fabrics are

- The tow or yarn running along the length of the fabric is called the *warp*. The warp direction is parallel to the long axis of the woven fabric.
- The tow or yarn running perpendicular to the warp is called the *fill* tow (also called the *weft* or the *woof* tow). The fill direction is perpendicular to the warp direction.

Some common fabric weaves are shown schematically in Figure 1.12. The *plain weave* (also called a *simple weave*) is shown in Figure 1.12a and is being

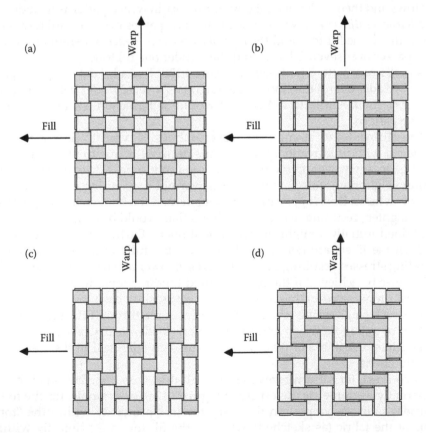

FIGURE 1.12
Some common woven fabrics used with high-performance fibers. (a) Plain weave pattern; (b) basket weave pattern; (c) four-harness (crowfoot) weave pattern; (d) 2×2 twill weave pattern. (From Donnet, J.-B. et al. *Carbon Fibers*, 3rd edition, Marcel Dekker, Inc., New York, 1998, ISBN 0-8247-0172-0.)

produced using the loom shown in Figure 1.11. The plain weave is the simplest fabric pattern available and is most commonly used. It is produced by repetitively weaving a given warp tow over one fill tow and under the next. The point at which a tow passes over/under another tow is called a *crossover point*. The plain weave pattern results in a very stable and firm fabric that exhibits minimum distortion (e.g., fiber slippage) during handling. A *basket weave*, shown in Figure 1.12b, is a variation of the simple weave pattern in which two (or more) warp tows pass over an equal number of fill tows, forming a rectangular pattern similar to a plain weave.

A family of woven fabric patterns known as *satin weaves* provides better *drape characteristics* than a plain weave. That is, a satin weave is more pliable and will more readily conform to complex-curved surfaces than plain weaves. The *four-harness satin weave* (also known as a *crowfoot* weave) is shown in Figure 1.12c. In this case, one warp tow passes over three adjacent fill tows and then under one fill tow. Similar satin weave patterns include the *five-harness satin weave*, wherein one warp tow passes over four fill tows and then under one fill tow, and the *eight-harness satin weave*, wherein one warp tow passes over seven fill tows and then under one fill tow.

A *twill weave* pattern in shown in Figure 1.12d. Two adjacent warp tow pass over two adjacent fill tow, forming a diagonal 2-over/2-under pattern known as a 2×2 twill. Similarly, a 4×4 twill would be based on a 4-over/4-under weave pattern.

The stiffness and strength of woven fabrics is typically less than that achieved with unidirectional fibers. This decrease is due to fiber waviness. That is, in any woven fabric the tow is required to pass over/under one (or more) neighboring tow(s) at each crossover point, resulting in pre-existing fiber waviness. Upon application of a tensile load the fibers within a ply tend to straighten, resulting in a lower stiffness than would be achieved if the ply contained initially straight unidirectional fibers. Further, due to the weave pattern the fibers are not allowed to straighten fully and are subjected to bending stresses, resulting in fiber failures at lower tensile loads than would otherwise be achieved if the ply contained unidirectional fibers. This effect is most pronounced in the case of plain weaves, since each tow passes over/under each neighboring tow. For plain weaves the through-thickness distribution of tow in the warp and fill directions is identical. Consequently, the strength and stiffness of plain weaves is usually identical in the warp and fill directions.

In contrast, for satin weaves the through-thickness distribution of tow is inherently asymmetric. Referring to Figure 1.12c, for example, for the four-harness satin weave pattern the warp tow are primarily within the "top" half of the fabric (as sketched), whereas the fill tow are primarily within the lower half. The asymmetric through-thickness distribution of tow causes a coupling between in-plane loading and bending deflections. That is, if a uniform tensile load is applied to the midplane of a single layer of a satin weave fabric, the fabric will not only stretch but will also deflect out-of-plane (i.e., bend). Similarly, the crossover points not symmetric with respect to either the warp or fill directions. This causes a coupling between in-plane loading and in-plane shear strain. That is, an in-plane shear strain is induced if a uniform tensile load is applied to a single layer of a satin weave fabric [8].

A woven fabric is in essence a 2D structure consisting of orthogonal warp and fill tows interlaced within a plane. Weaving or stitching several layers of a woven fabric together can produce a woven structure with a significant thickness. Structures produced in this fashion are called "3D weaves."

1.4.4 Braided Fabrics

Note from the preceding section that woven fabrics contain reinforcing tow in two orthogonal directions—namely, the warp and fill directions. In contrast, braided fabrics typically contain tow oriented in two (or more) nonorthogonal directions. Three common braiding patterns are shown in Figure 1.13. It is apparent from this figure that a braided fabric contains *bias* tows that intersect at a total included angle 2α. The angle α is called either the *braid angle* or the *bias angle*. Although the braid angle can be varied over a wide range, there is always some minimum and maximum possible value that depends on width of the tow and details of the braiding equipment used. Note that if $\alpha = 45°$ then the bias tow are in fact orthogonal and the braided fabric shown in Figure 1.13a,b are equivalent to a woven fabric. A braided fabric is described using the designation "$n \times n$," where n is the number of tows between crossover points. A 1×1 and 2×2 bias braided fabric is

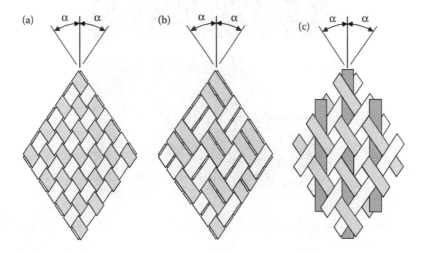

FIGURE 1.13
Some common braided fabrics used with high-performance fibers. (a) 1×1 bias braid pattern; (b) 2×2 bias braid pattern; (c) 1×1 triaxial braid pattern. (After Donnet, J.-B. et al. *Carbon Fibers*, 3rd edition, Marcel Dekker, Inc., New York, 1998, ISBN 0-8247-0172-0.)

shown in Figure 1.13a,b, respectively. A 1×1 triaxial-braided fabric is shown in Figure 1.13c. In this case, a third *axial* tow is present.

   Braided fabrics are produced in tubular form as shown in Figure 1.14. Tows are dispensed from two sets of roving spools, traveling in opposite directions on two outer circular races. The roving spools pass sequentially from one race to the other, producing the interwoven pattern that is readily apparent in Figure 1.14a. The tows are drawn toward the center of the braider and pass through a forming plate with circular opening. If a triaxial braid is to be produced then a third set of tows (not shown in Figure 1.14) is fed vertically through the center region of braider and interwoven with the bias tows. The surface texture of the braided tubes is dictated by the number and

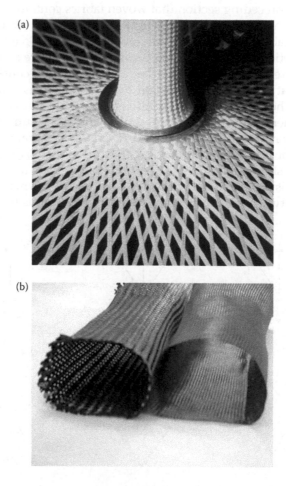

(a)

(b)

FIGURE 1.14
Braided fiber tubes. (a) Carriers passing through the central forming plate in a braider; (b) braided carbon fiber tubes of identical diameters, produced using a different number of tows and tow sizes. (Photo courtesy of A & P Technology: www.braider.com.)

size of the tows used. For example, two braided carbon tubes with the same diameter are shown in Figure 1.14b. The one on the left was produced using a 96-carrier braider and 12k carbon tow. In contrast, the tube on the right was produced using a 400-carrier braider and 1k carbon tow, resulting in a tube with substantially smoother surface texture. If desired, braided tubes may subsequently be slit lengthwise to form a flat braided fabric.

1.4.5 Pre-Impregnated Products or "Prepreg"

As is obvious from the preceding discussion, at some point during fabrication of a polymer composite the reinforcing fiber must be embedded within a polymeric matrix. One approach is to combine the fiber and resin during the manufacturing operation in which the final form of the composite structure is defined. Three manufacturing processes in which this approach is taken are filament winding (briefly described in Section 1.5.3), pultrusion (Section 1.5.4), and resin-transfer molding (Section 1.5.5).

An alternative approach is to combine the fiber and matrix in an intermediate step, resulting in an intermediate product. In this case either individual tow or a thin fabric of tow (which may be a unidirectional, woven, or braided fabric) are embedded within a polymeric matrix and delivered to the user in this form. Because the fibers have already been embedded within a polymeric matrix when delivered, the fibers are said to have been "preimpregnated" with resin, and products delivered in this condition are commonly known as "prepreg."

One method used to impregnate a large number of unidirectional tows with resin is sketched in Figure 1.15 [9]. As indicated, tows delivered from a

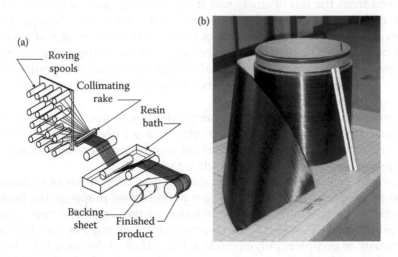

FIGURE 1.15
(a) Schematic representation of a "prepregger"; (b) a 12-in. wide roll of "prepreg tape."

large number of roving spools are arranged in a relatively narrow band. The tows are passed through a resin bath and then wound onto a roll. An inert backing sheet (also called a *scrim cloth*) is placed between layers on the roll to maintain a physical separation between layers and to aid during subsequent handling and processing. The tow/fibers are subjected to various surface pretreatments just prior to entering the resin bath. The pretreatments are proprietary but are intended to cause good wetting of the fiber by the resin, which ultimately helps ensure good adhesion between the fibers and polymer matrix in the cured composite. Products produced in this fashion are commonly known as "prepreg tape" (Figure 1.15b). Prepreg tape is available in width ranging from about 75 to 1220 mm (3–48 in.). Prepreg fabrics, produced using either woven or braided fabrics instead of unidirectional tows, are produced using similar techniques and are available in widths ranging from about 75 to 1220 mm.

A variety of fabrication methods have been developed based on the use of prepreg materials. A few such techniques will be described in Section 1.5.

The first commercially successful prepreg materials were based on B-staged epoxy resins. As discussed in Section 1.2, in the B-staged condition a thermoset resin has been partially polymerized, resulting in a relatively high viscosity, which aids in handling B-staged prepreg materials. However, prepreg must be kept at low temperatures until used, otherwise the resin continues to polymerize and slowly harden. This requires that the prepreg be shipped to the user in a refrigerated condition (for small amounts this is often accomplished using insulated shipping containers and dry ice). Further, the user must keep the stock of prepreg refrigerated until used. Typically storage temperatures are required to be −15°C (0°F) or below. In practice the prepreg material stock is removed from freezer, the amount of prepreg necessary is removed from the roll of stock, and the remaining stock is returned to the freezer. Hence, the cumulative "out-time" that a given roll of prepreg stock has experienced (i.e., total amount of time a roll has been out of a freezer) must be monitored and recorded. The need to store thermoset prepreg in a refrigerated condition and to maintain accurate records of cumulative out-time is a significant disadvantage, since these factors add significantly to the final cost of the composite structure.

Prepreg materials based on thermoplastic resins are also commercially available. In this case the polymeric matrix is a fully cured thermoplastic polymer, and hence the prepreg does not require refrigeration during shipping or storage, which is a distinct advantage.

Heat and/or pressure are applied during final fabrication of a composite based on prepreg materials. In the case of thermoset prepregs, the heat and pressure serve to complete polymerization of the polymeric resin, that is, the composite is "C-staged." For thermoplastic prepreg the objective is not necessarily to complete polymerization but rather to decrease the viscosity of the thermoplastic matrix so as to consolidate individual plies within the laminate.

1.5 Manufacturing Processes

A complete review of the many different manufacturing processes used to produce polymeric composite materials and structures is beyond the scope of this presentation. The most common manufacturing techniques will be briefly described here.

1.5.1 Layup Techniques

Many composites are produced using the tapes or fabrics discussed in the preceding section. These may be unidirectional tape, woven fabrics, or braided fabrics. These products are typically relatively thin. "Layup" simply refers to the process of stacking several layers together, much like a deck of cards. Stacking several layers together produces a laminate of significant thickness. The most direct method of producing a multi-ply composite laminate is to simply stack the desired number of layers of fabric by hand, referred to as "hand-layup" (see Figure 1.16). The layers may consist of either "dry" fabrics (i.e., fabrics which have not yet been impregnated with a resin) or prepreg materials. As will be discussed in later chapters, fiber angles are typically varied from one ply to the next, so as to insure adequate stiffness and strength in more than one direction.

Although hand-layup is simple and straightforward, it is labor intensive and, therefore, costly. It can also be very cumbersome if a large structure is being produced, such as a fuselage panel intended for use in a modern commercial aircraft. Therefore, various computer-controlled machines have been developed that automate the process of assembling the ply stack using prepreg materials. These include tow-placement and tape-placement machines (Figure 1.17). In either case, a roll (or rolls) of prepreg material is mounted on the head of a computer-controlled robot-arm or gantry.

FIGURE 1.16
Hand-layup of composite prepreg on a curved tool.

(a)    (b)

FIGURE 1.17
A computer-controlled tape-laying machine. (a) Producing a curved composite panel using the tool shown near the center of the image; (b) close-up via of the dispensing head.

The appropriate number of layers of prepreg is placed on a tool surface automatically and in the desired orientation. Although the capital costs of modern tow- or tape-placement machines may be very high, overall this approach is often less costly than hand-layup if production quantities are sufficiently high.

In the case of dry fabrics (which are usually either woven or braided fabrics), the stack must be impregnated with a low-viscosity polymeric resin following assembly of the fiber stack. Conceptually, this may be accomplished by pouring liquidous resin over the dry fiber stack, and using a squeegee or similar device to assist the resin to wet the fibers within the stack. This technique is commonly used in the recreational boat-building industry, for example. However, it is very difficult to insure uniform penetration of the resin and wetting of the fibers through the thickness of the stack, to insure that no air pockets remain trapped in the stack, and to avoid distortion of the fiber patterns while forcing resin into the fibrous assembly. There are also potential health issues associated with continually exposing workers to nonpolymerized resins. Hence, the technique of impregnating a dry fiber stack using hand-held tools such as squeegees is rarely employed in industries requiring low variability in stiffness and strength and/or high volumes, such as the aerospace or automotive industries, for example. Alternate methods of impregnating a dry fiber stack with resin have been developed, such as resin transfer molding (discussed in Section 1.5.5). These alternate techniques result in a composite with a much more uniform matrix volume fraction and almost no void content.

A major advantage of using prepreg materials, of course, is that the fibers have been impregnated with resin a priori. It is, therefore, much easier to maintain the desired matrix volume fraction and to avoid entrapped air-pockets. Further, prepreg material based on a B-staged thermoset are typically "tacky" (i.e., prepreg materials adhere to neighboring plies much like common masking tape), and hence once a given ply has been placed in the desired orientation it is less likely to move or be distorted relative to neighboring plies than is the case with dry fabrics.

1.5.2 Autoclave Process Cycles

Following layup (which may be accomplished using hand-layup, automatic tow- or tape-placement machines, or other techniques) the individual plies must be consolidated to form a solid laminate. Usually consolidation occurs through the application of pressure and heat. Whereas a simple hot press can be used for this purpose, applying pressure and heat using an *autoclave* produces highest-quality composites. An autoclave is simply a closed pressure vessel that can be used to apply a precisely controlled and simultaneous cycle of vacuum, pressure, and elevated temperature to the laminate during the consolidation process.

Although many variations exist, a typical assembly used to consolidate a laminate using an autoclave is shown in Figure 1.18. Some of the details of the assembly are as follows:

- The final shape of the composite is defined by a rigid *tool*. A simple flat tool is shown in Figure 1.18, but in practice the tool is rarely flat but instead mirrors the contour(s) desired in the final product (curved tools are shown in Figures 1.16 and 1.17, for example). Various materials may be used to produce the tool, including steel alloys, aluminum alloys, nickel alloys, ceramics, or composite materials.

- The tool surface is coated with a *release agent*. Various liquid or wax release agents are available which are either sprayed or wiped onto the surface. The purpose of the release agent is to prevent adherence between the tool and the polymeric matrix.

- A *peel ply* is placed next to both upper and lower surfaces of the composite laminate. The release ply does not develop a strong bond

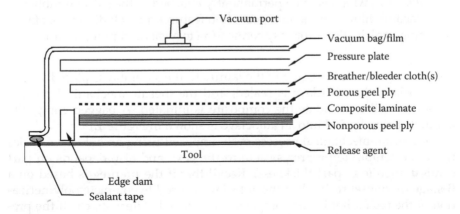

FIGURE 1.18
Typical assembly used to consolidate a polymeric composite laminate using an autoclave (expanded edge view).

to the composite, and hence can be easily removed following consolidation. The peel ply may be porous or nonporous. Porous peel plies allow resin to pass through the ply and be adsorbed by an adjacent bleeder/breather cloth (see below). Note that the surface texture of the consolidated laminate will be a mirror image of the peel ply used. For example, Teflon-coated porous glass fabrics are often used as peel plies, and these fabrics have a cloth-like surface texture. Hence, a composite laminate consolidated against such fabrics will exhibit a cloth-like surface texture as well.

- One or more layers of a *breather/bleeder cloth* is placed adjacent to the porous peel ply. The bleeder cloth has the texture of rather stiff cotton. Its purpose is to allow any gases released to be vented (hence the adjective *breather*).

- And also to adsorb any resin that passed through the porous peel ply (hence the adjective *bleeder*). The breather/bleeder is usually a glass, polyester, or jute cloth.

- An *edge dam* is placed around the periphery of the laminate. The edge dam is intended to maintain the position and resin content of the laminate edges.

- A *pressure plate* (also called a *caul plate*) is placed over the breather/bleeder cloth. The pressure plate insures a uniform distribution of pressures over the surface of the laminate.

- The entire assembly is sealed within a *vacuum film or bag*. Often this is a relatively thick (5 mm, say) layer of silicone rubber. Sealant tape is used to adhere the vacuum film to the tool surface, providing a pressure-tight seal around the periphery of the vacuum film.

- The volume within the vacuum bag is evacuated by means of a *vacuum port*, which is often permanently attached to the silicone rubber vacuum film. The vacuum port often features a quick-disconnect fitting, which allows for easy connection to a vacuum pump or line.

Following vacuum bagging of the laminate, the assembly is placed within an autoclave, the autoclave is sealed, and the thermo-mechanical process cycle that will consolidate the composite is initiated. A bagged composite laminate being loaded into an autoclave is shown in Figure 1.19.

The thermo-mechanical process cycle imposed using an autoclave varies from one composite prepreg system to the next, and also depends on part configuration (e.g., part thickness). Recall that if the prepreg is based on a B-stage thermoset resin, then the autoclave is used to complete polymerization of the resin, that is, the composite is C-staged. Alternatively, if the prepreg is based on a thermoplastic, the pressure and heat applied during the autoclave cycle softens the matrix and insures polymer flow across the ply interfaces. The laminate is then solidified upon cooling.

FIGURE 1.19
A vacuum bagged composite laminate being loaded into an autoclave.

A typical cure cycle, suitable for use with standard thermosetting resin systems such as epoxies, is as follows.

- Draw and hold a vacuum within the vacuum bag, resulting in a pressure of roughly 100 kPa (14.7 psi) applied to the laminate. The vacuum is typically maintained for about 30 min, and is intended to remove any entrapped air or volatiles, and to hold the laminate in place.
- While maintaining a vacuum, increase the temperature from room temperature to about 120°C (250°F), at a rate of about 2.8°C/min (5°F/min). Maintain this temperature for 30 min. During this 30-min dwell time any remaining air or other volatiles are removed.
- Increase the internal autoclave pressure from atmospheric to about 585 kPa (85 psi), at a rate of 21 kPa/min (3 psi/min). Release vacuum when autoclave pressure reaches 138 kPa (20 psi).
- Increase temperature from 120°C to 175°C (350°F), at a rate of about 2.8°C/min (5°F/min). Maintain at 175°C for 2 h. Polymerization of the thermosetting resin matrix is completed during this 2-h dwell.
- Cool to room temperature at a rate of about 2.8°C/min (5°F/min), release autoclave pressure, and remove cured laminate from the autoclave.

The polymerization process for many thermosetting resins, including epoxies, is an exothermic reaction. Consequently an inert gas (usually nitrogen) is used as the pressuring medium in most autoclave cure cycles, to avoid initiation of a fire within the autoclave chamber.

Process cycles used with thermoplastic prepregs are similar, except that higher temperatures (500°C or higher) are usually involved.

1.5.3 Filament Winding

Filament winding is an automated process in which tow is wound onto a mandrel at controlled position and orientation. A filament winder being used to produce a small pressure vessel is shown in Figure 1.20. During operation, the mandrel rotates about its axis, and a fiber carriage simultaneously moves in a controlled manner along the length of the mandrel. The angle at which fibers are placed on the mandrel surface is a function of the mandrel diameter, rate of mandrel rotation, and translational speed of the fiber carriage.

If dry tows are used, then the tow must pass through a liquid resin bath before being wound onto the mandrel. In this case the process is often referred to as "wet" filament winding. Often fiber tension provides sufficient compaction of the laminate, and so no additional external pressure is required. If a thermosetting resin that cures at room temperature is used, following completion of the winding operation the structure is simply left in the winder until polymerization is complete.

Of course, if prepreg tow is used then the tow is already impregnated with a resin and is not passed through a resin bath. This process is called "dry" filament winding. In this case heat and pressure are normally required to complete polymerization of the resin (in the case of a thermosetting polymer matrix) or to cause resin flow and consolidation (in the case of a thermoplastic polymer matrix). The appropriate heat and pressure are usually applied using an autoclave.

For simple wound shapes with open ends (such as cylindrical tubes) the mandrel is usually a simple solid cylinder whose surface has been coated with a release agent. In this case, the mandrel is forced out of the internal cavity after consolidation of the composite. Mandrel design and configuration become more complex when a shape with restricted openings at the

FIGURE 1.20
A filament winder being used to produce a pressure vessel.

ends is produced (such as the pressure tank shown in Figure 1.20). In these cases, the mandrel must somehow be removed after the part is consolidated. Several different types of mandrel designs are used in these cases, including

- *Soluble mandrels,* which are made from a material which can be dissolved in some fashion after the cure process is complete. In this approach the mandrel is cast and machined to the desired shape, the composite part is filament wound over the mandrel, the part is cured, and the mandrel is then simply dissolved. The wall of the composite structure must obviously have at least one opening, such that the dissolved (and now liquidous) mandrel material can be drained from the internal cavity. Soluble mandrels can be made from metallic alloys with suitably low-melting temperatures, eutectic salts, sand with water soluble binders, or various plasters.
- *Removable (or collapsible) mandrels,* which resemble giant 3D puzzles. That is, the entire mandrel can be taken apart piece-by-piece. The composite structure being wound must have at least one wall opening, which allows the mandrel pieces to be removed from the internal cavity after cure. Obviously, the mandrel is designed such that no single piece is larger than the available opening(s).
- *Inflatable mandrels,* which take on the desired shape when pressurized and then are simply deflated and removed after winding and consolidation.
- *Metal or polymer liners,* which are actually a modification of the inflatable mandrel concept. Liners can be described as metal or polymer "balloons," and remain in the filament wound vessel after cure. The liner does not contribute significantly to the strength or stiffness of the structure. In fact, the wall thickness of the liner is often so small that an internal pressure must be applied to the liner during the winding process to avoid buckling of the liner wall. Metal liners are almost always used in composite pressure vessels, where allowable leakage rates are very low, or in filament wound chemical storage tanks where corrosive liquids are stored.

1.5.4 Pultrusion

Pultrusion is a fabrication process in which continuous tows or fabrics impregnated with resin are pulled through a forming die, as shown schematically in Figure 1.21 [10]. If dry tow or fabric is used then the tow/fabric must pass through a resin bath prior to entering the forming die. In this case, the process is called "wet" pultrusion. If prepreg material is used then there is no need for a resin bath and the process is called "dry" pultrusion. The

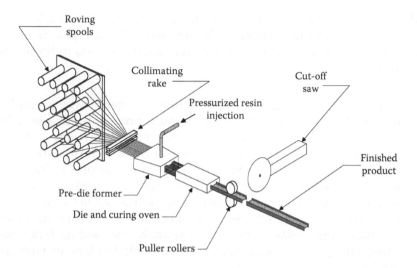

FIGURE 1.21
Sketch of a typical pultruder.

cross-sectional shape is defined by the die and is, therefore, constant along the length of the part. The principal attraction of pultrusion is that very high production rates are possible, as compared with other composite manufacturing techniques. Finished pultruded parts with various cross-sections are shown in Figure 1.22.

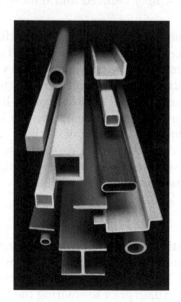

FIGURE 1.22
Finished pultruded parts.

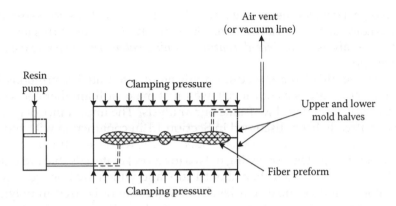

FIGURE 1.23
Summary of the resin-transfer molding process.

1.5.5 Resin Transfer Molding

In the resin transfer molding (RTM) process a dry fiber *preform* is placed within a cavity formed between two matched metal molds, as shown in schematically in Figure 1.23. The dry preform is often produced by stitching together woven or braided fabrics. Liquidous resin is forced into the cavity under pressure via a port located in the upper or lower mold halves. Air originally within the internal cavity (or other gases that evolve during cure of the resin) is allowed to escape via one or more air vents. Alternatively, a vacuum pump may be used to evacuate the internal cavity, which also assists in drawing the resin into the cavity. When a vacuum is used the process is one form of "vacuum-assisted resin transfer molding" (VARTM). Both the upper and the lower molds must be sufficiently rigid so as to resist the internal pressures applied and to maintain the desired shape of the internal cavity. Usually the closed molds are placed within a press, which provides a clamping pressure to assist in keeping the molds closed.

1.6 Scope of This Book

A broad overview of modern composite materials and manufacturing processes has been provided in preceding sections. It should be clear from this discussion that composite material systems is a multidisciplinary subject, involving topics drawn from polymer chemistry, fiber science, surface chemistry and adhesion, materials testing, structural analysis, and manufacturing techniques, to name a few. It is simply not possible to cover all of these topics in any depth in a single textbook. Accordingly, the material presented in this book represents a small fraction of the scientific and technological developments that have ultimately led to the successful use of modern composite material systems. Specifically, the focus of this text is the *structural* analysis of *laminated, continuous-fiber polymeric composite materials, and structures.*

Having identified the structural analysis of laminated continuous-fiber polymeric composites as our focus, we must make still another decision: at what *physical scale* should we frame our analysis? The importance of physical scale has already been discussed in Section 1.1 in conjunction with the very definition of a "composite material." Specifically, we have defined a composite as a material system consisting of two (or more) materials that are distinct at a physical scale greater than about 1 μm, and which are bonded together at the atomic and/or molecular levels. Fibers commonly used in polymeric composites possess diameters ranging from about 5 to 40 μm (Section 1.3). Therefore, we could perform a structural analysis at a physical scale comparable to the fiber diameter. Alternatively, laminated polymeric composites consist of well-defined layers (called plies) of fibers embedded in a polymeric matrix. The thickness of these layers range from about 0.125 to 0.250 mm (Section 1.4). We could, therefore, elect to begin a structural analysis at a physical scale comparable to the thickness of a single ply.

A distinction is drawn between structural analyses that begin at these two different physical scales. Analyses that are framed at a physical scale corresponding to the fiber diameter (or below) are classified as *micromechanics* analyses, whereas those framed at a physical scale corresponding to a single ply thickness (or above) are classified as *macromechanics* analyses. This distinction is comparable to the traditional distinction between *metallurgy* and *continuum mechanics*. That is, metallurgy typically involves the study of the crystalline nature of metals and metal alloys, and is, therefore, framed at a physical scale roughly corresponding to atomic dimensions. A metallurgist might attempt to predict Young's modulus* of a given metal alloy, based on knowledge of the constituent atoms and crystalline structure present in the alloy, for example. In contrast, continuum mechanics is formulated at a much larger physical scale, such that the existence of individual atoms is not perceptible. In continuum mechanics a metal or metal alloy is said to be "homogeneous," even though it actually consists of several different atomic species. A structural engineer wishing to apply a solution based on continuum mechanics would simply measure Young's modulus exhibited by the metal alloy of interest, rather than trying to predict it based on knowledge of the atomic crystalline structure.

In much the same way, composite micromechanics analyses are concerned with the predicting properties of composites based on the particular fiber and matrix materials involved, the spacing and orientation of the fibers, the adhesion (or lack thereof) between fiber and matrix, etc. For example,

---

* The definition of various material properties of interest to the structural engineer, such as Young's modulus, will be reviewed and discussed in greater detail in Chapter 3.

suppose that a unidirectional graphite-epoxy composite is to be produced by combining graphite fibers with a known Young's modulus (E_f) and an epoxy matrix with a known Young's modulus (E_m). An analysis framed at a physical scale corresponding to the fiber diameter, that is, a micromechanics analysis, is required to predict the Young's modulus that will be exhibited by the composite (E_c) formed using these two constituents.

In contrast, composite macromechanics analyses are framed at a physical scale corresponding to the ply thickness (or above). The existence or properties of individual fibers or the matrix material is not recognized (in a mathematical sense) in a macromechanics analysis. Instead, the ply is treated as a homogeneous layer whose properties are identical at all points, although they differ in different directions. Details of fiber or matrix type, fiber spacing, fiber orientation, etc., are represented in a macromechanics analysis only indirectly, via properties defined for the composite ply as a whole, rather than as properties of the individual constituents.

Finally then, the scope of this book is: macromechanics-based structural analysis of laminated, continuous-fiber, polymeric composites.

Micromechanics-based structural analyses will not be discussed in any detail. A simple micromechanics model that may be used to predict ply stiffness based on knowledge of fiber and matrix properties, called the *rule-of-mixtures*, will be developed in Section 3.6. However, the material devoted to micromechanics in this text is abbreviated and does not do justice to the many advances made in this area. The lack of emphasis on micromechanical topics is not meant to imply that such analyses are unimportant. Quite the contrary, micromechanics analyses are crucial during development of new composite material systems, since it is only through a detailed understanding of the behavior of composites at this physical scale that new and improved materials can be created. Micromechanics has been minimized herein simply due to space restrictions. The reader interested in learning more about micromechanics is referred to several excellent texts that cover this topic in greater detail, a few of which are [11–14].

References

1. *ASM Metals Reference Book*, 2nd edition, American Society of Metals, Metals Park, OH, 1983, ISBN 0-87170-156-1.
2. Rodriguez, F., *Principles of Polymer Systems*, 3rd edition, Hemisphere Pub. Co., New York, 1989, ISBN 0-89116-176-7.
3. Young, R.J., and Lovell, P.A., *Introduction to Polymers*, 2nd edition, Chapman & Hall Pub. Co., New York, 1989, ISBN 0-412-30630-1.
4. Strong, A.B., *Plastics: Materials and Processing*, 3rd edition, Prentice-Hall, Upper Saddle River, NJ, 2006, ISBN 0-13-021626-7.

5. *The Macrogalleria*, a website maintained by personnel at the University of Southern Mississippi and devoted to polymeric materials: http://pslc.ws/macrog/index.htm.
6. ASTM Standard C162-05 2010, *Standard Terminology of Glass and Glass Products*, ASTM International, West Conshohocken, PA, 2003, 10.1520/C0162-05R10, www.astm.org.
7. Donnet, J.-B., Wang, T.K., Peng, J.C.M., and Reboillat, M., *Carbon Fibers*, 3rd edition, Marcel Dekker, Inc., New York, 1998, ISBN 0-8247-0172-0.
8. Cox, B., and Flanagan, G., *Handbook of Analytical Methods for Textile Composites*, NASA Contractor Report 4750, NASA-Langley Res. Ctr., 1997.
9. Kalpakjian, S., *Manufacturing Processes for Engineering Materials*, 3rd edition, Addison-Wesley Longman, Inc., Menlo Park, CA, 1997, ISBN 0-201-82370-5.
10. Schwartz, M.M., *Composite Materials Handbook*, McGraw-Hill Book Co., New York, 1983, ISBN 0-07-055743-8.
11. Hyer, M.W., *Stress Analysis of Fiber-Reinforced Composite Materials*, McGraw-Hill Book Co., New York, 1998, ISBN 0-07-016700-1.
12. Herakovich, C.T., *Mechanics of Fibrous Composites*, John Wiley & Sons, New York, 1998, ISBN 0-471-10636-4.
13. Jones, R.M., *Mechanics of Composite Materials*, McGraw-Hill Book Co., New York, 1975, ISBN 0-07-032790-4.
14. Hull, D., *An Introduction to Composite Materials*, Cambridge University Press, Cambridge, UK, 1981. ISBN 0-521-23991-5.

2

Review of Force, Stress, and Strain Tensors

2.1 The Force Vector

Forces can be grouped into two broad categories: *surface* forces and *body* forces. Surface forces are those that act over a surface (as the name implies), and result from direct physical contact between two bodies. In contrast, body forces are those that act at a distance, and do not result from direct physical contact of one body with another. The force of gravity is the most common type of body force. In this chapter we are primarily concerned with surface forces, the effects of body forces (such as the weight of a structure) will be ignored.

A force is a three-dimensional (3D) vector. A force is defined by a *magnitude* and a *line of action*. In SI units, the magnitude of a force is expressed in *Newtons*, abbreviated N, whereas in English units the magnitude of a force is expressed in *pounds-force*, abbreviated lbf. A force vector $\overline{F}$ acting at a point P and referenced to a right-handed x–y–z coordinate system is shown in Figure 2.1. Components of $\overline{F}$ acting parallel to the x–y–z coordinate axes, F_x, F_y, and F_z, respectively, are also shown in the figure. The algebraic sign of each force component is defined in accordance with the algebraically positive direction of the corresponding coordinate axis. All force components shown in Figure 2.1 are algebraically positive, as each component "points" in the corresponding positive coordinate direction.

The reader is likely to have encountered several different ways of expressing force vectors in a mathematical sense. Three methods will be described here. The first is called *vector notation*, and involves the use of unit vectors. Unit vectors parallel to the x-, y-, and z-coordinate axes are typically labeled $\hat{i}$, $\hat{j}$, and $\hat{k}$, respectively, and by definition have a magnitude equal to unity. A force vector $\overline{F}$ is written in vector notation as follows:

$$\overline{F} = F_x\hat{i} + F_y\hat{j} + F_z\hat{k} \tag{2.1}$$

The magnitude of the force is given by

$$|\overline{F}| = \sqrt{F_x^2 + F_y^2 + F_z^2} \tag{2.2}$$

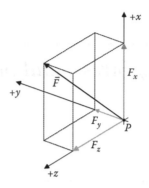

FIGURE 2.1
A force vector $\bar{F}$ acting at point P. Force components F_x, F_y, and F_z acting parallel to the x–y–z coordinate axes, respectively, are also shown.

A second method of defining a force vector is through the use of *indicial notation*. In this case, a subscript is used to denote individual components of the vectoral quantity:

$$\bar{F} = (F_x, F_y, F_z)$$

The subscript denotes the coordinate direction of each force component. One of the advantages of indicial notation is that it allows a shorthand notation to be used, as follows:

$$\bar{F} = F_i, \quad \text{where } i = x, y, \text{ or } z \tag{2.3}$$

Note that a *range* has been explicitly specified for the subscript "i" in Equation 2.3. That is, it is explicitly stated that the subscript i may take on values of x, y, or z. Usually, however, the range of a subscript(s) is not stated explicitly but rather is implied. For example, Equation 2.3 is normally written simply as

$$\bar{F} = F_i$$

where it is understood that the subscript i takes on values of x, y, and z.

The third approach is called *matrix notation*. In this case, individual components of the force vector are listed within braces in the form of a column array:

$$\bar{F} = \left\{ \begin{array}{c} F_x \\ F_y \\ F_z \end{array} \right\} \tag{2.4}$$

Indicial notation is sometimes combined with matrix notation as follows:

$$\bar{F} = \{F_i\} \tag{2.5}$$

2.2 Transformation of a Force Vector

One of the most common requirements in the study of mechanics is the need to describe a vector in more than one coordinate system. For example, suppose all components of a force vector F_i are known in one coordinate system (the x–y–z coordinate system, say) and it is desired to express this force vector in a second coordinate system (the x'–y'–z' coordinate system, say). To describe the force vector in the new coordinate system, we must calculate the components of the force parallel to the x', y', and z' axes—that is, we must calculate $F_{x'}$, $F_{y'}$, and $F_{z'}$. The process of relating force components in one coordinate system to those in another coordinate system is called *transformation* of the force vector. This terminology is perhaps unfortunate, in the sense that the force vector itself is not "transformed" but rather our *description* of the force vector transforms as we change from one coordinate system to another.

It can be shown [1,2] that the force components in the x'–y'–z' coordinate system ($F_{x'}$, $F_{y'}$, and $F_{z'}$) are related to the components in x–y–z coordinate system (F_x, F_y, F_z) according to:

$$F_{x'} = c_{x'x}F_x + c_{x'y}F_y + c_{x'z}F_z$$

$$F_{y'} = c_{y'x}F_x + c_{y'y}F_y + c_{y'z}F_z \tag{2.6a}$$

$$F_{z'} = c_{z'x}F_x + c_{z'y}F_y + c_{z'z}F_z$$

The terms c_{ij} that appear in Equation 2.6a are called *direction cosines*, and are to equal the cosine of the angle between the axes of the new and original coordinate systems. An angle of rotation is defined *from* the original x–y–z coordinate system *to* the new x'–y'–z' coordinate system. The algebraic sign of the angle of rotation is defined in accordance with the right-hand rule.

Equation 2.6a can be succinctly written using the summation convention as follows:

$$F_{i'} = c_{i'j}F_j \tag{2.6b}$$

Alternatively, these three equations can be written using matrix notation as

$$\begin{Bmatrix} F_{x'} \\ F_{y'} \\ F_{z'} \end{Bmatrix} = \begin{bmatrix} c_{x'x} & c_{x'y} & c_{x'z} \\ c_{y'x} & c_{y'y} & c_{y'z} \\ c_{z'x} & c_{z'y} & c_{z'z} \end{bmatrix} \begin{Bmatrix} F_x \\ F_y \\ F_z \end{Bmatrix} \tag{2.6c}$$

Note that although values of individual force components vary as we change from one coordinate system to another, the magnitude of the force vector (given by Equation 2.2) does not. The magnitude is independent of the coordinate system used, and is called an *invariant* of the force tensor.

Direction cosines relate unit vectors in the "new" and "old" coordinate systems. For example, a unit vector directed along the x'-axis (i.e., unit vector $\hat{i}\,'$) is related to unit vectors in the x–y–z coordinate system as follows:

$$\hat{i}\,' = c_{x'x}\,\hat{i} + c_{x'y}\,\hat{j} + c_{x'z}\,\hat{k} \tag{2.7}$$

As $\hat{i}\,'$ is a unit vector, then in accordance with Equation 2.2:

$$(c_{x'x})^2 + (c_{x'y})^2 + (c_{x'z})^2 = 1 \tag{2.8}$$

To this point we have referred to a force as a *vector*. A force vector can also be called a force *tensor*. The term "tensor" refers to any quantity that transforms in a physically meaningful way from one Cartesian coordinate system to another. The *rank* of a tensor equals the number of subscripts that must be used to describe the tensor. A force can be described using a single subscript, F_i, and therefore a force is said to be a *tensor of rank one*, or equivalently, *a first-order tensor*. Equation 2.6 is called the *transformation law for a first-order tensor*.

It is likely that the reader is already familiar with two other tensors: the *stress tensor*, σ_{ij}, and the strain tensor, ε_{ij}. The stress and strain tensors will be reviewed later in this chapter, but at this point it can be noted that *two* subscripts are used to describe stress and strain tensors. Hence, stress and strain tensors are said to be *tensors of rank two*, or equivalently, *second-order tensors*.

Example Problem 2.1

Given: All components of a force vector $\overline{F}$ are known in a given x–y–z coordinate system. It is desired to express this force in a new x''–y''–z'' coordinate system, where the x''–y''–z'' system is generated from the original x–y–z system by the following two rotations (see Figure 2.2):

- A rotation of θ-degrees about the original z-axis (which defines an intermediate x'–y'–z' coordinate system), followed by
- A rotation of β-degrees about the x'-axis (which defines the final x''–y''–z'' coordinate system)

PROBLEM

a. Determine the direction cosines $c_{i''j}$ relating the original x–y–z coordinate system to the new x''–y''–z'' coordinate system.
b. Obtain a general expression for the force vector $\overline{F}$ in the x''–y''–z'' coordinate system.

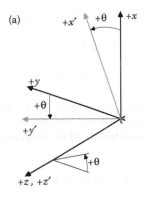

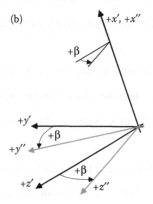

FIGURE 2.2
Generation of the $x''-y''-z''$ coordinate system from the $x-y-z$ coordinate system. (a) Rotation of θ-degrees about the original z-axis (which defines an intermediate $x'-y'-z'$ coordinate system); (b) rotation of β-degrees about the x'-axis (which defines the final $x''-y''-z''$ coordinate system).

    c. Calculate numerical values of the force vector $\bar{F}$ in the $x''-y''-z''$ coordinate system if $\theta = -20°$, $\beta = 60°$, and $F_x = 1000$ N, $F_y = 200$ N, $F_z = 600$ N.

SOLUTION

    a. One way to determine direction cosines $c_{i''j}$ is to rotate unit vectors. In this approach unit vectors are first rotated from the original $x-y-z$ coordinate system to the intermediate $x'-y'-z'$ coordinate system, and then from the $x'-y'-z'$ system to the final $x''-y''-z''$ coordinate system.

        Define a unit vector $\bar{I}$ that is aligned with the x-axis:

$$\bar{I} \equiv (1)\hat{i}$$

That is, vector $\bar{I}$ is a vector for which $I_x = 1$, $I_y = 0$, and $I_z = 0$. The vector $\bar{I}$ can be rotated to the intermediate $x'-y'-z'$ coordinate system using Equation 2.6:

$$I_{x'} = c_{x'x}I_x + c_{x'y}I_y + c_{x'z}I_z$$

$$I_{y'} = c_{y'x}I_x + c_{y'y}I_y + c_{y'z}I_z$$

$$I_{z'} = c_{z'x}I_x + c_{z'y}I_y + c_{z'z}I_z$$

The direction cosines associated with a transformation from the x–y–z coordinate system to the intermediate x'–y'–z' coordinate system can be determined by inspection (see Figure 2.2a), and are given by

$c_{x'x}$ = cosine(angle between x'- and x-axes) = $\cos\theta$

$c_{x'y}$ = cosine(angle between x'- and y-axes) = $\cos(90° - \theta)$ = $\sin\theta$

$c_{x'z}$ = cosine(angle between x'- and z-axes) = $\cos(90°)$ = 0

$c_{y'x}$ = cosine(angle between the y'- and x-axes) = $\cos(90° + \theta)$ = $-\sin\theta$

$c_{y'y}$ = cosine(angle between the y'- and y-axes) = $\cos\theta$

$c_{y'z}$ = cosine(angle between the y'- and z-axes) = $\cos(90°)$ = 0

$c_{z'x}$ = cosine(angle between the z'- and x-axes) = $\cos(90°)$ = 0

$c_{z'y}$ = cosine(angle between the z'- and y-axes) = $\cos(90°)$ = 0

$c_{z'z}$ = cosine(angle between the z'- and z-axes) = $\cos(0°)$ = 1

Using these direction cosines:

$$I_{x'} = c_{x'x}I_x + c_{x'y}I_y + c_{x'z}I_z = (\cos\theta)(1) + (\sin\theta)(0) + (0)(0) = \cos\theta$$

$$I_{y'} = c_{y'x}I_x + c_{y'y}I_y + c_{y'z}I_z = (-\sin\theta)(1) + (\cos\theta)(0) + (0)(0) = -\sin\theta$$

$$I_{z'} = c_{z'x}I_x + c_{z'y}I_y + c_{z'z}I_z = (0)(1) + (0)(0) + (1)(0) = 0$$

Therefore, in the x'–y'–z' coordinate system, the vector $\bar{I}$ is written:

$$\bar{I} = (\cos\theta)\hat{i}' + (-\sin\theta)\hat{j}'$$

Now define two additional unit vectors, one aligned with the original y-axis (vector $\bar{J}$) and one aligned with the original z-axis (vector $\bar{K}$); that is, let $\bar{J} = (1)\hat{j}$ and $\bar{K} = (1)\hat{k}$. Transforming these vectors to the x'–y'–z' coordinate system, again using the direction cosines listed above, results in

$$\bar{J} = (\sin\theta)\hat{i}' + (\cos\theta)\hat{j}'$$

$$\overline{K} = (1)\,\hat{k}'$$

We now rotate vectors $\overline{I}$, $\overline{J}$ and $\overline{K}$ from the *intermediate x'–y'–z'* coordinate system to the *final x"–y"–z"* coordinate system. The direction cosines associated with a transformation from the x'–y'–z' coordinate system to the final x"–y"–z" coordinate system are easily determined by inspection (see Figure 2.2b) and are given by

$$c_{x''x'} = 1 \quad c_{x''y'} = 0 \qquad c_{x''z'} = 0$$

$$c_{y''x'} = 0 \quad c_{y''y'} = \cos\beta \quad c_{y''z'} = \sin\beta$$

$$c_{z''x'} = 0 \quad c_{z''y'} = -\sin\beta \quad c_{z''z'} = \cos\beta$$

These direction cosines together with Equation 2.6 can be used to rotate the vector $\overline{I}$ from the intermediate x'–y'–z' coordinate system to the final x"–y"–z" coordinate system:

$$I_{x''} = c_{x''x'}I_{x'} + c_{x''y'}I_{y'} + c_{x''z'}I_{z'} = (1)(\cos\theta) + (0)(-\sin\theta) + (0)(0)$$

$$I_{x''} = \cos\theta$$

$$I_{y''} = c_{y''x'}I_{x'} + c_{y''y'}I_{y'} + c_{y''z'}I_{z'} = (0)(\cos\theta) + (\cos\beta)(-\sin\theta) + (\sin\beta)(0)$$

$$I_{y''} = -\cos\beta\,\sin\theta$$

$$I_{z''} = c_{z''x'}I_{x'} + c_{z''y'}I_{y'} + c_{z''z'}I_{z'} = (0)(\cos\theta) + (-\sin\beta)(-\sin\theta) + (\cos\beta)(0)$$

$$I_{z''} = \sin\beta\,\sin\theta$$

Therefore, in the final x"–y"–z" coordinate system, the vector $\overline{I}$ is written:

$$\overline{I} = (\cos\theta)\hat{i}'' + (-\cos\beta\sin\theta)\hat{j}'' + (\sin\beta\,\sin\theta)\hat{k}'' \qquad \text{(a)}$$

Recall that in the original x–y–z coordinate system $\overline{I}$ is simply a unit vector aligned with the original x-axis: $\overline{I} \equiv (1)\,\hat{i}$. Therefore, result (a) defines the direction cosines associated with the angle between the original x-axis and the final x"-, y"-, and z"-axes. That is

$$c_{x''x} = \cos\theta$$

$$c_{y''x} = -\cos\beta\,\sin\theta$$

$$c_{z''x} = \sin\beta\,\sin\theta$$

A similar procedure is used to rotate the unit vectors $\bar{J}$ and $\bar{K}$ from the intermediate x'–y'–z' coordinate system to the final x''–y''–z'' coordinate system. These rotations result in

$$\bar{J} = (\sin\,\theta)\hat{i}'' + (\cos\,\beta\cos\,\theta)\hat{j}'' + (-\sin\,\beta\,\cos\,\theta)\hat{k}'' \qquad \text{(b)}$$

$$\bar{K} = (0)\hat{i}'' + (\sin\,\beta)\hat{j}'' + (\cos\,\beta)\hat{k}'' \qquad \text{(c)}$$

As vector $\bar{J}$ is a unit vector aligned with the original y-axis, $\bar{J} = (1)\hat{j}$, result (b) defines the direction cosines associated with the angle between the original y-axis and the final x''-, y''-, and z''-axes:

$$c_{x''y} = \sin\,\theta$$

$$c_{y''y} = \cos\,\beta\,\cos\,\theta$$

$$c_{z''y} = -\sin\,\beta\,\cos\,\theta$$

Finally, result (c) defines the direction cosines associated with the angle between the original z-axis and the final x''-, y''-, and z''-axes:

$$c_{x''z} = 0$$

$$c_{y''z} = \sin\,\beta$$

$$c_{z''z} = \cos\,\beta$$

Assembling the preceding results, the set of direction cosines relating the original x–y–z coordinate system to the final x''–y''–z'' coordinate system can be written:

$$
\begin{bmatrix} c_{x''x} & c_{x''y} & c_{x''z} \\ c_{y''x} & c_{y''y} & c_{y''z} \\ c_{z''x} & c_{z''y} & c_{z''z} \end{bmatrix} = \begin{bmatrix} \cos\theta & \sin\theta & 0 \\ -\cos\beta\,\sin\theta & \cos\beta\,\cos\theta & \sin\beta \\ \sin\beta\,\sin\theta & -\sin\beta\,\cos\theta & \cos\beta \end{bmatrix}
$$

b. As direction cosines have been determined, transformation of force vector $\bar{F}$ can be accomplished using any version of Equation 2.6. For example, using matrix notation, Equation 2.6c:

$$
\begin{Bmatrix} F_{x''} \\ F_{y''} \\ F_{z''} \end{Bmatrix} = \begin{bmatrix} c_{x''x} & c_{x''y} & c_{x''z} \\ c_{y''x} & c_{y''y} & c_{y''z} \\ c_{z''x} & c_{z''y} & c_{z''z} \end{bmatrix} \begin{Bmatrix} F_x \\ F_y \\ F_z \end{Bmatrix}
$$

$$
= \begin{bmatrix} \cos\theta & \sin\theta & 0 \\ -\cos\beta\,\sin\theta & \cos\beta\,\cos\theta & \sin\beta \\ \sin\beta\,\sin\theta & -\sin\beta\,\cos\theta & \cos\beta \end{bmatrix} \begin{Bmatrix} F_x \\ F_y \\ F_z \end{Bmatrix}
$$

$$\begin{Bmatrix} F_{x''} \\ F_{y''} \\ F_{z''} \end{Bmatrix} = \begin{Bmatrix} (\cos\theta)F_x + (\sin\theta)F_y \\ (-\cos\beta\sin\theta)F_x + (\cos\beta\cos\theta)F_y + (\sin\beta)F_z \\ (\sin\beta\sin\theta)F_x + (-\sin\beta\cos\theta)F_y + (\cos\beta)F_z \end{Bmatrix}$$

c. Using the specified numerical values and the results of part (b):

$$\begin{Bmatrix} F_{x''} \\ F_{y''} \\ F_{z''} \end{Bmatrix} = \begin{Bmatrix} [\cos(-20°)](1000\ N) + [\sin(-20°)]200\ N \\ [-\cos(60°)(\sin -20°)](1000\ N) + [\cos(60°)\cos(-20°)](200\ N) \\ + [\sin(60°)](600\ N) \\ [\sin(60°)\sin(-20°)](1000\ N) + [-\sin(60°)\cos(-20°)](200\ N) \\ + (\cos60°)(600\ N) \end{Bmatrix}$$

$$\begin{Bmatrix} F_{x''} \\ F_{y''} \\ F_{z''} \end{Bmatrix} = \begin{Bmatrix} 873.1\ N \\ 784.6\ N \\ -159.0\ N \end{Bmatrix}$$

Using vector notation, $\overline{F}$ can now be expressed in the two different coordinate systems as

$$\overline{F} = (1000\ N)\hat{i} + (200\ N)\hat{j} + (600\ N)\hat{k}$$

or, equivalently

$$\overline{F} = (873.1\ N)\hat{i}'' + (784.6\ N)\hat{j}'' + (-159.0\ N)\hat{k}''$$

Where $\hat{i}, \hat{j}, \hat{k}$ and $\hat{i}'', \hat{j}'', \hat{k}''$ are unit vectors in the x–y–z and x''–y''–z'' coordinate systems, respectively. Force vector $\overline{F}$ drawn in the x–y–z and x''–y''–z'' coordinate systems is shown in Figure 2.3a and b, respectively. The two descriptions of $\overline{F}$ are entirely equivalent. A convenient way of (partially) verifying this equivalence is to calculate the magnitude of the original and transformed force vectors. As the magnitude is an invariant, it is independent of the coordinate system used to describe the force vector. Using Equation 2.2, the magnitude of the force vector in the x–y–z coordinate system is

$$|\overline{F}| = \sqrt{F_x^2 + F_y^2 + F_z^2} = \sqrt{(1000\,N)^2 + (200\,N)^2 + (600\,N)^2} = 1183\ N$$

The magnitude of the force vector in the x''–y''–z'' coordinate system is

$$|\overline{F}| = \sqrt{(F_{x''})^2 + (F_{y''})^2 + (F_{z''})^2} = \sqrt{(871.3\ N)^2 + (784.6\ N)^2 + (-159.0\ N)^2}$$

$$= 1183\ N\ \ (\text{agrees})$$

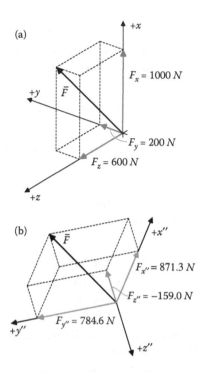

FIGURE 2.3
Force vector $\bar{F}$ drawn in two different coordinate systems. (a) Force vector $\bar{F}$ in the original x–y–z coordinate system; (b) force vector $\bar{F}$ in a new x''–y''–z'' coordinate system.

2.3 Normal Forces, Shear Forces, and Free-Body Diagrams

Force $\bar{F}$ acting at an angle to a planar surface is shown in Figure 2.4. As force is a vector, it can always be decomposed into two force components, a *normal* force component and a *shear* force component. The line-of-action of the normal force component is orthogonal to the surface, whereas the line-of-action of the shear force component is tangent to the surface.

Internal forces induced within a solid body by externally applied forces can be investigated with the aid of *free-body diagrams*. A simple example is shown in Figure 2.5, which shows a straight circular rod with constant diameter subjected to two external forces of equal magnitude (R) but opposite direction. The internal force ($\bar{F}_I$, say) induced at any cross-section of the rod can be investigated by making an imaginary cut along the plane of interest. Suppose an imaginary cut is made along plane a–a–a–a, which is perpendicular to the axis of the rod. The resulting free-body diagram for the lower half of the rod is shown in Figure 2.5a, where a x–y–z coordinate system has been assigned such that the x-axis is parallel to the rod axis, as shown. On the basis of this free-body diagram, it is concluded that an internal force $\bar{F}_I = (R)\hat{i} + (0)\hat{j} + (0)\hat{k}$ is induced at cross-section a–a. That

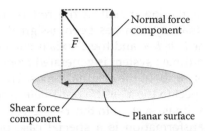

FIGURE 2.4
A force $\bar{F}$ acting at an angle to a planar surface.

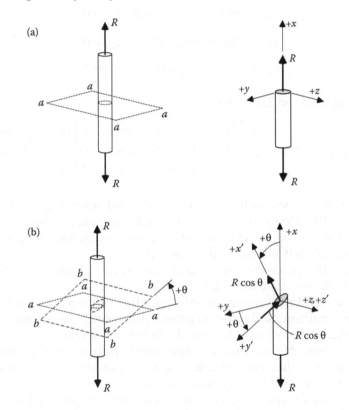

FIGURE 2.5
The use of free-body diagrams to determine internal forces acting on planes *a–a–a–a* and *b–b–b–b*. (a) Free-body diagram based on plane *a–a–a–a*, perpendicular to rod axis; (b) free-body diagram based on plane *b–b–b–b*, inclined at angle +θ to the rod axis.

is, only a normal force of magnitude R is induced at cross-section *a–a–a–a*, which has been defined to be perpendicular to the axis of the rod.

In contrast, the imaginary cut need not be made perpendicular to the axis of the rod. Suppose the imaginary cut is made along plane *b–b–b–b*, which is inclined at an angle of +θ with respect to the axis of the rod. The resulting free-body diagram for the lower half of the rod is shown in Figure 2.5b. A new x'–y'–z' coordinate system has been assigned so that the x'-axis is perpendicular to plane b–b–b–b and the z'-axis is coincident with the z-axis, that is, the x'–y'–z' coordinate system is generated from the x–y–z coordinate system by a rotation of $+\theta$ about the original z-axis. The internal force $\bar{F}_I$ can be expressed with respect to the x'–y'–z' coordinate system by transforming $\bar{F}_I$ from the x–y–z coordinate system to the x'–y'–z' coordinate system.

This coordinate transformation is a special case of the transformation considered in Example Problem 2.1. The direction cosines now become (with $\beta = 0°$):

$$c_{x'x} = \cos\theta \qquad c_{x'y} = \sin\theta \qquad c_{x'z} = 0$$

$$c_{y'x} = -\sin\theta \qquad c_{y'y} = \cos\theta \qquad c_{y'z} = 0$$

$$c_{z'x} = 0 \qquad c_{z'y} = 0 \qquad c_{z'z} = 1$$

Applying Equation 2.6, we have

$$\begin{Bmatrix} F_{x'} \\ F_{y'} \\ F_{z'} \end{Bmatrix} = \begin{bmatrix} c_{x'x} & c_{x'y} & c_{x'z} \\ c_{y'x} & c_{y'y} & c_{y'z} \\ c_{z'x} & c_{z'y} & c_{z'z} \end{bmatrix} \begin{Bmatrix} F_x \\ F_y \\ F_z \end{Bmatrix} = \begin{bmatrix} \cos\theta & \sin\theta & 0 \\ -\sin\theta & \cos\theta & 0 \\ 0 & 0 & 1 \end{bmatrix} \begin{Bmatrix} R \\ 0 \\ 0 \end{Bmatrix} = \begin{Bmatrix} (\cos\theta)R \\ (-\sin\theta)R \\ 0 \end{Bmatrix}$$

In the x'–y'–z' coordinate system, the internal force is $\bar{F}_I = (R\cos\theta)\hat{i}' - (R\sin\theta)\hat{j}' + (0)\hat{k}'$. Hence, by defining a coordinate system which is inclined to the axis of the rod, we conclude that *both* a normal force $(R\cos\theta)$ *and* a shear force $(-R\sin\theta)$ are induced in the rod.

Although the preceding discussion may seem simplistic, it has been included to demonstrate the following:

A specific coordinate system must be specified before a force vector can be defined in a mathematical sense. In general, the coordinate system is defined by the imaginary cut(s) used to form the free-body diagram.

All components of a force must be specified to fully define the force vector. Further, the individual components of a force change as the vector is transformed from one coordinate system to another.

These two observations are valid for all tensors, not just for force vectors. In particular, these observations hold in the case of stress and strain tensors, which will be reviewed in the following sections.

2.4 Definition of Stress

There are two fundamental types of stress: *normal* stress and *shear* stress. Both types of stress are defined as a force divided by the area over which it acts.

A general 3D solid body subjected to a system of external forces is shown in Figure 2.6a. It is assumed that the body is in static equilibrium, that is, it is assumed that the sum of all external forces is zero, $\Sigma \overline{F}_i = 0$. These external forces induce internal forces acting within the body. In general, the internal forces will vary in both magnitude and direction throughout the body. An illustration of the variation of internal forces along a single line within an internal plane is shown in Figure 2.6b. A small area (ΔA) isolated from this plane is shown in Figure 2.6c. Area ΔA is assumed to be "infinitesimally small." That

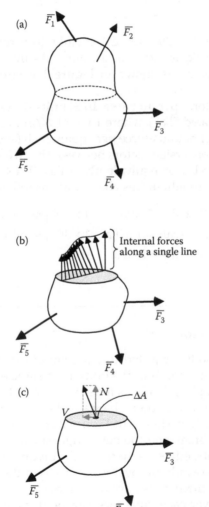

FIGURE 2.6
A solid 3-D body in equilibrium. (a) A solid 3-D body subject to external forces $\overline{F}_1 \rightarrow \overline{F}_5$; (b) variation of internal forces along an internal line; (c) internal force acting over infinitesimal area ΔA.

is, the area ΔA is small enough such that the internal forces acting over ΔA can be assumed to be of constant magnitude and direction. Therefore, the internal forces acting over ΔA can be represented by a force vector which can be decomposed into a normal force, N, and a shear force, V, as shown in Figure 2.6c.

Normal stress (usually denoted "σ"), and shear stress (usually denoted "τ") are defined as the force per unit area acting perpendicular and tangent to the area ΔA, respectively. That is,

$$\sigma \equiv \lim_{\Delta A \to 0} \frac{N}{\Delta A} \quad \tau \equiv \lim_{\Delta A \to 0} \frac{V}{\Delta A} \tag{2.9}$$

Note that by definition the area ΔA shrinks to zero: $\Delta A \to 0$. Stresses σ and τ are therefore said to exist "at a point." As internal forces generally vary from point-to-point (as shown in Figure 2.6), stresses also vary from point-to-point.

Stress has units of force per unit area. In SI units stress is reported in terms of "Pascals" (abbreviated "Pa"), where $1 \text{ Pa} = 1 \text{ N/m}^2$. In English units stress is reported in terms of pounds-force per square inch (abbreviated "psi"), that is, $1 \text{ psi} = 1 \text{ lbf/in.}^2$ Conversion factors between the two systems of measurement are $1 \text{ psi} = 6895 \text{ Pa}$, or equivalently, $1 \text{ Pa} = 0.1450 \times 10^{-3} \text{ psi}$. Common abbreviations used throughout this chapter are as follows:

$1 \times 10^3 \text{ Pa} = 1 \text{ kilo-Pascals} = 1 \text{ kPa}$ $1 \times 10^3 \text{ psi} = 1 \text{ kilo-psi} = 1 \text{ ksi}$

$1 \times 10^6 \text{ Pa} = 1 \text{ Mega-Pascals} = 1 \text{ MPa}$ $1 \times 10^6 \text{ psi} = 1 \text{ mega-psi} = 1 \text{ Msi}$

$1 \times 10^9 \text{ Pa} = 1 \text{ Giga-Pascals} = 1 \text{ GPa}.$

2.5 The Stress Tensor

A general 3D solid body subjected to a system of external forces is shown in Figure 2.7a. It is assumed that the body is in static equilibrium and that body forces are negligible, that is, is it assumed that the sum of all external forces is zero, $\Sigma F_i = 0$. A free-body diagram of an infinitesimally small cube removed from the body is shown in Figure 2.7b. The cube is referenced to a x–y–z coordinate system, and the cube edges are aligned with these axes. The lengths of the cube edges are denoted dx, dy, and dz. Although (in general) internal forces are induced over all six faces of the cube, for clarity the forces acting on only three faces have been shown.

The force acting over each cube face can be decomposed into a normal force component and two shear force components, as illustrated in Figure 2.7c. Although each force component could be identified with a single subscript (as force is a first-order tensor), for convenience two subscripts have been used. The first subscript identifies the face over which the force is distributed, whereas the second subscript identifies the direction in which the

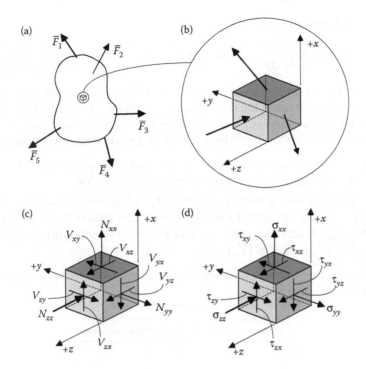

FIGURE 2.7
Free-body diagrams used to define stress induced in a solid body. (a) 3-D solid body in equilibrium; (b) infinitesimal cube removed from the solid body (internal forces acting on three faces shown); (c) normal force and two shear forces act over each face of the cube; (d) normal stress and two shear stresses act over each face of the cube.

force is oriented. For example, N_{xx} refers to a normal force component which is distributed over the x-face and which is acting parallel to the x-direction. Similarly, V_{zy} refers to a shear force distributed over the z-face which is acting parallel to the y-direction.

Three stress components can now be defined for each cube face, in accordance with Equation 2.9. For example, for the three faces of the infinitesimal element shown in Figure 2.7:

Stresses acting on the −x-face:

$$\sigma_{xx} = \lim_{dy,dz \to 0}\left(\frac{N_{xx}}{dy\,dz}\right) \quad \tau_{xy} = \lim_{dy,dz \to 0}\left(\frac{V_{xy}}{dy\,dz}\right) \quad \tau_{xz} = \lim_{dy,dz \to 0}\left(\frac{V_{xz}}{dy\,dz}\right)$$

Stresses acting on the −y-face:

$$\sigma_{yy} = \lim_{dx,dz \to 0}\left(\frac{N_{yy}}{dx\,dz}\right) \quad \tau_{yx} = \lim_{dx,dz \to 0}\left(\frac{V_{yx}}{dx\,dz}\right) \quad \tau_{yz} = \lim_{dx,dz \to 0}\left(\frac{V_{yz}}{dx\,dz}\right)$$

Stresses acting on the +z-face:

$$\sigma_{zz} = \lim_{dx,dy \to 0}\left(\frac{N_{zz}}{dx\,dy}\right) \quad \tau_{zx} = \lim_{dx,dy \to 0}\left(\frac{V_{zx}}{dx\,dy}\right) \quad \tau_{zy} = \lim_{dx,dy \to 0}\left(\frac{V_{zy}}{dx\,dy}\right)$$

As three force components (and therefore three stress components) exist on each of the six faces of the cube, it would initially appear that there are 18 independent force (stress) components. However, it is easily shown that for static equilibrium to be maintained (assuming body forces are negligible):

- Normal forces acting on opposite faces of the infinitesimal element must be of equal magnitude and opposite direction.
- Shear forces acting within a plane of the element must be orientated either "tip-to-tip" (e.g., forces V_{xz} and V_{zx} in Figure 2.7c) or "tail-to-tail" (e.g., forces V_{xy} and V_{yx}), and be of equal magnitude. That is $|V_{xy}| = |V_{yx}|$, $|V_{xz}| = |V_{zx}|$, $|V_{yz}| = |V_{zy}|$.

These restrictions reduce the number of independent force (stress) components from 18 to 6, as follows:

Independent Force Component(s)	Corresponding Stress Component(s)
N_{xx}	σ_{xx}
N_{yy}	σ_{yy}
N_{zz}	σ_{zz}
$V_{xy}\,(=V_{yx})$	$\tau_{xy}\,(=\tau_{yx})$
$V_{xz}\,(=V_{zx})$	$\tau_{xz}\,(=\tau_{zx})$
$V_{yz}\,(=V_{zy})$	$\tau_{yz}\,(=\tau_{zy})$

We must next define the algebraic sign convention we will use to describe individual stress components. The components acting on three faces of an infinitesimal element are shown in Figure 2.8. We first associate an algebraic sign with each face of the infinitesimal element. A cube face is *positive* if the outward unit normal of the face (i.e., the unit normal pointing away from the interior of the element) points in a positive coordinate direction; otherwise, the face is negative. For example, faces (*ABCD*) and (*ADEG*) in Figure 2.8 are a positive faces, whereas face (*DCFE*) is a negative face.

Having identified the positive and negative faces of the element, a stress component is *positive* if

- The stress component acts on a positive face and points in a positive coordinate direction, or if
- The stress component acts on a negative face and points in a negative coordinate direction.

If neither of these conditions is met, then the stress component is negative. This convention can be used to confirm that all stress components shown

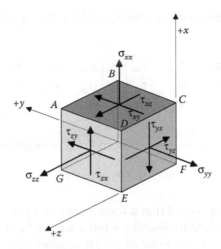

FIGURE 2.8
An infinitesimal stress element (all stress components shown in a positive sense).

in Figure 2.8 are algebraically positive. For example, to determine the algebraic sign of the normal stress σ_{xx} which acts on face $ABCD$ in Figure 2.8, note that (a) face $ABCD$ is positive, and (b) the normal stress σ_{xx} which acts on this face points in the positive x-direction. Therefore, σ_{xx} is positive. As a second example, the shear stress τ_{yz} which acts on cube face $DCFE$ is positive because (a) face $DCFE$ is a negative face, and (b) τ_{yz} points in the negative z-direction.

The preceding discussion shows that the *state of stress* at a point is defined by six components of stress: three normal stress components and three shear stress components. The state of stress is written using matrix notation as follows:

$$\begin{bmatrix} \sigma_{xx} & \tau_{xy} & \tau_{xz} \\ \tau_{yx} & \sigma_{yy} & \tau_{yz} \\ \tau_{zx} & \tau_{zy} & \sigma_{zz} \end{bmatrix} = \begin{bmatrix} \sigma_{xx} & \tau_{xy} & \tau_{xz} \\ \tau_{xy} & \sigma_{yy} & \tau_{yz} \\ \tau_{xz} & \tau_{yz} & \sigma_{zz} \end{bmatrix} \quad (2.10)$$

To express the state of stress using indicial notation we must first make the following change in notation:

$$\tau_{xy} \rightarrow \sigma_{xy}$$

$$\tau_{xz} \rightarrow \sigma_{xz}$$

$$\tau_{yx} \rightarrow \sigma_{yx}$$

$$\tau_{yz} \rightarrow \sigma_{yz}$$

$$\tau_{zx} \rightarrow \sigma_{zx}$$

$$\tau_{zy} \rightarrow \sigma_{zy}$$

With this change the matrix on the left side of the equality sign in Equation 2.10 becomes:

$$\begin{bmatrix} \sigma_{xx} & \tau_{xy} & \tau_{xz} \\ \tau_{yx} & \sigma_{yy} & \tau_{yz} \\ \tau_{zx} & \tau_{zy} & \sigma_{zz} \end{bmatrix} \rightarrow \begin{bmatrix} \sigma_{xx} & \sigma_{xy} & \sigma_{xz} \\ \sigma_{yx} & \sigma_{yy} & \sigma_{yz} \\ \sigma_{zx} & \sigma_{zy} & \sigma_{zz} \end{bmatrix}$$

which can be succinctly written using indicial notation as

$$\sigma_{ij}, \quad i,j = x, y, \text{ or } z \tag{2.11}$$

In Section 2.1, it was noted that a force vector is a *first-order tensor*, as only one subscript is required to describe a force tensor, F_i. From Equation 2.11 is it clear that stress is a *second-order tensor* (or equivalently, a *tensor of rank two*), as two subscripts are required to describe a state of stress.

Example Problem 2.2

Given: The stress element referenced to a x–y–z coordinate system and subject to the stress components shown in Figure 2.9.
Determine: Label all stress components, including algebraic sign.

SOLUTION

The magnitude and algebraic sign of each stress component is determined using the sign convention defined above. The procedure will

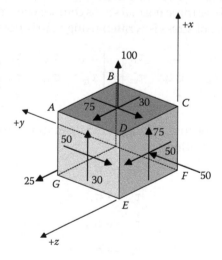

FIGURE 2.9
Stress components acting on an infinitesimal element (all stresses in MPa).

be illustrated using the stress components acting on face *DCFE*. First, note that face *DCFE* is a negative face, as an outward unit normal for this face points in the negative *y*-direction. The normal stress which acts on face *DCFE* has a magnitude of 50 MPa, and points in the positive *y*-direction. Hence, this stress component is negative and is labeled $\sigma_{yy} = -50$ MPa. One of the shear stress components acting on face *DCFE* has a magnitude of 75 MPa, and points in the positive *x*-direction. Hence, this stress component is also negative and is labeled $\tau_{yx} = -75$ MPa (or equivalently, $\tau_{xy} = -75$ MPa). Finally, the second shear force component acting on face *DCFE* has a magnitude of 50 MPa, and points in the positive *z*-direction. Hence, this component is labeled $\tau_{yz} = -50$ MPa (or equivalently, $\tau_{zy} = -50$ MPa).

Following this process for all faces of the element, the state of stress represented by the element shown in Figure 2.9 can be written:

$$\begin{bmatrix} \sigma_{xx} & \tau_{xy} & \tau_{xz} \\ \tau_{yx} & \sigma_{yy} & \tau_{yz} \\ \tau_{zx} & \tau_{zy} & \sigma_{zz} \end{bmatrix} = \begin{bmatrix} 100\,\text{MPa} & -75\,\text{MPa} & 30\,\text{MPa} \\ -75\,\text{MPa} & -50\,\text{MPa} & -50\,\text{MPa} \\ 30\,\text{MPa} & -50\,\text{MPa} & 25\,\text{MPa} \end{bmatrix}$$

2.6 Transformation of the Stress Tensor

In Section 2.5 the stress tensor was defined using a free-body diagram of an infinitesimal element removed from a 3D body in static equilibrium. This concept is again illustrated in Figure 2.10a, which shows the stress element referenced to an *x–y–z* coordinate system.

Now, the infinitesimal element need not be removed in the orientation shown in Figure 2.10a. An infinitesimal element removed from precisely the same point within the body but at a different orientation is shown in Figure 2.10b. This stress element is referenced to a new *x'–y'–z'* coordinate system. The state of stress at the point of interest is dictated by the external loads applied to the body, and is independent of the coordinate system used to describe it. Hence, the stress tensor referenced to the *x'–y'–z'* coordinate system is equivalent to the stress tensor referenced to the *x–y–z* coordinate system, although the direction and magnitude of individual stress components will differ.

The process of relating stress components in one coordinate system to those in another is called *transformation* of the stress tensor. This terminology is perhaps unfortunate, in the sense that the state of stress itself is not "transformed" but rather our *description* of the state of stress transforms as we change from one coordinate system to another.

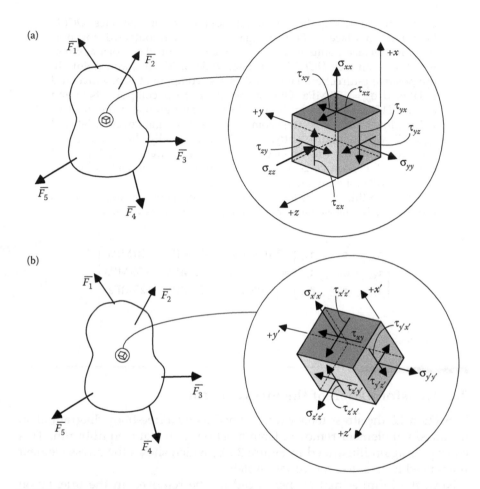

FIGURE 2.10
Infinitesimal elements removed from the same point within a 3D solid but in two different orientations. (a) Infinitesimal element referenced to the x–y–z coordinate system; (b) infinitesimal element referenced to the x'–y'–z' coordinate system.

It can be shown [1,2] that the stress components in the new x'–y'–z' coordinate system ($\sigma_{i'j'}$) are related to the components in the original x–y–z coordinate system (σ_{ij}) according to

$$\sigma_{i'j'} = c_{i'k}\, c_{j'l}\, \sigma_{kl} \quad \text{where} \quad i, j, k, l = x, y, z \qquad (2.12a)$$

Or, equivalently (using matrix notation):

$$[\sigma_{i'j'}] = [c_{i'j}][\sigma_{ij}][c_{i'j}]^{T}$$

where $[c_{i'\,j}]^T$ is the transpose of the direction cosine array. Writing in full matrix form:

$$
\begin{bmatrix} \sigma_{x'x'} & \sigma_{x'y'} & \sigma_{x'z'} \\ \sigma_{y'x'} & \sigma_{y'y'} & \sigma_{y'z'} \\ \sigma_{z'x'} & \sigma_{z'y'} & \sigma_{z'z'} \end{bmatrix} = \begin{bmatrix} c_{x'x} & c_{x'y} & c_{x'z} \\ c_{y'x} & c_{y'y} & c_{y'z} \\ c_{z'x} & c_{z'y} & c_{z'z} \end{bmatrix}
$$

$$
\times \begin{bmatrix} \sigma_{xx} & \sigma_{xy} & \sigma_{xz} \\ \sigma_{yx} & \sigma_{yy} & \sigma_{yz} \\ \sigma_{zx} & \sigma_{zy} & \sigma_{zz} \end{bmatrix} \begin{bmatrix} c_{x'x} & c_{y'x} & c_{z'x} \\ c_{x'y} & c_{y'y} & c_{z'y} \\ c_{x'z} & c_{y'z} & c_{z'z} \end{bmatrix} \quad (2.12b)
$$

As discussed in Section 2.2, the terms $c_{i'j}$ which appear in Equation 2.12a,b are *direction cosines* and equal the cosine of the angle between the axes of the $x–y–z$ and $x'–y'–z'$ coordinate systems. Recall that the algebraic sign of an angle of rotation is defined in accordance with the right-hand rule, and that angles are defined *from* the $x–y–z$ coordinate system *to* $x'–y'–z'$ coordinate system. Equation 2.12a,b is called the *transformation law for a second-order tensor*.

If an analysis is being performed with the aid of a digital computer, which nowadays is almost always the case, then matrix notation (Equation 2.12b) will most likely be used to transform a stress tensor from one coordinate system to another. Conversely, if a stress transformation is to be accomplished using hand calculations, then indicial notation (Equation 2.12a) may be the preferred choice. To apply Equation 2.12a, the stress component of interest is specified by selecting the appropriate values for subscripts i' and j', and then the terms on the right side of the equality are summed over the entire range of the remaining two subscripts, k and l. For example, suppose we wish write the relationship between $\sigma_{x'z'}$ and the stress components in the $x–y–z$ coordinate system in expanded form. We first specify that $i' = x'$ and $j' = z'$, and Equation 2.12a becomes:

$$
\sigma_{x'z'} = c_{x'k}\, c_{z'l}\, \sigma_{kl} \quad \text{where} \quad k, l = x, y, z
$$

We then sum all terms on the right side of the equality formed by cycling through the entire range of k and l. In expanded form, we have

$$
\sigma_{x'z'} = c_{x'x}\, c_{z'x}\, \sigma_{xx} + c_{x'x}\, c_{z'y}\, \sigma_{xy} + c_{x'x}\, c_{z'z}\, \sigma_{xz}
$$

$$
+ c_{x'y}\, c_{z'x}\, \sigma_{yx} + c_{x'y}\, c_{z'y}\, \sigma_{yy} + c_{x'y}\, c_{z'z}\, \sigma_{yz}
$$

$$
+ c_{x'z}\, c_{z'x}\, \sigma_{zx} + c_{x'z}\, c_{z'y}\, \sigma_{zy} + c_{x'z}\, c_{z'z}\, \sigma_{zz} \quad (2.13)
$$

Equations 2.12a and 2.12b show that the value of any individual stress component $\sigma_{i'j'}$ varies as the stress tensor is transformed from one coordinate system to another. However, it can be shown [1,2] that there are features of the *total* stress tensor that do not vary when the tensor is transformed from one coordinate system to another. These features are called the *stress invariants*. For a second-order tensor three independent stress invariants exist, and are defined as follows:

First stress invariant $= \Theta = \sigma_{ii}$ (2.14a)

Second stress invariant $= \Phi = \dfrac{1}{2}(\sigma_{ii}\sigma_{jj} - \sigma_{ij}\sigma_{ij})$ (2.14b)

Third stress invariant $= \Psi = \dfrac{1}{6}(\sigma_{ii}\sigma_{jj}\sigma_{kk} - 3\sigma_{ii}\sigma_{jk}\sigma_{jk} + 2\sigma_{ij}\sigma_{jk}\sigma_{ki})$ (2.14c)

Alternatively, by expanding these equations over the range $i, j, k = x, y, z$ and simplifying, the stress invariants can be written:

First stress invariant $= \Theta = \sigma_{xx} + \sigma_{yy} + \sigma_{zz}$ (2.15a)

Second stress invariant $= \Phi = \sigma_{xx}\sigma_{yy} + \sigma_{xx}\sigma_{zz} + \sigma_{yy}\sigma_{zz} - (\sigma_{xy}^2 + \sigma_{xz}^2 + \sigma_{yz}^2)$

(2.15b)

Third stress invariant $= \Psi = \sigma_{xx}\sigma_{yy}\sigma_{zz} - \sigma_{xx}\sigma_{yz}^2 - \sigma_{yy}\sigma_{xz}^2 - \sigma_{zz}\sigma_{xy}^2$

$$+ 2\sigma_{xy}\sigma_{xz}\sigma_{yz}$$ (2.15c)

The three stress invariants are conceptually similar to the magnitude of a force tensor. That is, the value of the three stress invariants is independent of the coordinate used to describe the stress tensor, just as the magnitude of a force vector is independent of the coordinate system used to describe the force. This invariance will be illustrated in the following Example Problem.

Example Problem 2.3

Given: A state of stress referenced to a x–y–z coordinate is known to be

$$\begin{bmatrix} \sigma_{xx} & \sigma_{xy} & \sigma_{xz} \\ \sigma_{yx} & \sigma_{yy} & \sigma_{yz} \\ \sigma_{zx} & \sigma_{zy} & \sigma_{zz} \end{bmatrix} = \begin{bmatrix} 50 & -10 & 15 \\ -10 & 25 & 30 \\ 15 & 30 & -5 \end{bmatrix} \text{(ksi)}$$

It is desired to express this state of stress in an x''–y''–z'' coordinate system, generated by the following two sequential rotations:

i. Rotation of $\theta = 20°$ about the original z-axis (which defines an intermediate x'–y'–z' coordinate system), followed by
ii. Rotation of $\beta = 35°$ about the x'-axis (which defines the final x''–y''–z'' coordinate system)

PROBLEM

a. Rotate the stress tensor to the x''–y''–z'' coordinate system.
b. Calculate the first, second, and third invariants of the stress tensor using (i) elements of the stress tensor referenced to the x–y–z coordinate system, σ_{ij}, and (ii) elements of the stress tensor referenced to the x''–y''–z'' coordinate system, $\sigma_{i''j''}$.

SOLUTION

a. General expressions for direction cosines relating the x–y–z and x''–y''–z'' coordinate systems were determined as a part of Example Problem 2.1. The direction cosines were found to be

$$c_{x''x} = \cos\theta$$
$$c_{x''y} = \sin\theta$$
$$c_{x''z} = 0$$
$$c_{y''x} = -\cos\beta\sin\theta$$
$$c_{y''y} = \cos\beta\cos\theta$$
$$c_{y''z} = \sin\beta$$
$$c_{z''x} = \sin\beta\sin\theta$$
$$c_{z''y} = -\sin\beta\cos\theta$$
$$c_{z''z} = \cos\beta$$

As in this problem $\theta = 20°$ and $\beta = 35°$, the numerical values of the direction cosines are

$$c_{x''x} = \cos(20°) = 0.9397$$
$$c_{x''y} = \sin(20°) = 0.3420$$
$$c_{x''z} = 0$$
$$c_{y''x} = -\cos(35°)\sin(20°) = -0.2802$$
$$c_{y''y} = \cos(35°)\cos(20°) = 0.7698$$
$$c_{y''z} = \sin(35°) = 0.5736$$
$$c_{z''x} = \sin(35°)\sin(20°) = 0.1962$$
$$c_{z''y} = -\sin(35°)\cos(20°) = -0.5390$$
$$c_{z''z} = \cos(35°) = 0.8192$$

Each component of the transformed stress tensor is now found through application of either Equation 2.12a or 2.12b. For example, if indicial notation is used stress component $\sigma_{x''z''}$ can be found using Equation 2.13:

$$\sigma_{x''z''} = c_{x''x}\, c_{z''x}\, \sigma_{xx} + c_{x''x}\, c_{z''y}\, \sigma_{xy} + c_{x''x}\, c_{z''z}\, \sigma_{xz}$$

$$+ c_{x''y}\, c_{z''x}\, \sigma_{yx} + c_{x''y}\, c_{z''y}\, \sigma_{yy} + c_{x''y}\, c_{z''z}\, \sigma_{yz}$$

$$+ c_{x''z}\, c_{z''x}\, \sigma_{zx} + c_{x''z}\, c_{z''y}\, \sigma_{zy} + c_{x''z}\, c_{z''z}\, \sigma_{zz}$$

$$\sigma_{x''z''} = (0.9397)(0.1962)(50\,\text{ksi}) + (0.9397)(-0.5390)(-10\,\text{ksi})$$

$$+ (0.9397)(0.8192)(15\,\text{ksi})$$

$$+ (0.3420)(0.1962)(-10\,\text{ksi}) + (0.3420)(-0.5390)(25\,\text{ksi})$$

$$+ (0.3420)(0.8192)(30\,\text{ksi})$$

$$+ (0)(0.1962)(15\,\text{ksi}) + (0)(-0.5390)(30\,\text{ksi}) + (0)(0.8192)(-5\,\text{ksi})$$

$$\sigma_{x''z''} = 28.95\,\text{ksi}$$

Alternatively, if matrix notation is used, then Equation 2.12b becomes:

$$
\begin{bmatrix} \sigma_{x''x''} & \sigma_{x''y''} & \sigma_{x''z''} \\ \sigma_{y''x''} & \sigma_{y''y''} & \sigma_{y''z''} \\ \sigma_{z''x''} & \sigma_{z''y''} & \sigma_{z''z''} \end{bmatrix} =
\begin{bmatrix} 0.9397 & 0.3420 & 0 \\ -0.2802 & 0.7698 & 0.5736 \\ 0.1962 & -0.5390 & 0.8192 \end{bmatrix}
\begin{bmatrix} 50 & -10 & 15 \\ -10 & 25 & 30 \\ 15 & 30 & -5 \end{bmatrix}
$$

$$
\times \begin{bmatrix} 0.9397 & -0.2802 & 0.1962 \\ 0.3420 & 0.7698 & -0.5390 \\ 0 & 0.5736 & 0.8192 \end{bmatrix}
$$

Completing the matrix multiplication indicated, there results:

$$
\begin{bmatrix} \sigma_{x''x''} & \sigma_{x''y''} & \sigma_{x''z''} \\ \sigma_{y''x''} & \sigma_{y''y''} & \sigma_{y''z''} \\ \sigma_{z''x''} & \sigma_{z''y''} & \sigma_{z''z''} \end{bmatrix} =
\begin{bmatrix} 40.65 & 1.113 & 28.95 \\ 1.113 & 43.08 & -10.60 \\ 28.95 & -10.60 & -13.72 \end{bmatrix} (\text{ksi})
$$

Notice that the value of $\sigma_{x''z''}$ determined through matrix multiplication is identical to that obtained using indicial notation, as previously described. The stress element is shown in the original and final coordinate systems in Figure 2.11.

b. The first, second, and third stress invariants will now be calculated using components of both σ_{ij} and $\sigma_{i''j''}$. It is expected

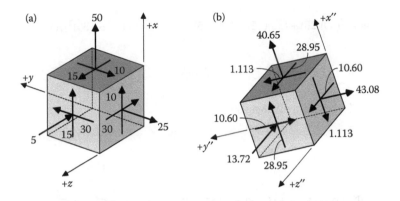

FIGURE 2.11
Stress tensor of Example Problem 2.3, referenced to two different coordinate systems (magnitude of all stress components in ksi). (a) Referenced to x–y–z coordinate system; (b) referenced to x''–y''–z'' coordinate system.

that identical values will be obtained, as the stress invariants are independent of coordinate system.

First Stress Invariant:
x–y–z coordinate system:

$$\Theta = \sigma_{ii} = \sigma_{xx} + \sigma_{yy} + \sigma_{zz}$$

$$\Theta = (50 + 25 - 5)^\circ \text{ksi}$$

$$\Theta = 70^\circ \text{ksi}$$

x''–y''–z'' coordinate system:

$$\Theta = \sigma_{i''i''} = \sigma_{x''x''} + \sigma_{y''y''} + \sigma_{z''z''}$$

$$\Theta = (40.65 + 43.08 - 13.72)$$

$$\Theta = 70 \text{ksi}$$

As expected, the first stress invariant is independent of coordinate system.

Second Stress Invariant:
x–y–z coordinate system:

$$\Phi = \frac{1}{2}\left(\sigma_{ii}\sigma_{jj} - \sigma_{ij}\sigma_{ij}\right) = \sigma_{xx}\sigma_{yy} + \sigma_{xx}\sigma_{zz} + \sigma_{yy}\sigma_{zz} - \left(\sigma_{xy}^2 + \sigma_{xz}^2 + \sigma_{yz}^2\right)$$

$$\Phi = \{(50)(25) + (50)(-5) + (25)(-5) - [(-10)^2 + (15)^2 + (30)^2]\}(\text{ksi})^2$$

$$\Phi = -350\,(\text{ksi})^2$$

$x''-y''-z''$ coordinate system:

$$\Phi = \frac{1}{2}\left(\sigma_{i''i''}\sigma_{j''j''} - \sigma_{i''j''}\sigma_{i''j''}\right)$$

$$\Phi = \sigma_{x''x''}\sigma_{y''y''} + \sigma_{x''x''}\sigma_{z''z''} + \sigma_{y''y''}\sigma_{z''z''} - \left(\sigma_{x''y''}^2 + \sigma_{x''z''}^2 + \sigma_{y''z''}^2\right)$$

$$\Phi = \{(40.65)(43.08) + (40.65)(-13.72) + (43.08)(-13.72)$$

$$- [(1.113)^2 + (28.95)^2 + (-10.60)^2]\}(\text{ksi})^2$$

$$\Phi = -350\,(\text{ksi})^2$$

As expected, the second stress invariant is independent of coordinate system.

Third Stress Invariant:

$x-y-z$ coordinate system:

$$\Psi = \frac{1}{6}\left(\sigma_{ii}\sigma_{jj}\sigma_{kk} - 3\sigma_{ii}\sigma_{jk}\sigma_{jk} + 2\sigma_{ij}\sigma_{jk}\sigma_{ki}\right)$$

$$\Psi = \sigma_{xx}\sigma_{yy}\sigma_{zz} - \sigma_{xx}\sigma_{yz}^2 - \sigma_{yy}\sigma_{xz}^2 - \sigma_{zz}\sigma_{xy}^2 + 2\sigma_{xy}\sigma_{xz}\sigma_{yz}$$

$$\Psi = [(50)(25)(-5) - (50)(30)^2 - (25)(15)^2 - (-5)(-10)^2 + 2(-10)(15)(30)](\text{ksi})^3$$

$$\Psi = -65375\,(\text{ksi})^3$$

$x''-y''-z''$ coordinate system:

$$\Psi = \frac{1}{6}\left(\sigma_{i''i''}\sigma_{j''j''}\sigma_{k''k''} - 3\sigma_{i''i''}\sigma_{j''k''}\sigma_{j''k''} + 2\sigma_{i''j''}\sigma_{j''k''}\sigma_{k''i''}\right)$$

$$\Psi = \sigma_{x''x''}\sigma_{y''y''}\sigma_{z''z''} - \sigma_{x''x''}\sigma_{y''z''}^2 - \sigma_{y''y''}\sigma_{x''z''}^2 - \sigma_{z''z''}\sigma_{x''y''}^2 + 2\sigma_{x''y''}\sigma_{x''z''}\sigma_{y''z''}$$

$$\Psi = [(40.65)(43.08)(-13.72) - (40.65)(-10.60)^2 - (43.08)(28.95)^2$$

$$- (-13.72)(1.113)^2 + 2(1.113)(28.95)(-10.60)](ksi)^3$$

$$\Psi = -65375\,(ksi)^3$$

As expected, the third stress invariant is independent of coordinate system.

2.7 Principal Stresses

It is always possible to rotate the stress tensor to a special coordinate system in which no shear stresses exist. This coordinate system is called the *principal stress coordinate system*, and the normal stresses that exist in this coordinate system are called *principal stresses*. Principal stresses can be used to predict failure of isotropic materials. Therefore, knowledge of the principal stresses and orientation of the principal stress coordinate system is of vital importance during design and analysis of isotropic metal structures.

This is not the case for anisotropic composite materials. Failure of composite material is *not* governed by principal stresses. Principal stresses are only of occasional interest to the composite engineer and are reviewed here only in the interests of completeness.

Principal stresses are usually denoted as σ_1, σ_2, and σ_3. However, in the study of composites, the labels "1," "2," and "3" are used to label a special coordinate system called the *principal material* coordinate system. Therefore, in this chapter the axes associated with the principal stress coordinate system will be labeled the "p_1," "p_2," and "p_3" axes, and the principal stresses will be denoted as σ_{p1}, σ_{p2}, and σ_{p3}.

Principal stresses may be related to stress components in an x–y–z coordinate system using the free-body diagram shown in Figure 2.12. It is assumed that plane ABC is one of the three principal planes (i.e., $n = 1, 2,$ or 3), and therefore, no shear stress exists on this plane. The line-of-action of principal stress σ_{pn} defines one axis of the principal stress coordinate system. The direction cosines between this principal axis and the x-, y-, and z-axes are c_{pnx}, c_{pny}, and c_{pnz}, respectively. The surface area of triangle ABC is denoted A_{ABC}. The normal force acting over triangle ABC therefore equals $(\sigma_{pn})(A_{ABC})$. The components of this normal force acting in the x-, y-, and z-directions equal $(c_{pnx})(\sigma_{pn})(A_{ABC})$, $(c_{pny})(\sigma_{pn})(A_{ABC})$ and $(c_{pnz})(\sigma_{pn})(A_{ABC})$, respectively.

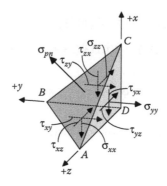

FIGURE 2.12

Free-body diagram used to relate strress components in the x–y–z coordinate system to a principal stress.

The area of the other triangular faces are given by

Area of triangle $ABD = (c_{pnx})(A_{ABC})$

Area of triangle $ACD = (c_{pny})(A_{ABC})$

Area of triangle $BCD = (c_{pnz})(A_{ABC})$

Summing forces in the x-direction and equating to zero, we obtain

$$c_{pnx}\sigma_{pn}A_{ABC} - \sigma_{xx}c_{pnx}A_{ABC} - \tau_{xy}c_{pny}A_{ABC} - \tau_{xz}c_{pnz}A_{ABC} = 0$$

which can be reduced and simplified to

$$(\sigma_{pn} - \sigma_{xx})c_{pnx} - \tau_{xy}c_{pny} - \tau_{xz}c_{pnz} = 0 \qquad (2.16a)$$

Similarly, summing forces in the y- and z-directions results in

$$-\tau_{xy}c_{pnx} + (\sigma_{pn} - \sigma_{yy})c_{pny} - \tau_{yz}c_{pnz} = 0 \qquad (2.16b)$$

$$-\tau_{xz}c_{pnx} - \tau_{yz}c_{pny} + (\sigma_{pn} - \sigma_{zz})c_{pnz} = 0 \qquad (2.16c)$$

Equation 2.16 represent three linear homogeneous equations which must be satisfied simultaneously. As direction cosines c_{pnx}, c_{pny}, and c_{pnz} must also satisfy Equation 2.8, and therefore cannot all equal zero, the solution can be obtained by requiring that the determinant of the coefficients of c_{pnx}, c_{pny}, and c_{pnz} equal zero:

$$\begin{vmatrix} (\sigma_{pn} - \sigma_{xx}) & -\tau_{xy} & -\tau_{xz} \\ -\tau_{xy} & (\sigma_{pn} - \sigma_{yy}) & -\tau_{yz} \\ -\tau_{xz} & -\tau_{yz} & (\sigma_{pn} - \sigma_{zz}) \end{vmatrix} = 0$$

Equating the determinant to zero results in the following cubic equation:

$$\sigma_{pn}^3 - \Theta\sigma_{pn}^2 + \Phi\sigma_{pn} - \Psi = 0 \qquad (2.17)$$

where Θ, Φ, and Ψ are the first, second, and third stress invariants, respectively, and have been previously listed as Equations 2.14 and 2.15. The three roots of this cubic equation represent the three principal stresses and may be found by application of the standard approach [3]. By convention, the principal stresses are numbered such that σ_{p1} is the algebraically greatest principal stress, whereas σ_{p3} is the algebraically least. That is, $\sigma_{p1} > \sigma_{p2} > \sigma_{p3}$.

Once the principal stresses are determined the three sets of direction cosines (which define the principal coordinate directions) are found by substituting the three principal stresses given by Equation 2.17 into Equation 2.16 in turn. As only two of Equation 2.16 are independent, Equation 2.8 is used as a third independent equation involving the three unknown constants, c_{pnx}, c_{pny} and c_{pnz}.

The process of finding principal stresses and direction cosines will be demonstrated in the following Example Problem.

Example Problem 2.4

Given: A state of stress referenced to an x–y–z coordinate is known to be:

$$\begin{bmatrix} \sigma_{xx} & \sigma_{xy} & \sigma_{xz} \\ \sigma_{yx} & \sigma_{yy} & \sigma_{yz} \\ \sigma_{zx} & \sigma_{zy} & \sigma_{zz} \end{bmatrix} = \begin{bmatrix} 50 & -10 & 15 \\ -10 & 25 & 30 \\ 15 & 30 & -5 \end{bmatrix} (\text{ksi})$$

PROBLEM

Find (a) the principal stresses and (b) the direction cosines that define the principal stress coordinate system.

SOLUTION

This is the same stress tensor considered in Example Problem 2.3. As a part of that problem the first, second, and third stress invariants were found to be

$$\Theta = 70 \text{ ksi}$$

$$\Phi = -350 \text{ (ksi)}^2$$

$$\Psi = -65375 \text{ (ksi)}^3$$

a. *Determining the Principal Stresses*: In accordance with Equation 2.17, the three principal stresses are the roots of the following cubic equation:

$$\sigma^3 - 70\sigma^2 - 350\sigma + 65375 = 0$$

The three roots of this equation represent the three principal stresses, and are given by

$$\sigma_{p1} = 54.21\,\text{ksi}, \quad \sigma_{p2} = 43.51\,\text{ksi}, \quad \text{and} \quad \sigma_{p3} = -27.72\,\text{ksi}$$

b. *Determining the Direction Cosines*: The first two of Equations 2.16 and 2.8 are used to form three independent equations in three unknowns. We have

$$(\sigma_{pn} - \sigma_{xx})c_{pnx} - \tau_{xy}c_{pny} - \tau_{xz}c_{pnz} = 0$$

$$-\tau_{xy}c_{pnx} + (\sigma_{pn} - \sigma_{yy})c_{pny} - \tau_{yz}c_{pnz} = 0$$

$$(c_{pnx})^2 + (c_{pny})^2 + (c_{pnz})^2 = 1$$

Direction cosines for σ_{p1}: The three independent equations become:

$$(54.21 - 50)c_{p1x} + 10c_{p1y} - 15c_{p1z} = 0$$

$$10c_{p1x} + (54.21 - 25)c_{p1y} - 30c_{p1z} = 0$$

$$(c_{p1x})^2 + (c_{p1y})^2 + (c_{p1z})^2 = 1$$

Solving simultaneously, we obtain:

$$c_{p1x} = -0.9726$$

$$c_{p1y} = 0.1666$$

$$c_{p1z} = -0.1620$$

Direction cosines for σ_{p2}: The three independent equations become:

$$(43.51 - 50)c_{p2x} + 10c_{p2y} - 15c_{p2z} = 0$$

$$10c_{p2x} + (43.51 - 25)c_{p2y} - 30c_{p2z} = 0$$

$$(c_{p2x})^2 + (c_{p2y})^2 + (c_{p2z})^2 = 1$$

Solving simultaneously, we obtain

$$c_{p2x} = -0.05466$$

$$c_{p2y} = -0.8416$$

$$c_{p2z} = -0.5738$$

Direction cosines for σ_{p3}: The three independent equations become:

$$(-27.72 - 50)c_{p3x} + 10c_{p3y} - 15c_{p3z} = 0$$

$$10c_{p3x} + (-27.72 - 25)c_{p3y} - 30c_{p3z} = 0$$

$$(c_{p3x})^2 + (c_{p3y})^2 + (c_{p3z})^2 = 1$$

Solving simultaneously, we obtain:

$$c_{p3x} = -0.8276$$

$$c_{p3y} = 0.2259$$

$$c_{p3z} = 0.5138$$

2.8 Plane Stress

A stress tensor is *always* defined by six components of stress: three normal stress components and three shear stress components. However, in practice a state of stress is often encountered in which the magnitudes of three stress components in one coordinate direction are known to be zero *a priori*. For example, suppose $\sigma_{zz} = \tau_{xz} = \tau_{yz} = 0$, as shown in Figure 2.13a. As the three remaining nonzero stress components (σ_{xx}, σ_{yy}, and τ_{xy}), all lie within the x–y plane, such a condition is called a state of *plane stress*. Plane stress conditions occur most often because of the geometry of the structure of interest. Specifically, the plane stress condition usually exists in relatively thin, plate-like structures. Examples include the web of an I-beam, the body panel of an automobile, or the skin of an airplane fuselage. In these instances the stresses induced normal to the plane of the structure are very small compared with those induced within the plane of the structure. Hence, the small out-of-plane stresses are assumed to be zero, and attention is focused on the relatively high stress components acting within the plane of the structure.

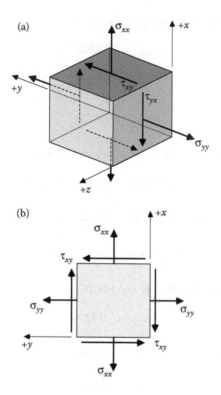

FIGURE 2.13
Stress elements subjected to a state of plane stress. (a) 3-D stress element subjected to a plane stress state (all stress components shown in a positive sense); (b) plane stress element drawn as a square rather than a cube (positive z-axis out of the plane of the figure; all stress components shown in a positive sense).

As laminated composites are often used in the form of thin plates or shells, the plane stress assumption is widely applicable in composite structures and will be used throughout most of the analyses discussed in this book. As the out-of-plane stresses are negligibly small, for convenience an infinitesimal stress element subjected to plane stress will usually be drawn as a square rather than a cube, as shown in Figure 2.13b.

Results discussed in earlier sections for general 3D state of stress will now be specialized for the plane stress condition. It will be assumed that the nonzero stresses lie in the x–y plane (i.e., $\sigma_{zz} = \tau_{xz} = \tau_{yz} = 0$). This allows the remaining components of stress to be written in the form of a column array, rather than a 3×3 array:

$$\begin{bmatrix} \sigma_{xx} & \tau_{xy} & 0 \\ \tau_{xy} & \sigma_{yy} & 0 \\ 0 & 0 & 0 \end{bmatrix} \rightarrow \begin{Bmatrix} \sigma_{xx} \\ \sigma_{yy} \\ \tau_{xy} \end{Bmatrix}$$

Note that when a plane stress state is described, stress *appears* to be a first-order tensor, as (apparently) only three components of stress (σ_{xx}, σ_{yy}, and τ_{xy}) need be specified to describe the state of stress. This is, of course, not the case. Stress is a second-order tensor in all instances, and six components of stress must *always* be specified to define a state of stress. When we invoke the plane stress assumption, we have simply assumed *a priori* that the magnitude of three stress components (σ_{zz}, τ_{xz}, and τ_{yz}) are zero.

Recall that either Equation 2.12a or Equation 2.12b governs the transformation of a stress tensor from one coordinate system to another. Equation 2.12b is repeated here for convenience:

$$
\begin{bmatrix} \sigma_{x'x'} & \sigma_{x'y'} & \sigma_{x'z'} \\ \sigma_{y'x'} & \sigma_{y'y'} & \sigma_{y'z'} \\ \sigma_{z'x'} & \sigma_{z'y'} & \sigma_{z'z'} \end{bmatrix} = \begin{bmatrix} c_{x'x} & c_{x'y} & c_{x'z} \\ c_{y'x} & c_{y'y} & c_{y'z} \\ c_{z'x} & c_{z'y} & c_{z'z} \end{bmatrix} \begin{bmatrix} \sigma_{xx} & \sigma_{xy} & \sigma_{xz} \\ \sigma_{yx} & \sigma_{yy} & \sigma_{yz} \\ \sigma_{zx} & \sigma_{zy} & \sigma_{zz} \end{bmatrix}
$$

$$
\times \begin{bmatrix} c_{x'x} & c_{y'x} & c_{z'x} \\ c_{x'y} & c_{y'y} & c_{z'y} \\ c_{x'z} & c_{y'z} & c_{z'z} \end{bmatrix} \qquad \text{(2.12b) (repeated)}
$$

When transformation of a plane stress tensor is considered, it will be assumed that the x'–y'–z' coordinate system is generated from the x–y–z system by a rotation θ about the z-axis. That is, the z- and z'-axes are coincident, as shown in Figure 2.14. In this case, the direction cosines are

$$c_{x'x} = \cos(\theta)$$

$$c_{x'y} = \cos(90° - \theta) = \sin(\theta)$$

$$c_{x'z} = \cos(90°) = 0$$

$$c_{y'x} = \cos(90° + \theta) = -\sin(\theta)$$

$$c_{y'y} = \cos(\theta)$$

$$c_{y'z} = \cos(90°) = 0$$

$$c_{z'x} = \cos(90°) = 0$$

$$c_{z'y} = \cos(90°) = 0$$

$$c_{z'z} = \cos(0°) = 1$$

If we now (a) substitute these direction cosines into Equation 2.12b, (b) label the shear stresses using the symbol using "τ" rather "σ,"

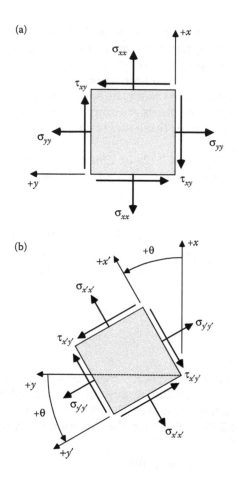

FIGURE 2.14
Transformation of a plane stress element from one coordinate system to another. (a) Plane stress element referenced to the *x–y–z* coordinate system; (b) plane stress element referenced to the *x′–y′–z′* coordinate system, oriented θ-degrees counter-clockwise from the *x–y–z* coordinate system.

and (c) note that $\sigma_{zz} = \tau_{xz} = \tau_{yz} = 0$ by assumption, then Equation 2.12b becomes:

$$
\begin{bmatrix} \sigma_{x'x'} & \tau_{x'y'} & \tau_{x'z'} \\ \tau_{y'x'} & \sigma_{y'y'} & \tau_{y'z'} \\ \tau_{z'x'} & \tau_{z'y'} & \sigma_{z'z'} \end{bmatrix} = \begin{bmatrix} \cos\theta & \sin\theta & 0 \\ -\sin\theta & \cos\theta & 0 \\ 0 & 0 & 1 \end{bmatrix} \begin{bmatrix} \sigma_{xx} & \tau_{xy} & 0 \\ \tau_{yx} & \sigma_{yy} & 0 \\ 0 & 0 & 0 \end{bmatrix}
$$

$$
\times \begin{bmatrix} \cos\theta & -\sin\theta & 0 \\ \sin\theta & \cos\theta & 0 \\ 0 & 0 & 1 \end{bmatrix}
$$

Completing the matrix multiplication indicated results in:

$$
\begin{bmatrix}
\sigma_{x'x'} & \tau_{x'y'} & \tau_{x'z'} \\
\tau_{y'x'} & \sigma_{y'y'} & \tau_{y'z'} \\
\tau_{z'x'} & \tau_{z'y'} & \sigma_{z'z'}
\end{bmatrix}
$$

$$
= \begin{bmatrix}
\cos^2\theta\,\sigma_{xx} + \sin^2\theta\,\sigma_{yy} + 2\cos\theta\sin\theta\,\tau_{xy} & -\cos\theta\sin\theta\,\sigma_{xx} + \cos\theta\sin\theta\,\sigma_{yy} + (\cos^2\theta - \sin^2\theta)\tau_{xy} & 0 \\
-\cos\theta\sin\theta\,\sigma_{xx} + \cos\theta\sin\theta\,\sigma_{yy} + (\cos^2\theta - \sin^2\theta)\tau_{xy} & \sin^2\theta\,\sigma_{xx} + \cos^2\theta\,\sigma_{yy} - 2\cos\theta\sin\theta\,\tau_{xy} & 0 \\
0 & 0 & 0
\end{bmatrix}
$$

(2.18)

As would be expected, the out-of-plane stresses are zero: $\sigma_{z'z'} = \tau_{x'z'} = \tau_{y'z'} = 0$. The remaining stress components are

$$
\sigma_{x'x'} = \cos^2(\theta)\sigma_{xx} + \sin^2(\theta)\sigma_{yy} + 2\cos(\theta)\sin(\theta)\tau_{xy}
$$

$$
\sigma_{y'y'} = \sin^2(\theta)\sigma_{xx} + \cos^2(\theta)\sigma_{yy} - 2\cos(\theta)\sin(\theta)\tau_{xy}
$$

(2.19)

$$
\tau_{x'y'} = -\cos(\theta)\sin(\theta)\sigma_{xx} + \cos(\theta)\sin(\theta)\sigma_{yy} + [\cos^2(\theta) - \sin^2(\theta)]\tau_{xy}
$$

Equation 2.19 can be written using matrix notation as

$$
\begin{Bmatrix}
\sigma_{x'x'} \\
\sigma_{y'y'} \\
\tau_{x'y'}
\end{Bmatrix}
=
\begin{bmatrix}
\cos^2(\theta) & \sin^2(\theta) & 2\cos(\theta)\sin(\theta) \\
\sin^2(\theta) & \cos^2(\theta) & -2\cos(\theta)\sin(\theta) \\
-\cos(\theta)\sin(\theta) & \cos(\theta)\sin(\theta) & \cos^2(\theta) - \sin^2(\theta)
\end{bmatrix}
\begin{Bmatrix}
\sigma_{xx} \\
\sigma_{yy} \\
\tau_{xy}
\end{Bmatrix}
\quad (2.20)
$$

It should be kept in mind that these results are valid only for a state of plane stress. More precisely, Equations 2.19 and 2.20 represent stress transformations within the *x–y* plane, *and are only valid if the z-axis is a principal stress axis.*
The 3×3 array that appears in Equation 2.20 is called the *transformation matrix*, and is abbreviated as [*T*]:

$$
[T] =
\begin{bmatrix}
\cos^2(\theta) & \sin^2(\theta) & 2\cos(\theta)\sin(\theta) \\
\sin^2(\theta) & \cos^2(\theta) & -2\cos(\theta)\sin(\theta) \\
-\cos(\theta)\sin(\theta) & \cos(\theta)\sin(\theta) & \cos^2(\theta) - \sin^2(\theta)
\end{bmatrix}
\quad (2.21)
$$

The stress invariants (given by Equation 2.14 or 2.15) are considerably simplified in the case of plane stress. Since by definition $\sigma_{zz} = \tau_{xz} = \tau_{yz} = 0$, the stress invariants become:

First stress invariant $= \Theta = \sigma_{xx} + \sigma_{yy}$

Second stress invariant $= \Phi = \sigma_{xx}\sigma_{yy} - \tau_{xy}^2$ (2.22)

Third stress invariant $= \Psi = 0$

The principal stresses equal the roots of the cubic equation previously listed as Equation 2.17. In the case of plane stress, this cubic equation becomes (since $\Psi = 0$):

$$\sigma^3 - \Theta\sigma^2 + \Phi\sigma = 0 \tag{2.23}$$

Obviously, one root of Equation 2.23 is $\sigma = 0$. This root corresponds to σ_{zz}, and for present purposes will be labeled σ_{p3} even though it may not be the algebraically least principal stress. Thus, in the case of plane stress the z-axis is a principal stress direction, and $\sigma_{zz} = \sigma_{p3} = 0$ is one of the three principal stresses. As the three principal stress directions are orthogonal, this implies that the remaining two principal stress directions must lie within the x–y plane.

Removing the known root from Equation 2.23, we have the following quadratic equation:

$$\sigma^2 - \Theta\sigma + \Phi = 0 \tag{2.24}$$

The two roots of this quadratic equation are found using the standard approach [3], and are given by

$$\sigma_{p1}, \sigma_{p2} = \frac{1}{2}\left[\Theta \pm \sqrt{\Theta^2 - 4\Phi}\right] \tag{2.25}$$

Substituting Equation 2.22 into 2.25 and simplifying, there results:

$$\sigma_{p1}, \sigma_{p2} = \frac{\sigma_{xx} + \sigma_{yy}}{2} \pm \sqrt{\left(\frac{\sigma_{xx} - \sigma_{yy}}{2}\right)^2 + \tau_{xy}^2} \tag{2.26}$$

The angle θ_p between the x-axis and either the p_1 or p_2 axis is given by

$$\theta_p = \frac{1}{2}\arctan\left(\frac{2\tau_{xy}}{\sigma_{xx} - \sigma_{yy}}\right) \tag{2.27}$$

Example Problem 2.5

Given: The plane stress element shown in Figure 2.15a.

PROBLEM

a. Rotate the stress element to a new coordinate system oriented
   15° clockwise from the *x*-axis, and redraw the stress element
   with all stress components properly oriented.

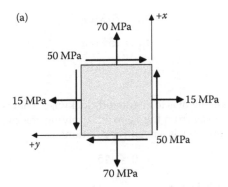

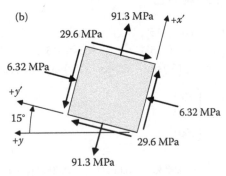

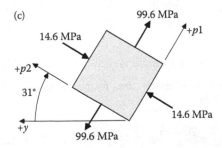

FIGURE 2.15
Plane stress elements associated with Example Problem 2.5. (a) Plane stress element in the
x–*y* coordinate system; (b) plane stress element in the *x'*–*y'* coordinate system; (c) plane stress
element in the principal stress coordinate system.

b. Determine the principal stresses and principal stress coordinate system, and redraw the stress element with the principal stress components properly oriented.

SOLUTION

a. The following components of stress are implied by the stress element shown (note that the shear stress is algebraically negative, in accordance with the sign convention discussed in Section 2.5):

$$\sigma_{xx} = 70\,\text{MPa}$$

$$\sigma_{yy} = 15\,\text{MPa}$$

$$\tau_{xy} = -50\,\text{MPa}$$

The stress element is to be rotated clockwise. That is, the $+x'$-axis is rotated away from the $+y$-axis. Applying the right-hand rule it is clear that this is a negative rotation:

$$\theta = -15°$$

Equation 2.20 becomes:

$$\begin{Bmatrix} \sigma_{x'x'} \\ \sigma_{y'y'} \\ \tau_{x'y'} \end{Bmatrix} = \begin{bmatrix} \cos^2(-15°) & \sin^2(-15°) & 2\cos(-15°)\sin(-15°) \\ \sin^2(-15°) & \cos^2(-15°) & -2\cos(-15°)\sin(-15°) \\ -\cos(-15°)\sin(-15°) & \cos(-15°)\sin(-15°) & \cos^2(-15°) - \sin^2(-15°) \end{bmatrix}$$

$$\times \begin{Bmatrix} 70 \\ 15 \\ -50 \end{Bmatrix}$$

$$\begin{Bmatrix} \sigma_{x'x'} \\ \sigma_{y'y'} \\ \tau_{x'y'} \end{Bmatrix} = \begin{bmatrix} 0.9330 & 0.0670 & -0.5000 \\ 0.0670 & 0.9330 & 0.5000 \\ 0.2500 & -0.2500 & 0.8660 \end{bmatrix} \begin{Bmatrix} 70 \\ 15 \\ -50 \end{Bmatrix} = \begin{Bmatrix} 91.3 \\ -6.32 \\ -29.6 \end{Bmatrix} \text{MPa}$$

The rotated stress element is shown in Figure 2.15b.

b. The principal stresses are found through application of Equation 2.26:

$$\sigma_{p1}, \sigma_{p2} = \frac{70 + 15}{2} \pm \sqrt{\left(\frac{70 - 15}{2}\right)^2 + (-50)^2} = 42.5 \pm 57.1\,\text{MPa}$$

$$\sigma_{p1} = 99.6\,\text{MPa}$$

$$\sigma_{p2} = -14.6\,\text{MPa}$$

The orientation of the principal stress coordinate system is given by Equation 2.27:

$$\theta_p = \frac{1}{2}\arctan\left(\frac{2(-50)}{70-15}\right) = -31°$$

Since θ_p is negative, the $+p1$-axis is oriented 31° clockwise from the x-axis. The stress element is shown in the principal stress coordinate system in Figure 2.15c.

2.9 Definition of Strain

All materials deform to some extent when subjected to external forces and/or environmental changes. In essence, the *state of strain* is a measure of the magnitude and orientation of the deformations induced by these effects. As in the case of stress, there are two types of strain: *normal* strain and *shear* strain.

The two types of strain can be visualized using the strain element shown in Figure 2.16. Imagine that a perfect square has been drawn on a flat surface of interest. Initially, angle $\angle ABC$ is exactly $\pi/2$ radians (i.e., initially $\angle ABC = 90°$), and sides AB and BC are of exactly equal lengths. Now suppose that some mechanism(s) causes the surface to deform. The mechanism(s) which cause the surface to deform need not be defined at this point, but might be external loading (i.e., stresses), a change in temperature, and/or (in the case of polymeric-based materials such as composites) the adsorption or desorption of water molecules. In any event, as the surface is deformed the initially square element drawn on the surface is deformed as well. As shown in Figure 2.16, point A moves to point A', and point C moves to point C'. It is

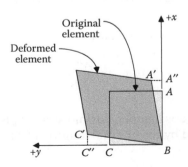

FIGURE 2.16
2-D element used to illustrate normal and shear strains (deformations are shown greatly exaggerated for clarity).

assumed that the element remains a parallelogram, that is, it is assumed that sides $A'B$ and $C'B$ remain straight lines after deformation. This assumption is valid if the element is *infinitesimally small*. In the present context "infinitesimally small" implies that lengths AB and CB are small enough such that the deformed element may be treated as a parallelogram.

Normal strain ε_{xx} is defined as the change in length of AB divided by the original length of AB:

$$\varepsilon_{xx} = \frac{\Delta AB}{AB} \tag{2.28}$$

The change in length AB is given by

$$\Delta AB = (A'B - AB)$$

From the figure it can be seen that the projection of length $A'B$ in the x-direction, that is length $A''B$, is given by

$$A''B = A'B\cos(\angle A'BA) \tag{2.29}$$

If we now assume that $\angle A'BA$ is "small" then we can invoke the *small angle approximation*,* which states that if $\angle A'BA$ is expressed in radians and is less than about 0.1745 radians (about 10°), then:

$$\sin(\angle A'BA) \approx \angle A'BA \quad \tan(\angle A'BA) \approx \angle A'BA \quad \cos(\angle A'BA) \approx 1$$

Based on the small angle approximation Equation 2.29 implies that $A''B \approx A'B$, and therefore that the change in length of AB is approximately given by

$$\Delta AB \approx (A''B - AB) = A''A$$

Equation 2.28 can now be written as:

$$\varepsilon_{xx} = \frac{A''A}{AB} \tag{2.30}$$

In an entirely analogous manner, normal strain ε_{yy} is defined as the change in length of CB divided by the original length of CB:

---

* The reader is encouraged to personally verify the "small angle approximation." For example, use a calculator to demonstrate that an angle of 5 degs equals 0.08727 radians, and that $\sin(0.08727\text{ rad}) = 0.08716$, $\tan(0.08727\text{ rad}) = 0.08749$ and $\cos(0.08727\text{ rad}) = 0.99619$. Therefore, in this example the small angle approximation results in a maximum error of less than 1%.

$$\varepsilon_{yy} = \frac{\Delta CB}{CB}$$

Based on the small angle approximation, the change in length of CB is approximately given by

$$\Delta CB = (C'B - CB) \approx C''C$$

and therefore

$$\varepsilon_{yy} = \frac{C''C}{CB} \qquad (2.31)$$

As before, the approximation for change in length CB is valid if angle $\angle C''BC$ is small.

Recall that the original element shown in Figure 2.16 was assumed to be perfectly square, and in particular that angle $\angle ABC$ is exactly $\pi/2$ radians (i.e., initially $\angle ABC = 90°$). *Engineering shear strain* is defined as the change in angle $\angle ABC$, expressed in radians:

$$\gamma_{xy} = \Delta(\angle ABC) = \angle A'BA + \angle C'BC \qquad (2.32)$$

The subscripts associated with a shear strain (e.g., subscripts "xy" in Equation 2.32) indicate that the shear strain represents the change in angle defined by line segments originally aligned with the x- and y-axes.

As discussed in the following sections, it is very convenient to describe a *state of strain* as a second-order tensor. To do so, we must use a slightly different definition of shear strain. Specifically, *tensoral shear strain* is defined as

$$\varepsilon_{xy} = \frac{1}{2}\gamma_{xy} \qquad (2.33)$$

As engineering shear strain has been defined as the total change in angle $\angle ABC$, tensoral shear strain is simply half this change in angle. The use of tensoral shear strain is convenient because it greatly simplifies the transformation of a state of strain from one coordinate system to another. However, the use of engineering shear strain is far more common in practice. In this chapter, tensoral shear strain will be used during initial mathematical manipulations of the strain tensor, but all final results will be converted to relations involving engineering shear strain.

Although strains are unit-less quantities, normal strains are usually reported in units of (length/length), and shear strains are usually reported in units of radians. The values of a strain is independent of the system of units used, for example, 1 (meter/meter) = 1 (inch/inch). Common abbreviations used throughout this chapter are as follows:

$$1 \times 10^{-6} \text{ meter/meter} = 1 \text{ micrometer/meter} = 1 \,\mu m/m = 1 \,\mu in./in.$$

$$1 \times 10^{-6} \text{ radians} = 1 \text{ microradians} = 1\mu rad$$

We must next define the algebraic sign convention used to describe individual strain components. The sign convention for normal strains is very straightforward and intuitive: a positive (or "tensile") normal strain is associated with an increase in length, while a negative (or "compressive") normal strain is associated with a decrease in length.

To define the algebraic sign of a shear strain, we first identify the algebraic sign of each face of the infinitesimal strain element (the algebraic sign of face was defined in Section 2.5). An algebraically positive shear strain corresponds to a *decrease* in the angle between two positive faces, or equivalently, to a *decrease* in the angle between two negative faces.

The above sign conventions can be used to confirm that all strains shown in Figure 2.16 are algebraically positive.

Example Problem 2.6

Given: The following two sets of strain components:
Set 1:

$$\varepsilon_{xx} = 1000 \,\mu m/m$$

$$\varepsilon_{yy} = -500 \,\mu m/m$$

$$\gamma_{xy} = 1500 \,\mu rad$$

Set 2:

$$\varepsilon_{xx} = 1000 \,\mu m/m$$

$$\varepsilon_{yy} = -500 \,\mu m/m$$

$$\gamma_{xy} = -1500 \,\mu rad$$

Determine: Prepare sketches (not to scale) of the deformed strain elements represented by the two sets of strain components.

SOLUTION

The required sketches are shown in Figure 2.17. Note that the only difference between the two sets of strain components is that in set 1 γ_{xy} is algebraically positive, whereas in set 2 γ_{xy} is algebraically negative.

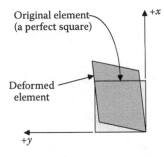
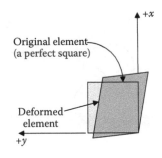

(a) Deformed strain element implied by:

$\varepsilon_{xx} = 1000\ \mu m/m$

$\varepsilon_{yy} = -500\ \mu m/m$

$\gamma_{xy} = 1500\ \mu rad$

(b) Deformed strain element implied by:

$\varepsilon_{xx} = 1000\ \mu m/m$

$\varepsilon_{yy} = -500\ \mu m/m$

$\gamma_{xy} = -1500\ \mu rad$

FIGURE 2.17
Strain elements associated with Example Problem 2.6 (not to scale).

2.10 The Strain Tensor

A general 3D solid body is shown in Figure 2.18a. An infinitesimally small cube isolated from an interior region of the body is shown in Figure 2.18b. The cube is referenced to an x–y–z coordinate system, and the cube edges are aligned with these axes.

Now assume that the body is subjected to some mechanism(s) which cause the body to deform. The mechanism(s) which causes this deformation need not be defined at this point, but might be external loading (i.e., stresses), a change in temperature, the adsorption or desorption of water molecules (in the case of polymeric-based materials such as composites), or any combination thereof.

As the entire body is deformed, the internal infinitesimal cube is deformed into a parallelepiped, as shown in Figure 2.18c. It can be shown [1,2] that the state of strain experienced by the cube can represented as a symmetric second-order tensor, involving six components of strain: three normal strains ($\varepsilon_{xx}, \varepsilon_{yy}, \varepsilon_{zz}$) and three tensoral shear strains ($\varepsilon_{xy}, \varepsilon_{xz}, \varepsilon_{yz}$). These six strain components are defined in the same manner as those discussed in the preceding section. Normal strains $\varepsilon_{xx}, \varepsilon_{yy}$, and ε_{zz} represent the change in length in the x-, y-, and z-directions, respectively. Tensoral shear strains $\varepsilon_{xy}, \varepsilon_{xz}$, and ε_{yz} represent the change in angle between cube edges initially aligned with the $(x$-, y-$)$, $(x$-, z-$)$, and $(y$-, z-$)$ axes, respectively. Using matrix notation the strain tensor may be written as

$$\begin{bmatrix} \varepsilon_{xx} & \varepsilon_{xy} & \varepsilon_{xz} \\ \varepsilon_{yx} & \varepsilon_{yy} & \varepsilon_{yz} \\ \varepsilon_{zx} & \varepsilon_{zy} & \varepsilon_{zz} \end{bmatrix} = \begin{bmatrix} \varepsilon_{xx} & \varepsilon_{xy} & \varepsilon_{xz} \\ \varepsilon_{xy} & \varepsilon_{yy} & \varepsilon_{yz} \\ \varepsilon_{xz} & \varepsilon_{yz} & \varepsilon_{zz} \end{bmatrix} \qquad (2.34)$$

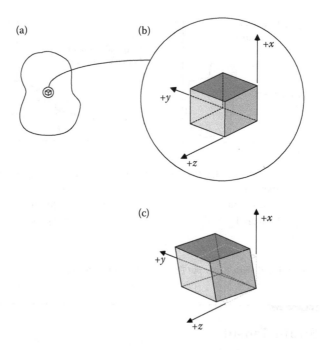

FIGURE 2.18
Infinitesimal element used to illustrate the strain tensor. (a) General 3-D solid body; (b) infinitesimal cube removed from the body, prior to deformation; (c) infinitesimal cube removed from the body, after deformation.

Alternatively, the strain tensor can be succinctly written using indicial notation as

$$\varepsilon_{ij}, \; i,j = \; x,y, \; \text{or} \; z \tag{2.35}$$

Note that if engineering shear strain is used, then Equation 2.34 becomes

$$\begin{bmatrix} \varepsilon_{xx} & \varepsilon_{xy} & \varepsilon_{xz} \\ \varepsilon_{xy} & \varepsilon_{yy} & \varepsilon_{yz} \\ \varepsilon_{xz} & \varepsilon_{yz} & \varepsilon_{zz} \end{bmatrix} = \begin{bmatrix} \varepsilon_{xx} & (\gamma_{xy}/2) & (\gamma_{xz}/2) \\ (\gamma_{xy}/2) & \varepsilon_{yy} & (\gamma_{yz}/2) \\ (\gamma_{xz}/2) & (\gamma_{yz}/2) & \varepsilon_{zz} \end{bmatrix}$$

If engineering shear strain is used, the strain tensor *cannot* be written using indicial notation (as in Equation 2.35), due to the 1/2 factor that appears in all off-diagonal positions.

In Section 2.1 it was noted that a force vector is a first-order tensor, as only one subscript is required to describe a force tensor, F_i. The fact that strain is a second-order tensor is evident from Equation 2.35, as two subscripts are necessary to describe a state of strain.

2.11 Transformation of the Strain Tensor

As both stress and strain are second-order tensors, transformation of the strain tensor from one coordinate system to another is analogous to transformation of the stress tensor, as discussed in Section 2.6. For example, it can be shown [1,2] that the strain components in the x'–y'–z' coordinate system ($\varepsilon_{i'j'}$) are related to the components in x–y–z coordinate system (ε_{ij}) according to

$$\varepsilon_{i'j'} = c_{i'k}\, c_{j'l}\, \varepsilon_{kl} \quad \text{where} \quad k, l = x, y, z \qquad (2.36a)$$

Alternatively, using matrix notation the strain tensor transforms according to

$$[\varepsilon_{i'j'}] = [c_{i'j}][\varepsilon_{ij}][c_{i'j}]^T$$

which expands as follows:

$$
\begin{bmatrix}
\varepsilon_{x'x'} & \varepsilon_{x'y'} & \varepsilon_{x'z'} \\
\varepsilon_{y'x'} & \varepsilon_{y'y'} & \varepsilon_{y'z'} \\
\varepsilon_{z'x'} & \varepsilon_{z'y'} & \varepsilon_{z'z'}
\end{bmatrix}
=
\begin{bmatrix}
c_{x'x} & c_{x'y} & c_{x'z} \\
c_{y'x} & c_{y'y} & c_{y'z} \\
c_{z'x} & c_{z'y} & c_{z'z}
\end{bmatrix}
\begin{bmatrix}
\varepsilon_{xx} & \varepsilon_{xy} & \varepsilon_{xz} \\
\varepsilon_{yx} & \varepsilon_{yy} & \varepsilon_{yz} \\
\varepsilon_{zx} & \varepsilon_{zy} & \varepsilon_{zz}
\end{bmatrix}
\begin{bmatrix}
c_{x'x} & c_{y'x} & c_{z'x} \\
c_{x'y} & c_{y'y} & c_{z'y} \\
c_{x'z} & c_{y'z} & c_{z'z}
\end{bmatrix}
$$

$$(2.36b)$$

The terms $c_{i'j}$ which appear in Equation 2.36a,b are direction cosines and equal the cosine of the angle between the axes of the x'–y'–z' and x–y–z coordinate systems.

As was the case for the stress tensor, there are certain features of the strain tensor that do not vary when the tensor is transformed from one coordinate system to another. These features are called the *strain invariants*. Three independent strain invariants exist, and are defined as follows:

$$\text{First strain invariant} = \Theta_\varepsilon = \varepsilon_{ii} \qquad (2.37a)$$

$$\text{Second strain invariant} = \Phi_\varepsilon = \frac{1}{2}(\varepsilon_{ii}\varepsilon_{jj} - \varepsilon_{ij}\varepsilon_{ij}) \qquad (2.37b)$$

$$\text{Third strain invariant} = \Psi_\varepsilon = \frac{1}{6}(\varepsilon_{ii}\varepsilon_{jj}\varepsilon_{kk} - 3\varepsilon_{ii}\varepsilon_{jk}\varepsilon_{jk} + 2\varepsilon_{ij}\varepsilon_{jk}\varepsilon_{ki}) \qquad (2.37c)$$

Alternatively, by expanding these equations over the range $i, j, k = x, y, z$ and simplifying, the strain invariants can be written as

$$\text{First strain invariant} = \Theta_\varepsilon = \varepsilon_{xx} + \varepsilon_{yy} + \varepsilon_{zz} \tag{2.38a}$$

$$\text{Second strain invariant} = \Phi_\varepsilon = \varepsilon_{xx}\varepsilon_{yy} + \varepsilon_{xx}\varepsilon_{zz} + \varepsilon_{yy}\varepsilon_{zz} - (\varepsilon_{xy}^2 + \varepsilon_{xz}^2 + \varepsilon_{yz}^2) \tag{2.38b}$$

$$\text{Third strain invariant} = \Psi_\varepsilon = \varepsilon_{xx}\varepsilon_{yy}\varepsilon_{zz} - \varepsilon_{xx}\varepsilon_{yz}^2 - \varepsilon_{yy}\varepsilon_{xz}^2 - \varepsilon_{zz}\varepsilon_{xy}^2 + 2\varepsilon_{xy}\varepsilon_{xz}\varepsilon_{yz} \tag{2.38c}$$

Example Problem 2.7

Given: A state of strain referenced to an $x-y-z$ coordinate is known to be

$$
\begin{bmatrix} \varepsilon_{xx} & \varepsilon_{xy} & \varepsilon_{xz} \\ \varepsilon_{yx} & \varepsilon_{yy} & \varepsilon_{yz} \\ \varepsilon_{zx} & \varepsilon_{zy} & \varepsilon_{zz} \end{bmatrix} =
\begin{bmatrix} 1000~\mu\text{m/m} & 500~\mu\text{rad} & 250~\mu\text{rad} \\ 500~\mu\text{rad} & 1500~\mu\text{m/m} & 750~\mu\text{rad} \\ 250~\mu\text{rad} & 750~\mu\text{rad} & 2000~\mu\text{m/m} \end{bmatrix}
$$

It is desired to express this state of strain in a $x''-y''-z''$ coordinate system, generated by

i. Rotation of $\theta = 20°$ about the original z-axis (which defines an intermediate $x'-y'-z'$ coordinate system), followed by
ii. Rotation of $\beta = 35°$ about the x'-axis (which defines the final $x''-y''-z''$ coordinate system)

(this coordinate transformation has been previously considered in Example Problem 2.1, and is shown in Figure 2.2).

PROBLEM

a. Rotate the strain tensor to the $x''-y''-z''$ coordinate system.
b. Calculate the first, second, and third invariants of the strain tensor using (i) elements of the strain tensor referenced to the $x-y-z$ coordinate system, ε_{ij}, and (ii) elements of the strain tensor referenced to the $x''-y''-z''$ coordinate system, $\varepsilon_{i''j''}$.

SOLUTION

a. General expressions for direction cosines relating the $x-y-z$ and $x''-y''-z''$ coordinate systems were determined as a part of Example Problem 2.1. Furthermore, numerical values for the particular rotation $\theta = 20°$ and $\beta = 35°$ were determined in Example Problem 2.3 and found to be:

$$c_{x''x} = \cos(20°) = 0.9397$$

$$c_{x''y} = \sin(20°) = 0.3420$$

$$c_{x''z} = 0$$

$$c_{y''x} = -\cos(35°)\sin(20°) = -0.2802$$

$$c_{y''y} = \cos(35°)\cos(20°) = 0.7698$$

$$c_{y''z} = \sin(35°) = 0.5736$$

$$c_{z''x} = \sin(35°)\sin(20°) = 0.1962$$

$$c_{z''y} = -\sin(35°)\cos(20°) = -0.5390$$

$$c_{z''z} = \cos(35°) = 0.8192$$

Each component of the transformed strain tensor can now be found through application of Equation 2.36a or Equation 2.36b. For example, setting $i' = x''$, $j' = x''$ and expanding Equation 2.36a, strain component $\varepsilon_{x''x''}$ is given by

$$\varepsilon_{x''x''} = c_{x''x}\, c_{x''x}\, \varepsilon_{xx} + c_{x''x}\, c_{x''y}\, \varepsilon_{xy} + c_{x''x}\, c_{x''z}\, \varepsilon_{xz}$$

$$+ c_{x''y}\, c_{x''x}\, \varepsilon_{yx} + c_{x''y}\, c_{x''y}\, \varepsilon_{yy} + c_{x''y}\, c_{x''z}\, \varepsilon_{yz}$$

$$+ c_{x''z}\, c_{x''x}\, \varepsilon_{zx} + c_{x''z}\, c_{x''y}\, \varepsilon_{zy} + c_{x''z}\, c_{x''z}\, \varepsilon_{zz}$$

$$\varepsilon_{x''x''} = (0.9397)(0.9397)(1000) + (0.9397)(0.3420)(500) + (0.9397)(0)(250)$$

$$+ (0.3420)(0.9397)(500) + (0.3420)(0.3420)(1500) + (0.3420)(0)(750)$$

$$+ (0)(0.9397)(250) + (0)(0.3420)(750) + (0)(0)(2000)$$

$$\varepsilon_{x''x''} = 1380\ \mu m/m$$

Alternatively, if matrix notation is used, then Equation 2.36b becomes

$$\begin{bmatrix} \varepsilon_{x''x''} & \varepsilon_{x''y''} & \varepsilon_{x''z''} \\ \varepsilon_{y''x''} & \varepsilon_{y''y''} & \varepsilon_{y''z''} \\ \varepsilon_{z''x''} & \varepsilon_{z''y''} & \varepsilon_{z''z''} \end{bmatrix} = \begin{bmatrix} 0.9397 & 0.3420 & 0 \\ -0.2802 & 0.7698 & 0.5736 \\ 0.1962 & -0.5390 & 0.8192 \end{bmatrix} \begin{bmatrix} 1000 & 500 & 250 \\ 500 & 1500 & 750 \\ 250 & 750 & 2000 \end{bmatrix}$$

$$\times \begin{bmatrix} 0.9397 & -0.2802 & 0.1962 \\ 0.3420 & 0.7698 & -0.5390 \\ 0 & 0.5736 & 0.8192 \end{bmatrix}$$

Completing the matrix multiplication indicated, there results:

$$
\begin{bmatrix}
\varepsilon_{x''x''} & \varepsilon_{x''y''} & \varepsilon_{x''z''} \\
\varepsilon_{y''x''} & \varepsilon_{y''y''} & \varepsilon_{y''z''} \\
\varepsilon_{z''x''} & \varepsilon_{z''y''} & \varepsilon_{z''z''}
\end{bmatrix}
=
\begin{bmatrix}
1380\ \mu\text{m/m} & 727\ \mu\text{rad} & 91\ \mu\text{rad} \\
727\ \mu\text{rad} & 1991\ \mu\text{m/m} & 625\ \mu\text{rad} \\
91\ \mu\text{rad} & 625\ \mu\text{rad} & 1129\ \mu\text{m/m}
\end{bmatrix}
$$

Notice that the value of $\varepsilon_{x''x''}$ determined through matrix multiplication is identical to that obtained using indicial notation, as expected.

b. The first, second, and third strain invariants will now be calculated using components of both ε_{ij} and $\varepsilon_{i''j''}$. It is expected that identical values will be obtained, as the strain invariants are independent of coordinate system.

First Strain Invariant

x–y–z coordinate system:

$$
\Theta_\varepsilon = \varepsilon_{ii} = \varepsilon_{xx} + \varepsilon_{yy} + \varepsilon_{zz}
$$

$$
\Theta_\varepsilon = (1000 + 1500 + 2000)\ \mu\text{m/m}
$$

$$
\Theta_\varepsilon = 4500\ \mu\text{m/m} = 0.004500\ \text{m/m}
$$

x''–y''–z'' coordinate system:

$$
\Theta_\varepsilon = \varepsilon_{i''i''} = \varepsilon_{x''x''} + \varepsilon_{y''y''} + \varepsilon_{z''z''}
$$

$$
\Theta_\varepsilon = (1380 + 1991 + 1129)\ \mu\text{m/m}
$$

$$
\Theta_\varepsilon = 4500\ \mu\text{m/m} = 0.004500\ \text{m/m}
$$

As expected, the first strain invariant is independent of coordinate system.

Second Strain Invariant

x–y–z coordinate system:

$$
\Phi_\varepsilon = \frac{1}{2}\left(\varepsilon_{ii}\varepsilon_{jj} - \varepsilon_{ij}\varepsilon_{ij}\right) = \varepsilon_{xx}\varepsilon_{yy} + \varepsilon_{xx}\varepsilon_{zz} + \varepsilon_{yy}\varepsilon_{zz} - \left(\varepsilon_{xy}^2 + \varepsilon_{xz}^2 + \varepsilon_{yz}^2\right)
$$

$$
\Phi_\varepsilon = \{(1000)(1500) + (1000)(2000) + (1500)(2000)
$$

$$
- [(500)^2 + (250)^2 + (750)^2]\}(\mu\text{m/m})^2
$$

$$
\Phi = 5.625 \times 10^6\ (\mu\text{m/m})^2 = 5.625 \times 10^{-6}\ (\text{m/m})^2
$$

x''–y''–z'' coordinate system:

$$\Phi_\varepsilon = \frac{1}{2}\left(\varepsilon_{i''i''}\varepsilon_{j''j''} - \varepsilon_{i''j''}\varepsilon_{i''j''}\right)$$

$$\Phi_\varepsilon = \varepsilon_{x''x''}\varepsilon_{y''y''} + \varepsilon_{x''x''}\varepsilon_{z''z''} + \varepsilon_{y''y''}\varepsilon_{z''z''} - \left(\varepsilon_{x''x''}^2 + \varepsilon_{x''z''}^2 + \varepsilon_{y''z''}^2\right)$$

$$\Phi_\varepsilon = \{(1380)(1991) + (1380)(1129) + (1991)(1129)$$
$$- [(727)^2 + (91)^2 + (625)^2]\}\,(\mu m/m)^2$$

$$\Phi_\varepsilon = 5.625 \times 10^6 \,(\mu m/m)^2 = 5.625 \times 10^{-6}(m/m)^2$$

As expected, the second strain invariant is independent of coordinate system.

Third Strain Invariant

x–y–z coordinate system:

$$\Psi_\varepsilon = \frac{1}{6}\left(\varepsilon_{ii}\varepsilon_{jj}\varepsilon_{kk} - 3\varepsilon_{ii}\varepsilon_{jk}\varepsilon_{jk} + 2\varepsilon_{ij}\varepsilon_{jk}\varepsilon_{ki}\right)$$

$$\Psi_\varepsilon = \varepsilon_{xx}\varepsilon_{yy}\varepsilon_{zz} - \varepsilon_{xx}\varepsilon_{yz}^2 - \varepsilon_{yy}\varepsilon_{xz}^2 - \varepsilon_{zz}\varepsilon_{xy}^2 + 2\varepsilon_{xy}\varepsilon_{xz}\varepsilon_{yz}$$

$$\Psi_\varepsilon = [(1000)(1500)(2000) - (1000)(750)^2 - (1500)(250)^2 - (2000)(500)^2$$
$$+ 2(500)(250)(750)]\,(\mu m/m)^3$$

$$\Psi = 2.031 \times 10^9 \,(\mu m/m)^3 = 2.031 \times 10^{-9}(m/m)^3$$

x''–y''–z'' coordinate system:

$$\Psi_\varepsilon = \frac{1}{6}\left(\varepsilon_{i''i''}\varepsilon_{j''j''}\varepsilon_{k''k''} - 3\varepsilon_{i''i''}\varepsilon_{j''k''}\varepsilon_{j''k''} + 2\varepsilon_{i''j''}\varepsilon_{j''k''}\varepsilon_{k''i''}\right)$$

$$\Psi_\varepsilon = \varepsilon_{x''x''}\varepsilon_{y''y''}\varepsilon_{z''z''} - \varepsilon_{x''x''}\varepsilon_{y''z''}^2 - \varepsilon_{y''y''}\varepsilon_{x''z''}^2 - \varepsilon_{z''z''}\varepsilon_{x''y''}^2 + 2\varepsilon_{x''y''}\varepsilon_{x''z''}\varepsilon_{y''z''}$$

$$\Psi_\varepsilon = [(1380)(1991)(1129) - (1380)(625)^2 - (1991)(91)^2 - (1129)(727)^2$$
$$+ 2(727)(91)(625)]\,(\mu m/m)^3$$

$$\Psi = 2.031 \times 10^9 (\mu m/m)^3 = 2.031 \times 10^{-9}(m/m)^3$$

As expected, the third stress invariant is independent of coordinate system.

2.12 Principal Strains

It is always possible to rotate the strain tensor to a special coordinate system in which no shear strains exist. This coordinate system is called the *principal strain coordinate system*, and the normal strains that exist in this coordinate system are called *principal strains*. Calculation of principal strains (and principal stresses) is important during the study of traditional isotropic materials and structures because principal strains or stresses can be used to predict failure of isotropic materials. This is not the case for anisotropic composite materials. Failure of composite material is *not* governed by principal strains or stresses. Principal strains are only of occasional interest to the composite engineer and are reviewed here only in the interests of completeness.

Principal strains are usually denoted ε_1, ε_2, and ε_3. However, in the study of composites the labels "1," "2," and "3" are used to refer to the principal material coordinate system rather than the directions of principal strain. Therefore, in this chapter the axes associated with the principal strain coordinate system will be labeled the "p1", "p2", and "p3" axes, and the principal strains will be denoted ε_{p1}, ε_{p2} and ε_{p3}.

As both stress and strain are second-order tensors, the principal strains can be found using an approach analogous to that used to find principal stresses. Specifically, it can be shown [1,2] that the principal strains must satisfy the following three simultaneous equations:

$$(\varepsilon_{pn} - \varepsilon_{xx})c_{pnx} - \varepsilon_{xy}c_{pny} - \varepsilon_{xz}c_{pnz} = 0 \tag{2.39a}$$

$$-\varepsilon_{xy}c_{pnx} + (\varepsilon_{pn} - \varepsilon_{yy})c_{pny} - \varepsilon_{yz}c_{pnz} = 0 \tag{2.39b}$$

$$-\varepsilon_{xz}c_{pnx} - \varepsilon_{yz}c_{pny} + (\varepsilon_{pn} - \varepsilon_{zz})c_{pnz} = 0 \tag{2.39c}$$

As direction cosines c_{pnx}, c_{pny}, and c_{pnz} must also satisfy Equation 2.8, and therefore cannot all equal zero, the solution can be obtained by requiring that the determinant of the coefficients of c_{pnx}, c_{pny}, and c_{pnz} equal zero:

$$\begin{vmatrix} (\varepsilon_{pn} - \varepsilon_{xx}) & -\varepsilon_{xy} & -\varepsilon_{xz} \\ -\varepsilon_{xy} & (\varepsilon_{pn} - \varepsilon_{yy}) & -\varepsilon_{yz} \\ -\varepsilon_{xz} & -\varepsilon_{yz} & (\varepsilon_{pn} - \varepsilon_{zz}) \end{vmatrix} = 0$$

Equating the determinant to zero results in the following cubic equation:

$$\varepsilon_{pn}^3 - \Theta_\varepsilon \varepsilon_{pn}^2 + \Phi_\varepsilon \varepsilon_{pn} - \Psi_\varepsilon = 0 \tag{2.40}$$

where Θ_ε, Φ_ε, and Ψ_ε are the first, second, and third strain invariants, respectively, and have been previously listed as Equations 2.37 and 2.38. The three roots of the cubic equation represent the three principal strains and may be found by application of the standard approach [3]. By convention, the principal stresses are numbered such that ε_{p1} is the algebraically greatest principal stress, whereas ε_{p3} is the algebraically least. That is, $\varepsilon_{p1} > \varepsilon_{p2} > \varepsilon_{p3}$.

Once the principal strains are determined, the three sets of direction cosines (which define the principal coordinate directions) are found by substituting the three principal strains given by Equation 2.40 into Equation 2.39 in turn. As only two of Equation 2.39 are independent, Equation 2.8 is used as a third independent equation involving the three unknown constants, c_{pnx}, c_{pny}, and c_{pnz}.

The process of finding principal strains and direction cosines will be demonstrated in the following Example Problem.

Example Problem 2.8

Given: A state of strain referenced to an *x–y–z* coordinate is known to be

$$\begin{bmatrix} \varepsilon_{xx} & \varepsilon_{xy} & \varepsilon_{xz} \\ \varepsilon_{yx} & \varepsilon_{yy} & \varepsilon_{yz} \\ \varepsilon_{zx} & \varepsilon_{zy} & \varepsilon_{zz} \end{bmatrix} = \begin{bmatrix} 1000\,\mu m/m & 500\,\mu rad & 250\,\mu rad \\ 500\,\mu rad & 1500\,\mu m/m & 750\,\mu rad \\ 250\,\mu rad & 750\,\mu rad & 2000\,\mu m/m \end{bmatrix}$$

PROBLEM

Find (a) the principal strains and (b) the direction cosines that define the principal strain coordinate system.

SOLUTION

This is the same strain tensor considered in Example Problem 2.7. As a part of that problem, the first, second, and third strain invariants were found to be

$$\Theta_\varepsilon = 0.004500\,m/m$$

$$\Phi = 5.625 \times 10^{-6}\,(m/m)^2$$

$$\Psi = 2.031 \times 10^{-9}\,(m/m)^3$$

a. *Determining the Principal Strains:* In accordance with Equation 2.40, the three principal strains are the roots of the following cubic equation:

$$\varepsilon_{pn}^3 - (0.004500)\varepsilon_{pn}^2 + (5.625 \times 10^{-6})\varepsilon_{pn} - (2.031 \times 10^{-9}) = 0$$

The three roots of this equation represent the three principal strains, and are given by

$$\varepsilon_{p1} = 2689\,\mu m/m, \quad \varepsilon_{p2} = 1160\,\mu m/m, \quad \text{and} \quad \varepsilon_{p3} = 651\,\mu m/m$$

b. *Determining the Direction Cosines:* The first two of Equations 2.39 and 2.8 will be used to form three independent equations in three unknowns. We have

$$(\varepsilon_{pn} - \varepsilon_{xx})c_{pnx} - \varepsilon_{xy}c_{pny} - \varepsilon_{xz}c_{pnz} = 0$$

$$-\varepsilon_{xy}c_{pnx} + (\varepsilon_{pn} - \varepsilon_{yy})c_{pny} - \varepsilon_{yz}c_{pnz} = 0$$

$$(c_{pnx})^2 + (c_{pny})^2 + (c_{pnz})^2 = 1$$

Direction cosines for ε_{p1}: The three independent equations become

$$(2689 - 1000)c_{p1x} - 500c_{p1y} - 250c_{p1z} = 0$$

$$-500c_{p1x} + (2689 - 1500)c_{p1y} - 750c_{p1z} = 0$$

$$(c_{p1x})^2 + (c_{p1y})^2 + (c_{p1z})^2 = 1$$

Solving simultaneously, we obtain

$$c_{p1x} = 0.2872$$

$$c_{p1y} = 0.5945$$

$$c_{p1z} = 0.7511$$

Direction cosines for ε_{p2}: The three independent equations become

$$(1160 - 1000)c_{p2x} - 500c_{p2y} - 250c_{p2z} = 0$$

$$-500c_{p2x} + (1160 - 1500)c_{p2y} - 750c_{p2z} = 0$$

$$(c_{p2x})^2 + (c_{p2y})^2 + (c_{p2z})^2 = 1$$

Solving simultaneously, we obtain

$$c_{p2x} = 0.5960$$

$$c_{p2y} = 0.5035$$

$$c_{p2z} = -0.6256$$

Direction cosines for ε_{p3}: The three independent equations become

$$(651 - 1000)c_{p3x} - 500c_{p3y} - 250c_{p3z} = 0$$

$$-500c_{p3x} + (651 - 1500)c_{p3y} - 750c_{p3z} = 0$$

$$(c_{p3x})^2 + (c_{p3y})^2 + (c_{p3z})^2 = 1$$

Solving simultaneously, we obtain

$$c_{p3x} = -0.7481$$

$$c_{p3y} = 0.6286$$

$$c_{p3z} = -0.2128$$

2.13 Strains within a Plane Perpendicular to a Principal Strain Direction

It has been seen that a strain tensor is defined by six components of strain: three normal strain components and three shear strain components. However, in practice there are circumstances in which it is known *a priori* that both shear strain components in one direction are zero: $\varepsilon_{xz} = \varepsilon_{yz} = 0$, say (or equivalently, $\gamma_{xz} = \gamma_{yz} = 0$). *This implies that the z-axis is a principal strain axis.* In these instances, we are primarily interested in the strains induced within the x–y plane, ε_{xx}, ε_{yy}, ε_{xy}. Two different circumstances are encountered in which it is known *a priori* that the z-axis is a principal strain axis.

In the first case, *all three out-of-plane strain components in the z-direction* are known *a priori* to equal zero. That is, it is known *a priori* that $\varepsilon_{zz} = \varepsilon_{xz} = \varepsilon_{yz} = 0$. Not only is the z-axis a principal strain axis in this case, but in addition the principal strain equals zero: $\varepsilon_{zz} = \varepsilon_{p3} = 0$. As the three remaining nonzero strain components (ε_{xx}, ε_{yy}, and ε_{xy}) all lie within the x–y plane, it is natural to call this condition a *state of plane strain*. Plane strain conditions occur most often because of the geometry of the structure of interest. Specifically, the plane strain condition usually exists in internal regions of very long (or very thick) structures. Examples include solid shafts or long dams. In these instances the strains induced along the long axis of the structure are often negligibly small compared with those induced within the transverse plane of the structure.

The second case in which the out-of-plane z-axis *may* be a principal axis is when a structure is subjected to a state of *plane stress*. As has been discussed in Section 2.8, the state of plane stress occurs most often in thin, plate-like structures. In this case, the z-axis is a principal strain axis, and ε_{zz} is again one of the principal strains. However, in this second case the out of plane normal strain does not, in general, equal zero: $\varepsilon_{zz} \neq 0$.

It is emphasized that a state of plane stress usually, *but not always*, causes a state of strain in which the z-axis is a principal strain axis. This point will be further discussed in Chapter 4. It will be seen that it is possible for a material to exhibit a coupling between in-plane stresses and out-of-plane shear strains. That is, in some cases stresses acting within the x–y plane (σ_{xx}, σ_{yy}, and/or τ_{xy}) can cause out-of-plane shear strains (ε_{xz} and/or ε_{yz}). In these instances, the out-of-plane z-axis is not a principal strain axis, even though the out-of-plane stresses all equal zero.

In any event, for present purposes assume that it is known *a priori* that the out-of-plane z-axis is a principal strain axis, and we are primarily interested in the strains induced within the x–y plane, ε_{xx}, ε_{yy}, and ε_{xy}. We will write these strains in the form of a column array, rather than a 3×3 array:

$$\begin{bmatrix} \varepsilon_{xx} & \varepsilon_{xy} & 0 \\ \varepsilon_{xy} & \varepsilon_{yy} & 0 \\ 0 & 0 & \varepsilon_{zz} \end{bmatrix} = \begin{bmatrix} \varepsilon_{xx} & (\gamma_{xy}/2) & 0 \\ (\gamma_{xy}/2) & \varepsilon_{yy} & 0 \\ 0 & 0 & \varepsilon_{zz} \end{bmatrix} \rightarrow \begin{Bmatrix} \varepsilon_{xx} \\ \varepsilon_{yy} \\ \varepsilon_{xy} \end{Bmatrix} = \begin{Bmatrix} \varepsilon_{xx} \\ \varepsilon_{yy} \\ \gamma_{xy}/2 \end{Bmatrix} \quad (2.41)$$

Note that ε_{zz} does not appear in the column array. This is not of concern in the case of plane strain, as in this case $\varepsilon_{zz} = 0$. However, in the case of plane stress it is important to remember that (in general) $\varepsilon_{zz} \neq 0$. Although in following chapters we will be primarily interested in strains induced within the x–y plane, the reader is advised to remember that an out-of-plane strain ε_{zz} is also induced by a state of plane stress.

The transformation of a general 3D strain tensor has already been discussed in Section 2.11. The relations presented there will now be simplified for the case of transformation of strains within a plane.

Recall that either Equation 2.36a or Equation2.36b governs the transformation of a strain tensor from one coordinate system to another. Equation 2.36b is repeated here for convenience:

$$\begin{bmatrix} \varepsilon_{x'x'} & \varepsilon_{x'y'} & \varepsilon_{x'z'} \\ \varepsilon_{y'x'} & \varepsilon_{y'y'} & \varepsilon_{y'z'} \\ \varepsilon_{z'x'} & \varepsilon_{z'y'} & \varepsilon_{z'z'} \end{bmatrix} = \begin{bmatrix} c_{x'x} & c_{x'y} & c_{x'z} \\ c_{y'x} & c_{y'y} & c_{y'z} \\ c_{z'x} & c_{z'y} & c_{z'z} \end{bmatrix} \begin{bmatrix} \varepsilon_{xx} & \varepsilon_{xy} & \varepsilon_{xz} \\ \varepsilon_{yx} & \varepsilon_{yy} & \varepsilon_{yz} \\ \varepsilon_{zx} & \varepsilon_{zy} & \varepsilon_{zz} \end{bmatrix} \begin{bmatrix} c_{x'x} & c_{y'x} & c_{z'x} \\ c_{x'y} & c_{y'y} & c_{z'y} \\ c_{x'z} & c_{y'z} & c_{z'z} \end{bmatrix}$$

$$(2.36b) \text{ (repeated)}$$

Assuming that the x'–y'–z' coordinate system is generated from the x–y–z system by a rotation θ about the z-axis, the direction cosines are

$$c_{x'x} = \cos(\theta)$$

$$c_{x'y} = \cos(90° - \theta) = \sin(\theta)$$

$$c_{x'z} = \cos(90°) = 0$$

$$c_{y'x} = \cos(90° + \theta) = -\sin(\theta)$$

$$c_{y'y} = \cos(\theta)$$

$$c_{y'z} = \cos(90°) = 0$$

$$c_{z'x} = \cos(90°) = 0$$

$$c_{z'y} = \cos(90°) = 0$$

$$c_{z'z} = \cos(0°) = 1$$

Substituting these direction cosines into Equation 2.36b and noting that by assumption $\varepsilon_{xz} = \varepsilon_{yz} = 0$, we have

$$
\begin{bmatrix} \varepsilon_{x'x'} & \varepsilon_{x'y'} & \varepsilon_{x'z'} \\ \varepsilon_{y'x'} & \varepsilon_{y'y'} & \varepsilon_{y'z'} \\ \varepsilon_{z'x'} & \varepsilon_{z'y'} & \varepsilon_{z'z'} \end{bmatrix} =
\begin{bmatrix} \cos\theta & \sin\theta & 0 \\ -\sin\theta & \cos\theta & 0 \\ 0 & 0 & 1 \end{bmatrix}
\begin{bmatrix} \varepsilon_{xx} & \varepsilon_{xy} & 0 \\ \varepsilon_{yx} & \varepsilon_{yy} & 0 \\ 0 & 0 & \varepsilon_{zz} \end{bmatrix}
\begin{bmatrix} \cos\theta & -\sin\theta & 0 \\ \sin\theta & \cos\theta & 0 \\ 0 & 0 & 1 \end{bmatrix}
$$

Completing the matrix multiplication indicated results in:

$$
\begin{bmatrix} \varepsilon_{x'x'} & \varepsilon_{x'y'} & \varepsilon_{x'z'} \\ \varepsilon_{y'x'} & \varepsilon_{y'y'} & \varepsilon_{y'z'} \\ \varepsilon_{z'x'} & \varepsilon_{z'y'} & \varepsilon_{z'z'} \end{bmatrix}
$$

$$
= \begin{bmatrix} \cos^2\theta \varepsilon_{xx} + \sin^2\theta \varepsilon_{yy} + 2\cos\theta\sin\theta \varepsilon_{xy} & -\cos\theta\sin\theta \varepsilon_{xx} + \cos\theta\sin\theta \varepsilon_{yy} + (\cos^2\theta - \sin^2\theta)\varepsilon_{xy} & 0 \\ -\cos\theta\sin\theta \varepsilon_{xx} + \cos\theta\sin\theta \varepsilon_{yy} + (\cos^2\theta - \sin^2\theta)\varepsilon_{xy} & \sin^2\theta \varepsilon_{xx} + \cos^2\theta \varepsilon_{yy} - 2\cos\theta\sin\theta \varepsilon_{xy} & 0 \\ 0 & 0 & \varepsilon_{zz} \end{bmatrix}
$$

As would be expected $\varepsilon_{x'z'} = \varepsilon_{y'z'} = 0$. The remaining strain components are:

$$\varepsilon_{x'x'} = \cos^2(\theta)\varepsilon_{xx} + \sin^2(\theta)\varepsilon_{yy} + 2\cos(\theta)\sin(\theta)\varepsilon_{xy}$$

$$\varepsilon_{y'y'} = \sin^2(\theta)\varepsilon_{xx} + \cos^2(\theta)\varepsilon_{yy} - 2\cos(\theta)\sin(\theta)\varepsilon_{xy}$$

$$\varepsilon_{x'y'} = -\cos(\theta)\sin(\theta)\varepsilon_{xx} + \cos(\theta)\sin(\theta)\varepsilon_{yy} + [\cos^2(\theta) + \sin^2(\theta)]\varepsilon_{xy} \tag{2.42}$$

$$\varepsilon_{z'z'} = \varepsilon_{zz}$$

Tensoral shear strains were used in Equation 2.36a,b so that rotation of the strain tensor could be accomplished using the normal transformation law for a second-order tensor. As engineering shear strains are commonly used in practice, we will convert our final results, Equation 2.42, to ones which involve engineering shear strain (γ_{xy}). Recall from Section 2.9 that $\varepsilon_{xy} = (1\backslash2)\gamma_{xy}$. Hence, to convert Equation 2.42, simply replace ε_{xy} with $(1\backslash2)\gamma_{xy}$ everywhere, resulting in:

$$\varepsilon_{x'x'} = \cos^2(\theta)\varepsilon_{xx} + \sin^2(\theta)\varepsilon_{yy} + \cos(\theta)\sin(\theta)\gamma_{xy}$$

$$\varepsilon_{y'y'} = \sin^2(\theta)\varepsilon_{xx} + \cos^2(\theta)\varepsilon_{yy} - \cos(\theta)\sin(\theta)\gamma_{xy} \tag{2.43}$$

$$\frac{\gamma_{x'y'}}{2} = -\cos(\theta)\sin(\theta)\varepsilon_{xx} + \cos(\theta)\sin(\theta)\varepsilon_{yy} + [\cos^2(\theta) + \sin^2(\theta)]\frac{\gamma_{xy}}{2}$$

$$\varepsilon_{z'z'} = \varepsilon_{zz}$$

Equation 2.43 relates the components of strain in two different coordinate systems within a single plane, and will be used extensively throughout the remainder of this chapter. It is important to remember that these equations are only valid *if the out-of-plane z-axis is a principal strain axis*.

The first three of Equation 2.43 can be written using matrix notation as

$$\left\{ \begin{array}{c} \varepsilon_{x'x'} \\ \varepsilon_{y'y'} \\ \dfrac{\gamma_{x'y'}}{2} \end{array} \right\} = \left[\begin{array}{ccc} \cos^2(\theta) & \sin^2(\theta) & 2\cos(\theta)\sin(\theta) \\ \sin^2(\theta) & \cos^2(\theta) & -2\cos(\theta)\sin(\theta) \\ -\cos(\theta)\sin(\theta) & \cos(\theta)\sin(\theta) & \cos^2(\theta) - \sin^2(\theta) \end{array} \right] \left\{ \begin{array}{c} \varepsilon_{xx} \\ \varepsilon_{yy} \\ \dfrac{\gamma_{xy}}{2} \end{array} \right\} \tag{2.44}$$

Compare Equation 2.44 with Equation 2.20. In particular, note that the transformation matrix, $[T]$, which was previously encountered during the discussion of plane stress in Section 2.8, also appears in Equation 2.44.

The strain invariants (given by Equation 2.37 or 2.38) are considerably simplified when the out-of-plane z-axis is a principal axis. As by definition $\varepsilon_{xz} = \gamma_{xz}/2 = \varepsilon_{yz} = \gamma_{yz}/2 = 0$, the strain invariants become:

First strain invariant = $\Theta_\varepsilon = \varepsilon_{xx} + \varepsilon_{yy} + \varepsilon_{zz}$

$$\text{Second strain invariant} = \Phi_\varepsilon = \varepsilon_{xx}\varepsilon_{yy} + \varepsilon_{xx}\varepsilon_{zz} + \varepsilon_{yy}\varepsilon_{zz} - \frac{\gamma_{xy}^2}{4} \qquad (2.45)$$

$$\text{Third strain invariant} = \Psi_\varepsilon = \varepsilon_{xx}\varepsilon_{yy}\varepsilon_{zz} - \varepsilon_{zz}\frac{\gamma_{xy}^4}{4}$$

The principal strains equal the roots of the cubic equation previously listed as Equation 2.40. Substituting Equation 2.45 into Equation 2.40, there results:

$$\varepsilon_{pn}^3 - (\varepsilon_{xx} + \varepsilon_{yy} + \varepsilon_{zz})\varepsilon_{pn}^2 + \left(\varepsilon_{xx}\varepsilon_{yy} + \varepsilon_{xx}\varepsilon_{zz} + \varepsilon_{yy}\varepsilon_{zz} - \frac{\gamma_{xy}^2}{4} \right)\varepsilon_{pn}$$

$$- \left(\varepsilon_{xx}\varepsilon_{yy}\varepsilon_{zz} - \varepsilon_{zz}\frac{\gamma_{xy}^2}{4} \right) = 0 \qquad (2.46)$$

One root of Equation 2.23 is $\varepsilon_{pn} = \varepsilon_{zz}$. For present purposes, this root will be labeled ε_{p3} even though it may not be the algebraically least principal strain. In the case of plane strain $\varepsilon_{p3} = \varepsilon_{zz} = 0$.

Removing the known root from Equation 2.46, we have the following quadratic equation:

$$\varepsilon_{pn}^2 - (\varepsilon_{xx} + \varepsilon_{yy})\varepsilon_{pn} + \left(\varepsilon_{xx}\varepsilon_{yy} - \frac{\gamma_{xy}^2}{4} \right) = 0$$

The two roots of this quadratic equation (i.e., the two remaining principal strains, ε_{p1} and ε_{p2}) may be found by application of the standard approach [3], and are given by

$$\varepsilon_{p1}, \varepsilon_{p2} = \frac{\varepsilon_{xx} + \varepsilon_{yy}}{2} \pm \sqrt{\left(\frac{\varepsilon_{xx} - \varepsilon_{yy}}{2} \right)^2 + \left(\frac{\gamma_{xy}}{2} \right)^2} \qquad (2.47)$$

The angle θ_{p_ε} between the x-axis and either the p1 or p2 axis is given by

$$\theta_{p_\varepsilon} = \frac{1}{2} \arctan\left(\frac{\gamma_{xy}}{\varepsilon_{xx} - \varepsilon_{yy}} \right) \qquad (2.48)$$

Example Problem 2.9

Given: A state of plane strain is known to consist of:

$$\varepsilon_{xx} = 500 \,\mu m/m$$

$$\varepsilon_{yy} = -1000 \,\mu m/m$$

$$\gamma_{xy} = -2500 \,\mu rad$$

PROBLEM

a. Prepare a rough sketch (not to scale) of the deformed strain element in the $x–y$ coordinate system.

b. Determine the strain components which correspond to an $x'–y'$ coordinate system, oriented 25° CCW from the $x–y$ coordinate system, and prepare a rough sketch (not to scale) of the deformed strain element in the $x'–y'$ coordinate system.

c. Determine the principal strain components that exist within the $x–y$ plane, and prepare a rough sketch (not to scale) of the deformed strain element in the principal strain coordinate system.

SOLUTION

a. A sketch showing the deformed strain element (not to scale) in the $x–y$ coordinate system is shown in Figure 2.19a. Note that
  • The length of the element side parallel to the x-axis has increased (corresponding to the tensile strain $\varepsilon_{xx} = 500 \,\mu m/m$).
  • The length of the element side parallel to the y-axis has decreased (corresponding to the compressive strain $\varepsilon_{yy} = -1000 \,\mu m/m$).
  • The angle defined by $x–y$ axes has increased (corresponding to the negative shear strain $\gamma_{xy} = -2500 \,\mu rad$).

b. Since the x'-axis is oriented 25° CCW from the x-axis, in accordance with the right-hand rule the angle of rotation is *positive*, that is, $\theta = +25°$. Substituting this angle and the given strain components in Equation 2.44:

$$\begin{Bmatrix} \varepsilon_{x'x'} \\ \varepsilon_{y'y'} \\ \dfrac{\gamma_{x'y'}}{2} \end{Bmatrix} = \begin{bmatrix} \cos^2(25°) & \sin^2(25°) & 2\cos(25°)\sin(25°) \\ \sin^2(25°) & \cos^2(25°) & -2\cos(25°)\sin(25°) \\ -\cos(25°)\sin(25°) & \cos(25°)\sin(25°) & \cos^2(25°) - \sin^2(25°) \end{bmatrix}$$

$$\times \begin{Bmatrix} 500 \\ -1000 \\ \dfrac{-2500}{2} \end{Bmatrix}$$

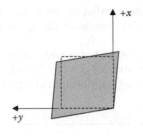

(a) Strain element in the x–y coordinate system:

$\varepsilon_{xx} = 500\ \mu\text{m/m}$
$\varepsilon_{yy} = -1000\ \mu\text{m/m}$
$\gamma_{xy} = -2500\ \mu\text{rad}$

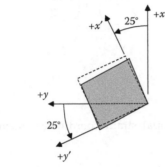

(b) Strain element in the x'–y' coordinate system:

$\varepsilon_{xx} = -725\ \mu\text{m/m}$
$\varepsilon_{yy} = 225\ \mu\text{m/m}$
$\gamma_{xy} = -1378\ \mu\text{rad}$

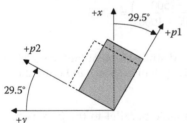

(c) Strain element in the principal strain coordinate system:

$\varepsilon_{p1} = 1208\ \mu\text{m/m}$
$\varepsilon_{p2} = -1708\ \mu\text{m/m}$

FIGURE 2.19
Strain elements associated with Example Problem 2.9 (all deformations shown greatly exaggerated for clarity).

Completing the matrix multiplication indicated results in

$$\left\{ \begin{array}{c} \varepsilon_{x'x'} \\ \varepsilon_{y'y'} \\ \dfrac{\gamma_{x'y'}}{2} \end{array} \right\} = \left\{ \begin{array}{c} -725\,\mu\text{m/m} \\ 225\,\mu\text{m/m} \\ -1378\,\mu\text{rad} \end{array} \right\}$$

A sketch showing the deformed strain element (not to scale) in the x'–y' coordinate system is shown in Figure 2.19b. Note that

- The length of the element side parallel to the x'-axis has decreased (corresponding to the compressive strain $\varepsilon_{x'x'} = -725\ \mu\text{m/m}$).

- The length of the element side parallel to the y'-axis has increased (corresponding to the tensile strain $\varepsilon_{y'y'} = 225$ μm/m).
- The angle defined by the x'–y' axes has increased (corresponding to the negative shear strain $\gamma_{x'y'} = -2756$ μrad).

c. The principal strains are found through application of Equation 2.47:

$$\varepsilon_{p1}, \varepsilon_{p2} = \frac{500 - 1000}{2} \pm \sqrt{\left(\frac{500 + 1000}{2}\right)^2 + \left(\frac{(-2500)}{2}\right)^2}$$

$$\varepsilon_{p1} = 1208\,\mu\text{m/m}$$

$$\varepsilon_{p2} = -1708\,\mu\text{m/m}$$

The orientation of the principal strain coordinate system is given by Equation 2.48:

$$\theta_{p\varepsilon} = \frac{1}{2}\arctan\left(\frac{-2500}{500 + 1000}\right) = -29.5°$$

A sketch showing the deformed strain element (not to scale) in the principal strain coordinate system is shown in Figure 2.19c. Note that

- The length of the element side parallel to the $p1$-axis has increased (corresponding to the tensile principal strain $\varepsilon_{p1} = 1208$ μm/m).
- The length of the element side parallel to the $p2$-axis has decreased (corresponding to the compressive principal strain $\varepsilon_{p2} = -1708$ μm/m).
- The angle defined by the principal strain axes has remained precisely $\pi/2$ radians (i.e., 90°), as in the principal strain coordinate system the shear strain is zero.

2.14 Relating Strains to Displacement Fields

Most analyses considered in later chapters begin with consideration of the *displacement fields* induced in the structure of interest. That is, mathematical expressions that describe the displacements induced at all points within a structure by external loading and/or environmental changes will be assumed or otherwise specified. Strains induced in the structure will then be inferred from these displacement fields.

In the most general case three displacement fields are involved. Specifically, displacements in the x-, y-, and z-directions, typically denoted as the u-, v-, and w-displacement fields, respectively. In general, all three displacement fields are functions of x, y, and z:

Displacements in the x-direction: $u = u(x,y,z)$

Displacements in the y-direction: $v = v(x,y,z)$

Displacements in the z-direction: $w = w(x,y,z)$.

However, if the out-of-plane z-axis is a principal strain axis, then u and v are (at most) functions of x and y only, while w is (at most) a function of z only. In this case:

Displacements in the x-direction: $u = u(x,y)$

Displacements in the y-direction: $v = v(x,y)$

Displacements in the z-direction: $w = w(z)$.

A detailed derivation of the relationship between displacements and strains is beyond the scope of this review, and the interested reader is referred to References 1, 2 for details. It can be shown that the relationship between displacement fields and the strain tensor depends upon the magnitude of *derivatives* of displacement fields (also called *displacement gradients*). If displacement gradients are arbitrarily large then the associated level of strain is said to be "finite," and each component of the strain tensor is related *nonlinearly* to displacement gradients as follows:

$$\varepsilon_{xx} = \frac{\partial u}{\partial x} + \frac{1}{2}\left[\left(\frac{\partial u}{\partial x}\right)^2 + \left(\frac{\partial v}{\partial x}\right)^2 + \left(\frac{\partial w}{\partial x}\right)^2\right]$$

$$\varepsilon_{yy} = \frac{\partial v}{\partial y} + \frac{1}{2}\left[\left(\frac{\partial u}{\partial y}\right)^2 + \left(\frac{\partial v}{\partial y}\right)^2 + \left(\frac{\partial w}{\partial y}\right)^2\right]$$

$$\varepsilon_{zz} = \frac{\partial w}{\partial z} + \frac{1}{2}\left[\left(\frac{\partial u}{\partial z}\right)^2 + \left(\frac{\partial v}{\partial z}\right)^2 + \left(\frac{\partial w}{\partial z}\right)^2\right]$$

$$\gamma_{xy} = \frac{\partial u}{\partial y} + \frac{\partial v}{\partial x} + \left(\frac{\partial u}{\partial x}\right)\left(\frac{\partial u}{\partial y}\right) + \left(\frac{\partial v}{\partial x}\right)\left(\frac{\partial v}{\partial y}\right) + \left(\frac{\partial w}{\partial x}\right)\left(\frac{\partial w}{\partial y}\right)$$

$$\gamma_{xz} = \frac{\partial u}{\partial z} + \frac{\partial w}{\partial x} + \left(\frac{\partial u}{\partial x}\right)\left(\frac{\partial u}{\partial z}\right) + \left(\frac{\partial v}{\partial x}\right)\left(\frac{\partial v}{\partial z}\right) + \left(\frac{\partial w}{\partial x}\right)\left(\frac{\partial w}{\partial z}\right)$$

$$\gamma_{yz} = \frac{\partial w}{\partial y} + \frac{\partial v}{\partial z} + \left(\frac{\partial u}{\partial y}\right)\left(\frac{\partial u}{\partial z}\right) + \left(\frac{\partial v}{\partial y}\right)\left(\frac{\partial v}{\partial z}\right) + \left(\frac{\partial w}{\partial y}\right)\left(\frac{\partial w}{\partial y}\right)$$

The expressions listed above define what is known as *Green's strain tensor* (also known as the *Lagrangian* strain tensor).

In most cases encountered in practice, however, displacement gradients are very small, and consequently the *products* of displacement gradients are negligibly small and can be discarded. For example, it can usually be assumed that

$$\left(\frac{\partial u}{\partial x}\right)^2 \approx 0 \qquad \left(\frac{\partial v}{\partial x}\right)^2 \approx 0$$

$$\left(\frac{\partial w}{\partial x}\right)^2 \approx 0 \qquad \left(\frac{\partial u}{\partial x}\right)\left(\frac{\partial u}{\partial y}\right) \approx 0, \quad \text{etc.}$$

When displacement gradients are very small the level of strain is said to be "infinitesimal," and each component of the strain tensor is *linearly* related to displacement gradients as follows:

$$\varepsilon_{xx} = \frac{\partial u}{\partial x} \qquad\qquad (2.49a)$$

$$\varepsilon_{yy} = \frac{\partial v}{\partial y} \qquad\qquad (2.49b)$$

$$\varepsilon_{zz} = \frac{\partial w}{\partial z} \qquad\qquad (2.49c)$$

$$\gamma_{xy} = \frac{\partial v}{\partial x} + \frac{\partial u}{\partial y} \qquad\qquad (2.49d)$$

$$\gamma_{xz} = \frac{\partial w}{\partial x} + \frac{\partial u}{\partial z} \qquad\qquad (2.49e)$$

$$\gamma_{yz} = \frac{\partial w}{\partial y} + \frac{\partial v}{\partial z} \qquad\qquad (2.49f)$$

For most analyses considered in later chapters we will assume that strains are infinitesimal and are related to displacement fields in accordance with Equation 2.49. The one exception occurs in Chapter 11, where it will be necessary to include nonlinear terms in the strain–displacement relationships.

At stated above, most analyses begin with consideration of the displacement fields induced in a structure of interest. Strains fields implied by these displacements are then calculated in accordance with Equation 2.49. This process insures that strain fields are consistent with displacements. Consider the opposite approach. Specifically, suppose that mathematical expressions for *strain fields* are assumed, perhaps on the basis of engineering judgment. In this case, it is possible that the *assumed* strain fields correspond to physically unrealistic displacement fields. For example, displacement fields inferred from assumed strain fields may imply that the solid body has voids and/or overlapping regions, a physically unrealistic circumstance. A system of six equations known as the *compatibility conditions* can be developed that guarantee that assumed expressions for the six components of strain do, in fact, correspond to physically reasonable displacement fields $u(x,y,z)$, $v(x,y,z)$, and $w(x,y,z)$. To develop the compatibility conditions, differentiate Equation 2.49d twice, once with respect to x and once with respect to y. We obtain

$$\frac{\partial^2 \gamma_{xy}}{\partial x \partial y} = \frac{\partial^3 u}{\partial x \partial y^2} + \frac{\partial^3 v}{\partial x^2 \partial y}$$

From Equations 2.49a,b is easily seen that

$$\frac{\partial^2 \varepsilon_{xx}}{\partial y^2} = \frac{\partial^3 u}{\partial x \partial y^2} \qquad \frac{\partial^2 \varepsilon_{yy}}{\partial x^2} = \frac{\partial^3 v}{\partial x^2 \partial y}$$

Combining these results, we see that assumed expressions for the strain components ε_{xx}, ε_{yy}, and γ_{xy} correspond to physically reasonable displacement fields (i.e., "are compatible") only if they satisfy:

$$\frac{\partial^2 \gamma_{xy}}{\partial x \partial y} = \frac{\partial^2 \varepsilon_{xx}}{\partial y^2} + \frac{\partial^2 \varepsilon_{yy}}{\partial x^2} \tag{2.50a}$$

Equation 2.50a is the first compatibility condition. Following a similar procedure using Equations 2.49e and 2.49f, we obtain

$$\frac{\partial^2 \gamma_{yz}}{\partial y \partial z} = \frac{\partial^2 \varepsilon_{yy}}{\partial z^2} + \frac{\partial^2 \varepsilon_{zz}}{\partial y^2} \tag{2.50b}$$

$$\frac{\partial^2 \gamma_{xz}}{\partial x \partial z} = \frac{\partial^2 \varepsilon_{xx}}{\partial z^2} + \frac{\partial^2 \varepsilon_{zz}}{\partial x^2} \tag{2.50c}$$

These are the second and third compatibility conditions. Next, the following expressions are obtained using Equations 2.49a through 2.49f), respectively:

$$\frac{\partial^3 u}{\partial x \partial y \partial z} = \frac{\partial^2 \varepsilon_{xx}}{\partial y \partial z}$$

$$\frac{\partial^3 u}{\partial x \partial y \partial z} = \frac{\partial^2 \gamma_{xy}}{\partial x \partial z} - \frac{\partial^3 v}{\partial x^2 \partial z}$$

$$\frac{\partial^3 u}{\partial x \partial y \partial z} = \frac{\partial^2 \gamma_{xz}}{\partial x \partial y} - \frac{\partial^3 w}{\partial x^2 \partial y}$$

$$\frac{\partial^3 w}{\partial x^2 \partial y} + \frac{\partial^3 v}{\partial x^2 \partial z} = \frac{\partial^2 \gamma_{yz}}{\partial x^2}$$

Combining these four expressions, we find that *assumed* expressions for strain components $\varepsilon_{xx}, \gamma_{xy}, \gamma_{xz},$ and γ_{yz} are compatible if

$$2\frac{\partial^2 \varepsilon_{xx}}{\partial y \partial z} = \frac{\partial}{\partial x}\left(\frac{\partial \gamma_{xy}}{\partial z} + \frac{\partial \gamma_{xz}}{\partial y} - \frac{\gamma_{yz}}{\partial x} \right) \qquad (2.50d)$$

This is the fourth compatibility condition. The final two compatibility conditions are developed using a similar process, and are given by

$$2\frac{\partial^2 \varepsilon_{yy}}{\partial x \partial z} = \frac{\partial}{\partial y}\left(\frac{\partial \gamma_{yz}}{\partial x} - \frac{\partial \gamma_{xz}}{\partial y} + \frac{\gamma_{xy}}{\partial z} \right) \qquad (2.50e)$$

$$2\frac{\partial^2 \varepsilon_{zz}}{\partial x \partial y} = \frac{\partial}{\partial z}\left(\frac{\partial \gamma_{yz}}{\partial x} + \frac{\partial \gamma_{xz}}{\partial y} - \frac{\gamma_{xy}}{\partial z} \right) \qquad (2.50f)$$

2.15　Computer Programs *3DROTATE* and *2DROTATE*

A review of force, stress, and strain tensors has been presented in this chapter. These concepts will be applied routinely in later chapters, as we develop a macromechanics-based analysis of structural composite materials and structures. It will be seen that transformation of stress and strain tensors is of particular importance. Indeed, nearly all analyses of composite materials and structures presented herein require multiple transformations of stress and strain tensors from one coordinate system to another.

Two computer programs, *3DROTATE* and *2DROTATE*, that can be used to perform transformations of force, stress, or strain tensors can be downloaded at no cost from the following website:

http://depts.washington.edu/amtas/computer.html

Program *3DROTATE* performs the calculations necessary to transform a force, stress, or strain tensor from the x–y–z coordinate system to the x'''–y'''–z''' coordinate system, where the x'''–y'''–z''' coordinate system is generated from the x–y–z coordinate system by (up to) three successive rotations. Derivation of the direction cosines that relate these two coordinate systems is left as a student exercise (see Homework Problem 2.2). Program *3DROTATE* also calculates the angles between the x–y–z and x'''–y'''–z''' coordinate axes, invariants of the force, stress, or strain tensors, and principal stresses and strains. All of the numerical results discussed in Example Problems 2.1, 2.3, and 2.7 can be obtained through the use of program *3DROTATE*.

The second program, *2DROTATE*, can be used to rotate stresses within a plane (as discussed in Section 2.8) and/or strains within a plane (as discussed in Section 2.13). It is important to remember that these transformations are only valid if the out-of-plane direction is a principal stress or principal strain axis. For the most part, thin plate-like composite structures will be considered in this book. Therefore, it can usually be assumed that the direction normal to the surface of the composite is a direction of principal stress or strain. Hence, most of the stress or strain transformations considered in this chapter involve rotations within a plane. Most of the numerical results discussed in Example Problems 2.5 and 2.9 can be obtained through the use of program *2DROTATE*.

HOMEWORK PROBLEMS

In the following problem statements the phrase "solve by hand" means that numerical solutions should be obtained using a pencil, paper, and nonprogrammable calculator. Solutions obtained by hand will then be compared with numerical results returned by appropriate computer programs. This process will insure understanding of the mathematical processes involved.

    2.1. Solve part (c) of Example Problem 2.1 by hand, based on the rotation angles listed below. In each case calculate that the magnitude of the transformed force vector. Confirm your calculations using program *3DROTATE*.

(a) $\theta = 60°$	$\beta = -45°$
(b) $\theta = 60°$	$\beta = 45°$
(c) $\theta = -60°$	$\beta = -45°$
(d) $\theta = -60°$	$\beta = 45°$
(e) $\theta = -45°$	$\beta = 60°$
(f) $\theta = -45°$	$\beta = -60°$
(g) $\theta = 45°$	$\beta = 60°$
(h) $\theta = 45°$	$\beta = -60°$

2.2. Consider an $x'''-y'''-z'''$ coordinate system, which is generated from an $x-y-z$ coordinate system by the following three rotations:

a. A rotation of θ-degrees about the original z-axis, which defines an intermediate $x'-y'-z'$ coordinate system (see Figure 2.2a).
b. A rotation of β-degrees about the x'-axis, which defines an intermediate $x''-y''-z''$ coordinate system (see Figure 2.2b).
c. A rotation of ψ-degrees about the y''-axis, which defines the final $x'''-y'''-z'''$ coordinate system.

Show that the $x'''-y'''-z'''$ and $x-y-z$ coordinate systems are related by the following direction cosines:

$$\begin{bmatrix} c_{x'''x} & c_{x'''y} & c_{x'''z} \\ c_{y'''x} & c_{y'''y} & c_{y'''z} \\ c_{z'''x} & c_{z'''y} & c_{z'''z} \end{bmatrix}$$

$$= \begin{bmatrix} \cos\psi\cos\theta - \sin\psi\sin\beta\sin\theta & \cos\psi\sin\theta + \sin\psi\sin\beta\cos\theta & -\sin\psi\cos\beta \\ -\cos\beta\sin\theta & \cos\beta\cos\theta & \sin\beta \\ \sin\psi\cos\theta + \cos\psi\sin\beta\sin\theta & \sin\psi\sin\theta - \cos\psi\sin\beta\cos\theta & \cos\psi\cos\beta \end{bmatrix}$$

2.3. The force vector discussed in Example Problem 2.1 is given by

$$\bar{F} = 1000\hat{i} + 200\hat{j} + 600\hat{k}$$

Using Equation 2.6c, express $\bar{F}$ in a new coordinate system defined by three successive rotations, as listed below, using the direction cosines listed in Problem 2.2. In each case compare the magnitude of the transformed force vector to the magnitudes calculated in

Example Problem 2.1. Solve these problems by hand, and then confirm your calculations using program *3DROTATE*.

(a) $\theta = 60°$	$\beta = -45°$	$\psi = 25°$
(b) $\theta = 60°$	$\beta = -45°$	$\psi = -25°$
(c) $\theta = 60°$	$\beta = -45°$	$\psi = 25°$
(d) $\theta = 60°$	$\beta = 45°$	$\psi = 25°$
(e) $\theta = -60°$	$\beta = -45°$	$\psi = 25°$

2.4. Solve Example Problem 2.3 by hand, except use the following rotation angles:

(a) $\theta = 20°$	$\beta = -35°$
(b) $\theta = -20°$	$\beta = 35°$
(c) $\theta = -20°$	$\beta = -35°$

Confirm your calculations using program *3DROTATE*.

2.5. Use Equation 2.12a to obtain an expression (in expanded form) for
the following stress component (in each case the expanded expression will be similar to Equation 2.13):

a. $\sigma_{x'x'}$
b. $\sigma_{x'y'}$
c. $\sigma_{y'y'}$
d. $\sigma_{y'z'}$
e. $\sigma_{z'z'}$

2.6. Use program *3DROTATE* to determine the stress invariants for
the stress tensor listed below, and compare to those determined in
Example Problem 2.3. (Note: this stress tensor is similar to the one
considered in Example Problem 2.3, except that the algebraic sign
of all three normal stresses has been reversed.):

$$\begin{bmatrix} \sigma_{xx} & \sigma_{xy} & \sigma_{xz} \\ \sigma_{yx} & \sigma_{yy} & \sigma_{yz} \\ \sigma_{zx} & \sigma_{zy} & \sigma_{zz} \end{bmatrix} = \begin{bmatrix} -50 & -10 & 15 \\ -10 & -25 & 30 \\ 15 & 30 & 5 \end{bmatrix} (\text{ksi})$$

2.7. Use program *3DROTATE* to determine the stress invariants for the
stress tensor listed below, and compare with those determined in
Example Problem 2.3. (Note: this stress tensor is similar to the one

considered in Example Problem 2.3, except that the algebraic sign of all three shear stresses has been reversed.):

$$\begin{bmatrix} \sigma_{xx} & \sigma_{xy} & \sigma_{xz} \\ \sigma_{yx} & \sigma_{yy} & \sigma_{yz} \\ \sigma_{zx} & \sigma_{zy} & \sigma_{zz} \end{bmatrix} = \begin{bmatrix} 50 & 10 & -15 \\ 10 & 25 & -30 \\ -15 & -30 & -5 \end{bmatrix} (ksi)$$

2.8. Use program *3DROTATE* to determine the stress invariants for the stress tensor listed below, and compare with those determined in Example Problem 2.3. (Note: this stress tensor is similar to the one considered in Example Problem 2.3, except that the algebraic sign of all stress components has been reversed.):

$$\begin{bmatrix} \sigma_{xx} & \sigma_{xy} & \sigma_{xz} \\ \sigma_{yx} & \sigma_{yy} & \sigma_{yz} \\ \sigma_{zx} & \sigma_{zy} & \sigma_{zz} \end{bmatrix} = \begin{bmatrix} -50 & 10 & -15 \\ 10 & -25 & -30 \\ -15 & -30 & 5 \end{bmatrix} (ksi)$$

2.9. Use program *3DROTATE* to determine the strain invariants for the strain tensor listed below, and compare with those determined in Example Problem 2.7. (Note: this strain tensor is similar to the one considered in Example Problem 2.7, except that the algebraic sign of all shear strain components has been reversed.):

$$\begin{bmatrix} \varepsilon_{xx} & \varepsilon_{xy} & \varepsilon_{xz} \\ \varepsilon_{yx} & \varepsilon_{yy} & \varepsilon_{yz} \\ \varepsilon_{zx} & \varepsilon_{zy} & \varepsilon_{zz} \end{bmatrix} = \begin{bmatrix} 1000\,\mu m/m & -500\,\mu rad & -250\,\mu rad \\ -500\,\mu rad & 1500\,\mu m/m & -750\,\mu rad \\ -250\,\mu rad & -750\,\mu rad & 2000\,\mu m/m \end{bmatrix}$$

2.10. Use program *3DROTATE* to determine the strain invariants for the strain tensor listed below, and compare with those determined in Example Problem 2.7. (Note: this strain tensor is similar to the one considered in Example Problem 2.7, except that the algebraic sign of all normal strain components has been reversed.):

$$\begin{bmatrix} \varepsilon_{xx} & \varepsilon_{xy} & \varepsilon_{xz} \\ \varepsilon_{yx} & \varepsilon_{yy} & \varepsilon_{yz} \\ \varepsilon_{zx} & \varepsilon_{zy} & \varepsilon_{zz} \end{bmatrix} = \begin{bmatrix} -1000\,\mu m/m & 500\,\mu rad & 250\,\mu rad \\ 500\,\mu rad & -1500\,\mu m/m & 750\,\mu rad \\ 250\,\mu rad & 750\,\mu rad & -2000\,\mu m/m \end{bmatrix}$$

2.11. Use program *3DROTATE* to determine the strain invariants for the strain tensor listed below, and compare with those determined in Example Problem 2.7. (Note: this strain tensor is similar to the one considered in Example Problem 2.7, except that the algebraic sign of all strain components has been reversed.):

$$
\begin{bmatrix} \varepsilon_{xx} & \varepsilon_{xy} & \varepsilon_{xz} \\ \varepsilon_{yx} & \varepsilon_{yy} & \varepsilon_{yz} \\ \varepsilon_{zx} & \varepsilon_{zy} & \varepsilon_{zz} \end{bmatrix} = \begin{bmatrix} -1000\,\mu m/m & -500\,\mu rad & -250\,\mu rad \\ -500\,\mu rad & -1500\,\mu m/m & -750\,\mu rad \\ -250\,\mu rad & -750\,\mu rad & -2000\,\mu m/m \end{bmatrix}
$$

References

1. Frederick, D., and Chang, T.S., *Continuum Mechanics*, Scientific Publishers, Cambridge, MA, 1972.
2. Fung, Y.C., *A First Course In Continuum Mechanics*, Prentice-Hall, Englewood Cliffs, NJ, 1969.
3. Consult any handbook of mathematical functions and tables, for example, *CRC Basic Mathematical Tables*, S.M. Selby, ed., The Chemical Company, Cleveland, OH, p. 59, 1970.

3

Material Properties

3.1 Material Properties of Anisotropic versus Isotropic Materials

The phrase *"material property"* refers to a measurable constant which is characteristic of a particular material, and which can be used to relate two disparate quantities of interest. Material properties are defined that describe the ability of a material to conduct electricity, to transmit (or reflect) visible light, to transfer heat, or to support mechanical loading, to name but a few. Material properties of interest herein are those used by engineers during the design of load-bearing composite structures. Two specific examples are *Young's modulus*, *E*, and *Poisson's ratio*, v. These familiar material properties, which will be reviewed and further discussed in the following section, are used to relate the stress and strain tensors.

The adjectives "anisotropic" and "isotropic" indicate whether a material exhibits a single value for a given material property. More specifically, if the properties of a material are *independent* of direction within the material, then the material is said to be *isotropic*. Conversely, if the material properties vary with direction within the material, then the material is said to be *anisotropic*.

To clarify this statement, suppose that three test specimens are machined from a large block at three different orientations, as shown in Figure 3.1. The geometry of the three specimens is assumed to be identical, so that the only difference between specimens is the original orientation of each specimen within the parent block. Now suppose that the axial stiffness (i.e., Young's modulus, *E*) is measured for each specimen. Young's modulus measured using specimen 1 will be denoted E_{xx}, that is, subscripts are used to indicate the original orientation of specimen 1 within the parent block. Similarly, Young's modulus measured using specimens 2 and 3 will be denoted E_{yy} and E_{zz}, respectively.

If the parent block consists of an isotropic material, then Young's modulus measured for each specimen will be identical (to within engineering accuracies):

$$\text{(for isotropic materials)} \quad E_{xx} = E_{yy} = E_{zz}$$

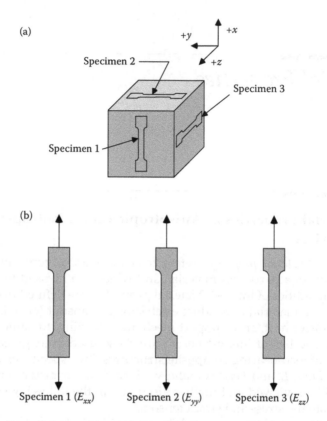

FIGURE 3.1
Illustration of method used to determine whether a material is isotropic or anisotropic. (a) Three specimens machined at different orientations from "parent" block; (b) tensile test of three individual specimens.

In this case, an identical value of Young's modulus is measured in the x-, y-, and z-directions, and is independent of direction within the material. In contrast, if the parent block is an anisotropic material, a different Young's modulus will, in general, be measured for each specimen:

$$\text{(for anisotropic materials)} \quad E_{xx} \neq E_{yy} \neq E_{zz}$$

In this case, the value of Young's modulus depends on the direction within the material the modulus is measured. For anisotropic materials, a similar dependence on direction can occur for any material property of interest (Poisson's ratio, thermal expansion coefficients, ultimate strengths, etc.).

It is the microstructural features of a material that determine whether it exhibits isotropic or anisotropic behavior. Consequently, to classify a given material as isotropic or anisotropic one must first define the physical scale of interest. For example, it is well known that metals and metal alloys are made up of individual *grains*, and that the atoms that exist within these grains are arranged in well-defined crystalline arrays. The most common crystalline arrays are the body-centered cubic (BCC), the face-centered cubic (FCC), or the hexagonal close-packed structure (HCP) [1]. Owing to the highly ordered atomic crystalline structures that exist within these arrays, an individual grain exhibits different properties in different directions and hence is anisotropic. That is to say, if material properties are defined at a physical scale on the order of a grain diameter or smaller, then all metals or metal alloys must be defined as anisotropic materials.

It does not necessarily follow, however, that a metal or metal alloy will exhibit anisotropic behavior at the *structural level*. This is because individual grains are typically very small, and the orientation of the atomic crystalline arrays usually varies randomly from one grain to the next. As a typical case, it is not uncommon for a steel alloy to exhibit an average grain diameter of 0.044 mm (0.0017 in.), or roughly 8200 grains/mm^3 (134×10^6 grains/in.^{3}) [1]. If the grains are randomly oriented, then at the structural level (at a physical scale >1 mm, say) the steel alloy will exhibit effectively isotropic properties even though the constituent grains are anisotropic. Conversely, if a significant percentage of grains are caused to be oriented by some mechanism (such as cold rolling, for example), then the same steel alloy will be anisotropic at the structural level.

Polymeric composites are anisotropic at the structural level, and the microstructural features that lead to this anisotropy are immediately apparent. Specifically, it is the uniform and symmetric orientation of the reinforcing fibers within a ply that leads to anisotropic behavior. As a simple example, suppose two specimens are machined from a thin unidirectional composite plate consisting of high-strength fibers embedded within a relatively flexible polymeric matrix, as shown in Figure 3.2a,b. Note that the coordinate system used to describe the plate has been labeled the 1–2–3 axes, and that fibers are arranged symmetrically about the 1–3 and 2–3 planes. The 1–2–3 coordinate system will henceforth be referred to as the *principal material coordinate system*. Referring to Figure 3.2a, specimen 1 is machined such that the fibers are aligned with the long axis of the specimen, whereas in specimen 2 the fibers are perpendicular to the axis of the specimen. Obviously, Young's modulus measured for these two specimens will be quite different. Specifically, the modulus measured for specimen 1 will approach that of the fibers, whereas the modulus measured for specimen 2 will approach that of the polymeric matrix. Therefore, for Figure 3.2a $E_{11} \gg E_{22}$ and Young's modulus varies with direction within the material, satisfying the definition of an anisotropic material.

The principal material coordinate system need not be aligned with the fibers, however. A typical example is shown in Figure 3.2b. In this case, the thin composite plate is formed using a braided fabric. As discussed in Section 3.1.4, braided fabrics contain fibers oriented in two (or more) nonorthogonal directions. The three principal material coordinate axes lie

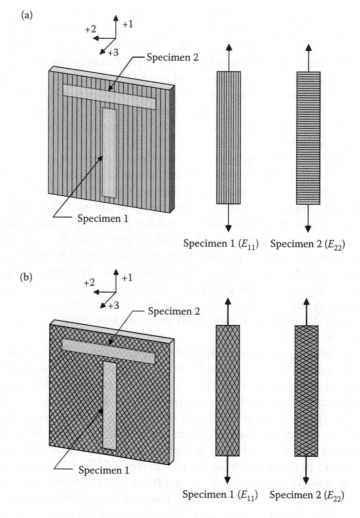

FIGURE 3.2
Illustration of the principal material coordinate system for thin composite laminates.
(a) Specimens machined from a thin unidirectional composite panel; (b) specimens machined
from a thin braided composite panel.

within planes that are symmetric with respect to the fiber array. As before,
the 1- and 2-axes lie within the plane of the plate, and the 3-axis is defined
normal to the plane of the plate.

One of the most unusual features of anisotropic materials is that they
can exhibit a coupling between normal stresses and shear strains, as well
as a coupling between shear stress and normal strains. A physical explanation of how this coupling occurs in the case of a unidirectional composite is
presented in Figure 3.3. A specimen in which the unidirectional fibers are

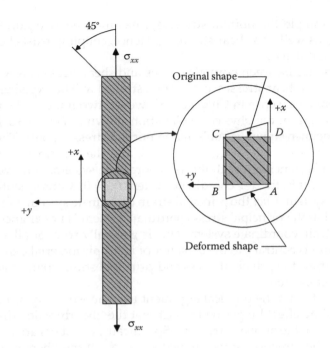

FIGURE 3.3
Forty five degrees off-axis composite specimen used to explain the origin of coupling effects.

oriented at an angle of 45° with respect to the *x*-axis is shown, and a small square element from the gage region is isolated. Since the element is initially square, and since in this example the fibers are defined to be at an angle of 45°, fibers are parallel to diagonal AC of the element. In contrast, fibers are perpendicular to diagonal BD. This implies that element is stiffer along diagonal AC than along diagonal BD.

Now assume that a tensile stress is applied, causing the square element (as well as the specimen as a whole) to deform. Since the stiffness is higher along diagonal AC than along diagonal BD, the length of diagonal BD is increased to a greater extent than that of diagonal AC. Hence, the initially square element deforms into a parallelogram, as shown in the figure. Note that:

- The length of the square element is increased in the *x*-direction (corresponding to a tensile strain, ε_{xx}).
- The length of the square element is decreased in the *y*-direction (corresponding to a compressive strain ε_{yy}, and associated with the Poisson effect).
- $\angle DAB$ is no longer $\pi/2$ radians, which indicates that a shear strain γ_{xy} has been induced.

In this example the normal stress, σ_{xx}, has induced two normal strains (ε_{xx} and ε_{yy}) as well as a shear strain, γ_{xy}. Hence, coupling exists between σ_{xx} and γ_{xy}, as stipulated.

The couplings between normal stresses and shear strains (as well as couplings between shear stresses and normal strains) will be explained in a formal mathematical sense in Chapter 4. However, two important conclusions can be drawn from the physical explanation shown in Figure 3.3. First, note that the specimen shown is subjected to normal stress σ_{xx} only. The in-plane principal stresses are $\sigma_{11} = \sigma_{xx}$, $\sigma_{22} = \sigma_{yy} = 0$, and the principal stress coordinate system is coincident with the x–y coordinate system. However, a shear *strain* does exist in the x–y coordinate system ($\gamma_{xy} \neq 0$). Consequently, the x–y coordinate system is *not* the principal strain coordinate system. We conclude, therefore, that the principal stress coordinate system is not aligned with the principal strain coordinate system. This is generally true for all anisotropic materials and is contrary to the behavior of isotropic materials, since for isotropic materials the principal stress and principal strain coordinate systems are always coincident.

Second, note that the physical argument used above to explain the origin of the coupling effect hinges on the fact that the fiber direction differs from the direction of the applied stress σ_{xx}. Specifically, the fibers are oriented 45° away from the direction of the applied stress σ_{xx}. If the fibers were aligned with either the x- or y-axis, then a coupling between σ_{xx} and γ_{xy} would not occur. We conclude that the unusual coupling effects exhibited by composites only occur *if stress and strain are referenced to a nonprincipal material coordinate system.*

Anisotropic materials are classified according to the number of *planes of symmetry* defined by the microstructure. The principal material coordinate axes lie within the planes of symmetry. For example, in the case of unidirectional composites three planes of symmetry can be defined: the 1–2 plane, the 1–3 plane, and the 2–3 plane. Composites fall within one of two classifications of anisotropic behavior. Specifically, composites are either *orthotropic* materials or *transversely isotropic* materials (the distinction between orthotropic and transversely isotropic materials will be further discussed in Section 3.3.2). During the composite structural analyses discussed in this and later chapters the composite will be called "anisotropic" if the coordinate system of reference is a nonprincipal material coordinate system. Use of the term "anisotropic" will, therefore, signal the possibility of couplings between normal stresses and shear strains and couplings between shear stresses and normal strains. If, instead, a structural analysis is referenced to the principal material coordinate system the composite will be called either orthotropic or transversely isotropic.

While many kinds of material properties may be defined, we are primarily interested in those properties used by structural engineers. The properties needed to perform a structural analysis of composite structures will be defined in the following sections. In each section, a general definition of the material property will be given, suitable for use with anisotropic materials. That is, the properties of a composite material when referenced to a nonprincipal material coordinate system will be discussed first. These general definitions will then be specialized to the principal material coordinate system, that is, they will be specialized for the case of orthotropic or transversely isotropic composites.

Typical values of the properties discussed in this chapter measured at room temperatures are listed for glass/epoxy, Kevlar/epoxy, and graphite/epoxy in Table 3.1. These properties do not represent the properties of any specific commercial composite material system, but rather should be viewed as typical values. Due to ongoing research and development activities within

TABLE 3.1

Typical Properties of Common Unidirectional Composites

Property	Glass/ Epoxy	Kevlar/ Epoxy	Graphite/ Epoxy
E_{11}	55 GPa	100 GPa	170 GPa
	(8.0 Msi)	(15 Msi)	(25 Msi)
E_{22}	16 GPa	6 GPa	10 GPa
	(2.3 Msi)	(0.90 Msi)	(1.5 Msi)
ν_{12}	0.28	0.33	0.30
G_{12}	7.6 GPa	2.1 GPa	13 GPa
	(1.1 Msi)	(0.30 Msi)	(1.9 Msi)
σ_{11}^{fT}	1050 MPa	1380 MPa	1500 MPa
	(150 ksi)	(200 ksi)	(218 ksi)
σ_{11}^{fC}	690 MPa	280 MPa	1200 MPa
	(100 ksi)	(40 ksi)	(175 ksi)
σ_{22}^{fT}	45 MPa	35 MPa	50 MPa
	(5.8 ksi)	(2.9 ksi)	(7.25 ksi)
σ_{22}^{fC}	120 MPa	105 MPa	100 MPa
	(16 ksi)	(15 ksi)	(14.5 ksi)
τ_{12}^{f}	40 MPa	40 MPa	90 MPa
	(4.4 ksi)	(4.0 ksi)	(13.1 ksi)
α_{11}	6.7 μm/m–°C	–3.6 μm/m–°C	–0.9 μm/m–°C
	(3.7 μin./in.–°F)	(–2.0 μin./in.–°F)	(–0.5 μin./in.–°F)
α_{22}	25 μm/m–°C	58 μm/m–°C	27 μm/m–°C
	(14 μin./in.–°F)	(32 μin./in.–°F)	(15 μin./in.–°F)
β_{11}	100 μm/m–%M	175 μm/m–%M	50 μm/m–%M
	(100 μin./in.–%M)	(175 μin./in.–%M)	(50 μin./in.–%M)
β_{22}	1200 μm/m–%M	1700 μm/m–%M	1200 μm/m–%M
	(1200 μin./in.–%M)	(1700 μin./in.–%M)	(1200 μin./in.–%M)
Ply Thickness	0.125 mm	0.125 mm	0.125 mm
	(0.005 in.)	(0.005 in.)	(0.005 in.)

the industry the properties of composites are improved more-or-less continuously. Therefore, the properties listed in Table 3.1 may not reflect those of currently available materials. The properties that appear in the table will be used in example and homework problems throughout the remainder of this book.

3.2 Material Properties That Relate Stress to Strain

Both stress and strain are second-order tensors, as discussed in Chapter 2. The materials properties used to relate the stress and strain tensors are inferred from experimental measurements. Conceptually, two different experimental approaches may be taken. In the first approach, the material of interest is subjected to a well-defined stress tensor, and components of the resulting strain tensor are measured. In the second approach, the material of interest is subjected to a well-defined strain tensor, and the components of the resulting stress tensor are measured. From an experimental standpoint it is far easier to impose a well-defined stress tensor than a well-defined strain tensor, and hence the first approach is almost always used in practice.

Recall that there are two fundamental types of stress components: normal stress and shear stress. As a consequence two fundamental types of tests are used to relate stress to strain. Specifically, a test that involves the application of a known normal stress component, and a test involving application of a known shear stress component. In either case a stress tensor is imposed in which five of the six stress components equal zero, and the resulting six components of the strain tensor are measured.

Tests that involve application of a known normal stress component are called *uniaxial tests*. In a typical case, a single normal stress component is applied to a test specimen (σ_{xx}, say), while insuring that the remaining five stress components are zero ($\sigma_{yy} = \sigma_{zz} = \tau_{xy} = \tau_{xz} = \tau_{yz} = 0$). The six components of strain caused by σ_{xx} are measured, which allows the calculation of various material properties relating a normal stress component to the strain tensor.

In contrast, tests that involve application of a known shear stress component are called *pure shear tests*. In a typical case a single shear stress component is applied to a test specimen (τ_{xy}, say), while insuring that the remaining five stress components are zero ($\sigma_{xx} = \sigma_{yy} = \sigma_{zz} = \tau_{xz} = \tau_{yz} = 0$). The six components of strain caused by τ_{xy} are measured, which allows the calculation of various material properties relating a shear stress component to the strain tensor.

A detailed description of experimental methods used to impose a specified state of stress is beyond the scope of the present discussion. It should be mentioned, however, that some of the stress states discussed below are very difficult to achieve in practice. For example, since composites are usually produced in the form of thin plate-like structures, it is very difficult to impose well-defined stresses acting normal to the plane of the composite, or to measure the strain components induced normal to the plane of the composite by a given state of stress.

3.2.1 Uniaxial Tests

Referring to Figure 3.1, suppose a uniaxial test is conducted using specimen 1, that is, material properties are measured in the x-direction. As the test proceeds stress σ_{xx} is increased from zero to some maximal level, and the components of strain induced as a result of this stress are measured. An idealized plot of strain data collected during a uniaxial test of an anisotropic material is shown in Figure 3.4, where it has been assumed that magnitude of stress is relatively low, such that a linear relationship exists between the stress component σ_{xx} and the resulting strains. Strains induced at high nonlinear stress levels, including failure stresses, will be considered in Section 3.3.5. *Note that for an anisotropic material stress σ_{xx} will induce all six components of strain: $\varepsilon_{xx}, \varepsilon_{yy}, \varepsilon_{zz}, \gamma_{xy}, \gamma_{xz},$ and γ_{yz}.* This is not the case for isotropic materials; a uniaxial stress σ_{xx} applied to an isotropic material will not induce any shear strains ($\gamma_{xy} = \gamma_{xz} = \gamma_{yz} = 0$), and further the transverse normal strains will be identical ($\varepsilon_{yy} = \varepsilon_{zz}$). Hence, for anisotropic material there is an

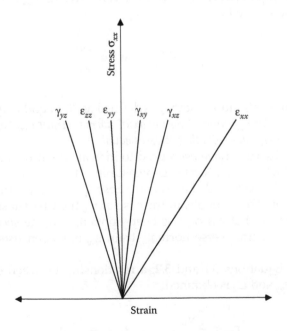

FIGURE 3.4
Idealized plot of the six strain components caused by the application a uniaxial stress σ_{xx}.

unusual *coupling* between normal stress and shear strain, which would not be expected based on previous experience with isotropic materials.

As would be expected, as the magnitude of σ_{xx} is increased, the magnitude of all resulting strain components is also increased. Since stress σ_{xx} causes six distinct components of strain for an anisotropic material, *six material properties must be defined in order to relate* σ_{xx} *to the resulting strains*.

Let us first consider material properties relating normal stress σ_{xx} to normal strains ε_{xx}, ε_{yy}, and ε_{zz}. The relationship between σ_{xx} and normal strain ε_{xx} is characterized by Young's modulus, E_{xx} (also called the "modulus of elasticity"):

$$E_{xx} \equiv \frac{\sigma_{xx}}{\varepsilon_{xx}} \tag{3.1}$$

Young's modulus is simply the slope of the σ_{xx} vs. ε_{xx} curve shown in Figure 3.4. In words, Young's modulus is defined as *"the normal stress σ_{xx} divided by the resulting normal strain ε_{xx}, with all other stress components equal zero."* Subscripts *"xx"* have been used to indicate the direction in which Young's modulus has been measured. Since we have restricted our attention to the linear region of the stress–strain curve, *Equation 3.1 is only valid at relatively low, linear stress levels.*

The relationship between the two transverse strains (ε_{yy} and ε_{zz}) and ε_{xx} is defined by Poisson's ratio:

$$\nu_{xy} \equiv \frac{-\varepsilon_{yy}}{\varepsilon_{xx}} \qquad \nu_{xz} \equiv \frac{-\varepsilon_{zz}}{\varepsilon_{xx}} \tag{3.2}$$

In words, Poisson's ratio ν_{xy} (or ν_{xz}) is defined as *"the negative of the transverse normal strain ε_{yy} (or ε_{zz}) divided by the axial normal strain ε_{xx}, both of which are induced by stress σ_{xx}, with all other stresses equal zero."*

As before, subscripts have been used to indicate the uniaxial stress condition under which Poisson's ratio is measured. The first subscript indicates the direction of stress, and the second subscript indicates the direction of transverse strain. For example, in the case of ν_{xy} the first subscript "x" indicates that a uniaxial stress σ_{xx} has been applied, and the second subscript "y" indicates that transverse normal strain ε_{yy} has been used to calculate Poisson's ratio.

Combining Equations 3.1 and 3.2, a relationship between σ_{xx} and transverse strains ε_{yy} and ε_{zz} is obtained:

$$\varepsilon_{yy} = -\frac{\nu_{xy}}{E_{xx}}\sigma_{xx} \qquad \varepsilon_{zz} = -\frac{\nu_{xz}}{E_{xx}}\sigma_{xx} \tag{3.3}$$

Now consider material properties relating normal strain ε_{xx} to shear strains γ_{xy}, γ_{xz}, and γ_{yz}. Material properties relating normal strains to shear strains were discussed by Lekhnitski [2] and are called *coefficients of mutual influence of the second kind*. In this chapter they will be denoted using the symbol "η," and are defined as follows:

$$\eta_{xx,xy} \equiv \frac{\gamma_{xy}}{\varepsilon_{xx}} \qquad \eta_{xx,xz} \equiv \frac{\gamma_{xz}}{\varepsilon_{xx}} \qquad \eta_{xx,yz} \equiv \frac{\gamma_{yz}}{\varepsilon_{xx}} \tag{3.4}$$

In words, the coefficient of mutual influence of the second kind $\eta_{xx,xy}$ (or $\eta_{xx,xz}$ or $\eta_{xx,yz}$) is defined as *"the shear strain γ_{xy} (or γ_{xz} or γ_{yz}) divided by the normal strain ε_{xx}, both of which are induced by normal stress σ_{xx}, when all other stresses equal zero."*

Subscripts have once again been used to indicate the stress condition under which the coefficient of mutual influence of the second kind is measured. The first set of subscripts indicates the direction of stress, and the second set of subscripts indicates the shear strain used to calculate the coefficient. For example, in the case of $\eta_{xx,xy}$ the first two subscripts "xx" indicates that a normal stress σ_{xx} has been applied, and the second two subscripts "xy" indicate that γ_{xy} has been used to calculate the coefficient.

Combining Equations 3.1 and 3.4, a relationship between σ_{xx} and shear strain γ_{xy}, γ_{xz}, or γ_{yz} is obtained:

$$\gamma_{xy} = \frac{\eta_{xx,xy}}{E_{xx}} \sigma_{xx} \qquad \gamma_{xz} = \frac{\eta_{xx,xz}}{E_{xx}} \sigma_{xx} \qquad \gamma_{yz} = \frac{\eta_{xx,yz}}{E_{xx}} \sigma_{xx} \tag{3.5}$$

Equations 3.1 through 3.5 define six properties measured in the x-direction, using specimen 1. Referring again to Figure 3.1, analogous results are obtained when properties are measured in the y- and z-directions, using specimens 2 and 3:

Properties Measured Using Specimen 2 (σ_{yy} Applied):

$$E_{yy} \equiv \frac{\sigma_{yy}}{\varepsilon_{yy}} \qquad \text{(or)} \qquad \varepsilon_{yy} = \frac{1}{E_{yy}} \sigma_{yy}$$

$$\nu_{yx} \equiv \frac{-\varepsilon_{xx}}{\varepsilon_{yy}} \qquad \text{(or)} \qquad \varepsilon_{xx} = -\frac{\nu_{yx}}{E_{yy}} \sigma_{yy} \tag{3.6}$$

$$\nu_{yz} \equiv \frac{-\varepsilon_{zz}}{\varepsilon_{yy}} \qquad \text{(or)} \qquad \varepsilon_{zz} = -\frac{\nu_{yz}}{E_{yy}} \sigma_{yy}$$

$$\eta_{yy,xy} \equiv \frac{\gamma_{xy}}{\varepsilon_{yy}} \quad \text{(or)} \quad \gamma_{xy} = \frac{\eta_{yy,xy}}{E_{yy}}\sigma_y$$

$$\eta_{yy,xz} \equiv \frac{\gamma_{xz}}{\varepsilon_{yy}} \quad \text{(or)} \quad \gamma_{xz} = \frac{\eta_{yy,xz}}{E_{yy}}\sigma_y$$

$$\eta_{yy,yz} \equiv \frac{\gamma_{yz}}{\varepsilon_{yy}} \quad \text{(or)} \quad \gamma_{yz} = \frac{\eta_{yy,yz}}{E_{yy}}\sigma_y$$

Properties Measured Using Specimen 3 (σ_{zz} Applied):

$$E_{zz} \equiv \frac{\sigma_{zz}}{\varepsilon_{zz}} \quad \text{(or)} \quad \varepsilon_{zz} = \frac{1}{E_{zz}}\sigma_{zz}$$

$$\nu_{zx} \equiv \frac{-\varepsilon_{xx}}{\varepsilon_{zz}} \quad \text{(or)} \quad \varepsilon_{xx} = -\frac{\nu_{zx}}{E_{zz}}\sigma_{zz}$$

$$\nu_{zy} \equiv \frac{-\varepsilon_{yy}}{\varepsilon_{zz}} \quad \text{(or)} \quad \varepsilon_{yy} = -\frac{\nu_{zy}}{E_{zz}}\sigma_{zz}$$

$$\eta_{zz,xy} \equiv \frac{\gamma_{xy}}{\varepsilon_{zz}} \quad \text{(or)} \quad \gamma_{xy} = \frac{\eta_{zz,xy}}{E_{zz}}\sigma_{zz} \qquad (3.7)$$

$$\eta_{zz,xz} \equiv \frac{\gamma_{xz}}{\varepsilon_{zz}} \quad \text{(or)} \quad \gamma_{xz} = \frac{\eta_{zz,xz}}{E_{zz}}\sigma_{zz}$$

$$\eta_{zz,yz} \equiv \frac{\gamma_{yz}}{\varepsilon_{zz}} \quad \text{(or)} \quad \gamma_{yz} = \frac{\eta_{zz,yz}}{E_{zz}}\sigma_{zz}$$

3.2.2 Pure Shear Tests

If a pure shear stress (τ_{xy}, say) is applied to an anisotropic material, six components of strain will be induced. An idealized plot of strain data collected during a pure shear test of an anisotropic material is shown schematically in Figure 3.5, where it is assumed that magnitude of shear stress is relatively low such that a linear relationship exists between the stress component τ_{xy} and the resulting strains. Strains induced at high nonlinear stress levels, including failure stresses, will be considered in Section 3.3.5. Once again, the stress–strain response of an anisotropic material differs markedly from that of an isotropic material. *Specifically, for an anisotropic material stress τ_{xy} will induce all six components of strain:* ε_{xx}, ε_{yy}, ε_{zz}, γ_{xy}, γ_{xz}, and γ_{yz}. If an isotropic material is subjected to a pure shear stress τ_{xy}, only one strain component is induced (γ_{xy}); all other strain components are zero ($\varepsilon_{xx} = \varepsilon_{yy} = \varepsilon_{zz} = \gamma_{xz} = \gamma_{yz} = 0$).

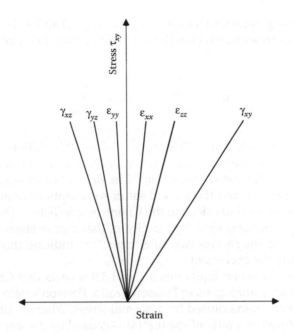

FIGURE 3.5
Idealized plot of the six strain components caused by the application a pure shear stress τ_{xy}.

Hence, for anisotropic material there is an unusual *coupling* between shear stress and normal strain, as well as an unusual coupling between shear stress in one plane (the *x–y* plane, say) and out-of-plane shear strains (γ_{xz} and γ_{yz}). Neither of these coupling effects occurs in isotropic materials.

As would be expected, as the magnitude of τ_{xy} is increased during the test the magnitude of the resulting strains is also increased. Since stress τ_{xy} causes six distinct components of strain, *six material properties must be defined in order to relate τ_{xy} to the resulting strains.*

Let us first consider material properties relating shear stress τ_{xy} to shear strains γ_{xy}, γ_{xz}, and γ_{yz}. The relationship between τ_{xy} and shear strain γ_{xy} is characterized by the shear modulus, G_{xy}:

$$G_{xy} \equiv \frac{\tau_{xy}}{\gamma_{xy}} \qquad (3.8)$$

In words, the shear modulus is defined as *"the shear stress τ_{xy} divided by the resulting shear strain γ_{xy}, with all other stress components equal zero."* Since we have restricted our attention to linear stress levels, *Equation 3.8 is only valid at relatively low, linear shear stress levels.*

The relationship between transverse strains (γ_{xz}, γ_{yz}) and γ_{xy} is characterized by Chentsov coefficients, which will be denoted using the symbol "μ" in this chapter:

$$\mu_{xy,xz} \equiv \frac{\gamma_{xz}}{\gamma_{xy}} \qquad \mu_{xy,yz} \equiv \frac{\gamma_{yz}}{\gamma_{xy}} \tag{3.9}$$

In words, the Chentsov coefficient $\mu_{xy,xz}$ (or $\mu_{xy,yz}$) is defined as *"the shear strain γ_{xz} (or γ_{yz}) divided by the shear strain γ_{xy}, both of which are induced by shear stress τ_{xy}, with all other stresses equal zero."* The first set of subscripts indicates the stress component, and the second set of subscripts indicates the out-of-plane shear strain used to calculate the Chentsov coefficient. For example, in the case of $\mu_{xy,xz}$ the subscripts *"xy"* indicates that a pure shear stress τ_{xy} has been applied, and the second two subscripts *"xz"* indicate that γ_{xz} has been used to calculate the coefficient.

A comparison between Equations 3.2 and 3.9 reveals that Chentsov coefficients are directly analogous to Poisson's ratio. Poisson's ratio is defined as a ratio of normal strains caused by a normal stress, whereas Chentsov coefficients are defined as a ratio of shear strains caused by a shear stress.

Combining Equations 3.8 and 3.9, a relationship between τ_{xy} and shear strains γ_{xz} or γ_{yz} is obtained:

$$\gamma_{xz} = \frac{\mu_{xy,xz}}{G_{xy}}\tau_{xy} \qquad \gamma_{yz} = \frac{\mu_{xy,yz}}{G_{xy}}\tau_{xy} \tag{3.10}$$

Finally, consider material properties relating shear stress τ_{xy} to normal strains ε_{xx}, ε_{yy}, and ε_{zz}. Material properties relating shear stress to normal strains were discussed by Lekhnitski [2] and are called *coefficients of mutual influence of the first kind*. In this chapter, they will be denoted using the symbol "η," and are defined as follows:

$$\eta_{xy,xx} \equiv \frac{\varepsilon_{xx}}{\gamma_{xy}} \qquad \eta_{xy,yy} \equiv \frac{\varepsilon_{yy}}{\gamma_{xy}} \qquad \eta_{xy,zz} \equiv \frac{\varepsilon_{zz}}{\gamma_{xy}} \tag{3.11}$$

In words, the coefficient of mutual influence of the first kind $\eta_{xy,xx}$ (or $\eta_{xy,yy}$ or $\eta_{xy,zz}$) is defined as *"the normal strain ε_{xx} (or ε_{yy} or ε_{zz}) divided by the shear strain γ_{xy}, both of which are induced by shear stress τ_{xy}, when all other stresses equal zero."* The first set of subscripts indicates the stress component applied, and the second set indicates the normal strain used to calculate the coefficient. For example, in the case of $\eta_{xy,xx}$ the first subscript *"xy"* indicates that shear stress τ_{xy} has been applied, and the second set *"xx"* indicate that ε_{xx} has been used to calculate the coefficient.

Combining Equations 3.8 and 3.11, a relationship between τ_{xy} and normal strain ε_{xx}, ε_{yy}, or ε_{zz} is obtained:

$$\varepsilon_{xx} = \frac{\eta_{xy,xx}}{G_{xy}}\tau_{xy} \qquad \varepsilon_{yy} = \frac{\eta_{xy,yy}}{G_{xy}}\tau_{xy} \qquad \varepsilon_{zz} = \frac{\eta_{xy,zz}}{G_{xy}}\tau_{xy} \qquad (3.12)$$

Equations 3.8 through 3.12 define six properties measured when a pure shear stress τ_{xy} is applied. Analogous material properties are defined during tests in which pure shears τ_{xz} or τ_{yz} are applied.

Properties Measured Using Pure Shear τ_{xz}:

$$G_{xz} \equiv \frac{\tau_{xz}}{\gamma_{xz}} \qquad (\text{or}) \qquad \gamma_{xz} = \frac{1}{G_{xz}}\tau_{xz}$$

$$\mu_{xz,xy} \equiv \frac{\gamma_{xy}}{\gamma_{xz}} \qquad (\text{or}) \qquad \gamma_{xy} = \frac{\mu_{xz,xy}}{G_{xz}}\tau_{xz}$$

$$\mu_{xz,yz} \equiv \frac{\gamma_{yz}}{\gamma_{xz}} \qquad (\text{or}) \qquad \gamma_{yz} = \frac{\mu_{xz,yz}}{G_{xz}}\tau_{xz}$$

$$\eta_{xz,xx} \equiv \frac{\varepsilon_{xx}}{\gamma_{xz}} \qquad (\text{or}) \qquad \varepsilon_{xx} = \frac{\eta_{xz,xx}}{G_{xz}}\tau_{xz} \qquad (3.13)$$

$$\eta_{xz,yy} \equiv \frac{\varepsilon_{yy}}{\gamma_{xz}} \qquad (\text{or}) \qquad \varepsilon_{yy} = \frac{\eta_{xz,yy}}{G_{xz}}\tau_{xz}$$

$$\eta_{xz,zz} \equiv \frac{\varepsilon_{zz}}{\gamma_{xz}} \qquad (\text{or}) \qquad \varepsilon_{zz} = \frac{\eta_{xz,zz}}{G_{xz}}\tau_{xz}$$

Properties Measured Using Pure Shear τ_{yz}:

$$G_{yz} \equiv \frac{\tau_{yz}}{\gamma_{yz}} \qquad (\text{or}) \qquad \gamma_{yz} = \frac{1}{G_{yz}}\tau_{yz}$$

$$\mu_{yz,xy} \equiv \frac{\gamma_{xy}}{\gamma_{yz}} \qquad (\text{or}) \qquad \gamma_{xy} = \frac{\mu_{yz,xy}}{G_{yz}}\tau_{yz} \qquad (3.14)$$

$$\mu_{yz,xz} \equiv \frac{\gamma_{xz}}{\gamma_{yz}} \qquad (\text{or}) \qquad \gamma_{xz} = \frac{\mu_{yz,xz}}{G_{yz}}\tau_{yz}$$

$$\eta_{yz,xx} \equiv \frac{\varepsilon_{xx}}{\gamma_{yz}} \quad \text{(or)} \quad \varepsilon_{xx} = \frac{\eta_{yz,xx}}{G_{yz}} \tau_{yz}$$

$$\eta_{yz,yy} \equiv \frac{\varepsilon_{yy}}{\gamma_{yz}} \quad \text{(or)} \quad \varepsilon_{yy} = \frac{\eta_{yz,yy}}{G_{yz}} \tau_{yz}$$

$$\eta_{yz,zz} \equiv \frac{\varepsilon_{zz}}{\gamma_{yz}} \quad \text{(or)} \quad \varepsilon_{zz} = \frac{\eta_{yz,zz}}{G_{yz}} \tau_{yz}$$

3.2.3 Specialization to Orthotropic and Transversely Isotropic Composites

As previously shown in Figure 3.2a,b, the principal material coordinate system is established by planes of symmetry within the reinforcing fiber architecture. Thus, for unidirectional composites the 1-axis is defined parallel to the fiber direction, the 2-axis is perpendicular to the fibers and lies within the plane of the composite, and the 3-axis is perpendicular to the fibers and lies out-of-plane. In other cases, such as the braided composite shown in Figure 3.2c,d, the 1–2–3 principal material coordinate system is not necessarily aligned with the fiber direction. For all composite fabrics based on continuous fibers and typically encountered in practice (i.e., unidirectional, woven, or braided fabrics) planes of symmetry and the principal material coordinate system are readily identified.

We will now consider those properties that are measured when the composite is referenced to the principal material coordinate system. It will be seen later that properties of an anisotropic composite (i.e., a composite referenced to a nonprincipal material coordinate system) can always be related to those measured relative to the 1–2–3 coordinate system. To simplify our discussion we will assume that the composite under consideration is a unidirectional composite, and hence that the 1–2–3 axes are parallel and perpendicular to the fibers.

A unidirectional composite subjected to uniaxial tensile stress σ_{11} is shown in Figure 3.6a. This specimen is commonly referred to as a [0°] laminate. The deformed shape of the element is also shown in the figure. Note that:

- The element has increased in length in the 1-direction, corresponding to a tensile strain ε_{11}.

- The element has decreased in width and thickness in the 2- and 3-directions, corresponding to compressive strains ε_{22} and ε_{33}, respectively.

- The deformed element is a rectangular parallelepiped. That is, due to the symmetric distribution of fibers with respect to the 1-, 2-, and 3-coordinate axes, in the deformed condition all angles remained $\pi/2$ radians (90°). Hence, all shear strains equal zero ($\gamma_{12} = \gamma_{13} = \gamma_{23} = 0$).

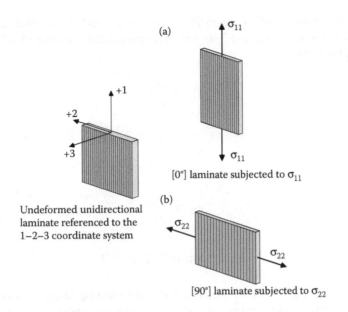

FIGURE 3.6
Deformations induced in a unidirectional composite by (a) uniaxial stress σ_{11}; (b) uniaxial stress σ_{22} (deformations exaggerated for clarity).

Applying Equations 3.1, 3.2, and 3.4, we have

$$E_{11} \equiv \frac{\sigma_{11}}{\varepsilon_{11}} \tag{3.15a}$$

$$\nu_{12} \equiv \frac{-\varepsilon_{22}}{\varepsilon_{11}} \tag{3.15b}$$

$$\nu_{13} \equiv \frac{-\varepsilon_{33}}{\varepsilon_{11}} \tag{3.15c}$$

$$\eta_{11,12} = \eta_{11,13} = \eta_{11,23} = 0 \tag{3.15d}$$

Since no shear strains are induced by σ_{11}, the coefficients of mutual influence of the second kind all equal zero. This is only true *when the composite is referenced to the principal material coordinate system*. That is, uniaxial stress acting in a nonprincipal coordinate system will cause a shear strain, as previously shown in Figure 3.3 for example. Therefore, the coefficients of mutual influence of the second kind do not equal zero for anisotropic composites, that is, if the composite is referenced to a nonprincipal material coordinate system. Methods of calculating composite material properties in nonprincipal coordinate systems will be presented in Chapter 4.

A unidirectional [90°] composite laminate subjected to uniaxial tensile stress σ_{22} is shown in Figure 3.6b. Material properties measured when stress σ_{22} is applied are

$$E_{22} \equiv \frac{\sigma_{22}}{\varepsilon_{22}} \tag{3.16a}$$

$$\nu_{21} \equiv \frac{-\varepsilon_{11}}{\varepsilon_{22}} \tag{3.16b}$$

$$\nu_{23} \equiv \frac{-\varepsilon_{33}}{\varepsilon_{22}} \tag{3.16c}$$

$$\eta_{22,12} = \eta_{22,13} = \eta_{22,23} = 0 \tag{3.16d}$$

Once again, due to the symmetrical distribution of fibers, stress σ_{22} does not induce any shear strains, so the coefficients of mutual influence of the second kind all equal zero.

As previously mentioned, due to the thin plate-like nature of composites it is difficult in practice to apply a well-defined out-of-plane uniaxial stress σ_{33}, or to measure the resulting normal strain induced in the out-of-plane direction, ε_{33}. Assuming these practical difficulties are overcome, the material properties measured when stress σ_{33} is applied are

$$E_{33} \equiv \frac{\sigma_{33}}{\varepsilon_{33}} \tag{3.17a}$$

$$\nu_{31} \equiv \frac{-\varepsilon_{11}}{\varepsilon_{33}} \tag{3.17b}$$

$$\nu_{32} \equiv \frac{-\varepsilon_{22}}{\varepsilon_{33}} \tag{3.17c}$$

$$\eta_{33,12} = \eta_{33,13} = \eta_{33,23} = 0 \tag{3.17d}$$

For unidirectional composites, both the 2- and 3-axes are defined to be perpendicular to the fibers, and hence properties measured in the 2- and 3-directions are typically similar in magnitude. In fact, if the distribution of fibers in the 2- and 3-directions is identical at the microlevel, then properties measured in these directions will be equal: $E_{22} = E_{33}$, $\nu_{12} = \nu_{13}$, $\nu_{21} = \nu_{31}$, and $\nu_{23} = \nu_{32}$. If this occurs, then the composite is classified as a *transversely isotropic material.* In contrast, if the distribution of fibers differs in the 2- and 3-directions, or if the composite under consideration is a woven or braided composite, then properties measured in the 2- and 3-directions will not be identical and the composite is classified as an *orthotropic material.*

Optical micrographs showing the fiber distribution in the 2–3 plane for a graphite-bismaleimide (IM7/5260) laminate produced using unidirectional plies at different orientations are shown in Figure 3.7. The fiber ends appear elliptical because they were inclined at an angle to the plane of the photograph (this is particularly noticeable in Figure 3.7c). It is apparent that a resin-rich zone exists between plies. In this case, the thickness of the resin-rich zone is about 10 μm. However, the thickness of the resin-rich zone varies from one laminate to the next and depends on the material system used, stacking sequence, and processing conditions used to produce the laminate. If the resin-rich zone is very thin (less than about 1/10 the ply thickness, say), and if the fibers are uniformly distributed within the interior of each ply, the composite will respond as a transversely isotropic material at the macrolevel. If these conditions do not exist (if the thickness of the resin-rich zone is an appreciable fraction of the ply thickness, or if the distribution of fibers in the 2- and 3-directions differ substantially), then E_{33} will differ from E_{22}, and the composite will respond as an orthotropic material. For the laminate shown in Figure 3.7 E_{33} would likely differ significantly from E_{22}, and so should be treated as an orthotropic material.

Let us now consider properties measured through application of a pure shear stress in the principal material coordinate system. A stress element representing a unidirectional composite subjected to a pure shear stress τ_{12} is shown in Figure 3.8. The deformed shape of the element is also shown. Note that:

- The angle originally defined by the 1–2 axes has decreased, corresponding to a positive shear strain γ_{12}.
- Due to the symmetric distribution of fibers with respect to the 1-, 2-, and 3-coordinate axes, in the deformed condition all remaining angles remained $\pi/2$ radians (90°). Hence, the remaining two shear strains equal zero ($\gamma_{13} = \gamma_{23} = 0$).
- The length, width, and thickness of the element has not changed, and hence all normal strain are zero ($\varepsilon_{11} = \varepsilon_{22} = \varepsilon_{33} = 0$).

Applying Equations 3.8, 3.9, and 3.10, we have

$$G_{12} \equiv \frac{\tau_{12}}{\gamma_{12}} \tag{3.18a}$$

$$\mu_{12,13} = \mu_{12,23} = 0 \tag{3.18b}$$

$$\eta_{12,11} = \eta_{12,22} = \eta_{12,33} = 0 \tag{3.18c}$$

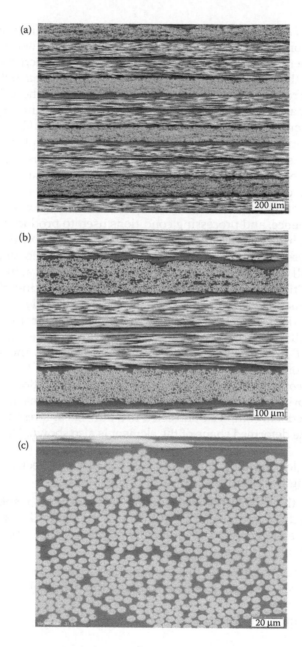

FIGURE 3.7
Optical micrographs of a graphite/bismaleimide (IM7/5260) composite laminate, taken at three different magnification levels. Note the resin-rich zone between plies. (a) Eleven adjacent plies; (b) six adjacent plies; (c) fibers in one ply.

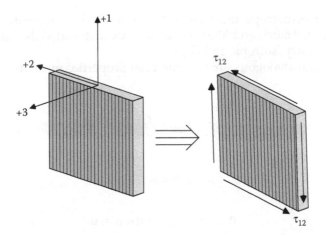

FIGURE 3.8
Deformations induced in a unidirectional composite by pure shear stress τ_{12} (deformations shown exaggerated for clarity).

Since only γ_{12} is induced by τ_{12}, the Chentsov coefficients as well as the coefficients of mutual influence of the first kind are all equal to zero. This is only true *when the composite is referenced to the principal material coordinate system*. That is, a shear stress acting in a nonprincipal material coordinate system will in general cause both normal strains and shear strains. Therefore, neither the Chentsov coefficients nor the coefficients of mutual influence of the first kind equal zero if the composite is referenced to a nonprincipal material coordinate system. Methods of calculating composite material properties in nonprincipal coordinate systems will be presented in Chapter 4.

Once again, due to the thin plate-like nature of composites, in practice it is difficult to apply well-defined out-of-plane shear stresses τ_{13} or τ_{23}, or to measure the resulting shear strains induced in the out-of-plane direction, γ_{13} or γ_{23}. Assuming these practical difficulties were overcome, the material properties measured when stress τ_{13} is applied are

$$G_{13} \equiv \frac{\tau_{13}}{\gamma_{13}} \tag{3.19a}$$

$$\mu_{13,12} = \mu_{13,23} = 0 \tag{3.19b}$$

$$\eta_{13,11} = \eta_{13,22} = \eta_{13,33} = 0 \tag{3.19c}$$

If the fibers are not uniformly distributed within the 2–3 plane, or if the composite is based on woven or braided fabrics, then the composite will behave as an orthotropic material and $G_{12} \neq G_{13}$. If the composite is based on a unidirectional fabric and fibers are uniformly distributed then the composite is transversely isotropic and $G_{12} = G_{13}$.

Following an analogous process, material properties measured when τ_{23} is applied are

$$G_{23} \equiv \frac{\tau_{23}}{\gamma_{23}} \qquad (3.20a)$$

$$\mu_{23,12} = \mu_{23,13} = 0 \qquad (3.20b)$$

$$\eta_{23,11} = \eta_{23,22} = \eta_{23,33} = 0 \qquad (3.20c)$$

A total of 12 material properties have been defined above for orthotropic or transversely isotropic composites: three Young's moduli (E_{11}, E_{22}, and E_{33}), six Poisson's ratios ($\nu_{12}, \nu_{13}, \nu_{21}, \nu_{23}, \nu_{31}$, and ν_{32}), and three shear moduli (G_{12}, G_{13}, and G_{23}). However, it will be seen later that for orthotropic composites only 9 of these 12 properties are independent, and for transversely isotropic composites only 5 of the 12 properties are independent. Therefore, only nine material properties must be measured to *fully* characterize the elastic response of orthotropic composites; for transversely isotropic composites only five material properties must be measured.

The number of material properties required in most practical engineering applications of composite is reduced further still. For reasons that will be explained later, it is usually appropriate to assume that a composite structure is subjected to a state of plane stress. Ultimately, this means that we only require material properties *in one plane*. Hence, while an orthotropic composite possesses nine distinct elastic material properties (and a transversely isotropic composites possesses five), in practice only four of these properties are ordinarily required: E_{11}, E_{22}, ν_{12}, and G_{12}. A brief summary of common experimental methods used to measure in-plane properties in tension is provided in Appendix A. Typical values for several composite material systems are listed in Table 3.1.

As a final comment, an often overlooked fact is that the elastic properties of composites usually differ in tension and compression (in fact, this is true for many materials, not just for composites). For example, for polymeric composites it is not uncommon for E_{22} measured in tension to differ by 10–15% from that measured in compression. Materials that exhibit this behavior are called "bi-modulus materials." Although it is possible to account for these differences during a structural analysis, for example, see Reference 3, the bi-modulus phenomenon is a significant complication and will not be accounted for herein. Throughout this and in later chapters it will be assumed that in-plane elastic properties E_{11}, E_{22}, and ν_{12} are identical in tension and compression. The reader should be aware that these differences usually exist, however. If in practice the measured response of a composite structure differs from the predicted behavior, the discrepancy may well be due to differences in elastic properties in tension versus compression.

3.3 Material Properties Relating Temperature to Strain

If an unconstrained anisotropic composite is subjected to a uniform change in temperature ΔT, six components of strain will be induced: $\varepsilon_{xx}^T, \varepsilon_{yy}^T, \varepsilon_{zz}^T, \gamma_{xy}^T, \gamma_{xz}^T,$ and γ_{yz}^T. The superscript "T" has been used to indicate that these strains are caused solely by a change in temperature. Note that three of these strains are *shear* strains; for anisotropic materials a change in temperature will, in general, cause shear strains to develop. Strains induced solely by a change in temperature are referred to as "free thermal strains" or simply "thermal strains." Properties that relate strains to temperature change are called *coefficients of thermal expansion.*

As previously discussed, it is the microstructural features of a material that determine whether it exhibits isotropic or anisotropic behavior. The contention that a change in temperature will induce shear strains may seem unusual (since isotropic materials do not exhibit such behavior), but can be easily explained in the case of unidirectional composites. An initially square unidirectional composite is shown in Figure 3.9, where it has been assumed that the fibers are oriented at an angle of 45° with respect to the x-axis. Since the composite is initially square, and since in this example the fibers are defined to be at an angle of 45°, fibers are parallel to diagonal AC and are perpendicular to diagonal BD. Now, the coefficient of thermal expansion exhibited by high-performance fibers is typically very low (or even slightly

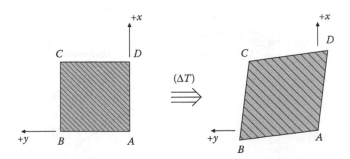

FIGURE 3.9
Deformations caused in a 45° unidirectional composite by a uniform change in temperature, ΔT (deformations shown exaggerated for clarity).

negative), whereas for most polymers it is relatively high. For example, the coefficient of thermal expansion of graphite fibers is about -1 μm/m–°C, whereas for epoxies it is on the order of 30 μm/m–°C. Therefore, assuming the composite shown in Figure 3.9 consists of a graphite–epoxy system, an increase in temperature will cause a slight *decrease* in the length of diagonal AC, but will cause a relatively large *increase* in the length of diagonal BD. Hence, the initially square composite deforms into a parallelogram, as shown in the figure. The fact that angle $\angle DAB$ has increased reveals that a shear strain γ_{xy} (in this case, a negative shear strain γ_{xy}) has been induced by the change in temperature ΔT. Hence, there is a coupling between the change in temperature and shear strains, as stipulated. Note that this physical explanation of the coupling between a uniform change in temperature and shear strain indicates that *this coupling only occurs if strain is referenced to a nonprincipal material coordinate system.*

An idealized plot of the six strain components induced in an anisotropic composite by a change in temperature is shown in Figure 3.10. As would be expected, as ΔT is increased the magnitude of all strain components also increases. For modest changes in temperature ($\Delta T < 200$°C, say) the change in temperature is linearly related to the resulting thermal strain components. That is, the slopes of the six strain vs ΔT curves shown in Figure 3.10 are constant at relatively low levels of ΔT. At high levels of ΔT the slopes of the curves typically increase. For polymeric composites, the temperature at which the curves become nonlinear is related to the glass-transition temperature, T_g, of the polymeric matrix (the glass-transition temperature

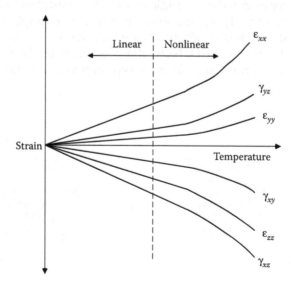

FIGURE 3.10
Idealized plot of the six strain components caused by a change in temperature, ΔT.

of a polymer is discussed in Section 3.1.2). Most composite structures are designed to operate at temperatures below the T_g, and hence we will focus our attention on the linear range shown in Figure 3.10.

Since a change in temperature causes six strains to develop for anisotropic composites, six coefficients of thermal expansion must be defined:

$$\alpha_{xx} \equiv \frac{\varepsilon_{xx}^T}{\Delta T} \qquad \alpha_{yy} \equiv \frac{\varepsilon_{yy}^T}{\Delta T} \qquad \alpha_{zz} \equiv \frac{\varepsilon_{zz}^T}{\Delta T}$$

$$\alpha_{xy} \equiv \frac{\gamma_{xy}^T}{\Delta T} \qquad \alpha_{xz} \equiv \frac{\gamma_{xz}^T}{\Delta T} \qquad \alpha_{yz} \equiv \frac{\gamma_{yz}^T}{\Delta T}$$

(3.21)

Since we have limited our discussion to the linear range shown in Figure 3.10, the coefficients of thermal expansion defined by Equation 3.21 equal the slopes of the corresponding strain vs ΔT curves within the linear range. These properties will henceforth be called *linear coefficients of thermal expansion*, and will be abbreviated "CTE's." In SI units, they are usually reported in terms of $\mu m/m$–°C or μrad/°C (for normal or shear strains, respectively). In English units they are usually reported in terms of $\mu in./in.$–°F or μrad/°F. A CTE can be converted from SI units into English units by multiplying by the factor 5/9. For example, a CTE of 15 $\mu m/m$–°C equals 8.3 $\mu in./in.$–°F.

Equations 3.21 can be easily rearranged and written in matrix form as follows:

$$\begin{bmatrix} \varepsilon_{xx}^T & \gamma_{xy}^T & \gamma_{xz}^T \\ \gamma_{yx}^T & \varepsilon_{yy}^T & \gamma_{yz}^T \\ \gamma_{zx}^T & \gamma_{zy}^T & \varepsilon_{zz}^T \end{bmatrix} = (\Delta T) \begin{bmatrix} \alpha_{xx} & \alpha_{xy} & \alpha_{xz} \\ \alpha_{xy} & \alpha_{yy} & \alpha_{yz} \\ \alpha_{xz} & \alpha_{yz} & \alpha_{zz} \end{bmatrix}$$

(3.22)

The reader should note that the strains caused by a change in temperature can be transformed from one coordinate system to another, in exactly the same way that mechanically induced strains are transformed. In particular, any of the strain transformation equations reviewed in Chapter 2 (e.g., Equations 2.36, 2.41, 2.43, 2.44, or 2.47) can be used to transform thermally induced strains from one coordinate system to another.

3.3.1 Specialization to Orthotropic and Transversely Isotropic Composites

We will now apply these general definitions to the case of a unidirectional composite referenced to the principal material coordinate system. It will be seen later that CTEs of an anisotropic composite (i.e., a composite referenced to a nonprincipal material coordinate system) can always be related to those measured in the 1–2–3 coordinate system. Although unidirectional

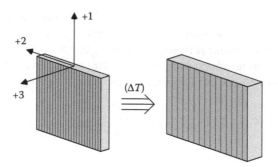

FIGURE 3.11
Deformations induced in a unidirectional composite by a change in temperature ΔT (deformations shown exaggerated for clarity).

composites are used in the following discussion, an equivalent discussion applies to woven or braided composites, when referenced to the principal material coordinate system.

A unidirectional composite subjected to a uniform change in temperature ΔT and referenced to the 1–2–3 coordinate system is shown in Figure 3.11. Since the fibers are distributed symmetrically with respect to the 1-, 2-, and 3-coordinate axes, a change in temperature does not cause a shear strain to develop. Hence, in the principal material coordinate system Equations 3.21 become

$$\alpha_{11} = \frac{\varepsilon_{11}^T}{\Delta T} \qquad \alpha_{22} = \frac{\varepsilon_{22}^T}{\Delta T} \qquad \alpha_{33} = \frac{\varepsilon_{33}^T}{\Delta T}$$

$$\alpha_{12} = \alpha_{13} = \alpha_{23} = 0$$

(3.23)

As before, if the fibers are distributed uniformly throughout the 2–3 plane, then $\alpha_{22} = \alpha_{33}$ and the composite will be transversely isotropic. If the fibers are not uniformly distributed, then $\alpha_{22} \neq \alpha_{33}$, and the composite will be orthotropic. Woven or braided composites are always orthotropic, since the distribution of fibers in the 2-direction differs substantially from that in the 3-direction. Representative CTEs for common unidirectional polymeric composites are included in Table 3.1.

3.4 Material Properties Relating Moisture Content to Strain

Water molecules diffuse into (or out of) the overall molecular structure of polymeric materials. Consequently the *moisture content* of polymeric-based materials, including polymeric composites, slowly varies as the relative humidity of the surrounding atmosphere varies. The moisture content of a polymer is usually expressed as a percentage by weight, and typically ranges from ~0% to a maximum of about ~3–5%.

From a structural point of view, the effects of a change in moisture content are analogous to those caused by a change in temperature. For example, a plot of strains as a function of moisture content would resemble Figure 3.10, except that ΔM would be plotted along the horizontal axis rather than temperature. Hence, if an unconstrained anisotropic composite is subjected to a uniform change in moisture content ΔM, then six components of strain may be induced: ε_{xx}^M, ε_{yy}^M, ε_{zz}^M, γ_{xy}^M, γ_{xz}^M, and γ_{yz}^M. The superscript "M" has been used to indicate that these strains are caused solely by a change in moisture content. Strains induced by a change in moisture content are sometimes referred to as *hygroscopic* strains, and can be just as large (or larger) than those associated with a change in temperature.

In this chapter, it will be assumed that strain is linearly related to changes in moisture content. Properties that relate strains to changes in moisture content will be called *linear coefficients of moisture expansion*, abbreviated as "CME's," and will be denoted using the symbol "β." Since a change in moisture content causes six strains to develop for anisotropic composites, six CME's must be defined:

$$\beta_{xx} \equiv \frac{\varepsilon_{xx}^M}{\Delta M} \qquad \beta_{yy} \equiv \frac{\varepsilon_{yy}^M}{\Delta M} \qquad \beta_{zz} \equiv \frac{\varepsilon_{zz}^M}{\Delta M}$$

$$\beta_{xy} \equiv \frac{\gamma_{xy}^M}{\Delta M} \qquad \beta_{xz} \equiv \frac{\gamma_{xz}^M}{\Delta M} \qquad \beta_{yz} \equiv \frac{\gamma_{yz}^M}{\Delta M} \tag{3.24}$$

The units of the CMEs are typically μm/m/%M or μrad/%M. Equations 3.24 can be easily rearranged and written in matrix form as follows:

$$\begin{bmatrix} \varepsilon_{xx}^M & \gamma_{xy}^M & \gamma_{xz}^M \\ \gamma_{yx}^M & \varepsilon_{yy}^M & \gamma_{yz}^M \\ \gamma_{zx}^M & \gamma_{zy}^M & \varepsilon_{zz}^M \end{bmatrix} = (\Delta M) \begin{bmatrix} \beta_{xx} & \beta_{xy} & \beta_{xz} \\ \beta_{xy} & \beta_{yy} & \beta_{yz} \\ \beta_{xz} & \beta_{yz} & \beta_{zz} \end{bmatrix} \tag{3.25}$$

A comparison between Equations 3.24 and 3.21, or between Equations 3.25 and 3.23, will reinforce the fact that strains induced by a change in moisture content are analogous (in a mathematical sense) to those caused by a change in temperature.

3.4.1 Specialization to Orthotropic and Transversely Isotropic Composites

In the principal material coordinate system, there is no coupling between a change in moisture content and shear strains. Hence, in the principal material coordinate system Equations 3.24 become

$$\beta_{11} \equiv \frac{\varepsilon_{11}^M}{\Delta M} \qquad \beta_{22} \equiv \frac{\varepsilon_{22}^M}{\Delta M} \qquad \beta_{33} \equiv \frac{\varepsilon_{33}^M}{\Delta M}$$

(3.26)

$$\beta_{12} = \beta_{13} = \beta_{23} = 0$$

If the composite is based on a unidirectional fabric and fibers are distributed uniformly throughout the 2–3 plane, then $\beta_{22} = \beta_{33}$ and the composite will be transversely isotropic. If the fibers are not uniformly distributed, or if the composite is based on woven or braided fabrics, then $\beta_{22} \neq \beta_{33}$ and the composite will be orthotropic. Typical CME's for several common polymeric composites are included in Table 3.1.

3.5 Material Properties Relating Stress or Strain to Failure

Fundamental material properties relating stress or strain to failure are measured during either a uniaxial test or a pure shear test. Strengths measured for unidirectional composites are discussed in this chapter. Later it will be seen that failure strengths are often measured for multiangle composite laminates as well. Discussion of the strengths of multiangle composite laminates will be deferred to Chapter 7.

Strengths of unidirectional composites are referenced to the principal material 1–2–3 coordinate system, and are measured parallel to the fibers (i.e., parallel to the 1-axis) and transverse to the fibers (parallel to the 2- and 3-axes). As would be anticipated, the strength of a composite in the 1-direction is determined primarily by the strength of the fibers, whereas strengths transverse to the fibers are determined primarily by the strength of the matrix.

A σ_{11} vs ε_{11} curve in tension is shown in Figure 3.12a. In the fiber direction unidirectional composites typically exhibit very linear behavior, even at stress levels approaching final fracture. The stress that causes fracture will be denoted σ_{11}^{fT} or σ_{11}^{fC}, depending on whether σ_{11} is tensile or compressive, respectively. Similarly, the strain at failure is denoted ε_{11}^{fT} or ε_{11}^{fC}.

A typical σ_{22} vs ε_{22} curve for advanced polymeric composites is shown in Figure 3.12b. As indicated, the stress–strain is usually only slightly nonlinear. In most cases the σ_{33} vs ε_{33} curve (not shown in Figure 3.12) is nearly identical to the σ_{22} vs ε_{22} curve. Due to the limited nonlinear response it is usually not necessary to define a composite yield stress for advanced composites.* In this chapter the stress and strain values present at the onset

---

* Discontinuous-fiber composites can exhibit a substantial nonlinear response prior to fracture, however, so for these materials it may be appropriate to define a yield strength, and to base subsequent predictions of structural failure on yield strength rather than fracture strength.

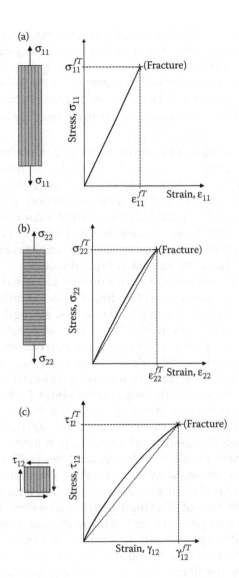

FIGURE 3.12
Idealized plots of stress–strain curves for unidirectional composite laminates. (a) Typical stress–strain curve for a unidirectional composite in the fiber direction; (b) typical stress–strain curve for a unidirectional composite transverse to the fiber direction; (c) typical shear stress–strain curve measured in the 1–2 coordinate system.

of failure in tension or compression will be denoted $(\sigma_{22}^{fT}, \varepsilon_{22}^{fT}, \sigma_{33}^{fT}, \varepsilon_{33}^{fT})$ or $(\sigma_{22}^{fC}, \varepsilon_{22}^{fC}, \sigma_{33}^{fC}, \varepsilon_{33}^{fC})$, respectively.

Finally, a typical shear stress–strain curve (i.e., a plot of τ_{12} vs γ_{12}, τ_{13} vs γ_{13}, or τ_{23} vs γ_{23}) for unidirectional composites is shown in Figure 3.12c. Although the shear response is usually slightly more nonlinear than that measured for normal stress in the 2- or 3-directions, the level of nonlinearity is low enough such that it is usually not necessary to define a shear yield strength. In the principal material coordinate system the shear response is insensitive to the algebraic sign of the shear stress. For example, in the 1–2 coordinate system an identical τ_{12} vs γ_{12} curve will be measured regardless of whether τ_{12} is positive or negative.* The shear stress and strain at fracture will be denoted τ_{12}^f and γ_{12}^f (or τ_{13}^f and γ_{13}^f, or τ_{23}^f and γ_{23}^f).

Owing to the thin plate-like nature of most composites it is often difficult in practice to apply well-defined out-of-plane stresses, or to measure the resulting out-of-plane strains. Hence, in practice the yield and fracture stresses associated with the out-of-plane 3-direction are rarely measured. Typical fracture strengths in the 1–2 plane for three common polymeric composites at room temperatures are included in Table 3.1. These properties do not represent the properties of any specific commercial composite material system, but rather should be viewed as typical values.

It is interesting to note that the *matrix*-dominated tensile strengths exhibited by polymeric composites are often lower than the failure strength of the polymeric matrix alone. For example, the tensile strength of a *nonreinforced* bulk epoxy is commonly about 70 MPa (10 ksi), whereas from Table 3.1 we see that graphite–epoxy typically possesses a matrix-dominated tensile failure strength on the order of 50 MPa (7.25 ksi). Even more pronounced is the reduction in tensile strain at fracture: for a nonreinforced bulk epoxy the tensile strain at fracture commonly ranges from about 1% to 5% (10,000 μm/m to 50,000 μm/m), whereas for graphite–epoxy the matrix-dominated tensile strain at fracture $\left(\varepsilon_{22}^{fT}\right)$ rarely exceeds about 0.7% (7000 μm/m). The relatively low matrix-dominated strengths exhibited by polymeric composites can be explained on the basis of micromechanics analyses [4–7]. Briefly, two factors lead to low matrix-dominated tensile strengths. The first is thermal stresses induced at the microlevel during cooldown from cure temperatures. The thermal expansion coefficient of most high-performance fibers is very low, and in fact is often slightly negative. In contrast, the thermal expansion coefficient of polymers is quite high and usually exceeds 30 μm/m–°C. Consequently during cooldown from cure temperatures the matrix is restrained from thermal contraction by the fibers, leading to self-equilibrating tensile stresses in the matrix and compressive stresses in the fibers. The second factor is the stress concentrating effect of the fibers. Since most advanced composites are produced with a fiber volume fraction of about 0.65, the tensile stresses induced within the matrix surrounding a fiber are not dictated strictly by the CTE mismatch between matrix and fiber but are also influenced by the presence of neighboring fibers. Together these two factors give rise to thermal stresses at the microlevel that are generally tensile in the matrix and

---

* However, in a general, nonprincipal material coordinate system the shear strength is sensitive to the algebraic sign of the shear stress. This important point will be further discussed in Sections 5.5 and 5.6.

compressive in the fibers. The magnitude of the thermal stresses induced in the fiber is very low relative to the fiber strength. However, the magnitude of the tensile stresses induced in the matrix represents a substantial fraction of the tensile strength of the matrix alone. Numerical micromechanics analyses based on the finite-element method have shown that thermal matrix stresses can often be 50–60% of the bulk matrix tensile strength. Hence, these thermal stresses are responsible for the low matrix-dominated tensile strengths and tensile strain at fracture exhibited by composites. They also explain in a qualitative sense why the magnitudes of matrix-dominated compressive strengths are invariably higher than matrix-dominated tensile strengths (i.e., $\sigma_{22}^{yC} > \sigma_{22}^{yT}$ and $\sigma_{22}^{fC} > \sigma_{22}^{fT}$).

3.6 Predicting Elastic Composite Properties Based on Constituents: The Rule of Mixtures

Various material properties exhibited by composites at the structural level have been described in preceding sections. These properties are usually simply *measured* for a composite material of interest. However, occasionally there is a need to *predict* composite material properties exhibited at the structural level. As a typical example, suppose a new high-performance graphite fiber has recently been developed, and properties of the fiber itself have been measured. Naturally, the structural engineer is interested in determining whether this new fiber will lead to improvements in composite material properties at the structural level. The potential improvement in properties can of course be evaluated directly, by embedding the new fiber in a polymeric matrix of interest and measuring the properties exhibited by the new composite material system. However, creating and testing the new material system in this fashion is time-consuming and expensive. A need exists to estimate the properties that will be provided by the new fiber, so as to justify the time and money that will be invested during development of the composite material system based on the new graphite fiber.

As discussed in Section 3.1.6, an analysis performed at a physical scale corresponding to the fiber diameter is classified as a *micromechanics* analysis. In the present instance we wish to use a micromechanics-based analysis to predict composite properties at the structural level. A simple micromechanics model that can be used to make this prediction is called the *rule-of-mixtures* and is developed as follows.

Consider the representative composite element shown in Figure 3.13. The element consists of unidirectional fibers embedded within a polymeric matrix. The principal material coordinate system, labeled the 1–2–3 coordinate system, is defined by the fiber direction. It is assumed that the fibers are evenly spaced, and that the matrix is perfectly bonded to the fiber.

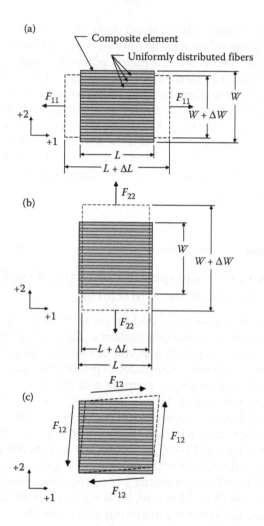

FIGURE 3.13
Composite element used to derive rule-of-mixtures equations. (a) Composite element deformed by a load F_{11}, acting parallel to the fiber direction; (b) composite element deformed by a load F_{22}, acting perpendicular to the fiber direction; (c) composite element deformed by shear load F_{12}.

If a force F_{11} is applied to the element as shown in Figure 3.13, the length of the element is increased by an amount ΔL and the width of the element is decreased by an amount ΔW. Force F_{11} is related to the average stress imposed in the 1-direction by

$$F_{11} = \sigma_{11}A$$

where A is the cross-sectional area of the element. The sum of forces present in the matrix and fibers must equal the total applied force, which implies:

$$\sigma_{11}A = \sigma_f A_f + \sigma_m A_m \tag{3.27}$$

where A_f is the total cross-sectional area of the fibers presented within the element and A_m is the cross-sectional area of the matrix. The strain in the 1-direction is associated with the change in length (ΔL), and is identical in fiber and matrix since the fiber and matrix are assumed to be perfectly bonded. That is,

$$\varepsilon_f = \varepsilon_m = \varepsilon_{11} = \frac{\Delta L}{L}$$

Stresses are assumed related to strains according to

$$\sigma_{11} = \varepsilon_{11}E_{11} \tag{3.28a}$$

$$\sigma_f = \varepsilon_f E_f = \varepsilon_{11}E_f \tag{3.28b}$$

$$\sigma_m = \varepsilon_m E_m = \varepsilon_{11}E_m \tag{3.28c}$$

The expressions for stresses σ_f and σ_m are only approximate. In reality, a triaxial state of stress is induced rather than a uniaxial stress state as implied by Equation 3.28b, 3.28c, due to the mismatch in fiber and matrix properties as well as the presence of adjacent fibers. Properly accounting for this (and other) complicating factors requires a rigorous analysis that is beyond the scope of the brief introduction presented here. Therefore, we will assume that fiber and matrix stresses are given by Equations 3.28b,c despite their shortcomings. Substituting these expressions into Equation 3.27 and rearranging, we find

$$E_{11} = E_f \frac{A_f}{A} + E_m \frac{A_m}{A}$$

This expression allows us to predict E_{11} based on properties of the constituents (E_f and E_m) and the area fractions of fiber and matrix, (A_f/A) and (A_m/A). If no voids are present, then

$$A = A_f + A_m$$

Usually rule of mixtures expressions are written in terms of volume fractions rather than area fractions. Volume fractions are given by

$$V_f = \frac{A_f}{A} \qquad V_m = \frac{A_m}{A} = \frac{A - A_f}{A} = (1 - V_f)$$

where V_f is the volume fraction of fibers and $V_m = (1 - V_f)$ is the volume fraction of matrix material. Consequently, the predicted value of E_{11} based on the rule of mixtures approach is given by

$$E_{11} = E_m + V_f(E_f - E_m) \tag{3.29}$$

Polymeric composites used in practice are typically produced with a fiber volume fraction V_f of about 0.65, although it can be lower ($V_f = 0.30$, say) depending on application and the manufacturing process used to consolidate the composite. Equation 3.29 shows that if $E_f \gg E_m$ (which is usually the case), then to a first approximation $E_{11} \approx V_f E_f$. The value of E_{11} is dictated primarily by the fiber modulus E_f and fiber volume fraction, V_f. E_{11} is, therefore, called a *fiber-dominated* property of the composite.

Now consider the Poisson effect exhibited by the composite element shown in Figure 3.13a. As per our normal definition, the average Poisson ratio is defined as the negative of the transverse normal strain (ε_{22}) divided by axial normal strain (ε_{11}), both of which are caused by σ_{11}:

$$\nu_{12} = \frac{-\varepsilon_{22}}{\varepsilon_{11}}$$

The transverse normal strain associated with the change in width of the entire element (ΔW) is given by

$$\varepsilon_{22} = \frac{\Delta W}{W}$$

The change in width can also be written as the sum of the change in width of the fibers present in the element, ΔW_f, and change in width of the matrix present, ΔW_m. These are approximated as follows:

$$\Delta W_f = -WV_f\nu_f\varepsilon_f = -WV_f\nu_f\varepsilon_{11}$$

$$\Delta W_m = -WV_m\nu_m\varepsilon_m = -WV_m\nu_m\varepsilon_{11}$$

where ν_f and ν_m are Poisson ratios of the fiber and matrix, respectively. Hence, the transverse strain is given by

$$\varepsilon_{22} = \frac{\Delta W}{W} = -\left[V_f\nu_f + (V_m)\nu_m\right]\varepsilon_{11}$$

Applying the definition of Poisson's ratio for the composite as a whole, we have:

$$v_{12} = \frac{-\varepsilon_{22}}{\varepsilon_{11}} = V_f v_f + V_m v_m$$

Noting as before that $V_m = (1 - V_f)$, the predicted value for Poisson's ratio v_{12} based on the rule of mixtures becomes

$$v_{12} = v_m - V_f(v_m - v_f) \tag{3.30}$$

Measurement of Poisson's ratio of the matrix material, v_m, is a straightforward matter. However, measuring Poisson's ratio of the fiber, v_f, is more difficult due to the small fiber diameters involved. Experimentally measured values of v_f are often unavailable, even for fibers widely used in practice. The data that are available imply that both v_m and v_f are algebraically positive, and also that $v_m > v_f$. Hence, Equation 3.29 implies that the composite Poisson ratio v_{12} varies linearly with fiber volume fraction V_f, and that Poisson's ratio of the composite is less than that of the matrix, $(v_{12} < v_m)$, since usually $(v_m - v_f) > 0$.

Assuming an identical fiber distribution in the 1–2 and 1–3 planes, then an identical analysis can be conducted to predict Poisson's ratio v_{13}, which will result in an identical expression: $v_{13} = v_{12}$.

Next, consider prediction of the transverse modulus E_{22} based on the rule-of-mixtures approach. A composite element subjected to a force applied perpendicular to the fibers, force F_{22}, is shown in Figure 3.13b. This force is related to the average stress imposed in the 2-direction by $F_{22} = \sigma_{22} A$. We assume that an identical and uniform stress σ_{22} is induced in *both* the fiber and matrix. Once again, this assumption is approximate; in reality a triaxial state of stress is induced in both the fiber and matrix. On the basis of this assumption the strains induced in the fiber and matrix perpendicular to the 1-axis are

$$\varepsilon_f = \frac{\sigma_{22}}{E_f}$$

$$\varepsilon_m = \frac{\sigma_{22}}{E_m}$$

The transverse length represented by the fibers present in the element equals $V_f W$, whereas the transverse length represented by the matrix equals $V_m W$. Hence, the change in width W caused by the application of σ_{22} is

$$\Delta W = (V_f W)\frac{\sigma_{22}}{E_f} + (V_m W)\frac{\sigma_{22}}{E_m}$$

Structural Analysis of Polymeric Composite Materials

The average transverse strain caused by σ_{22} is

$$\varepsilon_{22} = \frac{\Delta W}{W} = \sigma_{22}\left[\frac{V_f}{E_f} + \frac{V_m}{E_m}\right]$$

Young's modulus E_{22} as predicted by the rule of mixtures, therefore, becomes

$$E_{22} = \frac{\sigma_{22}}{\varepsilon_{22}} = \frac{1}{\left((V_f/E_f) + (V_m/E_m)\right)} = \frac{E_f E_m}{E_m V_f + E_f V_m}$$

As before, if no voids are present, then $V_m = (1 - V_f)$, and we obtain:

$$E_{22} = \frac{E_f E_m}{E_f - V_f(E_f - E_m)} \tag{3.31}$$

For most polymeric composite material systems $E_f \gg E_m$. Nevertheless, Equation 3.31 shows that E_{22} is dictated primarily by E_m, and is only modestly affected by the fiber modulus E_f. Indeed, even in the limit (i.e., as $E_f \to \infty$) the predicted value of E_{22} is only increased to

$$E_{22}\big|_{E_f \to \infty} = \frac{E_m}{1 - V_f}$$

Since V_f is usually about 0.65, this result shows that E_{22} is still less than 3 times the matrix modulus E_m, even if the composite is produced using a fiber whose stiffness is infinitely high ($E_f \to \infty$). E_{22} is, therefore, called a *matrix-dominated* property of the composite.

Assuming an identical fiber distribution in the 1–2 and 1–3 planes, then an identical analysis can be conducted to predict Young's modulus in the 3-direction, E_{33}, resulting in an identical expression. Hence, $E_{33} = E_{22}$. As before, E_{33} is dictated primarily by E_m and is a matrix-dominated property.

Now consider the shear modulus G_{12}. An element subjected to a pure shear force F_{12} is shown in Figure 3.13c. This force is related to the average shear stress according to $F_{12} = \tau_{12} A$. In a rule-of-mixture analysis it is assumed that an identical shear stress is induced in both the fibers and matrix regions. This assumption is approximate at best. Nevertheless, on the basis of this assumption the shear strains induced in fiber and matrix are given by

$$\gamma_f = \frac{\tau_{12}}{G_f}$$

$$\gamma_m = \frac{\tau_{12}}{G_m}$$

where G_f and G_m are the shear moduli of the fiber and matrix, respectively. The total shear strain is given by

$$\gamma_{12} = V_f\gamma_f + V_m\gamma_m = V_f\left(\frac{\tau_{12}}{G_f}\right) + V_m\left(\frac{\tau_{12}}{G_m}\right)$$

The shear modulus predicted by the rule of mixtures is then

$$G_{12} = \frac{\tau_{12}}{\gamma_{12}} = \frac{1}{\left((V_f/G_f)\right) + \left((V_m/G_m)\right)} = \frac{G_fG_m}{G_mV_f + G_fV_m}$$

Assuming no voids are present then $V_m = (1 - V_f)$, and we obtain:

$$G_{12} = \frac{G_fG_m}{G_f - V_f(G_f - G_m)} \quad\quad (3.32)$$

Comparing Equation 3.32 with Equation 3.31, it is seen that the shear modulus is related to fiber and matrix properties in a manner similar to E_{22} and E_{33}. The value of G_{12} is dictated primarily by the shear modulus of the matrix, G_m, and is considered a matrix-dominated property. Assuming an identical fiber distribution in the 1–2 and 1–3 planes, then $G_{13} = G_{12}$.

To summarize, the analysis presented above allows prediction of elastic moduli E_{11}, $v_{12} = v_{13}$, $E_{22} = E_{33}$, and $G_{12} = G_{13}$, based on knowledge of the fiber modulus E_f, matrix modulus E_m, and fiber volume fraction V_f. Although not presented here, a rule of mixture approach can also be used to predict thermal expansion coefficients α_1 and α_2, or moisture expansion coefficients β_1 and β_2.

The analysis presented above is only one of several micromechanics-based models that have been proposed. The rule of mixtures is certainly the simplest approach, but unfortunately is often the least accurate. In general, fiber-dominated properties E_{11} and $v_{12} = v_{13}$ are reasonably well predicted by Equations 3.29 and 3.30, respectively. However, matrix-dominated properties $E_{22} = E_{33}$ and $G_{12} = G_{13}$ are generally underpredicted by Equations 3.31 and 3.32. The accuracy of these predictions is not high because many important factors have not been accounted for. A partial listing of factors not accounted for include:

- The more-or-less random distribution and spacing of fibers present in a real composite
- The triaxial state of stress induced in both matrix and fiber due to the mismatch in fiber/matrix properties
- Differences in fiber distribution in the 1–2 and 1–3 planes
- The adhesion (or lack thereof) between fiber and matrix

- Variations in fiber cross-sections from one fiber to the next
- The presence of voids or other defects
- The anisotropic nature of many high-performance fibers (e.g., Young's modulus parallel and transverse to the long axis of the fiber usually differs)

A rigorous closed-form analytical solution that accounts for all of these factors is not available. Consequently, most advanced micromechanics analyses are performed numerically using finite-element methods. Since the primary objective of this book is to investigate composite materials at the structural (i.e., macroscopic) level, only the simple rule of mixtures is presented herein. The reader interested in learning more about micromechanics analyses is referred to the many excellent texts that discuss this topic in greater detail, a few of which are listed here as References 4–7.

HOMEWORK PROBLEMS

3.1. An anisotropic material is known to have the following elastic properties:

$E_{xx} = 100$ GPa	$E_{yy} = 200$ GPa	$E_{zz} = 75$ GPa
$\nu_{xy} = 0.20$	$\nu_{xz} = -0.25$	$\nu_{yz} = 0.60$
$\nu_{yx} = 0.40$	$\nu_{zx} = -0.1875$	$\nu_{zy} = 0.225$
$G_{xy} = 60$ GPa	$G_{xz} = 75$ GPa	$G_{yz} = 50$ GPa
$\eta_{xx,xy} = -0.30$	$\eta_{xx,xz} = 0.25$	$\eta_{xx,yz} = 0.30$
$\eta_{yy,xy} = 0.60$	$\eta_{yy,xz} = 0.75$	$\eta_{yy,yz} = 0.20$
$\eta_{zz,xy} = -0.20$	$\eta_{zz,xz} = -0.05$	$\eta_{zz,yz} = -0.15$
$\eta_{xy,xx} = -0.18$	$\eta_{xy,yy} = 0.18$	$\eta_{xy,zz} = -0.16$
$\eta_{xz,xx} = 0.19$	$\eta_{xz,yy} = 0.28$	$\eta_{xz,zz} = -0.05$
$\eta_{yz,xx} = 0.15$	$\eta_{yz,yy} = 0.05$	$\eta_{yz,zz} = -0.10$
$\mu_{xy,xz} = -0.10$	$\mu_{xy,yz} = -0.05$	$\mu_{xz,yz} = 0.10$
$\mu_{xz,xy} = -0.12$	$\mu_{yz,xy} = -0.042$	$\mu_{yz,xz} = 0.067$

a. What strains are induced if a uniaxial tensile stress $\sigma_{xx} = 300$ MPa is applied?

b. What strains are induced if a uniaxial tensile stress $\sigma_{yy} = 300$ MPa is applied?

c. What strains are induced if a uniaxial tensile stress $\sigma_{zz} = 300$ MPa is applied?

d. What strains are induced if a pure shear stress $\tau_{xy} = 100$ MPa is applied?

e. What strains are induced if a pure shear stress $\tau_{xz} = 100$ MPa is applied?

f. What strains are induced if a pure shear stress $\tau_{yz} = 100$ MPa is applied?

3.2. An orthotropic material is known to have the following elastic properties:

$E_{11} = 100$ GPa	$E_{22} = 200$ GPa	$E_{33} = 75$ GPa
$v_{12} = 0.20$	$v_{13} = -0.25$	$v_{23} = 0.60$
$v_{21} = 0.40$	$v_{31} = -0.19$	$v_{32} = 0.22$
$G_{12} = 60$ GPa	$G_{13} = 75$ GPar	$G_{23} = 50$ GPa

a. What strains are induced if a uniaxial tensile stress $\sigma_{11} = 300$ MPa is applied?

b. What strains are induced if a uniaxial tensile stress $\sigma_{22} = 300$ MPa is applied?

c. What strains are induced if a uniaxial tensile stress $\sigma_{33} = 300$ MPa is applied?

d. What strains are induced if a pure shear stress $\tau_{12} = 100$ MPa is applied?

e. What strains are induced if a pure shear stress $\tau_{13} = 100$ MPa is applied?

f. What strains are induced if a pure shear stress $\tau_{23} = 100$ MPa is applied?

3.3. A tensile specimen is machined from an anisotropic material. The specimen is referenced to an x–y–z coordinate system, as shown in Figure 3.14a. The cross-section of the specimen is initially a perfect 5 mm × 5 mm square. In addition, perfect 5 mm × 5 mm squares are drawn on the x–y and x–z surfaces of the specimen, as shown.

A uniaxial tensile stress $\sigma_{xx} = 700$ MPa is then applied, causing the specimen to deform as shown in Figure 3.14b–d. Determine the values of E_{xx}, v_{xy}, v_{xz}, $\eta_{xx,xy}$, $\eta_{xx,xz}$, and $\eta_{xx,yz}$ that correspond to these deformations.

3.4. Load versus strain data collected during two different composite tensile tests are shown in Tables 3.2 and 3.3.

Use linear regression to determine the following properties for this composite material:

a. Determine E_{11} and v_{12} using the data collected using the [0°] specimen (Table 3.2).

b. Determine E_{22} using the data collected using the [90°] specimen (Table 3.3).

c. In Chapter 4, it will be seen that $v_{21} = v_{12}(E_{22}/E_{11})$. Determine v_{21} for this composite material system.

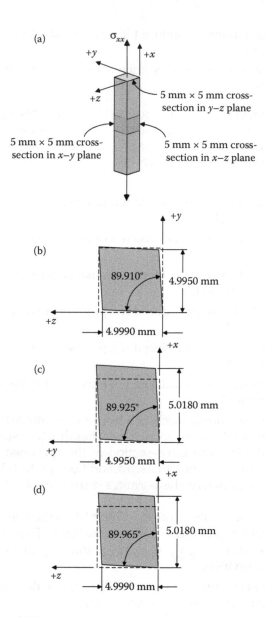

FIGURE 3.14
Tensile specimen described in Problem 3.4 (deformations shown exaggerated for clarity). (a) Tensile specimen machined from an anisotropic material; (b) change in cross section on y–z face; (c) change in dimensions on x–y face; (d) change in dimensions on x–z face.

TABLE 3.2

[0°] Specimen (Width = 1.251 in., Thick = 0.048 in.)

Load (lbf)	Axial Strain (μin./in.)	Trans Strain (μin./in.)
0	0	0
260	192	−61
630	454	−146
1220	860	−279
1910	1335	−433
2600	1807	−587
4100	2784	−930

TABLE 3.3

[90°] Specimen (Width = 1.254 in., Thick = 0.090 in.)

Load (lbf)	Axial Strain (μin./in.)
0	0
64	300
102	539
172	923
275	1489
385	2072

3.5. A thin tensile specimen is machined from a material with unknown properties. A "perfect" square with dimensions 5×5 mm^2 is drawn on one surface of the specimen, as shown in Figure 3.15. A tensile stress of 500 MPa is then applied, causing the square to deform. Determine E_{xx}, ν_{xy}, and $\eta_{xx,xy}$ for this material.

3.6. A perfect square with dimensions 1×1 mm^2 is drawn on the surface of a plate. The temperature of the plate is then uniformly increased by 300°C, causing the square to deform as shown in Figure 3.16. Determine the corresponding strains ε_{xx}, ε_{yy}, and γ_{xy}, and coefficients of thermal expansion α_{xx}, α_{yy}, and α_{xy}.

3.7. A new structural fiber with an elastic stiffness and Poisson ratio of 310 GPa and 0.33, respectively is available. Suppose new composite material system is developed by embedding this fiber in a polymer with elastic stiffness and Poisson ratio of 10 GPa and 0.45, respectively. Use the rule of mixtures to predict E_{11}, E_{22}, and ν_{12} of the new composite material system, assuming:

    a. A fiber volume fraction of 0.50
    b. A fiber volume fraction of 0.55
    c. A fiber volume fraction of 0.60

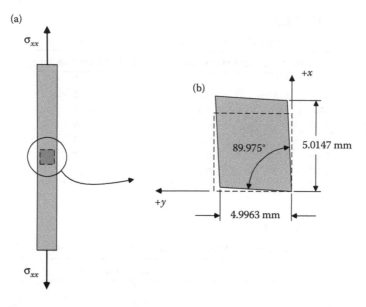

FIGURE 3.15
Tensile specimen described in Problem 3.6. (a) Tensile specimen; (b) deformed 5 mm × 5 mm square element (deformations shown exaggerated for clarity).

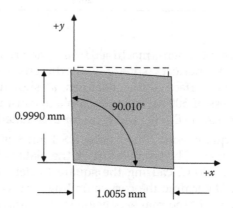

FIGURE 3.16
Deformed square described in Problem 3.7 (deformations shown exaggerated for clarity).

References

1. Dieter, G.E., *Mechanical Metallurgy*, McGraw-Hill, Inc., New York, ISBN0-07-016893-8, 1986.
2. Lekhnitski, S.G., *Theory of Elasticity of an Anisotropic Body*, Holden-Day, San Francisco, 1963.

3. Bert, C.W., Reddy, J.N., Reddy, V.S., and Chao, W.C., Analysis of thick rectangular plates laminated of bimodulus composite materials, *AIAA Journal*, 19(10), 1342–1349, 1981.
4. Hyer, M.W. and White. S.R., *Stress Analysis of Fiber-Reinforced Composite Materials*, Updated Edition, DEStech Publications, Lancaster, PA, ISBN 9781932078862, 2009.
5. Herakovich, C. T., *Mechanics of Fibrous Composites*, John Wiley & Sons, New York, ISBN 0-471-10636-4, 1998.
6. Hull, D., *An Introduction to Composite Materials*, Cambridge University Press, Cambridge, UK, ISBN 0-521-23991-5, 1981.
7. Gibson, R.F., *Principles of Composite Material Mechanics*, 2nd edition, Mechanical Engineering, 205, Marcel Dekker, Inc., New York, ISSN 0899-3858, 2007.

4

Elastic Response of Anisotropic Materials

4.1 Strains Induced by Stress: Anisotropic Materials

As discussed in Chapter 2, stress is a symmetric second-order tensor and can be written as σ_{ij}, where subscripts i and j take on values of x, y, and z. Alternatively, the stress tensor can be written using matrix notation as

$$\sigma_{ij} = \begin{bmatrix} \sigma_{xx} & \sigma_{xy} & \sigma_{xz} \\ \sigma_{yx} & \sigma_{yy} & \sigma_{yz} \\ \sigma_{zx} & \sigma_{zy} & \sigma_{zz} \end{bmatrix} \tag{4.1}$$

Similarly, strain is a symmetric second-order tensor, ε_{ij}, and can be written as

$$\varepsilon_{ij} = \begin{bmatrix} \varepsilon_{xx} & \varepsilon_{xy} & \varepsilon_{xz} \\ \varepsilon_{yx} & \varepsilon_{yy} & \varepsilon_{yz} \\ \varepsilon_{zx} & \varepsilon_{zy} & \varepsilon_{zz} \end{bmatrix} = \begin{bmatrix} (\varepsilon_{xx}) & (\gamma_{xy}/2) & (\gamma_{xz}/2) \\ (\gamma_{yx}/2) & (\varepsilon_{yy}) & (\gamma_{yz}/2) \\ (\gamma_{zx}/2) & (\gamma_{zy}/2) & (\varepsilon_{zz}) \end{bmatrix} \tag{4.2}$$

For any elastic solid, the strain and stress tensors are related as follows (assuming environmental factors such as temperature or moisture content remain constant):

$$\varepsilon_{ij} = S_{ijkl}\sigma_{kl} \tag{4.3}$$

All subscripts that appear in Equation 4.3 take on values of x, y, and z. Equation 4.3 is called *generalized Hooke's law*, and is valid for any elastic solid under constant environmental conditions. The strain and stress tensors are related via the fourth-order *compliance* tensor, S_{ijkl}. As strains are unit-less quantities, from Equation 4.3 it is seen that the units of S_{ijkl} are stress^{-1}, that is, either (1/Pa) or (1/psi).

As the compliance tensor is described using four subscripts, and since each subscript may take on three distinct value (e.g., x, y, or z), it would initially appear that $3^4 = 81$ independent terms appear within the compliance tensor. However, owing to symmetry of both the strain and stress tensors, it will be shown below that the compliance tensor consists of (at most) 36 material constants.

It will be very convenient to express Equation 4.3 using matrix notation. However, as S_{ijkl} is a fourth-order tensor (and hence can be viewed as having "four dimensions"), we cannot expand S_{ijkl} directly. To expand Equation 4.3, we must first define the components of stress and strain using *contracted notation*, as follows:

$$
\begin{array}{ll}
\varepsilon_{xx} \rightarrow \varepsilon_1 & \sigma_{xx} \rightarrow \sigma_1 \\
\varepsilon_{yy} \rightarrow \varepsilon_2 & \sigma_{yy} \rightarrow \sigma_2 \\
\varepsilon_{zz} \rightarrow \varepsilon_3 & \sigma_{zz} \rightarrow \sigma_3 \\
\gamma_{yz} = \gamma_{zy} \rightarrow \varepsilon_4 & \sigma_{yz} = \sigma_{zy} \rightarrow \sigma_4 \\
\gamma_{xz} = \gamma_{zx} \rightarrow \varepsilon_5 & \sigma_{xz} = \sigma_{zx} \rightarrow \sigma_5 \\
\gamma_{xy} = \gamma_{yx} \rightarrow \varepsilon_6 & \sigma_{xy} = \sigma_{yx} \rightarrow \sigma_6
\end{array}
\tag{4.4}
$$

Notice that symmetry of the strain and stress tensors ($\gamma_{yz} = \gamma_{zx}$, etc.) is embedded within the very definition of contracted notation. Also note that the shear strain components (ε_4, ε_5, and ε_6) represent engineering shear strains, rather than tensoral shear strains. On the basis of this change in notation, we can now write Equation 4.3 as

$$
\varepsilon_i = S_{ij}\sigma_j
\tag{4.5}
$$

where $i, j = 1$ to 6.

In contracted notation the strain and stress tensors are expressed with a single subscript (i.e., ε_i and σ_j), and hence in Equation 4.5 they *appear* to be first-order tensors. This is, of course, not the case. Both strain and stress are second-order tensors. We are able to write them using contracted notation only because they are both *symmetric* tensors. Similarly, contracted notation allows us to refer to individual components of the fourth-order compliance tensor expressed using only two subscripts. We will henceforth refer to S_{ij} as the *compliance matrix*, and use of contracted notation will be implied.

Expanding Equation 4.5, we have

$$
\begin{Bmatrix} \varepsilon_1 \\ \varepsilon_2 \\ \varepsilon_3 \\ \varepsilon_4 \\ \varepsilon_5 \\ \varepsilon_6 \end{Bmatrix} =
\begin{bmatrix}
S_{11} & S_{12} & S_{13} & S_{14} & S_{15} & S_{16} \\
S_{21} & S_{22} & S_{23} & S_{24} & S_{25} & S_{26} \\
S_{31} & S_{32} & S_{33} & S_{34} & S_{35} & S_{36} \\
S_{41} & S_{42} & S_{43} & S_{44} & S_{45} & S_{46} \\
S_{51} & S_{52} & S_{53} & S_{54} & S_{55} & S_{56} \\
S_{61} & S_{62} & S_{63} & S_{64} & S_{65} & S_{66}
\end{bmatrix}
\begin{Bmatrix} \sigma_1 \\ \sigma_2 \\ \sigma_3 \\ \sigma_4 \\ \sigma_5 \\ \sigma_6 \end{Bmatrix}
\tag{4.6}
$$

In contracted notation, the compliance matrix has six rows and six columns, so it is now clear that it consists of 36 independent material constants (at most), as previously stated. Furthermore, through a consideration of strain energy it can be shown that *the compliance matrix must itself be symmetric*. That is, all terms in symmetric off-diagonal positions must be equal:

$$
\begin{aligned}
&S_{21} = S_{12} \\
&S_{31} = S_{13} \quad S_{32} = S_{23} \\
&S_{41} = S_{14} \quad S_{42} = S_{24} \quad S_{43} = S_{34} \\
&S_{51} = S_{15} \quad S_{52} = S_{25} \quad S_{53} = S_{35} \quad S_{54} = S_{45} \\
&S_{61} = S_{16} \quad S_{62} = S_{26} \quad S_{63} = S_{36} \quad S_{64} = S_{46} \quad S_{65} = S_{56}
\end{aligned}
\tag{4.7}
$$

Hence, while the compliance matrix for an anisotropic composite consists of 36 material constants, *only 21 of these constants are independent.*

Substituting the original strain and stress terms (defined in Equation 4.4) into Equation 4.6, we have

$$
\begin{Bmatrix} \varepsilon_{xx} \\ \varepsilon_{yy} \\ \varepsilon_{zz} \\ \gamma_{yz} \\ \gamma_{xz} \\ \gamma_{xy} \end{Bmatrix}
=
\begin{bmatrix}
S_{11} & S_{12} & S_{13} & S_{14} & S_{15} & S_{16} \\
S_{21} & S_{22} & S_{23} & S_{24} & S_{25} & S_{26} \\
S_{31} & S_{32} & S_{33} & S_{34} & S_{35} & S_{36} \\
S_{41} & S_{42} & S_{43} & S_{44} & S_{45} & S_{46} \\
S_{51} & S_{52} & S_{53} & S_{54} & S_{55} & S_{56} \\
S_{61} & S_{62} & S_{63} & S_{64} & S_{65} & S_{66}
\end{bmatrix}
\begin{Bmatrix} \sigma_{xx} \\ \sigma_{yy} \\ \sigma_{zz} \\ \tau_{yz} \\ \tau_{xz} \\ \tau_{xy} \end{Bmatrix}
\tag{4.8}
$$

The individual constants that appear in the compliance matrix can be easily related to the material properties defined in Chapter 3 by invoking the *principal of superposition*. As we have restricted our attention to linear elastic behavior, an individual component of strain caused by several stress components acting simultaneously can be obtained by adding the strain caused by each stress component acting independently. For example, the strains ε_{xx} caused by each stress component independently are given by

σ_{xx} causes (from Equation 3.1):

$$\varepsilon_{xx} = \frac{1}{E_{xx}}\sigma_{xx}$$

σ_{yy} causes (from Equation 3.6):

$$\varepsilon_{xx} = -\frac{\nu_{yx}}{E_{yy}}\sigma_{yy}$$

σ_{zz} causes (from Equation 3.7):

$$\varepsilon_{xx} = -\frac{v_{zx}}{E_{zz}}\sigma_{zz}$$

τ_{yz} causes (from Equation 3.14):

$$\varepsilon_{xx} = \frac{\eta_{yz,xx}}{G_{yz}}\tau_{yz}$$

τ_{xz} causes (from Equation 3.13):

$$\varepsilon_{xx} = \frac{\eta_{xz,xx}}{G_{xz}}\tau_{xz}$$

τ_{xy} causes (from Equation 3.12):

$$\varepsilon_{xx} = \frac{\eta_{xy,xx}}{G_{xy}}\tau_{xy}$$

To determine the strain ε_{xx} induced if *all* stress components are acting simultaneously, simply add up the contribution to ε_{xx} caused by each stress individually, to obtain:

$$\varepsilon_{xx} = \left(\frac{1}{E_{xx}}\right)\sigma_{xx} + \left(\frac{-v_{yx}}{E_{yy}}\right)\sigma_{yy} + \left(\frac{-v_{zx}}{E_{zz}}\right)\sigma_{zz}$$

$$+ \left(\frac{\eta_{yz,xx}}{G_{yz}}\right)\tau_{yz} + \left(\frac{\eta_{xz,xx}}{G_{xz}}\right)\tau_{xz} + \left(\frac{\eta_{xy,xx}}{G_{xy}}\right)\tau_{xy} \qquad (4.9a)$$

Using an identical procedure, the remaining five strain components caused by an arbitrary combination of stresses are

$$\varepsilon_{yy} = \left(\frac{-v_{xy}}{E_{xx}}\right)\sigma_{xx} + \left(\frac{1}{E_{yy}}\right)\sigma_{yy} + \left(\frac{-v_{zy}}{E_{zz}}\right)\sigma_{zz}$$

$$+ \left(\frac{\eta_{yz,yy}}{G_{yz}}\right)\tau_{yz} + \left(\frac{\eta_{xz,yy}}{G_{xz}}\right)\tau_{xz} + \left(\frac{\eta_{xy,yy}}{G_{xy}}\right)\tau_{xy} \qquad (4.9b)$$

$$\varepsilon_{zz} = \left(\frac{-\nu_{xz}}{E_{xx}}\right)\sigma_{xx} + \left(\frac{-\nu_{yz}}{E_{yy}}\right)\sigma_{yy} + \left(\frac{1}{E_{zz}}\right)\sigma_{zz} + \left(\frac{\eta_{yz,zz}}{G_{yz}}\right)\tau_{yz} + \left(\frac{\eta_{xz,zz}}{G_{xz}}\right)\tau_{xz} + \left(\frac{\eta_{xy,zz}}{G_{xy}}\right)\tau_{xy}$$

(4.9c)

$$\gamma_{yz} = \left(\frac{\eta_{xx,yz}}{E_{xx}}\right)\sigma_{xx} + \left(\frac{\eta_{yy,yz}}{E_{yy}}\right)\sigma_{yy} + \left(\frac{\eta_{zz,yz}}{E_{zz}}\right)\sigma_{zz} + \left(\frac{1}{G_{yz}}\right)\tau_{yz} + \left(\frac{\mu_{xz,yz}}{G_{xz}}\right)\tau_{xz} + \left(\frac{\mu_{xy,yz}}{G_{xy}}\right)\tau_{xy}$$

(4.9d)

$$\gamma_{xz} = \left(\frac{\eta_{xx,xz}}{E_{xx}}\right)\sigma_{xx} + \left(\frac{\eta_{yy,xz}}{E_{yy}}\right)\sigma_{yy} + \left(\frac{\eta_{zz,xz}}{E_{zz}}\right)\sigma_{zz} + \left(\frac{\mu_{yz,xz}}{G_{yz}}\right)\tau_{yz} + \left(\frac{1}{G_{xz}}\right)\tau_{xz} + \left(\frac{\mu_{xy,xz}}{G_{xy}}\right)\tau_{xy}$$

(4.9e)

$$\gamma_{xy} = \left(\frac{\eta_{xx,xy}}{E_{xx}}\right)\sigma_{xx} + \left(\frac{\eta_{yy,xy}}{E_{yy}}\right)\sigma_{yy} + \left(\frac{\eta_{zz,xy}}{E_{zz}}\right)\sigma_{zz} + \left(\frac{\mu_{yz,xy}}{G_{yz}}\right)\tau_{yz} + \left(\frac{\mu_{xz,xy}}{G_{xz}}\right)\tau_{xz} + \left(\frac{1}{G_{xy}}\right)\tau_{xy}$$

(4.9f)

Equation 4.9 can be assembled in matrix form:

$$\begin{Bmatrix} \varepsilon_{xx} \\ \varepsilon_{yy} \\ \varepsilon_{zz} \\ \gamma_{yz} \\ \gamma_{xz} \\ \gamma_{xy} \end{Bmatrix} = \begin{bmatrix} \left(\dfrac{1}{E_{xx}}\right) & \left(\dfrac{-\nu_{yx}}{E_{yy}}\right) & \left(\dfrac{-\nu_{zx}}{E_{zz}}\right) & \left(\dfrac{\eta_{yz,xx}}{G_{yz}}\right) & \left(\dfrac{\eta_{xz,xx}}{G_{xz}}\right) & \left(\dfrac{\eta_{xy,xx}}{G_{xy}}\right) \\[2mm] \left(\dfrac{-\nu_{xy}}{E_{xx}}\right) & \left(\dfrac{1}{E_{yy}}\right) & \left(\dfrac{-\nu_{zy}}{E_{zz}}\right) & \left(\dfrac{\eta_{yz,yy}}{G_{yz}}\right) & \left(\dfrac{\eta_{xz,yy}}{G_{xz}}\right) & \left(\dfrac{\eta_{xy,yy}}{G_{xy}}\right) \\[2mm] \left(\dfrac{-\nu_{xz}}{E_{xx}}\right) & \left(\dfrac{-\nu_{yz}}{E_{yy}}\right) & \left(\dfrac{1}{E_{zz}}\right) & \left(\dfrac{\eta_{yz,zz}}{G_{yz}}\right) & \left(\dfrac{\eta_{xz,zz}}{G_{xz}}\right) & \left(\dfrac{\eta_{xy,zz}}{G_{xy}}\right) \\[2mm] \left(\dfrac{\eta_{xx,yz}}{E_{xx}}\right) & \left(\dfrac{\eta_{yy,yz}}{E_{yy}}\right) & \left(\dfrac{\eta_{zz,yz}}{E_{zz}}\right) & \left(\dfrac{1}{G_{yz}}\right) & \left(\dfrac{\mu_{xz,yz}}{G_{xz}}\right) & \left(\dfrac{\mu_{xy,yz}}{G_{xy}}\right) \\[2mm] \left(\dfrac{\eta_{xx,xz}}{E_{xx}}\right) & \left(\dfrac{\eta_{yy,xz}}{E_{yy}}\right) & \left(\dfrac{\eta_{zz,xz}}{E_{zz}}\right) & \left(\dfrac{\mu_{yz,xz}}{G_{yz}}\right) & \left(\dfrac{1}{G_{xz}}\right) & \left(\dfrac{\mu_{xy,xz}}{G_{xy}}\right) \\[2mm] \left(\dfrac{\eta_{xx,xy}}{E_{xx}}\right) & \left(\dfrac{\eta_{yy,xy}}{E_{yy}}\right) & \left(\dfrac{\eta_{zz,xy}}{E_{zz}}\right) & \left(\dfrac{\mu_{yz,xy}}{G_{yz}}\right) & \left(\dfrac{\mu_{xz,xy}}{G_{xz}}\right) & \left(\dfrac{1}{G_{xy}}\right) \end{bmatrix} \begin{Bmatrix} \sigma_{xx} \\ \sigma_{yy} \\ \sigma_{zz} \\ \tau_{yz} \\ \tau_{xz} \\ \tau_{xy} \end{Bmatrix}$$

(4.10)

By comparing Equations 4.8 and 4.10, it can be seen that the individual components of the compliance matrix are directly related to the material properties measured during uniaxial tests or pure shear tests:

$$S_{11} = \frac{1}{E_{xx}} \qquad S_{22} = \frac{1}{E_{yy}} \qquad S_{33} = \frac{1}{E_{zz}}$$

$$S_{44} = \frac{1}{G_{yz}} \qquad S_{55} = \frac{1}{G_{xz}} \qquad S_{66} = \frac{1}{G_{xy}}$$

$$S_{21} = S_{12} = \frac{-\nu_{xy}}{E_{xx}} = \frac{-\nu_{yx}}{E_{yy}} \qquad S_{31} = S_{13} = \frac{-\nu_{xz}}{E_{xx}} = \frac{-\nu_{zx}}{E_{zz}}$$

$$S_{32} = S_{23} = \frac{-\nu_{yz}}{E_{yy}} = \frac{-\nu_{zy}}{E_{zz}} \qquad S_{41} = S_{14} = \frac{\eta_{xx,yz}}{E_{xx}} = \frac{\eta_{yz,xx}}{G_{yz}}$$

$$S_{42} = S_{24} = \frac{\eta_{yy,yz}}{E_{yy}} = \frac{\eta_{yz,yy}}{G_{yz}} \qquad S_{43} = S_{34} = \frac{\eta_{zz,yz}}{E_{zz}} = \frac{\eta_{yz,zz}}{G_{yz}}$$

$$S_{51} = S_{15} = \frac{\eta_{xx,xz}}{E_{xx}} = \frac{\eta_{xz,xx}}{G_{xz}} \qquad S_{52} = S_{25} = \frac{\eta_{yy,xz}}{E_{yy}} = \frac{\eta_{xz,yy}}{G_{xz}}$$

$$S_{53} = S_{35} = \frac{\eta_{zz,xz}}{E_{zz}} = \frac{\eta_{xz,zz}}{G_{xz}} \qquad S_{54} = S_{45} = \frac{\mu_{yz,xz}}{G_{yz}} = \frac{\mu_{xz,yz}}{G_{xz}}$$

$$S_{61} = S_{16} = \frac{\eta_{xx,xy}}{E_{xx}} = \frac{\eta_{xy,xx}}{G_{xy}} \qquad S_{62} = S_{26} = \frac{\eta_{yy,xy}}{E_{yy}} = \frac{\eta_{xy,yy}}{G_{xy}}$$

$$S_{63} = S_{36} = \frac{\eta_{zz,xy}}{E_{zz}} = \frac{\eta_{xy,zz}}{G_{xy}} \qquad S_{64} = S_{46} = \frac{\mu_{yz,xy}}{G_{yz}} = \frac{\mu_{xy,yz}}{G_{xy}}$$

$$S_{65} = S_{56} = \frac{\mu_{xz,xy}}{G_{xz}} = \frac{\mu_{xy,xz}}{G_{xy}}$$

$$(4.11)$$

As the compliance matrix must be symmetric, Equations 4.11 show that many of the properties of anisotropic materials are related through the following *inverse relationships*:

$$\frac{\nu_{xy}}{E_{xx}} = \frac{\nu_{yx}}{E_{yy}}$$

$$\frac{\nu_{xz}}{E_{xx}} = \frac{\nu_{zx}}{E_{zz}} \qquad \frac{\nu_{yz}}{E_{yy}} = \frac{\nu_{zy}}{E_{zz}}$$

$$\frac{\eta_{xx,yz}}{E_{xx}} = \frac{\eta_{yz,xx}}{G_{yz}} \qquad \frac{\eta_{yy,yz}}{E_{yy}} = \frac{\eta_{yz,yy}}{G_{yz}} \qquad \frac{\eta_{zz,yz}}{E_{zz}} = \frac{\eta_{yz,zz}}{G_{yz}}$$

$$\frac{\eta_{xx,xz}}{E_{xx}} = \frac{\eta_{xz,xx}}{G_{xz}} \qquad \frac{\eta_{yy,xz}}{E_{yy}} = \frac{\eta_{xz,yy}}{G_{xz}} \qquad \frac{\eta_{zz,xz}}{E_{zz}} = \frac{\eta_{xz,zz}}{G_{xz}} \qquad \frac{\mu_{yz,xz}}{G_{yz}} = \frac{\mu_{xz,yz}}{G_{xz}}$$

$$\frac{\eta_{xx,xy}}{E_{xx}} = \frac{\eta_{xy,xx}}{G_{xy}} \qquad \frac{\eta_{yy,xy}}{E_{yy}} = \frac{\eta_{xy,yy}}{G_{xy}} \qquad \frac{\eta_{zz,xy}}{E_{zz}} = \frac{\eta_{xy,zz}}{G_{xy}} \qquad \frac{\mu_{yz,xy}}{G_{yz}} = \frac{\mu_{xy,yz}}{G_{xy}} \qquad \frac{\mu_{xz,xy}}{G_{xz}} = \frac{\mu_{xy,xz}}{G_{xy}}$$

$$(4.12)$$

The inverse relationships are very significant from an experimental point of view, as they dramatically reduce the number of tests that must be performed to determine the value of the many terms that appear within the compliance matrix of an anisotropic composite. If the compliance matrix was not symmetric, and hence if the inverse relationships did not exist, then 36 tests would be required to measure all components of the compliance matrix. The fact that the compliance matrix must be symmetric reduces the number of tests required to 21. This is still a large number of tests. Fortunately, as the principal material coordinate system of composites is readily apparent, the elastic properties of composites are usually measured relative to the principal material coordinate system rather than an arbitrary (nonprincipal) coordinate system. As discussed in the following section, this further reduces the number of tests required.

Thus far we have discussed Hooke's law in the form of "strain–stress" relationships. That is, given values of the components of stress we can calculate the resulting strains using Equation 4.8 or 4.10, for example. In practice, we are often interested in the reversed calculation. That is, a common circumstance is that the components of strain induced in a structure have been measured, and we wish to calculate the stresses that caused these strains. In this case we need a "stress–strain" form of Hooke's law. The stress–strain form of Hooke's law can be obtained by simply inverting Hooke's law given by Equation 4.5, resulting in

$$\sigma_i = C_{ij}\varepsilon_j \tag{4.13}$$

where $i, j = 1$ to 6, $C_{ij} = S_{ij}^{-1}$ is called the "stiffness matrix."*

In expanded form, Equation 4.13 is written:

$$
\begin{Bmatrix} \sigma_1 \\ \sigma_2 \\ \sigma_3 \\ \sigma_4 \\ \sigma_5 \\ \sigma_6 \end{Bmatrix}
=
\begin{bmatrix}
C_{11} & C_{12} & C_{13} & C_{14} & C_{15} & C_{16} \\
C_{21} & C_{22} & C_{23} & C_{24} & C_{25} & C_{26} \\
C_{31} & C_{32} & C_{33} & C_{34} & C_{35} & C_{36} \\
C_{41} & C_{42} & C_{43} & C_{44} & C_{45} & C_{46} \\
C_{51} & C_{52} & C_{53} & C_{54} & C_{55} & C_{56} \\
C_{61} & C_{62} & C_{63} & C_{64} & C_{65} & C_{66}
\end{bmatrix}
\begin{Bmatrix} \varepsilon_1 \\ \varepsilon_2 \\ \varepsilon_3 \\ \varepsilon_4 \\ \varepsilon_5 \\ \varepsilon_6 \end{Bmatrix}
\tag{4.14}
$$

The stiffness matrix is symmetric ($C_{21} = C_{12}$, $C_{31} = C_{13}$, etc.). The units of each stiffness term are the same as stress, either Pa or psi.

---

* The variable names assigned to the compliance and stiffness matrices in this chapter have evolved over many years, and are widely used within the structural mechanics community. The reader should note that, unfortunately, the symbol "S" is customarily used to refer to the "c"ompliance matrix, whereas the symbol "C" is customarily used to refer to the "s"tiffness matrix.

4.2 Strains Induced by Stress: Orthotropic and Transversely Isotropic Materials

Many of the unusual couplings between stress and strain exhibited by composites referenced to an arbitrary coordinate system do not occur if the stress and strain tensors are referenced to the principal material coordinate system. A material referenced to an arbitrary (nonprincipal) coordinate system is called "anisotropic," whereas if the same material is referenced to the principal material coordinate system it is called either an "orthotropic" or "transversely isotropic" material.

All of the following coupling terms are zero for orthotropic or transversely isotropic materials:

- Coefficients of mutual influence of the second kind:

$$\eta_{11,12} = \eta_{11,13} = \eta_{11,23} = \eta_{22,12} = \eta_{22,13} = \eta_{22,23} = \eta_{33,12} = \eta_{33,13} = \eta_{33,23} = 0$$

- Coefficients of mutual influence of the first kind:

$$\eta_{12,11} = \eta_{12,22} = \eta_{12,33} = \eta_{13,11} = \eta_{13,22} = \eta_{13,33} = \eta_{23,11} = \eta_{23,22} = \eta_{23,33} = 0$$

- Chentsov coefficients:

$$\mu_{12,13} = \mu_{12,23} = \mu_{13,12} = \mu_{13,23} = \mu_{23,12} = \mu_{23,13} = 0$$

Since these coupling terms do not exist, Hooke's law for orthotropic or transversely isotropic materials is simplified considerably relative to that of an anisotropic material. For an orthotropic material, Hooke's law becomes (compare with Equation 4.10)

$$
\begin{Bmatrix} \varepsilon_{11} \\ \varepsilon_{22} \\ \varepsilon_{33} \\ \gamma_{23} \\ \gamma_{13} \\ \gamma_{12} \end{Bmatrix} = \begin{Bmatrix} \sigma_{11} \\ \sigma_{22} \\ \sigma_{33} \\ \tau_{23} \\ \tau_{13} \\ \tau_{12} \end{Bmatrix} \begin{bmatrix} \left(\dfrac{1}{E_{11}}\right) & \left(\dfrac{-v_{21}}{E_{22}}\right) & \left(\dfrac{-v_{31}}{E_{33}}\right) & 0 & 0 & 0 \\ \left(\dfrac{-v_{12}}{E_{11}}\right) & \left(\dfrac{1}{E_{22}}\right) & \left(\dfrac{-v_{32}}{E_{33}}\right) & 0 & 0 & 0 \\ \left(\dfrac{-v_{13}}{E_{11}}\right) & \left(\dfrac{-v_{23}}{E_{22}}\right) & \left(\dfrac{1}{E_{33}}\right) & 0 & 0 & 0 \\ 0 & 0 & 0 & \left(\dfrac{1}{G_{23}}\right) & 0 & 0 \\ 0 & 0 & 0 & 0 & \left(\dfrac{1}{G_{13}}\right) & 0 \\ 0 & 0 & 0 & 0 & 0 & \left(\dfrac{1}{G_{12}}\right) \end{bmatrix} \begin{Bmatrix} \sigma_{11} \\ \sigma_{22} \\ \sigma_{33} \\ \tau_{23} \\ \tau_{13} \\ \tau_{12} \end{Bmatrix}
$$

(4.15)

Alternatively, Equation 4.15 may be written as

$$
\begin{Bmatrix} \varepsilon_{11} \\ \varepsilon_{22} \\ \varepsilon_{33} \\ \gamma_{23} \\ \gamma_{13} \\ \gamma_{12} \end{Bmatrix} = \begin{bmatrix} S_{11} & S_{12} & S_{13} & 0 & 0 & 0 \\ S_{12} & S_{22} & S_{23} & 0 & 0 & 0 \\ S_{13} & S_{23} & S_{33} & 0 & 0 & 0 \\ 0 & 0 & 0 & S_{44} & 0 & 0 \\ 0 & 0 & 0 & 0 & S_{55} & 0 \\ 0 & 0 & 0 & 0 & 0 & S_{66} \end{bmatrix} \begin{Bmatrix} \sigma_{11} \\ \sigma_{22} \\ \sigma_{33} \\ \tau_{23} \\ \tau_{13} \\ \tau_{12} \end{Bmatrix}
\tag{4.16}
$$

The fact that the compliance matrix must be symmetric ($S_{21} = S_{12}$, etc.) has been included in Equation 4.16. Each compliance term in Equation 4.16 are related to the more familiar engineering properties as follows:

$$
S_{11} = \frac{1}{E_{11}} \quad S_{22} = \frac{1}{E_{22}} \quad S_{33} = \frac{1}{E_{33}}
$$

$$
S_{44} = \frac{1}{G_{23}} \quad S_{55} = \frac{1}{G_{13}} \quad S_{66} = \frac{1}{G_{12}}
$$

$$
S_{21} = S_{12} = \frac{-\nu_{12}}{E_{11}} = \frac{-\nu_{21}}{E_{22}}
\tag{4.17}
$$

$$
S_{31} = S_{13} = \frac{-\nu_{13}}{E_{11}} = \frac{-\nu_{31}}{E_{33}}
$$

$$
S_{32} = S_{23} = \frac{-\nu_{23}}{E_{22}} = \frac{-\nu_{32}}{E_{33}}
$$

It can be seen that only nine independent material constants exist for an orthotropic material. The set of nine independent constants can be viewed as

$$
(S_{11},\ S_{22},\ S_{33},\ S_{44},\ S_{55},\ S_{66},\ S_{12},\ S_{13},\ \text{and}\ S_{23})
$$

or equivalently, as

$$
(E_{11},\ E_{22},\ E_{33},\ \nu_{12},\ \nu_{13},\ \nu_{23},\ G_{12},\ G_{13},\ \text{and}\ G_{23})
$$

Equation 4.16 is the strain–stress form of Hooke's law suitable for use with orthotropic materials. To obtain a stress–strain relationship Equation 4.16 is inverted, resulting in

$$
\begin{Bmatrix} \sigma_{11} \\ \sigma_{22} \\ \sigma_{33} \\ \tau_{23} \\ \tau_{13} \\ \tau_{12} \end{Bmatrix} = \begin{bmatrix} C_{11} & C_{12} & C_{13} & 0 & 0 & 0 \\ C_{12} & C_{22} & C_{23} & 0 & 0 & 0 \\ C_{13} & C_{23} & C_{33} & 0 & 0 & 0 \\ 0 & 0 & 0 & C_{44} & 0 & 0 \\ 0 & 0 & 0 & 0 & C_{55} & 0 \\ 0 & 0 & 0 & 0 & 0 & C_{66} \end{bmatrix} \begin{Bmatrix} \varepsilon_{11} \\ \varepsilon_{22} \\ \varepsilon_{33} \\ \gamma_{23} \\ \gamma_{13} \\ \gamma_{12} \end{Bmatrix}
\tag{4.18}
$$

Individual components within the stiffness matrix for an orthotropic material are related to the compliance terms as follows:

$$C_{11} = \frac{S_{22}S_{33} - S_{23}^2}{S} \quad C_{12} = \frac{S_{13}S_{23} - S_{12}S_{33}}{S} \quad C_{13} = \frac{S_{12}S_{23} - S_{13}S_{22}}{S}$$

$$C_{22} = \frac{S_{11}S_{33} - S_{13}^2}{S} \quad C_{23} = \frac{S_{12}S_{13} - S_{11}S_{23}}{S} \quad C_{33} = \frac{S_{11}S_{22} - S_{12}^2}{S} \tag{4.19}$$

$$C_{44} = \frac{1}{S_{44}} \quad\quad C_{55} = \frac{1}{S_{55}} \quad\quad C_{66} = \frac{1}{S_{66}}$$

where

$$S = S_{11}S_{22}S_{33} - S_{11}S_{23}^2 - S_{22}S_{13}^2 - S_{33}S_{12}^2 + 2S_{12}S_{13}S_{23}$$

Alternatively, the stiffness terms may be calculated using the elastic properties described in Section 3.2:

$$C_{11} = \frac{(E_{22} - v_{23}^2 E_{33})E_{11}^2}{\Omega} \quad\quad C_{12} = \frac{(v_{12}E_{22} + v_{13}v_{23}E_{33})l}{\Omega}$$

$$C_{13} = \frac{(v_{12}v_{23} + v_{13})E_{11}E_{22}E_{33}}{\Omega} \quad\quad C_{22} = \frac{(E_{11} - v_{13}^2 E_{33})E_{22}^2}{\Omega}$$

$$C_{23} = \frac{(v_{23}E_{11} + v_{12}v_{13}E_{22})E_{22}E_{33}}{\Omega} \quad C_{33} = \frac{(E_{11} - v_{12}^2 E_{22})E_{22}E_{33}}{\Omega} \tag{4.20}$$

$$C_{44} = G_{23} \quad C_{55} = G_{13} \quad C_{66} = G_{12}$$

where

$$\Omega = E_{11}E_{22} - v_{12}^2 E_{22}^2 - v_{13}^2 E_{22}E_{33} - v_{23}^2 E_{11}E_{33} - 2v_{12}v_{13}v_{23}E_{22}E_{33}$$

Hooke's law for transversely isotropic materials is simplified still further, as in this case $E_{22} = E_{33}$, $v_{12} = v_{13}$, $v_{21} = v_{31}$, $v_{23} = v_{32}$, and $G_{12} = G_{13}$. Also, it can be shown that for transversely isotropic materials $G_{23} = E_{22}/2(1 + v_{23})$. Hence, for transversely isotropic composites Equation 4.15 reduces to

$$\begin{Bmatrix} \varepsilon_{11} \\ \varepsilon_{22} \\ \varepsilon_{33} \\ \gamma_{23} \\ \gamma_{13} \\ \gamma_{12} \end{Bmatrix} = \begin{bmatrix} \left(\frac{1}{E_{11}}\right) & \left(\frac{-v_{21}}{E_{22}}\right) & \left(\frac{-v_{21}}{E_{22}}\right) & 0 & 0 & 0 \\ \left(\frac{-v_{12}}{E_{11}}\right) & \left(\frac{1}{E_{22}}\right) & \left(\frac{-v_{23}}{E_{22}}\right) & 0 & 0 & 0 \\ \left(\frac{-v_{12}}{E_{11}}\right) & \left(\frac{-v_{23}}{E_{22}}\right) & \left(\frac{1}{E_{22}}\right) & 0 & 0 & 0 \\ 0 & 0 & 0 & \left(\frac{2(1+v_{23})}{E_{22}}\right) & 0 & 0 \\ 0 & 0 & 0 & 0 & \left(\frac{1}{G_{12}}\right) & 0 \\ 0 & 0 & 0 & 0 & 0 & \left(\frac{1}{G_{12}}\right) \end{bmatrix} \begin{Bmatrix} \sigma_{11} \\ \sigma_{22} \\ \sigma_{33} \\ \tau_{23} \\ \tau_{13} \\ \tau_{12} \end{Bmatrix} \tag{4.21}$$

which may be written as

$$
\begin{Bmatrix} \varepsilon_{11} \\ \varepsilon_{22} \\ \varepsilon_{33} \\ \gamma_{23} \\ \gamma_{13} \\ \gamma_{12} \end{Bmatrix} =
\begin{bmatrix}
S_{11} & S_{12} & S_{12} & 0 & 0 & 0 \\
S_{12} & S_{22} & S_{23} & 0 & 0 & 0 \\
S_{12} & S_{23} & S_{22} & 0 & 0 & 0 \\
0 & 0 & 0 & 2(S_{22} - S_{23}) & 0 & 0 \\
0 & 0 & 0 & 0 & S_{66} & 0 \\
0 & 0 & 0 & 0 & 0 & S_{66}
\end{bmatrix}
\begin{Bmatrix} \sigma_{11} \\ \sigma_{22} \\ \sigma_{33} \\ \tau_{23} \\ \tau_{13} \\ \tau_{12} \end{Bmatrix}
\tag{4.22}
$$

Only five independent material constants exist for a transversely isotropic material. The set of five independent constants can be viewed as

$$(S_{11}, S_{22}, S_{66}, S_{12}, \text{ and } S_{23})$$

or equivalently, as

$$(E_{11}, E_{22}, v_{12}, v_{23}, \text{ and } G_{12})$$

Equation 4.22 is the strain–stress form of Hooke's law suitable for use with transversely isotropic composites. To obtain a stress–strain relationship, Equation 4.22 is inverted, resulting in

$$
\begin{Bmatrix} \sigma_{11} \\ \sigma_{22} \\ \sigma_{33} \\ \tau_{23} \\ \tau_{13} \\ \tau_{12} \end{Bmatrix} =
\begin{bmatrix}
C_{11} & C_{12} & C_{12} & 0 & 0 & 0 \\
C_{12} & C_{22} & C_{23} & 0 & 0 & 0 \\
C_{12} & C_{23} & C_{22} & 0 & 0 & 0 \\
0 & 0 & 0 & (C_{22} - C_{23})/2 & 0 & 0 \\
0 & 0 & 0 & 0 & C_{66} & 0 \\
0 & 0 & 0 & 0 & 0 & C_{66}
\end{bmatrix}
\begin{Bmatrix} \varepsilon_{11} \\ \varepsilon_{22} \\ \varepsilon_{33} \\ \gamma_{23} \\ \gamma_{13} \\ \gamma_{12} \end{Bmatrix}
\tag{4.23}
$$

Individual components within the stiffness matrix for a transversely isotropic composite are related to the compliance terms as follows:

$$
C_{11} = \frac{S_{22} + S_{33}}{\Omega} \qquad C_{12} = \frac{-S_{12}}{\Omega} \qquad C_{22} = \frac{S_{11}S_{22} - S_{12}^2}{\Omega(S_{22} - S_{23})}
$$

$$
C_{23} = \frac{S_{12}^2 - S_{11}S_{23}}{\Omega(S_{22} - S_{23})} \qquad C_{66} = \frac{1}{S_{66}}
\tag{4.24}
$$

where

$$\Omega = S_{11}(S_{22} + S_{23}) - 2S_{12}^2$$

Alternatively, the stiffness terms may be calculated using the elastic properties described in Section 3.2:

$$C_{11} = \frac{E_{11}^2(1 - v_{23})}{\Omega} \qquad C_{12} = \frac{v_{12}E_{11}E_{22}}{\Omega} \qquad C_{22} = \frac{E_{22}(E_{11} - v_{12}^2 E_{22})}{\Omega(1 + v_{23})}$$

$$C_{23} = \frac{E_{22}(v_{23}E_{11} + v_{12}^2 E_{22})}{\Omega(1 + v_{23})} \qquad C_{44} = \frac{C_{22} - C_{23}}{2} = \frac{E_{22}}{2(1 + v_{23})} \qquad C_{66} = G_{12}$$

(4.25)

where

$$\Omega = E_{11}(1 - v_{23}) - 2v_{12}^2 E_{22}$$

Example Problem 4.1

The properties of a composite material are known to be

$$E_{11} = 170 \text{ GPa} \quad E_{22} = 10 \text{ GPa} \quad E_{33} = 8 \text{ GPa}$$

$$v_{12} = 0.30 \qquad v_{13} = 0.35 \qquad v_{23} = 0.40$$

$$G_{12} = 13 \text{ GPa} \quad G_{13} = 10 \text{ GPa} \quad G_{23} = 8 \text{ GPa}$$

Note that nine distinct material properties have been specified, indicating that this composite material is orthotropic. Determine the strains caused by the following state of stress:

$$\begin{Bmatrix} \sigma_{11} \\ \sigma_{22} \\ \sigma_{33} \\ \tau_{23} \\ \tau_{13} \\ \tau_{12} \end{Bmatrix} = \begin{Bmatrix} 350 \text{ MPa} \\ 35 \text{ MPa} \\ 15 \text{ MPa} \\ 30 \text{ MPa} \\ 10 \text{ MPa} \\ 25 \text{ MPa} \end{Bmatrix}$$

SOLUTION

As the composite is orthotropic strains are calculated using Equation 4.16, where each term within the compliance matrix is calculated using Equation 4.17:

$$S_{11} = \frac{1}{E_{11}} = \frac{1}{170 \text{ GPa}} = \frac{5.88}{10^{12} \text{Pa}} \qquad S_{22} = \frac{1}{E_{22}} = \frac{1}{10 \text{ GPa}} = \frac{100.0}{10^{12} \text{ GPa}}$$

$$S_{33} = \frac{1}{E_{33}} = \frac{1}{8\,\text{GPa}} = \frac{125.0}{10^{12}\,\text{Pa}} \qquad S_{44} = \frac{1}{G_{23}} = \frac{1}{8\,\text{GPa}} = \frac{125.0}{10^{12}\,\text{Pa}}$$

$$S_{55} = \frac{1}{G_{13}} = \frac{1}{10\,\text{GPa}} = \frac{100}{10^{12}\,\text{Pa}} \qquad S_{66} = \frac{1}{G_{12}} = \frac{1}{13\,\text{GPa}} = \frac{76.9}{10^{12}\,\text{Pa}}$$

$$S_{21} = S_{12} = \frac{-v_{12}}{E_{11}} = \frac{-0.30}{170\,\text{GPa}} = \frac{-1.76}{10^{12}\,\text{Pa}}$$

$$S_{31} = S_{13} = \frac{-v_{13}}{E_{11}} = \frac{-0.35}{170\,\text{GPa}} = \frac{-2.06}{10^{12}\,\text{Pa}}$$

$$S_{32} = S_{23} = \frac{-v_{23}}{E_{22}} = \frac{-0.40}{10\,\text{GPa}} = \frac{-40.0}{10^{12}\,\text{Pa}}$$

In this case, Equation 4.16 becomes

$$
\begin{Bmatrix} \varepsilon_{11} \\ \varepsilon_{22} \\ \varepsilon_{33} \\ \gamma_{23} \\ \gamma_{13} \\ \gamma_{12} \end{Bmatrix}
=
\begin{bmatrix}
5.88 & -1.76 & -2.06 & 0 & 0 & 0 \\
-1.76 & 100.0 & -40.0 & 0 & 0 & 0 \\
-2.06 & -40.0 & 125.0 & 0 & 0 & 0 \\
0 & 0 & 0 & 125.0 & 0 & 0 \\
0 & 0 & 0 & 0 & 100.0 & 0 \\
0 & 0 & 0 & 0 & 0 & 76.9
\end{bmatrix}
\left(\frac{1}{10^{12}\,\text{Pa}}\right)
\begin{Bmatrix} 350 \\ 35 \\ 15 \\ 30 \\ 10 \\ 25 \end{Bmatrix}
(10^6\,\text{Pa})
$$

$$
\begin{Bmatrix} \varepsilon_{11} \\ \varepsilon_{22} \\ \varepsilon_{33} \\ \gamma_{23} \\ \gamma_{13} \\ \gamma_{12} \end{Bmatrix}
=
\begin{Bmatrix}
1966\ \mu\text{m/m} \\
2284\ \mu\text{m/m} \\
-246\ \mu\text{m/m} \\
3750\ \mu\text{rad} \\
1000\ \mu\text{rad} \\
1923\ \mu\text{rad}
\end{Bmatrix}
$$

Example Problem 4.2

The properties of a composite material are known to be

$$E_{11} = 25\ \text{Msi} \qquad\qquad E_{22} = E_{33} = 1.5\ \text{Msi}$$

$$v_{12} = v_{13} = 0.30 \qquad\qquad v_{23} = 0.40$$

$$G_{12} = G_{13} = 2.0\ \text{Msi}$$

Note that only five distinct material properties have been specified, indicating that this composite material is transversely isotropic. Determine the strains caused by the following state of stress:

$$\begin{Bmatrix} \sigma_{11} \\ \sigma_{22} \\ \sigma_{33} \\ \tau_{23} \\ \tau_{13} \\ \tau_{12} \end{Bmatrix} = \begin{Bmatrix} 50\,\text{ksi} \\ 5\,\text{ksi} \\ 2\,\text{ksi} \\ 4\,\text{ksi} \\ 1.5\,\text{ksi} \\ 3.5\,\text{ksi} \end{Bmatrix}$$

SOLUTION

As the composite is transversely isotropic, strains are calculated using Equation 4.21. Individual terms within the compliance matrix are

$$S_{11} = \frac{1}{E_{11}} = \frac{1}{25\,\text{Msi}} = \frac{40.0}{10^9\,\text{psi}}$$

$$S_{22} = S_{33} = \frac{1}{E_{22}} = \frac{1}{1.5\,\text{Msi}} = \frac{667}{10^9\,\text{psi}}$$

$$S_{44} = \frac{1}{G_{23}} = \frac{2(1 + \nu_{23})}{E_{22}} = \frac{2(1 + 0.40)}{1.5\,\text{Msi}} = \frac{1866}{10^9\,\text{psi}}$$

$$S_{55} = S_{66} = \frac{1}{G_{12}} = \frac{1}{2.0\,\text{Msi}} = \frac{500}{10^9\,\text{psi}}$$

$$S_{12} = S_{21} = S_{13} = S_{31} = \frac{-\nu_{12}}{E_{11}} = \frac{-0.30}{25\,\text{Msi}} = \frac{-12.0}{10^9\,\text{psi}}$$

$$S_{23} = S_{32} = \frac{-\nu_{23}}{E_{22}} = \frac{-0.40}{1.5\,\text{Msi}} = \frac{-266}{10^9\,\text{psi}}$$

Equation 4.16 becomes

$$\begin{Bmatrix} \varepsilon_{11} \\ \varepsilon_{22} \\ \varepsilon_{33} \\ \gamma_{23} \\ \gamma_{13} \\ \gamma_{12} \end{Bmatrix} = \begin{bmatrix} 40.0 & -12.0 & -12.0 & 0 & 0 & 0 \\ -12.0 & 667.0 & -266.0 & 0 & 0 & 0 \\ -12.0 & -266.0 & 667.0 & 0 & 0 & 0 \\ 0 & 0 & 0 & 800.0 & 0 & 0 \\ 0 & 0 & 0 & 0 & 500.0 & 0 \\ 0 & 0 & 0 & 0 & 0 & 500.0 \end{bmatrix} \left(\frac{1}{10^9\,\text{psi}} \right) \begin{Bmatrix} 50 \\ 5 \\ 2 \\ 4 \\ 1.5 \\ 3.5 \end{Bmatrix} (10^3\,\text{psi})$$

$$\begin{Bmatrix} \varepsilon_{11} \\ \varepsilon_{22} \\ \varepsilon_{33} \\ \gamma_{23} \\ \gamma_{13} \\ \gamma_{12} \end{Bmatrix} = \begin{Bmatrix} 1916\ \mu\text{in./in.} \\ 2203\ \mu\text{in./in.} \\ -596\ \mu\text{in./in.} \\ 3200\ \mu\text{rad} \\ 750\ \mu\text{rad} \\ 1750\ \mu\text{rad} \end{Bmatrix}$$

Example Problem 4.3

An orthotropic composite is subjected to a state of stress that causes the following state of strain:

$$\begin{Bmatrix} \varepsilon_{11} \\ \varepsilon_{22} \\ \varepsilon_{33} \\ \gamma_{23} \\ \gamma_{13} \\ \gamma_{12} \end{Bmatrix} = \begin{Bmatrix} -1500\ \mu\text{m/m} \\ 2000\ \mu\text{m/m} \\ 1000\ \mu\text{m/m} \\ -2500\ \mu\text{rad} \\ 500\ \mu\text{rad} \\ -2000\ \mu\text{rad} \end{Bmatrix}$$

Determine the stresses that caused these strains (use material properties listed in Example Problem 4.1).

SOLUTION

Since the composite is orthotropic, stresses are calculated using Equation 4.18. The stiffness matrix can be obtained by (a) inverting the compliance matrix determined as a part of Example Problem 4.1, (b) through the use of Equation 4.19, or (c) through the use of Equation 4.20. All three methods are entirely equivalent, and which procedure is selected for use is simply a matter of convenience. Equation 4.20 will be used in this example:

$$\Omega = E_{11}E_{22} - v_{12}^2 E_{22}^2 - v_{13}^2 E_{22}E_{33} - v_{23}^2 E_{11}E_{33} - 2v_{12}v_{13}v_{23}E_{22}E_{33}$$

$$\Omega = \big\{ (170)(10) - (0.30)^2(10)^2 - (0.35)^2(10)(8) - (0.40)^2(170)(8)$$

$$- 2(0.30)(0.35)(0.40)(10)(8) \big\} (\text{GPa})^2$$

$$\Omega = 1457 (\text{GPa})^2$$

$$C_{11} = \frac{(E_{22} - v_{23}^2 E_{33})E_{11}^2}{\Omega} = \frac{\big\{ (10\ \text{GPa}) - (0.40)^2(8\ \text{GPa}) \big\}(170\ \text{GPa})^2}{1457\ (\text{GPa})^2} = 172.98\ \text{GPa}$$

$$C_{12} = \frac{(\nu_{12}E_{22} + \nu_{13}\nu_{23}E_{33})E_{11}E_{22}}{\Omega}$$

$$= \frac{\{(0.30)(10\text{ GPa}) + (0.35)(0.40)(8\text{ GPa})\}(170\text{ GPa})(10\text{ GPa})}{1457(\text{GPa})^2} = 4.808\text{ GPa}$$

$$C_{13} = \frac{(\nu_{12}\nu_{23} + \nu_{13})E_{11}E_{22}E_{33}}{\Omega}$$

$$= \frac{\{(0.30)(0.40) + (0.35)\}(170\text{ GPa})(10\text{ GPa})(8\text{ GPa})}{1457(\text{GPa})^2} = 4.387\text{ GPa}$$

$$C_{22} = \frac{(E_{11} - \nu_{13}^2 E_{33})E_{22}^2}{\Omega} = \frac{\{(170\text{ GPa}) - (0.35)^2(8\text{ GPa})\}(10\text{ GPa})^2}{1457(\text{GPa})^2} = 11.602\text{ GPa}$$

$$C_{23} = \frac{(\nu_{23}E_{11} + \nu_{12}\nu_{13}E_{22})E_{22}E_{33}}{\Omega}$$

$$= \frac{\{(0.40)(170\text{ GPa}) + (0.30)(0.35)(10\text{ GPa})\}(10\text{ GPa})(8\text{ GPa})}{1457(\text{GPa})^2} = 3.792\text{ GPa}$$

$$C_{33} = \frac{(E_{11} - \nu_{12}^2 E_{22})E_{22}E_{33}}{\Omega}$$

$$= \frac{\{(170\text{ GPa}) - (0.30)^2(10\text{ GPa})\}(10\text{ GPa})(8\text{ GPa})}{1457(\text{GPa})^2} = 9.286\text{ GPa}$$

$$C_{44} = G_{23} = 8\text{ GPa} \quad C_{55} = G_{13} = 10\text{ GPa} \quad C_{66} = G_{12} = 13\text{ GPa}$$

Applying Equation 4.18, the stresses are

$$
\begin{Bmatrix} \sigma_{11} \\ \sigma_{22} \\ \sigma_{33} \\ \tau_{23} \\ \tau_{13} \\ \tau_{12} \end{Bmatrix} =
\begin{bmatrix}
172.98 & 4.808 & 4.387 & 0 & 0 & 0 \\
4.808 & 11.602 & 3.792 & 0 & 0 & 0 \\
4.387 & 3.792 & 9.286 & 0 & 0 & 0 \\
0 & 0 & 0 & 8.0 & 0 & 0 \\
0 & 0 & 0 & 0 & 10.0 & 0 \\
0 & 0 & 0 & 0 & 0 & 13.0
\end{bmatrix}(\text{GPa})
\begin{Bmatrix} -1500\ \mu\text{m/m} \\ 2000\ \mu\text{m/m} \\ 1000\ \mu\text{m/m} \\ -2500\ \mu\text{rad} \\ 500\ \mu\text{rad} \\ -2000\ \mu\text{rad} \end{Bmatrix}
$$

$$
\begin{Bmatrix} \sigma_{11} \\ \sigma_{22} \\ \sigma_{33} \\ \tau_{23} \\ \tau_{13} \\ \tau_{12} \end{Bmatrix} =
\begin{Bmatrix} -245.5\text{ MPa} \\ 19.78\text{ MPa} \\ 10.29\text{ MPa} \\ -20.0\text{ MPa} \\ 5.00\text{ MPa} \\ -26.0\text{ MPa} \end{Bmatrix}
$$

Example Problem 4.4

A transversely isotropic composite is subjected to a state of stress that causes the following state of strain:

$$\begin{Bmatrix} \varepsilon_{11} \\ \varepsilon_{22} \\ \varepsilon_{33} \\ \gamma_{23} \\ \gamma_{13} \\ \gamma_{12} \end{Bmatrix} = \begin{Bmatrix} -1250\ \mu\text{in./in.} \\ -1000\ \mu\text{in./in.} \\ -500\ \mu\text{in./in.} \\ -2500\ \mu\text{rad} \\ -1000\ \mu\text{rad} \\ 2000\ \mu\text{rad} \end{Bmatrix}$$

Determine the stresses that caused these strains (use material properties listed in Example Problem 4.2).

SOLUTION

As the composite is transversely isotropic, stresses are calculated using Equation 4.23. The stiffness matrix can be obtained by (a) inverting the compliance matrix determined as a part of Example Problem 4.2, (b) through the use of Equation 4.24, or (c) through the use of Equation 4.25. All three methods are entirely equivalent, and which procedure is selected for use is simply a matter of convenience. Equation 4.25 will be used in this example:

$$\Omega = E_{11}(1 - \nu_{23}) - 2\nu_{12}^2 E_{22}$$

$$\Omega = (25\text{ Msi})(1 - 0.40) - 2(0.30)^2(1.5\text{ Msi}) = 14.73\text{ Msi}$$

$$C_{11} = \frac{E_{11}^2(1 - \nu_{23})}{\Omega} = \frac{(25\text{ Msi})^2(1 - 0.40)}{14.73\text{ Msi}} = 25.46\text{ Msi}$$

$$C_{12} = \frac{\nu_{12}E_{11}E_{22}}{\Omega} = \frac{(0.30)(25\text{ Msi})(1.5\text{ Msi})}{14.73\text{ Msi}} = 0.7637\text{ Msi}$$

$$C_{22} = \frac{E_{22}(E_{11} - \nu_{12}^2 E_{22})}{\Omega(1 + \nu_{23})} = \frac{(1.5\text{ Msi})\{(25\text{ Msi}) - (0.30)^2(1.5\text{ Msi})\}}{14.73\text{ Msi}(1 + 0.40)} = 1.809\text{ Msi}$$

$$C_{23} = \frac{E_{22}(\nu_{23}E_{11} + \nu_{12}^2 E_{22})}{\Omega(1 + \nu_{23})} = \frac{(1.5\text{ Msi})\{(0.40)(25\text{ Msi}) + (0.30)^2(1.5\text{ Msi})\}}{14.73\text{ Msi}(1 + 0.40)}$$

$$= 0.7372\text{ Msi}$$

$$C_{44} = \frac{E_{22}}{2(1 + \nu_{23})} = \frac{(1.5 \text{ Msi})}{2(1 + 0.40)} = 0.5357 \text{ Msi}$$

$$C_{66} = G_{12} = 2.0 \text{ Msi}$$

Applying Equation 4.23, the stresses are

$$\begin{Bmatrix} \sigma_{11} \\ \sigma_{22} \\ \sigma_{33} \\ \tau_{23} \\ \tau_{13} \\ \tau_{12} \end{Bmatrix} = \begin{bmatrix} 25.46 & 0.7637 & 0.7637 & 0 & 0 & 0 \\ 0.7637 & 1.809 & 0.7372 & 0 & 0 & 0 \\ 0.7637 & 0.7372 & 1.809 & 0 & 0 & 0 \\ 0 & 0 & 0 & 0.5357 & 0 & 0 \\ 0 & 0 & 0 & 0 & 2.0 & 0 \\ 0 & 0 & 0 & 0 & 0 & 2.0 \end{bmatrix} (\text{Msi}) \begin{Bmatrix} -1250 \text{ μin./in.} \\ -1000 \text{ μin./in.} \\ -500 \text{ μin./in.} \\ -2500 \text{ μrad} \\ -1000 \text{ μrad} \\ 2000 \text{ μrad} \end{Bmatrix}$$

$$\begin{Bmatrix} \sigma_{11} \\ \sigma_{22} \\ \sigma_{33} \\ \tau_{23} \\ \tau_{13} \\ \tau_{12} \end{Bmatrix} = \begin{Bmatrix} -32.97 \text{ ksi} \\ -3.13 \text{ ksi} \\ -2.60 \text{ ksi} \\ -1.34 \text{ ksi} \\ -2.00 \text{ ksi} \\ 4.00 \text{ ksi} \end{Bmatrix}$$

4.3 Strains Induced by a Change in Temperature or Moisture Content

Material properties relating strains to a uniform change in temperature and a uniform change in moisture content were defined in Sections 3.3 and 3.4, respectively. For anisotropic materials strains caused by a change in temperature are given by Equation 3.22, repeated here for convenience:

$$\begin{bmatrix} \varepsilon_{xx}^T & \varepsilon_{xy}^T & \varepsilon_{xz}^T \\ \varepsilon_{yx}^T & \varepsilon_{yy}^T & \varepsilon_{yz}^T \\ \varepsilon_{zx}^T & \varepsilon_{zy}^T & \varepsilon_{zz}^T \end{bmatrix} = \Delta T \begin{bmatrix} \alpha_{xx} & \alpha_{xy} & \alpha_{xz} \\ \alpha_{yx} & \alpha_{yy} & \alpha_{yz} \\ \alpha_{zx} & \alpha_{zy} & \alpha_{zz} \end{bmatrix} \qquad (3.22)$$

Similarly, strains caused by a change in moisture content are given by Equation 3.25, repeated here for convenience:

$$\begin{bmatrix} \varepsilon_{xx}^M & \gamma_{xy}^M & \gamma_{xz}^M \\ \gamma_{yx}^M & \varepsilon_{yy}^M & \gamma_{yz}^M \\ \gamma_{zx}^M & \gamma_{zy}^M & \varepsilon_{zz}^M \end{bmatrix} = (\Delta M) \begin{bmatrix} \beta_{xx} & \beta_{xy} & \beta_{xz} \\ \beta_{xy} & \beta_{yy} & \beta_{yz} \\ \beta_{xz} & \beta_{yz} & \beta_{zz} \end{bmatrix} \qquad (3.25)$$

As before, the strain tensors must be symmetric. This allows the use of contracted notation, and hence Equations 3.22 and 3.25 can be written in the form of column arrays:

$$
\begin{Bmatrix} \varepsilon_{xx}^T \\ \varepsilon_{yy}^T \\ \varepsilon_{zz}^T \\ \gamma_{yz}^T \\ \gamma_{xz}^T \\ \gamma_{xy}^T \end{Bmatrix} = \Delta T \begin{Bmatrix} \alpha_{xx} \\ \alpha_{yy} \\ \alpha_{zz} \\ \alpha_{yz} \\ \alpha_{xz} \\ \alpha_{xy} \end{Bmatrix} \quad \text{(and)} \quad \begin{Bmatrix} \varepsilon_{xx}^M \\ \varepsilon_{yy}^M \\ \varepsilon_{zz}^M \\ \gamma_{yz}^M \\ \gamma_{xz}^M \\ \gamma_{xy}^M \end{Bmatrix} = \Delta M \begin{Bmatrix} \beta_{xx} \\ \beta_{yy} \\ \beta_{zz} \\ \beta_{yz} \\ \beta_{xz} \\ \beta_{xy} \end{Bmatrix} \tag{4.26}
$$

In the case of an orthotropic material $\alpha_{12} = \alpha_{13} = \alpha_{23} = 0$, and $\beta_{12} = \beta_{13} = \beta_{23} = 0$. In this case, Equation 4.26 becomes

$$
\begin{Bmatrix} \varepsilon_{11}^T \\ \varepsilon_{22}^T \\ \varepsilon_{33}^T \\ \gamma_{23}^T \\ \gamma_{13}^T \\ \gamma_{12}^T \end{Bmatrix} = \Delta T \begin{Bmatrix} \alpha_{11} \\ \alpha_{22} \\ \alpha_{33} \\ 0 \\ 0 \\ 0 \end{Bmatrix} \quad \text{(and)} \quad \begin{Bmatrix} \varepsilon_{11}^M \\ \varepsilon_{22}^M \\ \varepsilon_{33}^M \\ \gamma_{23}^M \\ \gamma_{13}^M \\ \gamma_{12}^M \end{Bmatrix} = \Delta M \begin{Bmatrix} \beta_{11} \\ \beta_{22} \\ \beta_{33} \\ 0 \\ 0 \\ 0 \end{Bmatrix} \tag{4.27}
$$

In addition to these simplifications, for a transversely isotropic material with symmetry in the 2–3 plane, $\alpha_{33} = \alpha_{22}$ and $\beta_{33} = \beta_{22}$. Hence, for a transversely isotropic material Equation 4.27 becomes:

$$
\begin{Bmatrix} \varepsilon_{11}^T \\ \varepsilon_{22}^T \\ \varepsilon_{33}^T \\ \gamma_{23}^T \\ \gamma_{13}^T \\ \gamma_{12}^T \end{Bmatrix} = \Delta T \begin{Bmatrix} \alpha_{11} \\ \alpha_{22} \\ \alpha_{22} \\ 0 \\ 0 \\ 0 \end{Bmatrix} \quad \text{(and)} \quad \begin{Bmatrix} \varepsilon_{11}^M \\ \varepsilon_{22}^M \\ \varepsilon_{33}^M \\ \gamma_{23}^M \\ \gamma_{13}^M \\ \gamma_{12}^M \end{Bmatrix} = \Delta M \begin{Bmatrix} \beta_{11} \\ \beta_{22} \\ \beta_{22} \\ 0 \\ 0 \\ 0 \end{Bmatrix} \tag{4.28}
$$

4.4 Strains Induced by Combined Effects of Stress, Temperature, and Moisture

The strains induced by stress under constant environmental conditions were discussed in Section 4.1, whereas strains induced by a uniform change in temperature or moisture content in the absence of stress were discussed in Section 4.3. We will now consider the strains induced if *all* of these mechanisms occur simultaneously. That is, we wish to consider the strains induced by the *combined* effects of stress, a uniform change in temperature, and a uniform change in moisture content. We will call this the total strain, and equals the sum of the strains induced by each of mechanism acting independently:

$$\varepsilon_{ij} = \varepsilon_{ij}^{\sigma} + \varepsilon_{ij}^{T} + \varepsilon_{ij}^{M} \tag{4.29}$$

The superscripts σ, T, and M used in Equation 4.29 indicate that the individual components of strain are caused by the application of stress, by a uniform change in temperature, and by a uniform change in moisture content, respectively.

Based on this assumption, for anisotropic materials the total strain is obtained by superimposing Equations 4.8 and 4.26:

$$\begin{Bmatrix} \varepsilon_{xx} \\ \varepsilon_{yy} \\ \varepsilon_{zz} \\ \gamma_{yz} \\ \gamma_{xz} \\ \gamma_{xy} \end{Bmatrix} = \begin{bmatrix} S_{11} & S_{12} & S_{13} & S_{14} & S_{15} & S_{16} \\ S_{21} & S_{22} & S_{23} & S_{24} & S_{25} & S_{26} \\ S_{31} & S_{32} & S_{33} & S_{34} & S_{35} & S_{36} \\ S_{41} & S_{42} & S_{43} & S_{44} & S_{45} & S_{46} \\ S_{51} & S_{52} & S_{53} & S_{54} & S_{55} & S_{56} \\ S_{61} & S_{62} & S_{63} & S_{64} & S_{65} & S_{66} \end{bmatrix} \begin{Bmatrix} \sigma_{xx} \\ \sigma_{yy} \\ \sigma_{zz} \\ \tau_{yz} \\ \tau_{xz} \\ \tau_{xy} \end{Bmatrix} + \Delta T \begin{Bmatrix} \alpha_{xx} \\ \alpha_{yy} \\ \alpha_{zz} \\ \alpha_{yz} \\ \alpha_{xz} \\ \alpha_{xy} \end{Bmatrix} + \Delta M \begin{Bmatrix} \beta_{xx} \\ \beta_{yy} \\ \beta_{zz} \\ \beta_{yz} \\ \beta_{xz} \\ \beta_{xy} \end{Bmatrix} \tag{4.30}$$

Equation 4.30 allows the prediction of the strains induced by the simultaneous effects of stress and uniform changes in temperature and/or moisture content. In practice, the inverse problem is often encountered. That is, a common circumstance is that the strains, the change in temperature, and the change in moisture content have been measured, and we wish to calculate stresses. This can be accomplished by inverting Equation 4.30 according to the laws of matrix algebra, resulting in

$$\begin{Bmatrix} \sigma_{xx} \\ \sigma_{yy} \\ \sigma_{zz} \\ \tau_{yz} \\ \tau_{xz} \\ \tau_{xy} \end{Bmatrix} = \begin{bmatrix} C_{11} & C_{12} & C_{13} & C_{14} & C_{15} & C_{16} \\ C_{21} & C_{22} & C_{23} & C_{24} & C_{25} & C_{26} \\ C_{31} & C_{32} & C_{33} & C_{34} & C_{35} & C_{36} \\ C_{41} & C_{42} & C_{43} & C_{44} & C_{45} & C_{46} \\ C_{51} & C_{52} & C_{53} & C_{54} & C_{55} & C_{56} \\ C_{61} & C_{62} & C_{63} & C_{64} & C_{65} & C_{66} \end{bmatrix} \begin{Bmatrix} \varepsilon_{xx} - \Delta T \alpha_{xx} - \Delta M \beta_{xx} \\ \varepsilon_{yy} - \Delta T \alpha_{yy} - \Delta M \beta_{yy} \\ \varepsilon_{zz} - \Delta T \alpha_{zz} - \Delta M \beta_{zz} \\ \gamma_{yz} - \Delta T \alpha_{yz} - \Delta M \beta_{yz} \\ \gamma_{xz} - \Delta T \alpha_{xz} - \Delta M \beta_{xz} \\ \gamma_{xy} - \Delta T \alpha_{xy} - \Delta M \beta_{xy} \end{Bmatrix} \tag{4.31}$$

where the stiffness matrix $C_{ij} = S_{ij}^{-1}$, as discussed in Section 4.1.

Following an analogous procedure, the strains induced in an orthotropic material by the combined effects of stress, a uniform change in temperature, and/or a uniform change in moisture content can be found by superimposing Equations 4.16 and 4.27:

$$
\begin{Bmatrix} \varepsilon_{11} \\ \varepsilon_{22} \\ \varepsilon_{33} \\ \gamma_{23} \\ \gamma_{13} \\ \gamma_{12} \end{Bmatrix} = \begin{bmatrix} S_{11} & S_{12} & S_{13} & 0 & 0 & 0 \\ S_{12} & S_{22} & S_{23} & 0 & 0 & 0 \\ S_{13} & S_{23} & S_{33} & 0 & 0 & 0 \\ 0 & 0 & 0 & S_{44} & 0 & 0 \\ 0 & 0 & 0 & 0 & S_{55} & 0 \\ 0 & 0 & 0 & 0 & 0 & S_{66} \end{bmatrix} \begin{Bmatrix} \sigma_{11} \\ \sigma_{22} \\ \sigma_{33} \\ \tau_{23} \\ \tau_{13} \\ \tau_{12} \end{Bmatrix} + \Delta T \begin{Bmatrix} \alpha_{11} \\ \alpha_{22} \\ \alpha_{33} \\ 0 \\ 0 \\ 0 \end{Bmatrix} + \Delta M \begin{Bmatrix} \beta_{11} \\ \beta_{22} \\ \beta_{33} \\ 0 \\ 0 \\ 0 \end{Bmatrix}
$$

(4.32)

Inverting Equation 4.32, we obtain

$$
\begin{Bmatrix} \sigma_{11} \\ \sigma_{22} \\ \sigma_{33} \\ \tau_{23} \\ \tau_{13} \\ \tau_{12} \end{Bmatrix} = \begin{bmatrix} C_{11} & C_{12} & C_{13} & 0 & 0 & 0 \\ C_{12} & C_{22} & C_{23} & 0 & 0 & 0 \\ C_{13} & C_{23} & C_{33} & 0 & 0 & 0 \\ 0 & 0 & 0 & C_{44} & 0 & 0 \\ 0 & 0 & 0 & 0 & C_{55} & 0 \\ 0 & 0 & 0 & 0 & 0 & C_{66} \end{bmatrix} \begin{Bmatrix} \varepsilon_{11} - \Delta T \alpha_{11} - \Delta M \beta_{11} \\ \varepsilon_{22} - \Delta T \alpha_{22} - \Delta M \beta_{22} \\ \varepsilon_{33} - \Delta T \alpha_{33} - \Delta M \beta_{33} \\ \gamma_{23} \\ \gamma_{13} \\ \gamma_{12} \end{Bmatrix}
$$

(4.33)

An implicit assumption in Equations 4.32 and 4.33 is that the strain tensor, stress tensor, and material properties are all referenced to the principal material coordinate system of the orthotropic material (i.e., the 1–2–3 coordinate system). If an orthotropic material is referenced to a nonprincipal material coordinate system then the relation between strain, stress, temperature, and moisture content is given by Equation 4.30 or 4.31.

Finally, the strains induced in a transversely isotropic material by the combined effects of stress, a uniform change in temperature, and/or a uniform change in moisture content can be found by superimposing Equations 4.22 and 4.28:

$$
\begin{Bmatrix} \varepsilon_{11} \\ \varepsilon_{22} \\ \varepsilon_{33} \\ \gamma_{23} \\ \gamma_{13} \\ \gamma_{12} \end{Bmatrix} = \begin{bmatrix} S_{11} & S_{12} & S_{12} & 0 & 0 & 0 \\ S_{12} & S_{22} & S_{23} & 0 & 0 & 0 \\ S_{12} & S_{23} & S_{22} & 0 & 0 & 0 \\ 0 & 0 & 0 & 2(S_{22} - S_{23}) & 0 & 0 \\ 0 & 0 & 0 & 0 & S_{66} & 0 \\ 0 & 0 & 0 & 0 & 0 & S_{66} \end{bmatrix} \begin{Bmatrix} \sigma_{11} \\ \sigma_{22} \\ \sigma_{33} \\ \tau_{23} \\ \tau_{13} \\ \tau_{12} \end{Bmatrix} + \Delta T \begin{Bmatrix} \alpha_{11} \\ \alpha_{22} \\ \alpha_{22} \\ 0 \\ 0 \\ 0 \end{Bmatrix} + \Delta M \begin{Bmatrix} \beta_{11} \\ \beta_{22} \\ \beta_{22} \\ 0 \\ 0 \\ 0 \end{Bmatrix}
$$

(4.34)

Inverting Equation 4.34, we have

$$
\begin{Bmatrix} \sigma_{11} \\ \sigma_{22} \\ \sigma_{33} \\ \tau_{23} \\ \tau_{13} \\ \tau_{12} \end{Bmatrix} =
\begin{bmatrix}
C_{11} & C_{12} & C_{12} & 0 & 0 & 0 \\
C_{12} & C_{22} & C_{23} & 0 & 0 & 0 \\
C_{12} & C_{23} & C_{22} & 0 & 0 & 0 \\
0 & 0 & 0 & (C_{22}-C_{23})/2 & 0 & 0 \\
0 & 0 & 0 & 0 & C_{66} & 0 \\
0 & 0 & 0 & 0 & 0 & C_{66}
\end{bmatrix}
\begin{Bmatrix} \varepsilon_{11} - \Delta T \alpha_{11} - \Delta M \beta_{11} \\ \varepsilon_{22} - \Delta T \alpha_{22} - \Delta M \beta_{22} \\ \varepsilon_{33} - \Delta T \alpha_{22} - \Delta M \beta_{22} \\ \gamma_{23} \\ \gamma_{13} \\ \gamma_{12} \end{Bmatrix}
$$

$$(4.35)$$

Once again, Equations 4.34 and 4.35 are valid only if referenced to the principal material coordinate system of the transversely isotropic material (i.e., the 1–2–3 coordinate system). If a transversely isotropic material is referenced to a nonprincipal material coordinate system then the relation between strain, stress, temperature, and moisture content is given by Equation 4.30 or 4.31.

HOMEWORK PROBLEMS

An anisotropic material with the following properties is considered in problems 4.1 through 4.4:

$E_{xx} = 100$ GPa	$E_{yy} = 200$ GPa	$E_{zz} = 75$ GPa
$\nu_{xy} = 0.20$	$\nu_{xz} = -0.25$	$\nu_{yz} = 0.60$
$G_{xy} = 60$ GPa	$G_{xz} = 75$ GPa	$G_{yz} = 50$ GPa
$\eta_{xx,xy} = -0.30$	$\eta_{xx,xz} = 0.25$	$\eta_{xx,yz} = 0.30$
$\eta_{yy,xy} = 0.60$	$\eta_{yy,xz} = 0.75$	$\eta_{yy,yz} = 0.20$
$\eta_{zz,xy} = -0.20$	$\eta_{zz,xz} = -0.05$	$\eta_{zz,yz} = -0.15$
$\mu_{xy,xz} = -0.10$	$\mu_{xy,yz} = -0.05$	$\mu_{xz,yz} = 0.10$
$\alpha_{xx} = 5\ \mu\text{m/m} - {}^{\circ}\text{C}$	$\alpha_{yy} = 10\ \mu\text{m/m} - {}^{\circ}\text{C}$	$\alpha_{zz} = 20\ \mu\text{m/m} - {}^{\circ}\text{C}$
$\alpha_{xy} = -5\ \mu\text{rad}/{}^{\circ}\text{C}$	$\alpha_{xz} = 15\ \mu\text{rad}/7{}^{\circ}\text{C}$	$\alpha_{yz} = 25\ \mu\text{rad}/{}^{\circ}\text{C}$
$\beta_{xx} = 300\ \mu\text{m/m} - \%M$	$\beta_{yy} = 60\ \mu\text{m/m} - \%M$	$\beta_{zz} = 1200\ \mu\text{m/m} - \%M$
$\beta_{xy} = 150\ \mu\text{rad}/\%M$	$\beta_{xz} = -1000\ \mu\text{rad}/\%M$	$\beta_{yz} = -350\ \mu\text{rad}/\%M$

4.1. Calculate the compliance matrix, S_{ij}.

4.2. Calculate the stiffness matrix, $C_{ij} = S_{ij}^{-1}$. (*Note*: perform this calculation using a suitable software package such as Maple, MATLAB®, Mathematica, etc.).

4.3. Consider the following stress tensor:

$$
\sigma_{ij} = \begin{bmatrix} \sigma_{xx} & \sigma_{xy} & \sigma_{xz} \\ \sigma_{yx} & \sigma_{yy} & \sigma_{yz} \\ \sigma_{zx} & \sigma_{zy} & \sigma_{zz} \end{bmatrix} = \begin{bmatrix} 75 & 10 & -25 \\ 10 & -90 & 30 \\ -25 & 30 & 25 \end{bmatrix} \text{(MPa)}
$$

a. Calculate the strains induced by this stress tensor, assuming no change in temperature or moisture content (i.e., assume $\Delta T = \Delta M = 0$).

b. Calculate the strains induced by this stress tensor and a temperature increase of 100°C (assume $\Delta M = 0$).

c. Calculate the strains induced by this stress tensor and a 2% increase in moisture content (assume $\Delta T = 0$).

d. Calculate the strains induced by this stress tensor, a temperature increase of 100°C, and a 2% increase in moisture content.

4.4. Consider the following strains:

$\varepsilon_{xx} = 1500 \ \mu m/m$	$\varepsilon_{yy} = -2000 \ \mu m/m$	$\varepsilon_{zz} = -1750 \ \mu m/m$
$\gamma_{xy} = 750 \ \mu rad$	$\gamma_{xz} = -500 \ \mu rad$	$\gamma_{yz} = 850 \ \mu rad$

a. Calculate the stress tensor that caused these strains, assuming no change in temperature or moisture content (i.e., assuming $\Delta T = \Delta M = 0$).

b. Calculate the stress tensor that caused these strains, if these strains were caused by the simultaneous effects of stress and a temperature decrease of 100°C (assume $\Delta M = 0$).

c. Calculate the stress tensor that caused these strains, if these strains were caused by the simultaneous effects of stress, a temperature decrease of 100°C, and a 2% increase in moisture content.

An orthotropic material with the following properties is considered in problems 4.5 through 4.8:

$E_{11} = 100 \ GPa$	$E_{22} = 200 \ GPa$	$E_{33} = 75 \ GPa$
$v_{12} = 0.20$	$v_{13} = -0.25$	$v_{23} = 0.60$
$G_{12} = 60 \ GPa$	$G_{13} = 75 \ GPa$	$G_{23} = 50 \ GPa$
$\alpha_{11} = 1 \ \mu m/m - °C$	$\alpha_{22} = 25 \ \mu m/m - °C$	$\alpha_{33} = 15 \ \mu m/m - °C$
$\beta_{11} = 100 \ \mu m/m - \%M$	$\beta_{22} = 600 \ \mu m/m - \%M$	$\beta_{33} = 1000 \ \mu m/m - \%M$

4.5. Calculate the compliance matrix, S_{ij}.

4.6. Calculate the stiffness matrix, C_{ij}.

4.7. Consider the following stress tensor:

$$\sigma_{ij} = \begin{bmatrix} \sigma_{11} & \sigma_{12} & \sigma_{13} \\ \sigma_{21} & \sigma_{22} & \sigma_{23} \\ \sigma_{31} & \sigma_{32} & \sigma_{33} \end{bmatrix} = \begin{bmatrix} 75 & 10 & -25 \\ 10 & -90 & 30 \\ -25 & 30 & 25 \end{bmatrix} (MPa)$$

   a. Calculate the strains induced by this stress tensor, assuming no change in temperature or moisture content (i.e., assume $\Delta T = \Delta M = 0$).

   b. Calculate the strains induced by this stress tensor and a temperature increase of 100°C (assume $\Delta M = 0$).

   c. Calculate the strains induced by this stress tensor and a 2% increase in moisture content (assume $\Delta T = 0$).

   d. Calculate the strains induced by this stress tensor, a temperature increase of 100°C, and a 2% increase in moisture content.

4.8. Consider the following strains:

$\varepsilon_{11} = -2000\ \mu m/m$	$\varepsilon_{22} = 3000\ \mu m/m$	$\varepsilon_{33} = 1500\ \mu m/m$
$\gamma_{12} = 750\ \mu rad$	$\gamma_{13} = -1000\ \mu rad$	$\gamma_{23} = 1250\ \mu rad$

   a. Calculate the stress tensor that caused these strains, assuming no change in temperature or moisture content (i.e., assuming $\Delta T = \Delta M = 0$).

   b. Calculate the stress tensor that caused these strains, if these strains were caused by the simultaneous effects of stress and a temperature decrease of 100°C (assume $\Delta M = 0$).

   c. Calculate the stress tensor that caused these strains, if these strains were caused by the simultaneous effects of stress, a temperature decrease of 100°C, and a 2% increase in moisture content.

5

Unidirectional Composite Laminates Subject to Plane Stress

5.1 Unidirectional Composites Referenced to the Principal Material Coordinate System

The strains induced in an orthotropic material subjected to a general 3-D stress tensor, a uniform change in temperature, and/or a uniform change in moisture content was described in Chapter 4. The strain response is summarized by Equation 4.32, repeated here for convenience:

$$
\begin{Bmatrix} \varepsilon_{11} \\ \varepsilon_{22} \\ \varepsilon_{33} \\ \gamma_{23} \\ \gamma_{13} \\ \gamma_{12} \end{Bmatrix} = \begin{bmatrix} S_{11} & S_{12} & S_{13} & 0 & 0 & 0 \\ S_{12} & S_{22} & S_{23} & 0 & 0 & 0 \\ S_{13} & S_{23} & S_{33} & 0 & 0 & 0 \\ 0 & 0 & 0 & S_{44} & 0 & 0 \\ 0 & 0 & 0 & 0 & S_{55} & 0 \\ 0 & 0 & 0 & 0 & 0 & S_{66} \end{bmatrix} \begin{Bmatrix} \sigma_{11} \\ \sigma_{22} \\ \sigma_{33} \\ \tau_{23} \\ \tau_{13} \\ \tau_{12} \end{Bmatrix} + \Delta T \begin{Bmatrix} \alpha_{11} \\ \alpha_{22} \\ \alpha_{33} \\ 0 \\ 0 \\ 0 \end{Bmatrix} + \Delta M \begin{Bmatrix} \beta_{11} \\ \beta_{22} \\ \beta_{33} \\ 0 \\ 0 \\ 0 \end{Bmatrix}
$$

$$(4.32) \text{ (repeated)}$$

Equation 4.32 is valid for any orthotropic material, as long as the strain tensor, stress tensor, and material properties are all referenced to the principal 1–2–3 coordinate system. Consider the strains induced by a state of plane stress. Assuming that $\sigma_{33} = \tau_{23} = \tau_{13} = 0$, Equation 4.32 becomes

$$
\begin{Bmatrix} \varepsilon_{11} \\ \varepsilon_{22} \\ \varepsilon_{33} \\ \gamma_{23} \\ \gamma_{13} \\ \gamma_{12} \end{Bmatrix} = \begin{bmatrix} S_{11} & S_{12} & S_{13} & 0 & 0 & 0 \\ S_{12} & S_{22} & S_{23} & 0 & 0 & 0 \\ S_{13} & S_{23} & S_{33} & 0 & 0 & 0 \\ 0 & 0 & 0 & S_{44} & 0 & 0 \\ 0 & 0 & 0 & 0 & S_{55} & 0 \\ 0 & 0 & 0 & 0 & 0 & S_{66} \end{bmatrix} \begin{Bmatrix} \sigma_{11} \\ \sigma_{22} \\ 0 \\ 0 \\ 0 \\ \tau_{12} \end{Bmatrix} + \Delta T \begin{Bmatrix} \alpha_{11} \\ \alpha_{22} \\ \alpha_{33} \\ 0 \\ 0 \\ 0 \end{Bmatrix} + \Delta M \begin{Bmatrix} \beta_{11} \\ \beta_{22} \\ \beta_{33} \\ 0 \\ 0 \\ 0 \end{Bmatrix}
$$

$$(5.1)$$

Note that Equation 5.1 shows that in the case of plane stress, the out-of-plane shear strains are always equal to zero ($\gamma_{23} = \gamma_{13} = 0$). It is customary to write the expressions for the remaining four strain components as follows:

$$
\begin{Bmatrix} \varepsilon_{11} \\ \varepsilon_{22} \\ \gamma_{12} \end{Bmatrix} = \begin{bmatrix} S_{11} & S_{12} & 0 \\ S_{12} & S_{22} & 0 \\ 0 & 0 & S_{66} \end{bmatrix} \begin{Bmatrix} \sigma_{11} \\ \sigma_{22} \\ \tau_{12} \end{Bmatrix} + \Delta T \begin{Bmatrix} \alpha_{11} \\ \alpha_{22} \\ 0 \end{Bmatrix} + \Delta M \begin{Bmatrix} \beta_{11} \\ \beta_{22} \\ 0 \end{Bmatrix}
\tag{5.2a}
$$

and

$$
\varepsilon_{33} = S_{13}\sigma_{11} + S_{23}\sigma_{22} + \Delta T \alpha_{33} + \Delta M \beta_{33}
\tag{5.2b}
$$

Equations 5.2a and 5.2b are called *reduced* forms of Hooke's law for an orthotropic composite. They are only valid for a state of plane stress, and are called "reduced" laws because we have reduced the allowable stress tensor from 3-dimensions to 2-dimensions. The 3×3 array in Equation 5.2a is called the *reduced compliance matrix*. Note that despite the reduction from 3- to 2-dimensions, we have retained the subscripts used in the original compliance matrix. For example, the element that appears in the (3,3) position of the reduced compliance matrix is labeled S_{66}. The definition of each compliance term is not altered by the reduction from 3- to 2-dimensions, and each term is still related to the more familiar engineering constants (E_{11}, E_{22}, ν_{12}, etc.) in accordance with Equations 4.17.

Inverting Equations 5.2a, we obtain:

$$
\begin{Bmatrix} \sigma_{11} \\ \sigma_{22} \\ \tau_{12} \end{Bmatrix} = \begin{bmatrix} Q_{11} & Q_{12} & 0 \\ Q_{12} & Q_{22} & 0 \\ 0 & 0 & Q_{66} \end{bmatrix} \begin{Bmatrix} \varepsilon_{11} - \Delta T \alpha_{11} - \Delta M \beta_{11} \\ \varepsilon_{22} - \Delta T \alpha_{22} - \Delta M \beta_{22} \\ \gamma_{12} \end{Bmatrix}
\tag{5.3}
$$

The 3×3 array that appears in Equation 5.3 is called the *reduced stiffness matrix*, and equals the inverse of the reduced compliance matrix:

$$
\begin{bmatrix} Q_{11} & Q_{12} & 0 \\ Q_{12} & Q_{22} & 0 \\ 0 & 0 & Q_{66} \end{bmatrix} \equiv \begin{bmatrix} S_{11} & S_{12} & 0 \\ S_{12} & S_{22} & 0 \\ 0 & 0 & S_{66} \end{bmatrix}^{-1}
\tag{5.4}
$$

Note that we are now using a different symbol to denote stiffness. That is, the original 3-D stiffness matrix was denoted "C_{ij}" (as in Equation 4.18, for example), whereas the reduced stiffness matrix is denoted "Q_{ij}." This change in notation is required because *individual members of the reduced stiffness matrix are not equal to the corresponding members in the original stiffness matrix.*

That is, $Q_{11} \neq C_{11}$, $Q_{12} \neq C_{12}$, $Q_{22} \neq C_{22}$, and $Q_{66} \neq C_{66}$. Relations between Q_{ij} and C_{ij} can be derived as follows. From Equation 4.18 it can be seen that for an orthotropic material subjected to an arbitrary state of stress:

$$\sigma_{11} = C_{11}\varepsilon_{11} + C_{12}\varepsilon_{22} + C_{13}\varepsilon_{33} \tag{5.5a}$$

and

$$\sigma_{33} = C_{13}\varepsilon_{11} + C_{23}\varepsilon_{22} + C_{33}\varepsilon_{33} \tag{5.5b}$$

However, the reduced stiffness matrix relates stress to strain *under conditions of plane stress*, by definition. Setting $\sigma_{33} = 0$ and solving Equation 5.5b for ε_{33}, we have

$$\varepsilon_{33} = \frac{-(C_{13}\varepsilon_{11} + C_{23}\varepsilon_{22})}{C_{33}} \tag{5.6}$$

The out-of-plane strain ε_{33} must be related to in-plane strains ε_{11} and ε_{22} in accordance with Equation 5.6, otherwise a state of plane stress does not exist in the composite. Substituting Equation 5.6 into 5.5a and simplifying, we have

$$\sigma_{11} = \left[\frac{C_{11}C_{33} - C_{13}^2}{C_{33}} \right]\varepsilon_{11} + \left[\frac{C_{12}C_{33} - C_{13}C_{23}}{C_{33}} \right]\varepsilon_{22} \tag{5.7}$$

On the other hand, from Equation 5.3 σ_{11} is given by (with $\Delta T = \Delta M = 0$):

$$\sigma_{11} = Q_{11}\varepsilon_{11} + Q_{12}\varepsilon_{22} \tag{5.8}$$

Comparing Equations 5.7 and 5.8, it is immediately apparent that

$$Q_{11} = \frac{C_{11}C_{33} - C_{13}^2}{C_{33}} \tag{5.9a}$$

$$Q_{12} = \frac{C_{12}C_{33} - C_{13}C_{23}}{C_{33}} \tag{5.9b}$$

Using a similar procedure it can be shown as

$$Q_{22} = \frac{C_{22}C_{33} - C_{23}^2}{C_{33}} \tag{5.9c}$$

$$Q_{66} = C_{66} \qquad (5.9d)$$

In essence, the definition of elements of the Q_{ij} matrix differs from those of the C_{ij} matrix because the Q_{ij} matrix is defined for *plane stress conditions only*, whereas C_{ij} can be used for any stress state.

Elements of the reduced stiffness matrix may be related to elements of the compliance matrix by either substituting Equations 4.19 into Equations 5.9, or by simply performing the matrix inversion indicated in Equation 5.4. In either case it will be found that

$$Q_{11} = \frac{S_{22}}{S_{11}S_{22} - S_{12}^2} \qquad Q_{12} = Q_{21} = \frac{-S_{12}}{S_{11}S_{22} - S_{12}^2}$$

$$Q_{22} = \frac{S_{11}}{S_{11}S_{22} - S_{12}^2} \qquad Q_{66} = \frac{1}{S_{66}} \qquad (5.10)$$

Alternatively, elements of the reduced stiffness matrix are related to more familiar engineering constants as follows:

$$Q_{11} = \frac{E_{11}^2}{E_{11} - v_{12}^2 E_{22}} \qquad Q_{12} = Q_{21} = \frac{v_{12}E_{11}E_{22}}{E_{11} - v_{12}^2 E_{22}}$$

$$Q_{22} = \frac{E_{11}E_{22}}{E_{11} - v_{12}^2 E_{22}} \qquad Q_{66} = G_{12} \qquad (5.11)$$

Equations 5.1 through 5.11 were developed assuming that the composite is an orthotropic material. Now consider the response of a transversely isotropic composite subjected to a state of plane stress. As before, we assume that $\sigma_{33} = \tau_{23} = \tau_{13} = 0$. From Equation 4.33 it is seen that the out-of-plane shear strains equal zero, ($\gamma_{23} = \gamma_{13} = 0$), and the remaining four strains can be written as

$$\begin{Bmatrix} \varepsilon_{11} \\ \varepsilon_{22} \\ \gamma_{12} \end{Bmatrix} = \begin{bmatrix} S_{11} & S_{12} & 0 \\ S_{12} & S_{22} & 0 \\ 0 & 0 & S_{66} \end{bmatrix} \begin{Bmatrix} \sigma_{11} \\ \sigma_{22} \\ \tau_{12} \end{Bmatrix} + \Delta T \begin{Bmatrix} \alpha_{11} \\ \alpha_{22} \\ 0 \end{Bmatrix} + \Delta M \begin{Bmatrix} \beta_{11} \\ \beta_{22} \\ 0 \end{Bmatrix} \qquad (5.12a)$$

and

$$\varepsilon_{33} = S_{12}\sigma_{11} + S_{23}\sigma_{22} + \Delta T \alpha_{22} + \Delta M \beta_{22} \qquad (5.12b)$$

Comparing Equations 5.12a with 5.2a, it is seen that relationship between in-plane strains (ε_{11}, ε_{22}, and γ_{12}) and in-plane stress components (σ_{11}, σ_{22}, and τ_{12})

are identical for orthotropic and transversely isotropic materials. In fact, identical results are obtained for the out-of-plane normal strain as well, since for a transversely isotropic material $S_{13} = S_{12}$, $\alpha_{33} = \alpha_{22}$, and $\beta_{33} = \beta_{22}$, and therefore Equation 5.2b is equivalent to Equation 5.12b. Consequently, Equation 5.3 can also be applied to a transversely isotropic material. Equations 5.9 are also still applicable, except that for a transversely isotropic material (with symmetry in the 2–3 plane) $C_{33} = C_{22}$ and $C_{13} = C_{12}$.

Inverting Equation 5.12a, we obtain:

$$\begin{Bmatrix} \sigma_{11} \\ \sigma_{22} \\ \tau_{12} \end{Bmatrix} = \begin{bmatrix} Q_{11} & Q_{12} & 0 \\ Q_{12} & Q_{22} & 0 \\ 0 & 0 & Q_{66} \end{bmatrix} \begin{Bmatrix} \varepsilon_{11} - \Delta T \alpha_{11} - \Delta M \beta_{11} \\ \varepsilon_{22} - \Delta T \alpha_{22} - \Delta M \beta_{22} \\ \gamma_{12} \end{Bmatrix}$$

Since this result is identical to Equation 5.3, it is seen once again that the relationship between in-plane strains (ε_{11}, ε_{22}, and γ_{12}) and in-plane stress components (σ_{11}, σ_{22}, and τ_{12}) are identical for orthotropic and transversely isotropic materials.

Example Problem 5.1

Determine the strains induced in a unidirectional graphite–epoxy composite subjected to the in-plane stresses shown in Figure 5.1. Assume $\Delta T = \Delta M = 0$, and use material properties listed in Table 3.1.

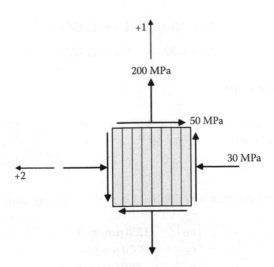

FIGURE 5.1
Unidirectional composite subjected to in-plane stresses (magnitudes of in-plane stresses shown).

SOLUTION

The magnitude of each stress component is indicated in Figure 5.1. On the basis of the sign conventions reviewed in Section 2.5, the algebraic sign of each stress component is

$$\sigma_{11} = 200\,\text{MPa} \quad \sigma_{22} = -30\,\text{MPa} \quad \tau_{12} = -50\,\text{MPa}$$

Strains can be calculated using either Equation 5.2a or Equation 5.12a. Since $\Delta T = \Delta M = 0$, we have

$$\begin{Bmatrix} \varepsilon_{11} \\ \varepsilon_{22} \\ \gamma_{12} \end{Bmatrix} = \begin{bmatrix} S_{11} & S_{12} & 0 \\ S_{12} & S_{22} & 0 \\ 0 & 0 & S_{66} \end{bmatrix} \begin{Bmatrix} \sigma_{11} \\ \sigma_{22} \\ \tau_{12} \end{Bmatrix}$$

Each term within the reduced compliance matrix is calculated using Equations 4.17:

$$\begin{Bmatrix} \varepsilon_{11} \\ \varepsilon_{22} \\ \gamma_{12} \end{Bmatrix} = \begin{bmatrix} \dfrac{1}{E_{11}} & \dfrac{-v_{12}}{E_{11}} & 0 \\ \dfrac{-v_{12}}{E_{11}} & \dfrac{1}{E_{22}} & 0 \\ 0 & 0 & \dfrac{1}{G_{12}} \end{bmatrix} \begin{Bmatrix} \sigma_{11} \\ \sigma_{22} \\ \tau_{12} \end{Bmatrix}$$

From Table 3.1:

$$E_{11} = 170\,\text{GPa} \quad E_{22} = 10\,\text{GPa}$$

$$v_{12} = 0.30 \quad G_{12} = 13\,\text{GPa}$$

Using these values:

$$\begin{Bmatrix} \varepsilon_{11} \\ \varepsilon_{22} \\ \gamma_{12} \end{Bmatrix} = \begin{bmatrix} 5.88 \times 10^{-12} & -1.76 \times 10^{-12} & 0 \\ -1.76 \times 10^{-12} & 100 \times 10^{-12} & 0 \\ 0 & 0 & 76.9 \times 10^{-12} \end{bmatrix} \left(\dfrac{1}{\text{Pa}}\right) \begin{Bmatrix} 200 \times 10^6 \\ -30 \times 10^6 \\ -50 \times 10^6 \end{Bmatrix} (\text{Pa})$$

Completing the matrix multiplication indicated, we obtain:

$$\begin{Bmatrix} \varepsilon_{11} \\ \varepsilon_{22} \\ \gamma_{12} \end{Bmatrix} = \begin{Bmatrix} 1230\,\mu\text{m/m} \\ -3350\,\mu\text{m/m} \\ -3850\,\mu\text{rad} \end{Bmatrix}$$

Example Problem 5.2

Determine the strains induced in a unidirectional graphite–epoxy composite subjected to

    a. The in-plane stresses shown in Figure 5.1
    b. A decrease in temperature $\Delta T = -155°C$
    c. An increase in moisture content $\Delta M = 0.5\%$

Use material properties listed in Table 3.1.

SOLUTION

This problem involves three different mechanisms that contribute to the total strain induced in the laminate: the applied stresses, the temperature change, and the change in moisture content. The total strains induced by all three mechanisms can be calculated using either Equation 5.2a or Equation 5.12a:

$$\begin{Bmatrix} \varepsilon_{11} \\ \varepsilon_{22} \\ \gamma_{12} \end{Bmatrix} = \begin{bmatrix} S_{11} & S_{12} & 0 \\ S_{12} & S_{22} & 0 \\ 0 & 0 & S_{66} \end{bmatrix} \begin{Bmatrix} \sigma_{11} \\ \sigma_{22} \\ \tau_{12} \end{Bmatrix} + \Delta T \begin{Bmatrix} \alpha_{11} \\ \alpha_{22} \\ 0 \end{Bmatrix} + \Delta M \begin{Bmatrix} \beta_{11} \\ \beta_{22} \\ 0 \end{Bmatrix}$$

Numerical values for the stresses and reduced compliance matrix are given in Example 5.1, and the linear thermal and moisture expansion coefficients for graphite–epoxy are (from Table 3.1):

$$\alpha_{11} = -0.9\,\mu m/m/°C$$

$$\alpha_{22} = 27\,\mu m/m/°C$$

$$\beta_{11} = 150\,\mu m/m/\%M$$

$$\beta_{22} = 4800\,\mu m/m/\%M$$

Hence, Equation 5.2a or Equation 5.12a becomes

$$\begin{Bmatrix} \varepsilon_{11} \\ \varepsilon_{22} \\ \gamma_{12} \end{Bmatrix} = \begin{bmatrix} 5.88 \times 10^{-12} & -1.76 \times 10^{-12} & 0 \\ -1.76 \times 10^{-12} & 100 \times 10^{-12} & 0 \\ 0 & 0 & 76.9 \times 10^{-12} \end{bmatrix} \begin{Bmatrix} 200 \times 10^{6} \\ -30 \times 10^{6} \\ -50 \times 10^{6} \end{Bmatrix}$$

$$+ (-155) \begin{Bmatrix} -0.9 \times 10^{-6} \\ 27 \times 10^{-6} \\ 0 \end{Bmatrix} + (0.5) \begin{Bmatrix} 150 \times 10^{-6} \\ 4800 \times 10^{-6} \\ 0 \end{Bmatrix}$$

Completing the matrix multiplication indicated, we obtain:

$$\begin{Bmatrix} \varepsilon_{11} \\ \varepsilon_{22} \\ \gamma_{12} \end{Bmatrix} = \begin{Bmatrix} 1230\,\mu m/m \\ -3350\,\mu m/m \\ -3850\,\mu rad \end{Bmatrix} + \begin{Bmatrix} 140\,\mu m/m \\ -4185\,\mu m/m \\ 0 \end{Bmatrix} + \begin{Bmatrix} 75\,\mu m/m \\ 2400\,\mu m/m \\ 0 \end{Bmatrix} = \begin{Bmatrix} 1445\,\mu m/m \\ -5135\,\mu m/m \\ -3850\,\mu rad \end{Bmatrix}$$

An implicit assumption in this problem is that the composite is free to expand or contract, as dictated by changes in temperature and/or moisture content. Consequently neither ΔT nor ΔM affect the state of stress, but rather affect only the state of strain. Conversely, if the composite is *not* free to expand or contract, then a change in temperature and/or moisture content *does* contribute to the state of stress, as illustrated in Example Problem 5.3.

Example Problem 5.3

A thin unidirectional graphite–epoxy composite laminate is firmly mounted within an infinitely rigid square frame, as shown in Figure 5.2. The coefficients of thermal and moisture expansion of the rigid frame equal zero.

The composite is initially stress-free. Subsequently, however, the composite/frame assembly is subjected to a decrease in temperature $\Delta T = -155°C$ and an increase in moisture content $\Delta M = 0.5\%$. Determine the stresses induced in the composite by this change in temperature and moisture content. Use material properties listed in Table 3.1 and ignore the possibility that the thin composite will buckle if compressive stresses occur.

SOLUTION

According to the problem statement, the frame is "infinitely rigid" and is made of a material whose thermal and moisture expansion coefficients

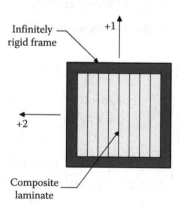

FIGURE 5.2
Unidirectional composite laminate mounted within an infinitely rigid frame.

equal zero. Consequently, the frame will retain its original shape, regardless of the temperature change, moisture change, or stresses imposed on the frame by the composite laminate. Furthermore, the composite is "firmly mounted" within the frame. Together, these stipulations imply that the *composite does not change shape*, even though changes in temperature and moisture content have occurred. Consequently, *the total strains experienced by the composite equal zero*:

$$\begin{Bmatrix} \varepsilon_{11} \\ \varepsilon_{22} \\ \gamma_{12} \end{Bmatrix} = \begin{Bmatrix} 0 \\ 0 \\ 0 \end{Bmatrix}$$

The stresses induced can be calculated using Equation 5.3:

$$\begin{Bmatrix} \sigma_{11} \\ \sigma_{22} \\ \tau_{12} \end{Bmatrix} = \begin{bmatrix} Q_{11} & Q_{12} & 0 \\ Q_{12} & Q_{22} & 0 \\ 0 & 0 & Q_{66} \end{bmatrix} \begin{Bmatrix} \varepsilon_{11} - \Delta T\alpha_{11} - \Delta M\beta_{11} \\ \varepsilon_{22} - \Delta T\alpha_{22} - \Delta M\beta_{22} \\ \gamma_{12} \end{Bmatrix}$$

The terms within the reduced stiffness matrix are calculated in accordance with Equations 5.11:

$$Q_{11} = \frac{E_{11}^2}{E_{11} - v_{12}^2 E_{22}} = \frac{(170\,\text{GPa})^2}{170\,\text{GPa} - (0.30)^2(10\,\text{GPa})} = 170.9\,\text{GPa}$$

$$Q_{12} = Q_{21} = \frac{v_{12}E_{11}E_{22}}{E_{11} - v_{12}^2 E_{22}} = \frac{(0.30)(170\,\text{GPa})(10\,\text{GPa})}{170\,\text{GPa} - (0.30)^2(10\,\text{GPa})} = 3.016\,\text{GPa}$$

$$Q_{22} = \frac{E_{11}E_{22}}{E_{11} - v_{12}^2 E_{22}} = \frac{(170\,\text{GPa})(10\,\text{GPa})}{170\,\text{GPa} - (0.30)^2(10\,\text{GPa})} = 10.05\,\text{GPa}$$

$$Q_{66} = G_{12} = 13\,\text{GPa}$$

Hence, in this case Equation 5.3 becomes

$$\begin{Bmatrix} \sigma_{11} \\ \sigma_{22} \\ \tau_{12} \end{Bmatrix} = \begin{bmatrix} 170.9 \times 10^9 & 3.016 \times 10^9 & 0 \\ 3.016 \times 10^9 & 10.05 \times 10^9 & 0 \\ 0 & 0 & 13.0 \times 10^9 \end{bmatrix} \begin{Bmatrix} 0 - (-155)(-0.9 \times 10^{-6}) - (0.5)(150 \times 10^{-6}) \\ 0 - (-155)(27 \times 10^{-6}) - (0.5)(4800 \times 10^{-6}) \\ 0 \end{Bmatrix}$$

$$\begin{Bmatrix} \sigma_{11} \\ \sigma_{22} \\ \tau_{12} \end{Bmatrix} = \begin{Bmatrix} -31.36\,\text{MPa} \\ 17.29\,\text{MPa} \\ 0 \end{Bmatrix}$$

The reader may initially view this example to be unrealistic. After all, the frame has been assumed to be "infinitely rigid." In reality, an infinitely rigid material (i.e., one for which $E \to \infty$) does not exist. Further, it is assumed that the thermal expansion coefficient for the frame is zero, which is also not valid for most real materials (the assumption that the moisture expansion coefficient is zero is true for metals and metal alloys).

Despite these unrealistic assumptions, this example problem illustrates a common occurrence in composite laminates. Specifically, most modern composites are cured at an elevated temperature (e.g., many graphite–epoxy systems are cured at 175°C ≈ 350°F). The composite can normally be considered to be stress- and strain-free *at the cure temperature*. After cure is complete, the composite is typically cooled to room temperatures, say 20°C ≈ 70°F, which corresponds to a temperature decrease of $\Delta T = -155°C \approx -280°F$. Also, the moisture content immediately after cure can usually be assumed to equal 0%, but will slowly increase following exposure to normal humidity levels over subsequent days or weeks. Although adsorption of moisture rarely causes a significant gain in weight (most composites adsorb a maximum of 1–3% moisture by weight, even if totally immersed in water), even this slight gain in moisture content can nevertheless cause significant strains to develop. If a ΔT and/or ΔM occur, and if the unidirectional composite is not free to expand or contract as dictated by these changes, then thermal- and/or moisture-induced stresses will develop.

Example Problem 5.4

A finite-element analysis of the U-shaped graphite–epoxy composite specimen shown in Figure 5.3 is performed using the commercial software package ANSYS. In-plane dimensions are shown in the figure.

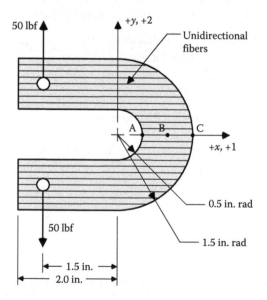

FIGURE 5.3
A 0° unidirectional composite U-shaped specimen; thickness = 0.160 in.

The specimen is produced using 32 plies and the thickness of each ply is 0.005 in., so total specimen thickness is 0.160 in. Fibers are aligned with the x-axis, so the principal material coordinate system (i.e., the 1–2 axes) is aligned with the x–y coordinate system. Points A, B, and C are defined on the specimen centerline (along $y = 0$), at $x = 0.5$, 1.0, and 1.5 in., respectively.

A transverse load $P = 50$ lbf is applied. Predicted in-plane stress and strain contours in the vicinity of the base of the U are shown in Figures 5.4 and 5.5, respectively. Also, the stresses and strains predicted along line ABC are plotted in Figures 5.6 and 5.7 (shear stress and strain are not plotted in these figures, since $\tau_{12} = \gamma_{12} = 0$ along line ABC). Confirm that the predicted stresses and strains at points A, B, and C are consistent with Equations 5.2 and 5.3. Ignore thermal and moisture effects.

SOLUTION

Numerical values for the in-plane stresses and strains at points A, B, and C, extracted from Figures 5.6 and 5.7, are as follows:

Point A: **Point B:**   **Point C:**

$$\begin{Bmatrix} \sigma_{11} \\ \sigma_{22} \\ \tau_{12} \end{Bmatrix} = \begin{Bmatrix} 0 \\ 5996\,\text{psi} \\ 0 \end{Bmatrix} \qquad \begin{Bmatrix} \sigma_{11} \\ \sigma_{22} \\ \tau_{12} \end{Bmatrix} = \begin{Bmatrix} 1313\,\text{psi} \\ 0 \\ 0 \end{Bmatrix} \qquad \begin{Bmatrix} \sigma_{11} \\ \sigma_{22} \\ \tau_{12} \end{Bmatrix} = \begin{Bmatrix} 0 \\ -3965\,\text{psi} \\ 0 \end{Bmatrix}$$

$$\begin{Bmatrix} \varepsilon_{11} \\ \varepsilon_{22} \\ \gamma_{12} \end{Bmatrix} = \begin{Bmatrix} -72\,\mu\text{in./in.} \\ 3997\,\mu\text{in./in.} \\ 0 \end{Bmatrix} \quad \begin{Bmatrix} \varepsilon_{11} \\ \varepsilon_{22} \\ \gamma_{12} \end{Bmatrix} = \begin{Bmatrix} 52\,\mu\text{in./in.} \\ -15\,\mu\text{in./in.} \\ 0 \end{Bmatrix} \quad \begin{Bmatrix} \varepsilon_{11} \\ \varepsilon_{22} \\ \gamma_{12} \end{Bmatrix} = \begin{Bmatrix} 48\,\mu\text{in./in.} \\ -2643\,\mu\text{in./in.} \\ 0 \end{Bmatrix}$$

Numerical values of the reduced compliance matrix for graphite–epoxy were calculated in Example Problem 5.1 using SI units. Recalculating the [S] matrix using English units, we have

$$[S] = \begin{bmatrix} \dfrac{1}{E_{11}} & \dfrac{-\nu_{12}}{E_{11}} & 0 \\ \dfrac{-\nu_{12}}{E_{11}} & \dfrac{1}{E_{22}} & 0 \\ 0 & 0 & \dfrac{1}{G_{12}} \end{bmatrix} = \begin{bmatrix} \dfrac{1}{25\,\text{Msi}} & \dfrac{-0.30}{25\,\text{Msi}} & 0 \\ \dfrac{-0.30}{25\,\text{Msi}} & \dfrac{1}{1.5\,\text{Msi}} & 0 \\ 0 & 0 & \dfrac{1}{1.9\,\text{Msi}} \end{bmatrix}$$

$$= \begin{bmatrix} 40.0 \times 10^{-9} & -12.0 \times 10^{-9} & 0 \\ -12.0 \times 10^{-9} & 667 \times 10^{-9} & 0 \\ 0 & 0 & 526 \times 10^{-9} \end{bmatrix} \left(\dfrac{1}{\text{psi}} \right)$$

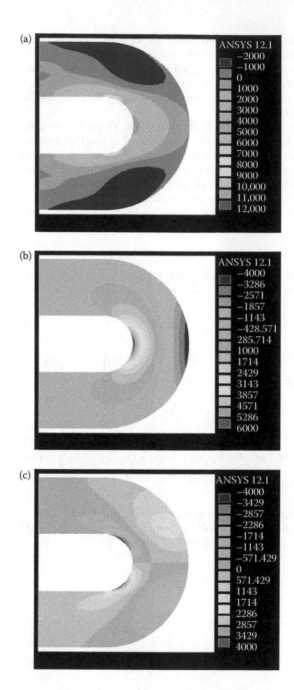

FIGURE 5.4
Predicted stress contours for the 0° unidirectional U-shaped specimen shown in Figure 5.3.
(a) σ_{11} contours; (b) σ_{22} contours; (c) τ_{12} contours.

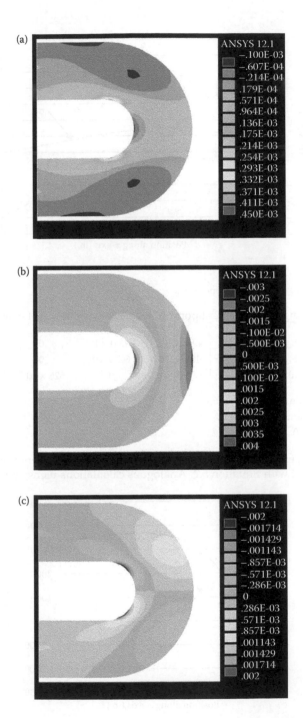

FIGURE 5.5
Predicted strain contours for the 0° unidirectional U-shaped specimen shown in Figure 5.3. (a) ε_{11} contours; (b) ε_{22} contours; (c) γ_{12} contours.

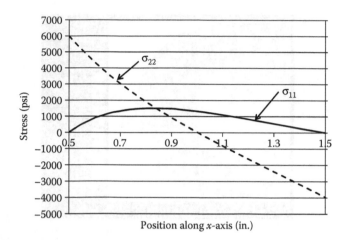

FIGURE 5.6
Plot of stresses σ_{11} and σ_{22} along line ABC.

Applying Equation 5.2 at point A, we have (with $\Delta T = \Delta M = 0$):

$$\begin{Bmatrix} \varepsilon_{11} \\ \varepsilon_{22} \\ \gamma_{12} \end{Bmatrix} = \begin{bmatrix} S_{11} & S_{12} & 0 \\ S_{12} & S_{22} & 0 \\ 0 & 0 & S_{66} \end{bmatrix} \begin{Bmatrix} \sigma_{11} \\ \sigma_{22} \\ \tau_{12} \end{Bmatrix} = \begin{bmatrix} 40.0 \times 10^{-9} & -12.0 \times 10^{-9} & 0 \\ -12.0 \times 10^{-9} & 667 \times 10^{-9} & 0 \\ 0 & 0 & 526 \times 10^{-9} \end{bmatrix} \left(\frac{1}{psi}\right) \begin{Bmatrix} 0 \\ 5996\,psi \\ 0 \end{Bmatrix}$$

$$\begin{Bmatrix} \varepsilon_{11} \\ \varepsilon_{22} \\ \gamma_{12} \end{Bmatrix} = \begin{Bmatrix} -72\,\mu in./in. \\ 3999\,\mu in./in. \\ 0 \end{Bmatrix}$$

Hence, except for minor round-off error the finite-element results are consistent with Equation 5.2. Analogous calculations using the stresses

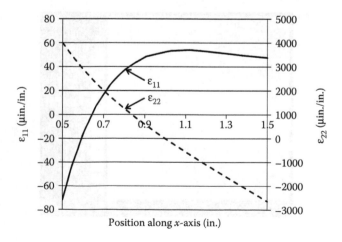

FIGURE 5.7
Plot of strains ε_{11} and ε_{22} along line ABC.

predicted at points B and C lead to comparable levels of agreement and are left as a student exercise (see Homework Problem 5.4).

Numerical values of the reduced stiffness matrix for graphite–epoxy were calculated in Example Problem 5.3 using SI units. Recalculating the [Q] matrix using English units, we have

$$Q_{11} = \frac{E_{11}^2}{E_{11} - v_{12}^2 E_{22}} = \frac{(25\,\text{Msi})^2}{25\,\text{Msi} - (0.30)^2 (1.5\,\text{Msi})} = 25.14\,\text{Msi}$$

$$Q_{12} = Q_{21} = \frac{v_{12} E_{11} E_{22}}{E_{11} - v_{12}^2 E_{22}} = \frac{(0.30)(25\,\text{Msi})(1.5\,\text{Msi})}{25\,\text{Msi} - (0.30)^2 (1.5\,\text{Msi})} = 0.4524\,\text{Msi}$$

$$Q_{22} = \frac{E_{11} E_{22}}{E_{11} - v_{12}^2 E_{22}} = \frac{(25\,\text{Msi})(1.5\,\text{Msi})}{25\,\text{Msi} - (0.30)^2 (1.5\,\text{Msi})} = 1.508\,\text{Msi}$$

$$Q_{66} = G_{12} = 1.900\,\text{Msi}$$

Applying Equation 5.3 at point A, we have (with $\Delta T = \Delta M = 0$):

$$\begin{Bmatrix} \sigma_{11} \\ \sigma_{22} \\ \tau_{12} \end{Bmatrix} = \begin{bmatrix} Q_{11} & Q_{12} & 0 \\ Q_{12} & Q_{22} & 0 \\ 0 & 0 & Q_{66} \end{bmatrix} \begin{Bmatrix} \varepsilon_{11} \\ \varepsilon_{22} \\ \gamma_{12} \end{Bmatrix} = \begin{bmatrix} 25.14 & 0.4524 & 0 \\ 0.4524 & 1.508 & 0 \\ 0 & 0 & 1.900 \end{bmatrix} (10^6\,\text{psi}) \begin{Bmatrix} -72 \\ 3997 \\ 0 \end{Bmatrix} (10^{-6}\,\text{in./in.})$$

$$\begin{Bmatrix} \sigma_{11} \\ \sigma_{22} \\ \tau_{12} \end{Bmatrix} = \begin{Bmatrix} -1.8\,\text{psi} \\ 5995\,\text{psi} \\ 0 \end{Bmatrix}$$

As before, the finite-element results are consistent with Equation 5.3, except for minor round-off error. Analogous calculations using the strains predicted at points B and C lead to comparable levels of agreement and are left as a student exercise (see Homework Problem 5.4).

The good agreement between finite-element analyses and Equations 5.2 and 5.3 is of course expected, since linear-elastic finite-element analyses are based, in part, on the orthotropic (or transversely isotropic) form of Hooke's Law presented in this section.

Note that the predicted stress and strain fields shown in Figures 5.4 and 5.5 exhibit symmetry about the x-axis, and that shear stress and strain are zero along the line of symmetry, line ABC. Symmetric stress and strain fields are expected for this case, due to the overall geometric symmetry of the specimen, the symmetric loading applied, *and* because the 1–2 principal material coordinate system is aligned with the axis of symmetry. Symmetry is not expected if any of these conditions are violated. In particular, symmetric stress and strain fields will *not* occur if the principal material coordinate system is not aligned with the axis of symmetry. This point will be illustrated in the following section, specifically in Example Problem 5.8.

5.2 Unidirectional Composites Referenced to an Arbitrary Coordinate System

A unidirectional composite referenced to two different coordinate systems is shown in Figure 5.8. In Figure 5.8a, the composite is referenced to the principal material coordinate system (i.e., the 1–2 coordinate system), and in this case either Equations 5.2 or 5.3 may be used to relate strains and stresses within the composite.

In Figure 5.8b, however, the composite is referenced to an arbitrary (nonprincipal) x–y coordinate system. This is often called an "off-axis" specimen, since the specimen is referenced to an arbitrary x–y coordinate system rather

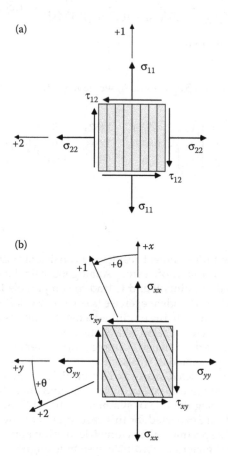

FIGURE 5.8

A unidirectional composite referenced to two different coordinate systems. (a) Unidirectional laminate referenced to the principal material coordinate system (the 1–2 axes); (b) unidirectional laminate referenced to an arbitrary x–y coordinate system (the 1–2 axes is oriented $+\theta°$ counterclockwise from the x–y axes).

than the principal 1–2 coordinate system. Suppose we wish to relate strains and stresses referenced to the x–y coordinate system. For example, suppose we know the stresses σ_{xx}, σ_{yy}, and τ_{xy}, as well as the material properties referenced to the principal material coordinate system (E_{11}, E_{22}, ν_{12}, etc.), and wish to calculate strains ε_{xx}, ε_{yy}, and γ_{xy}. In this case neither Equations 5.2 nor 5.3 can be used directly, since they require that the stresses and strains be referenced to the 1–2 coordinate system.

We can perform this calculation using a three-step process. Specifically, we can:

a. Transform the known stresses from the x–y coordinate system to the 1–2 coordinate system (using Equation 2.20, e.g.), which will give us the stress components σ_{11}, σ_{22}, and τ_{12} that correspond to the known values of σ_{xx}, σ_{yy}, and τ_{xy}.
b. Apply Equation 5.2a to obtain in-plane strains ε_{11}, ε_{22}, and γ_{12}.
c. Transform the calculated strains (ε_{11}, ε_{22}, and γ_{12}) from the 1–2 coordinate system back to the x–y coordinate system, finally obtaining the desired strains, ε_{xx}, ε_{yy}, and γ_{xy}.

This three-step process is a rigorously valid procedure. However, it will later be seen that during the analysis of a multiangle composite laminate the process of transforming strains/stresses from an arbitrary x–y coordinate system to the 1–2 coordinate system (and vice versa) must be performed for each ply in the laminate. Since this transformation is encountered so frequently it becomes cumbersome to apply the three-step process for every ply. Instead, it is very convenient to simply develop a form of reduced Hooke's law suitable for use in an *arbitrary* x–y coordinate system. In effect, we will *transform Hooke's law* from the 1–2 coordinate system to an arbitrary x–y coordinate system.

To simplify our discussion, assume for the moment that no change in environment occurs, that is, assume that $\Delta T = \Delta M = 0$. In this case Equation 5.2a becomes

$$\begin{Bmatrix} \varepsilon_{11} \\ \varepsilon_{22} \\ \gamma_{12} \end{Bmatrix} = \begin{bmatrix} S_{11} & S_{12} & 0 \\ S_{12} & S_{22} & 0 \\ 0 & 0 & S_{66} \end{bmatrix} \begin{Bmatrix} \sigma_{11} \\ \sigma_{22} \\ \tau_{12} \end{Bmatrix} \tag{5.13}$$

Equation 5.13 can be rewritten:

$$\begin{bmatrix} 1 & 0 & 0 \\ 0 & 1 & 0 \\ 0 & 0 & 2 \end{bmatrix} \begin{Bmatrix} \varepsilon_{11} \\ \varepsilon_{22} \\ \gamma_{12}/2 \end{Bmatrix} = \begin{bmatrix} S_{11} & S_{12} & 0 \\ S_{12} & S_{22} & 0 \\ 0 & 0 & S_{66} \end{bmatrix} \begin{Bmatrix} \sigma_{11} \\ \sigma_{22} \\ \tau_{12} \end{Bmatrix} \tag{5.14}$$

In Equation 5.14 we have employed the so-called "Reuter matrix" (named after the person who suggested this approach [1]) to, in effect, divide the shear strain by a factor of (1/2) within the strain array. Let:

$$[R] = \begin{bmatrix} 1 & 0 & 0 \\ 0 & 1 & 0 \\ 0 & 0 & 2 \end{bmatrix} \quad \text{(and)} \quad [S] = \begin{bmatrix} S_{11} & S_{12} & 0 \\ S_{12} & S_{22} & 0 \\ 0 & 0 & S_{66} \end{bmatrix}$$

so that Equation 5.14 can be written in the following abbreviated form:

$$[R]\begin{Bmatrix} \varepsilon_{11} \\ \varepsilon_{22} \\ \gamma_{12}/2 \end{Bmatrix} = [S]\begin{Bmatrix} \sigma_{11} \\ \sigma_{22} \\ \tau_{12} \end{Bmatrix}$$

The transformation of strains within a plane from a x–y coordinate system to another x'–y' coordinate system was discussed in Section 2.13. Adopting Equation 2.44 for our use here (i.e., using axes labels 1 and 2, rather than x' and y', respectively), we have

$$\begin{Bmatrix} \varepsilon_{11} \\ \varepsilon_{22} \\ \gamma_{12}/2 \end{Bmatrix} = \begin{bmatrix} \cos^2(\theta) & \sin^2(\theta) & 2\cos(\theta)\sin(\theta) \\ \sin^2(\theta) & \cos^2(\theta) & -2\cos(\theta)\sin(\theta) \\ -\cos(\theta)\sin(\theta) & \cos(\theta)\sin(\theta) & \cos^2(\theta)-\sin^2(\theta) \end{bmatrix} \begin{Bmatrix} \varepsilon_{xx} \\ \varepsilon_{yy} \\ \gamma_{xy}/2 \end{Bmatrix} = [T]\begin{Bmatrix} \varepsilon_{xx} \\ \varepsilon_{yy} \\ \gamma_{xy}/2 \end{Bmatrix}$$

Similarly, the transformation of stress within a plane was discussed in Section 2.8, and Equation 2.20 can be adopted as follows:

$$\begin{Bmatrix} \sigma_{11} \\ \sigma_{22} \\ \tau_{12} \end{Bmatrix} = \begin{bmatrix} \cos^2(\theta) & \sin^2(\theta) & 2\cos(\theta)\sin(\theta) \\ \sin^2(\theta) & \cos^2(\theta) & -2\cos(\theta)\sin(\theta) \\ -\cos(\theta)\sin(\theta) & \cos(\theta)\sin(\theta) & \cos^2(\theta)-\sin^2(\theta) \end{bmatrix} \begin{Bmatrix} \sigma_{xx} \\ \sigma_{yy} \\ \tau_{xy} \end{Bmatrix} = [T]\begin{Bmatrix} \sigma_{xx} \\ \sigma_{yy} \\ \tau_{xy} \end{Bmatrix}$$

As pointed out in Chapter 2, the identical transformation matrix, $[T]$, is used to relate strains and stresses in the two coordinate systems:

$$[T] = \begin{bmatrix} \cos^2(\theta) & \sin^2(\theta) & 2\cos(\theta)\sin(\theta) \\ \sin^2(\theta) & \cos^2(\theta) & -2\cos(\theta)\sin(\theta) \\ -\cos(\theta)\sin(\theta) & \cos(\theta)\sin(\theta) & \cos^2(\theta)-\sin^2(\theta) \end{bmatrix}$$

Inserting these transformation relationships into Equation 5.14, we have

$$[R][T]\begin{Bmatrix} \varepsilon_{xx} \\ \varepsilon_{yy} \\ \gamma_{xy}/2 \end{Bmatrix} = [S][T]\begin{Bmatrix} \sigma_{xx} \\ \sigma_{yy} \\ \tau_{xy} \end{Bmatrix} \qquad (5.15)$$

To simplify Equation 5.15, first multiply both sides of Equation 5.15 by the inverse of the Reuter matrix, $[R]^{-1}$, and then by the inverse of the transformation matrix, $[T]^{-1}$:

$$\begin{Bmatrix} \varepsilon_{xx} \\ \varepsilon_{yy} \\ \gamma_{xy}/2 \end{Bmatrix} = [T]^{-1}[R]^{-1}[S][T]\begin{Bmatrix} \sigma_{xx} \\ \sigma_{yy} \\ \tau_{xy} \end{Bmatrix} \qquad (5.16)$$

where

$$[R]^{-1} = \begin{bmatrix} 1 & 0 & 0 \\ 0 & 1 & 0 \\ 0 & 0 & \dfrac{1}{2} \end{bmatrix}$$

and

$$[T]^{-1} = \begin{bmatrix} \cos^2(\theta) & \sin^2(\theta) & -2\cos(\theta)\sin(\theta) \\ \sin^2(\theta) & \cos^2(\theta) & 2\cos(\theta)\sin(\theta) \\ \cos(\theta)\sin(\theta) & -\cos(\theta)\sin(\theta) & \cos^2(\theta) - \sin^2(\theta) \end{bmatrix} \qquad (5.17)$$

We next extract the factor of $(1/2)$ from the shear strain within the strain array using the $[R]^{-1}$ matrix:

$$[R]^{-1}\begin{Bmatrix} \varepsilon_{xx} \\ \varepsilon_{yy} \\ \gamma_{xy} \end{Bmatrix} = [T]^{-1}[R]^{-1}[S][T]\begin{Bmatrix} \sigma_{xx} \\ \sigma_{yy} \\ \tau_{xy} \end{Bmatrix} \qquad (5.18)$$

Multiplying both sides of Equation 5.18 by the $[R]$ matrix, we arrive at our final result:

$$\begin{Bmatrix} \varepsilon_{xx} \\ \varepsilon_{yy} \\ \gamma_{xy} \end{Bmatrix} = [R][T]^{-1}[R]^{-1}[S][T]\begin{Bmatrix} \sigma_{xx} \\ \sigma_{yy} \\ \tau_{xy} \end{Bmatrix} \qquad (5.19)$$

Equation 5.19 represents an "off-axis" version of Hooke's Law. That is, it relates the strains induced in the arbitrary x–y coordinate system (ε_{xx}, ε_{yy}, and γ_{xy}) to the stresses in the same x–y coordinate system (σ_{xx}, σ_{yy}, and τ_{xy}), via material properties referenced to the principal 1–2 coordinate system (represented by the [S] matrix) and the fiber angle θ (represented by the transformation matrix, [T]). Completing the matrix algebra indicated, we obtain:

$$\begin{Bmatrix} \varepsilon_{xx} \\ \varepsilon_{yy} \\ \gamma_{xy} \end{Bmatrix} = \begin{bmatrix} \bar{S} \end{bmatrix} \begin{Bmatrix} \sigma_{xx} \\ \sigma_{yy} \\ \tau_{xy} \end{Bmatrix} \tag{5.20}$$

where

$$[\bar{S}] = [R][T]^{-1}[R]^{-1}[S][T]$$

Equation 5.20 is known as *transformed, reduced Hooke's law*. It is called "transformed" because Equation 5.2 has been transformed from the 1–2 coordinate system to the x–y coordinate system, and "reduced" because we have reduced the allowable state of stress from 3-dimensions to 2-dimensions (i.e., Equation 5.20 is only valid for a plane stress state). Matrix $[\bar{S}]$ is called the *transformed reduced compliance matrix*.[*] In expanded form, Equation 5.20 is written as

$$\begin{Bmatrix} \varepsilon_{xx} \\ \varepsilon_{yy} \\ \gamma_{xy} \end{Bmatrix} = \begin{bmatrix} \bar{S}_{11} & \bar{S}_{12} & \bar{S}_{16} \\ \bar{S}_{21} & \bar{S}_{22} & \bar{S}_{26} \\ \bar{S}_{61} & \bar{S}_{62} & \bar{S}_{66} \end{bmatrix} \begin{Bmatrix} \sigma_{xx} \\ \sigma_{yy} \\ \tau_{xy} \end{Bmatrix} \tag{5.21}$$

where

$$\bar{S}_{11} = S_{11}\cos^4\theta + (2S_{12} + S_{66})\cos^2\theta\sin^2\theta + S_{22}\sin^4\theta$$

$$\bar{S}_{12} = \bar{S}_{21} = S_{12}(\cos^4\theta + \sin^4\theta) + (S_{11} + S_{22} - S_{66})\cos^2\theta\sin^2\theta$$

$$\bar{S}_{16} = \bar{S}_{61} = (2S_{11} - 2S_{12} - S_{66})\cos^3\theta\sin\theta - (2S_{22} - 2S_{12} - S_{66})\cos\theta\sin^3\theta$$

$$\bar{S}_{22} = S_{11}\sin^4\theta + (2S_{12} + S_{66})\cos^2\theta\sin^2\theta + S_{22}\cos^4\theta \tag{5.22}$$

$$\bar{S}_{26} = \bar{S}_{62} = (2S_{11} - 2S_{12} - S_{66})\cos\theta\sin^3\theta - (2S_{22} - 2S_{12} - S_{66})\cos^3\theta\sin\theta$$

$$\bar{S}_{66} = 2(2S_{11} + 2S_{22} - 4S_{12} - S_{66})\cos^2\theta\sin^2\theta + S_{66}(\cos^4\theta + \sin^4\theta)$$

---

[*] In common parlance the $[\bar{S}]$ matrix is often called the "S-bar" matrix.

Three important observations should be made regarding the transformed reduced compliance matrix. First, we have retained the original subscripts used in our earlier discussion of three-dimensional states of stress and strain. For example, the term that appears in the (3,3) position of the transformed reduced compliance matrix is labeled $\bar{S}_{66}$. Second, the $[\bar{S}]$ matrix is symmetric. Therefore, $\bar{S}_{21} = \bar{S}_{12}$, $\bar{S}_{61} = \bar{S}_{16}$, and $\bar{S}_{62} = \bar{S}_{26}$, as indicated in Equations 5.22. Third, the $[\bar{S}]$ matrix is fully populated. That is, (in general) none of the terms within the $[\bar{S}]$ matrix equal zero, in contrast to the $[S]$ matrix where four off-diagonal terms equal zero (see Equation 5.2a). This simply reveals the anisotropic nature of unidirectional composites. Following the convention adopted earlier, a unidirectional composite laminate referenced to the principal 1–2 coordinate system is referred to as an orthotropic (or transversely isotropic) material, whereas if the same material is referenced to an arbitrary (nonprincipal) x–y coordinate system it is called an anisotropic material. Since neither $\bar{S}_{16}$ nor $\bar{S}_{26}$ equal zero for an anisotropic composite, a coupling exists between shear stress and normal strains. For example, a shear stress τ_{xy} will cause normal strains ε_{xx} and ε_{yy} to occur, as indicated by Equation 5.21. This coupling does not occur for orthotropic or transversely isotropic composites (or for that matter, for isotropic materials).

Let us now include thermal and moisture strains in the transformed, reduced form of Hooke's law. In the 1–2 coordinate system in-plane thermal strains are given by

$$
\begin{Bmatrix} \varepsilon_{11}^T \\ \varepsilon_{22}^T \\ \gamma_{12}^T \end{Bmatrix} = \Delta T \begin{Bmatrix} \alpha_{11} \\ \alpha_{22} \\ 0 \end{Bmatrix}
\tag{5.23}
$$

As has been previously discussed, in the 1–2 coordinate system a uniform change in temperature does not produce a shear strain, that is, $\alpha_{12} = 0$. Now, thermally induced strains can be transformed from one coordinate system to another in exactly the same way as mechanically induced strains are transformed. That is, we can relate thermal strains in the 1–2 coordinate system to the x–y coordinate system using the transformation matrix:

$$
\begin{Bmatrix} \varepsilon_{11}^T \\ \varepsilon_{22}^T \\ \gamma_{12}^T/2 \end{Bmatrix} = \begin{bmatrix} \cos^2(\theta) & \sin^2(\theta) & 2\cos(\theta)\sin(\theta) \\ \sin^2(\theta) & \cos^2(\theta) & -2\cos(\theta)\sin(\theta) \\ -\cos(\theta)\sin(\theta) & \cos(\theta)\sin(\theta) & \cos^2(\theta) - \sin^2(\theta) \end{bmatrix} \begin{Bmatrix} \varepsilon_{xx}^T \\ \varepsilon_{yy}^T \\ \gamma_{xy}^T/2 \end{Bmatrix}
$$

Inverting this expression, we have:

$$\left\{\begin{array}{c} \varepsilon_{xx}^T \\ \varepsilon_{yy}^T \\ \gamma_{xy}^T/2 \end{array}\right\} = [T]^{-1} \left\{\begin{array}{c} \varepsilon_{11}^T \\ \varepsilon_{22}^T \\ 0 \end{array}\right\}$$

$$= \begin{bmatrix} \cos^2(\theta) & \sin^2(\theta) & -2\cos(\theta)\sin(\theta) \\ \sin^2(\theta) & \cos^2(\theta) & 2\cos(\theta)\sin(\theta) \\ \cos(\theta)\sin(\theta) & -\cos(\theta)\sin(\theta) & \cos^2(\theta)-\sin^2(\theta) \end{bmatrix} \left\{\begin{array}{c} \varepsilon_{11}^T \\ \varepsilon_{22}^T \\ \gamma_{12}^T/2 \end{array}\right\}$$

Substituting Equation 5.23 in this result, completing the matrix multiplication indicated, and simplifying the resulting expressions, we obtain:

$$\left\{\begin{array}{c} \varepsilon_{xx}^T \\ \varepsilon_{yy}^T \\ \gamma_{xy}^T \end{array}\right\} = \Delta T \left\{\begin{array}{c} \alpha_{xx} \\ \alpha_{yy} \\ \alpha_{xy} \end{array}\right\} \tag{5.24}$$

where

$$\alpha_{xx} = \alpha_{11}\cos^2(\theta) + \alpha_{22}\sin^2(\theta)$$

$$\alpha_{yy} = \alpha_{11}\sin^2(\theta) + \alpha_{22}\cos^2(\theta) \tag{5.25}$$

$$\alpha_{xy} = 2\cos(\theta)\sin(\theta)(\alpha_{11} - \alpha_{22})$$

In Section 5.6 we will define the properties of a unidirectional composite laminate when referenced to an arbitrary x–y coordinate system. As further discussed there, Equations 5.25 define the effective coefficients of thermal expansion for an anisotropic composite laminate. Note that for anisotropic composites a coupling exists between a uniform change in temperature and shear strain. That is, a change in temperature causes shear strain γ_{xy}^T as well as normal strains ε_{xx}^T and ε_{yy}^T.

In the 1–2 coordinate system the in-plane strains caused by a uniform change in moisture content are given by

$$\left\{\begin{array}{c} \varepsilon_{11}^M \\ \varepsilon_{22}^M \\ \gamma_{12}^M \end{array}\right\} = \Delta M \left\{\begin{array}{c} \beta_{11} \\ \beta_{22} \\ 0 \end{array}\right\} \tag{5.26}$$

As was the case for thermal strains, we wish to express strains induced by ΔM in an arbitrary x–y coordinate system. Using the identical procedure as before, moisture strains in the x–y coordinate system are given by

$$
\begin{Bmatrix} \varepsilon_{xx}^M \\ \varepsilon_{yy}^M \\ \gamma_{xy}^M \end{Bmatrix} = \Delta M \begin{Bmatrix} \beta_{xx} \\ \beta_{yy} \\ \beta_{xy} \end{Bmatrix}
\tag{5.27}
$$

where

$$
\begin{aligned}
\beta_{xx} &= \beta_{11}\cos^2(\theta) + \beta_{22}\sin^2(\theta) \\
\beta_{yy} &= \beta_{11}\sin^2(\theta) + \beta_{22}\cos^2(\theta) \\
\beta_{xy} &= 2\cos(\theta)\sin(\theta)(\beta_{11} - \beta_{22})
\end{aligned}
\tag{5.28}
$$

In Section 5.6 we will define the effective properties of a unidirectional composite laminate when referenced to an arbitrary x–y coordinate system. As further discussed there, Equations 5.28 define the effective coefficients of moisture expansion for an anisotropic composite laminate.

We can now calculate the strains induced by the combined effects of stress, a uniform change in temperature, and a uniform change in moisture content. Specifically, adding Equations 5.21, 5.24, and 5.27 together, we obtain:

$$
\begin{Bmatrix} \varepsilon_{xx} \\ \varepsilon_{yy} \\ \gamma_{xy} \end{Bmatrix} = \begin{bmatrix} \bar{S}_{11} & \bar{S}_{12} & \bar{S}_{16} \\ \bar{S}_{12} & \bar{S}_{22} & \bar{S}_{26} \\ \bar{S}_{16} & \bar{S}_{26} & \bar{S}_{66} \end{bmatrix} \begin{Bmatrix} \sigma_{xx} \\ \sigma_{yy} \\ \tau_{xy} \end{Bmatrix} + \Delta T \begin{Bmatrix} \alpha_{xx} \\ \alpha_{yy} \\ \alpha_{xy} \end{Bmatrix} + \Delta M \begin{Bmatrix} \beta_{xx} \\ \beta_{yy} \\ \beta_{xy} \end{Bmatrix}
\tag{5.29}
$$

Equation 5.29 allows calculation of the in-plane strains induced by any combination of in-plane stresses, a uniform change in temperature, and a uniform change in moisture content. If, instead, we have *measured* the total strains induced by a known ΔT, ΔM, and unknown in-plane stresses, we can calculate the stresses that caused these strains by inverting Equation 5.29 to obtain:

$$
\begin{Bmatrix} \sigma_{xx} \\ \sigma_{yy} \\ \tau_{xy} \end{Bmatrix} = \begin{bmatrix} \bar{Q}_{11} & \bar{Q}_{12} & \bar{Q}_{16} \\ \bar{Q}_{12} & \bar{Q}_{22} & \bar{Q}_{26} \\ \bar{Q}_{16} & \bar{Q}_{26} & \bar{Q}_{66} \end{bmatrix} \begin{Bmatrix} \varepsilon_{xx} - \Delta T\alpha_{xx} - \Delta M\beta_{xx} \\ \varepsilon_{yy} - \Delta T\alpha_{yy} - \Delta M\beta_{yy} \\ \gamma_{xy} - \Delta T\alpha_{xy} - \Delta M\beta_{xy} \end{Bmatrix}
\tag{5.30}
$$

where

$$
\bar{Q}_{11} = Q_{11}\cos^4\theta + 2(Q_{12} + 2Q_{66})\cos^2\theta\sin^2\theta + Q_{22}\sin^4\theta
$$

$$\overline{Q}_{12} = \overline{Q}_{21} = Q_{12}(\cos^4\theta + \sin^4\theta) + (Q_{11} + Q_{22} - 4Q_{66})\cos^2\theta\sin^2\theta$$

$$\overline{Q}_{16} = \overline{Q}_{61} = (Q_{11} - Q_{12} - 2Q_{66})\cos^3\theta\sin\theta - (Q_{22} - Q_{12} - 2Q_{66})\cos\theta\sin^3\theta$$

$$\overline{Q}_{22} = Q_{11}\sin^4\theta + 2(Q_{12} + 2Q_{66})\cos^2\theta\sin^2\theta + Q_{22}\cos^4\theta$$

$$\overline{Q}_{26} = \overline{Q}_{62} = (Q_{11} - Q_{12} - 2Q_{66})\cos\theta\sin^3\theta - (Q_{22} - Q_{12} - 2Q_{66})\cos^3\theta\sin\theta$$

$$\overline{Q}_{66} = (Q_{11} + Q_{22} - 2Q_{12} - 2Q_{66})\cos^2\theta\sin^2\theta + Q_{66}(\cos^4\theta + \sin^4\theta)$$

$$(5.31)$$

The $[\overline{Q}]$ matrix is called the *transformed, reduced stiffness matrix.** This name again reminds us that we have *transformed* Hooke's law from the 1–2 coordinate system to an arbitrary $x–y$ coordinate system, and that we have *reduced* our analysis to the 2-D plane stress case. Note that the $[\overline{Q}]$ matrix is (in general) fully populated. This reflects the anisotropic nature of unidirectional composites when referenced to a nonprincipal material coordinate system. Also, the $[\overline{Q}]$ matrix is symmetric, so that $\overline{Q}_{21} = \overline{Q}_{12}$, $\overline{Q}_{61} = \overline{Q}_{16}$, and $\overline{Q}_{62} = \overline{Q}_{26}$, as indicated in Equations 5.31.

The reader should note that the functional form of the equations that define elements of the transformed reduced stiffness matrix, that is, Equations 5.31, is not identical to the functional form of the equations defining elements of the transformed reduced compliance matrix, Equation 5.22. That is, Equations 5.31 cannot be transformed into Equations 5.22 by a simple substitution of S_{11} for Q_{11}, S_{12} for Q_{12}, S_{22} for Q_{22}, and so on. This difference in functional form is due to the fact that we have defined both the stiffness and compliance matrices in terms of engineering shear strains, rather than tensoral shear strains. The use of Equations 5.29 and 5.30 to solve simple problems involving off-axis composite laminates is demonstrated in Example Problems 5.5 through 5.8.

Example Problem 5.5

Determine the strains induced in the off-axis graphite–epoxy composite subjected to the in-plane stresses shown in Figure 5.9. Assume $\Delta T = \Delta M = 0$, and use material properties listed in Table 3.1.

SOLUTION

This problem is analogous to Example Problem 5.1, except that we are now considering the behavior of an off-axis composite. The magnitude of each stress component is indicated in Figure 5.9. Based on the sign conventions reviewed in Section 2.5, the algebraic sign of each stress component is

$$\sigma_{xx} = 200\,\text{MPa} \quad \sigma_{yy} = -30\,\text{MPa} \quad \tau_{xy} = -50\,\text{MPa}$$

---

* In common parlance the $[\overline{Q}]$ matrix is often called the "Q-bar" matrix.

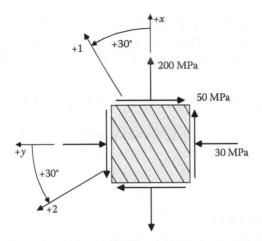

FIGURE 5.9
A 30° off-axis composite subjected to in-plane stresses (magnitudes of in-plane stresses shown).

Fiber angles are measured from the +x-axis to the +1-axis (or, equivalently, from the +y-axis to the +2-axis). In accordance with the right-hand rule, the fiber angle in Figure 5.9 is algebraically positive: $\theta = +30°$.

Strains are calculated using Equation 5.29. Since $\Delta T = \Delta M = 0$, we have

$$\begin{Bmatrix} \varepsilon_{xx} \\ \varepsilon_{yy} \\ \gamma_{xy} \end{Bmatrix} = \begin{bmatrix} \bar{S}_{11} & \bar{S}_{12} & \bar{S}_{16} \\ \bar{S}_{12} & \bar{S}_{22} & \bar{S}_{26} \\ \bar{S}_{16} & \bar{S}_{26} & \bar{S}_{66} \end{bmatrix} \begin{Bmatrix} \sigma_{xx} \\ \sigma_{yy} \\ \tau_{xy} \end{Bmatrix}$$

Recall that elements of the reduced compliance matrix, the [S] matrix, were calculated as a part of Example Problem 5.1. Each term within the transformed reduced compliance matrix, the [$\bar{S}$] matrix, can therefore be calculated via a straightforward application of Equations 5.22. Calculation of $\bar{S}_{11}$, for example, proceeds as follows:

$$\bar{S}_{11} = S_{11}\cos^4\theta + (2S_{12} + S_{66})\cos^2\theta\sin^2\theta + S_{22}\sin^4\theta$$

$$\bar{S}_{11} = (5.88 \times 10^{-12})\cos^4(30°)$$

$$+ \left\{2(-1.76 \times 10^{-12}) + 76.9 \times 10^{-12}\right\}\cos^2(30°)\sin^2(30°)$$

$$+ (100.0 \times 10^{-12})\sin^4(30°)$$

$$\bar{S}_{11} = 23.32 \times 10^{-12}(\text{Pa}^{-1})$$

Calculating the remaining elements of the [$\bar{S}$] matrix in similar fashion and applying Equation 5.29, we find:

$$
\begin{Bmatrix} \varepsilon_{xx} \\ \varepsilon_{yy} \\ \gamma_{xy} \end{Bmatrix} = \begin{bmatrix} 23.32 \times 10^{-12} & 4.327 \times 10^{-12} & -33.72 \times 10^{-12} \\ 4.327 \times 10^{-12} & 70.38 \times 10^{-12} & -47.79 \times 10^{-12} \\ -33.72 \times 10^{-12} & -47.79 \times 10^{-12} & 101.3 \times 10^{-12} \end{bmatrix} \left(\frac{1}{\mathrm{Pa}} \right) \begin{Bmatrix} 200 \times 10^{6} \\ -30 \times 10^{6} \\ -50 \times 10^{6} \end{Bmatrix} (\mathrm{Pa})
$$

Completing the matrix multiplication indicated, we obtain:

$$
\begin{Bmatrix} \varepsilon_{xx} \\ \varepsilon_{yy} \\ \gamma_{xy} \end{Bmatrix} = \begin{Bmatrix} 6220\,\mu\mathrm{m/m} \\ 1144\,\mu\mathrm{m/m} \\ -10375\,\mu\mathrm{rad} \end{Bmatrix}
$$

Example Problem 5.6

Determine the strains induced in the off-axis graphite–epoxy composite subjected to (a) the in-plane stresses shown in Figure 5.9, (b) a decrease in temperature $\Delta T = -155°C$, and (c) an increase in moisture content $\Delta M = 0.5\%$. Use material properties listed in Table 3.1.

SOLUTION

This problem involves three different mechanisms that contribute to the total strain induced in the laminate: the applied stresses, the temperature change, and the change in moisture content. The total strains induced by all three mechanisms can be calculated using Equation 5.29:

$$
\begin{Bmatrix} \varepsilon_{xx} \\ \varepsilon_{yy} \\ \gamma_{xy} \end{Bmatrix} = \begin{bmatrix} \bar{S}_{11} & \bar{S}_{12} & \bar{S}_{16} \\ \bar{S}_{12} & \bar{S}_{22} & \bar{S}_{26} \\ \bar{S}_{16} & \bar{S}_{26} & \bar{S}_{66} \end{bmatrix} \begin{Bmatrix} \sigma_{xx} \\ \sigma_{yy} \\ \tau_{xy} \end{Bmatrix} + \Delta T \begin{Bmatrix} \alpha_{xx} \\ \alpha_{yy} \\ \alpha_{xy} \end{Bmatrix} + \Delta M \begin{Bmatrix} \beta_{xx} \\ \beta_{yy} \\ \beta_{xy} \end{Bmatrix}
$$

Numerical values for the stresses and transformed reduced compliance matrix are given in Example 5.5. The linear thermal and moisture expansion coefficients for graphite–epoxy, referenced to the x–y coordinate system, are calculated using Equations 5.25 and 5.28, respectively.

The thermal expansion coefficients are

$\alpha_{xx} = \alpha_{11}\cos^2(\theta) + \alpha_{22}\sin^2(\theta) = (-0.9 \times 10^{-6})\cos^2(30°) + (27 \times 10^{-6})\sin^2(30°)$

$\alpha_{xx} = 6.1\,\mu\mathrm{m/m/°C}$

$\alpha_{yy} = \alpha_{11}\sin^2(\theta) + \alpha_{22}\cos^2(\theta) = (-0.9 \times 10^{-6})\sin^2(30°) + (27 \times 10^{-6})\cos^2(30°)$

$\alpha_{yy} = 20.0\,\mu\mathrm{m/m/°C}$

$\alpha_{xy} = 2\cos(\theta)\sin(\theta)(\alpha_{11} - \alpha_{22}) = 2\cos(30°)\sin(30°)(-0.9 \times 10^{-6} - 27 \times 10^{-6})$

$\alpha_{xy} = -24.2\,\mu\mathrm{rad/°C}$

The moisture expansion coefficients are

$$\beta_{xx} = \beta_{11}\cos^2(\theta) + \beta_{22}\sin^2(\theta) = (150 \times 10^{-6})\cos^2(30°) + (4800 \times 10^{-6})\sin^2(30°)$$

$$\beta_{xx} = 1313\,\mu m/m/\%M$$

$$\beta_{yy} = \beta_{11}\sin^2(\theta) + \beta_{22}\cos^2(\theta) = (150 \times 10^{-6})\sin^2(30°) + (4800 \times 10^{-6})\cos^2(30°)$$

$$\beta_{yy} = 3638\,\mu m/m/\%M$$

$$\beta_{xy} = 2\cos(\theta)\sin(\theta)(\beta_{11} - \beta_{22}) = 2\cos(30°)\sin(30°)(150 \times 10^{-6} - 4800 \times 10^{-6})$$

$$\beta_{xy} = -4027\,\mu rad/\%M$$

Hence, Equation 5.29 becomes

$$\begin{Bmatrix} \varepsilon_{xx} \\ \varepsilon_{yy} \\ \gamma_{xy} \end{Bmatrix} = \begin{bmatrix} 23.32 \times 10^{-12} & 4.327 \times 10^{-12} & -33.72 \times 10^{-12} \\ 4.327 \times 10^{-12} & 70.38 \times 10^{-12} & -47.79 \times 10^{-12} \\ -33.72 \times 10^{-12} & -47.79 \times 10^{-12} & 101.3 \times 10^{-12} \end{bmatrix} \begin{Bmatrix} 200 \times 10^6 \\ -30 \times 10^6 \\ -50 \times 10^6 \end{Bmatrix}$$

$$+ (-155)\begin{Bmatrix} 6.1 \times 10^{-6} \\ 20.0 \times 10^{-6} \\ -24.2 \times 10^{-6} \end{Bmatrix} + (0.5)\begin{Bmatrix} 1313 \times 10^{-6} \\ 3638 \times 10^{-6} \\ -4027 \times 10^{-6} \end{Bmatrix}$$

Completing the matrix multiplication indicated, we obtain:

$$\begin{Bmatrix} \varepsilon_{xx} \\ \varepsilon_{yy} \\ \gamma_{xy} \end{Bmatrix} = \begin{Bmatrix} 6220\,\mu m/m \\ 1144\,\mu m/m \\ -10375\,\mu rad \end{Bmatrix} + \begin{Bmatrix} -946\,\mu m/m \\ -3100\,\mu m/m \\ 3751\,\mu rad \end{Bmatrix} + \begin{Bmatrix} 657\,\mu m/m \\ 1819\,\mu m/m \\ -2014\,\mu rad \end{Bmatrix} = \begin{Bmatrix} 5931\,\mu m/m \\ -137\,\mu m/m \\ -8638\,\mu rad \end{Bmatrix}$$

An implicit assumption in this problem is that the composite is free to expand or contract, as dictated by changes in temperature and/or moisture content. Consequently neither ΔT nor ΔM affect the state of stress, but rather affect only the state of strain. Conversely, if the composite is *not* free to expand or contract, then a change in temperature and/or moisture content *does* contribute to the state of stress, as illustrated in Example Problem 5.7.

Example Problem 5.7

A thin off-axis graphite–epoxy composite laminate is firmly mounted within an infinitely rigid square frame, as shown in Figure 5.10. The coefficients of thermal and moisture expansion of the rigid frame equal zero.

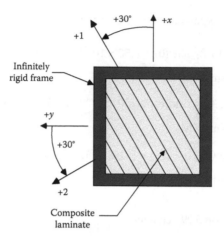

FIGURE 5.10
Off-axis composite laminate mounted within an infinitely rigid frame.

The composite is initially stress-free. Subsequently, however, the composite/frame assembly is subjected to a decrease in temperature $\Delta T = -155°C$ and an increase in moisture content $\Delta M = 0.5\%$. Determine the stresses induced in the composite by this change in temperature and moisture content. Ignore the possibility that the thin composite will buckle if compressive stresses occur.

SOLUTION

As discussed in Example Problem 5.3, the situation described in this problem is very often encountered in composite materials, despite the unrealistic assumptions regarding an "infinitely rigid" frame with zero thermal expansion coefficients. Since the problem states that the frame is "infinitely rigid" and is made of a material whose thermal and moisture expansion coefficients equal zero, the frame will retain its original shape. Since the off-axis composite is "firmly mounted" within the frame the *composite does not change shape*, even though changes in temperature and moisture content have occurred. Consequently, *the total strains experienced by the composite equal zero*:

$$\begin{Bmatrix} \varepsilon_{xx} \\ \varepsilon_{yy} \\ \gamma_{xy} \end{Bmatrix} = \begin{Bmatrix} 0 \\ 0 \\ 0 \end{Bmatrix}$$

The stresses induced can be calculated using Equation 5.30:

$$\begin{Bmatrix} \sigma_{xx} \\ \sigma_{yy} \\ \tau_{xy} \end{Bmatrix} = \begin{bmatrix} \overline{Q}_{11} & \overline{Q}_{12} & \overline{Q}_{16} \\ \overline{Q}_{12} & \overline{Q}_{22} & \overline{Q}_{26} \\ \overline{Q}_{16} & \overline{Q}_{26} & \overline{Q}_{66} \end{bmatrix} \begin{Bmatrix} \varepsilon_{xx} - \Delta T \alpha_{xx} - \Delta M \beta_{xx} \\ \varepsilon_{yy} - \Delta T \alpha_{yy} - \Delta M \beta_{yy} \\ \gamma_{xy} - \Delta T \alpha_{xy} - \Delta M \beta_{xy} \end{Bmatrix}$$

Recall that elements of the reduced stiffness matrix, the $[Q]$ matrix, were calculated as a part of Example Problem 5.3. Each term within the transformed reduced stiffness matrix, the $[\overline{Q}]$ matrix, can therefore be calculated via a straightforward application of Equations 5.31. Calculation of $\overline{Q}_{11}$, for example, proceeds as follows:

$$\overline{Q}_{11} = Q_{11}\cos^4\theta + 2(Q_{12} + 2Q_{66})\cos^2\theta\sin^2\theta + Q_{22}\sin^4\theta$$

$$\overline{Q}_{11} = \left(170.9 \times 10^9\right)\cos^4(30°) + 2\left\{3.016 \times 10^9 + 2\left(13.0 \times 10^9\right)\right\}\cos^2(30°)\sin^2(30°)$$

$$+ \left(10.05 \times 10^9\right)\sin^4(30°)$$

$$\overline{Q}_{11} = 107.6 \times 10^9\,\text{Pa}$$

Calculating the remaining elements of the $[\overline{Q}]$ matrix in a similar fashion, and using the thermal and moisture expansion coefficients referenced to the x–y coordinate system (calculated as a part of Example Problem 5.6) we find

$$\begin{Bmatrix} \sigma_{xx} \\ \sigma_{yy} \\ \tau_{xy} \end{Bmatrix} = \begin{bmatrix} 107.6 \times 10^9 & 26.1 \times 10^9 & 48.1 \times 10^9 \\ 26.1 \times 10^9 & 27.2 \times 10^9 & 21.5 \times 10^9 \\ 48.1 \times 10^9 & 21.5 \times 10^9 & 36.0 \times 10^9 \end{bmatrix}$$

$$\times \begin{Bmatrix} 0 - (-155)(6.1 \times 10^{-6}) - (0.5)(1313 \times 10^{-6}) \\ 0 - (-155)(20.0 \times 10^{-6}) - (0.5)(3638 \times 10^{-6}) \\ 0 - (-155)(-24.2 \times 10^{-6}) - (0.5)(-4027 \times 10^{-6}) \end{Bmatrix}$$

$$\begin{Bmatrix} \sigma_{xx} \\ \sigma_{yy} \\ \tau_{xy} \end{Bmatrix} = \begin{Bmatrix} 19.0\,\text{MPa} \\ -5.04\,\text{MPa} \\ 21.1\,\text{MPa} \end{Bmatrix}$$

Example Problem 5.8

A finite-element analysis of the U-shaped graphite–epoxy composite specimen shown in Figure 5.11 is performed using the commercial software package ANSYS. In-plane dimensions are shown in the figure. The specimen is produced using 32 plies and the thickness of each ply is 0.005 in, so total specimen thickness is 0.160 in. Fibers are inclined at $\theta = 45°$ with respect to the x-axis. Hence, the principal material coordinate system (i.e., the 1–2 axes) is *not* aligned with the overall specimen line of geometric symmetry and loading. Points A, B, and C are defined on the specimen centerline (along $y = 0$), at $x = 0.5$, 1.0, and 1.5 in., respectively.

A transverse load $P = 50$ lbf is applied. Predicted in-plane stress and strain contours in the vicinity of the base of the U are shown in Figures 5.12 and 5.13, respectively. Also, the stresses and strains predicted along line ABC are plotted in Figures 5.14 and 5.15. Confirm that the predicted

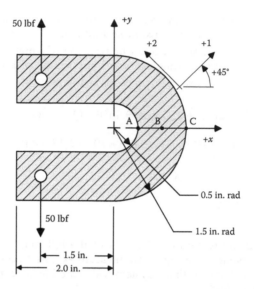

FIGURE 5.11
A 45° off-axis unidirectional composite U-shaped specimen; thickness = 0.160 in.

stresses and strains at points A, B, and C are consistent with Equations 5.29 and 5.30. Ignore thermal and moisture effects.

SOLUTION

Numerical values for the in-plane stresses and strains at points A, B, and C, extracted from Figures 5.14 and 5.15, are as follows:

| **Point A:** | **Point B:** | **Point C:** |

$$\begin{Bmatrix} \sigma_{xx} \\ \sigma_{yy} \\ \tau_{xy} \end{Bmatrix} = \begin{Bmatrix} 0 \\ 7506\,\text{psi} \\ 0 \end{Bmatrix} \qquad \begin{Bmatrix} \sigma_{xx} \\ \sigma_{yy} \\ \tau_{xy} \end{Bmatrix} = \begin{Bmatrix} 1031\,\text{psi} \\ -1149\,\text{psi} \\ -415\,\text{psi} \end{Bmatrix} \qquad \begin{Bmatrix} \sigma_{xx} \\ \sigma_{yy} \\ \tau_{xy} \end{Bmatrix} = \begin{Bmatrix} 0 \\ -1984\,\text{psi} \\ 0 \end{Bmatrix}$$

$$\begin{Bmatrix} \varepsilon_{xx} \\ \varepsilon_{yy} \\ \gamma_{xy} \end{Bmatrix} = \begin{Bmatrix} 291\,\mu\text{in./in.} \\ 2267\,\mu\text{in./in.} \\ -2348\,\mu\text{rad} \end{Bmatrix} \qquad \begin{Bmatrix} \varepsilon_{xx} \\ \varepsilon_{yy} \\ \gamma_{xy} \end{Bmatrix} = \begin{Bmatrix} 397\,\mu\text{in./in.} \\ -177\,\mu\text{in./in.} \\ -266\,\mu\text{rad} \end{Bmatrix} \qquad \begin{Bmatrix} \varepsilon_{xx} \\ \varepsilon_{yy} \\ \gamma_{xy} \end{Bmatrix} = \begin{Bmatrix} -76\,\mu\text{in./in.} \\ -598\,\mu\text{in./in.} \\ 618\,\mu\text{rad} \end{Bmatrix}$$

Calculating each term within the $[\bar{S}]$ matrix using Equations 5.22, with $\theta = 45°$ and based on the properties listed for graphite–epoxy in Table 3.1, the transformed reduced compliance matrix is found to be

$$\begin{bmatrix} \bar{S}_{11} & \bar{S}_{12} & \bar{S}_{16} \\ \bar{S}_{12} & \bar{S}_{22} & \bar{S}_{26} \\ \bar{S}_{16} & \bar{S}_{26} & \bar{S}_{66} \end{bmatrix} = \begin{bmatrix} 302.2 & 39.09 & -313.3 \\ 39.09 & 302.2 & -313.3 \\ -313.3 & -313.3 & 730.7 \end{bmatrix} \left(\frac{10^{-9}}{\text{psi}} \right)$$

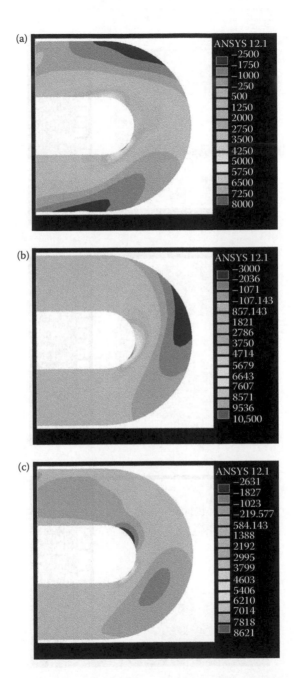

FIGURE 5.12
Predicted stress contours for the 45° off-axis unidirectional U-shaped specimen shown in Figure 5.8. (a) σ_{xx} contours; (b) σ_{yy} contours; (c) τ_{xy} contours.

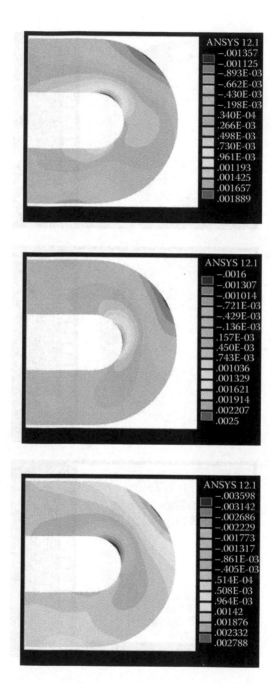

FIGURE 5.13
Predicted strain contours for the 45° off-axis unidirectional U-shaped specimen shown in Figure 5.8. (a) ε_{xx} contours; (b) ε_{yy} contours; (c) γ_{xy} contours.

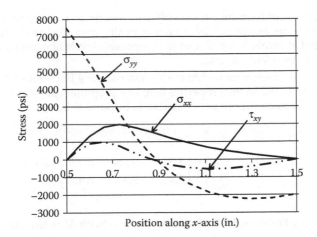

FIGURE 5.14
Plot of stresses σ_{xx}, σ_{yy}, and τ_{xy} along line ABC.

Applying Equation 5.29 at point A, we have (with $\Delta T = \Delta M = 0$):

$$\begin{Bmatrix} \varepsilon_{xx} \\ \varepsilon_{yy} \\ \gamma_{xy} \end{Bmatrix} = \begin{bmatrix} \bar{S}_{11} & \bar{S}_{12} & \bar{S}_{16} \\ \bar{S}_{12} & \bar{S}_{22} & \bar{S}_{26} \\ \bar{S}_{16} & \bar{S}_{26} & \bar{S}_{66} \end{bmatrix} \begin{Bmatrix} \sigma_{xx} \\ \sigma_{yy} \\ \tau_{xy} \end{Bmatrix} = \begin{bmatrix} 302.2 & 39.09 & -313.3 \\ 39.09 & 302.2 & -313.3 \\ -313.3 & -313.3 & 730.7 \end{bmatrix} \left(\frac{10^{-9}}{psi} \right) \begin{Bmatrix} 0 \\ 7506 \\ 0 \end{Bmatrix} psi$$

$$\begin{Bmatrix} \varepsilon_{11} \\ \varepsilon_{22} \\ \gamma_{12} \end{Bmatrix} = \begin{Bmatrix} 294 \ \mu in./in. \\ 2268 \ \mu in./in. \\ -2352 \ \mu rad \end{Bmatrix}$$

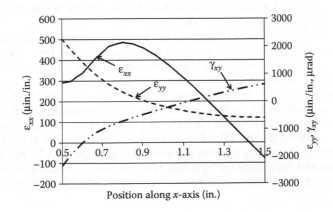

FIGURE 5.15
Plot of strains ε_{xx}, ε_{yy}, and γ_{xy} along line ABC.

Hence, except for minor round-off error the finite-element results are consistent with Equation 5.29. Analogous calculations using the stresses predicted at points B and C lead to comparable levels of agreement and are left as a student exercise.

Calculating each term within the $[\overline{Q}]$ matrix using Equations 5.31, with $\theta = 45°$ and based on the properties listed for graphite–epoxy in Table 3.1, the transformed reduced stiffness matrix is found to be

$$
\begin{bmatrix}
\overline{Q}_{11} & \overline{Q}_{12} & \overline{Q}_{16} \\
\overline{Q}_{12} & \overline{Q}_{22} & \overline{Q}_{26} \\
\overline{Q}_{16} & \overline{Q}_{26} & \overline{Q}_{66}
\end{bmatrix}
=
\begin{bmatrix}
8.787 & 4.987 & 5.907 \\
4.987 & 8.787 & 5.907 \\
5.907 & 5.907 & 6.435
\end{bmatrix} (10^6 \text{psi})
$$

Applying Equation 5.30 at point A, we have (with $\Delta T = \Delta M = 0$):

$$
\begin{Bmatrix}
\sigma_{xx} \\
\sigma_{yy} \\
\tau_{xy}
\end{Bmatrix}
=
\begin{bmatrix}
\overline{Q}_{11} & \overline{Q}_{12} & \overline{Q}_{16} \\
\overline{Q}_{12} & \overline{Q}_{22} & \overline{Q}_{26} \\
\overline{Q}_{16} & \overline{Q}_{26} & \overline{Q}_{66}
\end{bmatrix}
\begin{Bmatrix}
\varepsilon_{xx} \\
\varepsilon_{yy} \\
\gamma_{xy}
\end{Bmatrix}
$$

$$
=
\begin{bmatrix}
8.787 & 4.987 & 5.907 \\
4.987 & 8.787 & 5.907 \\
5.907 & 5.907 & 6.435
\end{bmatrix} (10^6 \text{ psi})
\begin{Bmatrix}
291 \\
2267 \\
-2348
\end{Bmatrix} (10^{-6} \text{ in./in.})
$$

$$
\begin{Bmatrix}
\sigma_{11} \\
\sigma_{22} \\
\tau_{12}
\end{Bmatrix}
=
\begin{Bmatrix}
-7.1\,\text{psi} \\
7502\,\text{psi} \\
0.73\,\text{psi}
\end{Bmatrix}
$$

As before, the finite-element results are consistent with Equation 5.30, except for minor round-off error. Analogous calculations using the strains predicted at points B and C lead to comparable levels of agreement and are left as a student exercise.

The good agreement between finite-element analyses and Equations 5.29 and 5.30 is of course expected, since linear-elastic finite-element analyses are based, in part, on the anisotropic form of Hooke's law presented in this chapter.

Recall that a 0° unidirectional U-specimen was considered in Example Problem 5.4. In that case, the predicted stress and strain fields exhibit symmetry about the *x*-axis. In contrast, a pronounced asymmetry is predicted for the 45° off-axis specimen considered in this example problem (compare Figures 5.4 and 5.12 and Figures 5.5 and 5.13). Stresses and strains are not symmetric in the present case, because the 1–2 principal material coordinate system is not aligned with the axis of symmetry.

5.3 Calculating Transformed Properties Using Material Invariants

The stresses and strains in a unidirectional composite referenced to an arbitrary x–y coordinate system may be related using either Equation 5.29 or 5.30. Equation 5.29 involves use of the transformed reduced compliance matrix, $[\bar{S}]$, and individual elements within the $[\bar{S}]$ matrix are calculated in accordance with Equation 5.22. Alternatively, Equation 5.30 involves use of the transformed reduced stiffness matrix, $[\bar{Q}]$, and individual elements of the $[\bar{Q}]$ matrix are calculated in accordance with Equations 5.31.

Now, both Equations 5.22 and 5.31 involve trigonometric functions raised to a power (i.e., $\sin^4\theta$, $\cos^4\theta$, $\cos\theta\sin^3\theta$, etc.). These equations can be simplified somewhat through the use of the following trigonometric identities:

$$\sin^4\theta = \frac{1}{8}(3 - 4\cos 2\theta + \cos 4\theta)$$

$$\cos^4\theta = \frac{1}{8}(3 + 4\cos 2\theta + \cos 4\theta)$$

$$\cos\theta\sin^3\theta = \frac{1}{8}(2\sin 2\theta - \sin 4\theta) \qquad (5.32)$$

$$\cos^2\theta\sin^2\theta = \frac{1}{8}(1 - \cos 4\theta)$$

$$\cos^3\theta\sin\theta = \frac{1}{8}(2\sin 2\theta + \sin 4\theta)$$

For example, substituting these identities into Equation 5.22 and simplifying results in:

$$\bar{S}_{11} = U_1^S + U_2^S\cos 2\theta + U_3^S\cos 4\theta$$

$$\bar{S}_{12} = \bar{S}_{21} = U_4^S - U_3^S\cos 4\theta$$

$$\bar{S}_{16} = \bar{S}_{61} = U_2^S\sin 2\theta + 2U_3^S\sin 4\theta$$

$$\bar{S}_{22} = U_1^S - U_2^S\cos 2\theta + U_3^S\cos 4\theta \qquad (5.33)$$

$$\bar{S}_{26} = \bar{S}_{62} = U_2^S\sin 2\theta - 2U_3^S\sin 4\theta$$

$$\bar{S}_{66} = U_5^S - 4U_3^S\cos 4\theta$$

The terms U_i^S which appear in Equation 5.33 are called *compliance invariants*, and are defined as follows:

$$U_1^S = \frac{1}{8}(3S_{11} + 3S_{22} + 2S_{12} + S_{66})$$

$$U_2^S = \frac{1}{2}(S_{11} - S_{22})$$

$$U_3^S = \frac{1}{8}(S_{11} + S_{22} - 2S_{12} - S_{66}) \tag{5.34}$$

$$U_4^S = \frac{1}{8}(S_{11} + S_{22} + 6S_{12} - S_{66})$$

$$U_5^S = \frac{1}{2}(S_{11} + S_{22} - 2S_{12} + S_{66})$$

The superscript "S" is used to indicate that these quantities are calculated using members of the reduced compliance matrix, $[S]$. They are called compliance invariants because they define elements of the compliance matrix that are *independent of coordinate system*. In this sense compliance invariants are analogous to stress and strain invariants, which were discussed in Sections 2.6 and 2.11, respectively.

In a similar manner, substituting the trigonometric identities listed as Equation 5.32 into Equation 5.31 and simplifying results in

$$Q_{11} = U_1^Q + U_2^Q \cos 2\theta + U_3^Q \cos 4\theta$$

$$\overline{Q}_{12} = \overline{Q}_{21} = U_4^Q - U_3^Q \cos 4\theta$$

$$\overline{Q}_{16} = \overline{Q}_{61} = \frac{1}{2}U_2^Q \sin 2\theta + U_3^Q \sin 4\theta$$

$$\overline{Q}_{22} = U_1^Q - U_2^Q \cos 2\theta + U_3^Q \cos 4\theta \tag{5.35}$$

$$\overline{Q}_{26} = \overline{Q}_{62} = \frac{1}{2}U_2^Q \sin 2\theta - U_3^Q \sin 4\theta$$

$$\overline{Q}_{66} = U_5^Q - U_3^Q \cos 4\theta$$

where the stiffness invariants, U_i^Q, are defined as

$$U_3^Q = \frac{1}{8}(Q_{11} + Q_{22} + 2Q_{12} + 4Q_{66})$$

$$U_4^Q = \frac{1}{8}(Q_{11} + Q_{22} + 6Q_{12} + 4Q_{66})$$

$$U_3^Q = \frac{1}{8}(Q_{11} + Q_{22} + 2Q_{12} + 4Q_{66})$$

$$U_4^Q = \frac{1}{8}(Q_{11} + Q_{22} + 6Q_{12} + 4Q_{66}) \tag{5.36}$$

$$U_4^Q = \frac{1}{8}(Q_{11} + Q_{22} + 6Q_{12} + 4Q_{66})$$

The superscript "Q" is used to indicate that these quantities are calculated using members of the reduced stiffness matrix, $[Q]$. The reader should note that the functional forms of the stiffness invariants defined in Equation 5.36 are not identical to those of the compliance invariants defined in Equation 5.34. The difference in functional form can be traced to the use of engineering shear strain rather than tensoral shear strain.

A comparison of Equations 5.22 and 5.33 reveals that use of compliance invariants does indeed simplify calculation of elements of the $[\bar{S}]$ matrix, although mathematically the two equations are entirely equivalent. Similarly, a comparison of Equations 5.35 and 5.31 shows that use of the stiffness invariants simplifies calculation of the terms within the $[\bar{Q}]$ matrix.

Example Problem 5.9

Use material invariants (i.e., Equations 5.33 and 5.34) to calculate the transformed reduced compliance matrix for a 30° graphite–epoxy laminate. Use material properties listed in Table 3.1.

SOLUTION

From Example Problem 5.1, the reduced compliance matrix for this material system is

$$\begin{bmatrix} S_{11} & S_{12} & 0 \\ S_{12} & S_{22} & 0 \\ 0 & 0 & S_{66} \end{bmatrix} = \begin{bmatrix} 5.88 \times 10^{-12} & -1.76 \times 10^{-12} & 0 \\ -1.76 \times 10^{-12} & 100.0 \times 10^{-12} & 0 \\ 0 & 0 & 76.9 \times 10^{-12} \end{bmatrix}\left(\frac{1}{Pa}\right)$$

The compliance invariants may be calculated using these values and Equations 5.34:

$$U_1^S = \frac{1}{8}(3S_{11} + 3S_{22} + 2S_{12} + S_{66})$$

$$= \frac{1}{8}\left[3(5.88 \times 10^{-12}) + 3(100 \times 10^{-12}) + 2(-1.76^{-12}) + 76.9 \times 10^{-12}\right] = 48.9 \times 10^{-12}$$

$$U_2^S = \frac{1}{2}(S_{11} - S_{22}) = \frac{1}{2}\left[5.88 \times 10^{-12} - 100.0 \times 10^{-12}\right] = -47.1 \times 10^{-12}$$

$$U_3^S = \frac{1}{8}(S_{11} + S_{22} - 2S_{12} - S_{66})$$

$$= \frac{1}{8}\left[5.88 \times 10^{-12} + 100.0 \times 10^{-12} - 2(-1.76 \times 10^{-12}) - 76.9 \times 10^{-12}\right] = 4.06 \times 10^{-12}$$

$$U_4^S = \frac{1}{8}(S_{11} + S_{22} + 6S_{12} - S_{66})$$

$$= \frac{1}{8}\left[5.88 \times 10^{-12} + 100.0 \times 10^{-12} + 6(-1.76 \times 10^{-12}) - 76.9 \times 10^{-12}\right] = 2.296 \times 10^{-12}$$

$$U_5^S = \frac{1}{2}(S_{11} + S_{22} - 2S_{12} + S_{66})$$

$$= \frac{1}{2}\left[5.88 \times 10^{-12} + 100.0 \times 10^{-12} - 2(-1.76 \times 10^{-12}) + 76.9 \times 10^{-12}\right] = 93.17 \times 10^{-12}$$

Using the first of Equations 5.33 and setting $\theta = 30°$:

$$\bar{S}_{11} = U_1^S + U_2^S \cos 2\theta + U_3^S \cos 4\theta$$

$$= (48.9 \times 10^{-12}) + (-47.1 \times 10^{-12})\cos(60°) + (4.06 \times 10^{-12})\cos(120°)$$

$$= 23.3 \times 10^{-12} \text{ Pa}^{-1}$$

Similarly, using the second of Equations 5.33 and setting $\theta = 30°$:

$$\bar{S}_{12} = \bar{S}_{21} = U_4^S - U_3^S \cos 4\theta$$

$$= (2.296 \times 10^{-12}) - (4.06 \times 10^{-12})\cos(120°)$$

$$= 4.33 \times 10^{-12} \text{Pa}^{-1}$$

The additional terms within the $[\bar{S}]$ matrix are calculated using the rest of Equations 5.33. A summary of our results is

$$[\bar{S}] = \begin{bmatrix} 23.3 \times 10^{-12} & 4.33 \times 10^{-12} & -33.7 \times 10^{-12} \\ 4.33 \times 10^{-12} & 70.4 \times 10^{-12} & -47.8 \times 10^{-12} \\ -33.7 \times 10^{-12} & -47.8 \times 10^{-12} & 101.3 \times 10^{-12} \end{bmatrix} \left(\frac{1}{\text{Pa}}\right)$$

Note that the $[\bar{S}]$ matrix is identical to that calculated in Example Problem 5.4.

5.4 Effective Elastic Properties of a Unidirectional Composite Laminate

The definitions of common engineering material properties were reviewed in Chapter 3. In this section these concepts will be used to define the "effective" properties of a unidirectional composite laminates referenced to an arbitrary x–y coordinate system.

Effective Properties Relating Stress to Strain: Consider the unidirectional composite laminate subjected to a uniaxial stress σ_{xx}, as shown in Figure 5.16. The strains induced in this laminate can be determined using Equation 5.29. Assuming $\Delta T = \Delta M = 0$, (and hence that thermal and moisture strains are zero), and also noting that by assumption $\sigma_{yy} = \tau_{xy} = 0$, Equation 5.29 becomes for this case:

$$\begin{Bmatrix} \varepsilon_{xx} \\ \varepsilon_{yy} \\ \gamma_{xy} \end{Bmatrix} = \begin{bmatrix} \bar{S}_{11} & \bar{S}_{12} & \bar{S}_{16} \\ \bar{S}_{12} & \bar{S}_{22} & \bar{S}_{26} \\ \bar{S}_{16} & \bar{S}_{26} & \bar{S}_{66} \end{bmatrix} \begin{Bmatrix} \sigma_{xx} \\ 0 \\ 0 \end{Bmatrix}$$

In-plane strains caused by uniaxial stress σ_{xx} are, therefore, given by

$$\varepsilon_{xx} = \bar{S}_{11}\sigma_{xx} \tag{5.37a}$$

$$\varepsilon_{yy} = \bar{S}_{12}\sigma_{xx} \tag{5.37b}$$

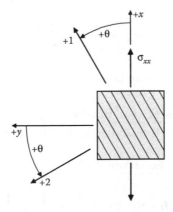

FIGURE 5.16
Unidirecitonal composite laminate subjected to uniaxial stress σ_{xx}.

$$\gamma_{xy} = \bar{S}_{16}\sigma_{xx} \tag{5.37c}$$

In Section 3.2, Young's modulus was defined as *"the normal stress σ_{xx} divided by the resulting normal strain ε_{xx}, with all other stress components equal zero."* Applying this definition to the unidirectional laminate shown in Figure 5.16, Young's modulus in the *x*-direction is given by

$$E_{xx} = \frac{\sigma_{xx}}{\varepsilon_{xx}} = \frac{\sigma_{xx}}{\bar{S}_{11}\sigma_{xx}} = \frac{1}{\bar{S}_{11}} \tag{5.38a}$$

Inserting the relation for $\bar{S}_{11}$ listed in Equation 5.22, we have

$$E_{xx} = \frac{1}{S_{11}\cos^4\theta + (2S_{12} + S_{66})\cos^2\theta\sin^2\theta + S_{22}\sin^4\theta} \tag{5.38b}$$

Since each of the compliance terms (S_{11}, S_{12}, etc.) can also be related to the more familiar engineering constants using Equations 4.17, Young's modulus can also be written:

$$E_{xx} = \frac{1}{\dfrac{\cos^4(\theta)}{E_{11}} + \left(\dfrac{1}{G_{12}} - \dfrac{2\nu_{12}}{E_{11}}\right)\cos^2(\theta)\sin^2(\theta) + \dfrac{\sin^4(\theta)}{E_{22}}} \tag{5.38c}$$

In Section 3.2, Poisson's ratio ν_{xy} was defined as *"the negative of the transverse normal strain ε_{yy} divided by the axial normal strain ε_{xx}, both of which are induced by stress σ_{xx}, with all other stresses equal zero."* Poisson's ratio for the unidirectional laminate shown in Figure 5.16 is, therefore, given by

$$\nu_{xy} = \frac{-\varepsilon_{yy}}{\varepsilon_{xx}} = \frac{-\bar{S}_{12}}{\bar{S}_{11}} \tag{5.39a}$$

Using Equations 5.22 this can be written as

$$\nu_{xy} = \frac{-(S_{12}(\cos^4\theta + \sin^4\theta) + (S_{11} + S_{22} - S_{66})\cos^2\theta\sin^2\theta}{S_{11}\cos^4\theta + (2S_{12} + S_{66})\cos^2\theta\sin^2\theta + S_{22}\sin^4\theta} \tag{5.39b}$$

Or, equivalently, using Equations 4.17:

$$\nu_{xy} = \frac{\dfrac{\nu_{12}}{E_{11}}(\cos^4\theta + \sin^4\theta) - \left(\dfrac{1}{E_{11}} + \dfrac{1}{E_{22}} - \dfrac{1}{G_{12}}\right)\cos^2\theta\sin^2\theta}{\dfrac{\cos^4(\theta)}{E_{11}} + \left(\dfrac{1}{G_{12}} - \dfrac{2\nu_{12}}{E_{11}}\right)\cos^2(\theta)\sin^2(\theta) + \dfrac{\sin^4(\theta)}{E_{22}}} \tag{5.39c}$$

In Section 3.2, the coefficient of mutual influence of the second kind $\eta_{xx,xy}$ was defined as "*the shear strain γ_{xy} divided by the normal strain ε_{xx}, both of which are induced by normal stress σ_{xx}, when all other stresses equal zero.*" For a unidirectional composite laminate $\eta_{xx,xy}$ is, therefore, given by

$$\eta_{xx,xy} = \frac{\gamma_{xy}}{\varepsilon_{xx}} = \frac{\overline{S}_{16}}{\overline{S}_{11}} \tag{5.40a}$$

which may be written as

$$\eta_{xx,xy} = \frac{(2S_{11} - 2S_{12} - S_{66})\cos^3\theta\sin\theta - (2S_{22} - 2S_{12} - S_{66})\cos\theta\sin^3\theta}{S_{11}\cos^4\theta + (2S_{12} + S_{66})\cos^2\theta\sin^2\theta + S_{22}\sin^4\theta} \tag{5.40b}$$

or, equivalently:

$$\eta_{xx,xy} = \frac{\left(\dfrac{2}{E_{11}} + \dfrac{2v_{12}}{E_{11}} - \dfrac{1}{G_{12}}\right)\cos^3(\theta)\sin(\theta) - \left(\dfrac{2}{E_{22}} + \dfrac{2v_{12}}{E_{11}} - \dfrac{1}{G_{12}}\right)\cos(\theta)\sin^2(\theta)}{\dfrac{\cos^4(\theta)}{E_{11}} + \left(\dfrac{1}{G_{12}} - \dfrac{2v_{12}}{E_{11}}\right)\cos^2(\theta)\sin^2(\theta) + \dfrac{\sin^4(\theta)}{E_{22}}} \tag{5.40c}$$

An identical procedure can be employed to define properties measured during a uniaxial test in which only σ_{yy} is applied. In this case Equation 5.29 becomes

$$\begin{Bmatrix} \varepsilon_{xx} \\ \varepsilon_{yy} \\ \gamma_{xy} \end{Bmatrix} = \begin{bmatrix} \overline{S}_{11} & \overline{S}_{12} & \overline{S}_{16} \\ \overline{S}_{12} & \overline{S}_{22} & \overline{S}_{26} \\ \overline{S}_{16} & \overline{S}_{26} & \overline{S}_{66} \end{bmatrix} \begin{Bmatrix} 0 \\ \sigma_{yy} \\ 0 \end{Bmatrix}$$

Inplane strains are

$$\varepsilon_{xx} = \overline{S}_{12}\sigma_{yy} \tag{5.41a}$$

$$\varepsilon_{yy} = \overline{S}_{22}\sigma_{yy} \tag{5.41b}$$

$$\gamma_{xy} = \overline{S}_{26}\sigma_{yy} \tag{5.41c}$$

These strains can be used to define the Young's modulus E_{yy}, Poisson's ratio v_{yx}, and coefficient of mutual influence of the second kind $\eta_{yy,xy}$:

$$E_{yy} = \frac{1}{\bar{S}_{22}}$$

(5.42a)

$$\nu_{yx} = \frac{-\bar{S}_{12}}{\bar{S}_{22}}$$

(5.42b)

$$\eta_{yy,xy} = \frac{\bar{S}_{26}}{\bar{S}_{22}}$$

(5.42c)

If desired, these relations can be expanded in terms of compliances referenced to the 1–2 coordinate system using Equations 5.22, or written in terms of measured engineering properties using Equations 4.17.

Next, consider the effective material properties measured during a pure shear test. A composite laminate subjected to pure shear stress τ_{xy} is shown in Figure 5.17. Assuming $\Delta T = \Delta M = 0$, Equation 5.29 becomes

$$\begin{Bmatrix} \varepsilon_{xx} \\ \varepsilon_{yy} \\ \gamma_{xy} \end{Bmatrix} = \begin{bmatrix} \bar{S}_{11} & \bar{S}_{12} & \bar{S}_{16} \\ \bar{S}_{12} & \bar{S}_{22} & \bar{S}_{26} \\ \bar{S}_{16} & \bar{S}_{26} & \bar{S}_{66} \end{bmatrix} \begin{Bmatrix} 0 \\ 0 \\ \tau_{xy} \end{Bmatrix}$$

Hence, the strains caused by a pure shear stress are given by

$$\varepsilon_{xx} = \bar{S}_{16}\tau_{xy}$$

(5.43a)

$$\varepsilon_{yy} = \bar{S}_{26}\tau_{xy}$$

(5.43b)

$$\gamma_{xy} = \bar{S}_{66}\tau_{xy}$$

(5.43c)

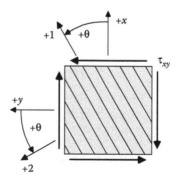

FIGURE 5.17
Unidirectional composite subjected to pure shear stress τ_{xy}.

In Section 3.2, the shear modulus was defined as *"the shear stress τ_{xy} divided by the resulting shear strain γ_{xy}, with all other stress components equal zero."* Applying this definition to the laminate shown in Figure 5.17, the shear modulus referenced to the x–y coordinate axes is given by

$$G_{xy} = \frac{\tau_{xy}}{\gamma_{xy}} = \frac{1}{\overline{S}_{66}} \tag{5.44}$$

As before, this expression can be expanded in terms of compliances referenced to the 1–2 coordinate system using Equations 5.22, or written in terms of measured engineering properties using Equations 4.17.

The coefficient of mutual influence of the first kind $\eta_{xy,xx}$ (or $\eta_{xy,yy}$) was defined as *"the normal strain ε_{xx} (or ε_{yy}) divided by the shear strain γ_{xy}, both of which are induced by shear stress τ_{xy}, when all other stresses equal zero."* For a unidirectional composite laminate the coefficient of mutual influence of the first kind $\eta_{xy,xx}$ is, therefore, given by

$$\eta_{xy,xx} = \frac{\varepsilon_{xx}}{\gamma_{xy}} = \frac{\overline{S}_{16}}{\overline{S}_{66}} \tag{5.45a}$$

While $\eta_{xy,yy}$ is given by

$$\eta_{xy,yy} = \frac{\varepsilon_{yy}}{\gamma_{xy}} = \frac{\overline{S}_{26}}{\overline{S}_{66}} \tag{5.45b}$$

Chentsov coefficients were defined in Section 3.2 as *"the shear strain γ_{xz} (or γ_{yz}) divided by the shear strain γ_{xy}, both of which are induced by shear stress τ_{xy}, with all other stresses equal zero."* For a thin composite laminate the principal material coordinate system lies within the plane of the laminate, and hence there is no coupling between a shear stress acting within the x–y plane (τ_{xy}) and out-of-plane shear strains (γ_{xz} or γ_{yz}). Consequently, Chentsov coefficients are always equal to zero for thin composite laminates.

Effective Properties Relate Temperature or Moisture Content to Strain: As discussed in Section 3.3, linear coefficients of thermal expansion are measured by determining the strains induced by a uniform change in temperature, and forming the following ratios:

$$\alpha_{xx} = \frac{\varepsilon_{xx}^{T}}{\Delta T} \quad \alpha_{yy} = \frac{\varepsilon_{yy}^{T}}{\Delta T} \quad \alpha_{xy} = \frac{\gamma_{xy}^{T}}{\Delta T} \tag{5.46}$$

The superscript *"T"* is included as a reminder that the strains involved are those caused by a change in temperature only. The strains induced in a unidirectional laminate subjected to a change in temperature can be determined using Equation 5.29. Assuming $\sigma_{xx} = \sigma_{yy} = \tau_{xy} = \Delta M = 0$, Equation 5.29 becomes

$$\begin{Bmatrix} \varepsilon_{xx} \\ \varepsilon_{yy} \\ \gamma_{xy} \end{Bmatrix} = \Delta T \begin{Bmatrix} \alpha_{xx} \\ \alpha_{yy} \\ \alpha_{xy} \end{Bmatrix} \tag{5.47}$$

Hence, the thermal expansion coefficients for a unidirectional laminate are given by Equations 5.25, repeated here for convenience:

$$\alpha_{xx} = \alpha_{11}\cos^2(\theta) + \alpha_{22}\sin^2(\theta)$$

$$\alpha_{yy} = \alpha_{11}\sin^2(\theta) + \alpha_{22}\cos^2(\theta) \tag{5.25} \text{ (repeated)}$$

$$\alpha_{xy} = 2\cos(\theta)\sin(\theta)(\alpha_{11} - \alpha_{22})$$

Similarly, the linear coefficient of moisture expansion is measured by determining the strains induced by a uniform change in moisture content, and forming the following ratios:

$$\beta_{xx} = \frac{\varepsilon_{xx}^M}{\Delta M} \quad \beta_{yy} = \frac{\varepsilon_{yy}^M}{\Delta M} \quad \beta_{xy} = \frac{\gamma_{xy}^M}{\Delta M} \tag{5.48}$$

The superscript "M" is included as a reminder that the strains involved are those caused by a change in moisture only. The strains induced in a unidirectional laminate subjected to a change in moisture content can be determined using Equation 5.29. Assuming $\sigma_{xx} = \sigma_{yy} = \tau_{xy} = \Delta T = 0$, Equation 5.29 becomes

$$\begin{Bmatrix} \varepsilon_{xx} \\ \varepsilon_{yy} \\ \gamma_{xy} \end{Bmatrix} = \Delta M \begin{Bmatrix} \beta_{xx} \\ \beta_{yy} \\ \beta_{xy} \end{Bmatrix} \tag{5.49}$$

Hence, the moisture expansion coefficients for a unidirectional laminate are given by Equations 5.28, repeated here for convenience:

$$\beta_{xx} = \beta_{11}\cos^2(\theta) + \beta_{22}\sin^2(\theta)$$

$$\beta_{yy} = \beta_{11}\sin^2(\theta) + \beta_{22}\cos^2(\theta) \tag{5.28} \text{ (repeated)}$$

$$\beta_{xy} = 2\cos(\theta)\sin(\theta)(\beta_{11} - \beta_{22})$$

Example Problem 5.10

Plot the effective properties listed below for a unidirectional $\theta°$ graphite–epoxy laminate, for all fiber angles ranging from $0° \le \theta \le 90°$:

   a. Effective Young's moduli, E_{xx} and E_{yy}
   b. Effective Poisson's ratio, v_{xy} and v_{yx}
   c. Effective shear modulus G_{xy}
   d. Coefficients of mutual influence of the first kind, $\eta_{xy,xx}$ and $\eta_{xy,yy}$
   e. Coefficients of mutual influence of the second kind, $\eta_{xx,xy}$ and $\eta_{yy,xy}$
   f. Coefficients of thermal expansion, α_{xx}, α_{yy} and α_{xy}
   g. Coefficients of moisture expansion, β_{xx}, β_{yy} and β_{xy}

Use material properties listed in Table 3.1.

SOLUTION

Plots of the effective elastic properties for unidirectional $\theta°$ graphite–epoxy laminates are presented in Figures 5.18 through 5.24.

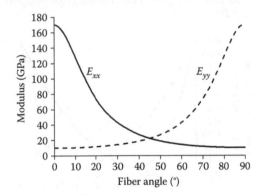

FIGURE 5.18
A plot of the effective Young's moduli E_{xx} and E_{yy} for unidirectional graphite–epoxy laminates and fiber angles ranging over $0° \leq \theta \leq 90°$.

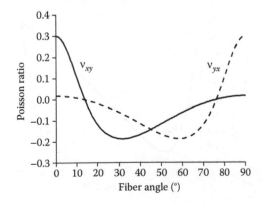

FIGURE 5.19
A plot of the effective Poisson ratios v_{xy} and v_{yx} for unidirectional graphite–epoxy laminates and fiber angles ranging over $0° \leq \theta \leq 90°$.

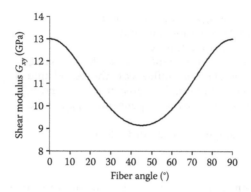

FIGURE 5.20
A plot of the effective shear modulus G_{xy} for unidirectional graphite–epoxy laminates and fiber angles ranging over $0° \leq 0 \leq 90°$.

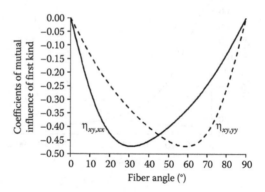

FIGURE 5.21
A plot of the effective coefficients of mutual influence of the first kind $\eta_{xy,xx}$ and $\eta_{xy,yy}$ for unidirectional graphite–epoxy laminates and fiber angles ranging over $0° \leq \theta \leq 90°$.

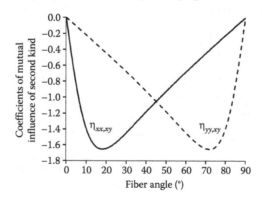

FIGURE 5.22
A plot of the effective coefficients of mutual influence of the second kind $\eta_{xx,xy}$ and $\eta_{yy,xy}$ for unidirectional graphite–epoxy laminates and fiber angles ranging over $0° \leq \theta \leq 90°$.

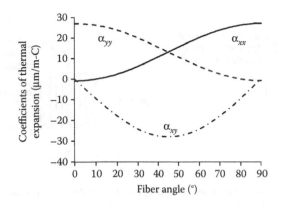

FIGURE 5.23
A plot of the effective coefficients of thermal expansion α_{xx}, α_{yy}, and α_{xy} for unidirectional graphite–epoxy laminates and fiber angles ranging over $0° \leq \theta \leq 90°$.

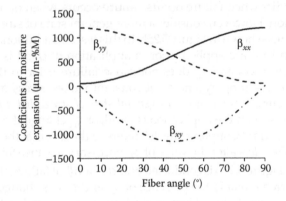

FIGURE 5.24
A plot of the effective coefficients of moisture expansion β_{xx}, β_{yy}, and β_{xy} for unidirectional graphite–epoxy laminates and fiber angles ranging over $0° \leq \theta \leq 90°$.

5.5 Failure of Unidirectional Composites Referenced to the Principal Material Coordinate System

The need for a "failure criteria" in engineering analysis and design is often misunderstood. In essence, the objective of any failure criterion is to account for potential *coupling effects* of individual stress components on the yielding and/or fracture phenomenon. This statement applies to both anisotropic *and* isotropic materials. To explain what is meant by the phrase "coupling effects between individual stress components," consider the two different tests of a unidirectional composite shown in Figure 5.25. A composite subjected to uniaxial stress σ_{11} is shown in Figure 5.25a. This is, of course, the very state of

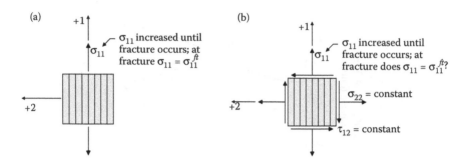

FIGURE 5.25
An illustration of what is meant by "coupling effects" of stress on the failure phenomenon.
(a) Uniaxial state of stress; (b) test in which a general state of plane stress exists.

stress used during measurement of the fundamental material strength σ_{11}^{fT} (as discussed in Section 3.5). In this case then, *we do not need to invoke any* failure criterion to predict when failure occurs: failure occurs when σ_{11} is increased to σ_{11}^{fT}, by definition. However, consider a more general state of stress, such as the state of plane stress shown in Figure 5.25b. In this test two additional components of stress (σ_{22} and τ_{12}) are applied prior to application of σ_{11}. It is assumed that the combination of σ_{22} and τ_{12} does not cause failure prior to the application of σ_{11}. While maintaining σ_{22} and τ_{12} at constant values, stress σ_{11} is increased until failure occurs. *It is for this more general state of stress that a failure criterion is required.* That is, does the application σ_{22} and/or τ_{12} change the value of σ_{11} at which failure occurs? "Coupling effects" refers to the fact that the application of σ_{22} and/or τ_{12} often *does* alter the value of σ_{11} necessary to cause failure. If the test represented by Figure 5.25b is conducted and $\sigma_{11} \neq \sigma_{11}^{ft}$ at failure, then a coupling effect has occurred—that is, the presence of σ_{22} and τ_{12} has changed the value of σ_{11} necessary to cause failure. Conversely, if the test depicted in Figure 5.25b is performed and $\sigma_{11} = \sigma_{11}^{ft}$ at failure, then no coupling has occurred.

Experimental measurements have shown that the coupling phenomenon is much more significant in some materials than in others. This is unfortunate, because it reveals that it is not possible to develop a "universal" failure criterion that can be applied to *all* materials. Further, there is no way of *predicting* a priori whether coupling effects are pronounced for a given material or not. For metals the general trend is that coupling effects are less pronounced in brittle materials (such as cast irons) than in ductile materials (such as aluminum alloys). The question as to whether this general trend holds in the case of composites is complicated by the fact that composites are (usually) brittle in the fiber direction but somewhat ductile transverse to the fiber. It is generally accepted that coupling effects do exist in composites, but despite extensive efforts no universally accepted composite failure theory has emerged.

Dozens of composite failure theories have been proposed, and reviews and direct comparisons of many of these have appeared in the literature [2,3]. A striking conclusion to be reached from these comparisons is that differing theories can give vastly different failure predictions. Further, these differences are not negligibly small—depending on stress state predictions may differ by 500% or more. *These major discrepancies serve to emphasize a very significant point*: at the present state of the art *no existing failure theory* can provide an accurate prediction for all possible states of stress. While a composite failure theory(ies) is always used during preliminary design and sizing of composite structures, the resulting failure predictions should be viewed only as a preliminary guide. During development prototype composite structures must *always* be thoroughly tested experimentally to insure that the structure can safely support the loads expected during service, particularly if failure of the structure will result in the loss of human life and/or substantial property damage.

In this section we will discuss three common composite failure criteria: the Maximum Stress Failure criterion, the Tsai–Hill Failure criterion, and the Tsai–Wu Failure criterion. As will be seen, the Maximum Stress criterion does not account for coupling effects, whereas potential stress coupling effects are accounted for in the Tsai–Hill and Tsai–Wu criterions.

5.5.1 The Maximum Stress Failure Criterion

According to this criterion a given state of stress will *not* cause failure of a unidirectional composite if *all* of the following nine inequalities are satisfied:

$$-1 * \sigma_{11}^{fC} < \sigma_{11} < \sigma_{11}^{fT}$$

(and)

$$-1 * \sigma_{22}^{fC} < \sigma_{22} < \sigma_{22}^{fT}$$

(and)

$$-1 * \sigma_{33}^{fC} < \sigma_{33} < \sigma_{33}^{fT} \tag{5.50}$$

(and)

$$|\tau_{12}| < \tau_{12}^{f}$$

(and)

$$|\tau_{13}| < \tau_{13}^{f}$$

(and)

$$|\tau_{23}| < \tau_{23}^{f}$$

According to the Maximum Stress Failure criterion, failure is predicted strictly on the basis of individual stress components. Thus, failure is assumed to be independent of any coupling effects between individual stress components.

In the case of plane stress ($\sigma_{33} = \tau_{13} = \tau_{23} = 0$), the Maximum Stress Failure criterion reduces to the following five inequalities:

$$-1 * \sigma_{11}^{fC} < \sigma_{11} < \sigma_{11}^{fT}$$

(and)

$$-1 * \sigma_{22}^{fC} < \sigma_{22} < \sigma_{22}^{fT} \tag{5.51}$$

(and)

$$|\tau_{12}| < \tau_{12}^{f}$$

The Maximum Stress Failure criterion is most commonly applied in the form of Equations 5.51, since in most cases an individual composite ply can be assumed to be in a state of plane stress.

5.5.2 The Tsai–Hill Failure Criterion

The von Mises yield criterion is widely used to predict yielding of isotropic metals and metal alloys.* In 1950 Hill proposed a modified version of the von Mises criterion for use with orthotropic metals [4]. Subsequently Tsai applied this method to predict failure of unidirectional polymeric composites [5], and the resulting theory is now known within the polymeric composites community as either the "Tsai–Hill" Failure criterion or as the "Quadratic" Failure criterion. For general 3-D states of stress, the Tsai–Hill criterion predicts that failure of an orthotropic composite will *not* occur if the following inequality is satisfied:

$$\frac{(\sigma_{11})^2}{(\sigma_{11}^{fT})^2} + \frac{(\sigma_{22})^2}{(\sigma_{22}^{fT})^2} + \frac{(\sigma_{33})^2}{(\sigma_{33}^{fT})^2} + \frac{(\tau_{23})^2}{(\tau_{23}^{f})^2} + \frac{(\tau_{13})^2}{(\tau_{13}^{f})^2} + \frac{(\tau_{12})^2}{(\tau_{12}^{f})^2}$$

$$-\sigma_{11}\sigma_{22}\left[\frac{1}{(\sigma_{11}^{fT})^2} + \frac{1}{(\sigma_{22}^{fT})^2} - \frac{1}{(\sigma_{33}^{fT})^2}\right] - \sigma_{11}\sigma_{33}\left[\frac{1}{(\sigma_{11}^{fT})^2} - \frac{1}{(\sigma_{22}^{fT})^2} + \frac{1}{(\sigma_{33}^{fT})^2}\right]$$

$$-\sigma_{22}\sigma_{33}\left[\frac{-1}{(\sigma_{11}^{fT})^2} + \frac{1}{(\sigma_{22}^{fT})^2} + \frac{1}{(\sigma_{33}^{fT})^2}\right] < 1 \tag{5.52}$$

---

* The von Mises yield criterion is mathematically equivalent to the "octahedral shear stress" and "distortional energy" yield criteria.

In the case of plane stress conditions ($\sigma_{33} = \tau_{13} = \tau_{23} = 0$), the Tsai–Hill criterion reduces to

$$\frac{(\sigma_{11})^2}{(\sigma_{11}^{fT})^2} + \frac{(\sigma_{22})^2}{(\sigma_{22}^{fT})^2} + \frac{(\tau_{12})^2}{(\tau_{12}^{f})^2} - \sigma_{11}\sigma_{22}\left[\frac{1}{(\sigma_{11}^{fT})^2} + \frac{1}{(\sigma_{22}^{fT})^2} - \frac{1}{(\sigma_{33}^{fT})^2}\right] < 1 \quad (5.53)$$

It is interesting to note that according to the Tsai–Hill failure criterion failure of orthotropic composites is sensitive to the out-of-plane strength term (σ_{33}^{fT}), even though plane stress conditions have been assumed ($\sigma_{33} = 0$).

If the composite is transversely isotropic (i.e., if $\sigma_{33}^{fT} = \sigma_{22}^{fT}$), then Equation 5.53 reduces to

$$\frac{(\sigma_{11})^2}{(\sigma_{11}^{fT})^2} + \frac{(\sigma_{22})^2}{(\sigma_{22}^{fT})^2} + \frac{(\tau_{12})^2}{(\tau_{12}^{f})^2} - \frac{\sigma_{11}\sigma_{22}}{(\sigma_{11}^{fT})^2} < 1 \quad (5.54)$$

A potential advantage of the Tsai–Hill failure criterion is that coupling effects between individual stress components are accounted for, in contrast with the Maximum Stress Failure Criteria. On the other hand, most composites exhibit significantly different failure strengths in tension and compression (as indicated in Table 3.1) and so a shortcoming of the Tsai–Hill failure criterion is that it does not directly account for these differences. That is, an implicit assumption of the Tsai–Hill criterion (as well as the original von Mises criterion) is that failure strengths in tension and compression have equal magnitudes. Hence, only tensile strengths σ_{11}^{fT}, σ_{22}^{fT}, and σ_{33}^{fT} appear in Equations 5.53 and 5.54. Differences in tensile and compressive strengths can be accounted for "artificially" in the Tsai–Hill criterion by using the appropriate compressive strength if the stress component involved is compressive. Suppose, for example, that a failure prediction is required for a transversely isotropic composite subjected to three stress components σ_{11}, σ_{22}, and τ_{12}, and also that σ_{11} is tensile but σ_{22} is compressive. In such a case the differences in tensile/compressive strengths can be accounted for by using the tensile strength in the 1-direction, σ_{11}^{fT}, but the compressive strength in the 2-direction, σ_{22}^{fC}.

While the Tsai–Hill criterion can be modified in this way to account for differences in tensile and compressive strengths, it would be ideal if a failure criterion were available that accounts for both coupling effects and differences in tensile and compressive strengths "automatically." One such criterion is the Tsai–Wu criterion, described in the next section.

5.5.3 The Tsai–Wu Failure Criterion

Tsai and Wu developed their criterion [6] by postulating that the strength of a unidirectional composite can be treated mathematically as a tensoral quantity, in much the same way as stress or strain tensors. For general 3-D states of stress, the Tsai–Wu criterion predicts that failure will *not* occur if the following inequality is satisfied:

$$X_1 \sigma_{11} + X_2 \sigma_{22} + X_3 \sigma_{33}$$

$$+ X_{11}\sigma_{11}^2 + X_{22}\sigma_{22}^2 + X_{33}\sigma_{33}^2 + X_{44}\tau_{23}^2 + X_{55}\tau_{13}^2 + X_{66}\tau_{12}^2$$

$$+ 2X_{12}\sigma_{11}\sigma_{22} + 2X_{13}\sigma_{11}\sigma_{33} + 2X_{23}\sigma_{22}\sigma_{33} < 1 \tag{5.55}$$

Most of the constants that appear in this inequality (i.e., X_1, X_2, X_3, X_{11}, etc.) can be determined based on fundamental strength measurements (i.e., σ_{11}^{fT}, σ_{11}^{fC}, σ_{22}^{fT}, σ_{22}^{fC}, etc.). First, consider a uniaxial strength measurement in which only stress σ_{11} is applied (i.e., a test in which $\sigma_{11} \neq 0$, $\sigma_{22} = \sigma_{33} = \tau_{23} = \tau_{13} = \tau_{12} = 0$). Stress σ_{11} is increased monotonically from zero until failure occurs. If σ_{11} is tensile, then at the moment of failure $\sigma_{11} = \sigma_{11}^{fT}$, and the Tsai–Wu criterion reduces to

$$X_1 \sigma_{11}^{fT} + X_{11}\left(\sigma_{11}^{fT}\right)^2 = 1$$

Conversely, if σ_{11} is compressive then at the moment of failure $\sigma_{11} = -\sigma_{11}^{fC}$ (where the measured compressive strength, σ_{11}^{fC}, is treated as an algebraically positive number), and the Tsai–Wu criterion reduces to

$$-X_1 \sigma_{11}^{fC} + X_{11}\left(-\sigma_{11}^{fC}\right)^2 = 1$$

Solving for X_1 and X_{11}, we find

$$X_1 = \frac{1}{\sigma_{11}^{fT}} - \frac{1}{\sigma_{11}^{fC}} \quad X_{11} = \frac{1}{\sigma_{11}^{fT}\sigma_{11}^{fC}} \tag{5.56}$$

Similarly, if failure is measured during two uniaxial stress tests in which only σ_{22} is applied we find

$$X_2 = \frac{1}{\sigma_{22}^{fT}} - \frac{1}{\sigma_{22}^{fC}} \quad X_{22} = \frac{1}{\sigma_{22}^{fT}\sigma_{22}^{fC}} \tag{5.57}$$

Using measurements obtained during two tests in which only σ_{33} is applied:

$$X_3 = \frac{1}{\sigma_{33}^{fT}} - \frac{1}{\sigma_{33}^{fC}} \quad X_{33} = \frac{1}{\sigma_{33}^{fT}\sigma_{33}^{fC}} \tag{5.58}$$

Three additional constants are determined using measured shear strengths:

$$X_{44} = \left(\frac{1}{\tau_{23}^f}\right)^2 \quad X_{55} = \left(\frac{1}{\tau_{13}^f}\right)^2 \quad X_{66} = \left(\frac{1}{\tau_{12}^f}\right)^2 \tag{5.59}$$

Only three coefficients remain to be determined, X_{12}, X_{13}, and X_{23}. Several methods of determining these coefficients have been suggested, but thus far no one technique has gained widespread acceptance. Two methods that have been proposed will be discussed here.

Conceptually, the most straightforward approach is through the use of additional biaxial testing. For example, X_{12} can be determined by conducting a biaxial test in which $\sigma_{11} = \sigma_{22} = \sigma$ and $\sigma_{33} = \tau_{23} = \tau_{13} = \tau_{12} = 0$. The magnitude of biaxial stresses (σ) is increased until failure occurs. At the moment of failure then, the stresses applied are $\sigma_{11} = \sigma_{22} = \sigma^f$ and $\sigma_{33} = \tau_{23} = \tau_{13} = \tau_{12} = 0$. Substituting these values into the Tsai–Wu criterion and solving for X_{12} results in

$$X_{12} = \frac{1}{2(\sigma^f)^2}\left[1 - (X_1 + X_2)\sigma^f - (X_{11} + X_{22})(\sigma^f)^2\right] \tag{5.60}$$

At least conceptually, X_{13} and X_{23} can also be determined in a similar manner. Two additional biaxial tests to failure would be required, where in one test $\sigma_{11} = \sigma_{33} = \sigma$ and in the second test $\sigma_{22} = \sigma_{33} = \sigma$. These data would then allow calculation of X_{13} and X_{23}, respectively. In practice, however, these tests would be very difficult to perform. Since composites are usually quite thin it is especially difficult to apply well-defined out-of-plane stress components (i.e., σ_{33}, τ_{13}, or τ_{23}). Hence, in most instances determining X_{13} or X_{23} in this manner is impractical.

A second approach is to assume that X_{12}, X_{13}, and X_{23} can be calculated as follows:

$$X_{12} = \frac{-1}{2}\sqrt{X_{11}X_{22}} = \frac{-1}{2\sqrt{\sigma_{11}^{fT}\sigma_{11}^{fC}\sigma_{22}^{fT}\sigma_{22}^{fC}}}$$

$$X_{13} = \frac{-1}{2}\sqrt{X_{11}X_{33}} = \frac{-1}{2\sqrt{\sigma_{11}^{fT}\sigma_{11}^{fC}\sigma_{33}^{fT}\sigma_{33}^{fC}}} \tag{5.61}$$

$$X_{23} = \frac{-1}{2}\sqrt{X_{22}X_{33}} = \frac{-1}{2\sqrt{\sigma_{22}^{fT}\sigma_{22}^{fC}\sigma_{33}^{fT}\sigma_{33}^{fC}}}$$

The basis of this approach is that, if Equations 5.61 are enforced and isotropic strengths are assumed (i.e., if $\sigma_{11}^{fT} = \sigma_{11}^{fC} = \sigma_{22}^{fT} = \sigma_{22}^{fC} = \sigma_{33}^{fT} = \sigma_{33}^{fC} = \sigma^f$, and $\tau_{12}^f = \tau_{13}^f = \tau_{23}^f = \sigma^f/\sqrt{3}$) then the Tsai–Wu criterion reduces to the original von Mises criterion for isotropic materials. This approach holds some intellectual appeal, since it "makes sense" that a failure criterion proposed for use with an orthotropic material should reduce to a well-known isotropic yield criterion if isotropic strengths are assumed. It is also a convenient assumption, since X_{12}, X_{13}, and X_{23} are now calculated using fundamental strength data and hence the need to perform any additional testing is avoided. However, there is little data available to assess the validity of these assumptions and so the accuracy of failure predictions obtained using this approach is unknown.

As discussed earlier, in most practical applications composites are subjected to a state of plane stress within the 1–2 plane. In this case the Tsai–Wu criterion reduces to

$$X_1\sigma_{11} + X_2\sigma_{22} + X_{11}\sigma_{11}^2 + X_{22}\sigma_{22}^2 + X_{66}\tau_{12}^2 + 2X_{12}\sigma_{11}\sigma_{22} < 1 \qquad (5.62)$$

Hence, in the plane-stress case six constants are involved, five of which can be calculated using readily available strength data (σ_{11}^{fT}, σ_{11}^{fC}, σ_{22}^{yT}, etc.). Only one problematic coefficient remains, X_{12}. This term can be determined using an off-axis specimen (which is, in effect, a biaxial test). For example, suppose a uniaxial stress σ_{xx} is applied to a unidirectional composite specimen in which the fibers are oriented at $\theta = 45°$ with respect to the direction of loading. Under these conditions the stresses in the 1–2 coordinate system are easily calculated:

$$\begin{Bmatrix} \sigma_{11} \\ \sigma_{22} \\ \tau_{12} \end{Bmatrix} = \begin{bmatrix} \cos^2(45°) & \sin^2(45°) & 2\cos(45°)\sin(45°) \\ \sin^2(45°) & \cos^2(45°) & -2\cos(45°)\sin(45°) \\ -\cos(45°)\sin(45°) & \cos(45°)\sin(45°) & \cos^2(45°) - \sin^2(45°) \end{bmatrix} \begin{bmatrix} \sigma_{xx} \\ 0 \\ 0 \end{bmatrix}$$

or

$$\begin{Bmatrix} \sigma_{11} \\ \sigma_{22} \\ \tau_{12} \end{Bmatrix} = \begin{Bmatrix} \sigma_{xx}\cos^2(45°) \\ \sigma_{xx}\sin^2(45°) \\ -\sigma_{xx}\cos(45°)\sin(45°) \end{Bmatrix} = \begin{Bmatrix} \sigma_{xx}/2 \\ \sigma_{xx}/2 \\ -\sigma_{xx}/2 \end{Bmatrix}$$

The strength of the 45° off-axis specimen is measured by increasing stress σ_{xx} until failure occurs. Denote the stress level at which failure occurs as $\sigma_{xx} = \sigma_{xx}^f$. At failure the ply stresses are $\sigma_{11} = \sigma_{22} = -\tau_{12} = \sigma_{xx}^f/2$. Substituting these stresses into Equation 5.62 and solving for X_{12}, we find

$$X_{12} = \frac{1}{2(\sigma_{xx}^f)^2}[4 - \sigma_{xx}^f\{2(X_1 + X_2) + \sigma_{xx}^f(X_{11} + X_{22} + X_{66})\}] \qquad (5.63)$$

While this example has been based on a 45° off-axis specimen, a similar approach can be used with any θ-deg off-axis specimen.

From an analytical standpoint the Tsai–Wu failure criterion is an improvement over the other two failure criteria considered. First, unlike the Maximum

Stress Failure Criteria, the coupling effects between individual stress components are accounted for in the Tsai–Wu criterion. Second, unlike the Tsai–Hill criterion, differences in tensile and compressive strengths are automatically and naturally accounted for via the X_1, X_{11}, X_2, X_{22}, X_3, and X_{33} terms.

5.6 Failure of Unidirectional Composites Referenced to an Arbitrary Coordinate System

In this section the three failure criteria introduced in Section 5.5 will be used to predict failure of unidirectional composites subjected to a state of plane stress, where stress components σ_{xx}, σ_{yy}, and τ_{xy} are referenced to an arbitrary x–y coordinate system. There are, of course, an infinite number of different combinations of σ_{xx}, σ_{yy}, and τ_{xy} that (collectively) define a state of plane stress. For illustrative purposes two simple stress states will be considered: first a state of uniaxial stress (i.e., $\sigma_{xx} \neq 0$, $\sigma_{yy} = \tau_{xy} = 0$), and second a state of pure shear ($\tau_{xy} \neq 0$, $\sigma_{xx} = \sigma_{yy} = 0$).

Numerical results for a unidirectional graphite–epoxy composite will be used to facilitate these comparisons. The following failure strengths are taken from Table 3.1 and are typical for graphite–epoxy at room temperature:

$$\sigma_{11}^{fT} = 1500 \, \text{MPa} \quad \sigma_{22}^{fT} = 50 \, \text{MPa} \quad \tau_{12}^{f} = 90 \, \text{MPa}$$

$$\sigma_{11}^{fC} = 1200 \, \text{MPa} \quad \sigma_{22}^{fC} = 100 \, \text{MPa}$$

5.6.1 Uniaxial Stress

An off-axis composite ply subjected to a uniaxial stress σ_{xx} has been previously shown in Figure 5.16. The stresses induced in the 1–2 coordinate system by stress σ_{xx} can be determined using Equation 2.20:

$$\begin{Bmatrix} \sigma_{11} \\ \sigma_{22} \\ \tau_{12} \end{Bmatrix} = \begin{bmatrix} \cos^2(\theta) & \sin^2(\theta) & 2\cos(\theta)\sin(\theta) \\ \sin^2(\theta) & \cos^2(\theta) & -2\cos(\theta)\sin(\theta) \\ -\cos(\theta)\sin(\theta) & \cos(\theta)\sin(\theta) & \cos^2(\theta) - \sin^2(\theta) \end{bmatrix} \begin{Bmatrix} \sigma_{xx} \\ 0 \\ 0 \end{Bmatrix}$$

or equivalently:

$$\sigma_{11} = \sigma_{xx} \cos^2(\theta)$$

$$\sigma_{22} = \sigma_{xx} \sin^2(\theta) \tag{5.64}$$

$$\tau_{12} = -\sigma_{xx} \cos(\theta)\sin(\theta)$$

5.6.1.1 Maximum Stress Criterion

Substituting Equations 5.64 into Equation 5.50, we obtain:

$$-\frac{\sigma_{11}^{fC}}{\cos^2(\theta)} < \sigma_{xx} < \frac{\sigma_{11}^{fT}}{\cos^2(\theta)}$$

(and)

$$\frac{-\sigma_{22}^{fC}}{\sin^2(\theta)} < \sigma_{xx} < \frac{\sigma_{22}^{fT}}{\sin^2(\theta)} \tag{5.65}$$

(and)

$$|\sigma_{xx}| < \frac{\tau_{12}^{f}}{\cos(\theta)\sin(\theta)}$$

According to the Maximum Stress criterion failure will *not* occur if these five inequalities are satisfied. The predicted tensile and compressive failure strengths, σ_{xx}^{fT} and σ_{xx}^{fC} respectively, for a $\theta°$–off-axis graphite–epoxy laminate are, therefore, the smallest values returned by the following expressions:*
Tensile strength:

$$\sigma_{xx}^{fT} = \frac{\sigma_{11}^{fT}}{\cos^2(\theta)} = \frac{1500\,\text{MPa}}{\cos^2(\theta)} \tag{5.66a}$$

$$\sigma_{xx}^{fT} = \frac{\sigma_{22}^{fT}}{\sin^2(\theta)} = \frac{50\,\text{MPa}}{\sin^2(\theta)} \tag{5.66b}$$

$$\sigma_{xx}^{fT} = \left|\frac{\tau_{12}^{f}}{\cos(\theta)\sin(\theta)}\right| = \left|\frac{90\,\text{MPa}}{\cos(\theta)\sin(\theta)}\right| \tag{5.66c}$$

Compressive strength:

$$\sigma_{xx}^{fC} = \frac{\sigma_{11}^{fC}}{\cos^2(\theta)} = \frac{1200\,\text{MPa}}{\cos^2(\theta)} \tag{5.67a}$$

$$\sigma_{xx}^{fC} = \frac{\sigma_{22}^{fC}}{\sin^2(\theta)} = \frac{100\,\text{MPa}}{\sin^2(\theta)} \tag{5.67b}$$

---

* As before, compressive strength is treated as an algebraically positive number.

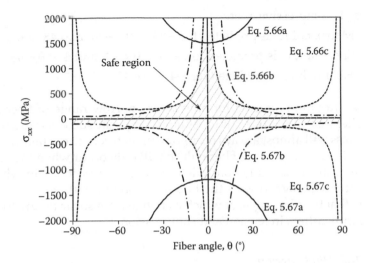

FIGURE 5.26
Failure envelope for a unidirectional graphite–epoxy laminate subjected to a uniaxial stress σ_{xx}, based on the Maximum Stress criterion.

$$\sigma_{xx}^{fC} = \left| \frac{\tau_{12}^f}{\cos(\theta)\sin(\theta)} \right| = \left| \frac{90\,\text{MPa}}{\cos(\theta)\sin(\theta)} \right| \qquad (5.67c)$$

Equations 5.66a, 5.66b, 5.66c and 5.67a, 5.67b, 5.67c were used to create the *failure envelope* for a unidirectional graphite–epoxy laminate shown in Figure 5.26. Equations 5.66 and 5.67 bound the safe region. The reader should note the following:

- The failure envelope shown in Figure 5.26 is valid for a *uniaxial state of stress only*. Specifically, Figure 5.26 is valid only if

$$\sigma_{xx} \neq 0$$

$$\sigma_{yy} = \sigma_{zz} = \tau_{xy} = \tau_{xz} = \tau_{yz} = 0$$

As will be seen later, failure envelopes for other states of stress differ substantially from Figure 5.26.
- The mode of failure depends on whether σ_{xx} is tensile or compressive, and also on the fiber angle, θ:

If σ_{xx} is tensile, then:
  - Matrix failure is predicted for: $-90° < \theta < -28°$ and $28° < \theta < 90°$
  - Shear failure is predicted for: $-28° < \theta < -3°$ and $3° < \theta < 28°$
  - Fiber failure is predicted for: $-3° < \theta < 3°$

If σ_{xx} is compressive, then:
- Matrix failure is predicted for: $-90° < \theta < -48°$ and $48° < \theta < 90°$
- Shear failure is predicted for: $-48° < \theta < -4°$ and $4° < \theta < 48°$
- Fiber failure is predicted for: $-4° < \theta < 4°$

Fiber failures are predicted for only a very narrow range of fiber angles. This implies that failure of a unidirectional composite subjected to a uniaxial state of stress will almost always occur due to matrix or shear failures, rather than fiber failure. In general, fiber failure will only occur when the composite is tested under carefully controlled laboratory conditions in which the uniaxial stress is aligned with the fiber direction to within a few degrees. Also note that the fiber angle at which a change from shear failure to matrix failure occurs differs in tension and compression.

5.6.1.2 Tsai–Hill Criterion

According to the Tsai–Hill criterion, failure of an orthotropic composite subjected to plane stress conditions is governed by Equation 5.53, whereas failure of a transversely isotropic composite is governed by Equation 5.54. For present purposes, assume the composite is transversely isotropic. Substituting Equations 5.64 into Equation 5.54, the Tsai–Hill failure criterion predicts that failure will not occur if the following inequality is satisfied:

$$\sigma_{xx} < \left\{ \frac{\cos^2(\theta)[\cos^2(\theta) - \sin^2(\theta)]}{(\sigma_{11}^{fT})^2} + \frac{\sin^4(\theta)}{(\sigma_{22}^{fT})^2} + \frac{\cos^2(\theta)\sin^2(\theta)}{(\tau_{12}^{f})^2} \right\}^{-1/2} \quad (5.68)$$

As previously noted, the Tsai–Hill criterion does not automatically account for differences in tensile and compressive strengths. A failure envelope for a unidirectional graphite–epoxy composite will be generated using tensile or compressive strengths, as appropriate. Thus, the tensile strength predicted by the Tsai–Hill criterion is

$$\sigma_{xx}^{fT} = \left\{ \frac{\cos^2(\theta)[\cos^2(\theta) - \sin^2(\theta)]}{\left(\sigma_{11}^{fT}\right)^2} + \frac{\sin^4(\theta)}{\left(\sigma_{22}^{fT}\right)^2} + \frac{\cos^2(\theta)\sin^2(\theta)}{\left(\tau_{12}^{f}\right)^2} \right\}^{-1/2}$$

Similarly, the compressive strength predicted by the Tsai–Hill criterion is

$$\sigma_{xx}^{fC} = \left\{ \frac{\cos^2(\theta)[\cos^2(\theta) - \sin^2(\theta)]}{\left(\sigma_{11}^{fC}\right)^2} + \frac{\sin^4(\theta)}{\left(\sigma_{22}^{fC}\right)^2} + \frac{\cos^2(\theta)\sin^2(\theta)}{\left(\tau_{12}^{f}\right)^2} \right\}^{-1/2}$$

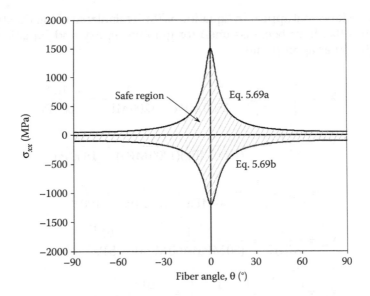

FIGURE 5.27
Failure envelope for a unidirectional graphite–epoxy laminate subjected to a uniaxial stress σ_{xx}, based on the Tsai–Hill criterion.

Substituting the strength values that have been assumed for graphite–epoxy, we have

$$\sigma_{xx}^{fT} = \left\{ \frac{\cos^2(\theta)[\cos^2(\theta) - \sin^2(\theta)]}{(1500\,\text{MPa})^2} + \frac{\sin^4(\theta)}{(50\,\text{MPa})^2} + \frac{\cos^2(\theta)\sin^2(\theta)}{(90\,\text{MPa})^2} \right\}^{-1/2} \quad (5.69a)$$

$$\sigma_{xx}^{fC} = \left\{ \frac{\cos^2(\theta)[\cos^2(\theta) - \sin^2(\theta)]}{(1200\,\text{MPa})^2} + \frac{\sin^4(\theta)}{(100\,\text{MPa})^2} + \frac{\cos^2(\theta)\sin^2(\theta)}{(90\,\text{MPa})^2} \right\}^{-1/2} \quad (5.69b)$$

A failure envelope based on Equations 5.69a, 5.69b is shown in Figure 5.27. As before, it is important to realize that this failure envelope is valid for a *uniaxial state of stress only*. Failure envelopes obtained using the Tsai–Hill criterion but for other states of stress differ substantially from Figure 5.27.

5.6.1.3 Tsai–Wu Criterion

The Tsai–Wu criterion for the case of plane stress is given by Equation 5.62. Substituting Equations 5.64 into Equation 5.62 we obtain:

$$(\sigma_{xx}^f)^2[X_{11}\cos^4(\theta) + X_{22}\sin^4(\theta) + \cos^2(\theta)\sin^2(\theta)(X_{66} + 2X_{12})]$$

$$+ \sigma_{xx}^f[X_1\cos^2(\theta) + X_2\sin^2(\theta)] - 1 = 0 \quad (5.70)$$

The constants that appear in Equation 5.70 are calculated using the strength properties that have been assumed for graphite–epoxy and Equations 5.56 through 5.59, as appropriate:

$$X_1 = \frac{1}{\sigma_{11}^{fT}} - \frac{1}{\sigma_{11}^{fC}} = \frac{1}{1500\,\text{MPa}} - \frac{1}{1200\,\text{MPa}} = \frac{-10^{-6}}{6000\,\text{Pa}}$$

$$X_{11} = \frac{1}{\sigma_{11}^{fT}\sigma_{11}^{fC}} = \frac{1}{(1500\,\text{MPa})(1200\,\text{MPa})} = \frac{10^{-15}}{1800\,\text{Pa}^2}$$

$$X_2 = \frac{1}{\sigma_{22}^{fT}} - \frac{1}{\sigma_{22}^{fC}} = \frac{1}{50\,\text{MPa}} - \frac{1}{100\,\text{MPa}} = \frac{10^{-6}}{100\,\text{Pa}}$$

$$X_{22} = \frac{1}{\sigma_{22}^{fT}\sigma_{22}^{fC}} = \frac{1}{(50\,\text{MPa})(100\,\text{MPa})} = \frac{10^{-15}}{5\,\text{Pa}^2}$$

$$X_{66} = \left(\frac{1}{\tau_{12}^{f}}\right)^2 = \left(\frac{1}{90\,\text{MPa}}\right)^2 = \frac{10^{-12}}{8100\,\text{Pa}^2}$$

As previously discussed, there is no widely accepted technique used to calculate X_{12}. For present purposes X_{12} will be calculated in accordance with Equation 5.61:

$$X_{12} = \frac{-1}{2}\sqrt{X_{11}X_{22}} = \frac{-1}{2}\sqrt{\left(\frac{10^{-15}}{1800\,\text{Pa}^2}\right)\left(\frac{10^{-15}}{5\,\text{Pa}^2}\right)} = \left(\frac{-\sqrt{10}}{600\,\text{Pa}^2}\right)(10^{-15})$$

Substituting these values into Equation 5.70, we obtain:

$$(\sigma_{xx}^{f})^2\left\{\frac{(10^{-15})\cos^4(\theta)}{1800\,\text{Pa}^2} + \frac{(10^{-15})\sin^4(\theta)}{5\,\text{Pa}^2} + \cos^2(\theta)\sin^2(\theta)\left[\frac{10^{-12}}{8100\,\text{Pa}^2} - \frac{(10^{-15})\sqrt{10}}{300\,\text{Pa}^2}\right]\right\}$$

$$+ \sigma_{xx}^{f}\left\{\frac{-10^{-6}\cos^2(\theta)}{6000\,\text{Pa}} + \frac{10^{-6}\sin^2(\theta)}{100\,\text{Pa}}\right\} - 1 = 0 \tag{5.71}$$

Equation 5.71 is a second-order polynomial in the unknown failure stress, σ_{xx}^{f}. For any given fiber angle θ, there will be two roots to this equation. The predicted tensile strength equals the algebraically positive root, whereas the predicted compressive strength equals the negative root. For example, for a fiber angle $\theta = 30°$, Equation 5.71 becomes

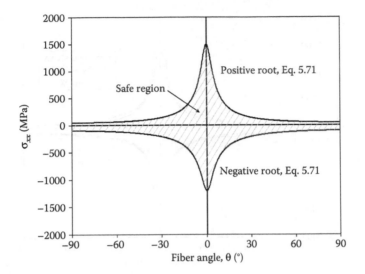

FIGURE 5.28
Failure envelope for a unidirectional graphite–epoxy laminate subjected to a uniaxial stress σ_{xx}, based on the Tsai–Wu criterion.

$$(\sigma_{xx}^f)^2 \left\{ \frac{3398 \times 10^{-20}}{\text{Pa}^2} \right\} + \sigma_{xx}^f \left\{ \frac{2375 \times 10^{-12}}{\text{Pa}} \right\} - 1 = 0$$

The two roots of this expression are found to be (140×10^6 Pa, -210×10^6 Pa). Hence, the strengths predicted by the Tsai–Wu criterion for a 30° graphite–epoxy specimen are

$$\sigma_{xx}^{fT} = 140\,\text{MPa}$$

$$\sigma_{xx}^{fC} = 210\,\text{MPa}$$

A failure envelope based on the Tsai–Wu criterion for a unidirectional graphite–epoxy laminate subjected to a uniaxial state of stress is shown in Figure 5.28. This figure is analogous to those obtained using the Maximum Stress criterion and the Tsai–Hill criterion (Figures 5.26 and 5.27, respectively). As before, it is important to realize that this failure envelope is valid for a *uniaxial state of stress only*. Failure envelopes based on the Tsai–Wu criterion but for other states of stress differ substantially from Figure 5.28.

5.6.1.4 Comparison

The failure envelopes for uniaxial stress obtained on the basis of the three failure criteria considered are compared directly in Figure 5.29, and an expanded view of the just first quadrant is presented in Figure 5.30. It is

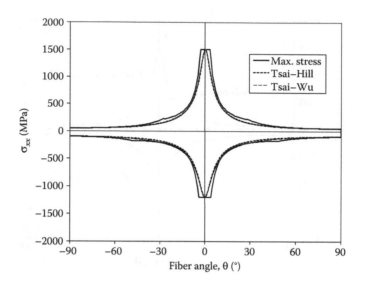

FIGURE 5.29
Comparison of the failure envelopes for a unidirectional graphite–epoxy laminate subjected to a uniaxial stress σ_{xx}, obtained using the Maximum Stress, Tsai–Hill, and Tsai–Wu failure criteria.

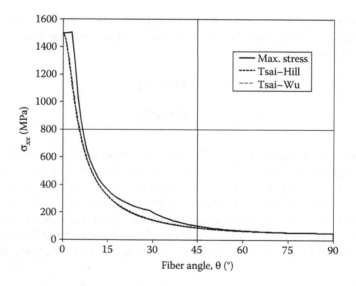

FIGURE 5.30
Comparison of the failure envelopes (first quadrant only) for a unidirectional graphite–epoxy laminate subjected to a uniaxial stress σ_{xx}, obtained using the Maximum Stress, Tsai–Hill, and Tsai–Wu failure criteria.

apparent that similar predictions are obtained on the basis of all three criteria. The most significant difference occurs at small fiber angles, where the magnitude of the failure stress predicted by the Maximum Stress criterion exceeds that predicted by either the Tsai–Hill or Tsai–Wu criterion. However, one should not conclude that the failure criterion described above *always* lead to similar predictions. In fact, depending on the state of stress considered the predicted failure envelopes may differ substantially. One stress state that exhibits this effect is the state of pure shear stress, considered in the following subsection.

5.6.2 Pure Shear Stress States

It was mentioned in Section 3.5 that the shear strength of composites is sensitive to the algebraic sign of the shear stress when referenced to a nonprincipal material coordinate system. This sensitivity is would not be expected based on previous experience with isotropic materials, since the shear strength of isotropic materials is not sensitive to algebraic sign. We are now in a position to explain this phenomenon. An off-axis composite ply subjected to a pure shear stress state is shown in Figure 5.31. The stresses induced in the 1–2 coordinate system can be determined using Equation 2.20:

$$
\begin{Bmatrix} \sigma_{11} \\ \sigma_{22} \\ \tau_{12} \end{Bmatrix} = \begin{bmatrix} \cos^2(\theta) & \sin^2(\theta) & 2\cos(\theta)\sin(\theta) \\ \sin^2(\theta) & \cos^2(\theta) & -2\cos(\theta)\sin(\theta) \\ -\cos(\theta)\sin(\theta) & \cos(\theta)\sin(\theta) & \cos^2(\theta)-\sin^2(\theta) \end{bmatrix} \begin{Bmatrix} 0 \\ 0 \\ \tau_{xy} \end{Bmatrix}
$$

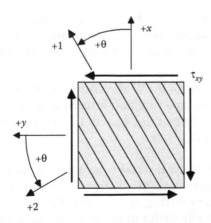

FIGURE 5.31
Unidirectional composite subjected to pure shear stress τ_{xy}.

or equivalently:

$$\sigma_{11} = 2\tau_{xy}\cos(\theta)\sin(\theta)$$

$$\sigma_{22} = -2\tau_{xy}\cos(\theta)\sin(\theta) \tag{5.72}$$

$$\tau_{12} = \tau_{xy}[\cos^2(\theta) - \sin^2(\theta)]$$

5.6.2.1 Maximum Stress Criterion

Substituting Equations 5.72 into Equations 5.51, we obtain:

$$\frac{-\sigma_{11}^{fC}}{2\cos(\theta)\sin(\theta)} < \tau_{xy} < \frac{\sigma_{11}^{fT}}{2\cos(\theta)\sin(\theta)} \tag{5.73a}$$

(and)

$$\frac{-\sigma_{22}^{fC}}{2\cos(\theta)\sin(\theta)} < \tau_{xy} < \frac{\sigma_{22}^{fT}}{2\cos(\theta)\sin(\theta)} \tag{5.73b}$$

(and)

$$|\tau_{xy}| < \frac{\tau_{12}^{f}}{\cos^2(\theta) - \sin^2(\theta)} \tag{5.73c}$$

Equations 5.73 will now be used to generate a failure envelope for a graphite–epoxy laminate subjected to pure shear. With reference to Equation 5.73, the following subtleties in these calculations should be noted:

- Over the range $0° < \theta < 90°$, both the cosine and sine functions return algebraically positive values. Consequently, for this range a positive shear stress τ_{xy} will induce a tensile value σ_{11} and a compressive value for σ_{22}.
- Over the range $-90° < \theta < 0°$ the cosine function returns a positive value whereas the sine function returns a negative value. Over this range a positive shear stress τ_{xy} will induce a compressive stress σ_{11} but tensile stress σ_{22}.

Of course, if a negative shear stress τ_{xy} is applied rather than a positive shear stress then the algebraic signs of all stress components are reversed. These subtleties are important during application of Equations 5.73, because composite strengths typically differ in tension and compression.

With these observations in mind, the following equations may be used to generate a failure envelope for a graphite–epoxy laminate subjected to pure shear, based on the strength properties previously listed:

Positive Shear Strengths:

- For $0° < \theta < 90°$:

$$\tau_{xy}^{(+f)} = \frac{\sigma_{11}^{fT}}{2\cos(\theta)\sin(\theta)} = \frac{1500\,\text{MPa}}{2\cos(\theta)\sin(\theta)} \tag{5.74a}$$

$$\tau_{xy}^{(+f)} = \frac{\sigma_{22}^{fC}}{2\cos(\theta)\sin(\theta)} = \frac{100\,\text{MPa}}{2\cos(\theta)\sin(\theta)} \tag{5.74b}$$

$$\tau_{xy}^{(+f)} = \frac{\tau_{12}^{f}}{\cos^2(\theta) - \sin^2(\theta)} = \frac{90\,\text{MPa}}{\cos^2(\theta) - \sin^2(\theta)} \tag{5.74c}$$

- For $-90° < \theta < 0°$:

$$\tau_{xy}^{(+f)} = \frac{\sigma_{11}^{fC}}{2\cos(\theta)\sin(\theta)} = \frac{1200\,\text{MPa}}{2\cos(\theta)\sin(\theta)} \tag{5.74d}$$

$$\tau_{xy}^{(+f)} = \frac{\sigma_{22}^{fT}}{2\cos(\theta)\sin(\theta)} = \frac{50\,\text{MPa}}{2\cos(\theta)\sin(\theta)} \tag{5.74e}$$

$$\tau_{xy}^{(+f)} = \frac{\tau_{12}^{f}}{\cos^2(\theta) - \sin^2(\theta)} = \frac{90\,\text{MPa}}{\cos^2(\theta) - \sin^2(\theta)} \tag{5.74f}$$

Negative Shear Strengths:

- For $0° < \theta < 90°$:

$$\tau_{xy}^{(-f)} = \frac{\sigma_{11}^{fC}}{2\cos(\theta)\sin(\theta)} = \frac{1200\,\text{MPa}}{2\cos(\theta)\sin(\theta)} \tag{5.75a}$$

$$\tau_{xy}^{(-f)} = \frac{\sigma_{22}^{fT}}{2\cos(\theta)\sin(\theta)} = \frac{50\,\text{MPa}}{2\cos(\theta)\sin(\theta)} \tag{5.75b}$$

$$\tau_{xy}^{(-f)} = \frac{\tau_{12}^{f}}{\cos^2(\theta) - \sin^2(\theta)} = \frac{90\,\text{MPa}}{\cos^2(\theta) - \sin^2(\theta)} \tag{5.75c}$$

- For $-90° < \theta < 0°$:

$$\tau_{xy}^{(-f)} = \frac{\sigma_{11}^{fT}}{2\cos(\theta)\sin(\theta)} = \frac{1500\,\text{MPa}}{2\cos(\theta)\sin(\theta)} \tag{5.75d}$$

$$\tau_{xy}^{(-f)} = \frac{\sigma_{22}^{fC}}{2\cos(\theta)\sin(\theta)} = \frac{100\,\text{MPa}}{2\cos(\theta)\sin(\theta)} \tag{5.75e}$$

$$\tau_{xy}^{(-f)} = \frac{\tau_{12}^{f}}{\cos^2(\theta) - \sin^2(\theta)} = \frac{75\,\text{MPa}}{\cos^2(\theta) - \sin^2(\theta)} \tag{5.75f}$$

Equations 5.74 and 5.75 were used to create the failure envelope shown in Figure 5.32. Note the following:

- The failure envelope shown in Figure 5.32 is valid for a *pure shear stress state only*. Failure envelopes for other states of stress differ substantially (for example, compare Figures 5.26 and 5.32, both of which are based on the Maximum Stress failure criterion).
- None of the curves shown in Figure 5.32 are associated with fiber failure. The magnitudes of the critical shear stress values returned by Equations 5.74a, 5.74d, 5.75a, or 5.75b are all large enough that they do not appear in Figure 5.32 due to the scale used for the vertical axis. These results indicate that failure of a unidirectional composite

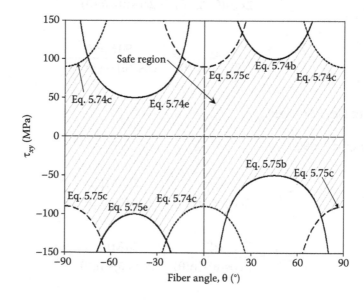

FIGURE 5.32
Failure envelope for a unidirectional graphite–epoxy laminate subjected to a pure shear stress τ_{xy}, based on the Maximum Stress criterion.

laminate subjected to a pure shear stress state will *never* occur due to fiber failure.

- The shear strength of an off-axis unidirectional composite depends on algebraic sign of the shear stress. For example, a 45° specimen is predicted to have a positive shear strength of 100 MPa and a negative shear strength of 50 MPa.

5.6.2.2 Tsai–Hill Criterion

Substituting Equations 5.72 into Equation 5.54 we obtain:

$$\tau_{xy}^2\left\{4\cos^2(\theta)\sin^2(\theta)\left[\frac{2}{(\sigma_{11}^{fT})^2}+\frac{1}{(\sigma_{22}^{fT})^2}-\frac{1}{2(\tau_{12}^f)^2}\right]+\frac{\cos^4(\theta)+\sin^4(\theta)}{(\tau_{12}^f)^2}\right\}<1 \quad (5.76)$$

Equating the left-hand side to unity and solving for τ_{xy}:

$$\tau_{xy}^f=\left\{\frac{1}{4\cos^2(\theta)\sin^2(\theta)\left[\dfrac{2}{(\sigma_{11}^{fT})^2}+\dfrac{1}{(\sigma_{22}^{fT})^2}-\dfrac{1}{2(\tau_{12}^f)^2}\right]+\dfrac{\cos^4(\theta)+\sin^4(\theta)}{(\tau_{12}^f)^2}}\right\}^{1/2} \quad (5.77)$$

Recall that the Tsai–Hill criterion does not automatically account for differences in tensile and compressive stresses. Therefore, to predict shear strengths using Equation 5.77 the failure strengths used must be selected according to whether σ_{11} and σ_{22} are positive or negative.

Positive Shear Strengths:

- For $-90°<\theta<0°$, σ_{11} is negative while σ_{22} is positive, therefore, use σ_{11}^{fC} and σ_{22}^{fT}
- For $0°<\theta<90°$, σ_{11} is positive and σ_{22} is negative, therefore, use σ_{11}^{fT} and σ_{22}^{fC}

Negative Shear Strengths:

- For $-90°<\theta<0°$, σ_{11} positive while σ_{22} is negative, therefore, use σ_{11}^{fT} and σ_{22}^{fC}
- For $0°<\theta<90°$, σ_{11} is negative and σ_{22} is positive, therefore, use σ_{11}^{fC} and σ_{22}^{fT}

A failure envelope based on these failure strengths and Equation 5.77 is shown in Figure 5.33.

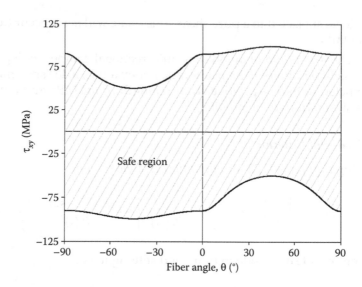

FIGURE 5.33
Failure envelope for a unidirectional graphite–epoxy laminate subjected to a pure shear stress τ_{xy}, based on the Tsai–Hill criterion.

5.6.2.3 Tsai–Wu Criterion

Substituting Equations 5.72 into Equation 5.62 we obtain:

$$\tau_{xy}^2 \left\{ X_{66}[\cos^4(\theta) + \sin^4(\theta)] + 2\cos^2(\theta)\sin^2(\theta)[2X_{11} + 2X_{22} - 4X_{12} - X_{66}] \right\}$$
$$+ 2\tau_{xy}\cos(\theta)\sin(\theta)[X_1 - X_2] < 1 \tag{5.78}$$

Numerical values for constants X_1, X_2, X_{11}, and so on, were calculated in Section 5.6.1.3 (based on strengths assumed for graphite–epoxy). Substituting these values in the left-hand side of Equation 5.78 and equating to unity, we obtain:

$$\tau_{xy}^2 \left\{ \frac{10^{-12}}{5625\,\text{Pa}^2}[\cos^4(\theta) + \sin^4(\theta)] + \cos^2(\theta)\sin^2(\theta)\left[\frac{67 \times 10^{-15}}{150\,\text{Pa}^2} - \frac{(\sqrt{10})(10^{-15})}{75\,\text{Pa}^2} \right] \right\}$$
$$- \tau_{xy}\cos(\theta)\sin(\theta)\left[\frac{61 \times 10^{-6}}{3000} \right] = 1 \tag{5.79}$$

Equation 5.79 is a second-order polynomial in the unknown shear failure stress, τ_{xy}^f. For a given fiber angle, θ, there are two roots to this equation. The predicted positive shear strength, τ_{xy}^{fp}, equals the algebraically positive root,

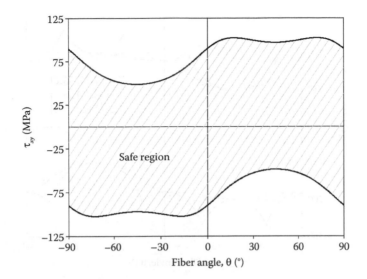

FIGURE 5.34
Failure envelope for a unidirectional graphite–epoxy laminate subjected to a pure shear stress τ_{xy}, based on the Tsai–Wu criterion.

whereas the predicted negative shear strength, τ_{xy}^{fN}, equals the negative root. For example, for a fiber angle $\theta = 30°$, Equation 5.79 becomes

$$(\tau_{xy}^{f})^2 \left\{ \frac{189.2 \times 10^{-18}}{Pa^2} \right\} - \tau_{xy}^{f} \left\{ \frac{88.05 \times 10^{-10}}{Pa} \right\} - 1 = 0$$

The two roots of this expression are found to be $(99.6 \times 10^6 \ Pa, -53.1 \times 10^6 \ Pa)$. Hence the shear strengths predicted by the Tsai–Wu criterion for a 30° graphite–epoxy laminate are

$$\tau_{xy}^{(+f)} = 99.6 \, MPa$$

$$\tau_{xy}^{(-f)} = 53.1 \, MPa$$

A failure envelope based on the Tsai–Wu criterion for a unidirectional graphite–epoxy laminate subjected to a pure shear stress state is shown in Figure 5.34. This figure is analogous to those obtained using the Maximum Stress criterion and the Tsai–Hill criterion (Figures 5.32 and 5.33, respectively).

5.6.2.4 Comparisons

The failure envelops for pure shear stress obtained on the basis of the three failure criteria considered are compared directly in Figure 5.35, and an expanded view of the just the first and fourth quadrants is presented in Figure 5.36. The difference between predictions obtained using the three

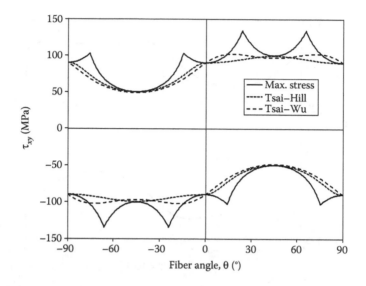

FIGURE 5.35
Comparison of the failure envelopes for a unidirectional graphite–epoxy laminate subjected to pure shear stress τ_{xy}, obtained using the Maximum Stress, Tsai–Hill, and Tsai–Wu failure criteria.

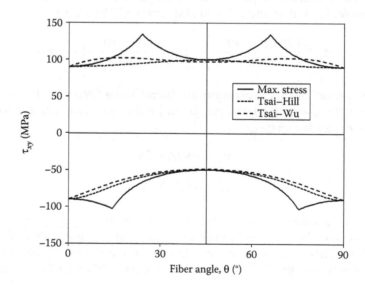

FIGURE 5.36
Comparison of the failure envelopes (first and fourth quadrants only) for a unidirectional graphite–epoxy laminate subjected to pure shear stress τ_{xy}, obtained using the Maximum Stress, Tsai–Hill, and Tsai–Wu failure criteria.

failure criteria is more striking in pure shear than was the case in uniaxial stress (e.g., compare Figures 5.36 and 5.30). Predictions on the basis of the Tsai–Hill and Tsai–Wu criteria are similar, although some difference exists. The Maximum Stress criterion predicts local maximums in positive shear strength near fiber angles of $\theta = 26°$ and 66°, and a local maximums in negative shear strengths near $\theta = 10°$ and 76°. These maximums are associated with a change in predicted failure mode, from a shear failure mode to a matrix failure mode (or vice versa).

5.7 Computer Programs *UNIDIR* and *UNIFAIL*

The results derived in this chapter will be used extensively throughout the remainder of this book. As is already abundantly clear, the calculations associated with any thermo-mechanical analysis of an anisotropic composite are tedious and time consuming if performed using a hand calculator. Consequently, most composite analyses are performed with the aid of a digital computer.

Two computer programs are available to supplement the material presented in this chapter: *UNIDIR* and *UNIFAIL*. These programs can be downloaded at no cost from the following website:

http://depts.washington.edu/amtas/computer.html

The analyses that can be performed with the aid of these programs will be discussed in the following subsections. Both programs require the user to provide various numerical values required during the calculations performed. The user must define these values using a consistent set of units. For example, program *UNIDIR* requires the user to input elastic modulii, thermal expansion coefficients, and moisture expansion coefficients for the composite material system of interest. Using the properties listed in Table 3.1 and based on the SI system of units, the following numerical values would be input for graphite–epoxy:

$$E_{11} = 170 \times 10^9 \, \text{Pa} \qquad\qquad E_{22} = 10 \times 10^9 \, \text{Pa} \; \nu_{12} = 0.30$$

$$G_{12} = 13 \times 10^9 \, \text{Pa}$$

$$\alpha_{11} = -0.9 \times 10^{-6} \, \text{m/m} - °\text{C} \qquad \alpha_{11} = 27.0 \times 10^{-6} \, \text{m/m} - °\text{C}$$

$$\beta_{11} = 150.0 \times 10^{-6} \, \text{m/m} - \%\text{M} \quad \beta_{22} = 4800 \times 10^{-6} \, \text{m/m} - \%\text{M}$$

If the analysis requires the user to input numerical values for stresses, then stresses must be input in *Pascals* (*not* in MPa). A typical value would be $\sigma_{xx} = 200 \times 10^6$ Pa. If, instead, the analysis requires the user to input numerical values for strains, then strains must be input in m/m (*not* in μm/m). A typical value would be $\varepsilon_{xx} = 2000 \times 10^{-6}$ m/m = 0.002000 m/m. All temperatures would be input in °C.

In contrast, if the English system of units were used, then the following numerical values would be input for the same graphite–epoxy material system:

$$E_{11} = 25.0 \times 10^6 \text{psi} \qquad E_{22} = 1.5 \times 10^6 \text{psi}$$

$$\nu_{12} = 0.30 \qquad G_{12} = 1.9 \times 10^6 \text{ psi}$$

$$\alpha_{11} = -0.5 \times 10^{-6} \text{in./in.} - °F \qquad \alpha_{11} = 15 \times 10^{-6} \text{in./in.} - °F$$

$$\beta_{11} = 150.0 \times 10^{-6} \text{in./in.} - \%M \quad \beta_{22} = 4800 \times 10^{-6} \text{in./in.} - \%M$$

If the analysis requires the user to input numerical values for stresses, then stresses must be input in *psi* (*not* in ksi). A typical value would be $\sigma_{xx} = 30{,}000$ psi. If, instead, the analysis requires the user to input numerical values for strains, then strains must be input in in./in. (*not* in μin./in.). A typical value would be $\varepsilon_{xx} = 2000 \times 10^{-6}$ in./in = 0.002000 in./in. All temperatures would be input in °F.

5.7.1 Program *UNIDIR*

Program *UNIDIR* may be used to predict the elastic behavior of unidirectional composites, and is based on the material presented in Sections 5.1 through 5.4. Two different types of analyses may be performed. The program may be used either to:

- Calculate total strains ($\varepsilon_{xx}, \varepsilon_{yy}, \gamma_{xy}$) caused by a specified combination of stresses ($\sigma_{xx}, \sigma_{yy}, \tau_{xy}$), a uniform temperature change (ΔT), and a uniform change in moisture content (ΔM). Calculations performed as a part of Example Problems 5.1, 5.2, 5.5, and 5.6 are typical of this type of analysis.

or

- Calculate stresses ($\sigma_{xx}, \sigma_{yy}, \tau_{xy}$), caused by a specified combination of total strains ($\varepsilon_{xx}, \varepsilon_{yy}, \gamma_{xy}$), a uniform temperature change (ΔT), and a uniform change in moisture content (ΔM). Calculations performed as a part of Example Problems 5.3 and 5.7 are typical of this type of analysis.

The program also determines the effective properties of a unidirectional composite, based on the definitions described in Section 5.4. An implicit assumption in these calculations is that *all* material properties (E_{11}, α_{11}, β_{11}, etc.) input by the user correspond to the temperature and moisture content dictated by ΔT and ΔM.

5.7.2 Program *UNIFAIL*

Program *UNIFAIL* may be used to obtain failure predictions for unidirectional composites based on the Maximum Stress, Tsai–Hill, or Tsai–Wu failure criterion introduced in Section 5.5. Two different types of analyses may be performed. The program may be used either to:

- Calculate predicted uniaxial and shear strengths of a unidirectional laminate with a specified fiber angle, θ.

or

- Generate a data file that can subsequently be used to produce failure envelopes for unidirectional composites subjected to several types of plane stress conditions.

Note that program *UNIFAIL* itself does not create a failure envelope. Rather, the program creates a file (named *Envelop.txt*) that contains the stress(es) predicted to cause failure of a unidirectional composite as a function of fiber angle, based on the particular failure criterion specified by the user. A failure envelope may then be created using a second software package to "import" the data generated by program *UNIFAIL* and then plotting failure stress versus fiber angle. For example, any of the failure envelopes presented in Section 5.6 may be easily re-created in this way. As before, an implicit assumption is that failure strengths (σ_{11}^{fT}, α_{22}^{yT}, etc.) input by the user correspond to the values exhibited by the composite at the temperature and moisture content of interest.

HOMEWORK PROBLEMS

Notes:

a. In the following problems the phrase "by hand calculation" means that solutions are to be obtained using a calculator (or equivalent), pencil and paper.

b. Computer programs UNIDIR and/or UNIFAIL are referenced in many of the following problems. As described in Section 5.7, these programs can be downloaded from the following website:

http://depts.washington.edu/amtas/computer.html

5.1. Calculate the reduced compliance matrix for the materials listed below, first by hand calculation and then using program UNIDIR. Use material properties listed in Table 3.1.

   a. Glass/epoxy
   b. Kevlar/epoxy
   c. Graphite/epoxy

5.2. Calculate the reduced stiffness matrix for the materials listed below, first by hand calculation and then using program UNIDIR. Use material properties listed in Table 3.1.

   a. Glass/epoxy
   b. Kevlar/epoxy
   c. Graphite/epoxy

5.3. Confirm that:

$$Q_{22} = \frac{C_{22}C_{33} - C_{23}^2}{C_{33}}$$

5.4. Referring to Example Problem 5.4, confirm that the predicted stresses and strains at points B and C are consistent with Equations 5.2 and 5.3.

5.5. A thin unidirectional glass/epoxy composite laminate is simultaneously subjected to a uniform temperature change $\Delta T = -175°C$, an increase in moisture content $\Delta M = 0.5\%$, and the following in-plane stresses:

$$\sigma_{11} = 350\,\text{MPa}$$

$$\sigma_{22} = 40\,\text{MPa}$$

$$\tau_{12} = 60\,\text{MPa}$$

Determine the resulting strains (ε_{11}, ε_{22}, and γ_{12}), first by hand calculation and then using program UNIDIR. Use material properties listed in Table 3.1.

5.6. Repeat Problem 5.5 for a unidirectional Kevlar/epoxy composite laminate.

5.7. Repeat Problem 5.5 for a unidirectional graphite/epoxy composite laminate.

5.8. A thin unidirectional glass/epoxy composite laminate is simultaneously subjected to a uniform temperature change $\Delta T = -275°F$ and an unknown plane stress state. The following strains are measured as a result (moisture content remains constant):

$$\varepsilon_{11} = -1250\,\mu\text{in./in.}$$

$$\varepsilon_{22} = 2000\,\mu\text{in./in.}$$

$$\gamma_{12} = 0$$

Determine the stresses (σ_{11}, σ_{22}, and τ_{12}), first by hand calculation and then using program UNIDIR. Use material properties listed in Table 3.1.

5.9. Repeat Problem 5.8 for a unidirectional Kevlar/epoxy composite laminate.

5.10. Repeat Problem 5.8 for a unidirectional graphite/epoxy composite laminate.

5.11. A square unidirectional glass/epoxy composite laminate with dimensions 1 m × 1 m is clamped between an infinitely rigid frame, as shown in Figure 5.37. Material properties are listed in Table 3.1. Initially the clamped composite is stress-free, but the temperature is subsequently decreased by 100°C. The thermal expansion coefficient of the rigid walls is zero, so the rigid walls do not expand or contract.

 a. Calculate the stresses ($\sigma_{11}, \sigma_{22}, \tau_{12}$) induced by this change in temperature, first by hand calculation and then using program UNIDIR.

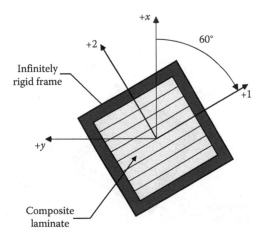

FIGURE 5.37
Clamped composite laminate considered in Problems 5.10, 5.11, and 5.12.

b. Predict whether the composite will fail, based on the Maximum Stress Failure criterion.

5.12. Repeat Problem 5.11 for a unidirectional Kevlar/epoxy composite laminate.

5.13. Repeat Problem 5.11 for a unidirectional graphite/epoxy composite laminate.

5.14. A $1\,\text{m} \times 1\,\text{m}$ square unidirectional glass/epoxy composite laminate is placed within a cavity defined by four rigid walls, as shown in Figure 5.38. An initial gap of 0.050 mm exists between all edges of the ply and the walls. The composite ply adsorbs 1.5% moisture, causing the ply to expand and completely fill the cavity. Temperature remains constant. The rigid walls do not adsorb moisture, and hence do not expand or contract. Assuming the ply does not buckle, calculate the stresses (σ_{11}, σ_{22}, τ_{12}) caused by the change in moisture content.

5.15. Repeat Problem 5.13 for a unidirectional Kevlar/epoxy composite laminate.

5.16. Repeat Problem 5.13 for a unidirectional graphite/epoxy composite laminate.

5.17. A perfectly square unidirectional glass/epoxy composite laminate is mounted in a frame consisting of four infinitely rigid frame members, as shown in Figure 5.39a. The frame members are pinned at each corner. Since the laminate is perfectly square, the angle defined by corners *ABC* is initially 90° (precisely).

A force *F* is then applied to two diagonal corners, as shown in Figure 5.39b. After *F* is applied angle *ABC* is measured to be

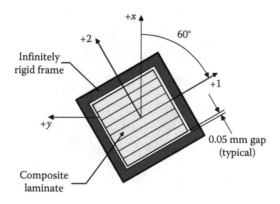

FIGURE 5.38
A composite laminate placed within a cavity defined by four rigid walls (considered in Problems 5.13, 5.14, and 5.15).

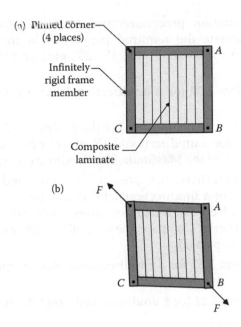

FIGURE 5.39
Unidirectional composite laminate considered in Problems 5.16, 5.17, and 5.18. (a) Panel before loading; (b) panel after loading (deformations shown greatly exaggerated for clarity).

89.50° (precisely). Both temperature and moisture content remain constant. What stresses (σ_{11}, σ_{22}, τ_{12}) are induced in the panel?

5.18. Repeat Problem 5.16 for a unidirectional Kevlar/epoxy composite laminate.

5.19. Repeat Problem 5.16 for a unidirectional graphite/epoxy composite laminate.

5.20. Referring to Example Problem 5.8, confirm that the predicted stresses and strains at points B and C are consistent with Equations 5.29 and 5.30.

5.21. Create a plot of the following effective properties for a unidirectional glass-epoxy composite, with fiber angles ranging from −90° to +90°.

   a. E_{xx} and E_{yy}

   b. v_{xy} and v_{yx}

   c. G_{xy}

   d. $\eta_{xy,xx}$ and $\eta_{xy,yy}$

   e. $\eta_{xx,xy}$ and $\eta_{yy,xy}$

   f. α_{xx}, α_{yy}, and α_{xy}

   g. β_{xx}, β_{yy}, and β_{xy}

Suggested solution procedure: use program *UNIDIR* (repeatedly) to calculate the required properties in increments of 5° (i.e., calculate for $\theta = 0°$, 5°, 10°, 15°, 20°, etc.) and then plot these calculations.

5.22. Repeat Problem 5.20 for a unidirectional Kevlar/epoxy composite laminate.

5.23. a. Using hand calculation, predict the positive and negative shear strengths for a unidirectional 45° glass/epoxy composite laminate, based on the Maximum Stress Failure criterion.

   b. Use program *UNIFAIL* to predict the positive and negative shear strengths for a unidirectional 45° glass/epoxy composite laminate, based on the Maximum Stress, Tsai–Hill, and Tsai–Wu Failure criteria. Compare these results with your calculations obtained in part (a).

5.24. Repeat Problem 5.22 for a unidirectional Kevlar/epoxy composite laminate.

5.25. Repeat Problem 5.22 for a unidirectional graphite/epoxy composite laminate.

5.26. *On the same graph*, plot failure envelopes for a unidirectional glass/epoxy composite laminate, for the following two conditions:

   a. Unidirectional stress: σ_{xx}, $\sigma_{yy} = \tau_{xy} = 0$

   b. Biaxial normal stress: $\sigma_{yy} = \sigma_{xx}/10$, $\tau_{xy} = 0$

   Use the Tsai–Hill failure criterion. Suggested solution procedure: use program *UNIFAIL* (twice) to generate data files corresponding to the specified loading conditions, and then plot these data files.

5.27. Repeat Problem 5.25 for a unidirectional Kevlar/epoxy composite laminate.

5.28. Repeat Problem 5.25 for a unidirectional graphite/epoxy composite laminate.

5.29. Using program *UNIFAIL* and the properties listed in Table 3.1, prepare plots similar to Figure 5.29 for

   a. Glass-epoxy

   b. Kevlar-epoxy

5.30. Using program *UNIFAIL* and the properties listed in Table 3.1, prepare plots similar to Figure 5.35 for

   a. Glass-epoxy

   b. Kevlar-epoxy

References

1. Reuter, R.C., Concise property transformation relations for an anisotropic lamina, *Journal of Composite Materials*, April, 270–272, 1971.
2. Soden, P.D., Hinton, M.J., and Kaddour, A.S., A comparison of the predictive capabilities of current failure theories for composite laminates, *Composite Science and Technology*, 58, 1225–1254, 1998.
3. Hinton, M.J., Kaddour, A.S., and Soden, P.D., *Failure Criteria in Fibre Reinforced Polymer Composites: The World-Wide Failure Exercise*, Elsevier, Amsterdam, 2004.
4. Hill, R., *The Mathematical Theory of Plasticity*, Oxford University Press, 1950.
5. Tsai, S.W., Strength theories of filamentary structures, *Fundamental Aspects of Fiber Reinforced Plastic Composites*, R.T. Schwartz and H.S. Schwartz, eds., Wiley Interscience, New York, pp. 3–11, 1968.
6. Tsai, S.W. and Wu, E.M., A general theory of strength for anisotropic materials, *Journal of Composite Materials*, January, 58–80, 1971.

6

Thermomechanical Behavior of Multiangle Composite Laminates

6.1 Definition of a "Thin Plate" and Allowable Plate Loadings

A "thin plate" with in-plane dimensions a and b and thickness t is shown schematically in Figure 6.1. The plate can be considered "thin" if the plate thickness is less than about one-tenth the in-plane dimensions, that is, if $t < a/10$ and $t < b/10$. A x–y–z coordinate system is defined as indicated. Note that the origin of the x–y–z coordinate system is positioned at the geometric center of the plate, such that the *midplane* (or *midsurface*) of the plate lies within the plane $z = 0$. Consequently, the plate exists within the space defined by the planes $z = -t/2$ and $z = +t/2$.

We will assume that the thin plate is subjected to plane stress conditions. Therefore, we will only consider plate loadings that result in a plane stress state within the plate. Furthermore, in this chapter we will only consider *uniformly distributed* loads. That is, we will assume that the loads are constant and uniformly distributed along the edge of the plate. The more general case in which loads vary along the edge of the plate will be considered in Chapter 10.

Six types of uniformly distributed loads that give rise to plane stress conditions within the x–y plane are shown in Figure 6.2. All load components are shown in an algebraically positive sense. Since the line-of-action of all load vectors shown in Figure 6.2 lie within the x–y plane, these load components are called *in-plane loads*.

First consider load components N_{xx}, N_{yy}, and N_{xy}. Two subscripts are used to identify these load components. The algebraic sense of each component is interpreted in a manner analogous to that previously used to identify the algebraic sense of individual stress components (discussed in Section 2.5). That is, the first subscript indicates the face of the plate a given load acts upon, while the second subscript indicates the line of action of the load. A positive load is one that:

- Acts on a positive face and points in a positive coordinate direction, or
- Acts on a negative face and points in a negative coordinate direction.

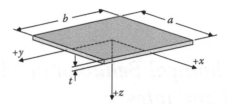

FIGURE 6.1
A thin plate with in-plane dimensions a and b and thickness t.

The algebraic sense of normal loads N_{xx} and N_{yy} (Figure 6.2a) is readily apparent and intuitive: a positive (tensile) normal load is one that tends to cause the plate to stretch. The algebraic sense of shear loads N_{xy} and N_{yx} (Figure 6.2b) is not as immediately apparent, but application of the sign convention just described will confirm that the shear loads shown in Figure 6.2b are indeed positive. That is, two of the shear loads shown are acting on a positive face and point in a positive coordinate direction, whereas two of the shear loads shown act on a negative face and point in a negative coordinate direction. Since individual load components are not allowed to vary spatially (i.e., loads are assumed to be constant and uniformly distributed along each edge of the plate), static equilibrium requires that the shear loads acting along the x- and y-edge be oriented tip-to-tip and tail-to-tail, as shown in Figure 6.2b. Furthermore, the magnitude of the shear loads must

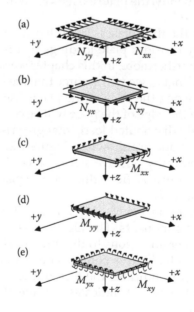

FIGURE 6.2
Schematic of allowable plate loadings. (a) In-plane normal forces N_{xx} and N_{yy}; (b) in-plane shear forces N_{xy} and N_{yx}; (c) bending moment M_{xx}; (d) bending moment M_{yy}; (e) in-plane torques M_{xy} and M_{yx}.

be identical, $|N_{xy}| = |N_{yx}|$. These requirements are also analogous to those of shear stresses acting on an infinitesimal stress element, as discussed in Section 2.5. It is emphasized that N_{xx}, N_{yy}, and N_{xy} are all defined as *distributed* loads, expressed in units of (force/plate length), such as N/m or lbf/in.

The remaining loads shown in Figure 6.2 (M_{xx}, M_{yy}, and M_{xy}) are bending moments (or torques) distributed along the edge of the plate. Load components M_{xx} and M_{yy} are uniformly distributed bending moments acting along the x- and y-edge of the plate, respectively, as shown in Figure 6.2c and d, respectively. These loads are shown in an algebraically positive sense. The subscripts assigned to M_{xx} and M_{yy} may seem puzzling, since (for example) M_{xx} represents a bending moment acting about the y-axis. However, M_{xx} is directly related to the distribution of σ_{xx} through the thickness of the plate, as will be shown below. Since M_{xx} arises due to the distribution of σ_{xx} (or vice versa), it is customary to use the same subscripts for both entities. Unfortunately, the convention used to assign an algebraic sign to distributed bending moments varies from one author to the next. The sign convention used herein is most commonly used in the study of composite plates. An algebraically positive distributed bending moment is defined as one that tends to cause tensile stresses in the positive z-face of the plate and compressive stresses in the negative z-face. Referring to Figure 6.2c and noting that the positive z-direction is downwards as drawn, it is seen that M_{xx} tends to cause tensile stresses in the positive z-face (i.e., the lower face) of the plate. Hence M_{xx} is positive as drawn. A similar observation holds for M_{yy}, shown in Figure 6.2d.

Finally, loads M_{xy} and M_{yx} are defined as uniformly distributed in-plane moments (or torques) acting along neighboring edges of the plate, as shown in Figure 6.2e. It will be shown below that M_{xy} and M_{yx} are directly related to in-plane shear stresses τ_{xy} and τ_{yx}, respectively. Since M_{xy} and M_{yx} arise due to the distribution of τ_{xy} and τ_{yx}, it can be shown that for static equilibrium to be maintained, $|M_{xy}| = |M_{yx}|$. An algebraically positive distributed torque is defined as one that tends to cause a positive shear stress in the positive z-face (i.e., the lower face) of the plate.

Recall that the units of an applied moment or torque are force–length, such as N–m, or lbf–in. Since M_{xx}, M_{yy}, and M_{xy} all represent *uniformly distributed moments* acting along the plate edge, they are all expressed in units of force–length/plate length, such as N–m/m or lbf–in./in.

It is expected that most readers will have considered the behavior of isotropic *prismatic beams* during earlier studies. It is, therefore, instructive to contrast the definitions just given for a thin plate, as well as the loads applied thereon, to those encountered in fundamental beam theory. As previously shown in Figure 6.1, a "thin plate" is defined as a structure whose thickness, t, is much less than the in-plane dimensions, a and b. That is, $t \ll a,b$. In contrast, a beam is a structure for which two dimensions are small compared to the third. Hence, a beam can be described as a structure in which one of the in-plane dimensions, width b say, is of the same order as the thickness t. Hence, the thin plate shown in Figure 6.1 is "converted" to a beam if we allow $b \approx t \ll a$. In this way, we

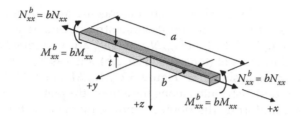

FIGURE 6.3
A prismatic beam with rectangular cross-section (compare with Figure 6.1).

describe a beam with rectangular cross-section $b \times t$ and length a, as shown in Figure 6.3. The beam is called "prismatic" if the cross-section remains constant along the length of the beam, that is, if b and t remain constant along length a.

Regarding the description of applied loading, in the case of thin plates all loading conditions are specified in terms of *distributed* loads, as described in preceding sections. For example, in SI units N_{xx} is expressed in terms of N/m, while M_{xx} is expressed in terms of N–m/m. In contrast, in fundamental beam theory *point* loads are often specified. For a beam a normal load is expressed in terms of N or lbf, while bending moments (or torques) are expressed in terms of N–m or lbf–ft. Loads that correspond to N_{xx} and M_{xx}, when applied to a beam, have also been shown in Figure 6.3. They have been denoted as N_{xx}^b and M_{xx}^b, where the superscript "b" is used to denote that these loads are defined in the sense traditionally used in beam theory and, therefore, have different units than the distributed loads used in plate theory. Since the width of the beam is b, the two load definitions are related according to $N_{xx}^b = bN_{xx}$ and $M_{xx}^b = bM_{xx}$. As mentioned above, the sign convention used to define an algebraically positive bending moment varies from one author to the next. The bending moment applied to the beam shown in Figure 6.3 is considered to be positive according to the sign convention used throughout this chapter. This corresponds to the convention most commonly used in the study of composite plates. However, according to the sign convention used in many textbooks devoted to fundamental beam theory, the bending moment shown in Figure 6.3 would be considered to be negative. Hence, the sign convention used to describe bending moments in this and other textbooks devoted to composites differs from the sign convention used in many textbooks devoted to beam theory. The reader must simply be aware of this potential source of confusion, and carefully note which convention has been used when comparing the results described in this chapter to those developed elsewhere.

Let us now return to the topic of thin plates. We wish to relate the external distributed loads applied to the plate to the resulting internal stresses. An edge view of a plate loaded only by distributed load N_{xx} and moment M_{xx} is shown in Figure 6.4a. A free-body diagram of a section of the plate is shown in Figure 6.4b. The free-body diagram has been drawn showing the distributed load N_{xx} and distributed moment M_{xx} on the left-hand side, and the resulting internal

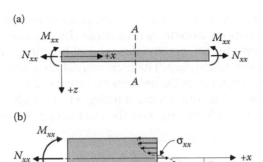

FIGURE 6.4
Edge view of a thin plate subjected to loads N_{xx} and M_{xx} only. (a) Edge view of a plate loaded by external loads N_{xx} and M_{xx}; (b) a free-body diagram of the plate (using a magnified scale), based on a cut at section A-A and showing both the externally applied loads and the resulting internal stress σ_{xx}.

stress σ_{xx} on the right-hand side. At this point, the through-thickness distribution of σ_{xx} is unknown and is shown as a dashed line. The plate is assumed to be in static equilibrium. Therefore, $\Sigma F = 0$, and the force per unit width associated with the unknown distribution of stress σ_{xx} through the thickness of the plate must be exactly balanced by the distributed load N_{xx}. Let the free-body diagram have a width of "1," and consider an incremental strip of height dz. The cross-sectional area dA of this strip is $dA = (1)(dz)$. The incremental force dF_{xx} associated with the stress acting over this thin strip is $dF_{xx} = (dN_{xx})$ $(1) = (\sigma_{xx})\, dA = \sigma_{xx}\, dz$. We can now relate the total distributed force N_{xx} acting on the left-hand side of the free-body diagram to the distribution of σ_{xx} acting on the right-hand side by simply "adding up" the forces acting over all incremental strips; that is, we integrate over the thickness of the plate:

$$N_{xx} = \int_{-t/2}^{+t/2} dN_{xx} = \int_{-t/2}^{t/2} \sigma_{xx}\, dz \tag{6.1a}$$

In an entirely equivalent manner, we can relate distributed forces N_{yy} and N_{xy} to stresses σ_{yy} and τ_{xy}, respectively:

$$N_{yy} = \int_{-t/2}^{t/2} \sigma_{yy}\, dz \tag{6.1b}$$

$$N_{xy} = \int_{-t/2}^{t/2} \tau_{xy}\, dz \tag{6.1c}$$

Now consider moment M_{xx}. As before, the plate is assumed to be in static equilibrium, and hence moments acting about the y-axis must sum to zero: $\Sigma M_y = 0$. Again consider an incremental strip of height dz, which is located a distance z from the midsurface. The incremental distributed moment dM_{xx} contributed by N_{xx} acting over the incremental strip is $dM_{xx} = (dN_{xx})(z) = (\sigma_{xx})(z)(dz)$. We can obtain the total moment acting on the right-hand side of the free-body diagram by integrating over the thickness of the plate:

$$M_{xx} = \int_{-t/2}^{t/2} \sigma_{xx} z \, dz \qquad (6.2a)$$

In an entirely equivalent manner, we can relate moments M_{yy} and M_{xy} to stresses σ_{yy} and τ_{xy}, respectively:

$$M_{yy} = \int_{-t/2}^{t/2} \sigma_{yy} z \, dz \qquad (6.2b)$$

$$M_{xy} = \int_{-t/2}^{t/2} \tau_{xy} z \, dz \qquad (6.2c)$$

Equations 6.1 and 6.2 show that the uniformly distributed loads and moments applied to the plate edge are directly related to the stresses within the plate. Distributed loads N_{xx}, N_{yy}, and N_{xy} are commonly called *stress resultants*, and moments M_{xx}, M_{yy}, and M_{xy} are commonly called *moment resultants*.

6.2 Plate Deformations: The Kirchhoff Hypothesis

Let us now consider the deformation of a thin flat plate. Figure 6.5 represents a (magnified) edge view of the plate in both the "initial" and "deformed" positions. The positive x-direction is to the right, the positive y-direction is out of the plane of the figure, and the positive z-direction is downward, which is consistent with the original definition shown in Figure 6.1. Although we will eventually apply our results to a composite laminate, for the moment we will not consider the existence of individual plies, and, therefore, the ply interfaces are not shown in the figure. If the flat plate is loaded and/or is subjected to a change in environment, it will be deformed and (in general) will change length and become curved, as shown in the figure. We will base our analysis on the *Kirchhoff hypothesis*, which states that a *straight line which*

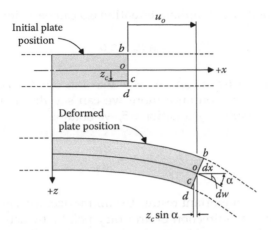

FIGURE 6.5
Initial and deformed positions of a flat plate (deformations exaggerated for clarity).

is initially perpendicular to the midplane of the plate remains straight and perpendicular to the midplane of the plate after deformation. For example, let us consider straight line *b–o–d*. This line is shown in Figure 6.5, in the sketch of *both* the initial *and* deformed positions of the plate. In accordance with the Kirchhoff hypothesis, line *b–o–d* has been drawn perpendicular to the midplane in *both* the initial *and* deformed positions. In effect, we have assumed that out-of-plane shear strains are zero ($\gamma_{xz} = \gamma_{yz} = 0$), which is equivalent to saying that we have assumed that the z-axis is a *principal strain axis*.

We are interested in describing the displacement of an arbitrary point *c*, which is located within the thickness of the plate at some distance z_c from the midsurface. Point *c* is shown in Figure 6.5, and lies along line *b–o–d*. Denote the displacement of point *o* in the x-direction as distances u_o. From the figure it can be seen that the distance point *c* has moved in the x-direction, u_c, is approximately given by

$$u_c \cong u_o - z_c \sin \alpha \tag{6.3}$$

where α is the angle formed by the plate midplane and the x-axis in the deformed condition. Equation 6.3 is approximate because we have ignored any change in plate thickness, that is, we have ignored any change in distance z_c which may have occurred during deformation of the plate. If we now further assume that angle α is small, then we can simplify Equation 6.3 using the *small angle approximation*, which states that if α is expressed in radians and is less than about 0.1745 rad (about 10°), then:

$$\sin \alpha \cong \alpha \qquad \tan \alpha \cong \alpha \qquad \cos \alpha \cong 1 \tag{6.4}$$

On the basis of this assumption, Equation 6.3 can be written as

$$u_c \cong u_o - z_c\alpha \tag{6.5}$$

Now, from Figure 6.5 it can be seen that $\tan \alpha = dw/dx$. Applying the small-angle approximation once more we can say that $\tan \alpha \cong \alpha \cong dw/dx$. Substituting this result into Equation 6.5, we obtain:

$$u_c = u_o - z_c\frac{dw}{dx} \tag{6.6}$$

Equation 6.6 is an important result. To summarize, we have expressed the displacement in the x-direction of arbitrary point c (which we have called distance u_c) as a function of the displacement in the x-direction of a point on the plate midsurface (distance u_o), the position of point c with respect to the midsurface (length z_c), and the slope of the plate midsurface (dw/dx).

We will also require an expression for the displacement of point c in the y-direction. We will denote the displacement of point c in the y-direction as v_c. Using a procedure that is entirely equivalent to that just described, it can be shown:

$$v_c = v_o - z_c\frac{dw}{dy} \tag{6.7}$$

Equations 6.6 and 6.7 represent the *displacements* of point c in the x- and y-directions, respectively, and follow directly from the Kirchhoff hypothesis. We can now determine the *infinitesimal in-plane strains* at point c, in accordance with Equation 2.49:

$$\varepsilon_{xx} = \frac{\partial u}{\partial x} \quad \varepsilon_{yy} = \frac{\partial v}{\partial y} \quad \gamma_{xy} = \frac{\partial v}{\partial x} + \frac{\partial u}{\partial y} \tag{6.8}$$

Substituting Equations 6.6 and 6.7 into Equation 6.8, we obtain the following expressions for the strains induced at point c:

$$\varepsilon_{xx}^c = \frac{\partial u_o}{\partial x} - z_c\frac{\partial^2 w}{\partial x^2}$$

$$\varepsilon_{yy}^c = \frac{\partial v_o}{\partial y} - z_c\frac{\partial^2 w}{\partial y^2} \tag{6.9}$$

$$\gamma_{xy}^c = \frac{\partial u_o}{\partial y} + \frac{\partial v_o}{\partial x} - 2z_c\frac{\partial^2 w}{\partial x\partial y}$$

Let

$$\varepsilon_{xx}^o = \frac{\partial u_o}{\partial x} \tag{6.10a}$$

$$\varepsilon_{yy}^o = \frac{\partial v_o}{\partial y} \tag{6.10b}$$

$$\gamma_{xy}^o = \frac{\partial u_o}{\partial y} + \frac{\partial v_o}{\partial x} \tag{6.10c}$$

$$\kappa_{xx} = -\frac{\partial^2 w}{\partial x^2} \tag{6.10d}$$

$$\kappa_{yy} = -\frac{\partial^2 w}{\partial y^2} \tag{6.10e}$$

$$\kappa_{xy} = -2\frac{\partial^2 w}{\partial x \partial y} \tag{6.10f}$$

where ε_{xx}^o, ε_{yy}^o, and γ_{xy}^o are the in-plane strains which exist *at the midplane* of the plate. The terms κ_{xx}, κ_{yy}, and κ_{xy} are called *midplane curvatures*, and represent the rate of change of the slope of the midplane of the plate.

The reader is likely to have encountered the concept of midplane curvatures during earlier studies of fundamental beam theory. Unfortunately, the algebraic sign convention used to define curvatures varies from one author to the next. The sign convention used throughout this chapter and defined by Equations 6.10 is most commonly used in the study of composite plates. However, in many textbooks devoted to beam theory curvatures are defined using the opposite sign convention. For example, in beam theory curvature κ_{xx} is often defined as $\kappa_{xx} = +\partial^2 w/dx^2$, rather than $\kappa_{xx} = -\partial^2 w/dx^2$ as indicated above. Also, in some textbooks devoted to plate theory κ_{xy} is defined as $\kappa_{xy} = +\partial^2 w/dx\,dy$, rather than as $\kappa_{xy} = -2\partial^2 w/dx\,dy$ as indicated in Equation 6.10. These unfortunate deviations from one author to the next have developed over many years, and a universal agreement on algebraic signs or even the fundamental definition of κ_{xy} are not likely to occur for the foreseeable future. The reader must simply be aware of these potential sources of confusion, and carefully note the convention has been used when comparing the results described in this chapter to those developed elsewhere.

Note from Figure 6.5 that point c is located at an arbitrary distance z from the neutral surface. We will discontinue the use of the subscript "c" in Equations 6.9. Substituting Equations 6.10 into Equation 6.9, we obtain:

$$\varepsilon_{xx} = \varepsilon_{xx}^o + z\kappa_{xx}$$

$$\varepsilon_{yy} = \varepsilon_{yy}^o + z\kappa_{yy} \qquad (6.11)$$

$$\gamma_{xy} = \gamma_{xy}^o + z\kappa_{xy}$$

Equations 6.11 can be conveniently written in matrix form as

$$\begin{Bmatrix} \varepsilon_{xx} \\ \varepsilon_{yy} \\ \gamma_{xy} \end{Bmatrix} = \begin{Bmatrix} \varepsilon_{xx}^o \\ \varepsilon_{yy}^o \\ \gamma_{xy}^o \end{Bmatrix} + z \begin{Bmatrix} \kappa_{xx} \\ \kappa_{yy} \\ \kappa_{xy} \end{Bmatrix} \qquad (6.12)$$

Equation 6.12 is the primary result we require for present purposes from classical thin-plate theory. It allows us to calculate the in-plane strains (ε_{xx}, ε_{yy}, γ_{xy}) induced at any position z through the thickness of the plate, based on the midplane strains (ε_{xx}^o, ε_{yy}^o, γ_{xy}^o) and midplane curvatures (κ_{xx}, κ_{yy}, κ_{xy}). Note that this result is based strictly on the Kirchhoff hypothesis. We have made no assumptions regarding the mechanism(s) that *caused* the flat plate to deform. Hence, Equation 6.12 is valid if the plate is deformed by a change in temperature, a change in moisture content, by externally applied mechanical loads, or by any combination thereof. Also, we have made no assumptions regarding material properties. Equation 6.12 is, therefore, valid for isotropic, transversely isotropic, orthotropic, or anisotropic thin plates.

Example Problem 6.1

A thin plate with a thickness of 1 mm is subjected to mechanical loads, a change in temperature, and a change in moisture content. Strain gages are used to measure the surface strains induced in the plate. They are found to be

at z = −t/2 = −0.5 mm: $\varepsilon_{xx} = 250$ µm/m, $\varepsilon_{yy} - 1500$ µm/m, $\gamma_{xy} = 1000$ µrad

at z = +t/2 = +0.5 mm: $\varepsilon_{xx} = -250$ µm/m, $\varepsilon_{yy} - 1100$ µm/m, $\gamma_{xy} = 800$ µrad

What midplane strains and curvatures are induced in the plate?

SOLUTION

To solve this problem, we simply apply Equation 6.12 to both surfaces of the plate. For example, using the measured strains for ε_{xx} we have:

at z = −t/2 = −0.0005 m: $\varepsilon_{xx} = 250\,\mu m/m = \varepsilon_{xx}^o - (0.0005)\kappa_{xx}$

at z = +t/2 = +0.0005 m: $\varepsilon_{xx} = -250\,\mu m/m = \varepsilon_{xx}^o + (0.0005)\kappa_{xx}$

Solving simultaneously, we find:

$$\varepsilon_{xx}^o = 0\ \mu m/m \qquad \kappa_{xx} = -0.50\ rad/m$$

Using a similar approach using the measured values for ε_{yy} and γ_{xy}, we find:

$$\varepsilon_{yy}^o = -1300\ \mu m/m \qquad \kappa_{yy} = 0.40\ rad/m$$

$$\gamma_{xy}^o = 900\ \mu rad \qquad \kappa_{xy} = -0.20\ rad/m$$

6.3 Principal Curvatures

In Section 6.2, we invoked the Kirchhoff hypothesis, where it is assumed that a straight line which is initially perpendicular to the midplane of the plate remains straight and perpendicular to the midplane after deformation. The Kirchhoff hypothesis has ultimately allowed us to calculate the in-plane strains referenced to the x–y coordinate system (ε_{xx}, ε_{yy}, and γ_{xy}) induced at any position z through the thickness of a thin plate, using either Equations 6.11 or 6.12. These equations are valid for any combination of midplane strains (ε_{xx}^o, ε_{yy}^o, and γ_{xy}^o) and midplane curvatures (κ_{xx}, κ_{yy}, and κ_{xy}).

In this section, we will consider a special case. Specifically, we will consider a state of deformation in which the midplane strains are zero: $\varepsilon_{xx}^o = \varepsilon_{yy}^o = \gamma_{xy}^o = 0$. In this special case, Equation 6.12 becomes

$$\begin{Bmatrix} \varepsilon_{xx} \\ \varepsilon_{yy} \\ \gamma_{xy} \end{Bmatrix} = z \begin{Bmatrix} \kappa_{xx} \\ \kappa_{yy} \\ \kappa_{xy} \end{Bmatrix} \tag{6.13}$$

This state of deformation is known as *pure bending*. When a thin plate is in a state of pure bending all midplane strains are zero ($\varepsilon_{xx}^o = \varepsilon_{yy}^o = \gamma_{xy}^o = 0$), and the midplane of the plate is called the *neutral surface*.

Equation 6.13 gives the in-plane strains in pure bending referenced to the x–y coordinate system. Referring to Figure 6.6, suppose we wish to express these strains relative to a new x'–y' coordinate system, obtained by rotating through angle α about the z-axis. As noted in the preceding section, the Kirchhoff hypothesis implies $\gamma_{xz} = \gamma_{yz} = 0$. Therefore, *the z-axis is a principal strain axis.* Consequently, we can rotate in-plane strains from the x–y

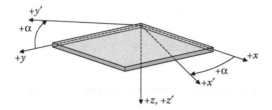

FIGURE 6.6
In-plane coordinate system x'–y', obtained by rotating through ange α about the z-axis.

coordinate system to the new x'–y'' coordinate system using Equation 2.44 (developed in Section 2.13), repeated here for convenience:

$$
\left\{
\begin{array}{c}
\varepsilon_{x'x'} \\
\varepsilon_{y'y'} \\
\dfrac{\gamma_{x'y'}}{2}
\end{array}
\right\}
=
\left[
\begin{array}{ccc}
\cos^2(\alpha) & \sin^2(\alpha) & 2\cos(\alpha)\sin(\alpha) \\
\sin^2(\alpha) & \cos^2(\alpha) & -2\cos(\alpha)\sin(\alpha) \\
-\cos(\alpha)\sin(\alpha) & \cos(\alpha)\sin(\alpha) & \cos^2(\alpha)-\sin^2(\alpha)
\end{array}
\right]
\left\{
\begin{array}{c}
\varepsilon_{xx} \\
\varepsilon_{yy} \\
\dfrac{\gamma_{xy}}{2}
\end{array}
\right\}
$$

(2.44) (repeated)

Substituting Equation 6.13 into Equation 2.44, we can write

$$
\left\{
\begin{array}{c}
\varepsilon_{x'x'} \\
\varepsilon_{y'y'} \\
\gamma_{x'y'}
\end{array}
\right\}
= z
\left\{
\begin{array}{c}
\kappa_{x'x'} \\
\kappa_{y'y'} \\
\kappa_{x'y'}
\end{array}
\right\}
\tag{6.14}
$$

where

$$
\left\{
\begin{array}{c}
\kappa_{x'x'} \\
\kappa_{y'y'} \\
\dfrac{\kappa_{x'y'}}{2}
\end{array}
\right\}
=
\left[
\begin{array}{ccc}
\cos^2(\alpha) & \sin^2(\alpha) & 2\cos(\alpha)\sin(\alpha) \\
\sin^2(\alpha) & \cos^2(\alpha) & -2\cos(\alpha)\sin(\alpha) \\
-\cos(\alpha)\sin(\alpha) & \cos(\alpha)\sin(\alpha) & \cos^2(\alpha)-\sin^2(\alpha)
\end{array}
\right]
\left\{
\begin{array}{c}
\kappa_{xx} \\
\kappa_{yy} \\
\dfrac{\kappa_{xy}}{2}
\end{array}
\right\}
\tag{6.15}
$$

Note that midplane curvatures in the x'–y' coordinate system are related to curvatures in the x–y coordinate system by means of the familiar transformation matrix $[T]$. This reveals that midplane curvatures can be treated as a second-order tensor, and can be transformed from one coordinate system to another in exactly the same way as the strain tensor (or stress tensor) is transformed.

The in-plane principal strains and the orientation of the principal strain coordinate system can also be determined using Equations 2.47 and 2.48, respectively, repeated here for convenience:

$$\varepsilon_{p_1}, \varepsilon_{p_2} = \frac{\varepsilon_{xx} + \varepsilon_{yy}}{2} \pm \sqrt{\left(\frac{\varepsilon_{xx} - \varepsilon_{yy}}{2}\right)^2 + \left(\frac{\gamma_{xy}}{2}\right)^2} \qquad \text{(2.47) (repeated)}$$

$$\theta_{p_\varepsilon} = \frac{1}{2} \arctan\left(\frac{\gamma_{xy}}{\varepsilon_{xx} - \varepsilon_{yy}}\right) \qquad \text{(2.48) (repeated)}$$

Substituting Equation 6.13 into Equation 2.47, we find that for pure bending the in-plane principal strains are given by

$$\varepsilon_{p_1}, \varepsilon_{p_2} = z\left[\frac{(\kappa_{xx} + \kappa_{yy})}{2} \pm \sqrt{\left(\frac{(\kappa_{xx} - \kappa_{yy})}{2}\right)^2 + \left(\frac{\kappa_{xy}}{2}\right)^2}\right] \qquad (6.16)$$

Substituting Equation 6.13 into Equation 2.48, we find that the orientation of the principal strain coordinate system is given by

$$\theta_{p_\varepsilon} = \frac{1}{2} \arctan\left(\frac{\kappa_{xy}}{\kappa_{xx} - \kappa_{yy}}\right) \qquad (6.17)$$

Noting that κ_{xx}, κ_{yy}, and κ_{xy} are midplane values, Equation 6.16 shows that principal strains are linear functions of z. In contrast, Equation 6.17 shows that, for the case of pure bending, the orientation of the principal strain coordinate system is constant and does not vary with through-thickness position, even though the magnitudes of the principal strains do vary with z.

A simplified expression for the principal strains is obtained by writing Equation 6.16 as

$$\varepsilon_{p1} = z\kappa_{p1}$$
$$\varepsilon_{p2} = z\kappa_{p2} \qquad (6.18)$$

where κ_{p1} and κ_{p2} are called *principal curvatures* and are given by

$$\kappa_{p1}, \kappa_{p2} = \frac{(\kappa_{xx} + \kappa_{yy})}{2} \pm \sqrt{\left(\frac{(\kappa_{xx} - \kappa_{yy})}{2}\right)^2 + \left(\frac{\kappa_{xy}}{2}\right)^2} \qquad (6.19)$$

For the case of pure bending, the principal curvatures occur in the same coordinate system as the principal strains. Hence, Equation 6.17 gives the orientation of the coordinate system in which the principal curvatures exist. Since shear strain is zero in the principal strains coordinate system, $\kappa_{p1p2} = 0$ as well.

A physical interpretation of the preceding results can be obtained through sketches of deformed strain elements parallel to the x–y plane, as was done in Section 2.13 (in particular, refer to Example Problem 2.9). A thin plate referenced to a x–y–z coordinate system is shown in Figure 6.7a. A rectangular 3-D element cut out of this plate by two pairs of planes parallel to the x–z and y–z planes is also shown. The dimensions of the element in the x- and y-directions are dx and dy, respectively, while the height of the element equals the plate thickness t. Assuming this plate is subjected to pure bending, then the strains induced at any position z, relative to the x–y coordinate system, can be calculated using Equation 6.13.

Consider as representative examples three 2-D strain elements parallel to the x–y plane and located at positions defined by

- $z = +t/2$ (element i–j–k–l, shown in Figure 6.7b)
- $z = 0$ (element e–f–g–h, shown in Figure 6.7c)
- $z = -t/2$ (element a–b–c–d, shown in Figure 6.7d)

In each case, we imagine a 2-D strain element whose sides are parallel to the x- and y-axis prior to deformation. As the plate is deformed the length of the

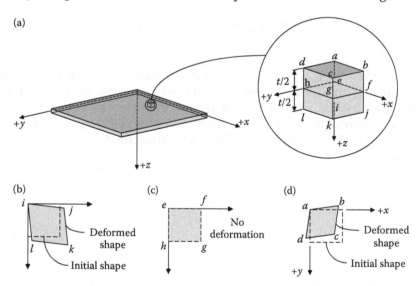

FIGURE 6.7
Illustration of strains induced at the three through-thickness positions $z = -t/2$, 0, and $+t/2$ by pure bending (deformations shown exaggerated for clarity). (a) A 3-D strain element removed from a thin plate referenced to a x–y–z coordinate system; (b) 2-D strain element at $z = +t/2$; (c) 2-D strain element at $z = 0$; (d) 2-D strain element at $z = -t/2$.

element sides increase or decrease, in accordance with the algebraic sign of strains ε_{xx} and ε_{yy}, and the angle between adjacent face of the element changes from $\pi/2$ radians (i.e., $90°$), in accordance with the algebraic sign of γ_{xy}.

First consider strain element i–j–k–l, located at $z = +t/2$ (Figure 6.7b). For a state of pure bending the strains induced at this through-thickness are given by Equation 6.13:

$$\varepsilon_{xx}\big|_{z=t/2} = t\kappa_{xx}/2$$

$$\varepsilon_{yy}\big|_{z=t/2} = t\kappa_{yy}/2$$

$$\gamma_{xy}\big|_{z=t/2} = t\kappa_{xy}/2$$

Assume for illustrative purposes that all curvatures are positive (κ_{xx}, κ_{yy}, $\kappa_{xy} > 0$). This implies that all strains induced at $z = +t/2$ are algebraically positive. A sketch of a deformed element that corresponds to these assumptions is shown (not to scale) in Figure 6.7b. In the deformed condition, the lengths of the element sides have increased, since ε_{xx} and ε_{yy} are positive, and angle j–i–l has decreased, since γ_{xy} is positive.

Now consider strain element e–f–g–h, located at the midplane of the plate, $z = 0$. Since we have assumed a state of pure bending, the strains at the midplane are zero, and consequently element e–f–g–h is not deformed, as shown in Figure 6.7c.

Finally, consider strain element a–b–c–d, located at $z = -t/2$ (Figure 6.7d). Using Equations 6.13 the strains induced at this position are

$$\varepsilon_{xx}\big|_{z=-t/2} = -t\kappa_{xx}/2$$

$$\varepsilon_{yy}\big|_{z=-t/2} = -t\kappa_{yy}/2$$

$$\gamma_{xy}\big|_{z=-t/2} = -t\kappa_{xy}/2$$

Since we have already assumed that all midplane curvatures are positive, these results show that all strains induced at $z = -t/2$ are algebraically negative. A sketch of the deformed element that corresponds to this condition is shown (not to scale) in Figure 6.7d. Note that in this case, the lengths of the element sides have *decreased*, since ε_{xx} and ε_{yy} are negative, and angle b–a–d has *increased*, since γ_{xy} is negative.

The deformed 2-D strains elements shown in Figure 6.7b,c,d are assembled to create a sketch of the entire 3-D element in Figure 6.8. Note that, in accordance with the Kirchhoff hypothesis, the four line segments that define the vertical edges of the element (line segments a–e–i, b–f–j, c–g–k, and d–h–l) remain straight lines after deformation. However, the transverse planes *are*

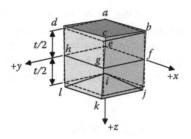

FIGURE 6.8
3-D strain element assembled from the 2-D deformed elements shown in Figure 6.7b,c,d (original shape shown in dashed lines; deformations shown exaggerated for clarity).

no longer plane after deformation. For example, plane *b–j–k–c* has been *twisted* during deformation of the plate. Inspection of Figures 6.7b,c,d and 6.8 reveal that transverse planes do not remain plane after deformation due to curvature κ_{xy}. That is, if $\kappa_{xy} \neq 0$, shear strain γ_{xy} varies with through-thickness position z, in accordance with Equation 6.13. It is this through-thickness variation in γ_{xy} that leads to twisting of the transverse planes. For this reason κ_{xy} is known as the *twist curvature*.

We will now repeat this process for a rectangular 3-D element referenced to the principal strain coordinate system, as shown in Figure 6.9a. Once again we

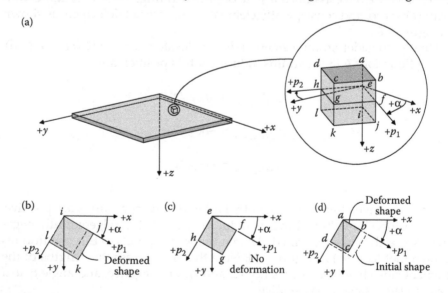

FIGURE 6.9
Illustration of principal strains induced at the three through-thickness positions $z = -t/2$, 0, and $+t/2$ by pure bending; compare with Figure 6.7 (deformations shown exaggerated for clarity). (a) A 3-D strain element removed from a thin plate referenced to the principal strain coordinate system; (b) 2-D strain element at $z = +t/2$; (c) 2-D strain element at $z = 0$; (d) 2-D strain element at $z = -t/2$.

assume the plate is subjected to pure bending. The principal strains induced at any position z can, therefore, be calculated using Equation 6.18. We consider three 2-D strain elements located at through-thickness positions $z = -t/2$, 0, and $+t/2$. A 2-D sketch of the deformed strain elements at these three positions is shown in Figure 6.9b,c,d. Since the element is aligned with the principal strain coordinate system no shear strain is induced in any element, that is, all corner angles equal $\pi/2$ radians before and after deformation. Assuming for illustrative purposes that both principal strains are positive ($\kappa_{p_1}, \kappa_{p_2} > 0$), the principal strains induced at $z = +t/2$ are tensile (Figure 6.9b), whereas the principal strains induced at $z = -t/2$ are compressive (Figure 6.9d). No deformations occur at $z = 0$ since we have assumed pure bending and the midplane is, therefore, the neutral surface. The deformed 2-D strain elements shown in Figures 6.9b,c,d are assembled to create a sketch of the deformed 3-D element in Figure 6.10. As before, the four line segments that define the vertical edges of the element (line segments $a–e–i$, $b–f–j$, $c–g–k$, and $d–h–l$) remain straight lines after deformation. However, in contrast to Figure 6.8, the *planes* in which these line segment lie *remain plane* after deformation. Twisting of these transverse planes does not occur. When referenced to the coordinate system in which the principal curvatures exist, the transverse planes of the strain element simply *rotate* about the neutral surface.

A summary of the results presented in this section is as follows. We have found that curvatures can be treated as second-order tensors, and can be rotated from one coordinate system to another using the same process as that used to transform the strain or stress tensors. For a thin plate governed by the Kirchhoff hypothesis, midplane curvatures transform according to Equation 6.15. In general, three midplane curvatures are induced in a thin plate: κ_{xx}, κ_{yy}, and κ_{xy}. Curvature κ_{xy} is called the twist curvature because it represents a twisting of a plane transverse to the midplane of the plate. The principal curvatures, κ_{p1} and κ_{p2}, are the maximum and minimum curvatures, respectively, induced at a given point in a plate. Equation 6.17 gives the orientation of the coordinate system in which the principal curvatures exist, and no twisting occurs in this coordinate system (the twist curvature equals zero in the

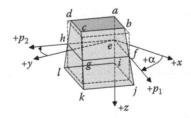

FIGURE 6.10
3-D strain element assembled from 2-D deformed elements referenced to the principal strain coordinate system and shown in Figure 6.9b,c,d; compare with Figure 6.8 (deformations shown greatly exaggerated for clarity).

principal coordinate system). For a thin plate in pure bending, the orientation of the principal strain coordinate system is constant through the thickness of the plate, and the principal curvatures are induced in this coordinate system.

The reader should note that the results in this section are valid for the special case of pure bending. Some of the results presented above are not valid for the case of general nonuniform plate bending. For example, if the midplane is not the neutral surface (i.e., if $\varepsilon_{xx}^o, \varepsilon_{yy}^o, \gamma_{xy}^o \neq 0$) then it can be shown that the orientation of the principal strain coordinate system is not constant but rather varies as a function of z. However, even in this more general case midplane curvatures transform according to Equation 6.15, and principal curvatures are given by Equation 6.19.

A more detailed discussion of principal strains and curvatures under general conditions will not be presented, since these topics are not of immediate interest. The results presented in this section for pure bending will be applied in Chapter 8, where the topic of composite beams is considered.

6.4 Standard Methods of Describing Composite Laminates

A magnified edge view of a thin composite laminate that contains "n" plies is shown in Figure 6.11. The figure is similar to the edge view of a thin plate shown in Figure 6.4, except that now the ply interface positions are shown. The thickness of ply "k" will be denoted t_k. The origin of the $x–z$ axes lies at the geometric midsurface of the laminate, and so the outer surfaces of the laminate exist at $z = -t/2$ and $z = +t/2$, where t equals the total thickness of the laminate. The total thickness of the laminate equals the sum of all ply

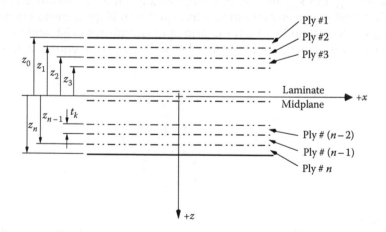

FIGURE 6.11
An edge view of an n-ply laminate showing ply interface positions.

thicknesses: $t = \sum_{k=1}^{n} t_k$. Note that a ply interface does not necessarily exist at the midplane of the laminate, as indicated in Figure 6.11.

We will require a method of specifying the coordinate position of each ply interface with respect to the laminate midplane. By convention, we will denote the coordinate position of the outermost laminate surface in the *negative* z-direction as position "z_0," that is, $z_0 \equiv -t/2$. Note that z_0 is always an algebraically negative number. The coordinate position of the interface between plies 1 and 2 is denoted "z_1," and $z_1 = z_0 + t_1$. Similarly, the coordinate position of the interface between plies 2 and 3 is denoted "z_2," and $z_2 = z_1 + t_2$, and so on. For an *n*-ply laminate, the outermost surface of the laminate in the positive z-direction is labeled z_n ; obviously, $z_n = +t/2$. Note that in all cases z_n is an algebraically positive number. Also note that the total thickness of the laminate equals $(z_n - z_0)$, and the thickness of an individual ply k is $t_k = (z_k - z_k-1)$. For example, the thickness of ply 2 is $t_2 = (z_2 - z_1)$.

We also need a method of consistently describing the *stacking sequence* of a composite laminate. That is, we need to develop a method of indicating the orientation of the principal material coordinate system of each ply with respect to the x-axis, and the order in which they appear. As discussed in previous chapters, a ply may contain unidirectional fibers, or may consist of a woven or braided fabric. In the case of fabric plies there are two or more fiber directions present within each ply, but the orientation of the principal material coordinate system is always evident due to the symmetric pattern of the fiber architecture. For simplicity, in the following discussion, it will be assumed that all plies are composed of unidirectional fibers. In this case, the angle between the principal material coordinate system and the x-axis is equivalent to the angle between the fibers and the x-axis. Hence, in the discussion to follow we will simply refer to the "fiber angle" in each ply. It should be understood that this angle actually refers to the orientation of the principal material coordinate system. This terminology is adopted simply because the phrase "fiber angle" is more concise than the phrase "principal material coordinate system angle."

To describe the stacking sequence of a laminate we list fiber angles within square brackets. The fiber angle (in degrees) of ply number 1 is listed first, followed by the fiber angle of ply 2, ply 3, and so on. Each fiber angle is separated by a slash, "/". For example, a four-ply laminate consisting of plies with fiber angles of 0°, 45°, –20°, and 90° is shown in Figure 6.12a. This laminate is denoted $[0/45/-20/90]_T$. The subscript "T" has been used to indicate that the "total" laminate has been described, that is, a fiber angle is listed for all plies within the laminate within the square brackets.

In practice, it is common to encounter laminates with 10, 20, 30, or (in unusual cases) even hundreds of plies. In such cases it becomes very tedious to list all fiber angles within the laminate. Fortunately, for many reasons (some of which will be described later in this chapter) composite laminates are usually designed with some systematic pattern of fiber angles, which allows us to abbreviate the listing of ply fiber angles which appear within the laminate. It is easiest to introduce these abbreviations with a series of examples.

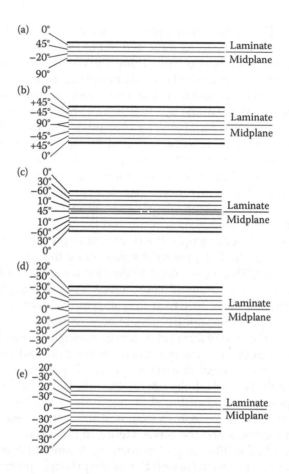

FIGURE 6.12
Edge view of several composite laminates illustrating abbreviations used to describe stacking sequences. (a) $[0/45/{-}20/90]_T$; (b) $[0/{\pm}45/90]_s$; (c) $[0/30/{-}60/10/\overline{45}]_s$; (d) $[(20/{-}30)_s/0]_s$; (e) $[(20/{-}30)_2/0]_s$.

Consider the 8-ply laminate shown in Figure 6.12b. In this case, the fiber angles are (starting from ply #1): $0°$, $+45°$, $-45°$, $90°$, $90°$, $-45°$, $+45°$, and $0°$. This is an example of a *symmetric laminate*, since the fiber angles are symmetric about the laminate midplane. This laminate is denoted $[0/\pm45/90]_s$. The subscript "s" indicates that the four fiber angles listed appear symmetrically about the midplane, and hence a total of 8 plies exist within the laminate, even though only four angles are listed.

A 9-ply laminate containing fiber angles $0°$, $30°$, $-60°$, $10°$, $45°$, $10°$, $-60°$, $30°$, and $0°$ is shown in Figure 6.12c. This laminate is symmetric about the geometric midplane, but because an odd number of plies are present the midplane passes through the center of the 45-degree ply (ply number 5). This laminate is denoted $[0/30/{-}60/10/\overline{45}]_s$. That is, a bar is used to indicate that the midplane passes through the 45-degree ply, and hence "$4\frac{1}{2}$" plies exist symmetrically about the midplane of this laminate.

A 10-ply laminate containing fiber angles 20°, –30°, –30°, 20°, 0°, 0°, 20°, –30°, –30°, and 20° is shown in Figure 6.12d. This laminate is symmetric about the midplane, but also contains a symmetric pattern within both halves of the laminate. In this case, the laminate is denoted $[(20/-30)_s/0]_s$. The subscript "s" appears twice, first to indicate that fiber angles 20° and –30° appear symmetrically within one-half of the laminate, and the second to indicate that the entire laminate is symmetric about the midplane.

A final example is the 10-ply laminate shown in Figure 6.12e. In this case, the fiber angles are 20°, –30°, 20°, –30°, 0°, 0°, –30°, 20°, –30°, and 20°. This laminate is denoted $[(20/-30)_2/0]_s$ where the subscript "2" indicates that the fiber pattern listed within the parentheses occurs twice. Note that this laminate is similar but not identical to that shown in Figure 6.12d.

Example Problem 6.2

A $[(0/\pm30/90)_2/45/\overline{20}]_s$ laminate is fabricated using a graphite–epoxy material system. Each ply has a thickness of 0.125 mm. Determine the number of plies in the laminate, the total laminate thickness, and the z-coordinate of each ply interface.

SOLUTION

An ordered listing of all fiber angles that appear in the laminate is as follows:

$$[0°,30°,-30°,90°,0°,30°,-30°,90°,45°,20°,45°,90°,-30°,30°,0°,90°,-30°,30°,0°]$$

↑   ↑   ↑

ply #1   ply # 10   ply # 19

(midplane)

The laminate contains a total of 19 plies, the fiber angles appear symmetrically about the midplane of the laminate, and the midplane passes through the center of the 20-degree ply. Since all plies are made of the same composite material system, they all have the same thickness. The total laminate thickness is, therefore, $t = 19 \ (0.125 \text{ mm}) = 2.375 \text{ mm}$.

Ply interface positions are

$z_0 = -t/2 = -1.1875 \text{ mm}$ $z_1 = z_0 + t_1 = -1.0625 \text{ mm}$ $z_2 = z_1 + t_2 = -0.9375 \text{ mm}$

$z_3 = z_2 + t_3 = -0.8125 \text{ mm}$ $z_4 = z_3 + t_4 = -0.6875 \text{ mm}$ $z_5 = z_4 + t_5 = -0.5625 \text{ mm}$

$z_6 = z_5 + t_6 = -0.4375 \text{ mm}$ $z_7 = z_6 + t_7 = -0.3125 \text{ mm}$ $z_8 = z_7 + t_8 = -0.1875 \text{ mm}$

$z_9 = z_8 + t_9 = -0.0625 \text{ mm}$ $z_{10} = z_9 + t_{10} = 0.0625 \text{ mm}$ $z_{11} = z_{10} + t_{11} = 0.1875 \text{ mm}$

$z_{12}=z_{11}+t_{12}=0.3125$ mm $z_{13}=z_{12}+t_{13}=0.4375$ mm $z_{14}=z_{13}+t_{14}=0.5625$ mm

$z_{15}=z_{14}+t_{15}=0.6875$ mm $z_{16}=z_{15}+t_{16}=0.8125$ mm $z_{17}=z_{16}+t_{17}=0.9375$ mm

$z_{18}=z_{17}+t_{18}=1.0625$ mm $z_{19}=z_{18}+t_{19}=1.1875$ mm

Note that the total laminate thickness equals the difference between z_{19} and z_0, as expected: $t = z_{19} - z_0 = 1.1875$ mm $- (-1.1875$ mm$) = 2.375$ mm.

6.5 Calculating Ply Strains and Stresses

The theory developed to this point allows calculation of the elastic strains and stresses present at any through-thickness position within a multiangle composite laminate subjected to known midplane strains and curvatures. A summary of how strains and stresses are calculated is as follows:

Laminate description: A composite laminate is described by specifying:

- The laminate stacking sequence (i.e., the number of plies within a laminate and the fiber angles of each ply)
- The material properties and thickness of each ply

Note that the plies are not necessarily all of the same material type. For example, some plies within a laminate may be of graphite–epoxy while others may be glass-epoxy.

Once the stacking sequence and thickness of each ply has been specified, the total laminate thickness and interface positions throughout the laminate may be determined, as previously illustrated in Figure 6.11 and Example Problem 6.2. Also, the transformed reduced stiffness matrix, $[\bar{Q}]$, can be calculated for each ply in accordance with Equations 5.31.

Ply Strains: Strains are calculated using the laminate strains and curvatures, $(\varepsilon_{xx}^o, \varepsilon_{yy}^o, \gamma_{xy}^o)$ and $(\kappa_{xx}, \kappa_{yy}, \kappa_{xy})$ respectively, in accordance with the Kirchhoff Hypothesis (Equation 6.12). For example, the strains induced in a distance z_k from the laminate midplane, are

$$\left\{ \begin{array}{c} \varepsilon_{xx} \\ \varepsilon_{yy} \\ \gamma_{xy} \end{array} \right\}\Bigg|_{z=z_k} = \left\{ \begin{array}{c} \varepsilon_{xx}^o \\ \varepsilon_{yy}^o \\ \gamma_{xy}^o \end{array} \right\} + z_k \left\{ \begin{array}{c} \kappa_{xx} \\ \kappa_{yy} \\ \kappa_{xy} \end{array} \right\}$$

Note that these strains are referenced to the x–y coordinate system. If desired, these strains can be rotated from the x–y coordinate system to the "local" 1–2 coordinate system for each ply (defined by the ply fiber angle) using Equation 2.44:

$$\left\{ \begin{matrix} \varepsilon_{11} \\ \varepsilon_{22} \\ \gamma_{12}/2 \end{matrix} \right\}\Bigg|_{z=z_k} = [T]_k \left\{ \begin{matrix} \varepsilon_{xx} \\ \varepsilon_{yy} \\ \gamma_{xy}/2 \end{matrix} \right\}\Bigg|_{z=z_k} = \begin{bmatrix} \cos^2\theta & \sin^2\theta & 2\cos\theta\sin\theta \\ \sin^2\theta & \cos^2\theta & -2\cos\theta\sin\theta \\ -\cos\theta\sin\theta & \cos\theta\sin\theta & \cos^2\theta - \sin^2\theta \end{bmatrix}_k$$

$$\times \left\{ \begin{matrix} \varepsilon_{xx} \\ \varepsilon_{yy} \\ \gamma_{xy}/2 \end{matrix} \right\}\Bigg|_{z=z_k}$$

Ply Stresses: Once ply strains are determined, ply stresses are calculated using Hooke's Law, as discussed in Section 5.2. For example, the stresses induced at a distance z_k from the laminate midplane are calculated using Equation 5.30:

$$\left\{ \begin{matrix} \sigma_{xx} \\ \sigma_{yy} \\ \tau_{xy} \end{matrix} \right\}\Bigg|_{z=z_k} = \begin{bmatrix} \overline{Q}_{11} & \overline{Q}_{12} & \overline{Q}_{16} \\ \overline{Q}_{12} & \overline{Q}_{22} & \overline{Q}_{26} \\ \overline{Q}_{16} & \overline{Q}_{26} & \overline{Q}_{66} \end{bmatrix}_{z=z_k} \left\{ \begin{matrix} \varepsilon_{xx} - \Delta T \alpha_{xx} - \Delta M \beta_{xx} \\ \varepsilon_{yy} - \Delta T \alpha_{yy} - \Delta M \beta_{yy} \\ \gamma_{xy} - \Delta T \alpha_{xy} - \Delta M \beta_{xy} \end{matrix} \right\}\Bigg|_{z=z_k}$$

Note that the material properties used in this calculation (specifically, $[\overline{Q}]$, α_{ij}, and β_{ij}), are properties of the ply that exists at position z_k, and in particular are functions of the ply fiber angle, θ. Since fiber angle generally varies from one ply to the next, these material properties also vary from one ply to the next.

If desired, stresses can be rotated from the x–y coordinate system to the "local" 1–2 coordinate system for each ply using Equation 2.20:

$$\left\{ \begin{matrix} \sigma_{11} \\ \sigma_{22} \\ \tau_{12} \end{matrix} \right\}\Bigg|_{z=z_k} = [T]_{z=z_k} \left\{ \begin{matrix} \sigma_{xx} \\ \sigma_{yy} \\ \tau_{xy} \end{matrix} \right\}\Bigg|_{z=z_k}$$

A numerical example that illustrates these calculations is presented in the following Example Problem.

Example Problem 6.3

Assume that the panel considered in Example Problem 6.1 is actually an eight-ply $[0/30/90/-30]_s$ graphite–epoxy laminate. Assume the laminate was initially flat and stress-free, that is, ignore possible preexisting stresses/strains due to temperature and/or moisture changes.

Determine the strains and stresses induced at each ply interface. Use material properties listed in Table 3.1, and assume the thickness of each ply is 0.125 mm.

SOLUTION

From Example Problem 6.1, the midplane strains and curvatures are

$$\varepsilon_{xx}^o = 0\,\mu m/m \qquad \kappa_{xx} = -0.50\,rad/m$$

$$\varepsilon_{yy}^o = -1300\,\mu m/m \quad \kappa_{yy} = 0.40\,rad/m$$

$$\gamma_{xy}^o = 900\,\mu rad \qquad \kappa_{xy} = -0.20\,rad/m$$

To determine ply interface positions, first note that the total laminate thickness is

$$t = (8\ plies)(0.125\ mm) = 1.0\ mm = 0.001\ m$$

A total of nine ply interface positions must be determined, since there are eight plies in the laminate. Following the numbering scheme discussed in Section 6.4, and referring to Figure 6.11, ply interface positions are

$$z_0 = -t/2 = -(0.00100\ m)/2 = -0.000500\ m$$

$$z_1 = z_0 + t_1 = -0.000500\ m + 0.000125\ m = -0.000375\ m$$

$$z_2 = z_1 + t_2 = -0.000375\ m + 0.000125\ m = -0.000250\ m$$

$$z_3 = z_2 + t_3 = -0.000250\ m + 0.000125\ m = -0.000125\ m$$

$$z_4 = z_3 + t_4 = -0.000125\ m + 0.000125\ m = 0.000000\ m$$

$$z_5 = z_4 + t_5 = 0.000000\ m + 0.000125\ m = 0.000125\ m$$

$$z_6 = z_5 + t_6 = 0.000125\ m + 0.000125\ m = 0.000250\ m$$

$$z_7 = z_6 + t_7 = 0.000250\ m + 0.000125\ m = 0.000375\ m$$

$$z_8 = z_7 + t_7 = 0.000375\ m + 0.000125\ m = 0.000500\ m$$

Strain Calculations: Strains are calculated using Equation 6.12, and can be determined at any through-thickness position. Usually, strains of greatest interest are those induced at the ply interface locations. For example, strains present at the outer surface of ply #1 (i.e., strains present at $z_o = -0.000500$ m) are

$$\left\{\begin{matrix} \varepsilon_{xx} \\ \varepsilon_{yy} \\ \gamma_{xy} \end{matrix}\right\}\Bigg|_{z=z_0} = \left\{\begin{matrix} \varepsilon_{xx}^o \\ \varepsilon_{yy}^o \\ \gamma_{xy}^o \end{matrix}\right\} + z_0 \left\{\begin{matrix} \kappa_{xx} \\ \kappa_{yy} \\ \kappa_{xy} \end{matrix}\right\} = \left\{\begin{matrix} 0 \\ -1300 \times 10^{-6} \text{ m/m} \\ 900 \times 10^{-6} \text{ m/m} \end{matrix}\right\} + (-0.000500 \text{ m}) \left\{\begin{matrix} -0.50 \text{ rad/m} \\ 0.40 \text{ rad/m} \\ -0.20 \text{ rad/m} \end{matrix}\right\}$$

$$\left\{\begin{matrix} \varepsilon_{xx} \\ \varepsilon_{yy} \\ \gamma_{xy} \end{matrix}\right\}\Bigg|_{z=z_0} = \left\{\begin{matrix} 250 \text{ }\mu\text{m/m} \\ -1500 \text{ }\mu\text{m/m} \\ 1000 \text{ }\mu\text{rad} \end{matrix}\right\}$$

Similarly, strains present at the interface between plies #1 and #2 (i.e., strains present at $z_1 = -0.000375$ m) are

$$\left\{\begin{matrix} \varepsilon_{xx} \\ \varepsilon_{yy} \\ \gamma_{xy} \end{matrix}\right\}\Bigg|_{z=z_1} = \left\{\begin{matrix} \varepsilon_{xx}^o \\ \varepsilon_{yy}^o \\ \gamma_{xy}^o \end{matrix}\right\} + z_1 \left\{\begin{matrix} \kappa_{xx} \\ \kappa_{yy} \\ \kappa_{xy} \end{matrix}\right\} = \left\{\begin{matrix} 0 \\ -1300 \times 10^{-6} \text{ m/m} \\ 900 \times 10^{-6} \text{ m/m} \end{matrix}\right\} + (-0.000375 \text{ m}) \left\{\begin{matrix} -0.50 \text{ rad/m} \\ 0.40 \text{ rad/m} \\ -0.20 \text{ rad/m} \end{matrix}\right\}$$

$$\left\{\begin{matrix} \varepsilon_{xx} \\ \varepsilon_{yy} \\ \gamma_{xy} \end{matrix}\right\}\Bigg|_{z=z_1} = \left\{\begin{matrix} 188 \text{ }\mu\text{m/m} \\ -1450 \text{ }\mu\text{m/m} \\ 975 \text{ }\mu\text{rad} \end{matrix}\right\}$$

Strains present at all remaining interfaces are calculated in exactly the same fashion.

Strains calculated at all ply interfaces are summarized in Table 6.1, and are plotted in Figure 6.13. Note that all three strain components (ε_{xx}, ε_{yy}, and γ_{xy}) are predicted to be linearly distributed through the plate thickness. This linear distribution is a direct consequence of the Kirchhoff hypothesis, which is a good approximation as long as the plate is "thin." In fact, identical strain distributions would be predicted for any thin plate subjected to the midplane strains and curvatures specified in Example Problem 6.1. For example, we would predict the identical strains if an aluminum plate were under consideration rather than a laminated composite plate.

The strains listed in Table 6.1 and plotted in Figure 6.13 are referenced to the global x–y coordinate system. As will be seen, knowledge of ply strains referenced to the local 1–2 coordinate system (defined by the fiber angle within each ply) is often required. Transformation of the strain tensor from one coordinate system to another was reviewed in Chapter 2, and in particular strains can be rotated from the x–y coordinate system to the 1–2 coordinate system using Equation 2.44. In practice, strains are usually calculated at both the "top" and "bottom" interface for each ply. Example calculations for plies 1 and 2 are listed below:

TABLE 6.1

Ply Interface Strains in a $[0/{-30}/90/30]_s$ Graphite–Epoxy Laminate Subjected to the Midplane Strains and Curvatures Discussed in Example Problem 6.1

z-Coordinate (mm)	ε_{xx} (μm/m)	ε_{yy} (μm/m)	γ_{xy} (μ Radians)
−0.500	250	−1500	1000
−0.375	188	−1450	975
−0.250	125	−1400	950
−0.125	62	−1350	925
0.0	0	−1300	900
0.125	−62	−1250	875
0.250	−125	−1200	850
0.375	−188	−1150	825
0.500	−250	−1100	800

Note: Strains are referenced to the x–y coordinate system.

Ply 1: Since $\theta_1 = 0°$, the x–y and 1–2 coordinate systems are coincident, and, therefore, the description of the strain tensor is identical in both coordinate systems. This can be confirmed through application of Equation 2.44:

Top interface:

$$
\left\{ \begin{array}{c} \varepsilon_{11} \\ \varepsilon_{22} \\ \gamma_{12}/2 \end{array} \right\}^{Ply1}_{z=z0} =
\left[\begin{array}{ccc}
\cos^2\theta_1 & \sin^2\theta_1 & 2\cos\theta_1\sin\theta_1 \\
\sin^2\theta_1 & \cos^2\theta_1 & -2\cos\theta_1\sin\theta_1 \\
-\cos\theta_1\sin\theta_1 & \cos\theta_1\sin\theta_1 & \cos^2\theta_1 - \sin^2\theta_1
\end{array} \right]
\left\{ \begin{array}{c} \varepsilon_{xx} \\ \varepsilon_{yy} \\ \gamma_{xy}/2 \end{array} \right\}^{Ply1}_{z=z0}
$$

$$
\left\{ \begin{array}{c} \varepsilon_{11} \\ \varepsilon_{22} \\ \gamma_{12}/2 \end{array} \right\}^{Ply1}_{z=z0} =
\left[\begin{array}{ccc}
\cos^2(0°) & \sin^2(0°) & 2\cos(0°)\sin(0°) \\
\sin^2(0°) & \cos^2(0°) & -2\cos(0°)\sin(0°) \\
-\cos(0°)\sin(0°) & \cos(0°)\sin(0°) & \cos^2(0°) - \sin^2(0°)
\end{array} \right]
$$

$$
\times \left\{ \begin{array}{c} 250\ \mu m/m \\ -1500\ \mu m/m \\ 1000\ \mu rad/2 \end{array} \right\}^{Ply1}_{z=z0}
$$

$$
\left\{ \begin{array}{c} \varepsilon_{11} \\ \varepsilon_{22} \\ \gamma_{12}/2 \end{array} \right\}^{Ply1}_{z=z0} =
\left[\begin{array}{ccc}
1 & 0 & 0 \\
0 & 1 & 0 \\
0 & 0 & 1
\end{array} \right]
\left\{ \begin{array}{c} 250\ \mu m/m \\ -1500\ \mu m/m \\ 1000\ \mu rad/2 \end{array} \right\}^{Ply1}_{z=z0}
$$

$$
\left\{ \begin{array}{c} \varepsilon_{11} \\ \varepsilon_{22} \\ \gamma_{12} \end{array} \right\}^{Ply1}_{z=z0} =
\left\{ \begin{array}{c} 250\ \mu m/m \\ -1500\ \mu m/m \\ 1000\ \mu rad \end{array} \right\}^{Ply1}_{z=z0}
$$

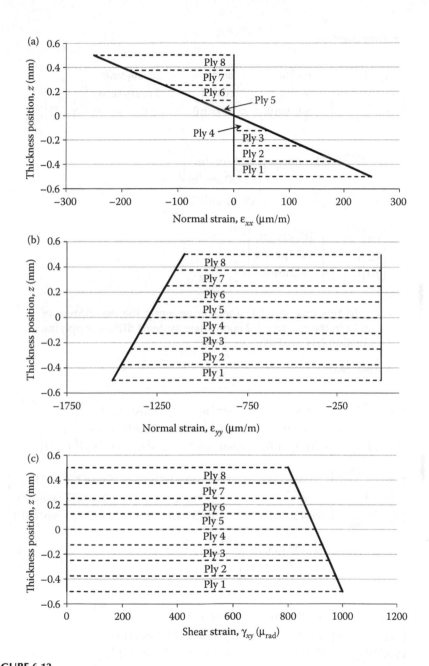

FIGURE 6.13
Through-thickness strain plots implied by the midplane strains and curvatures discussed in Example Problem 6.1. Strains referenced to the x–y coordinate system. (a) Normal strain ε_{xx}; (b) normal strain ε_{yy}; (c) shear strain γ_{xy}.

Bottom interface:

$$\left\{\begin{array}{c} \varepsilon_{11} \\ \varepsilon_{22} \\ \gamma_{12}/2 \end{array}\right\}^{Ply1}_{z=z1} = \begin{bmatrix} \cos^2\theta_1 & \sin^2\theta_1 & 2\cos\theta_1\sin\theta_1 \\ \sin^2\theta_1 & \cos^2\theta_1 & -2\cos\theta_1\sin\theta_1 \\ -\cos\theta_1\sin\theta_1 & \cos\theta_1\sin\theta_1 & \cos^2\theta_1-\sin^2\theta_1 \end{bmatrix} \left\{\begin{array}{c} \varepsilon_{xx} \\ \varepsilon_{yy} \\ \gamma_{xy}/2 \end{array}\right\}^{Ply1}_{z=z1}$$

$$\left\{\begin{array}{c} \varepsilon_{11} \\ \varepsilon_{22} \\ \gamma_{12}/2 \end{array}\right\}^{Ply1}_{z=z1} = \begin{bmatrix} 1 & 0 & 0 \\ 0 & 1 & 0 \\ 0 & 0 & 1 \end{bmatrix} \left\{\begin{array}{c} 188\,\mu m/m \\ -1450\,\mu m/m \\ (975\,\mu rad)/2 \end{array}\right\}^{Ply1}_{z=z1}$$

$$\left\{\begin{array}{c} \varepsilon_{11} \\ \varepsilon_{22} \\ \gamma_{12} \end{array}\right\}^{Ply1}_{z=z1} = \left\{\begin{array}{c} 188\,\mu m/m \\ -1450\,\mu m/m \\ 975\,\mu rad \end{array}\right\}^{Ply1}_{z=z1}$$

Ply 2: In this case, $\theta_2 = 30°$ and consequently the description of strain in the x–y and 1–2 coordinate systems differs. Applying Equation 2.44, we have:

Top interface:

$$\left\{\begin{array}{c} \varepsilon_{11} \\ \varepsilon_{22} \\ \gamma_{12}/2 \end{array}\right\}^{Ply2}_{z=z1} = \begin{bmatrix} \cos^2\theta_2 & \sin^2\theta_2 & 2\cos\theta_2\sin\theta_2 \\ \sin^2\theta_2 & \cos^2\theta_2 & -2\cos\theta_2\sin\theta_2 \\ -\cos\theta_2\sin\theta_2 & \cos\theta_2\sin\theta_2 & \cos^2\theta_2-\sin^2\theta_2 \end{bmatrix} \left\{\begin{array}{c} \varepsilon_{xx} \\ \varepsilon_{yy} \\ \gamma_{xy}/2 \end{array}\right\}^{Ply2}_{z=z1}$$

$$\left\{\begin{array}{c} \varepsilon_{11} \\ \varepsilon_{22} \\ \gamma_{12}/2 \end{array}\right\}^{Ply2}_{z=z1} = \begin{bmatrix} \cos^2(30°) & \sin^2(30°) & 2\cos(30°)\sin(30°) \\ \sin^2(30°) & \cos^2(30°) & -2\cos(30°)\sin(30°) \\ -\cos(30°)\sin(30°) & \cos(30°)\sin(30°) & \cos^2(30°)-\sin^2(30°) \end{bmatrix}$$

$$\times \left\{\begin{array}{c} 188\,\mu m/m \\ -1450\,\mu m/m \\ (975\,\mu rad)/2 \end{array}\right\}^{Ply2}_{z=z1}$$

$$\left\{\begin{array}{c} \varepsilon_{11} \\ \varepsilon_{22} \\ \gamma_{12}/2 \end{array}\right\}^{Ply2}_{z=z1} = \begin{bmatrix} 0.750 & 0.250 & 0.866 \\ 0.250 & 0.750 & -0.866 \\ -0.433 & 0.433 & 0.500 \end{bmatrix} \left\{\begin{array}{c} 188\,\mu m/m \\ -1450\,\mu m/m \\ (975\,\mu rad)/2 \end{array}\right\}^{Ply2}_{z=z1}$$

$$\left\{\begin{array}{c} \varepsilon_{11} \\ \varepsilon_{22} \\ \gamma_{12} \end{array}\right\}^{Ply2}_{z=z1} = \left\{\begin{array}{c} 200\,\mu m/m \\ -1463\,\mu m/m \\ -931\,\mu rad \end{array}\right\}^{Ply2}_{z=z1}$$

Bottom interface:

$$\left\{\begin{matrix} \varepsilon_{11} \\ \varepsilon_{22} \\ \gamma_{12}/2 \end{matrix}\right\}\Bigg|_{z=z_2}^{Ply2} = \begin{bmatrix} \cos^2\theta_2 & \sin^2\theta_2 & 2\cos\theta_2\sin\theta_2 \\ \sin^2\theta_2 & \cos^2\theta_2 & -2\cos\theta_2\sin\theta_2 \\ -\cos\theta_2\sin\theta_2 & \cos\theta_2\sin\theta_2 & \cos^2\theta_2-\sin^2\theta_2 \end{bmatrix} \left\{\begin{matrix} \varepsilon_{xx} \\ \varepsilon_{yy} \\ \gamma_{xy}/2 \end{matrix}\right\}\Bigg|_{z=z_2}^{Ply2}$$

$$\left\{\begin{matrix} \varepsilon_{11} \\ \varepsilon_{22} \\ \gamma_{12}/2 \end{matrix}\right\}\Bigg|_{z=z_2}^{Ply2} = \begin{bmatrix} \cos^2(30°) & \sin^2(30°) & 2\cos(30°)\sin(30°) \\ \sin^2(30°) & \cos^2(30°) & -2\cos(30°)\sin(30°) \\ -\cos(30°)\sin(30°) & \cos(30°)\sin(30°) & \cos^2(30°)-\sin^2(30°) \end{bmatrix}$$

$$\times \left\{\begin{matrix} 125\,\mu m/m \\ -1400\,\mu m/m \\ (950\,\mu rad)/2 \end{matrix}\right\}\Bigg|_{z=z_2}^{Ply2}$$

$$\left\{\begin{matrix} \varepsilon_{11} \\ \varepsilon_{22} \\ \gamma_{12}/2 \end{matrix}\right\}\Bigg|_{z=z_2}^{Ply2} = \begin{bmatrix} 0.750 & 0.250 & 0.866 \\ 0.250 & 0.750 & -0.866 \\ -0.433 & 0.433 & 0.500 \end{bmatrix} \left\{\begin{matrix} 125\,\mu m/m \\ -1400\,\mu m/m \\ (950\,\mu rad)/2 \end{matrix}\right\}\Bigg|_{z=z_2}^{Ply2}$$

$$\left\{\begin{matrix} \varepsilon_{11} \\ \varepsilon_{22} \\ \gamma_{12} \end{matrix}\right\}\Bigg|_{z=z_2}^{Ply2} = \left\{\begin{matrix} 155\,\mu m/m \\ -1430\,\mu m/m \\ -846\,\mu rad \end{matrix}\right\}\Bigg|_{z=z_2}^{Ply2}$$

Ply strains referenced to the local 1–2 coordinate systems at all interface locations are summarized in Table 6.2 and plotted in Figure 6.14. Comparing Figures 6.13 and 6.14, it is apparent that the through-thickness strain distributions no longer appear to be continuous when referenced to the ply 1–2 coordinate systems. This is illusionary, in the sense that strains appear to be discontinuous only because the coordinate system used to describe the through-thickness strain is varied discontinuously at each ply interface.

Stress calculations: Since strains are now known at all ply interface positions, we can calculate stresses at these locations using Equation 5.30. Note that we have assumed that temperature and moisture content both remain constant in this example, so $\Delta T = \Delta M = 0$. During these calculations we will require the transformed reduced stiffness matrix for each ply. Using graphite–epoxy material properties (from Table 3.1), Equations 5.11 and 5.31, we find:

For 0°plies:

$$\left[\overline{Q}\right]_{0°plies} = \begin{bmatrix} 170.9\times10^9 & 3.016\times10^9 & 0 \\ 3.016\times10^9 & 10.05\times10^9 & 0 \\ 0 & 0 & 13.00\times10^9 \end{bmatrix}(Pa)$$

288 Structural Analysis of Polymeric Composite Materials

TABLE 6.2

Ply Interface Strains in a $[0/-30/90/30]_s$ Graphite–Epoxy Laminate Subjected to the Midplane Strains and Curvatures Discussed in Example Problem 6.1

Ply Number	z-Coordinate (mm)	ε_{11} (μm/m)	ε_{22} (μm/m)	γ_{12} (μrad)
Ply	−0.500	250	−1500	1000
1	−0.375	188	−1450	975
Ply	−0.375	200	−1463	−931
2	−0.250	155	−1430	−846
Ply	−0.250	−1400	125	−950
3	−0.125	−1350	63	−925
Ply	−0.125	−691	−596	1686
4	0.000	−715	−585	1576
Ply	0.000	−715	−585	1576
5	0.125	−738	−574	1466
Ply	0.125	−1250	−62	−875
6	0.250	−1200	−125	−850
Ply	0.250	−26	−1299	−506
7	0.375	−71	−1267	−421
Ply	0.375	−188	−1150	825
8	0.500	−250	−1100	800

Note: Strains are referenced to the 1–2 coordinate system local to individual plies.

For 30° plies:

$$\left[\overline{Q}\right]_{30°\,plies} = \begin{bmatrix} 107.6 \times 10^9 & 26.06 \times 10^9 & 48.3 \times 10^9 \\ 26.06 \times 10^9 & 27.22 \times 10^9 & 21.52 \times 10^9 \\ 48.3 \times 10^9 & 21.52 \times 10^9 & 36.05 \times 10^9 \end{bmatrix} (Pa)$$

For 90° plies:

$$\left[\overline{Q}\right]_{90°\,plies} = \begin{bmatrix} 10.05 \times 10^9 & 3.016 \times 10^9 & 0 \\ 3.016 \times 10^9 & 170.9 \times 10^9 & 0 \\ 0 & 0 & 13.00 \times 10^9 \end{bmatrix} (Pa)$$

For −30° plies:

$$\left[\overline{Q}\right]_{-30°\,plies} = \begin{bmatrix} 107.6 \times 10^9 & 26.06 \times 10^9 & -48.3 \times 10^9 \\ 26.06 \times 10^9 & 27.22 \times 10^9 & -21.52 \times 10^9 \\ -48.3 \times 10^9 & -21.52 \times 10^9 & 36.05 \times 10^9 \end{bmatrix} (Pa)$$

Stresses present at the outer surface of ply #1 (i.e., strains present at $z_o = -0.000500$ m) can now be calculated:

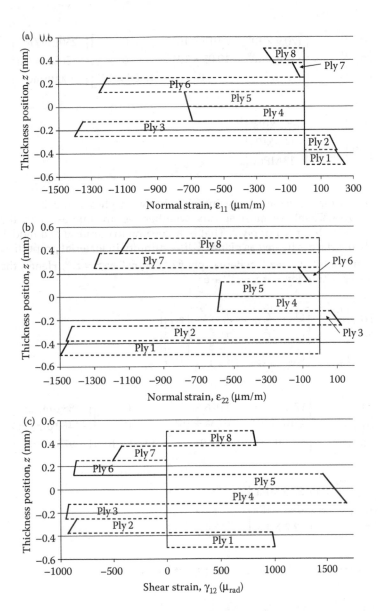

FIGURE 6.14

Through-thickness strain plots implied by the midplane strains and curvatures discussed in Example Problem 6.1. Strains referenced to the 1–2 coordinate system. (a) Normal strain ε_{11}; (b) normal strain ε_{22}; (c) shear strain γ_{12}.

$$
\left\{\begin{array}{c} \sigma_{xx} \\ \sigma_{yy} \\ \tau_{xy} \end{array}\right\}_{z=z_0}^{Ply\ 1} = \left[\begin{array}{ccc} \overline{Q}_{11} & \overline{Q}_{12} & \overline{Q}_{16} \\ \overline{Q}_{12} & \overline{Q}_{22} & \overline{Q}_{26} \\ \overline{Q}_{16} & \overline{Q}_{26} & \overline{Q}_{66} \end{array}\right]_{z=z_0}^{Ply1} \left\{\begin{array}{c} \varepsilon_{xx} \\ \varepsilon_{yy} \\ \gamma_{xy} \end{array}\right\}_{z=z_0}
$$

$$\left\{ \begin{matrix} \sigma_{xx} \\ \sigma_{yy} \\ \tau_{xy} \end{matrix} \right\}_{z=z_0}^{Ply\ 1} = \begin{bmatrix} 170.9 \times 10^9 & 3.016 \times 10^9 & 0 \\ 3.016 \times 10^9 & 10.05 \times 10^9 & 0 \\ 0 & 0 & 13.00 \times 10^9 \end{bmatrix} \left\{ \begin{matrix} 250\,\mu m/m \\ -1500\,\mu m/m \\ 1000\,\mu rad \end{matrix} \right\}$$

$$\left\{ \begin{matrix} \sigma_{xx} \\ \sigma_{yy} \\ \tau_{xy} \end{matrix} \right\}_{z=z_0}^{Ply\ 1} = \left\{ \begin{matrix} 38.2\,MPa \\ -14.3\,MPa \\ 13\,MPa \end{matrix} \right\}$$

To calculate stresses at the interface between plies 1 and 2 (i.e., at $z_1 = -0.000375$ m) we must specify whether we are interested in the stresses within ply 1 or ply 2. That is, according to our idealized model, a ply interface is treated as plane of discontinuity in material properties. Ply #1 "ends" at $z = z_1^{(-)}$, whereas ply #2 "begins" at $z = z_1^{(+)}$. Hence, the stresses within ply 1 at $z = z_1^{(-)}$ are

$$\left\{ \begin{matrix} \sigma_{xx} \\ \sigma_{yy} \\ \tau_{xy} \end{matrix} \right\}_{z=z_1}^{Ply\ 1} = \begin{bmatrix} \bar{Q}_{11} & \bar{Q}_{12} & \bar{Q}_{16} \\ \bar{Q}_{12} & \bar{Q}_{22} & \bar{Q}_{26} \\ \bar{Q}_{16} & \bar{Q}_{26} & \bar{Q}_{66} \end{bmatrix}_{z=z_1}^{Ply1} \left\{ \begin{matrix} \varepsilon_{xx} \\ \varepsilon_{yy} \\ \gamma_{xy} \end{matrix} \right\}_{z=z_1}$$

$$\left\{ \begin{matrix} \sigma_{xx} \\ \sigma_{yy} \\ \tau_{xy} \end{matrix} \right\}_{z=z_1}^{Ply\ 1} = \begin{bmatrix} 170.9 \times 10^9 & 3.016 \times 10^9 & 0 \\ 3.016 \times 10^9 & 10.05 \times 10^9 & 0 \\ 0 & 0 & 13.00 \times 10^9 \end{bmatrix} \left\{ \begin{matrix} 188\,\mu m/m \\ -1450\,\mu m/m \\ 975\,\mu rad \end{matrix} \right\}$$

$$\left\{ \begin{matrix} \sigma_{xx} \\ \sigma_{yy} \\ \tau_{xy} \end{matrix} \right\}_{z=z_1}^{Ply\ 1} = \left\{ \begin{matrix} 27.8\,MPa \\ -14.0\,MPa \\ 12.7\,MPa \end{matrix} \right\}$$

The stresses within ply 2 (a 30° ply) at $z = z_1^{(+)}$ are

$$\left\{ \begin{matrix} \sigma_{xx} \\ \sigma_{yy} \\ \tau_{xy} \end{matrix} \right\}_{z=z_1}^{Ply\ 2} = \begin{bmatrix} \bar{Q}_{11} & \bar{Q}_{12} & \bar{Q}_{16} \\ \bar{Q}_{12} & \bar{Q}_{22} & \bar{Q}_{26} \\ \bar{Q}_{16} & \bar{Q}_{26} & \bar{Q}_{66} \end{bmatrix}_{z=z_1}^{Ply2} \left\{ \begin{matrix} \varepsilon_{xx} \\ \varepsilon_{yy} \\ \gamma_{xy} \end{matrix} \right\}_{z=z_1}$$

$$\left\{ \begin{matrix} \sigma_{xx} \\ \sigma_{yy} \\ \tau_{xy} \end{matrix} \right\}_{z=z_1}^{Ply\ 2} = \begin{bmatrix} 107.6 \times 10^9 & 26.06 \times 10^9 & 48.3 \times 10^9 \\ 26.06 \times 10^9 & 27.22 \times 10^9 & 21.52 \times 10^9 \\ 48.3 \times 10^9 & 21.52 \times 10^9 & 36.05 \times 10^9 \end{bmatrix} \left\{ \begin{matrix} 188\,\mu m/m \\ -1450\,\mu m/m \\ 975\,\mu rad \end{matrix} \right\}$$

TABLE 6.3

Ply Interface Stresses in a $[0/-30/90/30]_s$ Graphite–Epoxy Laminate Subjected to the Midplane Strains and Curvatures Discussed in Example Problem 6.1

Ply Number	z-Coordinate (mm)	σ_{xx} (MPa)	σ_{yy} (MPa)	τ_{xy} (MPa)	Θ (MPa)	Φ (MPa)²
Ply	−0.500	38.2	−14.3	13.0	23.9	−715
1	−0.375	27.8	−14.0	12.7	13.8	−550
Ply	−0.375	29.3	−13.6	13.0	15.7	−567
2	−0.250	22.7	−14.4	10.1	8.3	−429
Ply	−0.250	−2.97	−239	12.4	−242	556
3	−0.125	−3.44	−231	12.0	−234	651
Ply	−0.125	−73.0	−55.0	59.4	−128	487
4	0.000	−77.2	−54.7	60.4	−132	575
Ply	0.000	−77.2	−54.7	60.4	−132	575
5	0.125	−81.4	−54.5	61.4	−136	666
Ply	0.125	−4.40	−214.	11.4	−218	812
6	0.250	−4.90	−205	11.0	−210	884
Ply	0.250	−3.82	−17.6	−1.20	−21.4	65.8
7	0.375	−10.4	−18.4	−4.03	−28.8	175
Ply	0.375	−35.5	−12.1	10.7	−47.6	315
8	0.500	−46.0	−11.8	10.4	−57.8	435

Note: Stresses are referenced to the x–y coordinate system.

$$\left. \begin{Bmatrix} \sigma_{xx} \\ \sigma_{yy} \\ \tau_{xy} \end{Bmatrix} \right|_{z=z_1}^{Ply\ 2} = \begin{Bmatrix} 29.5\ \text{MPa} \\ -13.6\ \text{MPa} \\ 13.0\ \text{MPa} \end{Bmatrix}$$

Stresses are calculated at all remaining ply interfaces in exactly the same fashion. Ply interface stresses are summarized in Table 6.3, and are plotted in Figure 6.15. Obviously, stresses are *not* linearly distributed through the thickness of the laminate, even when referenced to the global x–y coordinate system. In general, all stress components exhibit a sudden discontinuous change at all ply interface positions. The abrupt change in stresses at ply interfaces is due to the discontinuous change in the $[Q]$ matrix from one ply to the next. In turn, the discontinuous change in $[Q]$ occurs because the fiber angle (in general) changes from one ply to the next. Indeed, in this example problem the same fiber angle occurs in only two adjacent plies (namely, plies 4 and 5, both of which have a fiber angle of −30°), and inspection of Figure 6.15 shows that the interface between plies 4 and 5 is the only interface for which the stresses do not change abruptly.

It has been mentioned that the linear *strain* distributions shown in Figure 6.13 would be the same for any thin-plate, regardless of the material the plate is made of. The same statement cannot be made for *stress*

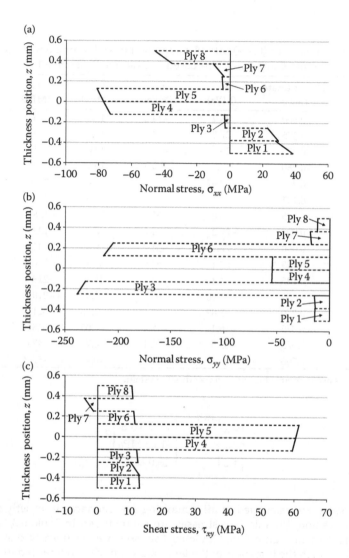

FIGURE 6.15
Through-thickness stress plots implied by the midplane strains and curvatures discussed in
Example Problem 6.1. Stresses referenced to the x–y coordinate system stress. (a) Normal stress
σ_{xx}; (b) normal stress σ_{yy}; (c) shear stress τ_{xy}.

distributions. In general, through-thickness stress distributions for isotropic plates (e.g., an isotropic aluminum plate) are linear and continuous,
unless high nonlinear (plastic) stresses occur, in which case the stress distribution may not be linear but will nevertheless be continuous. In contrast, the stress distributions in laminated composite plates are usually
discontinuous. The only conditions under which a linear and continuous
stress distribution is encountered is when (a) the laminate is subjected to

elastic stress/strain levels and (b) when the $[\overline{Q}]$ matrix does not vary from one ply to the next, that is, for unidirectional laminates in which the fiber angle does not vary from one ply to the next.

Knowledge of ply stresses referenced to the local 1–2 coordinate system (defined by the fiber angle within each ply) is often required. Transformation of the stress tensor from one coordinate system to another was reviewed in Chapter 2, and in particular stresses can be rotated from the x–y coordinate system to the 1–2 coordinate system using Equation 2.20. Typically stresses are calculated at both the "top" and "bottom" interfaces for all plies. For example, rotation of the ply stresses that exist within ply 2 at the interface between plies 1 and 2 (i.e., at $z = z_1 = -0.375$ mm) proceeds as follows:

$$\begin{Bmatrix} \sigma_{11} \\ \sigma_{22} \\ \tau_{12} \end{Bmatrix}\Bigg|_{z=z1}^{Ply2} = \begin{bmatrix} \cos^2\theta_2 & \sin^2\theta_2 & 2\cos\theta_2\sin\theta_2 \\ \sin^2\theta_2 & \cos^2\theta_2 & -2\cos\theta_2\sin\theta_2 \\ -\cos\theta_2\sin\theta_2 & \cos\theta_2\sin\theta_2 & \cos^2\theta_2 - \sin^2\theta_2 \end{bmatrix} \begin{Bmatrix} \sigma_{xx} \\ \sigma_{yy} \\ \tau_{xy} \end{Bmatrix}\Bigg|_{z=z1}^{Ply2}$$

$$\begin{Bmatrix} \sigma_{11} \\ \sigma_{22} \\ \tau_{12} \end{Bmatrix}\Bigg|_{z=z1}^{Ply2} = \begin{bmatrix} \cos^2(30°) & \sin^2(30°) & 2\cos(30°)\sin(30°) \\ \sin^2(30°) & \cos^2(30°) & -2\cos(30°)\sin(30°) \\ -\cos(30°)\sin(30°) & \cos(30°)\sin(30°) & \cos^2(30°) - \sin^2(30°) \end{bmatrix}$$

$$\times \begin{Bmatrix} 29.3\ \text{MPa} \\ -13.6\ \text{MPa} \\ 13.0\ \text{MPa} \end{Bmatrix}\Bigg|_{z=z1}^{Ply2}$$

$$\begin{Bmatrix} \sigma_{11} \\ \sigma_{22} \\ \tau_{12} \end{Bmatrix}\Bigg|_{z=z1}^{Ply2} = \begin{bmatrix} 0.750 & 0.250 & 0.866 \\ 0.250 & 0.750 & -0.866 \\ -0.433 & 0.433 & 0.500 \end{bmatrix} \begin{Bmatrix} 29.3\ \text{MPa} \\ -13.6\ \text{MPa} \\ 13.0\ \text{MPa} \end{Bmatrix}\Bigg|_{z=z1}^{Ply2}$$

$$\begin{Bmatrix} \sigma_{11} \\ \sigma_{22} \\ \tau_{12} \end{Bmatrix}\Bigg|_{z=z1}^{Ply2} = \begin{Bmatrix} 29.8\ \text{MPa} \\ -14.1\ \text{MPa} \\ -12.1\ \text{MPa} \end{Bmatrix}\Bigg|_{z=z1}^{Ply2}$$

Ply interface stresses referenced to local 1–2 coordinate systems are summarized in Table 6.4 and plotted in Figure 6.16. Once again, stresses are not linearly distributed through the thickness of the laminate, but instead exhibit a sudden discontinuous change at all ply interface positions.

Stress invariants can be used to confirm that the ply stresses referenced to the x–y coordinate system, listed in Table 6.3, are equivalent to the ply stresses referenced to the 1–2 coordinate system, listed in Table 6.4. The concept of "stress invariants" was discussed in Chapter 2. The

TABLE 6.4

Ply Interface Stresses in a $[0/-30/90/30]_s$ Graphite–Epoxy Laminate Subjected to the Midplane Strains and Curvatures Discussed in Example Problem 6.1

Ply Number	z-Coordinate (mm)	σ_{11} (MPa)	σ_{22} (MPa)	τ_{12} (MPa)	Θ (MPa)	Φ (MPa)^2
Ply	−0.500	38.2	−14.3	13.0	23.9	−715
1	−0.375	27.8	−14.0	12.7	13.8	−550
Ply	−0.375	29.8	−14.1	−12.1	15.7	−567
2	−0.250	22.2	−13.9	−11.0	8.3	−429
Ply	−0.250	−239	−2.97	−12.4	−242	556
3	−0.125	−231	−3.44	−12.0	−234	651
Ply	−0.125	−120	−8.08	21.9	−128	487
4	0.000	−124	−8.04	20.5	−132	575
Ply	0.000	−124	−8.04	20.5	−132	575
5	0.125	−128	−8.00	19.1	−136	666
Ply	0.125	−214	−4.40	−11.4	−218	812
6	0.250	−205	−4.88	−11.0	−210	884
Ply	0.250	−8.31	−13.1	−6.58	−21.4	65.8
7	0.375	−15.9	−13.0	−5.47	−28.8	175
Ply	0.375	−35.5	−12.1	10.7	−47.6	315
8	0.500	−46.0	−11.8	10.4	−57.8	435

Note: Stresses are referenced to the 1–2 coordinate system.

stress invariants for the case of plane stress are given by Equation 2.22, repeated here for convenience:

First Stress Invariant $= \Theta = \sigma_{xx} + \sigma_{yy}$

Second Stress Invariant $= \Phi = \sigma_{xx}\sigma_{yy} - \tau_{xy}^2$ (2.22) (repeated)

Third Stress Invariant $= \Psi = 0$

For plane stress conditions the third stress invariant always equals zero, and so Ψ cannot be used to evaluate whether two plane stress states are equivalent. The first and second stress invariants, Θ and Φ respectively, have been calculated using the ply stress components referenced to both the x–y and 1–2 coordinate systems. Values calculated for Θ and Φ are included in the last two columns of both Tables 6.3 and 6.4. Identical values are obtained in all cases, indicating the equivalence of the ply stress states described using the two different coordinate systems.

6.6 Classical Lamination Theory

Stress and moment resultants were introduced in Section 6.1. As was discussed, a thin plate subjected to any combination of stress and moment resultants will experience a state of plane stress. Deformations of a thin plate

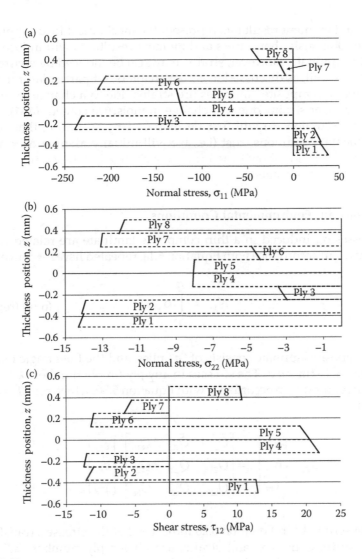

FIGURE 6.16
Through-thickness stress plots implied by the midplane strains and curvatures discussed in Example Problem 6.1. Stresses referenced to the 1–2 coordinate system. (a) Normal stress σ_{11}; (b) normal stress σ_{22}; (c) shear stress τ_{12}.

were then considered in Section 6.2. There the Kirchhoff hypothesis was invoked, which allows us to calculate the in-plane strains induced at any location through the thickness of a thin plate. In this section we will combine the material presented in Sections 6.1 and 6.2, as well as certain material presented in Chapter 5. This will lead to the ability to relate stress and moment resultants to the resulting strains (and hence stresses) induced within a thin composite plate. This combination of analysis tools is commonly known as classical lamination theory, often abbreviated simply as CLT.

Stress and moment resultants represent the mechanical loads applied to a laminate. Obviously then, stress and moment resultants will induce strains within the laminate. However, strains may also be induced by environmental factors as well, as discussed in earlier chapters. Of particular importance for polymeric composite laminates are strains due to a change in temperature (ΔT) and/or strains due to a change in moisture content (ΔM). To simplify our discussion we will first develop CLT by assuming that constant environmental conditions exist (i.e., we will initially assume $\Delta T = \Delta M = 0$). We will then consider how to account for a change in temperature and/or a change in moisture content.

6.6.1 Constant Environmental Conditions

The stresses σ_{xx} induced in a thin composite laminate are related to stress resultant N_{xx} in accordance with Equation 6.1a, repeated here for convenience:

$$N_{xx} = \int_{-t/2}^{t/2} \sigma_{xx}dz \qquad \text{(6.1a) (repeated)}$$

The composite laminate consists of "n" plies, and the fiber angle may vary from one ply to the next. The stresses in any ply (in ply number "k," say) are related to ply strains in accordance with Equation 5.30, which for $\Delta T = \Delta M = 0$ becomes

$$\begin{Bmatrix} \sigma_{xx} \\ \sigma_{yy} \\ \tau_{xy} \end{Bmatrix}_k = \begin{bmatrix} \overline{Q}_{11} & \overline{Q}_{12} & \overline{Q}_{16} \\ \overline{Q}_{21} & \overline{Q}_{22} & \overline{Q}_{26} \\ \overline{Q}_{61} & \overline{Q}_{62} & \overline{Q}_{66} \end{bmatrix}_k \begin{Bmatrix} \varepsilon_{xx} \\ \varepsilon_{yy} \\ \gamma_{xy} \end{Bmatrix}_k \qquad (6.20)$$

The subscript "k" in Equation 6.20 indicates that the stresses, transformed reduced stiffness matrix, and strains are all for ply number "k," where $1 \leq k \leq n$, and $n =$ number of plies in the laminate.

From Equation 6.20, the stress σ_{xx} induced in ply k is

$$(\sigma_{xx})_k = \left\{ \overline{Q}_{11}\varepsilon_{xx} + \overline{Q}_{12}\varepsilon_{yy} + \overline{Q}_{16}\gamma_{xy} \right\}_k$$

Substituting this relationship into Equation 6.1a, we have

$$N_{xx} = \int_{-t/2}^{t/2} \left\{ \overline{Q}_{11}\varepsilon_{xx} + \overline{Q}_{12}\varepsilon_{yy} + \overline{Q}_{16}\gamma_{xy} \right\}_k dz \qquad (6.21)$$

The strains induced in ply k can be related to the midplane strains and curvatures via the Kirchhoff hypothesis, in accordance with Equation 6.11 or 6.12. Substituting Equation 6.11, 6.12 into Equation 6.21, we obtain:

$$N_{xx} = \int_{-t/2}^{t/2} \left\{ \overline{Q}_{11}\varepsilon_{xx}^o + \overline{Q}_{12}\varepsilon_{yy}^o + \overline{Q}_{16}\gamma_{xy}^o + z\overline{Q}_{11}\kappa_{xx} + z\overline{Q}_{12}\kappa_{yy} + z\overline{Q}_{16}\kappa_{xy} \right\} dz \quad (6.22)$$

We cannot integrate Equation 6.22 directly, because the integrand is a discontinuous function of z. That is, the transformed reduced stiffness terms $\overline{Q}_{11}, \overline{Q}_{12}$, and $\overline{Q}_{16}$ are all directly related to the ply material properties and fiber angle θ (see Equations 5.31). Since the ply material and/or fiber angle may change from one ply to the next, the transformed reduced stiffness terms also change, and hence are discontinuous functions of z. Note, however, that the midplane strains and curvatures are not functions of z, but instead are constants for a given laminate. Hence, they may be brought out from under the integral sign. Equation 6.15 can, therefore, be broken into six individual integrals:

$$N_{xx} = \varepsilon_{xx}^o \int_{-t/2}^{t/2} \left\{ \overline{Q}_{11} \right\}_k dz + \varepsilon_{yy}^o \int_{-t/2}^{t/2} \left\{ \overline{Q}_{12} \right\}_k dz + \gamma_{xy}^0 \int_{-t/2}^{t/2} \left\{ \overline{Q}_{16} \right\}_k dz$$

$$+ \kappa_{xx} \int_{-t/2}^{t/2} z\left\{ \overline{Q}_{11} \right\}_k dz + \kappa_{yy} \int_{-t/2}^{t/2} z\left\{ \overline{Q}_{12} \right\}_k dz + \kappa_{xy} \int_{-t/2}^{t/2} z\left\{ \overline{Q}_{16} \right\}_k dz \quad (6.23)$$

Since the transformed stiffness terms are constant over each ply thickness, each of the six integrals in Equation 6.23 can be evaluated in a "piecewise" fashion:

$$N_{xx} = \varepsilon_{xx}^o \left\{ \left(\overline{Q}_{11} \right)_1 \int_{z_0}^{z_1} dz + \left(\overline{Q}_{11} \right)_2 \int_{z_1}^{z_2} dz + \left(\overline{Q}_{11} \right)_3 \int_{z_2}^{z_3} dz + \cdots + \left(\overline{Q}_{11} \right)_{n-1} \int_{z_{n-2}}^{z_{n-1}} dz + \left(\overline{Q}_{11} \right)_n \int_{z_{n-1}}^{z_n} dz \right\}$$

$$+ \varepsilon_{yy}^o \left\{ \left(\overline{Q}_{12} \right)_1 \int_{z_0}^{z_1} dz + \left(\overline{Q}_{12} \right)_2 \int_{z_1}^{z_2} dz + \left(\overline{Q}_{12} \right)_3 \int_{z_2}^{z_3} dz + \cdots + \left(\overline{Q}_{12} \right)_{n-1} \int_{z_{n-2}}^{z_{n-1}} dz + \left(\overline{Q}_{12} \right)_n \int_{z_{n-1}}^{z_n} dz \right\}$$

$$+ \gamma_{xy}^o \left\{ \left(\overline{Q}_{16} \right)_1 \int_{z_0}^{z_1} dz + \left(\overline{Q}_{16} \right)_2 \int_{z_1}^{z_2} dz + \left(\overline{Q}_{16} \right)_3 \int_{z_2}^{z_3} dz + \cdots + \left(\overline{Q}_{16} \right)_{n-1} \int_{z_{n-2}}^{z_{n-1}} dz + \left(\overline{Q}_{16} \right)_n \int_{z_{n-1}}^{z_n} dz \right\}$$

$$+\kappa_{xx}\left\{\left(\overline{Q}_{11}\right)_1\int_{z_0}^{z_1}z\,dz+\left(\overline{Q}_{11}\right)_2\int_{z_1}^{z_2}z\,dz+\left(\overline{Q}_{11}\right)_3\int_{z_2}^{z_3}z\,dz+\cdots+\left(\overline{Q}_{11}\right)_{n-1}\int_{z_{n-2}}^{z_{n-1}}z\,dz+\left(\overline{Q}_{11}\right)_n\int_{z_{n-1}}^{z_n}z\,dz\right\}$$

$$+\kappa_{yy}\left\{\left(\overline{Q}_{12}\right)_1\int_{z_0}^{z_1}z\,dz+\left(\overline{Q}_{12}\right)_2\int_{z_1}^{z_2}z\,dz+\left(\overline{Q}_{12}\right)_3\int_{z_2}^{z_3}z\,dz+\cdots+\left(\overline{Q}_{12}\right)_{n-1}\int_{z_{n-2}}^{z_{n-1}}z\,dz+\left(\overline{Q}_{12}\right)_n\int_{z_{n-1}}^{z_n}z\,dz\right\}$$

$$+\kappa_{xy}\left\{\left(\overline{Q}_{16}\right)_1\int_{z_0}^{z_1}z\,dz+\left(\overline{Q}_{16}\right)_2\int_{z_1}^{z_2}z\,dz+\left(\overline{Q}_{16}\right)_3\int_{z_2}^{z_3}z\,dz+\cdots+\left(\overline{Q}_{16}\right)_{n-1}\int_{z_{n-2}}^{z_{n-1}}z\,dz+\left(\overline{Q}_{16}\right)_n\int_{z_{n-1}}^{z_n}z\,dz\right\}$$

$$(6.24)$$

Although Equation 6.24 may appear daunting, closer inspection reveals that evaluation of Equation 6.24 is actually a simple matter. All integrals that appear in Equation 6.24 are one of the following two forms, both of which are easily evaluated:

$$\int_{z_{k-1}}^{z_k}dz=(z_k-z_{k-1})$$

or

$$\int_{z_{k-1}}^{z_k}z\,dz=\frac{1}{2}(z_k^2-z_{k-1}^2)$$

Hence, evaluating all integrals that appear in Equation 6.24, we obtain:

$$N_{xx}=\varepsilon_{xx}^o\left\{\left(\overline{Q}_{11}\right)_1[z_1-z_0]+\left(\overline{Q}_{11}\right)_2[z_2-z_1]+\left(\overline{Q}_{11}\right)_3[z_3-z_2]+\cdots+\left(\overline{Q}_{11}\right)_n\right.$$

$$\times[z_n-z_{n-1}]\}+\varepsilon_{yy}^o\left\{\left(\overline{Q}_{12}\right)_1[z_1-z_0]+\left(\overline{Q}_{12}\right)_2[z_2-z_1]+\left(\overline{Q}_{12}\right)_3[z_3-z_2]+\cdots\right.$$

$$+\left(\overline{Q}_{12}\right)_n[z_n-z_{n-1}]\}+\gamma_{xy}^o\left\{\left(\overline{Q}_{16}\right)_1[z_1-z_0]+\left(\overline{Q}_{16}\right)_2[z_2-z_1]+\left(\overline{Q}_{16}\right)_3\right.$$

$$\times[z_3-z_2]+\cdots+\left(\overline{Q}_{16}\right)_n[z_n-z_{n-1}]\}+\frac{1}{2}\kappa_{xx}\left\{\left(\overline{Q}_{11}\right)_1[z_1^2-z_0^2]+\left(\overline{Q}_{11}\right)_2\right.$$

$$\times [z_2^2 - z_1^2] + \left(\overline{Q}_{11}\right)_3 [z_3^2 - z_2^2] + \cdots + \left(\overline{Q}_{11}\right)_n [z_n^2 - z_{n-1}^2]\} + \frac{1}{2}\kappa_{yy}\{\left(\overline{Q}_{12}\right)_1 [z_1^2 - z_0^2]$$

$$\times [z_1^2 - z_0^2] + \left(\overline{Q}_{12}\right)_2 [z_2^2 - z_1^2] + \left(\overline{Q}_{12}\right)_3 [z_3^2 - z_2^2] + \cdots + \left(\overline{Q}_{12}\right)_n [z_n^2 - z_{n-1}^2]\}$$

$$+ \frac{1}{2}\kappa_{xy}\{\left(\overline{Q}_{16}\right)_1 + \left(\overline{Q}_{16}\right)_2 [z_2^2 - z_1^2] + \left(\overline{Q}_{16}\right)_3 [z_3^2 - z_2^2] + \cdots + \left(\overline{Q}_{16}\right)_n [z_n^2 - z_{n-1}^2]\}$$

$$\tag{6.25}$$

Equation 6.25 can be simplified substantially by defining the following terms:

$$A_{11} = \{\left(\overline{Q}_{11}\right)_1 [z_1 - z_0] + \left(\overline{Q}_{11}\right)_2 [z_2 - z_1] + \left(\overline{Q}_{11}\right)_3 [z_3 - z_2] + \cdots + \left(\overline{Q}_{11}\right)_n [z_n - z_{n-1}]\}$$

$$A_{12} = \{\left(\overline{Q}_{12}\right)_1 [z_1 - z_0] + \left(\overline{Q}_{12}\right)_2 [z_2 - z_1] + \left(\overline{Q}_{12}\right)_3 [z_3 - z_2] + \cdots + \left(\overline{Q}_{12}\right)_n [z_n - z_{n-1}]\}$$

$$A_{16} = \{\left(\overline{Q}_{16}\right)_1 [z_1 - z_0] + \left(\overline{Q}_{16}\right)_2 [z_2 - z_1] + \left(\overline{Q}_{16}\right)_3 [z_3 - z_2] + \cdots + \left(\overline{Q}_{16}\right)_n [z_n - z_{n-1}]\}$$

$$B_{11} = \frac{1}{2}\{\left(\overline{Q}_{11}\right)_1 [z_1^2 - z_0^2] + \left(\overline{Q}_{11}\right)_2 [z_2^2 - z_1^2] + \left(\overline{Q}_{11}\right)_3 [z_3^2 - z_2^2] + \cdots + \left(\overline{Q}_{11}\right)_n [z_n^2 - z_{n-1}^2]\}$$

$$B_{12} = \frac{1}{2}\{\left(\overline{Q}_{12}\right)_1 [z_1^2 - z_0^2] + \left(\overline{Q}_{12}\right)_2 [z_2^2 - z_1^2] + \left(\overline{Q}_{12}\right)_3 [z_3^2 - z_2^2] + \cdots + \left(\overline{Q}_{12}\right)_n [z_n^2 - z_{n-1}^2]\}$$

$$B_{16} = \frac{1}{2}\{\left(\overline{Q}_{16}\right)_1 [z_1^2 - z_0^2] + \left(\overline{Q}_{16}\right)_2 [z_2^2 - z_1^2] + \left(\overline{Q}_{16}\right)_3 [z_3^2 - z_2^2] + \cdots + \left(\overline{Q}_{16}\right)_n [z_n^2 - z_{n-1}^2]\}$$

With these definitions Equations 6.25, become:

$$N_{xx} = A_{11}\varepsilon_{xx}^0 + A_{12}\varepsilon_{yy}^0 + A_{16}\gamma_{xy}^0 + B_{11}\kappa_{xx} + B_{12}\kappa_{yy} + B_{16}\kappa_{xy} \tag{6.26a}$$

Following an entirely analogous procedure for stress resultants N_{yy} and N_{xy}, it can be shown:

$$N_{yy} = A_{21}\varepsilon_{xx}^0 + A_{22}\varepsilon_{yy}^0 + A_{26}\gamma_{xy}^0 + B_{21}\kappa_{xx} + B_{22}\kappa_{yy} + B_{26}\kappa_{xy} \tag{6.26b}$$

$$N_{xy} = A_{61}\varepsilon_{xx}^0 + A_{62}\varepsilon_{yy}^0 + A_{66}\gamma_{xy}^0 + B_{61}\kappa_{xx} + B_{62}\kappa_{yy} + B_{66}\kappa_{xy} \tag{6.26c}$$

where

$$A_{ij} = \sum_{k=1}^{n} \{\overline{Q}_{ij}\}_k (z_k - z_{k-1}) \tag{6.27a}$$

$$B_{ij} = \frac{1}{2}\sum_{k=1}^{n}\left\{\overline{Q}_{ij}\right\}_k (z_k^2 - z_{k-1}^2) \qquad (6.27b)$$

and $i, j = 1, 2,$ or 6. Since subscripts i and j may take on one of three values, both A_{ij} and B_{ij} can be written as 3×3 matrices. Also, recall that the transformed reduced stiffness matrix is symmetric (see Equations 5.31). Hence, both A_{ij} and B_{ij} are also symmetric:

$$A_{ij} = \begin{bmatrix} A_{11} & A_{12} & A_{16} \\ A_{21} & A_{22} & A_{26} \\ A_{61} & A_{62} & A_{66} \end{bmatrix} = \begin{bmatrix} A_{11} & A_{12} & A_{16} \\ A_{12} & A_{22} & A_{26} \\ A_{16} & A_{22} & A_{66} \end{bmatrix}$$

$$B_{1j} = \begin{bmatrix} B_{11} & B_{12} & B_{16} \\ B_{21} & B_{22} & B_{26} \\ B_{61} & B_{62} & B_{66} \end{bmatrix} = \begin{bmatrix} B_{11} & B_{12} & B_{16} \\ B_{12} & B_{22} & B_{26} \\ B_{16} & B_{26} & B_{66} \end{bmatrix}$$

Equations 6.26 can be written in matrix form as follows:

$$\begin{Bmatrix} N_{xx} \\ N_{yy} \\ N_{xy} \end{Bmatrix} = \begin{bmatrix} A_{11} & A_{12} & A_{16} & B_{11} & B_{12} & B_{16} \\ A_{12} & A_{22} & A_{26} & B_{12} & B_{22} & B_{26} \\ A_{16} & A_{26} & A_{66} & B_{16} & B_{26} & B_{66} \end{bmatrix} \begin{Bmatrix} \varepsilon_{xx}^o \\ \varepsilon_{yy}^o \\ \gamma_{xy}^o \\ \kappa_{xx} \\ \kappa_{yy} \\ \kappa_{xy} \end{Bmatrix} \qquad (6.28)$$

To summarize our results to this point, Equation 6.28 relates the stress resultants applied to a composite laminate to the resulting midplane strains and curvatures via the A_{ij} and B_{ij} matrices. The values of each term within the A_{ij} and B_{ij} matrices depend on the material properties and fiber angle of each ply (i.e., they depend on terms within the $\overline{Q}_{ij}$ matrix) as well as the stacking sequence (i.e., the distance z_k of each ply from the laminate midplane), in accordance with Equations 6.27. In practice then, if the midplane strains and curvatures induced in a laminate under constant environmental conditions are measured, then the stress resultants that caused these strains and curvatures can be calculated using Equation 6.28.

This entire process must now be repeated for the moment resultants. The stresses σ_{xx} induced in a thin composite laminate are related to moment resultant M_{xx} in accordance with Equation 6.2a, repeated here for convenience:

$$M_{xx} = \int\limits_{-t/2}^{t/2} \sigma_{xx} z \, dz \qquad \text{(6.2a) (repeated)}$$

Substituting the expression for σ_{xx} from Equation 6.20, we obtain:

$$M_{xx} = \int\limits_{-t/2}^{t/2} \left\{ \overline{Q}_{11}\varepsilon_{xx} + \overline{Q}_{12}\varepsilon_{yy} + \overline{Q}_{16}\gamma_{xy} \right\}_k z \, dz \qquad (6.29)$$

Each strain that appears in Equation 6.29 can be related to the midplane strains and curvatures via the Kirchhoff hypothesis. Hence, substituting either Equation 6.11, 6.12, we have

$$M_{xx} = \int\limits_{-t/2}^{t/2} \left\{ z\overline{Q}_{11}\varepsilon_{xx}^o + z\overline{Q}_{12}\varepsilon_{yy}^o + z\overline{Q}_{16}\gamma_{xy}^o + z^2\overline{Q}_{11}\kappa_{xx} + z^2\overline{Q}_{12}\kappa_{yy} + z^2\overline{Q}_{16}\kappa_{xy} \right\} dz$$

$$(6.30)$$

Equation 6.30 is similar to Equation 6.22. Once again, this integral cannot be evaluated directly, because the integrand is a discontinuous function of z. However, (a) noting that the midplane strains and curvatures are not functions of z and can be brought outside the integral sign, and then (b) evaluating the integral in a "piecewise" fashion through the thickness of the laminate, we obtain:

$$M_{xx} = \varepsilon_{xx}^o \left\{ \left(\overline{Q}_{11}\right)_1 \int\limits_{z_0}^{z_1} z \, dz + \left(\overline{Q}_{11}\right)_2 \int\limits_{z_1}^{z_2} z \, dz + \left(\overline{Q}_{11}\right)_3 \int\limits_{z_2}^{z_3} z dz + \cdots + \left(\overline{Q}_{11}\right)_{n-1} \right.$$

$$\left. \times \int\limits_{z_{n-2}}^{z_{n-1}} z \, dz + \left(\overline{Q}_{11}\right)_n \int\limits_{z_{n-1}}^{z_n} z \, dz \right\} + \varepsilon_{yy}^o \left\{ \left(\overline{Q}_{12}\right)_1 \int\limits_{z_0}^{z_1} z \, dz + \left(\overline{Q}_{12}\right)_2 \right.$$

$$\left. \times \int\limits_{z_1}^{z_2} z \, dz + \left(\overline{Q}_{12}\right)_3 \int\limits_{z_2}^{z_3} z \, dz + \cdots + \left(\overline{Q}_{12}\right)_{n-1} \int\limits_{z_{n-2}}^{z_{n-1}} z \, dz + \left(\overline{Q}_{12}\right)_n \int\limits_{z_{n-1}}^{z_n} z \, dz \right\}$$

$$+ \gamma_{xy}^o \left\{ \left(\overline{Q}_{16}\right)_1 \int\limits_{z_0}^{z_1} z \, dz + \left(\overline{Q}_{16}\right)_2 \int\limits_{z_1}^{z_2} z \, dz + \left(\overline{Q}_{16}\right)_3 \int\limits_{z_2}^{z_3} z \, dz + \cdots + \left(\overline{Q}_{16}\right)_{n-1} \right.$$

$$\left. \times \int\limits_{z_{n-2}}^{z_{n-1}} z \, dz + \left(\overline{Q}_{16}\right)_n \int\limits_{z_{n-1}}^{z_n} z \, dz \right\} + \kappa_{xx} \left\{ \left(\overline{Q}_{11}\right)_1 \int\limits_{z_0}^{z_1} z^2 \, dz + \left(\overline{Q}_{11}\right)_2 \int\limits_{z_1}^{z_2} z^2 \, dz + \left(\overline{Q}_{11}\right)_3 \right.$$

$$\times \int_{z_2}^{z_3} z^2 \, dz + \cdots + \left(\overline{Q}_{11}\right)_{n-1} \int_{z_{n-2}}^{z_{n-1}} z^2 \, dz + \left(\overline{Q}_{11}\right)_{n} \int_{z_{n-1}}^{z_n} z^2 \, dz \Bigg\}$$

$$+ \kappa_{yy} \Bigg\{ \left(\overline{Q}_{12}\right)_{1} \int_{z_0}^{z_1} z^2 \, dz + \left(\overline{Q}_{12}\right)_{2} \int_{z_1}^{z_2} z^2 \, dz + \left(\overline{Q}_{12}\right)_{3} \int_{z_2}^{z_3} z^2 \, dz + \cdots + \left(\overline{Q}_{12}\right)_{n-1}$$

$$\times \int_{z_{n-2}}^{z_{n-1}} z^2 \, dz + \left(\overline{Q}_{12}\right)_{n} \int_{z_{n-1}}^{z_n} z^2 \, dz \Bigg\} + \kappa_{xy} \Bigg\{ \left(\overline{Q}_{16}\right)_{1} \int_{z_0}^{z_1} z^2 \, dz + \left(\overline{Q}_{16}\right)_{2} \int_{z_1}^{z_2} z^2 \, dz$$

$$+ \left(\overline{Q}_{16}\right)_{3} \int_{z_2}^{z_3} z^2 \, dz + \cdots + \left(\overline{Q}_{16}\right)_{n-1} \int_{z_{n-2}}^{z_{n-1}} z^2 \, dz + \left(\overline{Q}_{16}\right)_{n} \int_{z_{n-1}}^{z_n} z^2 \, dz \Bigg\} \tag{6.31}$$

The piecewise integrals that appear in Equation 6.31 are of one of the following two forms, both of which are easily evaluated:

$$\int_{z_{k-1}}^{z_k} z \, dz = \frac{1}{2}(z_k^2 - z_{k-1}^2)$$

or

$$\int_{z_{k-1}}^{z_k} z^2 \, dz = \frac{1}{3}(z_k^3 - z_{k-1}^3)$$

Hence, evaluating all integrals, we obtain:

$$M_{xx} = \frac{1}{2}\varepsilon_{xx}^{o} \Big\{ \left(\overline{Q}_{11}\right)_{1} [z_1^2 - z_0^2] + \left(\overline{Q}_{11}\right)_{2} [z_2^2 - z_1^2] + \left(\overline{Q}_{11}\right)_{3} [z_3^2 - z_2^2] + \cdots$$

$$+ \left(\overline{Q}_{11}\right)_{n} [z_n^2 - z_{n-1}^2] \Big\} + \frac{1}{2}\varepsilon_{yy}^{o} \Big\{ \left(\overline{Q}_{12}\right)_{1} [z_1^2 - z_0^2] + \left(\overline{Q}_{12}\right)_{2} [z_2^2 - z_1^2]$$

$$+ \left(\overline{Q}_{12}\right)_{3} [z_3^2 - z_2^2] + \cdots + \left(\overline{Q}_{12}\right)_{n} [z_n^2 - z_{n-1}^2] \Big\} + \frac{1}{2}\gamma_{xy}^{o} \Big\{ \left(\overline{Q}_{16}\right)_{1} [z_1^2 - z_0^2]$$

$$+ \left(\overline{Q}_{16}\right)_{2} [z_2^2 - z_1^2] + \left(\overline{Q}_{16}\right)_{3} [z_3^2 - z_2^2] + \cdots + \left(\overline{Q}_{16}\right)_{n} [z_n^2 - z_{n-1}^2] \Big\}$$

The 6×6 array that appears in Equation 6.36 is called the "*ABD* matrix." Since each of the individual matrices which make up the total *ABD* matrix are themselves symmetric (e.g., $A_{12} = A_{21}$, $B_{12} = B_{21}$, $D_{12} = D_{21}$, etc.), the entire *ABD* matrix is also symmetric.

It should be noted that the above results are applicable to thin laminates fabricated using any combination of ply materials. Since the A_{ij}, B_{ij}, and D_{ij} matrices are each calculated based on a summation over all plies, and individual ply properties (represented by the $\overline{Q}_{ij}$ matrix) are embedded within these summations, both ply material type and fiber angle can vary from one ply to the next. Hence, the *ABD* matrix for any thin plate can be calculated using Equations 6.27a, 6.27b, and 6.34. For example, the *ABD* matrix for "hybrid" laminates (i.e., laminates fabricated using two different pre-preg material systems) are calculated using Equations 6.27a, 6.27b, and 6.34.

The A_{ij} matrix relates in-plane stress resultants to in-plane midplane strains. For this reason, the A_{ij} terms are called *extensional stiffnesses*. Similarly, the matrix relates moment resultants to midplane curvatures, and elements within the D_{ij} matrix are, therefore, called *bending stiffnesses*. The B_{ij} matrix relates in-plane stress resultants to midplane curvatures, and also relates moment resultants to the in-plane midplane strains. The B_{ij} terms are called *coupling stiffnesses*. For an isotropic plate the coupling stiffnesses are always zero.

The stress and moment resultants can be thought of as "stress-like" quantities, since they are directly related to the stresses through the thickness of the laminate via Equations 6.1 and 6.2. On the other hand, the midplane strains and curvatures are "strain-like" quantities, since they can be used to calculate the strains at any position through the thickness of the laminate via Equations 6.11 and 6.12. Hence, Equation 6.36 relates "stress-like" quantities to "strain-like" quantities, and in this sense can be thought of as "Hooke's law" for a composite laminate.

Equation 6.36 is in convenient form if we measure midplane strains and curvatures and wish to calculate the stress and moment resultants that caused these strains and curvatures. Suppose, instead, that the stress and moment resultants are known and we wish to calculate the midplane strain and curvatures that will be caused by these known loads. In this case, we must invert Equation 6.36, to obtain a relationship of the form:

$$\left\{\begin{array}{c} \varepsilon^o \\ \kappa \end{array}\right\} = \left[\begin{array}{cc} A & B \\ B & D \end{array}\right]^{-1} \left\{\begin{array}{c} N \\ M \end{array}\right\} \tag{6.37}$$

In this book, the inverse of the *ABD* matrix will be labeled the *abd* matrix:

$$\left[\begin{array}{cc} a & b \\ b & d \end{array}\right] = \left[\begin{array}{cc} A & B \\ B & D \end{array}\right]^{-1}$$

Methods of inverting the [*ABD*] matrix analytically are discussed in several composite texts, including References 1–3. However, in practice the *ABD* matrix is most often inverted numerically with the aid of a digital computer, since many commercial software packages (e.g., MATLAB®, Maple, Mathematica, etc.) are available nowadays that can invert a 6×6 matrix routinely.

Written out in full, Equation 6.37 is

$$
\begin{Bmatrix} \varepsilon_{xx}^o \\ \varepsilon_{yy}^o \\ \gamma_{xy}^o \\ \kappa_{xx} \\ \kappa_{yy} \\ \kappa_{xy} \end{Bmatrix} =
\begin{bmatrix}
a_{11} & a_{12} & a_{16} & b_{11} & b_{12} & b_{16} \\
a_{12} & a_{22} & a_{26} & b_{21} & b_{22} & b_{26} \\
a_{16} & a_{26} & a_{66} & b_{61} & b_{62} & b_{66} \\
b_{11} & b_{21} & b_{61} & d_{11} & d_{12} & d_{16} \\
b_{12} & b_{22} & b_{62} & d_{12} & d_{22} & d_{26} \\
b_{16} & b_{26} & b_{66} & d_{16} & d_{26} & d_{66}
\end{bmatrix}
\begin{Bmatrix} N_{xx} \\ N_{yy} \\ N_{xy} \\ M_{xx} \\ M_{yy} \\ M_{xy} \end{Bmatrix}
\tag{6.38}
$$

The reader should carefully inspect the subscripts used in Equation 6.38. Note that the [*abd*] matrix is symmetric. Furthermore, the individual 3×3 matrices that appear in the upper left-hand quadrant and lower right-hand quadrant of the [*abd*] matrix, a_{ij} and d_{ij} respectively, are also symmetric. However, the 3×3 matrix that appears in the upper right-hand quadrant is not symmetric ($b_{12} \neq b_{21}, b_{16} \neq b_{61}$, and $b_{26} \neq b_{62}$). Also, the 3×3 matrix in the lower left-hand quadrant is the *transpose* of the 3×3 matrix that appears in the upper right-hand quadrant.

Example Problem 6.4

Determine the [*ABD*] and [*abd*] matrices for a $[30/0/90]_T$ graphite–epoxy laminate. Use material properties listed for graphite–epoxy in Table 3.1, and assume each ply has a thickness of 0.125 mm.

SOLUTION

An edge view of the laminate is shown in Figure 6.17. The total laminate thickness $t = 3(0.125 \text{ mm}) = 0.375$ mm. Since all three plies are of the same material, the thickness of each ply is identical: $t_1 = t_2 = t_3 = 0.125$ mm. Note that since an odd number of plies are used, the origin of the $x–y–z$ coordinate system exists at the midplane of ply 2. The ply interface coordinates can be calculated as

$$z_0 = -t/2 = -(0.375 \text{ mm})/2 = -0.1875 \text{ mm} = -0.0001875 \text{ m}$$

$$z_1 = z_0 + t_1 = -0.1875 \text{ mm} + 0.125 \text{ mm} = -0.0625 \text{ mm} = -0.0000625 \text{ m}$$

$$z_2 = z_1 + t_2 = -0.0625 \text{ mm} + 0.125 \text{ mm} = 0.0625 \text{ mm} = 0.0000625 \text{ m}$$

$$z_3 = z_2 + t_3 = 0.0625 \text{ mm} + 0.125 \text{ mm} = 0.1875 \text{ mm} = 0.0001875 \text{ m}$$

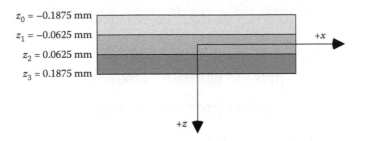

FIGURE 6.17
Edge view of the $[30/0/90]_T$ laminate considered in Example Problem 6.4.

We will also require the transformed reduced stiffness matrix for each ply. Elements of the $[\bar{Q}]_k$ matrices are calculated using Equations 5.31* and are equal to
For ply #1 (the 30° ply):

$$
[\bar{Q}]_{30°\text{ply}} = \begin{bmatrix} \bar{Q}_{11} & \bar{Q}_{12} & \bar{Q}_{16} \\ \bar{Q}_{12} & \bar{Q}_{22} & \bar{Q}_{26} \\ \bar{Q}_{16} & \bar{Q}_{26} & \bar{Q}_{66} \end{bmatrix} = \begin{bmatrix} 107.6 \times 10^9 & 26.06 \times 10^9 & 48.13 \times 10^9 \\ 26.06 \times 10^9 & 27.22 \times 10^9 & 21.52 \times 10^9 \\ 48.13 \times 10^9 & 21.52 \times 10^9 & 36.05 \times 10^9 \end{bmatrix} (\text{Pa})
$$

For ply #2 (the 0° ply):

$$
[\bar{Q}]_{0°\text{ply}} = \begin{bmatrix} \bar{Q}_{11} & \bar{Q}_{12} & \bar{Q}_{16} \\ \bar{Q}_{12} & \bar{Q}_{22} & \bar{Q}_{26} \\ \bar{Q}_{16} & \bar{Q}_{26} & \bar{Q}_{66} \end{bmatrix} = \begin{bmatrix} 170.9 \times 10^9 & 3.016 \times 10^9 & 0 \\ 3.016 \times 10^9 & 10.05 \times 10^9 & 0 \\ 0 & 0 & 13.00 \times 10^9 \end{bmatrix} (\text{Pa})
$$

For ply #3 (the 90° ply):

$$
[\bar{Q}]_{90°\text{ply}} = \begin{bmatrix} \bar{Q}_{11} & \bar{Q}_{12} & \bar{Q}_{16} \\ \bar{Q}_{12} & \bar{Q}_{22} & \bar{Q}_{26} \\ \bar{Q}_{16} & \bar{Q}_{26} & \bar{Q}_{66} \end{bmatrix} = \begin{bmatrix} 10.05 \times 10^9 & 3.016 \times 10^9 & 0 \\ 3.016 \times 10^9 & 170.9 \times 10^9 & 0 \\ 0 & 0 & 13.00 \times 10^9 \end{bmatrix} (\text{Pa})
$$

We can now calculate each member of the A_{ij}, B_{ij}, and D_{ij} matrices, in accordance with Equations 6.27a, 6.27b, and 6.34, respectively.

- Using Equation 6.27a, element A_{11} is calculated as follows:

$$
A_{11} = \sum_{k=1}^{3} \{\bar{Q}_{11}\}_k (z_k - z_{k-1})
$$

* The $[\bar{Q}]$ matrix for a 30° graphite–epoxy ply was calculated as a part of Example Problem 5.6.

$$A_{11} = \left\{\overline{Q}_{11}\right\}_1 (z_1 - z_0) + \left\{\overline{Q}_{11}\right\}_2 (z_2 - z_1) + \left\{\overline{Q}_{11}\right\}_3 (z_3 - z_2)$$

$$A_{11} = \left\{107.6 \times 10^9\right\}(-.0000625 + 0.0001875)$$

$$+ \left\{170.9 \times 10^9\right\}(0.0000625 + 0.0000625)$$

$$+ \left\{10.05 \times 10^9\right\}(0.0001875 - 0.0000625)$$

$$A_{11} = 36.07 \times 10^6 \text{ Pa-m}$$

The remaining elements of the A_{ij} matrix are found in similar fashion:

$$A_{ij} = \begin{bmatrix} 36.07 & 4.012 & 6.016 \\ 4.012 & 26.02 & 2.690 \\ 6.016 & 2.690 & 7.756 \end{bmatrix} \times 10^6 \text{(Pa-m)}$$

- Using Equations 6.27b, element B_{11} is calculated as follows:

$$B_{11} = \frac{1}{2} \sum_{k=1}^{3} \left\{\overline{Q}_{11}\right\}_k (z_k^2 - z_{k-1}^2)$$

$$B_{11} = \frac{1}{2}\left[\left\{\overline{Q}_{11}\right\}_1 (z_1^2 - z_0^2) + \left\{\overline{Q}_{11}\right\}_2 (z_2^2 - z_1^2) + \left\{\overline{Q}_{11}\right\}_3 (z_3^2 - z_2^2)\right]$$

$$B_{11} = \frac{1}{2}\left[\left\{107.6 \times 10^9\right\}\left\{(-.0000625)^2 - (-0.0001875)^2\right\}\right.$$

$$+ \left\{170.9 \times 10^9\right\}\left\{(0.0000625)^2 - (-0.0000625)^2\right\}$$

$$\left. + \left\{10.05 \times 10^9\right\}\left\{(0.0001875)^2 - (0.0000625)^2\right\}\right]$$

$$B_{11} = -1.524 \times 10^3 \text{ Pa} - \text{m}^2$$

The remaining elements of the B_{ij} matrix are found in similar fashion:

$$B_{ij} = \begin{bmatrix} -1.524 & -0.3601 & -0.7521 \\ -0.3601 & 2.245 & -0.3362 \\ -0.7521 & -0.3362 & -0.3601 \end{bmatrix} \times 10^3 \text{ (Pa-m}^2)$$

In passing, in this example, it *appears* that B_{12} is numerically equal to B_{66}. This is not true, in general. In this problem the apparent numerical equivalence is due to the fact that only 4 significant digits have been listed. Nevertheless, for laminates produced using a single material system it is often (but not always) the case that $B_{12} \approx B_{66}$. This common occurrence can be traced to the fact the functional form and magnitude of $\overline{Q}_{12}$ and $\overline{Q}_{66}$ are similar (see Equations 5.31). Since B_{12} and B_{66} are directly related to $\overline{Q}_{12}$ and $\overline{Q}_{66}$, respectively, their values are often nearly identical. Also, in the next section it will be seen that *all* elements within the B_{ij} matrix are zero for *symmetric* laminates. Hence, for symmetric laminates these two terms are, in fact, numerically equal, that is, $B_{12} = B_{66} = 0$ for symmetric laminates.

- Using Equations 6.34, element D_{11} is calculated as follows:

$$D_{11} = \frac{1}{3} \sum_{k=1}^{3} \{\overline{Q}_{11}\}_k (z_k^3 - z_{k-1}^3)$$

$$D_{11} = \frac{1}{3}\Big[\{\overline{Q}_{11}\}_1 (z_1^3 - z_0^3) + \{\overline{Q}_{11}\}_2 (z_2^3 - z_1^3) + \{\overline{Q}_{11}\}_3 (z_3^3 - z_2^3)\Big]$$

$$D_{11} = \frac{1}{3}\Big[\{107.6 \times 10^9\}\{(-.0000625)^3 - (-0.0001875)^3\}$$

$$+ \{170.9 \times 10^9\}\{(0.0000625)^3 - (-0.0000625)^3\}$$

$$+ \{10.05 \times 10^9\}\{(0.0001875)^3 - (0.0000625)^3\}\Big]$$

$$D_{11} = 0.2767 \text{ Pa–m}^3$$

The remaining elements of the D_{ij} matrix are found in similar fashion:

$$D_{ij} = \begin{bmatrix} 0.2767 & 0.0620 & 0.1018 \\ 0.0620 & 0.4208 & 0.0455 \\ 0.1018 & 0.0455 & 0.1059 \end{bmatrix} (\text{Pa} - \text{m}^3)$$

The [*ABD*] matrix can now be assembled:

$$[ABD] = \begin{bmatrix} 36.07 \times 10^6 & 4.012 \times 10^6 & 6.016 \times 10^6 & -1524 & -360.1 & -752.1 \\ 4.012 \times 10^6 & 26.02 \times 10^6 & 2.690 \times 10^6 & -360.1 & 2245 & -336.2 \\ 6.016 \times 10^6 & 2.690 \times 10^6 & 7.756 \times 10^6 & -752.1 & -336.2 & -360.1 \\ -1524 & -360.1 & -752.1 & 0.2767 & 0.0620 & 0.1018 \\ -360.1 & 2245 & -336.2 & 0.0620 & 0.4208 & 0.0455 \\ -752.1 & -336.2 & -360.1 & 0.1018 & 0.0455 & 0.1059 \end{bmatrix}$$

The [*abd*] matrix is obtained by inverting the [*ABD*] matrix, and is found to be

$$[abd] = \begin{bmatrix} 3.757 \times 10^{-8} & -1.964 \times 10^{-9} & -1.038 \times 10^{-8} & 1.440 \times 10^{-4} \\ -1.964 \times 10^{-9} & 1.037 \times 10^{-7} & -4.234 \times 10^{-8} & -1.866 \times 10^{-5} \\ -1.038 \times 10^{-8} & -4.234 \times 10^{-8} & 2.004 \times 10^{-7} & 3.661 \times 10^{-4} \\ 1.440 \times 10^{-4} & -1.866 \times 10^{-5} & 3.661 \times 10^{-4} & 7.064 \\ 3.905 \times 10^{-6} & -6.361 \times 10^{-4} & 3.251 \times 10^{-4} & -3.122 \times 10^{-2} \\ 8.513 \times 10^{-5} & 4.628 \times 10^{4} & -1.851 \times 10^{-5} & -4.572 \end{bmatrix}$$

$$\begin{bmatrix} 3.905 \times 10^{-6} & 8.513 \times 10^{-5} \\ -6.361 \times 10^{-4} & 4.628 \times 10^{4} \\ 3.251 \times 10^{-4} & -1.851 \times 10^{-5} \\ -3.122 \times 10^{-2} & -4.572 \\ 6.429 & -3.620 \\ -3.620 & 17.41 \end{bmatrix}$$

Example Problem 6.5

A $[30/0/90]_T$ graphite–epoxy laminate is subjected to the following stress and moment resultants:

$N_{xx} = 50$ kN/m $N_{yy} = -10$ kN/m $N_{xy} = 0$ N/m

$M_{xx} = 1$ N–m/m $M_{yy} = -1$ N–m/m $M_{xy} = 0$ N–m/m

Determine the following quantities caused by these stress and moment resultants:

a. Midplane strains and curvatures
b. Ply strains relative to the x–y coordinate system
c. Ply stresses relative to the x–y coordinate system

Use material properties listed for graphite–epoxy in Table 3.1 and assume each ply has a thickness of 0.125 mm.

SOLUTION

Note that this is the same laminate considered in Example Problem 6.4. An edge view of the laminate appears in Figure 6.17.

a. *Midplane strains and curvatures:* The [abd] matrix for this laminate was calculated as a part of Example Problem 6.4. Hence, midplane strains and curvature may be obtained through application of Equation 6.38, which becomes

$$
\begin{Bmatrix} \varepsilon_{xx}^o \\ \varepsilon_{yy}^o \\ \gamma_{xy}^o \\ \kappa_{xx} \\ \kappa_{yy} \\ \kappa_{xy} \end{Bmatrix} =
\begin{bmatrix}
3.757 \times 10^{-8} & -1.964 \times 10^{-9} & -1.038 \times 10^{-8} & 1.440 \times 10^{-4} \\
-1.964 \times 10^{-9} & 1.037 \times 10^{-7} & -4.234 \times 10^{-8} & -1.866 \times 10^{-5} \\
-1.038 \times 10^{-8} & -4.234 \times 10^{-8} & 2.004 \times 10^{-7} & 3.661 \times 10^{-4} \\
1.440 \times 10^{-4} & -1.866 \times 10^{-5} & 3.661 \times 10^{-4} & 7.064 \\
3.905 \times 10^{-6} & -6.361 \times 10^{-4} & 3.251 \times 10^{-4} & -3.122 \times 10^{-2} \\
8.513 \times 10^{-5} & 4.628 \times 10^{4} & -1.851 \times 10^{-5} & -4.572
\end{bmatrix}
$$

$$
\begin{bmatrix}
3.905 \times 10^{-6} & 8.513 \times 10^{-5} \\
-6.361 \times 10^{-4} & 4.628 \times 10^{4} \\
3.251 \times 10^{-4} & -1.851 \times 10^{-5} \\
-3.122 \times 10^{-2} & -4.572 \\
6.429 & -3.620 \\
-3.620 & 17.41
\end{bmatrix}
\begin{Bmatrix} 50 \times 10^{3} \\ -10 \times 10^{3} \\ 0 \\ 1 \\ -1 \\ 0 \end{Bmatrix}
$$

Completing this matrix multiplication, we obtain:

$$
\begin{Bmatrix} \varepsilon_{xx}^o \\ \varepsilon_{yy}^o \\ \gamma_{xy}^o \\ \kappa_{xx} \\ \kappa_{yy} \\ \kappa_{xy} \end{Bmatrix} =
\begin{Bmatrix}
2039 \ \mu m/m \\
-518 \ \mu m/m \\
-55 \ \mu rad \\
14.48 \ m^{-1} \\
0.096 \ m^{-1} \\
-1.323 \ m^{-1}
\end{Bmatrix}
$$

b. *Ply strains relative to the x–y coordinate system:* Ply strains may now be calculated using Equation 6.12. For example, strains present at the outer surface of ply #1 (i.e., strains present at $z_o = -0.0001875$ m) are

$$
\begin{Bmatrix} \varepsilon_{xx} \\ \varepsilon_{yy} \\ \gamma_{xy} \end{Bmatrix}_{z=z0} =
\begin{Bmatrix} \varepsilon_{xx}^o \\ \varepsilon_{yy}^o \\ \gamma_{xy}^o \end{Bmatrix} + z_0 \begin{Bmatrix} \kappa_{xx} \\ \kappa_{yy} \\ \kappa_{xy} \end{Bmatrix} =
\begin{Bmatrix} 2038 \times 10^{-6} \ m/m \\ -518 \times 10^{-6} \ m/m \\ -55 \times 10^{-6} \ m/m \end{Bmatrix} + (-0.0001875 \ m) \begin{Bmatrix} 14.48 \ rad/m \\ 0.096 \ rad/m \\ -1.328 \ rad/m \end{Bmatrix}
$$

$$
\begin{Bmatrix} \varepsilon_{xx} \\ \varepsilon_{yy} \\ \gamma_{xy} \end{Bmatrix}_{z=z0} =
\begin{Bmatrix} -677 \ \mu m/m \\ -536 \ \mu m/m \\ 194 \ \mu rad \end{Bmatrix}
$$

Strains calculated at the remaining ply interface positions are summarized in Table 6.5.

TABLE 6.5

Ply Interface Strains in a [30/0/90] Graphite–Epoxy Laminate Caused by the Stress and Moment Resultants Specified in Example Problem 6.5

z-Coordinate (mm)	ε_{xx} (μm/m)	ε_{yy} (μm/m)	γ_{xy} (μRadians)
−0.1875	−677	−536	194
−0.0625	1133	−524	28
0.0625	2943	−512	−137
0.1875	4753	−500	−303

Note: Strains are referenced to the *x–y* coordinate system.

c. *Ply stresses relative to the x–y coordinate system:* The $[\overline{Q}]$ matrix for all plies was calculated as a part of Example Problem 6.4. Ply stresses may now be calculated using Equation 5.30, with $\Delta T = \Delta M = 0$. The stresses present at the outer surface of ply #1 (i.e., at $z = z_0$) are

$$
\left\{ \begin{matrix} \sigma_{xx} \\ \sigma_{yy} \\ \tau_{xy} \end{matrix} \right\}^{Ply\ 1}_{z=z0} = \begin{bmatrix} \overline{Q}_{11} & \overline{Q}_{12} & \overline{Q}_{16} \\ \overline{Q}_{12} & \overline{Q}_{22} & \overline{Q}_{26} \\ \overline{Q}_{16} & \overline{Q}_{26} & \overline{Q}_{66} \end{bmatrix}^{Ply1}_{z=z0} \left\{ \begin{matrix} \varepsilon_{xx} \\ \varepsilon_{yy} \\ \gamma_{xy} \end{matrix} \right\}^{Ply1}_{z=z0}
$$

$$
\left\{ \begin{matrix} \sigma_{xx} \\ \sigma_{yy} \\ \tau_{xy} \end{matrix} \right\}^{Ply\ 1}_{z=z0} = \begin{bmatrix} 107.6 \times 10^9 & 26.06 \times 10^9 & 48.13 \times 10^9 \\ 26.06 \times 10^9 & 27.22 \times 10^9 & 21.52 \times 10^9 \\ 48.13 \times 10^9 & 21.52 \times 10^9 & 36.05 \times 10^9 \end{bmatrix} \left\{ \begin{matrix} -677 \times 10^{-6} \\ -536 \times 10^{-6} \\ 194 \times 10^{-6} \end{matrix} \right\}
$$

$$
\left\{ \begin{matrix} \sigma_{xx} \\ \sigma_{yy} \\ \tau_{xy} \end{matrix} \right\}^{Ply1}_{z=z0} = \left\{ \begin{matrix} -77.5\ \text{MPa} \\ -28.1\ \text{MPa} \\ -37.1\,\text{MPa} \end{matrix} \right\}
$$

Stresses calculated at remaining ply interface positions are summarized in Table 6.6.

6.6.2 Including Changes in Environmental Conditions

Recall that we simplified the analysis leading up to Equation 6.36 by assuming $\Delta T = \Delta M = 0$. We will now consider how to predict the behavior of a laminate subjected to a change in temperature and/or moisture content as well as external mechanical loads.

To begin, the stresses in any ply (in ply number "*k*," say) are related to ply strains in accordance with Equation 5.30:

TABLE 6.6

Ply Interface Stresses in a $[30/0/90]_T$ Graphite–Epoxy Laminate Caused by the Stress and Moment Resultants Specified in Example Problem 6.5

Ply Number	z-Coordinate (mm)	σ_{xx} (MPa)	σ_{yy} (MPa)	τ_{xy} (MPa)
Ply	−0.1875	−77.5	−28.1	−37.1
1	−0.0625	109.7	15.9	44.3
Ply	−0.0625	192.1	−1.85	0.366
2	0.0625	501.5	3.73	−1.78
Ply	0.0625	28.0	−78.6	−1.78
3	0.1875	46.3	−71.1	−3.93

Note: Stresses are referenced to the x–y coordinate system.

$$\left\{ \begin{matrix} \sigma_{xx} \\ \sigma_{yy} \\ \tau_{xy} \end{matrix} \right\}_k = \begin{bmatrix} \overline{Q}_{11} & \overline{Q}_{12} & \overline{Q}_{16} \\ \overline{Q}_{21} & \overline{Q}_{22} & \overline{Q}_{26} \\ \overline{Q}_{61} & \overline{Q}_{62} & \overline{Q}_{66} \end{bmatrix}_k \left\{ \begin{matrix} \varepsilon_{xx} - \Delta T \alpha_{xx} - \Delta M \beta_{xx} \\ \varepsilon_{yy} - \Delta T \alpha_{yy} - \Delta M \beta_{yy} \\ \gamma_{xy} - \Delta T \alpha_{xy} - \Delta M \beta_{xy} \end{matrix} \right\}_k \qquad \text{(5.30) (repeated)}$$

Stress σ_{xx} in ply k is given by

$$\sigma_{xx} = \overline{Q}_{11}\{\varepsilon_{xx} - \Delta T \alpha_{xx} - \Delta M \beta_{xx}\} + \overline{Q}_{12}\{\varepsilon_{yy} - \Delta T \alpha_{yy} - \Delta M \beta_{xx}\}$$
$$+ \overline{Q}_{16}\{\gamma_{xy} - \Delta T \alpha_{xy} - \Delta M \beta_{xy}\} \qquad (6.39)$$

Stress resultant N_{xx} is related to σ_{xx} via Equation 6.1a. Substituting Equation 6.39 into Equation 6.1a, we have

$$N_{xx} = \int_{-t/2}^{t/2} \left\{ \overline{Q}_{11}\varepsilon_{xx} + \overline{Q}_{12}\varepsilon_{yy} + \overline{Q}_{16}\gamma_{xy} \right\}_k dz$$

$$- \Delta T \int_{-t/2}^{t/2} \left\{ \overline{Q}_{11}\alpha_{xx} + \overline{Q}_{12}\alpha_{yy} + \overline{Q}_{16}\gamma_{xy} \right\}_k dz$$

$$- \Delta M \int_{-t/2}^{t/2} \left\{ \overline{Q}_{11}\beta_{xx} + \overline{Q}_{12}\beta_{yy} + \overline{Q}_{16}\beta_{xy} \right\}_k dz \qquad (6.40)$$

The first integral on the right-hand side of the equality sign is identical to Equation 6.21, and after evaluation (using the same techniques as previously described) will result in Equation 6.26a. The second and third integrals were not previously encountered, since they involve ΔT and ΔM, which were previously assumed to equal zero. Using methods similar to those used previously, it can be shown that the second integral may be written as

$$\Delta T \int_{-t/2}^{t/2} \left\{ \overline{Q}_{11}\alpha_{xx} + \overline{Q}_{12}\alpha_{yy} + \overline{Q}_{16}\gamma_{xy} \right\}_k dz = \Delta T \sum_{k=1}^{n} \left\{ \left[\overline{Q}_{11}\alpha_{xx} + \overline{Q}_{12}\alpha_{yy} + \overline{Q}_{16}\alpha_{xy} \right]_k \right.$$

$$\left. \times \left[z_k - z_{k-1} \right] \right\}$$

This quantity is called a *thermal stress resultant*, and will be denoted N_{xx}^T. That is,

$$N_{xx}^T \equiv \Delta T \sum_{k=1}^{n} \left\{ \left[\overline{Q}_{11}\alpha_{xx} + \overline{Q}_{12}\alpha_{yy} + \overline{Q}_{16}\alpha_{xy} \right]_k \left[z_k - z_{k-1} \right] \right\} \qquad (6.41a)$$

Similarly, the third integral in Equation 6.40 can be evaluated to give the *moisture stress resultant*, denoted N_{xx}^M:

$$N_{xx}^M \equiv \Delta M \sum_{k=1}^{n} \left\{ \left[\overline{Q}_{11}\beta_{xx} + \overline{Q}_{12}\beta_{yy} + \overline{Q}_{16}\beta_{xy} \right]_k \left[z_k - z_{k-1} \right] \right\} \qquad (6.42a)$$

Hence, after evaluating all integrals, Equation 6.40 may be written as

$$N_{xx} = A_{11}\varepsilon_{xx}^o + A_{12}\varepsilon_{yy}^o + A_{16}\gamma_{xy}^o + B_{11}\kappa_{xx} + B_{12}\kappa_{yy} + B_{16}\kappa_{xy} - N_{xx}^T - N_{xx}^M \qquad (6.43)$$

This result should be compared to Equation 6.26a. It is easily seen that the inclusion of temperature and/or moisture changes in our analysis has resulted in the addition of two new terms (N_{xx}^T and N_{xx}^M), but otherwise our earlier results remain unchanged.

If an analogous procedure is now followed for the remaining stress and moment resultants, using Equations 6.1b, 6.1c and Equations 6.2a, 6.2b, 6.2c, five additional thermal stress/moment resultants and five additional moisture stress/moment resultants will be identified, as follows:

$$N_{yy}^T \equiv \Delta T \sum_{k=1}^{n} \left\{ \left[\overline{Q}_{12}\alpha_{xx} + \overline{Q}_{22}\alpha_{yy} + \overline{Q}_{26}\alpha_{xy} \right]_k \left[z_k - z_{k-1} \right] \right\} \qquad (6.41b)$$

$$N_{xy}^T \equiv \Delta T \sum_{k=1}^{n} \left\{ \left[\overline{Q}_{16}\alpha_{xx} + \overline{Q}_{26}\alpha_{yy} + \overline{Q}_{66}\alpha_{xy} \right]_k \left[z_k - z_{k-1} \right] \right\} \qquad (6.41c)$$

$$M_{xx}^{T} \equiv \frac{\Delta T}{2} \sum_{k=1}^{n} \left\{ \left[\overline{Q}_{11}\alpha_{xx} + \overline{Q}_{12}\alpha_{yy} + \overline{Q}_{16}\alpha_{xy} \right]_{k} \left[z_{k}^{2} - z_{k-1}^{2} \right] \right\} \tag{6.41d}$$

$$M_{yy}^{T} \equiv \frac{\Delta T}{2} \sum_{k=1}^{n} \left\{ \left[\overline{Q}_{12}\alpha_{xx} + \overline{Q}_{22}\alpha_{yy} + \overline{Q}_{26}\alpha_{xy} \right]_{k} \left[z_{k}^{2} - z_{k-1}^{2} \right] \right\} \tag{6.41e}$$

$$M_{xy}^{T} \equiv \frac{\Delta T}{2} \sum_{k=1}^{n} \left\{ \left[\overline{Q}_{16}\alpha_{xx} + \overline{Q}_{26}\alpha_{yy} + \overline{Q}_{66}\alpha_{xy} \right]_{k} \left[z_{k}^{2} - z_{k-1}^{2} \right] \right\} \tag{6.41f}$$

$$N_{yy}^{M} \equiv \Delta M \sum_{k=1}^{n} \left\{ \left[\overline{Q}_{12}\beta_{xx} + \overline{Q}_{22}\beta_{yy} + \overline{Q}_{26}\beta_{xy} \right]_{k} \left[z_{k} - z_{k-1} \right] \right\} \tag{6.42b}$$

$$N_{xy}^{M} \equiv \Delta M \sum_{k=1}^{n} \left\{ \left[\overline{Q}_{16}\beta_{xx} + \overline{Q}_{26}\beta_{yy} + \overline{Q}_{66}\beta_{xy} \right]_{k} \left[z_{k} - z_{k-1} \right] \right\} \tag{6.42c}$$

$$M_{xx}^{M} \equiv \frac{\Delta M}{2} \sum_{k=1}^{n} \left\{ \left[\overline{Q}_{11}\beta_{xx} + \overline{Q}_{12}\beta_{yy} + \overline{Q}_{16}\beta_{xy} \right]_{k} \left[z_{k}^{2} - z_{k-1}^{2} \right] \right\} \tag{6.42d}$$

$$M_{yy}^{M} \equiv \frac{\Delta M}{2} \sum_{k=1}^{n} \left\{ \left[\overline{Q}_{12}\beta_{xx} + \overline{Q}_{22}\beta_{yy} + \overline{Q}_{26}\beta_{xy} \right]_{k} \left[z_{k}^{2} - z_{k-1}^{2} \right] \right\} \tag{6.42e}$$

$$M_{xy}^{M} \equiv \frac{\Delta M}{2} \sum_{k=1}^{n} \left\{ \left[\overline{Q}_{16}\beta_{xx} + \overline{Q}_{26}\beta_{yy} + \overline{Q}_{66}\beta_{xy} \right]_{k} \left[z_{k}^{2} - z_{k-1}^{2} \right] \right\} \tag{6.42f}$$

Finally, the response of a composite laminate subjected to mechanical loads, a change in temperature, and a change in moisture content can be written in a form similar to Equation 6.36:

$$\begin{Bmatrix} N_{xx} \\ N_{yy} \\ N_{xy} \\ M_{xx} \\ M_{yy} \\ M_{xy} \end{Bmatrix} = \begin{bmatrix} A_{11} & A_{12} & A_{16} & B_{11} & B_{12} & B_{16} \\ A_{12} & A_{22} & A_{26} & B_{12} & B_{22} & B_{26} \\ A_{16} & A_{26} & A_{66} & B_{16} & B_{26} & B_{66} \\ B_{11} & B_{12} & B_{16} & D_{11} & D_{12} & D_{16} \\ B_{12} & B_{22} & B_{26} & D_{12} & D_{22} & D_{26} \\ B_{16} & B_{26} & B_{66} & D_{16} & D_{26} & D_{66} \end{bmatrix} \begin{Bmatrix} \varepsilon_{xx}^{o} \\ \varepsilon_{yy}^{o} \\ \gamma_{xy}^{o} \\ \kappa_{xx} \\ \kappa_{yy} \\ \kappa_{xy} \end{Bmatrix} - \begin{Bmatrix} N_{xx}^{T} \\ N_{yy}^{T} \\ N_{xy}^{T} \\ M_{xx}^{T} \\ M_{yy}^{T} \\ M_{xy}^{T} \end{Bmatrix} - \begin{Bmatrix} N_{xx}^{M} \\ N_{yy}^{M} \\ N_{xy}^{M} \\ M_{xx}^{M} \\ M_{yy}^{M} \\ M_{xy}^{M} \end{Bmatrix}$$

$$\tag{6.44}$$

Equation 6.44 will sometimes be abbreviated as

$$\begin{Bmatrix} N \\ M \end{Bmatrix} = \begin{bmatrix} A & B \\ B & D \end{bmatrix} \begin{Bmatrix} \varepsilon^o \\ \kappa \end{Bmatrix} - \begin{Bmatrix} N^T \\ M^T \end{Bmatrix} - \begin{Bmatrix} N^M \\ M^M \end{Bmatrix}$$

Equation 6.44 is comparable to Equation 6.36, except we have now included the effects due to a change in temperature and/or moisture content. Equation 6.44 can be viewed as "Hooke's Law" for a composite laminate, in the sense that it may be used to relate stress-like quantities (i.e., stress and moment resultants) to strain-like quantities (i.e., midplane strains and curvatures). Inverting Equation 6.44, we obtain:

$$\begin{Bmatrix} \varepsilon_{xx}^o \\ \varepsilon_{yy}^o \\ \gamma_{xy}^o \\ \kappa_{xx} \\ \kappa_{yy} \\ \kappa_{xy} \end{Bmatrix} = \begin{bmatrix} a_{11} & a_{12} & a_{16} & b_{11} & b_{12} & b_{16} \\ a_{12} & a_{22} & a_{26} & b_{21} & b_{22} & b_{26} \\ a_{16} & a_{26} & a_{66} & b_{61} & b_{62} & b_{66} \\ b_{11} & b_{21} & b_{61} & d_{11} & d_{12} & d_{16} \\ b_{12} & b_{22} & b_{62} & d_{12} & d_{22} & d_{26} \\ b_{16} & b_{26} & b_{66} & d_{16} & d_{26} & d_{66} \end{bmatrix} \begin{Bmatrix} N_{xx} + N_{xx}^T + N_{xx}^M \\ N_{yy} + N_{yy}^T + N_{yy}^M \\ N_{xy} + N_{xy}^T + N_{xy}^M \\ M_{xx} + M_{xx}^T + M_{xx}^M \\ M_{yy} + M_{yy}^T + M_{yy}^M \\ M_{xy} + M_{xy}^T + M_{xy}^M \end{Bmatrix} \quad (6.45)$$

where, as before:

$$\begin{bmatrix} a & b \\ b & d \end{bmatrix} = \begin{bmatrix} A & B \\ B & D \end{bmatrix}^{-1}$$

A subtlety embedded within the preceding discussion is the fact that most composites are subjected to a significant state of stress *prior to the application of any external mechanical loading*. That is, most modern composite material systems are cured at an elevated temperature (common cure temperatures are either 120°C or 175°C), and are nominally stress-free *at the cure temperature*. Once the polymerization process is complete the composite is cooled to room temperatures (20°C, say) and consequently the composite experiences a uniform change in temperature of $\Delta T = -100°C$ or $-155°C$ *during cooldown*. In general, this change in temperature results in thermal stress and/or moment resultants to develop, causing thermal stresses within all plies of the laminate. These thermal stresses can be quite high, and contribute toward failure of the laminate.*

---

* Determination of the "stress-free temperature" is actually more complex than is implied here. It is true that thermal stresses begin to develop as cooldown begins, but since polymeric materials exhibit viscoelastic characteristics at these elevated temperatures the matrix will creep, initially relieving thermal stresses somewhat. As temperature is decreased further the viscoelastic nature of the matrix is rapidly decreased, and thermal stresses develop as

A further complicating factor is related to measurement of strains. In most practical situations strain measurement devices (e.g., resistance foil strain gages) are bonded to a composite material or structure *after cooldown to room temperature*. Hence, in practice the zero-reference state of a measurement device mounted on a laminate does not necessarily correspond to the stress-free (or strain-free) state of the composite. This complication will be further explored in Chapter 7. At this point it will simply be noted that a significant difficulty arises when prediction of nonlinear behavior (or more generally, the prediction of composite failure) is required based on measured laminate strains.

Example Problem 6.6

A $[30/0/90]_T$ graphite–epoxy laminate is cured at 175°C and then cooled to room temperature (20°C). Determine:

a. Midplane strains and curvatures
b. Ply strains relative to the x–y coordinate system
c. Ply stresses relative to the x–y coordinate system

which are induced during cooldown. Use material properties listed for graphite–epoxy in Table 3.1, assume each ply has a thickness of 0.125 mm, and assume no change in moisture content (i.e., assume $\Delta M = 0$).

SOLUTION

Note that this is the same laminate considered in Sample Problem 6.4. A side view of the laminate appears in Figure 6.17.

a. *Midplane strains and curvatures:* The laminate has experienced a change in temperature $\Delta T = (20-175) = -155°C$, and consequently is subjected thermal stress and moment resultants. However, no external loads are applied and there has been no change in moisture content, therefore the stress and moment resultants and the moisture stress and moment results are zero:

$$\begin{Bmatrix} N_{xx} \\ N_{yy} \\ N_{xy} \\ M_{xx} \\ M_{yy} \\ M_{xy} \end{Bmatrix} = \begin{Bmatrix} N_{xx}^M \\ N_{yy}^M \\ N_{xy}^M \\ M_{xx}^M \\ M_{yy}^M \\ M_{xy}^M \end{Bmatrix} = \begin{Bmatrix} 0 \\ 0 \\ 0 \\ 0 \\ 0 \\ 0 \end{Bmatrix}$$

described. A second factor as that all polymers exhibit some shrinkage during the polymerization process (see Section 1.2), and this shrinkage results in additional stresses similar to thermal stresses. As a rule-of-thumb the stress-free temperature is often estimated to be 10–25°C below the final cure temperature. Nevertheless, this complication will be ignored in this text; it will be assumed that the final cure temperature defines the stress-free temperature.

The effective thermal expansion coefficients for each ply are calculated using Equation 5.25, repeated here for convenience:

$$\alpha_{xx} = \alpha_{11} \cos^2 (\theta) + \alpha_{22} \sin^2 (\theta)$$

$$\alpha_{yy} = \alpha_{11} \sin^2 (\theta) + \alpha_{22} \cos^2 (\theta) \qquad \text{(5.25) (repeated)}$$

$$\alpha_{xy} = 2 \cos (\theta) \sin (\theta) (\alpha_{11} - \alpha_{22})$$

From Table 3.1, the thermal expansion coefficients for graphite–epoxy (relative to the 1–2 coordinate system) are $\alpha_{11} = -0.9$ μm/m – °C and $\alpha_{22} = 27$ μm/m – °C. Therefore:

For ply #1 (the 30°ply):

$$\alpha_{xx}^{(1)} = (-0.9\,\mu m/m - °C)\cos^2(30°) + (27\,\mu m/m - °C)\sin^2(30°) = 6.08\ \mu m/m - °C$$

$$\alpha_{yy}^{(1)} = (-0.9\,\mu m/m - °C)\sin^2(30°) + (27\,\mu m/m - °C)\cos^2(30°) = 20.0\ \mu m/m - °C$$

$$\alpha_{xy}^{(1)} = 2\cos(30)\sin(30)[(-0.9 - 27)\mu m/m - °C] = -24.2\ \mu rad/°C$$

For ply #2 (the 0°ply):

$$\alpha_{xx}^{(2)} = (-0.9\,\mu m/m - °C)\cos^2(0°) + (27\,\mu m/m - °C)\sin^2(0°) = -0.9\ \mu m/m - °C$$

$$\alpha_{yy}^{(2)} = (-0.9\,\mu m/m - °C)\sin^2(0°) + (27\,\mu m/m - °C)\cos^2(0°) = 27.0\ \mu m/m - °C$$

$$\alpha_{xy}^{(2)} = 2\cos(0°)\sin(0°)[(-0.9 - 27)\,\mu m/m - °C] = 0\ \mu rad/°C$$

For ply #3 (the 90°ply):

$$\alpha_{xx}^{(3)} = (-0.9\,\mu m/m - °C)\cos^2(90°) + (27\,\mu m/m - °C)\sin^2(90°) = 27.0\ \mu m/m-°C$$

$$\alpha_{yy}^{(3)} = (-0.9\,\mu m/m - °C)\sin^2(90°) + (27\,\mu m/m - °C)\cos^2(90°) = -0.9\ \mu m/m-°C$$

$$\alpha_{xy}^{(3)} = 2\cos(90°)\sin(90°)[(-0.9 - 27)\,\mu m/m - °C] = 0\ \mu rad/°C$$

Both the ply interface positions as well as the $\overline{Q}_{ij}$ matrices for each ply were calculated as a part of Example Problem 6.4. Hence, we now have all the information needed to calculate the thermal stress and moment resultants, using Equations 6.41. For example, Equation 6.41a is evaluated as follows:

$$N_{xx}^T \equiv \Delta T \sum_{k=1}^{3} \left\{ \left[\overline{Q}_{11}\alpha_{xx} + \overline{Q}_{12}\alpha_{yy} + \overline{Q}_{16}\alpha_{xy} \right]_k [z_k - z_{k-1}] \right\}$$

$$N_{xx}^T = \Delta T \left\{ \left(\left[\overline{Q}_{11}\alpha_{xx} + \overline{Q}_{12}\alpha_{yy} + \overline{Q}_{16}\alpha_{xy} \right]_1 [z_1 - z_0] \right) \right.$$

$$+ \left(\left[\overline{Q}_{11}\alpha_{xx} + \overline{Q}_{12}\alpha_{yy} + \overline{Q}_{16}\alpha_{xy} \right]_2 [z_2 - z_1] \right)$$

$$\left. + \left(\left[\overline{Q}_{11}\alpha_{xx} + \overline{Q}_{12}\alpha_{yy} + \overline{Q}_{16}\alpha_{xy} \right]_3 [z_3 - z_2] \right) \right\}$$

$$N_{xx}^T = (-155) \left\{ \left(\left[(107.6 \times 10^9)(6.08 \times 10^{-6}) + (26.06 \times 10^9)(20.0 \times 10^{-6}) \right. \right. \right.$$

$$+ (48.13 \times 10^9)(-24.2 \times 10^{-6}) \right] \times \left[(-0.0625 + 0.1875) \times 10^{-3} \right] \Big)$$

$$+ \left(\left[(170.9 \times 10^9)(-0.9 \times 10^{-6}) + (3.016 \times 10^9)(27.0 \times 10^{-6}) + (0)(0) \right] \right.$$

$$\times \left[(0.0625 + 0.0625) \times 10^{-3} \right] \Big) + \left(\left[(10.05 \times 10^9)(27 \times 10^{-6}) + (3.016 \times 10^9) \right. \right.$$

$$\times (-0.9 \times 10^{-6}) + (0)(0) \right] \left[(0.1875 - 0.0625) \times 10^{-3} \right] \Big) \Big\}$$

$$N_{xx}^T = 4060 \text{ N/m}$$

The remaining thermal stress and moment resultants are calculated in similar fashion, resulting in

$$\begin{Bmatrix} N_{xx}^T \\ N_{yy}^T \\ N_{xy}^T \\ M_{xx}^T \\ M_{yy}^T \\ M_{xy}^T \end{Bmatrix} = \begin{Bmatrix} -4060 \text{ N/m} \\ -7360 \text{ N/m} \\ 2860 \text{ N/m} \\ -0.62 \text{ N} - \text{m/m} \\ 0.62 \text{ N} - \text{m/m} \\ -0.36 \text{ N} - \text{m/m} \end{Bmatrix}$$

We can now calculate midplane strains and curvature using Equation 6.45, which becomes*

---

* The [abd] matrix for a [30/0/90]_T graphite–epoxy laminate was calculated in Sample Problem 6.3.

$$
\begin{Bmatrix} \varepsilon_{xx}^o \\ \varepsilon_{yy}^o \\ \gamma_{xy}^o \\ \kappa_{xx} \\ \kappa_{yy} \\ \kappa_{xy} \end{Bmatrix} = \begin{bmatrix} 3.757 \times 10^{-8} & -1.964 \times 10^{-9} & -1.038 \times 10^{-8} & 1.440 \times 10^{-4} \\ -1.964 \times 10^{-9} & 1.037 \times 10^{-7} & -4.234 \times 10^{-8} & -1.866 \times 10^{-5} \\ -1.038 \times 10^{-8} & -4.234 \times 10^{-8} & 2.004 \times 10^{-7} & 3.661 \times 10^{-4} \\ 1.440 \times 10^{-4} & -1.866 \times 10^{-5} & 3.661 \times 10^{-4} & 7.064 \\ 3.905 \times 10^{-6} & -6.361 \times 10^{-4} & 3.251 \times 10^{-4} & -3.122 \times 10^{-2} \\ 8.513 \times 10^{-5} & 4.628 \times 10^{4} & -1.851 \times 10^{-5} & -4.572 \end{bmatrix}
$$

$$
\begin{matrix} 3.905 \times 10^{-6} & 8.513 \times 10^{-5} \\ -6.361 \times 10^{-4} & 4.628 \times 10^{4} \\ 3.251 \times 10^{-4} & -1.851 \times 10^{-5} \\ -3.122 \times 10^{-2} & -4.572 \\ 6.429 & -3.620 \\ -3.620 & 17.41 \end{matrix} \begin{Bmatrix} -4060 \\ -7360 \\ 2860 \\ -0.62 \\ 0.62 \\ -0.36 \end{Bmatrix}
$$

$$
\begin{Bmatrix} \varepsilon_{xx}^o \\ \varepsilon_{yy}^o \\ \gamma_{xy}^o \\ \kappa_{xx} \\ \kappa_{yy} \\ \kappa_{xy} \end{Bmatrix} = \begin{Bmatrix} -285\ \mu m/m \\ -1424\ \mu m/m \\ 908\ \mu rad \\ -2.16\ m^{-1} \\ 10.9\ m^{-1} \\ -9.4\ m^{-1} \end{Bmatrix}
$$

b. *Ply strains relative to the x–y coordinate system:* Ply strains may now be calculated using Equation 6.12. For example, strains present at the outer surface of ply #1 (i.e., strains present at $z_o = -0.0001875$ m) are

$$
\begin{Bmatrix} \varepsilon_{xx} \\ \varepsilon_{yy} \\ \gamma_{xy} \end{Bmatrix}\Bigg|_{z=z0} = \begin{Bmatrix} \varepsilon_{xx}^o \\ \varepsilon_{yy}^o \\ \gamma_{xy}^o \end{Bmatrix} + z_0 \begin{Bmatrix} \kappa_{xx} \\ \kappa_{yy} \\ \kappa_{xy} \end{Bmatrix} = \begin{Bmatrix} -285 \times 10^{-6}\ m/m \\ -1424 \times 10^{-6}\ m/m \\ 908 \times 10^{-6}\ m/m \end{Bmatrix} + (-0.0001875\,m) \begin{Bmatrix} -2.16\ rad/m \\ 10.9\ rad/m \\ -9.4\ rad/m \end{Bmatrix}
$$

$$
\begin{Bmatrix} \varepsilon_{xx} \\ \varepsilon_{yy} \\ \gamma_{xy} \end{Bmatrix}\Bigg|_{z=z0} = \begin{Bmatrix} 120\ \mu m/m \\ -3468\ \mu m/m \\ 2672\ \mu rad \end{Bmatrix}
$$

Strains calculated at the remaining ply interface positions are summarized in Table 6.7.

c. *Ply stresses relative to the x–y coordinate system:* Ply stresses may now be calculated using Equation 5.30, with $\Delta M = 0$. The stresses present at the outer surface of ply #1 are (i.e., at $z = z_0$):

TABLE 6.7

Ply Interface Strains in a [30/0/90] Graphite–Epoxy Laminate Caused by a Cooldown from 175°C to 20°C

z-Coordinate (mm)	ε_{xx} (μm/m)	ε_{yy} (μm/m)	γ_{xy} (μRadians)
−0.1875	120	−3468	2672
−0.0625	−150	−2100	1500
0.0625	−420	−750	320
0.1875	−690	620	−860

Note: Strains are referenced to the x–y coordinate system.

$$\left\{ \begin{matrix} \sigma_{xx} \\ \sigma_{yy} \\ \tau_{xy} \end{matrix} \right\}^{Ply\,1}_{z=z0} = \begin{bmatrix} \overline{Q}_{11} & \overline{Q}_{12} & \overline{Q}_{16} \\ \overline{Q}_{12} & \overline{Q}_{22} & \overline{Q}_{26} \\ \overline{Q}_{16} & \overline{Q}_{26} & \overline{Q}_{66} \end{bmatrix}^{Ply1}_{z=z0} \left\{ \begin{matrix} \varepsilon_{xx} - \Delta T \alpha_{xx} \\ \varepsilon_{yy} - \Delta T \alpha_{yy} \\ \gamma_{xy} - \Delta T \alpha_{xy} \end{matrix} \right\}_{z=z0}$$

$$\left\{ \begin{matrix} \sigma_{xx} \\ \sigma_{yy} \\ \tau_{xy} \end{matrix} \right\}^{Ply\,1}_{z=z0} = \begin{bmatrix} 107.6 \times 10^9 & 26.06 \times 10^9 & 48.13 \times 10^9 \\ 26.06 \times 10^9 & 27.22 \times 10^9 & 21.52 \times 10^9 \\ 48.13 \times 10^9 & 21.52 \times 10^9 & 36.05 \times 10^9 \end{bmatrix}$$

$$\times \left\{ \begin{matrix} [(120) - (-155)(6.08)] \times 10^{-6} \\ [(-3468) - (-155)(20.0)] \times 10^{-6} \\ [(2672) - (-155)(-24.2)] \times 10^{-6} \end{matrix} \right\}$$

$$\left\{ \begin{matrix} \sigma_{xx} \\ \sigma_{yy} \\ \tau_{xy} \end{matrix} \right\}^{Ply1}_{z=z0} = \left\{ \begin{matrix} 53 \text{ MPa} \\ -5.2 \text{ MPa} \\ 4.8 \text{ MPa} \end{matrix} \right\}$$

Stresses calculated at the remaining plies and ply interface positions are summarized in Table 6.8.

TABLE 6.8

Ply Interface Stresses in a [30/0/90] Graphite–Epoxy Laminate Caused by a Cooldown from 175°C to 20°C

Ply Number	z-Coordinate (mm)	σ_{xx} (MPa)	σ_{yy} (MPa)	τ_{xy} (MPa)
Ply	−0.1875	53	−5.2	4.8
1	−0.0625	3.1	−0.53	−21
Ply	−0.0625	−43	20	19
2	0.0625	−85	33	4.1
Ply	0.0625	35	−140	4.1
3	0.1875	37	92	−11

Note: Stresses are referenced to the x–y coordinate system.

322 *Structural Analysis of Polymeric Composite Materials*

Example Problem 6.7

A $[30/0/90]_T$ graphite–epoxy laminate is cured at 175°C and cooled to room temperature (20°C). Initially the moisture content of the laminate is zero. However, the laminate is subjected to a humid environment for several weeks, resulting in an increase of moisture content of 0.5% (by weight). Determine:

a. Midplane strains and curvatures
b. Ply strains relative to the x–y coordinate system
c. Ply stresses relative to the x–y coordinate system

which are present following the increase in moisture content. Use material properties listed for graphite–epoxy in Table 3.1, and assume each ply has a thickness of 0.125 mm.

SOLUTION

Note that this is the same laminate considered in Sample Problem 6.6, and the midplane strains and curvatures, ply strains, and ply stresses that will be induced immediately upon cooldown by the change in temperature have already been calculated. These quantities will all be modified due to the slow diffusion of water molecules into the epoxy matrix.

a. *Midplane strains and curvatures:* The laminate has experienced a change in moisture content $\Delta M = +0.5\%$, and consequently is subjected moisture stress and moment resultants. The effective moisture expansion coefficients for each ply are calculated using Equation 5.28, repeated here for convenience:

$$\beta_{xx} = \beta_{11} \cos^2(\theta) + \beta_{22} \sin^2(\theta)$$

$$\beta_{yy} = \beta_{11} \sin^2(\theta) + \beta_{22} \cos^2(\theta) \qquad \text{(5.28) (repeated)}$$

$$\beta_{xy} = 2\cos(\theta) \sin(\theta) (\beta_{11} - \beta_{22})$$

From Table 3.1, the moisture expansion coefficients for graphite–epoxy (relative to the 1–2 coordinate system) are $\beta_{11} = 150$ μm/m–%M and $\beta_{22} = 4800$ μm/m–%M. Therefore,

For ply #1 (the 30°ply):

$$\beta_{xx}^{(1)} = (150\,\mu m/m - \%M)\cos^2(30°) + (4800\,\mu m/m - \%M)\sin^2(30°)$$

$$= 1310\,\mu m/m - \%M$$

$$\beta_{yy}^{(1)} = (150\,\mu m/m - \%M)\sin^2(30°) + (4800\,\mu m/m - \%M)\cos^2(30°)$$

$$= 3640\,\mu m/m - \%M$$

$$\beta_{xy}^{(1)} = 2\cos(30)\sin(30)[(150 - 4800)\,\mu m/m - \%M] = -4030\,\mu rad/\%M$$

For ply #2 (the 0°ply):

$$\beta_{xx}^{(2)} = (150\ \mu\mathrm{m/m} - \%M)\cos^2(0°) + (4800\ \mu\mathrm{m/m} - \%M)\sin^2(0°)$$

$$= 150\ \mu\mathrm{m/m} - \%M$$

$$\beta_{yy}^{(2)} = (150\ \mu\mathrm{m/m} - \%M)\sin^2(0°) + (4800\ \mu\mathrm{m/m} - \%M)\cos^2(0°)$$

$$= 4800\ \mu\mathrm{m/m} - \%M$$

$$\beta_{xy}^{(2)} = 2\cos(0°)\sin(0°)[(150 - 4800)\ \mu\mathrm{m/m} - \%M] = 0\ \mu\mathrm{rad}/\%M$$

For ply #3 (the 90°ply):

$$\beta_{xx}^{(3)} = (150\ \mu\mathrm{m/m} - \%M)\cos^2(90°) + (4800\ \mu\mathrm{m/m} - \%M)\sin^2(90°)$$

$$= 4800\ \mu\mathrm{m/m} - \%M$$

$$\beta_{yy}^{(3)} = (150\ \mu\mathrm{m/m} - \%M)\sin^2(90°) + (4800\ \mu\mathrm{m/m} - \%M)\cos^2(90°)$$

$$= 150\ \mu\mathrm{m/m} - \%M$$

$$\beta_{xy}^{(3)} = 2\cos(90°)\sin(90°)[(150 - 4800)\ \mu\mathrm{m/m} - \%M] = 0\ \mu\mathrm{rad}/\%M$$

Both the ply interface positions as well as the $[\overline{Q}]$ matrices for each ply were calculated as a part of Example Problem 6.4. Hence, we now have all the information needed to calculate the moisture stress and moment resultants, using Equations 6.42. For example, Equation 6.42a is evaluated as follows:

$$N_{xx}^M \equiv \Delta M \sum_{k=1}^{3} \left\{ \left[\overline{Q}_{11}\beta_{xx} + \overline{Q}_{12}\beta_{yy} + \overline{Q}_{16}\beta_{xy} \right]_k [z_k - z_{k-1}] \right\}$$

$$N_{xx}^M = \Delta M \left\{ \left(\left[\overline{Q}_{11}\beta_{xx} + \overline{Q}_{12}\beta_{yy} + \overline{Q}_{16}\beta_{xy} \right]_1 [z_1 - z_0] \right) + \left(\left[\overline{Q}_{11}\beta_{xx} + \overline{Q}_{12}\beta_{yy} + \overline{Q}_{16}\beta_{xy} \right]_2 \right. \right.$$

$$\left. \left. \times [z_2 - z_1] + \left(\left[\overline{Q}_{11}\beta_{xx} + \overline{Q}_{12}\beta_{yy} + \overline{Q}_{16}\alpha_{xy} \right]_3 [z_3 - z_2] \right) \right\} \right.$$

$$N_{xx}^M = (+0.5) \left\{ \left(\left[(107.6 \times 10^9)(1312 \times 10^{-6}) + (26.06 \times 10^9)(3638 \times 10^{-6}) \right. \right. \right.$$

$$\left. + (48.13 \times 10^9)(-4027 \times 10^{-6}) \right] \left[(-0.0625 + 0.1875) \times 10^{-3} \right] + \left(\left[(170.9 \times 10^9) \right. \right.$$

$$\left. \times (150 \times 10^{-6}) + (3.016 \times 10^9)(4800 \times 10^{-6}) + (0)(0) \right] \left[(0.0625 + 0.0625) \times 10^{-3} \right] \right\}$$

$$\left. + \left(\left[(10.05 \times 10^9)(4800 \times 10^{-6}) + (3.016 \times 10^9)(150 \times 10^{-6}) + (0)(0) \right] \right) \right.$$

$$\left. \times \left[(0.1875 - 0.0625) \times 10^{-3} \right] \right) \right\}$$

$$N_{xx}^M = 8190 \text{ N/m}$$

The remaining moisture stress and moment resultants are calculated in similar fashion, resulting in

$$\begin{Bmatrix} N_{xx}^M \\ N_{yy}^M \\ N_{xy}^M \\ M_{xx}^M \\ M_{yy}^M \\ M_{xy}^M \end{Bmatrix} = \begin{Bmatrix} 8190 \text{ N/m} \\ 8460 \text{ N/m} \\ -233 \text{ N/m} \\ 0.05 \text{ N-m/m} \\ -0.05 \text{ N-m/m} \\ 0.03 \text{ N-m/m} \end{Bmatrix}$$

We can now calculate midplane strains and curvatures using Equation 6.45, which becomes*

$$\begin{Bmatrix} \varepsilon_{xx}^o \\ \varepsilon_{yy}^o \\ \gamma_{xy}^o \\ \kappa_{xx} \\ \kappa_{yy} \\ \kappa_{xy} \end{Bmatrix} = \begin{bmatrix} 3.757 \times 10^{-8} & -1.964 \times 10^{-9} & -1.038 \times 10^{-8} & 1.440 \times 10^{-4} \\ -1.964 \times 10^{-9} & 1.037 \times 10^{-7} & -4.234 \times 10^{-8} & -1.866 \times 10^{-5} \\ -1.038 \times 10^{-8} & -4.234 \times 10^{-8} & 2.004 \times 10^{-7} & 3.661 \times 10^{-4} \\ 1.440 \times 10^{-4} & -1.866 \times 10^{-5} & 3.661 \times 10^{-4} & 7.064 \\ 3.905 \times 10^{-6} & -6.361 \times 10^{-4} & 3.251 \times 10^{-4} & -3.122 \times 10^{-2} \\ 8.513 \times 10^{-5} & 4.628 \times 10^4 & -1.851 \times 10^{-5} & -4.572 \end{bmatrix}$$

$$\begin{matrix} 3.905 \times 10^{-6} & 8.513 \times 10^{-5} \\ -6.361 \times 10^{-4} & 4.628 \times 10^4 \\ 3.251 \times 10^{-4} & -1.851 \times 10^{-5} \\ -3.122 \times 10^{-2} & -4.572 \\ 6.429 & -3.620 \\ -3.620 & 17.41 \end{matrix} \begin{bmatrix} -4060 + 8190 \\ -7360 + 8460 \\ 2860 - 233 \\ -0.62 + 0.05 \\ 0.62 - 0.05 \\ -0.36 + 0.03 \end{bmatrix}$$

$$\begin{Bmatrix} \varepsilon_{xx}^o \\ \varepsilon_{yy}^o \\ \gamma_{xy}^o \\ \kappa_{xx} \\ \kappa_{yy} \\ \kappa_{xy} \end{Bmatrix} = \begin{Bmatrix} 18 \text{ μm/m} \\ -509 \text{ μm/m} \\ 420 \text{ μrad} \\ -1.0 \text{ m}^{-1} \\ 5.0 \text{ m}^{-1} \\ -4.4 \text{ m}^{-1} \end{Bmatrix}$$

b. *Ply strains relative to the x–y coordinate system:* Ply strains may now be calculated using Equation 6.12. For example, strains

---

* The [*abd*] matrix for a [30/0/90]$_T$ graphite–epoxy laminate was calculated in Sample Problem 6.3, and the thermal stress and moment resultants were calculated in Sample Problem 6.5.

present at the outer surface of ply #1 (i.e., strains present at $z_o = -0.0001875\ m$) are

$$
\left\{\begin{array}{c} \varepsilon_{xx} \\ \varepsilon_{yy} \\ \gamma_{xy} \end{array}\right\}\bigg|_{z=z0} = \left\{\begin{array}{c} \varepsilon_{xx}^o \\ \varepsilon_{yy}^o \\ \gamma_{xy}^o \end{array}\right\} + z_0 \left\{\begin{array}{c} \kappa_{xx} \\ \kappa_{yy} \\ \kappa_{xy} \end{array}\right\} = \left\{\begin{array}{c} 18 \times 10^{-6}\ m/m \\ -509 \times 10^{-6}\ m/m \\ 420 \times 10^{-6}\ m/m \end{array}\right\} + (-0.0001875\,m) \left\{\begin{array}{c} -1.0\ rad/m \\ 5.0\ rad/m \\ -4.4\ rad/m \end{array}\right\}
$$

$$
\left\{\begin{array}{c} \varepsilon_{xx} \\ \varepsilon_{yy} \\ \gamma_{xy} \end{array}\right\}\bigg|_{z=z0} = \left\{\begin{array}{c} 206\ \mu m/m \\ -1450\ \mu m/m \\ 1240\ \mu rad \end{array}\right\}
$$

Strains calculated at the remaining ply interface positions are summarized in Table 6.9.

c. *Ply stresses relative to the x–y coordinate system:* Ply stresses may now be calculated using Equation 5.30. The stresses present at the outer surface of ply #1 are (i.e., at $z = z_0$):

$$
\left\{\begin{array}{c} \sigma_{xx} \\ \sigma_{yy} \\ \tau_{xy} \end{array}\right\}^{Ply\ 1}_{z=z0} = \left[\begin{array}{ccc} \overline{Q}_{11} & \overline{Q}_{12} & \overline{Q}_{16} \\ \overline{Q}_{12} & \overline{Q}_{22} & \overline{Q}_{26} \\ \overline{Q}_{16} & \overline{Q}_{26} & \overline{Q}_{66} \end{array}\right]^{Ply1}_{z=z0} \left\{\begin{array}{c} \varepsilon_{xx} - \Delta T\alpha_{xx} - \Delta M\beta_{xx} \\ \varepsilon_{yy} - \Delta T\alpha_{yy} - \Delta M\beta_{yy} \\ \gamma_{xy} - \Delta T\alpha_{xy} - \Delta M\beta_{xy} \end{array}\right\}\bigg|_{z=z0}
$$

$$
\left\{\begin{array}{c} \sigma_{xx} \\ \sigma_{yy} \\ \tau_{xy} \end{array}\right\}^{Ply\ 1}_{z=z0} = \left[\begin{array}{ccc} 107.6 \times 10^9 & 26.06 \times 10^9 & 48.13 \times 10^9 \\ 26.06 \times 10^9 & 27.22 \times 10^9 & 21.52 \times 10^9 \\ 48.13 \times 10^9 & 21.52 \times 10^9 & 36.05 \times 10^9 \end{array}\right]
$$

$$
\left\{\begin{array}{c} [(206) - (-155)(6.08) - (0.5)(1312)] \times 10^{-6} \\ [(-1450) - (-155)(20.0) - (0.5)(3638)] \times 10^{-6} \\ [(1240) - (-155)(-24.2) - (0.5)(-4027)] \times 10^{-6} \end{array}\right\}
$$

TABLE 6.9

Ply Interface Strains in a [30/0/90] Graphite–Epoxy Laminate Caused by the Combined Effects of Cooldown from 175°C to 20°C and an Increase in Moisture Content of +0.5%

z-Coordinate (mm)	ε_{xx} ($\mu m/m$)	ε_{yy} ($\mu m/m$)	γ_{xy} (μ Radians)
−0.1875	206	−1450	1240
−0.0625	80	−820	690
0.0625	−44	−190	150
0.1875	−170	440	−400

Note: Strains are referenced to the *x–y* coordinate system.

TABLE 6.10

Ply Interface Stresses in a [30/0/90] Graphite–Epoxy Laminate Caused by the Combined Effects of Cooldown from 175°C to 20°C and an Increase in Moisture Content of +0.5%

Ply Number	z-Coordinate (mm)	σ_{xx} (MPa)	σ_{yy} (MPa)	τ_{xy} (MPa)
Ply	−0.1875	25	−2.2	2.2
1	−0.0625	1.4	−0.24	−9.9
Ply	−0.0625	−20	9.3	9.0
2	0.0625	−40	15	1.9
Ply	0.0625	16	−65	1.9
3	0.1875	17	43	−5.2

Note: Stresses are referenced to the *x*–*y* coordinate system.

$$\left\{ \begin{matrix} \sigma_{xx} \\ \sigma_{yy} \\ \tau_{xy} \end{matrix} \right\}^{|Ply1}_{\,z=z0} = \left\{ \begin{matrix} 25\text{ MPa} \\ -2.4\text{ MPa} \\ 2.2\text{ MPa} \end{matrix} \right\}$$

Stresses calculated at the remaining plies and ply interface positions are summarized in Table 6.10.

A comparison of the results obtained in Example Problems 6.6 and 6.7 leads to the following observation: the initial ply stresses and strains caused by cooldown from cure temperatures to room temperatures are partially relieved by the subsequent adsorption of moisture. Although the interaction between temperature and moisture effects obviously depends on the details of the situation (material properties involved, stacking sequence, magnitudes of ΔT and ΔM, etc.), this observation is often true. That is, the thermal stresses predicted to develop in a multiangle laminate during cooldown are usually predicted to be relieved somewhat by subsequent adsorption of moisture.

6.7 Simplifications due to Stacking Sequence

Equations 6.44 and 6.45 summarize the response of a multiangle composite laminate due to the combined effects of uniform mechanical loading, uniform changes in temperature, and/or uniform changes in moisture content. The primary objective of this section is to show that these equations may be substantially simplified through proper selection of the laminate stacking sequence. Before these simplifications are discussed, however, it is illustrative to consider the *simplest* case of all—specifically, let us consider Equations 6.44 and 6.45 when applied to a plate of total thickness *t* made from an *isotropic* material.

Recall that for isotropic materials all properties are independent of direction. For present purposes, let:

$$E_{11} = E_{22} = E$$

$$\nu_{12} = \nu_{21} = \nu$$

$$G_{12} = G$$

$$\alpha_{11} = \alpha_{22} = \alpha$$

$$\beta_{11} = \beta_{22} = \beta$$

Also recall that for isotropic materials only 2 of the elastic modulii are independent. That is,

$$G = \frac{E}{2(1 + \nu)}$$

If these interrelations between material properties are enforced, then Equation 6.44 reduces to

$$
\begin{Bmatrix} N_{xx} \\ N_{yy} \\ N_{xy} \\ M_{xx} \\ M_{yy} \\ M_{xy} \end{Bmatrix} =
\begin{bmatrix}
A_{11} & A_{12} & 0 & 0 & 0 & 0 \\
A_{12} & A_{11} & 0 & 0 & 0 & 0 \\
0 & 0 & \left(\dfrac{A_{11} - A_{12}}{2}\right) & 0 & 0 & 0 \\
0 & 0 & 0 & D_{11} & D_{12} & 0 \\
0 & 0 & 0 & D_{12} & D_{11} & 0 \\
0 & 0 & 0 & 0 & 0 & \dfrac{(D_{11} - D_{12})}{2}
\end{bmatrix}
\begin{Bmatrix} \varepsilon^o_{xx} \\ \varepsilon^o_{yy} \\ \gamma^o_{xy} \\ \kappa_{xx} \\ \kappa_{yy} \\ \kappa_{xy} \end{Bmatrix}
- \begin{Bmatrix} N^T \\ N^T \\ 0 \\ 0 \\ 0 \\ 0 \end{Bmatrix}
- \begin{Bmatrix} N^M \\ N^M \\ 0 \\ 0 \\ 0 \\ 0 \end{Bmatrix}
$$

$$(6.46)$$

where

$$A_{11} = \frac{E\,t}{1 - \nu^2} \qquad A_{12} = \nu\, A_{11} = \frac{\nu\, E\, t}{1 - \nu^2}$$

$$D_{11} = D_{22} = \frac{E\, t^3}{12\,(1 - \nu^2)} \qquad D_{12} = \nu\, D_{11} = \frac{\nu\, E\, t^3}{12\,(1 - \nu^2)}$$

$$D_{66} = \frac{(D_{11} - D_{12})}{2} = \frac{E t^3}{24\,(1 + \nu)}$$

$$N^T = \Delta T \left[\frac{E t \alpha}{(1 - v)} \right] \qquad N^M = \Delta M \left[\frac{E t \beta}{(1 - v)} \right]$$

The constant D_{11} is often called the *flexural rigidity* of an isotropic plate. Taking the inverse of Equation 6.46, we find:

$$\begin{Bmatrix} \varepsilon_{xx}^o \\ \varepsilon_{yy}^o \\ \gamma_{xy}^o \\ \kappa_{xx} \\ \kappa_{yy} \\ \kappa_{xy} \end{Bmatrix} = \begin{bmatrix} a_{11} & a_{12} & 0 & 0 & 0 & 0 \\ a_{12} & a_{11} & 0 & 0 & 0 & 0 \\ 0 & 0 & 2(a_{11} - a_{12}) & 0 & 0 & 0 \\ 0 & 0 & 0 & d_{11} & d_{12} & 0 \\ 0 & 0 & 0 & d_{12} & d_{11} & 0 \\ 0 & 0 & 0 & 0 & 0 & 2(d_{11} - d_{12}) \end{bmatrix} \begin{Bmatrix} N_{xx} + N^T + N^M \\ N_{yy} + N^T + N^M \\ N_{xy} \\ M_{xx} \\ M_{yy} \\ M_{xy} \end{Bmatrix}$$

$$(6.47)$$

where

$$a_{11} = \frac{1}{A_{11}(1 - v^2)} = \frac{1}{E t}$$

$$a_{12} = -v a_{11} \frac{-v}{A_{11}(1 - v^2)} = \frac{-v}{E t}$$

$$d_{11} = d_{22} = \frac{1}{D_{11}(1 - v^2)} = \frac{12}{E t^3}$$

$$d_{12} = -v d_{11} = \frac{-v}{D_{11}(1 - v^2)} = \frac{-12v}{E t^3}$$

$$d_{66} = 2(d_{11} - d_{12}) = \frac{24(1 + v)}{E t^3}$$

Comparing Equations 6.44, 6.45 with Equations 6.46, 6.47, it is apparent that multiangle composite laminates may exhibit unusual coupling effects, as compared to the more familiar behavior of isotropic plates. For example, referring to Equation 6.45 it can be seen that application of a normal stress resultant N_{xx} will (in general) induce a midplane shear strain γ_{xy}^o and curvatures κ_{xx}, κ_{yy}, and κ_{xy} due to the presence of the a_{16}, b_{11}, b_{12}, and b_{16} terms, respectively. Physically, these means that a uniform in-plane uniaxial loading will cause in-plane shear strains as well as out-of-plane curvatures in a composite plate (i.e., the plate will bend). These coupling do not exist for isotropic panels, as indicated by Equation 6.47.

Unusual couplings between thermal resultants and laminate strains and curvatures also exist, and are immediately apparent in practice. As previously discussed most modern composite material systems are cured at elevated temperatures and are subsequently cooled to room temperatures. Therefore, thermal stress and moment resultants (N_{ij}^T and M_{ij}^T, respectively) develop during cooldown. Equation 6.45 shows that the thermal stress resultants N_{ij}^T and M_{ij}^T will cause curvatures κ_{xx}, κ_{yy}, and κ_{xy} to develop upon cooldown. Physically, this means that (in general) a composite laminate that is flat at the elevated cure temperature will bend and warp as it is cooled to room temperature. Coupling effects due to moisture stress and moment resultants, N_{ij}^M and M_{ij}^M, are analogous to those associated with thermal stress and moment resultants. Thus, even if a composite laminate is cured and used at room temperature (so that $N_{ij}^T = M_{ij}^T = 0$), the laminate may still bend or warp if the surrounding humidity causes the moisture content of the laminate to change with time.

Since coupling effects greatly complicate the design of composite structures, it is of interest to determine whether these coupling effects can be reduced or eliminated. It will be seen that it is indeed possible to reduce or eliminate many of these coupling effects through proper selection of the laminate stacking sequence. Common stacking sequences used to eliminate coupling effects are described in separate sections below.

6.7.1 Symmetric Laminates

A symmetric laminate is one that possesses both geometric and material symmetry about the midplane. In a symmetric laminate plies located symmetrically about the laminate midplane are of the same material, have the same thickness, and have the same fiber angle. Several examples of symmetric stacking sequences have been previously shown in Figure 6.12. For a symmetric n-ply laminate, the material and fiber angle used in ply 1 is identical to that used in ply n, the material and fiber angle used in ply 2 is identical to that used ply n-1, and so on.

Use of a symmetric stacking sequence result in three major simplifications to Equations 6.44 and 6.45. Specifically, for a symmetric laminate:

- All coupling stiffnesses equal zero ($B_{ij} = 0$).
- All thermal moment resultants equal zero ($M_{ij}^T = 0$).
- All moisture moment resultants equal zero ($M_{ij}^M = 0$).

To demonstrate that coupling stiffnesses are zero for a symmetric laminate, consider the coupling stiffness B_{11}. From Equation 6.27b, B_{11} is given by

$$B_{11} = \frac{1}{2}\sum_{k=1}^{n}\left\{\overline{Q}_{11}\right\}_k (z_k^2 - z_{k-1}^2)$$

In expanded form, B_{11} is given by

$$B_{11} = \frac{1}{2}\left\{\left(\overline{Q}_{11}\right)_1 [z_1^2 - z_0^2] + \left(\overline{Q}_{11}\right)_2 [z_2^2 - z_1^2] + \left(\overline{Q}_{11}\right)_3 [z_3^2 - z_2^2] + \cdots \right.$$
$$\left. +\left(\overline{Q}_{11}\right)_{n-2} [z_{n-2}^2 - z_{n-3}^2] + \left(\overline{Q}_{11}\right)_{n-1} [z_{n-1}^2 - z_{n-2}^2] + \left(\overline{Q}_{11}\right)_n [z_n^2 - z_{n-1}^2] \right\} \quad (6.48)$$

Since the laminate is assumed to be symmetric, it must be that

$$\left(\overline{Q}_{11}\right)_1 = \left(\overline{Q}_{11}\right)_n$$
$$\left(\overline{Q}_{11}\right)_2 = \left(\overline{Q}_{11}\right)_{n-1}$$
$$\left(\overline{Q}_{11}\right)_3 = \left(\overline{Q}_{11}\right)_{n-2} \quad (6.49a)$$
$$\vdots$$

etc.

Also, due to symmetry the ply interface positions are located symmetrically about the midplane, and hence (recalling that $z_0 < 0$ and $z_n > 0$):

$$z_0 = -z_n$$
$$z_1 = -z_{n-1}$$
$$z_2 = -z_{n-2} \quad (6.49b)$$
$$\vdots$$

etc.

Together, the relations listed as Equations 6.49 imply that for any symmetric laminate:

$$\left(\overline{Q}_{11}\right)_1 [z_1^2 - z_0^2] = -\left(\overline{Q}_{11}\right)_n [z_n^2 - z_{n-1}^2]$$
$$\left(\overline{Q}_{11}\right)_2 [z_2^2 - z_1^2] = -\left(\overline{Q}_{11}\right)_{n-1} [z_{n-1}^2 - z_{n-2}^2]$$
$$\left(\overline{Q}_{11}\right)_3 [z_3^2 - z_2^2] = -\left(\overline{Q}_{11}\right)_{n-2} [z_{n-2}^2 - z_{n-3}^2] \quad (6.50)$$
$$\vdots$$

etc.

Substituting Equation 6.50 into Equation 6.48 it is seen that $B_{11} = 0$. Similar results may be demonstrated for all other coupling stiffnesses, and hence $B_{ij} = 0$ for any symmetric laminate, as stated.

To demonstrate that thermal moment resultants are zero for a symmetric laminate, consider the thermal moment resultant M_{xx}^T. From Equation 6.42d, M_{xx}^T is given by

$$M_{xx}^T \equiv \frac{\Delta T}{2} \sum_{k=1}^{n} \left\{ \left[\bar{Q}_{11}\alpha_{xx} + \bar{Q}_{12}\alpha_{yy} + \bar{Q}_{16}\alpha_{xy} \right]_k \left[z_k^2 - z_{k-1}^2 \right] \right\}$$

In expanded form, M_{xx}^T is given by

$$
\begin{aligned}
M_{xx}^T = \frac{\Delta T}{2} &\left\{ \left[\bar{Q}_{11}\alpha_{xx} + \bar{Q}_{12}\alpha_{yy} + \bar{Q}_{16}\alpha_{xy} \right]_1 \left[z_1^2 - z_0^2 \right] \right. \\
&+ \left[\bar{Q}_{11}\alpha_{xx} + \bar{Q}_{12}\alpha_{yy} + \bar{Q}_{16}\alpha_{xy} \right]_2 \left[z_2^2 - z_1^2 \right] \\
&+ \left[\bar{Q}_{11}\alpha_{xx} + \bar{Q}_{12}\alpha_{yy} + \bar{Q}_{16}\alpha_{xy} \right]_3 \left[z_3^2 - z_2^2 \right] \\
&+ \cdots + \left[\bar{Q}_{11}\alpha_{xx} + \bar{Q}_{12}\alpha_{yy} + \bar{Q}_{16}\alpha_{xy} \right]_{n-2} \left[z_{n-2}^2 - z_{n-3}^2 \right] \\
&+ \left[\bar{Q}_{11}\alpha_{xx} + \bar{Q}_{12}\alpha_{yy} + \bar{Q}_{16}\alpha_{xy} \right]_{n-1} \left[z_{n-1}^2 - z_{n-2}^2 \right] \\
&\left. + \left[\bar{Q}_{11}\alpha_{xx} + \bar{Q}_{12}\alpha_{yy} + \bar{Q}_{16}\alpha_{xy} \right]_n \left[z_n^2 - z_{n-1}^2 \right] \right\}
\end{aligned}
\tag{6.51}
$$

Since the laminate is assumed to be symmetric it must be that

$$
\left[\bar{Q}_{11}\alpha_{xx} + \bar{Q}_{12}\alpha_{yy} + \bar{Q}_{16}\alpha_{xy} \right]_1 \left[z_1^2 - z_0^2 \right] = -\left[\bar{Q}_{11}\alpha_{xx} + \bar{Q}_{12}\alpha_{yy} + \bar{Q}_{16}\alpha_{xy} \right]_n
$$
$$
\times \left[z_n^2 - z_{n-1}^2 \right]
$$
$$
\left[\bar{Q}_{11}\alpha_{xx} + \bar{Q}_{12}\alpha_{yy} + \bar{Q}_{16}\alpha_{xy} \right]_2 \left[z_2^2 - z_1^2 \right] = -\left[\bar{Q}_{11}\alpha_{xx} + \bar{Q}_{12}\alpha_{yy} + \bar{Q}_{16}\alpha_{xy} \right]_{n-1}
$$
$$
\times \left[z_{n-1}^2 - z_{n-2}^2 \right]
$$
$$
\left[\bar{Q}_{11}\alpha_{xx} + \bar{Q}_{12}\alpha_{yy} + \bar{Q}_{16}\alpha_{xy} \right]_3 \left[z_3^2 - z_2^2 \right] = -\left[\bar{Q}_{11}\alpha_{xx} + \bar{Q}_{12}\alpha_{yy} + \bar{Q}_{16}\alpha_{xy} \right]_{n-2}
$$
$$
\times \left[z_{n-2}^2 - z_{n-3}^2 \right]
$$

$$\vdots$$

$$\tag{6.52}$$

Substituting Equation 6.52 into Equation 6.51, it is seen that $M_{xx}^T = 0$ for a symmetric laminate. Similar results may be demonstrated for M_{yy}^T and M_{xy}^T, and hence all thermal moment resultants equal zero for any symmetric laminate, as stated. Therefore, a symmetric laminate will not bend or warp when subjected to a uniform change in temperature. In particular, a symmetric laminate will not warp or bend during cooldown from the cure temperature.

An identical procedure may be used to prove that all moisture moment resultants are zero for a symmetric laminate.

In summary then, for symmetric laminates Equations 6.44 and 6.45 reduce to

$$
\begin{Bmatrix} N_{xx} \\ N_{yy} \\ N_{xy} \\ M_{xx} \\ M_{yy} \\ M_{xy} \end{Bmatrix} = \begin{bmatrix} A_{11} & A_{12} & A_{16} & 0 & 0 & 0 \\ A_{12} & A_{22} & A_{26} & 0 & 0 & 0 \\ A_{16} & A_{26} & A_{66} & 0 & 0 & 0 \\ 0 & 0 & 0 & D_{11} & D_{12} & D_{16} \\ 0 & 0 & 0 & D_{12} & D_{22} & D_{26} \\ 0 & 0 & 0 & D_{16} & D_{26} & D_{66} \end{bmatrix} \begin{Bmatrix} \varepsilon_{xx}^o \\ \varepsilon_{yy}^o \\ \gamma_{xy}^o \\ \kappa_{xx} \\ \kappa_{yy} \\ \kappa_{xy} \end{Bmatrix} - \begin{Bmatrix} N_{xx}^T \\ N_{yy}^T \\ N_{xy}^T \\ 0 \\ 0 \\ 0 \end{Bmatrix} - \begin{Bmatrix} N_{xx}^M \\ N_{yy}^M \\ N_{xy}^M \\ 0 \\ 0 \\ 0 \end{Bmatrix}
$$

$$(6.53a)$$

and

$$
\begin{Bmatrix} \varepsilon_{xx}^o \\ \varepsilon_{yy}^o \\ \gamma_{xy}^o \\ \kappa_{xx} \\ \kappa_{yy} \\ \kappa_{xy} \end{Bmatrix} = \begin{bmatrix} a_{11} & a_{12} & a_{16} & 0 & 0 & 0 \\ a_{12} & a_{22} & a_{26} & 0 & 0 & 0 \\ a_{16} & a_{26} & a_{66} & 0 & 0 & 0 \\ 0 & 0 & 0 & d_{11} & d_{12} & d_{16} \\ 0 & 0 & 0 & d_{12} & d_{22} & d_{26} \\ 0 & 0 & 0 & d_{16} & d_{26} & d_{66} \end{bmatrix} \begin{Bmatrix} N_{xx} + N_{xx}^T + N_{xx}^M \\ N_{yy} + N_{yy}^T + N_{yy}^M \\ N_{xy} + N_{xy}^T + N_{xy}^M \\ M_{xx} \\ M_{yy} \\ M_{xy} \end{Bmatrix}
$$

$$(6.53b)$$

Owing to these dramatic simplifications symmetric laminates are almost always used in practice. In those rare circumstances in which the use of a nonsymmetric laminate is required for some reason, it is best to place the nonsymmetric ply (or plies) at or near the laminate midplane, which will minimize the coupling stiffnesses and thermal and moisture moment resultants.

It is noted in passing that unidirectional composite laminates, $[\theta]_n$, are symmetric and hence $B_{ij} = M_{ij}^T = M_{ij}^M = 0$ for unidirectional laminates. In addition, for the case of $[0°]_n$ or $[90°]_n$ laminates, $A_{16} = A_{26} = D_{16} = D_{26} = N_{xy}^T = N_{xy}^M = 0$.

6.7.2 Cross-Ply Laminates

Composite laminates that contain plies with fiber angles of 0° or 90° *only* are called "cross-ply" laminates. Inspection of Equation 5.31 reveals that for any $0°$ or $90°$ ply, $\overline{Q}_{16} = \overline{Q}_{26} = 0$. From Equations 6.27 and 6.34 it is seen that $A_{16}, A_{26}, B_{16}, B_{26}, D_{16},$ and D_{26} all involve a summation of terms involving $\overline{Q}_{16}$ and $\overline{Q}_{26}$, and hence all of these terms also equal zero for cross-ply laminates. Furthermore, from Equation 5.25 it is seen that $\alpha_{xy} = 0$ for any $0°$ or $90°$ ply, and from Equation 5.38 that $\beta_{xy} = 0$ for any $0°$ or $90°$ ply. From Equations 6.42c,f and 6.43c,f it is seen that $N_{xy}^T = M_{xy}^T = N_{xy}^M = M_{xy}^M = 0$ for cross-ply laminates (since $\overline{Q}_{16} = \overline{Q}_{26} = \alpha_{xy} = \beta_{xy} = 0$). For cross-ply laminates then, Equations 6.44 and 6.45 reduce to

$$
\begin{Bmatrix} N_{xx} \\ N_{yy} \\ N_{xy} \\ M_{xx} \\ M_{yy} \\ M_{xy} \end{Bmatrix} = \begin{bmatrix} A_{11} & A_{12} & 0 & B_{11} & B_{12} & 0 \\ A_{12} & A_{22} & 0 & B_{12} & B_{22} & 0 \\ 0 & 0 & A_{66} & 0 & 0 & B_{66} \\ B_{11} & B_{12} & 0 & D_{11} & D_{12} & 0 \\ B_{12} & B_{22} & 0 & D_{12} & D_{22} & 0 \\ 0 & 0 & B_{66} & 0 & 0 & D_{66} \end{bmatrix} \begin{Bmatrix} \varepsilon_{xx}^o \\ \varepsilon_{yy}^o \\ \gamma_{xy}^o \\ \kappa_{xx} \\ \kappa_{yy} \\ \kappa_{xy} \end{Bmatrix} - \begin{Bmatrix} N_{xx}^T \\ N_{yy}^T \\ 0 \\ M_{xx}^T \\ M_{yy}^T \\ 0 \end{Bmatrix} - \begin{Bmatrix} N_{xx}^M \\ N_{yy}^M \\ 0 \\ M_{xx}^M \\ M_{yy}^M \\ 0 \end{Bmatrix}
$$

$$(6.54a)$$

and

$$
\begin{Bmatrix} \varepsilon_{xx}^o \\ \varepsilon_{yy}^o \\ \gamma_{xy}^o \\ \kappa_{xx} \\ \kappa_{yy} \\ \kappa_{xy} \end{Bmatrix} = \begin{bmatrix} a_{11} & a_{12} & 0 & b_{11} & b_{12} & 0 \\ a_{12} & a_{22} & 0 & b_{21} & b_{22} & 0 \\ 0 & 0 & a_{66} & 0 & 0 & b_{66} \\ b_{11} & b_{21} & 0 & d_{11} & d_{12} & 0 \\ b_{12} & b_{22} & 0 & d_{12} & d_{22} & 0 \\ 0 & 0 & b_{66} & 0 & 0 & d_{66} \end{bmatrix} \begin{Bmatrix} N_{xx} + N_{xx}^T + N_{xx}^M \\ N_{yy} + N_{yy}^T + N_{yy}^M \\ N_{xy} \\ M_{xx} + M_{xx}^T + M_{xx}^M \\ M_{yy} + M_{yy}^T + M_{yy}^M \\ M_{xy} \end{Bmatrix}
$$

$$(6.54b)$$

If a cross-ply laminate is *also* symmetric, then all remaining coupling stiffnesses equal zero, as well as all remaining thermal and moisture moment resultants. Hence, for symmetric cross-ply laminates Equation 6.54 are further simplified to

$$
\begin{Bmatrix} N_{xx} \\ N_{yy} \\ N_{xy} \\ M_{xx} \\ M_{yy} \\ M_{xy} \end{Bmatrix} = \begin{bmatrix} A_{11} & A_{12} & 0 & 0 & 0 & 0 \\ A_{12} & A_{22} & 0 & 0 & 0 & 0 \\ 0 & 0 & A_{66} & 0 & 0 & 0 \\ 0 & 0 & 0 & D_{11} & D_{12} & 0 \\ 0 & 0 & 0 & D_{12} & D_{22} & 0 \\ 0 & 0 & 0 & 0 & 0 & D_{66} \end{bmatrix} \begin{Bmatrix} \varepsilon_{xx}^o \\ \varepsilon_{yy}^o \\ \gamma_{xy}^o \\ \kappa_{xx} \\ \kappa_{yy} \\ \kappa_{xy} \end{Bmatrix} - \begin{Bmatrix} N_{xx}^T \\ N_{yy}^T \\ 0 \\ 0 \\ 0 \\ 0 \end{Bmatrix} - \begin{Bmatrix} N_{xx}^M \\ N_{yy}^M \\ 0 \\ 0 \\ 0 \\ 0 \end{Bmatrix}
$$

$$(6.55a)$$

and

$$
\begin{Bmatrix} \varepsilon_{xx}^o \\ \varepsilon_{yy}^o \\ \gamma_{xy}^o \\ \kappa_{xx} \\ \kappa_{yy} \\ \kappa_{xy} \end{Bmatrix} = \begin{bmatrix} a_{11} & a_{12} & 0 & 0 & 0 & 0 \\ a_{12} & a_{22} & 0 & 0 & 0 & 0 \\ 0 & 0 & a_{66} & 0 & 0 & 0 \\ 0 & 0 & 0 & d_{11} & d_{12} & 0 \\ 0 & 0 & 0 & d_{12} & d_{22} & 0 \\ 0 & 0 & 0 & 0 & 0 & d_{66} \end{bmatrix} \begin{Bmatrix} N_{xx} + N_{xx}^T + N_{xx}^M \\ N_{yy} + N_{yy}^T + N_{yy}^M \\ N_{xy} \\ M_{xx} \\ M_{yy} \\ M_{xy} \end{Bmatrix} \quad (6.55b)
$$

Note from Equations 6.55a, 6.55b that symmetric cross-ply laminates *do not exhibit any coupling stiffnesses*. That is,

$$ A_{16} = A_{26} = D_{16} = D_{26} = B_{ij} = 0 $$

or, equivalently,

$$ a_{16} = a_{26} = d_{16} = d_{26} = b_{ij} = 0 $$

Laminates that do not possess these coupling stiffnesses are called *specially orthotropic laminates*. The fact that symmetric cross-ply laminates are specially orthotropic will become an important factor in Chapters 9 and 10, where the mechanical response of such laminates subjected to varying loads will be considered.

6.7.3 Balanced Laminates

A laminate is "balanced" if for every ply with fiber angle θ there exists a second ply whose fiber angle is $-\theta$. The two plies must be otherwise identical, that is, they must be composed of the same material and have the same thickness. Inspection of Equation 5.31 reveals that the $\bar{Q}_{16}$ and $\bar{Q}_{26}$ terms for these balanced plies will always be of equal magnitude but opposite algebraic sign, that is, $\bar{Q}_{16}\big|_{\theta} = -\bar{Q}_{16}\big|_{-\theta}$ and $\bar{Q}_{26}\big|_{\theta} = -\bar{Q}_{26}\big|_{-\theta}$. Consequently, from Equation 6.27a, $A_{16} = A_{26} = 0$ for a balanced laminate. Further, from Equations 5.25 and 5.28 it is seen that α_{xy} and β_{xy} for the two balanced plies plies will always be of equal magnitude but opposite algebraic sign, that is, $\alpha_{xy}\big|_{\theta} = -\alpha_{xy}\big|_{-\theta}$ and $\beta_{xy}\big|_{\theta} = -\beta_{xy}\big|_{-\theta}$. Consequently, from Equations 6.42c and 6.43c, $N_{xy}^T = N_{xy}^M = 0$ for a balanced laminate. Equations 6.44, 6.45 are, therefore, simplified to

$$
\begin{Bmatrix} N_{xx} \\ N_{yy} \\ N_{xy} \\ M_{xx} \\ M_{yy} \\ M_{xy} \end{Bmatrix} =
\begin{bmatrix}
A_{11} & A_{12} & 0 & B_{11} & B_{12} & B_{16} \\
A_{12} & A_{22} & 0 & B_{12} & B_{22} & B_{26} \\
0 & 0 & A_{66} & B_{16} & B_{26} & B_{66} \\
B_{11} & B_{12} & B_{16} & D_{11} & D_{12} & D_{16} \\
B_{12} & B_{22} & B_{26} & D_{12} & D_{22} & D_{26} \\
B_{16} & B_{26} & B_{66} & D_{16} & D_{26} & D_{66}
\end{bmatrix}
\begin{Bmatrix} \varepsilon_{xx}^{o} \\ \varepsilon_{yy}^{o} \\ \gamma_{xy}^{o} \\ \kappa_{xx} \\ \kappa_{yy} \\ \kappa_{xy} \end{Bmatrix}
- \begin{Bmatrix} N_{xx}^{T} \\ N_{yy}^{T} \\ 0 \\ M_{xx}^{T} \\ M_{yy}^{T} \\ M_{xy}^{T} \end{Bmatrix}
- \begin{Bmatrix} N_{xx}^{M} \\ N_{yy}^{M} \\ 0 \\ M_{xx}^{M} \\ M_{yy}^{M} \\ M_{xy}^{M} \end{Bmatrix}
$$

$$(6.56a)$$

and

$$
\begin{Bmatrix} \varepsilon_{xx}^{o} \\ \varepsilon_{yy}^{o} \\ \gamma_{xy}^{o} \\ \kappa_{xx} \\ \kappa_{yy} \\ \kappa_{xy} \end{Bmatrix} =
\begin{bmatrix}
a_{11} & a_{12} & 0 & b_{11} & b_{12} & b_{16} \\
a_{12} & a_{22} & 0 & b_{21} & b_{22} & b_{26} \\
0 & 0 & a_{66} & b_{61} & b_{62} & b_{66} \\
b_{11} & b_{21} & b_{61} & d_{11} & d_{12} & d_{16} \\
b_{12} & b_{22} & b_{62} & d_{12} & d_{22} & d_{26} \\
b_{16} & b_{26} & b_{66} & d_{16} & d_{26} & d_{66}
\end{bmatrix}
\begin{Bmatrix} N_{xx} + N_{xx}^{T} + N_{xx}^{M} \\ N_{yy} + N_{yy}^{T} + N_{yy}^{M} \\ N_{xy} \\ M_{xx} + M_{xx}^{T} + M_{xx}^{M} \\ M_{yy} + M_{yy}^{T} + M_{yy}^{M} \\ M_{xy} + M_{xy}^{T} + M_{xy}^{M} \end{Bmatrix}
$$

$$(6.56b)$$

If a balanced laminate is *also* symmetric, then Equations 6.56 are simplified to

$$
\begin{Bmatrix} N_{xx} \\ N_{yy} \\ N_{xy} \\ M_{xx} \\ M_{yy} \\ M_{xy} \end{Bmatrix} =
\begin{bmatrix}
A_{11} & A_{12} & 0 & 0 & 0 & 0 \\
A_{12} & A_{22} & 0 & 0 & 0 & 0 \\
0 & 0 & A_{66} & 0 & 0 & 0 \\
0 & 0 & 0 & D_{11} & D_{12} & D_{16} \\
0 & 0 & 0 & D_{12} & D_{22} & D_{26} \\
0 & 0 & 0 & D_{16} & D_{26} & D_{66}
\end{bmatrix}
\begin{Bmatrix} \varepsilon_{xx}^{o} \\ \varepsilon_{yy}^{o} \\ \gamma_{xy}^{o} \\ \kappa_{xx} \\ \kappa_{yy} \\ \kappa_{xy} \end{Bmatrix}
- \begin{Bmatrix} N_{xx}^{T} \\ N_{yy}^{T} \\ 0 \\ 0 \\ 0 \\ 0 \end{Bmatrix}
- \begin{Bmatrix} N_{xx}^{M} \\ N_{yy}^{M} \\ 0 \\ 0 \\ 0 \\ 0 \end{Bmatrix}
$$

$$(6.57a)$$

and

$$
\begin{Bmatrix} \varepsilon_{xx}^{o} \\ \varepsilon_{yy}^{o} \\ \gamma_{xy}^{o} \\ \kappa_{xx} \\ \kappa_{yy} \\ \kappa_{xy} \end{Bmatrix} =
\begin{bmatrix}
a_{11} & a_{12} & 0 & 0 & 0 & 0 \\
a_{12} & a_{22} & 0 & 0 & 0 & 0 \\
0 & 0 & a_{66} & 0 & 0 & 0 \\
0 & 0 & 0 & d_{11} & d_{12} & d_{16} \\
0 & 0 & 0 & d_{12} & d_{22} & d_{26} \\
0 & 0 & 0 & d_{16} & d_{26} & d_{66}
\end{bmatrix}
\begin{Bmatrix} N_{xx} + N_{xx}^{T} + N_{xx}^{M} \\ N_{yy} + N_{yy}^{T} + N_{yy}^{M} \\ N_{xy} \\ M_{xx} \\ M_{yy} \\ M_{xy} \end{Bmatrix}
$$

$$(6.57b)$$

6.7.4 Balanced Angle-Ply Laminates

The definition of a "balanced" laminate was given in the preceding subsection. A balanced *angle-ply* laminate is really just a special class of balanced laminates; it is discussed in a separate subsection because of an additional simplification that occurs.

All plies in an *angle-ply laminate* have a fiber angle of the same *magnitude*. That is, all plies within an angle-ply laminate have a fiber angle of either $+\theta$ or $-\theta$, where the value θ *is the same for all plies*. In general, for an angle-ply laminate the number of plies with fiber angle $+\theta$ may differ from the number of plies with fiber angle $-\theta$. However, a *balanced* angle-ply laminate must have an equal number of plies with fiber angle $+\theta$ and $-\theta$, so as to satisfy the preceding definition of a balanced laminate. Carefully note the distinction between a balanced laminate and balanced angle-ply laminate. A balanced laminate may involve more than one "distinct" fiber angle. For example, a $[35/65/-35/-65]_T$ laminate is balanced (although not symmetric), and $A_{16} = A_{26} = 0$ for this laminate. In contrast, a balanced angle-ply laminate may involve only one "distinct" angle. A stacking sequence of either $[35/-35/35/-35]_T$ or $[65/65/-65/-65]_T$ result in balanced angle-ply laminates, for example.

The following simplifications occur for a balanced angle-ply laminate:

$$B_{11} = B_{22} = B_{66} = M_{xx}^T = M_{yy}^T = M_{xx}^M = M_{yy}^M = 0$$

In addition, the simplifications that exist for any balanced laminate ($A_{16} = A_{26} = N_{xy}^T = N_{xy}^M = 0$) also occur for a balanced angle-ply laminate. Equations 6.44, 6.45 are, therefore, simplified to

$$\begin{Bmatrix} N_{xx} \\ N_{yy} \\ N_{xy} \\ M_{xx} \\ M_{yy} \\ M_{xy} \end{Bmatrix} = \begin{bmatrix} A_{11} & A_{12} & 0 & 0 & 0 & B_{16} \\ A_{12} & A_{22} & 0 & 0 & 0 & B_{26} \\ 0 & 0 & A_{66} & B_{16} & B_{26} & 0 \\ 0 & 0 & B_{16} & D_{11} & D_{12} & D_{16} \\ 0 & 0 & B_{26} & D_{12} & D_{22} & D_{26} \\ B_{16} & B_{26} & 0 & D_{16} & D_{26} & D_{66} \end{bmatrix} \begin{Bmatrix} \varepsilon_{xx}^o \\ \varepsilon_{yy}^o \\ \gamma_{xy}^o \\ \kappa_{xx} \\ \kappa_{yy} \\ \kappa_{xy} \end{Bmatrix} - \begin{Bmatrix} N_{xx}^T \\ N_{yy}^T \\ 0 \\ 0 \\ 0 \\ M_{xy}^T \end{Bmatrix} - \begin{Bmatrix} N_{xx}^M \\ N_{yy}^M \\ 0 \\ 0 \\ 0 \\ M_{xy}^M \end{Bmatrix}$$

$$(6.58a)$$

and

$$\begin{Bmatrix} \varepsilon_{xx}^o \\ \varepsilon_{yy}^o \\ \gamma_{xy}^o \\ \kappa_{xx} \\ \kappa_{yy} \\ \kappa_{xy} \end{Bmatrix} = \begin{bmatrix} a_{11} & a_{12} & 0 & 0 & 0 & b_{16} \\ a_{12} & a_{22} & 0 & 0 & 0 & b_{26} \\ 0 & 0 & a_{66} & b_{61} & b_{62} & b_{66} \\ 0 & 0 & b_{61} & d_{11} & d_{12} & d_{16} \\ 0 & 0 & b_{62} & d_{12} & d_{22} & d_{26} \\ b_{16} & b_{26} & 0 & d_{16} & d_{26} & d_{66} \end{bmatrix} \begin{Bmatrix} N_{xx} + N_{xx}^T + N_{xx}^M \\ N_{yy} + N_{yy}^T + N_{yy}^M \\ N_{xy} \\ M_{xx} \\ M_{yy} \\ M_{xy} + M_{xy}^T + M_{xy}^M \end{Bmatrix} \qquad (6.58b)$$

If a balanced angle-ply laminate is *also* symmetric, then Equation 6.58 are simplified still further to

$$
\begin{Bmatrix} N_{xx} \\ N_{yy} \\ N_{xy} \\ M_{xx} \\ M_{yy} \\ M_{xy} \end{Bmatrix} = \begin{bmatrix} A_{11} & A_{12} & 0 & 0 & 0 & 0 \\ A_{12} & A_{22} & 0 & 0 & 0 & 0 \\ 0 & 0 & A_{66} & 0 & 0 & 0 \\ 0 & 0 & 0 & D_{11} & D_{12} & D_{16} \\ 0 & 0 & 0 & D_{12} & D_{22} & D_{26} \\ 0 & 0 & 0 & D_{16} & D_{26} & D_{66} \end{bmatrix} \begin{Bmatrix} \varepsilon^o_{xx} \\ \varepsilon^o_{yy} \\ \gamma^o_{xy} \\ \kappa_{xx} \\ \kappa_{yy} \\ \kappa_{xy} \end{Bmatrix} - \begin{Bmatrix} N^T_{xx} \\ N^T_{yy} \\ 0 \\ 0 \\ 0 \\ 0 \end{Bmatrix} - \begin{Bmatrix} N^M_{xx} \\ N^M_{yy} \\ 0 \\ 0 \\ 0 \\ 0 \end{Bmatrix}
$$

(6.59a)

(and)

$$
\begin{Bmatrix} \varepsilon^o_{xx} \\ \varepsilon^o_{yy} \\ \gamma^o_{xy} \\ \kappa_{xx} \\ \kappa_{yy} \\ \kappa_{xy} \end{Bmatrix} = \begin{bmatrix} a_{11} & a_{12} & 0 & 0 & 0 & 0 \\ a_{12} & a_{22} & 0 & 0 & 0 & 0 \\ 0 & 0 & a_{66} & 0 & 0 & 0 \\ 0 & 0 & 0 & d_{11} & d_{12} & d_{16} \\ 0 & 0 & 0 & d_{12} & d_{22} & d_{26} \\ 0 & 0 & 0 & d_{16} & d_{26} & d_{66} \end{bmatrix} \begin{Bmatrix} N_{xx} + N^T_{xx} + N^M_{xx} \\ N_{yy} + N^T_{yy} + N^M_{yy} \\ N_{xy} \\ M_{xx} \\ M_{yy} \\ M_{xy} \end{Bmatrix}
$$

(6.59b)

6.7.5 Quasi-Isotropic Laminates

A quasi-isotropic laminate is one that satisfies the following conditions:

- Three or more distinct fiber angles must be present within a laminate. The number of distinct fiber angles will be denoted "*m*," and hence if a laminate is quasi-isotropic then $m \geq 3$.
- The *m* distinct fiber angles must appear at equal increments of $(180/m)$ degrees.
- An equal number of plies must be present at each of the *m* distinct fiber angles.

It can be shown (see Reference 2) that if these three conditions are met then members of the A_{ij} matrix for the laminate are related as follows:

$$ A_{11} = A_{22} \qquad A_{66} = \frac{1}{2}(A_{11} - A_{12}) \qquad A_{16} = A_{26} = 0 $$

Now, these same relations between extensional stiffness also hold for an isotropic plate (see Equation 6.46). Hence, laminates that satisfy the above conditions are called *quasi-isotropic laminates*. Also, for a quasi-isotropic laminate $N^T_{xy} = M^T_{xy} = N^M_{xy} = M^M_{xy} = 0$. In this case, Equations 6.44 and 6.45 reduce to

$$
\begin{Bmatrix} N_{xx} \\ N_{yy} \\ N_{xy} \\ M_{xx} \\ M_{yy} \\ M_{xy} \end{Bmatrix} = \begin{bmatrix} A_{11} & A_{12} & 0 & B_{11} & B_{12} & B_{16} \\ A_{12} & A_{11} & 0 & B_{12} & B_{22} & B_{26} \\ 0 & 0 & \left(\dfrac{A_{11}-A_{12}}{2}\right) & B_{16} & B_{26} & B_{66} \\ B_{11} & B_{12} & B_{16} & D_{11} & D_{12} & D_{16} \\ B_{12} & B_{22} & B_{26} & D_{12} & D_{22} & D_{26} \\ B_{16} & B_{26} & B_{66} & D_{16} & D_{26} & D_{66} \end{bmatrix} \begin{Bmatrix} \varepsilon^o_{xx} \\ \varepsilon^o_{yy} \\ \gamma^o_{xy} \\ \kappa_{xx} \\ \kappa_{yy} \\ \kappa_{xy} \end{Bmatrix} - \begin{Bmatrix} N^T_{xx} \\ N^T_{yy} \\ 0 \\ M^T_{xx} \\ M^T_{yy} \\ 0 \end{Bmatrix} - \begin{Bmatrix} N^M_{xx} \\ N^M_{yy} \\ 0 \\ M^M_{xx} \\ M^M_{yy} \\ 0 \end{Bmatrix}
$$

$$(6.60\text{a})$$

$$
\begin{Bmatrix} \varepsilon^o_{xx} \\ \varepsilon^o_{yy} \\ \gamma^o_{xy} \\ \kappa_{xx} \\ \kappa_{yy} \\ \kappa_{xy} \end{Bmatrix} = \begin{bmatrix} a_{11} & a_{12} & 0 & b_{11} & b_{12} & b_{16} \\ a_{12} & a_{11} & 0 & b_{21} & b_{22} & b_{26} \\ 0 & 0 & 2(a_{11}-a_{12}) & b_{61} & b_{62} & b_{66} \\ b_{11} & b_{21} & b_{61} & d_{11} & d_{12} & d_{16} \\ b_{12} & b_{22} & b_{62} & d_{12} & d_{22} & d_{26} \\ b_{16} & b_{26} & b_{66} & d_{16} & d_{26} & d_{66} \end{bmatrix} \begin{Bmatrix} N_{xx} + N^T_{xx} + N^M_{xx} \\ N_{yy} + N^T_{yy} + N^M_{yy} \\ N_{xy} \\ M_{xx} + M^T_{xx} + M^M_{xx} \\ M_{yy} + M^T_{yy} + M^M_{yy} \\ M_{xy} \end{Bmatrix}
$$

$$(6.60\text{b})$$

The simplest possible quasi-isotropic laminate contains three plies ($m = 3$), oriented at equal increments of $(180°/3) = 60°$. For example, $[0/60/{-}60]_T$ or $[60/0/{-}60]_T$ laminates are quasi-isotropic. Probably the most common quasi-isotropic laminate involves four distinct fiber angles ($m = 4$). These laminates must have ply angles oriented at increments of $(180°/4) = 45°$. Typical stacking sequences in this case are $[0/45/90/{-}45]_T$ or $[45/0/{-}45/90]_T$. Although the extensional stiffnesses for these laminates are "quasi-isotropic," they still exhibit coupling stiffnesses (i.e., $B_{ij} \neq 0$) and furthermore the bending stiffnesses *are not* isotropic ($D_{11} \neq D_{22}$, for example).

If a quasi-isotropic laminate is *also* symmetric ($[0/45/90/{-}45]_s$, for example) then the coupling stiffnesses $B_{ij} = 0$, and all remaining thermal moment resultants and moisture moment resultants equal zero. Hence, Equations 6.60a, 6.60b are simplified still further:

$$
\begin{Bmatrix} N_{xx} \\ N_{yy} \\ N_{xy} \\ M_{xx} \\ M_{yy} \\ M_{xy} \end{Bmatrix} = \begin{bmatrix} A_{11} & A_{12} & 0 & 0 & 0 & 0 \\ A_{12} & A_{11} & 0 & 0 & 0 & 0 \\ 0 & 0 & \left(\dfrac{A_{11}-A_{12}}{2}\right) & 0 & 0 & 0 \\ 0 & 0 & 0 & D_{11} & D_{12} & D_{16} \\ 0 & 0 & 0 & D_{12} & D_{22} & D_{26} \\ 0 & 0 & 0 & D_{16} & D_{26} & D_{66} \end{bmatrix} \begin{Bmatrix} \varepsilon^o_{xx} \\ \varepsilon^o_{yy} \\ \gamma^o_{xy} \\ \kappa_{xx} \\ \kappa_{yy} \\ \kappa_{xy} \end{Bmatrix} - \begin{Bmatrix} N^T_{xx} \\ N^T_{yy} \\ 0 \\ 0 \\ 0 \\ 0 \end{Bmatrix} - \begin{Bmatrix} N^M_{xx} \\ N^M_{yy} \\ 0 \\ 0 \\ 0 \\ 0 \end{Bmatrix}
$$

$$(6.61\text{a})$$

$$\begin{Bmatrix} \varepsilon_{xx}^o \\ \varepsilon_{yy}^o \\ \gamma_{xy}^o \\ \kappa_{xx} \\ \kappa_{yy} \\ \kappa_{xy} \end{Bmatrix} = \begin{bmatrix} a_{11} & a_{12} & 0 & 0 & 0 & 0 \\ a_{12} & a_{11} & 0 & 0 & 0 & 0 \\ 0 & 0 & 2(a_{11} - a_{12}) & 0 & 0 & 0 \\ 0 & 0 & 0 & d_{11} & d_{12} & d_{16} \\ 0 & 0 & 0 & d_{12} & d_{22} & d_{26} \\ 0 & 0 & 0 & d_{16} & d_{26} & d_{66} \end{bmatrix} \begin{Bmatrix} N_{xx} + N_{xx}^T + N_{xx}^M \\ N_{yy} + N_{yy}^T + N_{yy}^M \\ N_{xy} \\ M_{xx} \\ M_{yy} \\ M_{xy} \end{Bmatrix}$$

(6.61b)

Again, note that only the in-plane extensional stiffnesses are "isotropic"; the bending stiffnesses for a symmetric quasi-isotropic laminate are not related in the same manner as in an isotropic plate.

It is noted in passing that symmetric quasi-isotropic laminates are not specially orthotropic, in contrast with symmetric cross-ply laminates. That is, for symmetric quasi-isotropic laminates $D_{16}, D_{26}, d_{16}, d_{26} \neq 0$.

6.8 Summary of CLT Calculations

At this point it is helpful to summarize the calculations steps involved in a composites analysis based on classical lamination theory. Two different analysis requirements are commonly encountered. In the first case the *loads applied to the composite laminate* are known, and the objective of the analysis is to determine the ply stresses and strains which will be induced by these loads. This situation is usually encountered during the design process. For example, suppose the wing of a new airplane is being designed, and the skin is to be fabricated using a composite laminate. Typically, the loads that must be supported by the skin, temperatures extremes that will be encountered in service, and the moisture content that may develop over long times are known (having been calculated on the basis of separate analyses), and are supplied as input data to the design engineer. The objective is to design the laminate such that the ply stresses and strains induced by the given mechanical and environmental loads will remain within safe limits.

A second case is when the *strains induced in a laminate* are known, and the objective of the analysis is to determine the loads that caused these strains. This situation is commonly encountered during *stress analysis* of an existing structure or prototype. For example, suppose a prototype of the wing described in the preceding paragraph has been built, and a stress analyst is assigned to evaluate its performance. During the evaluation process the entire wing structure is subjected to various aerodynamic maneuvers and/ or other loading conditions expected to be encountered during the service life of the aircraft. The strains, temperatures, and changes in moisture content are all measured. The objective is now to deduce the loads that caused these measured strains, as well as to determine the strains and stresses induced within individual plies.

A slightly different calculation path is used in these two different scenarios. The calculation procedures will be summarized in the following two subsections.

6.8.1 A CLT Analysis When Loads Are Known

The calculation steps followed when the applied loads are known are summarized below:

1. Define the problem:
   a. Specify the number of different ply materials to be used in the laminate
   b. Specify the elastic properties for each ply material used (i.e., specify E_{11}, E_{22}, v_{12}, G_{12}, α_{11}, α_{22}, β_{11}, and β_{22} for each ply material)
   c. Specify the laminate description (i.e., specify the number of plies n and the ply material and fiber angle for each ply)
   d. Specify the mechanical and environmental loads applied to the laminate (specify N_{xx}, N_{yy}, N_{xy}, M_{xx}, M_{yy}, M_{xy}, ΔT, and ΔM)
2. Calculate the ABD matrix:
   a. Calculate the reduced stiffness matrix, Q_{ij}, for each material used in the laminate (using Equation 5.11)
   b. Calculate the transformed reduced stiffness matrix, $\overline{Q}_{ij}$, for each ply based on the appropriate reduced stiffness matrix and ply fiber angle (using Equation 5.31)
   c. Calculate the A_{ij}, B_{ij}, and D_{ij} matrices, using Equations 6.27a, 6.27b, and 6.34, respectively
   d. Assemble the ABD matrix
3. Calculate the inverse of the ABD matrix: $abd = ABD^{-1}$
4. Calculate the thermal and moisture stress and moment resultants:
   a. Calculate the effective thermal and moisture expansion coefficients for each ply, using Equations 5.25 and 5.28, respectively
   b. Calculate the thermal and moisture stress and moment resultants using Equations 6.41 and 6.42, respectively
5. Calculate the midplane strains and curvatures induced in the laminate, using Equation 6.45.
6. For each ply:
   a. Calculate the ply strains in the x–y coordinate system, using Equation 6.12. Strains may be calculated at any desired position

z, but most often they are calculated at the ply interface positions, that is, at positions $z_0, z_1, z_2, z_3, \ldots, z_n$. Strains are also often transformed to the local 1–2 coordinate system, defined by the fiber angle within each ply.

b. Calculate the ply stresses in the x–y coordinate system, using Equation 5.30. Once again, while stresses may be calculated at any desired position z, most often they are calculated at the ply interface positions, that is, at positions $z_0, z_1, z_2, z_3, \ldots, z_n$. Stresses are also often transformed to the local 1–2 coordinate system, defined by the fiber angle within each ply.

6.8.2 A CLT Analysis When Midplane Strains and Curvatures Are Known

The calculation steps followed when midplane strains and curvatures are known are summarized below:

1. Define the problem:
   a. Specify the number of different ply materials to be used in the laminate.
   b. Specify the elastic properties for each ply material used (i.e., specify E_{11}, E_{22}, ν_{12}, G_{12}, α_{11}, α_{22}, β_{11}, and β_{22} for each ply material).
   c. Specify the laminate description (i.e., specify the number of plies n and the ply material and fiber angle for each ply).
   d. Specify the midplane strains and curvatures and environmental loads experienced by the laminate (specify $\varepsilon_{xx}^o, \varepsilon_{yy}^o, \gamma_{xy}^o,, \kappa_{xx}, \kappa_{yy}, \kappa_{xy}, \Delta T,$ and ΔM).

2. Calculate the ABD matrix:
   a. Calculate the reduced stiffness matrix, Q_{ij}, for each material used in the laminate (using Equations 5.11).
   b. Calculate the transformed reduced stiffness matrix, $\overline{Q}_{ij}$, for each ply based on the appropriate reduced stiffness matrix and ply fiber angle (using Equations 5.31).
   c. Calculate the A_{ij}, B_{ij}, and D_{ij} matrices, using Equations 6.27a, 6.27b, and 6.34, respectively.
   d. Assemble the ABD matrix.

3. Calculate the thermal and moisture stress and moment resultants:
   a. Calculate the effective thermal and moisture expansion coefficients for each ply, using Equations 5.25 and 5.28, respectively.
   b. Calculate the thermal and moisture stress and moment resultants using Equations 6.41 and 6.42, respectively.

4. Calculate the applied stress and moment resultants, using Equation 6.44.

5. For each ply:

   a. Calculate the ply strains in the x–y coordinate system, using Equation 6.12. Strains may be calculated at any desired position z, but most often they are calculated at the ply interface positions, that is, at positions $z_0, z_1, z_2, z_3, \ldots, z_n$. Strains are also often transformed to the local 1–2 coordinate system, defined by the fiber angle within each ply.

   b. Calculate the ply stresses in the x–y coordinate system, using Equation 5.30. Once again, while stresses may be calculated at any desired position z, most often they are calculated at the ply interface positions, that is, at positions $z_0, z_1, z_2, z_3, \ldots, z_n$. Stresses are also often transformed to the local 1–2 coordinate system, defined by the fiber angle within each ply.

6.9 Effective Properties of a Composite Laminate

The definitions of common engineering material properties were reviewed in Chapter 3. In this section, these concepts will be used to define the "effective" properties of multiangle composite laminates. Before we begin this discussion it is pertinent to note that most of the properties defined in Chapter 3 do not take into account the peculiar coupling effects exhibited by general composite laminates. As a typical example, consider the definition of the thermal expansion coefficient. As discussed in Section 3.3, the thermal expansion coefficient is defined as the ratio of a thermal strain divided by the temperature change that caused that strain. Six different thermal expansion coefficients may be defined for an anisotropic material (see Equation 3.21). For example, if a uniform temperature change (ΔT) causes a normal strain in the x-direction (ε_{xx}^T), then the thermal expansion coefficient associated with the x-direction is

$$\alpha_{xx} = \frac{\varepsilon_{xx}^T}{\Delta T}$$

However, this definition does not anticipate the out-of-plane coupling effects exhibited by general composite laminates. That is, for a general composite laminate a temperature change ΔT causes *both* midplane strain ε_{xx}^o *and* midplane curvature κ_{xx}, and consequently ε_{xx}^T varies through the laminate thickness. In this case then, what strain should be used to calculate α_{xx}?

A reasonable approach is to define α_{xx} based on the thermal strain induced at the laminate midplane, $\alpha_{xx} = \varepsilon_{xx}^o / \Delta T$, regardless of whether or not a curvature is induced as well. A second thermal property can then be defined as the ratio of midplane curvature to temperature change, $(\kappa_{xx}/\Delta T)$.

While several new properties representing unusual coupling effects can be defined in this manner, this topic will not be pursued here. Recall that coupling effects occur in composite laminates with arbitrary stacking sequences because $B_{ij} \neq 0$. Symmetric laminates are almost always used in practice so as to insure that $B_{ij} = 0$. Consequently these couplings are rarely encountered in practice. The complications due to out-of-plane coupling effects present in general laminates will be ignored in this section. The effective properties for *symmetric* composite laminates are discussed in the following subsections.

6.9.1 Effective Properties Relating Stress to Strain

The standard definitions of material properties used to relate stress to strain were discussed in Section 3.2. Three types of properties are measured during uniaxial tests: Young's modulus, Poisson's ratios, and coefficients of mutual influence of the second kind. Although these properties are usually measured using uniaxial tests, they can also be measured based on strains induced by *pure bending*. Properties measured for isotropic materials during uniaxial tests are identical to those measured during pure bending tests. For example, for an isotropic material the value of Young's measured during a uniaxial test is identical to that measured in pure bending. However, for composite materials the properties measured during uniaxial tests differ substantially from those measured during pure bending tests. Young's modulus of a composite as measured during a uniaxial test is substantially different than that measured in pure bending, for example. Therefore, we must distinguish between the two. In the following discussion those properties measured through application of in-plane loads are called "extensional" properties. In contrast, properties during a pure bending test are called "flexural" properties.

6.9.1.1 Extensional Properties

Consider a symmetric composite laminate subjected to uniaxial loading N_{xx}, as shown in Figure 6.18. The in-plane strains induced in this laminate can be determined using classical lamination theory, in accordance with Equation 6.45. Assuming $\Delta T = \Delta M = 0$, (and hence that thermal and moisture stress and moment resultants are zero), and also noting that by definition $N_{yy} = N_{xy} = M_{xx} = M_{yy} = M_{xy} = 0$, Equation 6.45 becomes (for symmetric laminates):

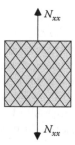

FIGURE 6.18
A symmetric composite laminate subjected to uniaxial load N_{xx}.

$$
\begin{Bmatrix} \varepsilon_{xx}^o \\ \varepsilon_{yy}^o \\ \gamma_{xy}^o \\ \kappa_{xx} \\ \kappa_{yy} \\ \kappa_{xy} \end{Bmatrix} = \begin{bmatrix} a_{11} & a_{12} & a_{16} & 0 & 0 & 0 \\ a_{12} & a_{22} & a_{26} & 0 & 0 & 0 \\ a_{16} & a_{26} & a_{66} & 0 & 0 & 0 \\ 0 & 0 & 0 & d_{11} & d_{12} & d_{16} \\ 0 & 0 & 0 & d_{12} & d_{22} & d_{26} \\ 0 & 0 & 0 & d_{16} & d_{26} & d_{66} \end{bmatrix} \begin{Bmatrix} N_{xx} \\ 0 \\ 0 \\ 0 \\ 0 \\ 0 \end{Bmatrix}
$$

The midplane strains caused by uniaxial loading are, therefore, given by

$$\varepsilon_{xx}^o = a_{11}N_{xx} \tag{6.62a}$$

$$\varepsilon_{yy}^o = a_{12}N_{xx} \tag{6.62b}$$

$$\gamma_{xy}^o = a_{16}N_{xx} \tag{6.62c}$$

Recalling that N_{xx} is defined as a constant load/unit plate length (with units of N/m or lbf/in.), the effective (or nominal) normal stress $\bar{\sigma}_{xx}$ applied to the laminate is given by $\bar{\sigma}_{xx} = N_{xx}/t$, where t is the total laminate thickness.

In Section 3.2, Young's modulus was defined as *"the normal stress σ_{xx} divided by the resulting normal strain ε_{xx}, with all other stress components equal zero."*

Applying the standard definition to the laminate shown in Figure 6.18, the effective extensional Young's modulus in the x-direction is given by

$$\bar{E}_{xx}^{ex} = \frac{\bar{\sigma}_{xx}}{\varepsilon_{xx}^o} = \frac{(N_{xx}/t)}{(a_{11}N_{xx})} = \frac{1}{ta_{11}} \tag{6.63}$$

The superscript *"ex"* is used to denote that this property is Young's modulus measured in extension.

In Section 3.2, Poisson's ratio v_{xy} was defined as *"the negative of the transverse normal strain ε_{yy} divided by the axial normal strain ε_{xx}, both of which are induced by stress σ_{xx}, with all other stresses equal zero."* The effective Poisson's ratio in extension for the laminate shown in Figure 6.18 is given by

$$\overline{v}_{xy}^{ex} = \frac{-\varepsilon_{yy}^o}{\varepsilon_{xx}^o} = \frac{-a_{12}N_{xx}}{a_{11}N_{xx}} = \frac{-a_{12}}{a_{11}} \tag{6.64}$$

Once again, a superscript *"ex"* has been used to indicate that this Poisson's ratio is measured in extension.

The coefficient of mutual influence of the second kind $\eta_{xx,xy}$ was defined as *"the shear strain γ_{xy} divided by the normal strain ε_{xx}, both of which are induced by normal stress σ_{xx}, when all other stresses equal zero."* For a composite laminate the effective coefficient of mutual influence of the second kind $\overline{\eta}_{xx,xy}^{ex}$ is, therefore, given by

$$\overline{\eta}_{xx,xy}^{ex} = \frac{\gamma_{xy}^o}{\varepsilon_{xx}^o} = \frac{a_{16}N_{xx}}{a_{11}N_{xx}} = \frac{a_{16}}{a_{11}} \tag{6.65}$$

An identical procedure can be employed to define properties measured during a uniaxial test in which only $\overline{\sigma}_{yy} = N_{yy}/t$ is applied. In this case, Equation 6.45 becomes

$$\begin{Bmatrix} \varepsilon_{xx}^o \\ \varepsilon_{yy}^o \\ \gamma_{xy}^o \\ \kappa_{xx} \\ \kappa_{yy} \\ \kappa_{xy} \end{Bmatrix} = \begin{bmatrix} a_{11} & a_{12} & a_{16} & 0 & 0 & 0 \\ a_{12} & a_{22} & a_{26} & 0 & 0 & 0 \\ a_{16} & a_{26} & a_{66} & 0 & 0 & 0 \\ 0 & 0 & 0 & d_{11} & d_{12} & d_{16} \\ 0 & 0 & 0 & d_{12} & d_{22} & d_{26} \\ 0 & 0 & 0 & d_{16} & d_{26} & d_{66} \end{bmatrix} \begin{Bmatrix} 0 \\ N_{yy} \\ 0 \\ 0 \\ 0 \\ 0 \end{Bmatrix}$$

Midplane strains induced are therefore:

$$\varepsilon_{xx}^o = a_{12}N_{yy} \tag{6.66a}$$

$$\varepsilon_{yy}^o = a_{22}N_{yy} \tag{6.66b}$$

$$\gamma_{xy}^o = a_{26}N_{yy} \tag{6.66c}$$

These strains can be used to define the effective Young's modulus $\overline{E}_{yy}^{ex}$, Poisson's ratio $\overline{\nu}_{yx}^{ex}$, and coefficient of mutual influence of the second kind $\overline{\eta}_{yy,xy}^{ex}$:

$$\overline{E}_{yy}^{ex} = \frac{1}{t a_{22}} \tag{6.67a}$$

$$\overline{\nu}_{yx}^{ex} = \frac{-a_{12}}{a_{22}} \tag{6.67b}$$

$$\overline{\eta}_{yy,xy}^{ex} = \frac{a_{26}}{a_{22}} \tag{6.67c}$$

Next, consider the effective material properties measured during a pure shear test. A symmetric composite laminate subjected to pure shear loading N_{xy} is shown in Figure 6.19. Since the load N_{xy} is applied within the plane of the laminate, properties measured during a pure shear test are included in this discussion of "extensional" properties, even though there is no common counterpart measured in a test involving out-of-plane loading. Assuming $\Delta T = \Delta M = 0$, Equation 6.45 becomes

$$\begin{Bmatrix} \varepsilon_{xx}^o \\ \varepsilon_{yy}^o \\ \gamma_{xy}^o \\ \kappa_{xx} \\ \kappa_{yy} \\ \kappa_{xy} \end{Bmatrix} = \begin{bmatrix} a_{11} & a_{12} & a_{16} & 0 & 0 & 0 \\ a_{12} & a_{22} & a_{26} & 0 & 0 & 0 \\ a_{16} & a_{26} & a_{66} & 0 & 0 & 0 \\ 0 & 0 & 0 & d_{11} & d_{12} & d_{16} \\ 0 & 0 & 0 & d_{12} & d_{22} & d_{26} \\ 0 & 0 & 0 & d_{16} & d_{26} & d_{66} \end{bmatrix} \begin{Bmatrix} 0 \\ 0 \\ N_{xy} \\ 0 \\ 0 \\ 0 \end{Bmatrix}$$

Hence, the midplane strains caused by pure shear loading are given by

$$\varepsilon_{xx}^o = a_{16} N_{xy} \tag{6.68a}$$

$$\varepsilon_{yy}^o = a_{26} N_{xy} \tag{6.68b}$$

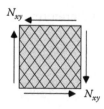

FIGURE 6.19
A symmetric composite laminate subjected to shear load N_{xy}.

$$\gamma^0_{xy} = a_{66}N_{xy} \tag{6.68c}$$

The effective (or nominal) shear stress $\bar{\tau}_{xy}$ applied to the laminate is given by $\bar{\tau}_{xy} = N_{xy}/t$, where t is the total laminate thickness.

In Section 3.2 the shear modulus was defined as *"the shear stress τ_{xy} divided by the resulting shear strain γ_{xy}, with all other stress components equal zero."* Applying this definition to the laminate shown in Figure 6.19, the effective shear modulus referenced to the x–y coordinate axes is given by

$$\bar{G}_{xy} = \frac{\bar{\tau}_{xy}}{\gamma^0_{xy}} = \frac{(N_{xy}/t)}{(a_{66}N_{xy})} = \frac{1}{ta_{66}} \tag{6.69}$$

The coefficient of mutual influence of the first kind $\eta_{xy,xx}$ (or $\eta_{xy,yy}$) was defined as *"the normal strain ε_{xx} (or ε_{yy}) divided by the shear strain γ_{xy}, both of which are induced by shear stress τ_{xy}, when all other stresses equal zero."* For a composite laminate the effective coefficient of mutual influence of the first kind $\bar{\eta}_{xy,xx}$ is, therefore, given by

$$\bar{\eta}_{xy,xx} = \frac{\varepsilon^0_{xx}}{\gamma^0_{xy}} = \frac{a_{16}N_{xx}}{a_{66}N_{xy}} = \frac{a_{16}}{a_{66}} \tag{6.70a}$$

While $\bar{\eta}_{xy,yy}$ is given by

$$\bar{\eta}_{xy,yy} = \frac{\varepsilon^0_{yy}}{\gamma^0_{xy}} = \frac{a_{26}}{a_{66}} \tag{6.70b}$$

Recall that during the derivation of classical lamination theory we assumed a state of plane stress exists within a thin composite laminate. This assumption implies that out-of-plane shear strains (γ^0_{xz} and γ^0_{yz}) are equal to zero. Consequently, Chentsov coefficients, which were defined in Section 3.2, are always equal to zero for thin composite laminates.

The effective properties of a laminate obey the inverse relations followed by any anisotropic plate, which are defined in Equation 4.13. In particular,

$$\frac{\bar{\nu}^{ex}_{xy}}{\bar{E}^{ex}_{xx}} = \frac{\bar{\nu}^{ex}_{yx}}{\bar{E}^{ex}_{yy}} \qquad \frac{\bar{\eta}^{ex}_{xx,xy}}{\bar{E}^{ex}_{xx}} = \frac{\bar{\eta}_{xy,xx}}{\bar{G}_{xy}} \qquad \frac{\bar{\eta}^{ex}_{yy,xy}}{\bar{E}^{ex}_{yy}} = \frac{\bar{\eta}_{xy,yy}}{\bar{G}_{xy}}$$

Cross-ply, balanced, and balanced angle-ply laminates were discussed in Sections 6.7.2, 6.7.3, and 6.7.4, respectively. Recall that $a_{16} = a_{26} = 0$ for these types of stacking sequences. Therefore, coefficients of mutual influence of the first and second kind always equal zero for cross-ply, balanced, and balanced angle-ply laminates. Finally, in the case of a quasi-isotropic laminate (discussed in Section 6.7.4) it will be found that the effective in-plane moduli $\overline{E}_{xx}^{ex} = \overline{E}_{yy}^{ex}$, $\overline{v}_{xy}^{ex} = \overline{v}_{yx}^{ex}$, and $\overline{G}_{xy}$ are related in the same manner as for an isotropic plate:

$$\overline{G}_{xy} = \frac{\overline{E}_{xx}^{ex}}{2(1 + \overline{v}_{xy}^{ex})}$$

6.9.1.2 Flexural Properties

All of the effective properties described in the preceding subsection are determined by subjecting the composite laminate to a load whose line of action lies within the plane of the plate and measuring the resulting strains. Properties measured in this way are called effective extensional properties. In this section we consider properties measured by applying a pure bending moment and measuring the resulting strains. Properties measured under a state of pure bending are called effective *flexural* properties. Effective flexural properties do not follow from the fundamental definitions of material properties reviewed in Section 3.2. Rather, effective flexural properties represent the response of a structure subjected to bending, and are defined in analogy to those exhibited by an isotropic plate. As discussed in Section 6.7, the bending stiffnesses (i.e., the D_{ij} matrix) for an isotropic plate are given by

$$D_{11} = D_{22} = \frac{Et^3}{12(1 - v^2)}$$

$$D_{12} = v\,D_{11} = \frac{v\,Et^3}{12(1 - v^2)}$$

$$D_{66} = \frac{(D_{11} - D_{12})}{2} = \frac{Et^3}{24(1 + v)}$$

Also, for an isotropic plate the inverse of the D_{ij} matrix (i.e., the d_{ij} matrix) is given by

$$d_{11} = d_{22} = \frac{1}{D_{11}(1 - v^2)} = \frac{12}{Et^3}$$

$$d_{12} = -vd_{11} = \frac{-v}{D_{11}(1 - v^2)} = \frac{-12v}{Et^3}$$

$$d_{66} = 2(d_{11} - d_{12}) = \frac{24(1 + v)}{Et^3}$$

Thus, for isotropic plates both the D_{ij} and d_{ij} matrices can be calculated directly from material properties E and v and the plate thickness t. Given these results one might anticipate that the D_{ij} and d_{ij} matrices for a composite laminate could be calculated in a similar manner, using the total laminate thickness t and effective extensional properties $\overline{E}^{ex}_{xx}$, $\overline{E}^{ex}_{yy}$, $\overline{v}^{ex}_{xy}$, and $\overline{v}^{ex}_{yx}$. This is not the case. The bending stiffnesses of a composite laminate (i.e., the D_{ij} matrix) cannot, in general, be calculated on the basis of effective extensional properties. For example, in general

$$ D_{11} \neq \frac{\overline{E}^{ex}_{xx} t^3}{12 \left[1 - \left(\overline{v}^{ex}_{xy} \right)^2 \right]} $$

for a composite laminate. In essence, this cannot be done because the flexural properties of a laminate are dictated in part by the laminate stacking sequence. In contrast, the effective extensional properties are *independent* of the stacking sequence. Hence, effective flexural properties cannot be directly related to effective extensional properties.

As stated above, effective flexural properties are defined for a state of pure bending. Note from the above discussion that for an isotropic plate:

$$ d_{11} = \frac{12}{E t^3} \qquad d_{12} = -v d_{11} $$

The effective Young's modulii and Poisson ratios exhibited by a composite laminate in flexure, denoted $\overline{E}^{fl}_{xx}$, $\overline{E}^{fl}_{yy}$, $\overline{v}^{fl}_{xy}$, and $\overline{v}^{fl}_{yx}$, can, therefore, be defined as follows:

$$ \overline{E}^{fl}_{xx} = \frac{12}{d_{11} t^3} \qquad \overline{E}^{fl}_{yy} = \frac{12}{d_{22} t^3} $$

$$ \overline{v}^{fl}_{xy} = \frac{-d_{12}}{d_{11}} \qquad \overline{v}^{fl}_{yx} = \frac{-d_{12}}{d_{22}} $$

Although the effective flexural Young's modulii and Poisson ratios differ from the corresponding effective extensional properties, direct substitution will shown that the flexural properties obey the inverse relations:

$$ \frac{\overline{v}^{fl}_{xy}}{\overline{E}^{fl}_{xx}} = \frac{\overline{v}^{fl}_{yx}}{\overline{E}^{fl}_{yy}} $$

It is also interesting to note that $d_{11} \neq d_{22}$ for a quasi-isotropic laminate. Consequently $\overline{E}^{fl}_{xx} \neq \overline{E}^{fl}_{yy}$, $\overline{v}^{fl}_{xy} \neq \overline{v}^{fl}_{yx}$, even for a quasi-isotropic laminate.

The effective flexural coefficient of mutual influence of the second kind, $\overline{\eta}^{fl}_{xx,xy}$ can be defined as *"the midplane shear strain γ^o_{xx} divided by the midplane normal strain ε^o_{xx}, both of which are induced by moment resultant M_{xx}, when all other stress resultants equal zero."* For a composite laminate the effective flexural coefficient of mutual influence of the second kind $\overline{\eta}^{fl}_{xx,xy}$ is, therefore, given by

$$\overline{\eta}^{fl}_{xx,xy} = \frac{\gamma^o_{xy}}{\varepsilon^o_{xx}} = \frac{d_{16}M_{xx}}{d_{11}M_{xx}} = \frac{d_{16}}{d_{11}}$$

An identical procedure can be employed to define the effective flexural coefficient of mutual influence of the second kind $\overline{\eta}^{fl}_{yy,xy}$, a property measured during a test in which only M_{yy} is applied:

$$\overline{\eta}^{fl}_{yy,xy} = \frac{\gamma^o_{xy}}{\varepsilon^o_{yy}} = \frac{d_{26}M_{yy}}{d_{22}M_{yy}} = \frac{d_{26}}{d_{22}}$$

It is mentioned in passing that flexural coefficients of mutual influence of the second kind are not encountered during study of isotropic plates, since $d_{16} = d_{26} = 0$ for isotropic plates. The effective flexural properties will be very useful during the study of composite beams, considered in Chapter 8.

6.9.2 Effective Properties Relating Temperature or Moisture Content to Strain

As discussed in Section 3.3, linear coefficients of thermal expansion are measured by determining the strains induced by a uniform change in temperature, and forming the following ratios:

$$\alpha_{xx} = \frac{\varepsilon^T_{xx}}{\Delta T} \qquad \alpha_{yy} = \frac{\varepsilon^T_{yy}}{\Delta T} \qquad \alpha_{xy} = \frac{\gamma^T_{xy}}{\Delta T} \tag{6.71}$$

The superscript *"T"* is included as a reminder that the strains involved are those caused by a change in temperature only. The midplane strains and curvatures induced by a uniform temperature change may be calculated according to Equation 6.45, which becomes (for a symmetric laminate and for $\Delta M = N_{xx} = N_{yy} = N_{xy} = M_{xx} = M_{yy} = M_{xy} = 0$):

$$
\begin{Bmatrix} \varepsilon^o_{xx} \\ \varepsilon^o_{yy} \\ \gamma^o_{xy} \\ \kappa_{xx} \\ \kappa_{yy} \\ \kappa_{xy} \end{Bmatrix}
=
\begin{bmatrix}
a_{11} & a_{12} & a_{16} & 0 & 0 & 0 \\
a_{12} & a_{22} & a_{26} & 0 & 0 & 0 \\
a_{16} & a_{26} & a_{66} & 0 & 0 & 0 \\
0 & 0 & 0 & d_{11} & d_{12} & d_{16} \\
0 & 0 & 0 & d_{12} & d_{22} & d_{26} \\
0 & 0 & 0 & d_{16} & d_{26} & d_{66}
\end{bmatrix}
\begin{Bmatrix} N^T_{xx} \\ N^T_{yy} \\ N^T_{xy} \\ 0 \\ 0 \\ 0 \end{Bmatrix}
\tag{6.72}
$$

Substituting the midplane strains indicated by Equation 6.72 into Equations 6.71, the effective linear thermal expansion coefficients for a general laminate are

$$\bar{\alpha}_{xx} = \frac{1}{\Delta T}\left[a_{11}N_{xx}^T + a_{12}N_{yy}^T + a_{16}N_{xy}^T\right]$$

$$\bar{\alpha}_{yy} = \frac{1}{\Delta T}\left[a_{12}N_{xx}^T + a_{22}N_{yy}^T + a_{26}N_{xy}^T\right] \qquad (6.73a)$$

$$\bar{\alpha}_{xy} = \frac{1}{\Delta T}\left[a_{16}N_{xx}^T + a_{26}N_{yy}^T + a_{66}N_{xy}^T\right]$$

It is noted in passing that these results are for a symmetric laminate and can, therefore, be inverted to give:

$$N_{xx}^T = \Delta T\left[A_{11}\bar{\alpha}_{xx} + A_{12}\bar{\alpha}_{yy} + A_{16}\bar{\alpha}_{xy}\right]$$

$$N_{yy}^T = \Delta T\left[A_{12}\bar{\alpha}_{xx} + A_{22}\bar{\alpha}_{yy} + A_{26}\bar{\alpha}_{xy}\right] \qquad (6.73b)$$

$$N_{xy}^T = \Delta T\left[A_{16}\bar{\alpha}_{xx} + A_{26}\bar{\alpha}_{yy} + A_{66}\bar{\alpha}_{xy}\right]$$

The effective linear coefficient of moisture expansion is measured by determining the strains induced by a uniform change in moisture content, and forming the following ratios:

$$\beta_{xx} = \frac{\varepsilon_{xx}^M}{\Delta M} \qquad \beta_{yy} = \frac{\varepsilon_{yy}^M}{\Delta M} \qquad \beta_{xy} = \frac{\gamma_{xy}^M}{\Delta M} \qquad (6.74)$$

The superscript "M" in the above equations is included as a reminder that the strains involved are those caused by a change in moisture only. The strains induced by a change in moisture content (only) may be calculated according to Equation 6.45, which becomes (for a symmetric laminate and for $\Delta T = N_{xx} = N_{yy} = N_{xy} = M_{xx} = M_{yy} = M_{xy} = 0$):

$$\begin{Bmatrix} \varepsilon_{xx}^o \\ \varepsilon_{yy}^o \\ \gamma_{xy}^o \\ \kappa_{xx} \\ \kappa_{yy} \\ \kappa_{xy} \end{Bmatrix} = \begin{bmatrix} a_{11} & a_{12} & a_{16} & 0 & 0 & 0 \\ a_{12} & a_{22} & a_{26} & 0 & 0 & 0 \\ a_{16} & a_{26} & a_{66} & 0 & 0 & 0 \\ 0 & 0 & 0 & d_{11} & d_{12} & d_{16} \\ 0 & 0 & 0 & d_{12} & d_{22} & d_{26} \\ 0 & 0 & 0 & d_{16} & d_{26} & d_{66} \end{bmatrix} \begin{Bmatrix} N_{xx}^M \\ N_{yy}^M \\ N_{xy}^M \\ 0 \\ 0 \\ 0 \end{Bmatrix} \qquad (6.75)$$

Substituting the midplane strains indicated by Equation 6.75 into Equations 6.74, the effective linear moisture expansion coefficients for a general laminate are

$$\bar{\beta}_{xx} = \frac{1}{\Delta M}\left[a_{11}N_{xx}^M + a_{12}N_{yy}^M + a_{16}N_{xy}^M\right]$$

$$\bar{\beta}_{yy} = \frac{1}{\Delta M}\left[a_{12}N_{xx}^M + a_{22}N_{yy}^M + a_{26}N_{xy}^M\right] \qquad (6.76a)$$

$$\bar{\beta}_{xy} = \frac{1}{\Delta M}\left[a_{16}N_{xx}^M + a_{26}N_{yy}^M + a_{66}N_{xy}^M\right]$$

Since these results are for a symmetric laminate they be inverted to give

$$N_{xx}^M = \Delta M\left[A_{11}\bar{\beta}_{xx} + A_{12}\bar{\beta}_{yy} + A_{16}\bar{\beta}_{xy}\right]$$

$$N_{yy}^M = \Delta M\left[A_{12}\bar{\beta}_{xx} + A_{22}\bar{\beta}_{yy} + A_{26}\bar{\beta}_{xy}\right] \qquad (6.76b)$$

$$N_{xy}^M = \Delta M\left[A_{16}\bar{\beta}_{xx} + A_{26}\bar{\beta}_{yy} + A_{66}\bar{\beta}_{xy}\right]$$

Example Problem 6.8

Determine the effective extensional and flexural moduli, thermal expansion coefficients, and moisture expansion coefficients for a $[30/0/90]_s$ graphite–epoxy laminate. Use material properties listed for graphite–epoxy in Table 3.1, and assume each ply has a thickness of 0.125 mm.

SOLUTION

As described in this section, effective moduli are calculated using various elements of the $[abd]$ matrix. A six-ply symmetric laminate is considered in this problem. The total laminate thickness is $t = 6(0.000125 \text{ m}) = 0.000750 \text{ m}$. Using methods discussed in Section 6.6, the $[ABD]$ matrix is determined to be

$$[ABD] = \begin{bmatrix} 72.2 \times 10^6 & 8.02 \times 10^6 & 12.0 \times 10^6 & 0 & 0 & 0 \\ 8.02 \times 10^6 & 52.0 \times 10^6 & 5.38 \times 10^6 & 0 & 0 & 0 \\ 12.0 \times 10^6 & 5.38 \times 10^6 & 15.5 \times 10^6 & 0 & 0 & 0 \\ 0 & 0 & 0 & 4.23 & 0.676 & 1.19 \\ 0 & 0 & 0 & 0.676 & 0.988 & 0.532 \\ 0 & 0 & 0 & 1.19 & 0.532 & 1.03 \end{bmatrix}$$

Since the laminate is symmetric, all elements of the B_{ij} matrix are zero, as expected. We obtain the $[abd]$ by inverting the $[ABD]$ numerically:

$$[abd] = \begin{bmatrix} 16.0 \times 10^{-9} & -1.23 \times 10^{-9} & -12.0 \times 10^{-9} & 0 & 0 & 0 \\ -1.23 \times 10^{-9} & 20.0 \times 10^{-9} & -6.0 \times 10^{-9} & 0 & 0 & 0 \\ -12.0 \times 10^{-9} & -6.0 \times 10^{-9} & 75.8 \times 10^{-9} & 0 & 0 & 0 \\ 0 & 0 & 0 & 3.51 \times 10^{-1} & -2.92 \times 10^{-2} & -3.92 \times 10^{-1} \\ 0 & 0 & 0 & -2.92 \times 10^{-2} & 1.408 & -6.96 \times 10^{-1} \\ 0 & 0 & 0 & -3.92 \times 10^{-1} & -6.96 \times 10^{-1} & 1.79 \end{bmatrix}$$

The effective extensional modulii of the laminate can now be calculated using Equations 6.63 through 6.70:

$$\bar{E}_{xx}^{ex} = \frac{1}{ta_{11}} = \frac{1}{(0.000750)(16.0 \times 10^{-9})} = 83.3\,\text{GPa}$$

$$\bar{E}_{yy}^{ex} = \frac{1}{ta_{22}} = \frac{1}{(0.000750)(20 \times 10^{-9})} = 66.7\,\text{GPa}$$

$$\bar{G}_{xy} = \frac{1}{ta_{66}} = \frac{1}{(0.000750)(75.8 \times 10^{-9})} = 17.6\,\text{GPa}$$

$$\bar{v}_{xy}^{ex} = \frac{-a_{12}}{a_{11}} = \frac{-(-1.23 \times 10^{-9})}{16.0 \times 10^{-9}} = 0.077$$

$$\bar{v}_{yx}^{ex} = \frac{-a_{12}}{a_{22}} = \frac{-(1.23 \times 10^{-9})}{20 \times 10^{-9}} = 0.061$$

$$\bar{\eta}_{xx,xy}^{ex} = \frac{a_{16}}{a_{11}} = \frac{-12 \times 10^{-9}}{16 \times 10^{-9}} = -0.75$$

$$\bar{\eta}_{yy,xy}^{ex} = \frac{a_{26}}{a_{22}} = \frac{-6.0 \times 10^{-9}}{20 \times 10^{-9}} = -0.30$$

$$\bar{\eta}_{xy,xx}^{ex} = \frac{a_{16}}{a_{66}} = \frac{-12.0 \times 10^{-9}}{75.8 \times 10^{-9}} = -0.16$$

$$\bar{\eta}_{xy,yy}^{ex} = \frac{a_{26}}{a_{66}} = \frac{-6.0 \times 10^{-9}}{75.8 \times 10^{-9}} = -0.079$$

Effective flexural properties are found to be

$$\bar{E}_{xx}^{fl} = \frac{12}{d_{11}t^3} = \frac{12}{(0.351)(0.000750)^3} = 81.0\,\text{GPa}$$

$$\bar{E}_{yy}^{fl} = \frac{12}{d_{22}t^3} = \frac{12}{(1.408)(0.000750)^3} = 20.2\,\text{GPa}$$

$$\bar{v}_{xy}^{fl} = \frac{-d_{12}}{d_{11}} = \frac{2.92 \times 10^{-2}}{0.351} = 0.083$$

$$\bar{v}_{yx}^{fl} = \frac{-d_{12}}{d_{22}} = \frac{2.92 \times 10^{-2}}{1.408} = 0.021$$

$$\bar{\eta}_{xx,xy}^{fl} = \frac{d_{16}}{d_{11}} = \frac{-0.392}{0.351} = -1.12$$

$$\bar{\eta}_{yy,xy}^{fl} = \frac{d_{26}}{d_{22}} = \frac{-0.696}{1.408} = -0.49$$

Note that the values of extensional properties are quite different from analogous flexural properties.

The thermal stress resultants associated with a given change in temperature must be determined in order to calculate the effective thermal expansion coefficients. Numerically speaking, any change in temperature can be used, but for present purposes a unit change in temperature will be assumed (i.e., $\Delta T = 1$). Using Equations 6.41, the thermal stress resultants associated with $\Delta T = 1$ are

$$N_{xx}^T\big|_{\Delta T=1} = 52.3\,\text{N/m} \qquad N_{yy}^T\big|_{\Delta T=1} = 94.9\,\text{N/m}$$

$$N_{xy}^T\big|_{\Delta T=1} = -36.9\,\text{N/m}$$

The effective thermal expansion coefficient $\bar{\alpha}_{xx}$ can now be calculated in accordance with Equation 6.73a:

$$\bar{\alpha}_{xx} = \frac{1}{\Delta T}\left[a_{11}N_{xx}^T + a_{12}N_{yy}^T + a_{16}N_{xy}^T\right]$$

$$\bar{\alpha}_{xx} = \frac{1}{(1)}\left[16.0 \times 10^{-9}(52.3) - 1.23 \times 10^{-9}(94.9) - 12.0 \times 10^{-9}(-36.9)\right]$$

$$\bar{\alpha}_{xx} = 1.16\,\mu\text{m/m–°C}$$

Using an equivalent procedure:

$$\bar{\alpha}_{yy} = 2.06\,\mu\text{m/m–°C}$$

$$\bar{\alpha}_{xy} = -4.00\,\mu\text{rad/°C}$$

Finally, moisture stress resultants associated with a given change in moisture content must be determined in order to calculate the effective moisture expansion coefficients. Numerically speaking, any change moisture content can be used, but for present purposes a unit change in content will be assumed (i.e., $\Delta M = 1$). Using Equations 6.42, the moisture stress resultants associated with $\Delta M = 1$ are

$$N_{xx}^M \Big|_{\Delta M=1} = 32800 \text{ N/m}$$

$$N_{yy}^M \Big|_{\Delta M=1} = 33800 \text{ N/m} \qquad N_{xy}^M \Big|_{\Delta M=1} = -930 \text{ N/m}$$

Applying Equations 6.76a, we find

$$\bar{\beta}_{xx} = 494 \ \mu\text{m/m} - \%M$$

$$\bar{\beta}_{yy} = 643 \ \mu\text{m/m} - \%M \qquad \bar{\beta}_{xy} = -667 \ \mu\text{rad}/\%M$$

6.10 Transformation of the *ABD* Matrix

The A_{ij}, B_{ij}, and D_{ij} matrices all involve a summation over the thickness of the laminate in accordance with Equations 6.27a, 6.27b, and 6.34, repeated here for convenience:

$$A_{ij} = \sum_{k=1}^{n} \left\{ \overline{Q}_{ij} \right\}_k (z_k - z_{k-1}) \qquad \text{(6.27a) (repeated)}$$

$$B_{ij} = \frac{1}{2} \sum_{k=1}^{n} \left\{ \overline{Q}_{ij} \right\}_k (z_k^2 - z_{k-1}^2) \qquad \text{(6.27b) (repeated)}$$

$$D_{ij} = \frac{1}{3} \sum_{k=1}^{n} \left\{ \overline{Q}_{ij} \right\}_k (z_k^3 - z_{k-1}^3) \qquad \text{(6.34) (repeated)}$$

As discussed in Section 5.2, $\overline{Q}_{ij}$ may be calculated using Equations 5.31, which involves use of the Q_{ij} matrix and various trigonometric functions raised to a power (e.g., $\cos^4 \theta$, $\cos^2 \theta \sin^2 \theta$, $\sin^4 \theta$, etc.). Alternatively as discussed in Section 5.3, $\overline{Q}_{ij}$ may be calculated using Equations 5.35, which involves use of material invariants U_i^Q and trigonometric functions, whose arguments involve fiber angle multiplied by a constant (i.e., $\cos 2\theta$, $\cos 4\theta$, $\sin 2\theta$, or $\sin 4\theta$).

Although either approach is mathematically equivalent, in some circumstances the use of material invariants (Equation 5.35) can be advantageous. Specifically, *if all plies within the laminate are of the same material type*, then use of Equation 5.35 leads to the ability to easily transform the *ABD* matrix from one coordinate system to another.

To aid in our development, define the following "geometry factors," which are related to the fiber angles and ply interface positions:

$$V_0^A = \sum_{k=1}^{n} (z_k - z_{k-1}) = t$$

$$V_1^A = \sum_{k=1}^{n} \cos 2\theta_k \, (z_k - z_{k-1})$$

$$V_2^A = \sum_{k=1}^{n} \sin 2\theta_k \, (z_k - z_{k-1}) \tag{6.77}$$

$$V_3^A = \sum_{k=1}^{n} \cos 4\theta_k \, (z_k - z_{k-1})$$

$$V_4^A = \sum_{k=1}^{n} \sin 4\theta_k \, (z_k - z_{k-1})$$

$$V_0^B = \frac{1}{2} \sum_{k=1}^{n} (z_k^2 - z_{k-1}^2) = 0$$

$$V_1^B = \frac{1}{2} \sum_{k=1}^{n} \cos 2\theta_k \, (z_k^2 - z_{k-1}^2)$$

$$V_2^B = \frac{1}{2} \sum_{k=1}^{n} \sin 2\theta_k \, (z_k^2 - z_{k-1}^2) \tag{6.78}$$

$$V_3^B = \frac{1}{2} \sum_{k=1}^{n} \cos 4\theta_k \, (z_k^2 - z_{k-1}^2)$$

$$V_4^B = \sum_{k=1}^{n} \sin 4\theta_k \, (z_k^2 - z_{k-1}^2)$$

$$V_0^D = \frac{1}{3} \sum_{k=1}^{n} (z_k^3 - z_{k-1}^3) = \frac{t^3}{12}$$

$$V_1^D = \frac{1}{3} \sum_{k=1}^{n} \cos 2\theta_k \, (z_k^3 - z_{k-1}^3)$$

$$V_2^D = \frac{1}{3} \sum_{k=1}^{n} \sin 2\theta_k \ (z_k^3 - z_{k-1}^3)$$

$$V_3^D = \frac{1}{3} \sum_{k=1}^{n} \cos 4\theta_k \ (z_k^3 - z_{k-1}^3) \tag{6.79}$$

$$V_4^D = \sum_{k=1}^{n} \sin 4\theta_k \ (z_k^3 - z_{k-1}^3)$$

Next, consider the steps necessary to calculate the A_{11} term. In this case, Equation 6.27a becomes

$$A_{11} = \sum_{k=1}^{n} \{\overline{Q}_{11}\}_k \ (z_k - z_{k-1}) \tag{6.80}$$

Using the invariant formulation the expression for $\overline{Q}_{11}$ listed in Equations 5.35 may be substituted, resulting in

$$A_{11} = \sum_{k=1}^{n} \{U_1^Q + U_2^Q \cos 2\theta + U_3^Q \cos 4\theta\}_k \ (z_k - z_{k-1}) \tag{6.81}$$

where U_1^Q, U_2^Q, and U_3^Q are the stiffness invariants defined in Equation 5.36. If all plies within the laminate are composed of the same material, then stiffness invariants are constant for the laminate and only fiber angle θ varies from one ply to the next. Therefore, Equation 6.81 can be rewritten as

$$A_{11} = U_1^Q \sum_{k=1}^{n} (z_k - z_{k-1}) + U_2^Q \sum_{k=1}^{n} \cos 2\theta_k (z_k - z_{k-1}) + U_3^Q \sum_{k=1}^{n} \cos 4\theta_k (z_k - z_{k-1})$$

$$\tag{6.82}$$

The geometry factors V_0^A, V_1^A, and V_3^A appear in Equation 6.82, and hence the equation may be written as

$$A_{11} = U_1^Q V_0^A + U_2^Q V_1^A + U_3^Q V_3^A \tag{6.83a}$$

Following an identical procedure, the remaining elements of the A_{ij} matrix are given by

$$A_{22} = U_1^Q V_0^A - U_2^Q V_1^A + U_3^Q V_3^A \tag{6.83b}$$

$$A_{12} = A_{21} = U_4^Q V_0^A - U_3^Q V_3^A \tag{6.83c}$$

$$A_{66} = U_5^Q V_0^A - U_3^Q V_3^A \tag{6.83d}$$

$$A_{16} = A_{61} = \frac{1}{2} U_2^Q V_2^A + U_3^Q V_4^A \tag{6.83e}$$

$$A_{26} = A_{62} = \frac{1}{2} U_2^Q V_2^A - U_3^Q V_4^A \tag{6.83f}$$

Analogous procedures can be used to calculate members of the B_{ij} and D_{ij} matrices. It can be shown that elements of the B_{ij} matrix are given by

$$B_{11} = U_1^Q V_0^B + U_2^Q V_1^B + U_3^Q V_3^B \tag{6.84a}$$

$$B_{22} = U_1^Q V_0^B - U_2^Q V_1^B + U_3^Q V_3^B \tag{6.84b}$$

$$B_{12} = B_{21} = U_4^Q V_0^B - U_3^Q V_3^B \tag{6.84c}$$

$$B_{66} = U_5^Q V_0^B - U_3^Q V_3^B \tag{6.84d}$$

$$B_{16} = B_{61} = \frac{1}{2} U_2^Q V_2^B + U_3^Q V_4^B \tag{6.84e}$$

$$B_{26} = B_{62} = \frac{1}{2} U_2^Q V_2^B - U_3^Q V_4^B \tag{6.84f}$$

Members of the D_{ij} matrix are given by

$$D_{11} = U_1^Q V_0^D + U_2^Q V_1^D + U_3^Q V_3^D \tag{6.85a}$$

$$D_{22} = U_1^Q V_0^D - U_2^Q V_1^D + U_3^Q V_3^D \tag{6.85b}$$

$$D_{12} = D_{21} = U_4^Q V_0^D - U_3^Q V_3^D \tag{6.85c}$$

$$D_{66} = U_5^Q V_0^D - U_3^Q V_3^D \tag{6.85d}$$

$$D_{16} = D_{61} = \frac{1}{2} U_2^Q V_2^D + U_3^Q V_4^D \tag{6.85e}$$

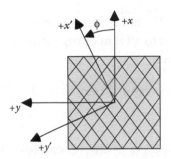

FIGURE 6.20
A general composite laminate, showing the "original" x–y coordinate system and the "new" x'–y' coordinate system.

$$D_{26} = D_{62} = \frac{1}{2}U_2^Q V_2^D - U_3^Q V_4^D \qquad (6.85f)$$

Let us now consider transformation of the *ABD* matrix from one coordinate system to another. A multiangle composite laminate referenced to a x–y coordinate system is shown in Figure 6.20. It is assumed that the *ABD* matrix for this laminate has been calculated and is known, based on fiber angles referenced to the x–y coordinate system. Now suppose that a different coordinate system is of interest, the x'–y' coordinate system, orientated φ degrees counter-clockwise from the original x–y coordinate system. The transformed stiffness matrices referenced to the x'–y' coordinate system will be labeled $\overline{A}_{ij}$, $\overline{B}_{ij}$, and $\overline{D}_{ij}$.

Inspection of Figure 6.20 shows that a ply with fiber angle θ_k relative to the original x-axis will form an angle $(\theta_k - \varphi)$ relative to the x'-axis. Element $\overline{A}_{11}$ may, therefore, be calculated using Equation 6.83 by substituting $(\theta_k - \varphi)$ for θ_k. Hence,

$$\overline{A}_{11} = U_1^Q \sum_{k=1}^{n}(z_k - z_{k-1}) + U_2^Q \sum_{k=1}^{n}[\cos 2(\theta_k - \varphi)](z_k - z_{k-1})$$

$$+ U_3^Q \sum_{k=1}^{n}[\cos 4(\theta_k - \varphi)](z_k - z_{k-1}) \qquad (6.86)$$

Recalling the general trigonometric identity:

$$\cos(\alpha - \beta) = \cos\alpha\cos\beta + \sin\alpha\sin\beta \qquad (6.87)$$

By utilizing his identity, Equation 6.86 can be written

$$\overline{A}_{11} = U_1^Q V_0^A + U_2^Q\left\{V_1^A\cos(2\varphi) + V_2^A\sin(2\varphi)\right\} + U_3^Q\left\{V_3^A\cos(4\varphi) + V_4^A\sin(4\varphi)\right\}$$

$$(6.88a)$$

The geometry factors V_i^A have been defined in Equation 6.77. A similar procedure can be applied to all remaining elements of the $\overline{A}_{ij}$ matrix as well as the $\overline{B}_{ij}$, and $\overline{D}_{ij}$ matrices:

$$\overline{A}_{22} = U_1^Q V_0^A - U_2^Q \left\{ V_1^A \cos(2\varphi) + V_2^A \sin(2\varphi) \right\} + U_3^Q \left\{ V_3^A \cos(4\varphi) + V_4^A \sin(4\varphi) \right\}$$

(6.88b)

$$\overline{A}_{12} = \overline{A}_{21} = U_4^Q V_0^A - U_3^Q \left\{ V_3^A \cos(4\varphi) + V_4^A \sin(4\varphi) \right\}$$

(6.88c)

$$\overline{A}_{66} = U_5^Q V_0^A - U_3^Q \left\{ V_3^A \cos(4\varphi) + V_4^A \sin(4\varphi) \right\}$$

(6.88d)

$$\overline{A}_{16} = \overline{A}_{61} = \frac{1}{2} U_2^Q \left\{ V_2^A \cos(2\varphi) - V_1^A \sin(2\varphi) \right\} + U_3^Q \left\{ V_4^A \cos(4\varphi) - V_3^A \sin(4\varphi) \right\}$$

(6.88e)

$$\overline{A}_{26} = \overline{A}_{62} = \frac{1}{2} U_2^Q \left\{ V_2^A \cos(2\varphi) - V_1^A \sin(2\varphi) \right\} - U_3^Q \left\{ V_4^A \cos(4\varphi) - V_3^A \sin(4\varphi) \right\}$$

(6.88f)

$$\overline{B}_{11} = U_1^Q V_0^B + U_2^Q \left\{ V_1^B \cos(2\varphi) + V_2^B \sin(2\varphi) \right\} + U_3^Q \left\{ V_3^B \cos(4\varphi) + V_4^B \sin(4\varphi) \right\}$$

(6.89a)

$$\overline{B}_{22} = U_1^Q V_0^B - U_2^Q \left\{ V_1^B \cos(2\varphi) + V_2^B \sin(2\varphi) \right\} + U_3^Q \left\{ V_3^B \cos(4\varphi) + V_4^B \sin(4\varphi) \right\}$$

(6.89b)

$$\overline{B}_{12} = \overline{B}_{21} = U_4^Q V_0^B - U_3^Q \left\{ V_3^B \cos(4\varphi) + V_4^B \sin(4\varphi) \right\}$$

(6.89c)

$$\overline{B}_{66} = U_5^Q V_0^B - U_3^Q \left\{ V_3^B \cos(4\varphi) + V_4^B \sin(4\varphi) \right\}$$

(6.89d)

$$\overline{B}_{16} = \overline{B}_{61} = \frac{1}{2} U_2^Q \left\{ V_2^B \cos(2\varphi) - V_1^B \sin(2\varphi) \right\} + U_3^Q \left\{ V_4^B \cos(4\varphi) - V_3^B \sin(4\varphi) \right\}$$

(6.89e)

$$\overline{B}_{26} = \overline{B}_{62} = \frac{1}{2} U_2^Q \left\{ V_2^B \cos(2\varphi) - V_1^B \sin(2\varphi) \right\} - U_3^Q \left\{ V_4^B \cos(4\varphi) - V_3^B \sin(4\varphi) \right\}$$

(6.89f)

$$\overline{D}_{11} = U_1^Q V_0^D + U_2^Q \left\{ V_1^D \cos(2\varphi) + V_2^D \sin(2\varphi) \right\} + U_3^Q \left\{ V_3^D \cos(4\varphi) + V_4^D \sin(4\varphi) \right\}$$

(6.90a)

$$\overline{D}_{22} = U_1^Q V_0^D - U_2^Q \left\{ V_1^D \cos(2\varphi) + V_2^D \sin(2\varphi) \right\} + U_3^Q \left\{ V_3^D \cos(4\varphi) + V_4^D \sin(4\varphi) \right\}$$

(6.90b)

$$\overline{D}_{12} = \overline{D}_{21} = U_4^Q V_0^D - U_3^Q \left\{ V_3^D \cos(4\varphi) + V_4^D \sin(4\varphi) \right\}$$ (6.90c)

$$\overline{D}_{66} = U_5^Q V_0^D - U_3^Q \left\{ V_3^D \cos(4\varphi) + V_4^D \sin(4\varphi) \right\}$$ (6.90d)

$$\overline{D}_{16} = \overline{D}_{61} = \frac{1}{2} U_2^Q \left\{ V_2^D \cos(2\varphi) - V_1^D \sin(2\varphi) \right\} + U_3^Q \left\{ V_4^D \cos(4\varphi) - V_3^D \sin(4\varphi) \right\}$$

(6.90e)

$$\overline{D}_{26} = \overline{D}_{62} = \frac{1}{2} U_2^Q \left\{ V_2^D \cos(2\varphi) - V_1^D \sin(2\varphi) \right\} - U_3^Q \left\{ V_4^D \cos(4\varphi) - V_3^D \sin(4\varphi) \right\}$$

(6.90f)

It is again emphasized that all new results presented in this section (in particular, Equations 6.82 through 6.90) are valid *only if all plies within the laminate are of the same material type.* In many cases, this is a severe restriction. These equations cannot be used to calculate the *ABD* matrix for a hybrid composite laminate, for example. On the other hand, an advantage of the invariant approach is the ability to easily rotate the ABD matrix from one coordinate system to another using Equations 6.88 through 6.90), so in the proper circumstances the invariant approach is convenient.

6.11 Computer Program *CLT*

The computer program *CLT* is available to supplement the material presented in this chapter. This program can be downloaded at no cost from the following website:

http://depts.washington.edu/amtas/computer.html

Program *CLT* can be used to recreate the numerical results discussed in the Example Problems presented so far in this chapter. In essence, the calculation steps described in Section 6.8 are implemented in this program. The user may select two different analysis paths. One analysis path corresponds to the case in which the loads applied to the laminate are specified, whereas the second corresponds to the case in which midplane strains and curvatures applied to the laminate are specified. In either case, the user must provide various numerical values required during the calculations performed. The user must define these values using a consistent set of units. For example, the user must input elastic modulii, thermal expansion coefficients, and moisture expansion coefficients for the composite material system(s) of interest. Using the properties listed in Table 3.1 and based on the SI system of units, the following numerical values would be input for graphite–epoxy:

$$E_{11} = 170 \times 10^9 \text{ Pa} \qquad E_{22} = 10 \times 10^9 \text{ Pa} \qquad \nu_{12} = 0.30 \qquad G_{12} = 13 \times 10^9 \text{ Pa}$$

$$\alpha_{11} = -0.9 \times 10^{-6} \text{ m/m} - {}^\circ\text{C} \qquad \alpha_{11} = 27.0 \times 10^{-6} \text{ m/m} - {}^\circ\text{C}$$

$$\beta_{11} = 150.0 \times 10^{-6} \text{ m/m} - \%M \qquad \beta_{22} = 4800 \times 10^{-6} \text{ m/m} - \%M$$

If the analysis requires the user to input numerical values for stress and moment resultants, then stress resultants must be input in N/m, and moment resultants must be input in N–m/m. Typical value would be $N_{xx} = 150 \times 10^3$ N/m and $M_{xx} = 5$ N–m/m. If, instead, the analysis requires the user to input numerical values for mid-plane strains and curvatures, then strains must be input in m/m (not in μm/m) and curvatures must be input in m^{-1}. Typical value would be $\varepsilon_{xx}^o = 2000 \times 10^{-6}$ m/m $= 0.002000$ m/m and $\kappa_{xx} = 0.5$ m^{-1}. All temperatures would be input in °C. Ply thicknesses must be input in meters (not milli-meters). A typical value would be $t_k = 0.000125$ *m* (corresponding to a ply thickness of 0.125 mm).

In contrast, if the English system of units were used, then the following numerical values would be input for the same graphite–epoxy material system:

$$E_{11} = 25.0 \times 10^6 \text{ psi} \qquad E_{22} = 1.5 \times 10^6 \text{ psi} \qquad \nu_{12} = 0.30 \qquad G_{12} = 1.9 \times 10^6 \text{ psi}$$

$$\alpha_{11} = -0.5 \times 10^{-6} \text{ in./in.–}{}^\circ\text{F} \qquad \alpha_{11} = 15 \times 10^{-6} \text{ in./in.–}{}^\circ\text{F}$$

$$\beta_{11} = 150.0 \times 10^{-6} \text{ in./in.–}\%M \qquad \beta_{22} = 4800 \times 10^{-6} \text{ in./in.–}\%M$$

Stress resultants must be input in lbf/in., and moment resultants must be input in lbf-in./in. Typical value would be $N_{xx} = 1000$ lbf/in. and $M_{xx} = 1$ lbf-in./in. If, instead, the analysis requires the user to input numerical values for midplane strains and curvatures, then strains must be input in in./in. and curvatures must be input in in^{-1}. Typical value would be $\varepsilon_{xx}^o = 2000 \times 10^{-6}$ in./in. $= 0.002000$ in./in. and $\kappa_{xx} = 0.01$ in^{-1}. All temperatures would be input in °F. Ply thicknesses must be input in inches. A typical value would be $t_k = 0.005$ in.

6.12 Comparing Classical Lamination Theory and Finite-Element Analyses

CLT is limited to a flat plate with constant thickness.* Obviously, a structure of practical interest is rarely a flat plate. Aircraft, automobiles, marine vessels, sporting goods, wind turbines, and so on, are largely composed of singly or doubly curved plates or shells, with varying thicknesses and cross-sections, and may feature bolt/rivet holes, cut-outs, reinforcing ribs/stiffeners, or other geometric complexities. CLT cannot be directly applied to study such structures. Instead, structural analyses of these more complex composite parts are usually performed by means of a finite-element (FE) analyses.

The objective of this section is to compare results from typical FE analyses to CLT. Note that a detailed discussion of the finite-element method itself is beyond the scope of this presentation. Dozens of excellent textbooks devoted to the finite-element method are available, and several have appeared in which the finite-element method is applied specifically to advanced composite materials and structures; References 5–8 are representative examples.

The FE analyses described here were completed using ANSYS, a widely used commercial finite-element software packages. Comparable results can be obtained using any FE analysis routine, as long as the element used allows for nonisotropic material properties. Discussion will center on the graphite–epoxy composite specimen shown in Figure 6.21. Note that this same specimen geometry and loading were considered for the case of *unidirectional* composites in Example Problems 5.4 and 5.8. The specimen shown in Figure 6.21 is produced using a multiangle stacking sequence; specifically, a 32-ply quasi-isotropic $[(0/45/90/-45)_4]_s$ stacking sequence. Two basic FE modeling approaches can be used. In the first approach the specimen is modeled on a *ply-by-ply* basis, whereas in the second approach the FE analysis is performed based on the *effective laminate* properties. These two methods are illustrated in the following two Example Problems.

Example Problem 6.9

The graphite–epoxy composite specimen shown in Figure 6.21 was modeled on a ply-by-ply basis, using the SHELL99 element featured in the ANSYS v 12 element library. This is an eight-noded element with six degrees of freedom at each node (the six degrees of freedom represent

---

* Although not presented in this chapter, it is possible to develop a CLT-type analysis for simple composite shells or tubes with constant thicknesses. See, for example, Reference 4.

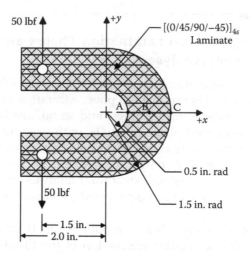

FIGURE 6.21

A quasi-isotropic [(0/45/90/−45)₄]ₛ U-shaped specimen; thickness = 0.160 in.

translations in the x-, y-, and z-directions and rotations about the x-, y-, and z-axes). The SHELL99 element is called a "layered" element. That is, to define the element the user must specify the number of through-thickness plies, the material type for each ply, and fiber angle of each ply. After a solution is obtained the user must specify the ply for which the stresses and/or strains are to be output.

In this example, the laminate is produced using a 32-ply quasi-isotropic [(0/45/90/−45)₄]ₛ stacking sequence, so 32 layers at these fiber angles are defined for each element. Ply properties listed in Table 3.1 for graphite–epoxy are assumed. The thickness of each ply is 0.005 in., so the total element thickness is 0.160 in. Points A, B, and C are defined on the specimen centerline (along $y = 0$), at $x = 0.5$, 1.0, and 1.5 in., respectively.

No bending will occur since the laminate is symmetric and the line of action of the applied load lies within the midplane of the specimen. Consequently the predicted strain contours are identical for all plies, regardless of ply fiber angle. Predicted in-plane strains are shown in Figure 6.22. Note that strain contours exhibit symmetry about the x-axis (i.e., about line ABC), and also that shear strains are zero along this line of symmetry. A symmetric strain field is expected in this case, due to the overall geometric symmetry of the specimen, the symmetric loading applied, *and* because a symmetric and balanced stacking sequence is used.

In contrast to predicted strain contours, predicted stress contours change dramatically from one ply to the next. Stress contours for 0°, 45°, 90°, and −45° plies are shown in Figures 6.23 through 6.26, respectively. Stress contours within the 0° and 90° plies (Figures 6.23 and 6.25) are both symmetric about line ABC, although the magnitude of stresses differs markedly. Shear stress is zero along line ABC for both of these

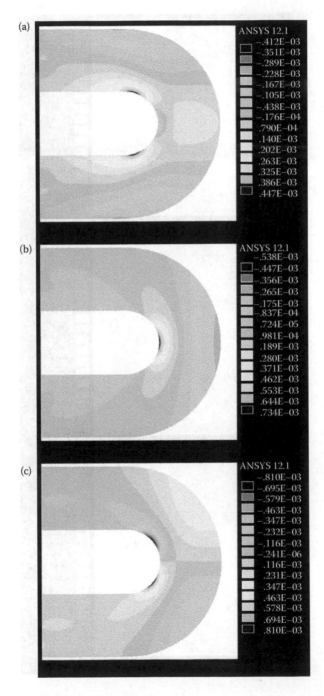

FIGURE 6.22

Predicted strain contours for all plies within the quasi-isotropic $[(0/45/90/-45)_4]_s$ U-shaped specimen shown in Figure 6.21. (a) ε_{xx} contours; (b) ε_{yy} contours; (c) γ_{xy} contours.

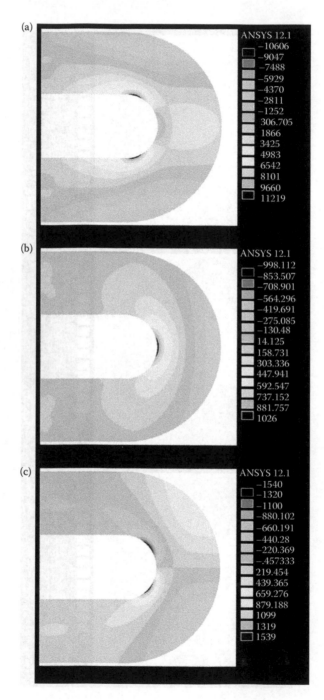

FIGURE 6.23
Predicted stress contours for all 0° plies within the quasi-isotropic $[(0/45/90/-45)_4]_s$ U-shaped specimen shown in Figure 6.21. (a) σ_{xx} contours; (b) σ_{yy} contours; (c) τ_{xy} contours.

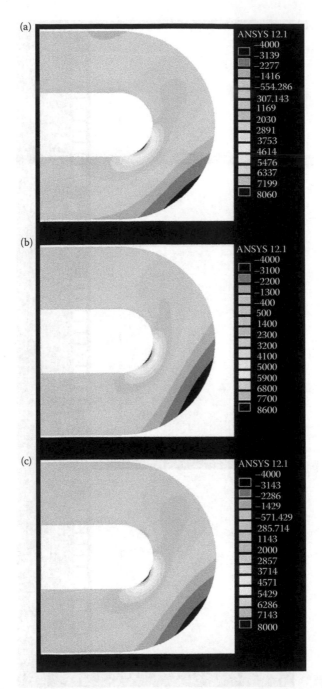

FIGURE 6.24
Predicted stress contours for all 45° plies within the quasi-isotropic $[(0/45/90/-45)_4]_s$ U-shaped specimen shown in Figure 6.21. (a) σ_{xx} contours; (b) σ_{yy} contours; (c) τ_{xy} contours.

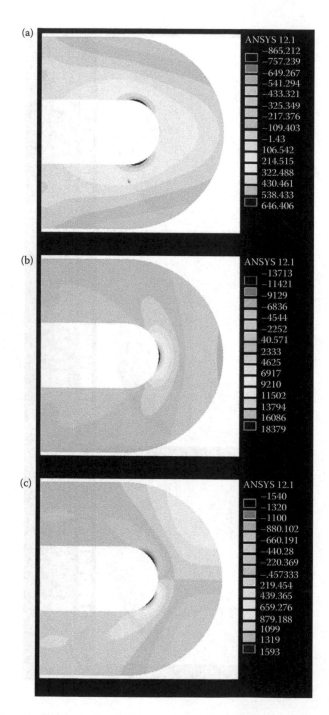

FIGURE 6.25
Predicted stress contours for all 90° plies within the quasi-isotropic $[(0/45/90/-45)_4]_s$ U-shaped specimen shown in Figure 6.21. (a) σ_{xx} contours; (b) σ_{yy} contours; (c) τ_{xy} contours.

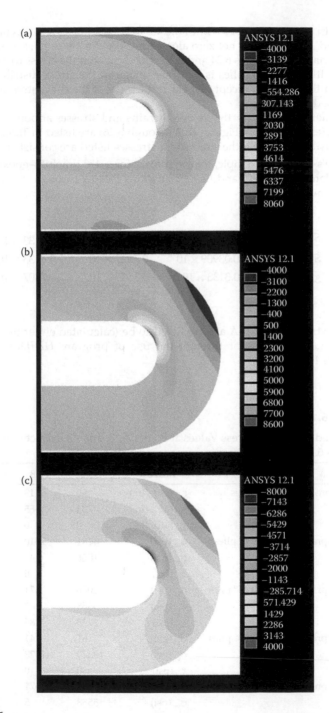

FIGURE 6.26
Predicted stress contours for all −45° plies within the quasi-isotropic [(0/45/90/−45)₄]ₛ U-shaped specimen shown in Figure 6.21. (a) σ_{xx} contours; (b) σ_{yy} contours; (c) τ_{xy} contours.

plies. In contrast, stresses within the 45° and −45° plies are not symmetric, and shear stress is not zero along line ABC for these plies. Careful examination of Figures 6.24 and 6.26 will show that the stress distribution within the −45° plies is a mirror image of the stress distribution within the 45° plies, except that algebraic signs of the shear stresses are reversed.

Numerical values for the in-plane strains and stresses at points A, B, and C (extracted from Figures 6.23 through 6.26) are listed in Table 6.11. It is easily shown that the strains and stresses listed are consistent with Hooke's Law. For example, the $[\overline{S}]$ matrix for a 45° graphite–epoxy ply (calculated either by hand using Equations 5.22 or through the use of program *UNDIR*) is

$$\begin{bmatrix} \overline{S}_{11} & \overline{S}_{12} & \overline{S}_{16} \\ \overline{S}_{21} & \overline{S}_{22} & \overline{S}_{26} \\ \overline{S}_{61} & \overline{S}_{62} & \overline{S}_{66} \end{bmatrix} = \begin{bmatrix} 0.3022 \times 10^{-6} & 0.03909 \times 10^{-6} & -0.3133 \times 10^{-6} \\ 0.03909 \times 10^{-6} & 0.3022 \times 10^{-6} & -0.3133 \times 10^{-6} \\ -0.3133 \times 10^{-6} & -0.3133 \times 10^{-6} & 0.7307 \times 10^{-6} \end{bmatrix} \left(\frac{1}{psi} \right)$$

The strains at point A are predicted to be (calculated either by hand using Equation 5.29, or through the use of program *UNIDIR*, with $\Delta T = \Delta M = 0$):

TABLE 6.11

Predicted Strain and Stress Values at Points A, B, and C, Extracted from Figures 6.22 through 6.26

	Variable	Pt A	Pt B	Pt C
Strains predicted for all plies	ε_{xx} (μin./in.)	−180	113	77
	ε_{yy} (μin./in.)	734	−65	−312
	γ_{xy} (μin./in.)	0	0	0
Stresses predicted for all 0° plies	σ_{xx} (psi)	−4181	2810	1796
	σ_{yy} (psi)	1026	−46	−436
	τ_{xy} (psi)	0	0	0
Stresses predicted for all 45° plies	σ_{xx} (psi)	2086	671	−880
	σ_{yy} (psi)	5558	−3	−2359
	τ_{xy} (psi)	3278	286	−1389
Stresses predicted for all 90° plies	σ_{xx} (psi)	61	141	−25
	σ_{yy} (psi)	18379	−1573	−7811
	τ_{xy} (psi)	0	0	0
Stresses predicted for all −45° plies	σ_{xx} (psi)	2084	671	−880
	σ_{yy} (psi)	5558	−5	−2359
	τ_{xy} (psi)	−3278	−286	1389

$$\begin{Bmatrix} \varepsilon_{xx} \\ \varepsilon_{yy} \\ \gamma_{xy} \end{Bmatrix} = \begin{bmatrix} \bar{S}_{11} & \bar{S}_{12} & \bar{S}_{16} \\ \bar{S}_{12} & \bar{S}_{22} & \bar{S}_{26} \\ \bar{S}_{16} & \bar{S}_{26} & \bar{S}_{66} \end{bmatrix} \begin{Bmatrix} \sigma_{xx} \\ \sigma_{yy} \\ \tau_{xy} \end{Bmatrix} = \begin{bmatrix} 0.3022 \times 10^{-6} & 0.03909 \times 10^{-6} & -0.3133 \times 10^{-6} \\ 0.03909 \times 10^{-6} & 0.3022 \times 10^{-6} & -0.3133 \times 10^{-6} \\ -0.3133 \times 10^{-6} & -0.3133 \times 10^{-6} & 0.7307 \times 10^{-6} \end{bmatrix} \begin{Bmatrix} 2086 \\ 5558 \\ 3278 \end{Bmatrix}$$

$$\begin{Bmatrix} \varepsilon_{xx} \\ \varepsilon_{yy} \\ \gamma_{xy} \end{Bmatrix} = \begin{Bmatrix} -179\ \mu\text{in./in.} \\ 734\ \mu\text{in./in.} \\ 0\ \mu\text{rad} \end{Bmatrix}$$

Hence, except for minor round-off error the finite-element results listed in Table 6.11 for the 45° ply at point A are consistent with Equation 5.29.

Alternatively, the $[\bar{Q}]$ matrix for a 45° graphite–epoxy ply (calculated either by hand using Equations 5.31, or through the use of program *UNDIR*) is

$$\begin{bmatrix} \bar{Q}_{11} & \bar{Q}_{12} & \bar{Q}_{16} \\ \bar{Q}_{21} & \bar{Q}_{22} & \bar{Q}_{26} \\ \bar{Q}_{61} & \bar{Q}_{62} & \bar{Q}_{66} \end{bmatrix} = \begin{bmatrix} 8.787 \times 10^6 & 4.987 \times 10^6 & 5.907 \times 10^6 \\ 4.987 \times 10^6 & 8.787 \times 10^6 & 5.907 \times 10^6 \\ 5.907 \times 10^6 & 5.907 \times 10^6 & 6.435 \times 10^6 \end{bmatrix} (\text{psi})$$

The stresses at point A are then predicted to be (calculated either by hand using Equation 5.30, or through the use of program *UNIDIR*, with $\Delta T = \Delta M = 0$):

$$\begin{Bmatrix} \sigma_{xx} \\ \sigma_{yy} \\ \tau_{xy} \end{Bmatrix}\Bigg|_{45°\text{plies}} = \begin{bmatrix} \bar{Q}_{11} & \bar{Q}_{12} & \bar{Q}_{16} \\ \bar{Q}_{12} & \bar{Q}_{22} & \bar{Q}_{26} \\ \bar{Q}_{16} & \bar{Q}_{26} & \bar{Q}_{66} \end{bmatrix}_{45°\text{plies}} \begin{Bmatrix} \varepsilon_{xx} \\ \varepsilon_{yy} \\ \gamma_{xy} \end{Bmatrix}_{45°\text{plies}}$$

$$\begin{Bmatrix} \sigma_{xx} \\ \sigma_{yy} \\ \tau_{xy} \end{Bmatrix}\Bigg|_{45°\text{plies}} = \begin{bmatrix} 8.787 \times 10^6 & 4.987 \times 10^6 & 5.907 \times 10^6 \\ 4.987 \times 10^6 & 8.787 \times 10^6 & 5.907 \times 10^6 \\ 5.907 \times 10^6 & 5.907 \times 10^6 & 6.435 \times 10^6 \end{bmatrix}_{45°\text{plies}} \begin{Bmatrix} -180 \times 10^{-6} \\ 734 \times 10^{-6} \\ 0 \end{Bmatrix}$$

$$\begin{Bmatrix} \sigma_{xx} \\ \sigma_{yy} \\ \tau_{xy} \end{Bmatrix}\Bigg|_{45°\text{plies}} = \begin{Bmatrix} 2079 \\ 5552 \\ 3272 \end{Bmatrix} (\text{psi})$$

As before, except for minor round-off error the finite-element results listed in Table 6.11 for the 45° ply at point A are consistent with Equation 5.30. Further evaluation of the stresses and strains predicted via the ANSYS analysis are left as a student exercise (see Homework Problem 6.23).

Example Problem 6.10

The graphite–epoxy composite specimen shown in Figure 6.21 was modeled using effective *laminate* properties. In this case, the PLANE183 element featured in the ANSYS v 12 element library was used. This is an eight-node element suitable for use in plane problems, which also allows the user to input orthotropic material properties. The element features 2 degrees of freedom at each node (translations in the x-, y-directions). To use this element the user must input the *effective properties of the laminate*, rather than ply properties and an associated stacking sequence.

Methods to calculate the effective properties of a laminate were discussed in Section 6.9, and the results developed there are implemented in program *CLT*. In this case, the effective extensional Young's modulus, major Poisson ratio, and shear modulus for a $[(0/45/90/-45)_4]_s$ graphite–epoxy laminate are needed. These can be calculated using Equations 6.63, 6.64, and 6.67 (or through the use of program *CLT*), using ply properties given in Table 3.1. The effective properties are found to be

$$\bar{E}_{xx}^{ex} = \bar{E}_{yy}^{ex} = 10.4\,\text{Msi}$$

$$\bar{v}_{xy}^{ex} = 0.246$$

$$\bar{G}_{xy} = 4.17\,\text{Msi}$$

The predicted strain distributions based on the PLANE183 element and these effective properties are virtually identical to those predicted using the SHELL99 element, as previously shown in Figure 6.22. However, the predicted stress distributions provided by the two analyses differ, since *effective* stresses are calculated using the PLANE183 element, whereas *ply* stresses are calculated using the layered SHELL99 element.

Contours of effective laminate stresses are shown in Figure 6.27. Although the distribution of effective stresses resembles the distribution of ply stresses in the 0° and 90° plies, the magnitudes are quite different (compare Figures 6.23, 6.25, and 6.27).

Numerical values of strains and effective stresses at points A, B, and C (extracted from Figures 6.22 and 6.27) are listed in Table 6.12. It can be shown that the strains and effective stresses predicted by the FE analysis are consistent with CLT by applying Equation 6.38. Note that stress resultants are related to effective stresses as follows:

$$N_{xx} = t\,\bar{\sigma}_{xx}$$

$$N_{yy} = t\,\bar{\sigma}_{yy}$$

$$N_{xy} = t\,\bar{\tau}_{xy}$$

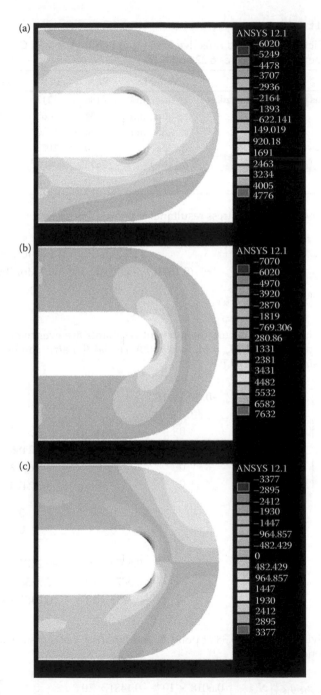

FIGURE 6.27
Predicted *effective* stress contours for the quasi-isotropic $[(0/45/90/-45)_4]_s$ U-shaped specimen shown in Figure 6.21. (a) $\bar{\sigma}_{xx}$ contours; (b) $\bar{\sigma}_{yy}$ contours; (c) $\bar{\tau}_{xy}$ contours.

TABLE 6.12

Predicted Strains and Effective Stresses at Points A, B, and C, Extracted from Figures 6.22 and 6.27

	Variable	Pt A	Pt B	Pt C
Strains predicted for all plies	ε_{xx} (μin./in.)	−180	113	77
	ε_{yy} (μin./in.)	734	−65	−312
	γ_{xy} (μin./in.)	0	0	0
Effective stresses	$\bar{\sigma}_{xx}$ (psi)	0	1073	0
	$\bar{\sigma}_{yy}$ (psi)	7632	−407	−3241
	$\bar{\tau}_{xy}$ (psi)	0	0	0

So, for example, the stress resultants at point B are

$$
\left\{\begin{matrix} N_{xx} \\ N_{yy} \\ N_{xy} \end{matrix}\right\}_{pt\,B} = t \left\{\begin{matrix} \bar{\sigma}_{xx} \\ \bar{\sigma}_{yy} \\ \bar{\tau}_{xy} \end{matrix}\right\}_{pt\,B} = (0.160\,\text{in})\left\{\begin{matrix} 1073\ \text{psi} \\ -407\ \text{psi} \\ 0 \end{matrix}\right\} = \left\{\begin{matrix} 171.68\ \text{lbf/in} \\ -65.12\ \text{lbf/in} \\ 0 \end{matrix}\right\}
$$

Since no bending occurs the moment resultants are everywhere zero for this problem (i.e., $M_{xx} = M_{yy} = M_{xy} = 0$). The [abd] matrix can be calculated using program CLT:

$$
[abd] = \begin{bmatrix}
0.6018 \times 10^{-6} & -0.1481 \times 10^{-6} & 0 & 0 \\
-0.1481 \times 10^{-6} & 0.6018 \times 10^{-6} & 0 & 0 \\
0 & 0 & 1.500 \times 10^{-6} & 0 \\
0 & 0 & 0 & 0.2486 \times 10^{-3} \\
0 & 0 & 0 & -0.6052 \times 10^{-4} \\
0 & 0 & 0 & -0.2470 \times 10^{-4}
\end{bmatrix}
$$

$$
\begin{bmatrix}
0 & 0 \\
0 & 0 \\
0 & 0 \\
-0.6052 \times 10^{-4} & -0.2470 \times 10^{-4} \\
0.3070 \times 10^{-3} & -0.3236 \times 10^{-4} \\
-0.3235 \times 10^{-4} & 0.7482 \times 10^{-3}
\end{bmatrix}
$$

Applying Equation 6.38 at point B, and ignoring variables associated with bending, we have:

$$
\left\{\begin{matrix} \varepsilon_{xx}^o \\ \varepsilon_{yy}^o \\ \gamma_{xy}^o \end{matrix}\right\} = \begin{bmatrix} a_{11} & a_{12} & a_{16} \\ a_{12} & a_{22} & a_{26} \\ a_{16} & a_{26} & a_{66} \end{bmatrix}\left\{\begin{matrix} N_{xx} \\ N_{yy} \\ N_{xy} \end{matrix}\right\} = \begin{bmatrix} 0.6018 \times 10^{-6} & -0.1481 \times 10^{-6} & 0 \\ -0.1481 \times 10^{-6} & 0.6018 \times 10^{-6} & 0 \\ 0 & 0 & 1.500 \times 10^{-6} \end{bmatrix}\left\{\begin{matrix} 171.68 \\ -65.12 \\ 0 \end{matrix}\right\}
$$

$$\begin{Bmatrix} \varepsilon_{xx}^{o} \\ \varepsilon_{yy}^{o} \\ \gamma_{xy}^{o} \end{Bmatrix} = \begin{Bmatrix} 113\,\mu\text{in./in.} \\ -65\,\mu\text{in./in.} \\ 0 \end{Bmatrix}$$

It is seen that the strains predicted by the FE analysis and listed in Table 6.12 are consistent with Equation 6.38. Further evaluation of the consistency between the ANSYS analysis and CLT are left as a student exercise (see Homework Problems 6.24).

6.13 Free Edge Stresses

Although it may seem non-intuitive, in-plane loads applied to a flat multi-angle laminate cause both in-plane *and* out-of-plane stresses near any free edge. This is depicted schematically in Figure 6.28, which shows that a simple uniaxial in-plane load N_{xx} will cause a 2-D state of plane stress at interior regions of a flat laminate, but a 3-D state of stress near a free edge.* As will be seen, free edge stresses occur whenever there is a mismatch in material properties of adjacent plies.

Free edge stresses are extremely important and often initiate composite failures. A tensile out-of-plane normal stress σ_{zz} may initiate delamination failures, for example. Since CLT is based on the plane stress assumption it cannot be used to calculate or predict 3-D free edge stresses. However, a CLT analysis can be used to explain *why* 3-D stresses occur near a free edge. In the following subsection a CLT analysis of a simple 3-ply laminate is derived and used to explain the origins of interlaminar stresses.

Calculating free edge stresses for general laminates used in practice, which often involve a large number of plies and/or two or more material systems, requires advanced numerical analysis methods based on finite element or finite difference techniques that are beyond the scope of this book. A summary of typical free edge stresses predicted for general laminates based on these methods is presented in a concluding subsection.

6.13.1 The Origins of Free Edge Stresses

Consider a *hybrid* three-ply $[0°]_3$ laminate subjected to uniaxial tensile load N_{xx} (only), as shown in Figure 6.29a. This simple laminate is assumed to consist of three 0° plies and made of two different materials. Material

---

* A free edge is defined as any unloaded edge of the laminate; examples include the unloaded edge of a tensile specimen or the edges of lightening holes and other cutouts in a composite structure.

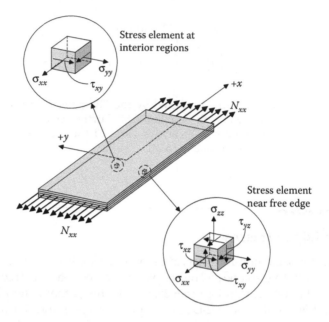

FIGURE 6.28
Stress elements illustrating the 3-D stress state induced near free edges in general composite laminates.

"A" is used in plies 1 and 3, whereas material "B" is used in ply 2. Assume that the only difference between materials A and B is that the value of Poisson's ratio for material A is greater than for material B. That is, assume $E_{11}^{(A)} = E_{11}^{(B)} = E_{11}$, $E_{22}^{(A)} = E_{22}^{(B)} = E_{22}$, $G_{12}^{(A)} = G_{12}^{(B)} = G_{12}$, but $\nu_{12}^{(A)} > \nu_{12}^{(B)}$. Note that in this example the coefficient of mutual influence of the second kind equals zero for all three plies, since $\theta = 0$ and the 1–2 axes are coincident with the x–y axis for all three plies. Hence, an in-plane normal load does not tend to cause a shear strain. We assume that the thickness of each ply is identical ($t_1 = t_2 = t_3 = t_p$, say), so that the total laminate thickness is $t = 3t_p$.

Based strictly on physical reasoning, it is clear that all plies will experience a tensile axial stress and strain, since the overall loading is tensile. Note that all three plies have the same cross-sectional area and the same axial stiffness, $E_{11}^{(A)} = E_{11}^{(B)} = E_{11}$. Therefore it would seem that each individual ply will support one-third of the overall loading. In other words, it would seem that the axial stress should be the same in each ply (we will later see that this is not quite true).

If the three plies are *firmly bonded together* then due to the Poisson effect the tensile load N_{xx} will cause a uniform contraction in the transverse y-direction, and the same compressive transverse strain will occur in each ply:

$$\varepsilon_{yy}\big|_{ply\ 1} = \varepsilon_{yy}\big|_{ply\ 2} = \varepsilon_{yy}\big|_{ply\ 3}.$$

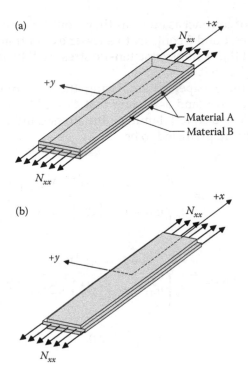

FIGURE 6.29
A hybrid 3-ply 0° laminate subjected to a uniaxial tensile loading, used to illustrate the origins of free edge stresses. (a) Plies 1 and 3 are composed of material A, while ply 2 is composed of material B; (b) deformed shape if plies could freely expand or contract and $v_{12}^A > v_{12}^B$ (deformations shown greatly exaggerated).

On the other hand, *suppose the three plies were not bonded together*. In this case, an individual ply would be free to contract as dictated by its' own Poisson's ratio. The transverse contractions that would occur if the three plies were not bonded together are shown schematically (and highly exaggerated) in Figure 6.29b. Note that plies 1 and 3 would contract to a greater extent than ply 2, since it has been assumed that Poisson's ratio of plies 1 and 3 is greater than that of ply 2: $v_{12}^{(A)} > v_{12}^{(B)}$. Thus, *if the plies were not bonded together* then the magnitude of the compressive transverse strains induced in plies 1 and 3 would be greater than that in ply 2: $|\varepsilon_{yy}|_{\text{plies } 1, 3} > |\varepsilon_{yy}|_{\text{ply } 2}$. Also, *if the three plies were not bonded together* then the transverse stress in each ply would equal zero: $\sigma_{yy}|_{\text{ply } 1} = \sigma_{yy}|_{\text{ply } 2} = \sigma_{yy}|_{\text{ply } 3} = 0$.

Since the three plies *are* bonded together the three plies will contract by an equal amount. Ply 2 is, therefore, forced to contract to a greater extent than it would if it were not bonded to neighboring plies 1 and 3. Based on this reasoning we expect that a transverse compressive stress will be induced in ply 2, $\sigma_{yy}|_{\text{ply } 2} < 0$, since this ply is forced by neighboring plies 1 and 3 to contract to a greater extent than it otherwise would. In contrast, plies 1 and 3 do not contract as much as they would if they were not bonded to ply 2. Since plies 1 and 3 contract to a lesser extent that they would otherwise, we expect that a transverse tensile stress will be induced in plies 1 and 3: $\sigma_{yy}|_{ply\,1}, \sigma_{yy}|_{ply\,3} > 0$.

CLT confirms these expectations. Using the CLT modeling approach described in Section 6.6 (and assuming that $\Delta T = \Delta M = 0$) it can be shown[*] that for this simple hybrid $[0°]_3$ laminate the stresses induced in the interior regions of each ply are predicted to be

$$\sigma_{xx}\big|_{plies\,1,3} = \frac{\left\{3E_{11} - E_{22}v_{12}^{(B)}\left[2v_{12}^{(B)} + v_{12}^{(A)}\right]\right\}N_{xx}}{t_p\left\{9E_{11} - E_{22}\left[\left\{v_{12}^{(A)}\right\} + 2\left\{v_{12}^{(B)}\right\}\right]^2\right\}} \tag{6.91a}$$

$$\sigma_{yy}\big|_{plies\,1,3} = \frac{\left[v_{12}^{(A)} - v_{12}^{(B)}\right]E_{22}N_{xx}}{t_p\left\{9E_{11} - E_{22}\left[\left\{v_{12}^{(A)}\right\} + 2\left\{v_{12}^{(B)}\right\}\right]^2\right\}} \tag{6.91b}$$

$$\tau_{xy}\big|_{plies\,1,3} = 0 \tag{6.91c}$$

$$\sigma_{xx}\big|_{ply\,2} = \frac{\left\{3E_{11} - E_{22}v_{12}^{(A)}\left[2v_{12}^{(B)} + v_{12}^{(A)}\right]\right\}N_{xx}}{t_p\left\{9E_{11} - E_{22}\left[\left\{v_{12}^{(A)}\right\} + 2\left\{v_{12}^{(B)}\right\}\right]^2\right\}} \tag{6.92a}$$

$$\sigma_{yy}\big|_{ply\,2} = \frac{-2\left[v_{12}^{(A)} - v_{12}^{(B)}\right]E_{22}N_{xx}}{t_p\left\{9E_{11} - E_{22}\left[\left\{v_{12}^{(A)}\right\} + 2\left\{v_{12}^{(B)}\right\}\right]^2\right\}} = -2\sigma_{yy}\big|_{plies\,1,3} \tag{6.92b}$$

$$\tau_{xy}\big|_{ply\,2} = 0 \tag{6.92c}$$

Comparing Equations 6.91a and 6.92a, it is interesting to note that since $v_{12}^{(A)} > v_{12}^{(B)}$ the axial stress σ_{xx} induced in plies 1 and 3 is not precisely equal to that induced in ply 2. Hence, the applied load is not shared equally among the three plies, even though each ply has the same axial stiffness and is subjected to the same axial strain. Of greater immediate interest, however, is that Equations 6.91b and 6.92b predict that a transverse

[*] Since this is a relatively simple laminate involving only 0° plies with equal thicknesses, Equations 6.91 and 6.92 can be easily derived in closed form, beginning with Equation 6.28.

tensile stress σ_{yy} is predicted in plies 1 and 3, and a compressive trans-verse stress is predicted in ply 2, $[v_{12}^{(A)} - v_{12}^{(B)}] > 0$. This result is in complete agreement with what we expect based strictly on physical reasoning, as previously discussed. Hence, we conclude that *a uniaxial load in the x-direc-tion (N_{xx}) is predicted to cause a transverse normal stress in the y-direction.* The transverse stress induced in ply 2 is predicted to be compressive (as expected), with a magnitude twice as high as the tensile transverse stress induced in plies 1 and 3. In this example, the transverse stresses occur *solely because of the mismatch in Poisson's ratio.* That is, if $v_{12}^{(A)} = v_{12}^{(B)}$ then from Equations 6.91b and 6.92b the transverse stress in all three plies becomes $\sigma_{yy} = 0$.

The physical reasoning described above is straightforward, and the expected algebraic sign of the transverse stresses expected in each ply agrees with CLT calculations. Nevertheless, *a transverse stress σ_{yy} cannot possibly exist at the free edge of the laminate, since no transverse load N_{yy} is applied to the laminate.* We are faced with a quandary: on the one hand predictions based on the CLT analysis are reasonable and agree with physical reasoning, and yet stresses are predicted that cannot possibly exist at the free edge.

A free-body diagram of an individual ply can now be used to understand why free edge stresses develop, and why they are three-dimensional. A cross-section within the y–z plane of the three-ply lami-nate, and a free body diagram of a section removed from ply 3, is shown in Figure 6.30. As per the preceding discussion, a transverse force F_{yy} is induced at interior regions of the ply, and is drawn on the left face of the free-body diagram. The magnitude of this force equals the stress σ_{yy} multiplied by the area over which it acts: $F_{yy} = \sigma_{yy} (t_p) (dx)$. An equili-brating normal force F_{yy} cannot exist on the unloaded free surface (the right side of the free-body diagram). Therefore, a distributed equilibrat-ing shear force V_{zy} *must* develop on the interior surface of ply three (i.e., on the "lower" surface of ply 3, as shown in Figure 6.30), so as to main-tain force equilibrium in the y-direction (i.e., to insure that $\Sigma F_y = 0$). The fact that a shear force V_{zy} is present implies, of course, that a shear stress $\tau_{yz} = \tau_{zy}$ is present near the free edge. The fact that shear force V_{zy} (and shear stress τ_{zy}) must be present does not tell us how this force/stress is distributed, however. Since no shear force can exist *at* the free edge ($V_{zy} = 0$ at $y = b$), V_{zy} can only be nonzero at interior regions of ply 3. The distribu-tion and magnitudes of V_{zy} and τ_{zy} depend on details of loading and lami-nate stacking sequence and must be calculated for each individual case, as discussed later.

Note from Figure 6.30 that the normal force acting on the left side, F_{yy}, is acting through the centroid of the free-body diagram, whereas the shear force on the right side, V_{zy}, is present on the lower surface only. Therefore, these two forces are not colinear, and consequently they induce a bending moment about the x-axis. The only way that this bending moment can be

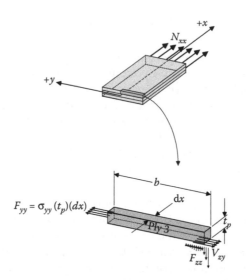

FIGURE 6.30
A free-body diagram of a section removed from ply 3 of the hybrid 3-ply laminate shown in Figure 6.29.

reacted so as to satisfy moment equilibrium (i.e., so as to maintain $\Sigma M_x = 0$) is if a force acting normal to the x–y plane (F_{zz}) *also* develops. The fact that a normal force F_{zz} is present implies, of course, that a normal stress σ_{zz} is present. Once again, while the equations of equilibrium dictate that a normal force F_{zz} (and associated normal stress σ_{zz}) must be present, they cannot be used to determine how this force/stress is distributed spatially. Since the only external force applied to the laminate is N_{xx}, the internal force F_{zz} (or equivalently, the normal stress σ_{zz}) must be self-equilibrating. That is, forces in the z-direction must sum to zero: $\Sigma F_z = \int_{y=0}^{b} \sigma_{zz}\,dxdy = 0$. This implies that σ_{zz} must undergo a change in algebraic sign in regions near the free edge. If, for example, σ_{zz} is tensile at the free edge $y = b$, then it must become compressive at some interior distance from the free edge, so as to maintain static equilibrium.

For the simple hybrid $[0°]_3$ laminate considered here the coefficient of mutual influence of the second kind $\eta_{xx,xy} = 0$ for all plies, since all fiber angles are 0°. Therefore the uniaxial load N_{xx} does not cause in-plane shear stresses (τ_{xy}) to occur in any ply. However, $\eta_{xx,xy} \neq 0$ if $\theta \neq 0°$ or 90°. Therefore, for more general laminates $\eta_{xx,xy}$ may not equal zero for all plies, and in these cases the uniaxial loading N_{xx} will cause both σ_{yy} and τ_{xy} at interior regions of the laminate. In these cases the state of stress near a free edge will involves all six components of stress (σ_{xx}, σ_{yy}, σ_{zz}, τ_{yz}, τ_{xz}, and τ_{xy}).

This simple example illustrated what was stated earlier: free edges stresses develop due to a mismatch in effective material properties of adjacent plies. In this case, free edge stresses develop due to the mismatch in Poisson ratios.

6.13.2 Analytical and Numerical Studies of Free Edge Stresses

Free edge stresses have a profound influence on initial delaminations and final fracture of multi-angle composite laminates, and have, therefore, been the topic of a great many research papers. The existence of free edge strains in multi-angle composite laminates has been well documented via experimental measurements (e.g., References 9–12). Methods used to *predict* free edge stresses may be roughly grouped into numerical solutions (e.g., References 12–18) and approximate analytical solutions (e.g., References 19–29). A detailed discussion of the numerical methods and/or approximate analytical tools that have been used is beyond the scope of this book. The results of these analyses will simply be summarized, and the interested reader is referred to the original references for additional details as needed. Review papers are also available (e.g., Reference 30).

Pipes and Pagano presented the first three-dimensional numerical study of free edge stresses in 1970 [13]. Their analysis was based on the method of finite differences. They investigated free edge stresses induced in symmetric graphite–epoxy angle-ply laminates having the general stacking sequence $[\theta/-\theta]_s$ and subjected to a remote uniform axial extension, which implies that at remote locations the laminate is subjected to a uniform axial tensile strain ε_{xx}. They found that, at points well away from a free edge, stresses in individual plies were well-predicted by CLT. In particular, their analysis showed that at interior regions a plane stress condition exists within all plies within the laminate. For arbitrary ply "k," and at points away from the free edge, only in-plane stresses σ_{xx}, σ_{yy}, and τ_{xy} were predicted and all out-of-plane stresses were zero; $\sigma_{zz} = \tau_{xz} = \tau_{yz} = 0$. In contrast, at points near a free edge a 3-D state of stress involving all 6 components of stress was predicted.

Pipes and Pagano performed a convergence study wherein finite-difference analyses based on successively finer and finer grid patterns were used. They found that the magnitudes of two stress components, σ_{zz} and τ_{xz}, did not converge to a constant value but rather continued to increase as grid density was increased. Based on this observation they speculated that σ_{zz} and τ_{xz} were singular. The singular nature of σ_{zz} and τ_{xz} could not proven definitively, however, due to the numerical nature of their study.

The pioneering study of Pipes and Pagano demonstrated that a 3-dimensional state of stress generally exists within a narrow region near an unloaded edge of $[\theta/-\theta]_s$ graphite–epoxy laminates subjected to a remote uniform axial strain ε_{xx}. Many numerical analyses of free edge stresses based on the use of the finite-element technique appeared following the initial Pipes and Pagano study; References 14–18 are a few representative examples. A growing consensus that a singular stress field does exist at the free edge developed. That is, as numerical analyses became more and more refined, allowing greater mesh densities to be used to study the free edge problem, the magnitude of predicted free edge stresses became larger and larger. The fact that individual stress components (in particular, the values of σ_{zz} and especially τ_{xz}) did not converge to a stable value as mesh densities were increased near the free edge is indicative of a singular stress field, although this conclusion cannot be proven rigorously on the basis of numerical studies alone.

Wang and Choi [23, 24] and Zwiers, Ting, and Spilker [25] eventually confirmed the singular nature of free edge stress fields on the basis of analytical studies. The approach taken by Wang and Choi is based on the use of complex-variable stress potentials, which is an analytical approach originated primarily by Lekhnitski [26]. In contrast, the approach taken by Zwiers, Ting, and Spilker is based on the use of complex-variable displacement potentials, which is an analytical approach originated primarily by Stroh [26]. Although either approach may be taken, most analytical studies of free edge stresses (e.g., References 23,24,27–29) have been based on the use of Lekhnitski's complex stress potentials. These analytical studies have shown that stresses induced near a free edge are singular in nature, and can be expressed in general form as

$$\sigma_{ij} = f(y,z)r^{-\delta} = \frac{f(y,z)}{r^\delta} \tag{6.93}$$

where $f(y,z)$ represents a generalized function that depends on laminate stacking sequence, material properties, and applied loading, and r represents the distance from the intersection between the interface between two plies and the free edge. It can be shown [23,24] that the exponent "δ" depends on both material properties as well as the difference in fiber angles between adjacent plies in the laminate. Also, δ must be a positive number bounded by $0 < \delta < 1$. Hence, as the free edge is approached (i.e., as $r \to 0$), the fraction $1/r^\delta \to \infty$ and in accordance with Equation 6.93 the stresses σ_{ij} are predicted to be infinitely high. That is, free edge stresses are singular.

Four summary comments are as follows:

- Since free edge stresses are due to any mismatch in material properties of adjacent plies, the distribution of free edge stresses depends on the material(s) and stacking sequence used to create the laminate. The number of different combinations of materials and stacking sequences is infinitely large, so it is not possible to tabulate solutions for free edge stresses; a new analysis is required for each new combination.

- The effective material properties of a ply are a function of fiber angle. Consequently, free edge stresses are, in general, minimized if the difference in fiber angles of adjacent plies is minimized. For example, free edge stresses in a quasi-isotropic laminate based on a [0/45/90/−45]$_s$ stacking sequence would be expected to be lower than those induced in [0/±45/90]$_s$ laminate, since in the former case the maximum difference in adjacent ply fiber angles is 45°, whereas in the latter case the maximum difference is 90°.

- Most analyses that have appeared in the literature have focused on free edge stresses caused in rectangular test specimens subjected to uniaxial tensile loading. In practice composite structures are required to support combinations of tensile, compressive, and shear loading. Practical composite structures are rarely flat and rectangular. Free edge stresses may occur at along curved edges, bolt holes, cutouts, or other geometric complexities. Therefore results of studies that have appeared in the literature may not be directly applicable during the design of a composite structure, even if the same material and stacking sequence is involved.

- As discussed by Herakovich et al. [12], the predicted singular nature of free edge stresses is in reality an artifact of available structural analysis methods. That is, all of the analyses described herein are based on the assumption of linear-elastic material behavior, and furthermore all analyses assume that a distinct interface exists between plies. Neither of these assumptions is rigorously true. A distinct interface does not exist in real composite laminates. While it is a certainty that high stress levels and stress gradients are induced near a free edge, in reality high stress levels cause a nonlinear material response, and, therefore, the magnitude of free edge stresses are large but finite. This observation is of course borne out by experiment, since if free edge stresses were truly singular then a composite laminate would delaminate immediately upon application of any external load, regardless of magnitude.

6.13.3 Typical Numerical Results

The finite-difference analysis described by Pipes and Pagano in [13] will be used to illustrate typical free edge stresses. This study was devoted to symmetric angle-ply laminates having the general stacking sequence $[\theta/-\theta]_s$, as shown in Figure 6.31. The thickness of each ply was labeled h_o. In this book the total laminate thickness has been denoted "t," so the total laminate thickness in this case is $t = 4h_o$. The width of the laminate is labeled $2b$. The following properties were assumed during the analysis:

$$E_{11} = 138 \text{ GPa (20 Msi)}$$

$$E_{22} = E_{33} = 14.5 \text{ GPa (2.1 Msi)}$$

$$G_{12} = G_{23} = G_{13} = 5.86 \text{ GPa (0.85 Msi)}$$

$$\nu_{12} = \nu_{23} = \nu_{31} = 0.21$$

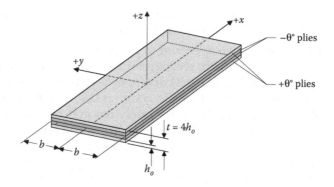

FIGURE 6.31
Description of the $[\theta/-\theta]_s$ laminate studied by Pipes and Pagano. (Adapted from Pipes, R.B. and Pagano, N.J., *Journal of Composite Materials*, 4, 538–548, 1970.)

Using these properties the normalized stresses predicted by a standard CLT analysis for the 45° plies in a $[45/-45]_s$ laminate subjected to axial stress are as follows[*]:

$$\frac{\sigma_{xx}}{\varepsilon_{xx}} = 20.4\,\text{GPa}\ (2.96\,\text{Msi})$$

$$\sigma_{yy} = 0$$

$$\frac{\tau_{xy}}{\varepsilon_{xx}} = 7.93\,\text{GPa}\ (1.15\,\text{Msi})$$

Since CLT is based on the assumption of plane stress, for a CLT calculation the out-of-plane stress components are zero by assumption: $\sigma_{zz} = \tau_{xz} = \tau_{yz} = 0$.

As already pointed out, free edge stresses occur when a mismatch in material properties of adjacent plies exist. In this case, the only mismatch that occurs is in the coefficient of mutual influence. That is, it can be shown (using program CLT, for example) that $\overline{\eta}_{xx,xy}^{ex}$ exhibited by the $+\theta$ plies is equal in magnitude but opposite in sign to that of the $-\theta$ plies. The effective Young's modulus and Poisson ratios are identical:

$$\overline{\eta}_{xx,xy}^{ex}\Big|_{+\theta^\circ} = (-1) * \overline{\eta}_{xx,xy}^{ex}\Big|_{-\theta^\circ}$$

$$\overline{E}_{xx}^{ex}\Big|_{+\theta^\circ} = \overline{E}_{xx}^{ex}\Big|_{-\theta^\circ}$$

$$\overline{E}_{yy}^{ex}\Big|_{+\theta^\circ} = \overline{E}_{yy}^{ex}\Big|_{-\theta^\circ}$$

$$\overline{\nu}_{xy}^{ex}\Big|_{+\theta^\circ} = \overline{\nu}_{xy}^{ex}\Big|_{-\theta^\circ}$$

---

[*] These values can be easily confirmed using program *CLT*.

Hence, the free edge stresses calculated during the Pipes and Pagano study are due solely to the mismatch in $\overline{\eta}^{ex}_{xx,xy}$ of adjacent plies.

Interlaminar stresses for a $[45/-45]_s$ laminate with $b/h_o = 8$ (i.e., for a laminate whose width is four times as great as it's thickness), at the through-thickness position $z = h_o^+$ (i.e., at a through-thickness position just within the $+45°$ply), are presented in Figures 6.32 and 6.33. The six individual stress components are plotted in these figures against normalized position y/b within the laminate. Referring to Figure 6.31, the position $y/b = 0$ corresponds to the centerline of the laminate, while point $y/b = 1.0$ corresponds to the free edge. A brief summary of each stress component is as follows:

- *Normal stress* σ_{xx} (Figure 6.32): At points nearthe laminate centerline the normalized stress predicted by the finite-difference analysis is $\sigma_{xx}/\varepsilon_{xx} = 2.96$, in agreement with the CLT analysis. As the free edge is approached the axial stress is predicted to increase very slightly, reaching a maximum normalized value of about 3.05 Msi at $y/b \approx 0.77$. It then decreases to a finite value of around 2.3 Msi at the free edge.

- *Shear stress* τ_{xy} (Figure 6.32): At points near the laminate centerline the normalized in-plane shear stress predicted by the finite-difference analysis is $\tau_{xy}/\varepsilon_{xx} = 1.15$, in agreement with the CLT analysis. As the free edge is approached the predicted normalized in-plane shear stress $\tau_{xy}/\varepsilon_{xx}$ increases very slightly from the value predicted by CLT, reaching

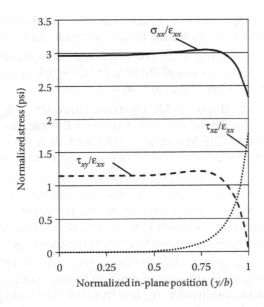

FIGURE 6.32

Normalized stresses $\sigma_{xx}/\varepsilon_{xx}$, $\tau_{xy}/\varepsilon_{xx}$, and $\tau_{xz}/\varepsilon_{xx}$, induced in a $[45/-45]_s$ laminate by a uniaxial tensile load.

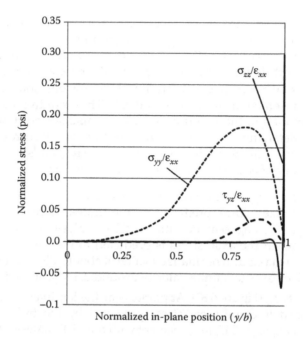

FIGURE 6.33
Normalized stresses $\sigma_{yy}/\varepsilon_{xx}$, $\sigma_{zz}/\varepsilon_{xx}$ and $\tau_{yz}/\varepsilon_{xx}$, induced in a $[45/-45]_s$ laminate by a uniaxial tensile load.

a maximum value of around 1.2 Msi at $y/b \approx 0.77$. It then decreases to zero at the free edge, as it must to satisfy equilibrium at that point.

- *Shear stress* τ_{xz} (Figure 6.32): The predicted normalized out-of-plane shear stress $\tau_{xz}/\varepsilon_{zz}$ is essentially zero for $y/b < 0.5$, in agreement with CLT, and then increases rapidly as $y/b \rightarrow 1.0$.

- *Normal stress* σ_{yy} (Figure 6.33): The normalized stress $\sigma_{yy}/\varepsilon_{xx}$ is essentially zero $y/b < 0.5$, in agreement with CLT. It reaches a maximum value of around 0.18 Msi at $y/b \approx 0.85$, and then decreases back to zero at the free edge, as it must to satisfy equilibrium at that point.

- *Normal stress* σ_{zz} (Figure 6.33): Normalized $\sigma_{zz}/\varepsilon_{xx}$ is restricted to an exceptionally narrow region very near the free edge; $\sigma_{zz}/\varepsilon_{xx}$ is essentially zero for $y/b < 0.90$. However, $\sigma_{zz}/\varepsilon_{xx}$ increases very rapidly as $y/b \rightarrow 1$. As will be discussed in Chapter 7, σ_{zz} is often responsible for the initiation of delamination failures at the free edge.

- *Shear stress* τ_{yz} (Figure 6.33): The predicted normalized out-of-plane shear stress $\tau_{yz}/\varepsilon_{zz}$ is almost non-existent throughout the entire width of the laminate. Values of $\tau_{yz}/\varepsilon_{zz}$ less than 0.05 Msi were predicted for $y/b > 0.75$, and then decrease to zero at the free edge, as it must to satisfy equilibrium at that point.

As a rule of thumb, the width of the region over which appreciable free edge stresses develop is equal to the total thickness of the laminate, t.

It is again emphasized that these are not general results. Rather, these stress distributions are predicted for the specific case of a $[45/-45]_s$ graphite–epoxy laminate subjected to an axial tensile load and based on the ply properties listed above. Different stress distributions would be predicted if a different stacking sequence were involved, if the laminate were subject to different loading conditions, and/or if different material properties were used.

HOMEWORK PROBLEMS

Notes:

a. In the following problems, the phrase "by hand calculation" means that solutions are to be obtained using a calculator, pencil and paper.

b. The computer programs *UNIDIR* and *CLT* are referenced in some of the following problems. As described in Section 6.11, these programs can be downloaded from the following website:

http://depts.washington.edu/amtas/computer.html

6.1. Three-element strain gage rosettes are mounted on opposite sides of a $[0/\pm30]_s$ graphite–epoxy laminate, as shown in Figure 6.34. An individual ply has a thickness of 0.005 in. The laminate is then subjected to an unknown system of forces. The strains measured by strain gage rosette 1 are

$$\varepsilon_{xx} = 2000 \ \mu\text{in./in.} \qquad \varepsilon_{yy} = 500 \ \mu\text{in./in.} \qquad \gamma_{xy} = -1000 \ \mu\text{rad}$$

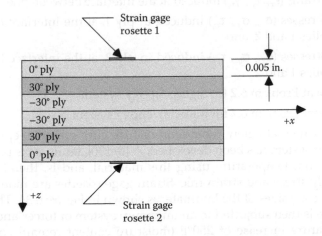

FIGURE 6.34
Edge view of a $[0°/\pm30°]_s$ composite laminate.

Similarly, the strains measured by strain gage rosette 2 are

$$\varepsilon_{xx} = 3000 \text{ } \mu\text{in./in.} \quad \varepsilon_{yy} = -2000 \text{ } \mu\text{in./in.} \quad \gamma_{xy} = 1000 \text{ } \mu\text{rad}$$

Determine by hand calculation:

a. Midplane strains and curvatures induced in the laminate, and

b. Strains (ε_{xx}, ε_{yy}, γ_{xy}) induced at the interface between plies 1 and 2.

6.2. Imagine that a new room-temperature cure graphite–epoxy pre-preg system has been developed. A $[0/\pm30]_s$ laminate is produced at room temperature using this material, and is, therefore, initially stress-and strain free. Strain gage rosettes are mounted on opposite sides of the laminate, as shown in Figure 6.34. The laminate is then subjected to an unknown system of forces, while temperature and moisture content remain constant. The strains measured by rosette 1 are

$$\varepsilon_{xx} = 500 \text{ } \mu\text{in./in.} \quad \varepsilon_{yy} = 1000 \text{ } \mu\text{in./in.} \quad \gamma_{xy} = 750 \text{ } \mu\text{rad}$$

Similarly, the strains measured by rosette 2 are

$$\varepsilon_{xx} = -1000 \text{ } \mu\text{in./in.} \quad \varepsilon_{yy} = -2000 \text{ } \mu\text{in./in.} \quad \gamma_{xy} = 750 \text{ } \mu\text{rad}$$

Obtain numerical values for all elements of the $[\bar{Q}]$ matrices for plies 1 and 2 using program *UNIDIR* and properties listed in Table 3.1. Assume an individual ply has a thickness of 0.005 in. Then determine the following by hand calculation:

a. Midplane strains and curvatures induced in the laminate,

b. Strains (ε_{xx}, ε_{yy}, γ_{xy}) induced at the interface between plies 1 and 2,

c. Stresses (σ_{xx}, σ_{yy}, τ_{xy}) induced in ply 1, at the interface between plies 1 and 2, and

d. Stresses (σ_{xx}, σ_{yy}, τ_{xy}) induced in ply 2, at the interface between plies 1 and 2.

6.3. Repeat Problem 6.2 for a glass/epoxy system.

6.4. Repeat Problem 6.2 for a Kevlar/epoxy system.

6.5. Imagine that a new room-temperature cure graphite–epoxy pre-preg system has been developed. A $[0/\pm30]_s$ laminate is produced at room temperature using this material, and is, therefore, initially stress-and strain free. Strain gage rosettes are mounted on opposite sides of the laminate, as shown in Figure 6.34. The laminate is then subjected to an unknown system of forces and a temperature increase of 250°F (moisture content remain constant). The strains measured by rosette 1 are

$$\varepsilon_{xx} = 500 \text{ } \mu\text{in./in.} \quad \varepsilon_{yy} = 1000 \text{ } \mu\text{in./in.} \quad \gamma_{xy} = 750 \text{ } \mu\text{rad}$$

Similarly, the strains measured by rosette 2 are

$$\varepsilon_{xx} = -1000 \ \mu in./in. \quad \varepsilon_{yy} = -2000 \ \mu in./in. \quad \gamma_{xy} = 750 \ \mu rad$$

Obtain numerical values for all elements of the $[\overline{Q}]$ matrices for plies 1 and 2 using program *UNIDIR* and properties listed in Table 3.1. Assume an individual ply has a thickness of 0.005 in. Then determine the following by hand calculation:

a. Midplane strains and curvatures induced in the laminate,

b. Strains ($\varepsilon_{xx}, \varepsilon_{yy}, \gamma_{xy}$) induced at the interface between plies 1 and 2,

c. Stresses ($\sigma_{xx}, \sigma_{yy}, \tau_{xy}$) induced in ply 1, at the interface between plies 1 and 2, and

d. Stresses ($\sigma_{xx}, \sigma_{yy}, \tau_{xy}$) induced in ply 2, at the interface between plies 1 and 2.

6.6. Repeat Problem 6.5 for a glass/epoxy system.

6.7. Repeat Problem 6.5 for a Kevlar/epoxy system.

6.8. An engineer is designing a structure that involves a $[0/\mp 10/90]_s$ graphite/epoxy composite laminate that will be cured at 350°F. During service the structure must support a load of 1000 lb$_f$, and will experience a temperature of 150°F in a dry environment. Based on the cure temperature and expected service conditions the engineer predicts that the laminate will experience the following midplane strains and curvatures:

$$\varepsilon_{xx}^o = 1500 \ \mu in./in.$$

$$\varepsilon_{yy}^o = 2200 \ \mu in./in.$$

$$\gamma_{xy}^o = -1000 \ \mu rad$$

$$\kappa_{xx} = \kappa_{yy} = \kappa_{xy} = 0$$

Obtain numerical values for all elements of the $[\overline{Q}]$ matrix for ply 3 using program *UNIDIR* and properties listed in Table 3.1. Assume an individual ply has a thickness of 0.005 in. Then determine the stresses induced in ply 3 relative to the 1–2 coordinate system by hand calculation.

6.9. Repeat Problem 6.8 for a glass/epoxy structure.

6.10. Repeat Problem 6.8 for a Kevlar/epoxy structure.

6.11. A $[0/\pm 30/90]_s$ graphite/epoxy laminate is cured at 350°F and then cooled to room temperature (70°F). Use hand calculation to determine the following thermal resultants induced during cooling (use program *UNIDIR* and properties listed in Table 3.1 to determine elements of the $[\overline{Q}]$ matrix as necessary):

a. N_{xx}^T

b. N_{yy}^T

c. N_{xy}^T

d. M_{xx}^T

e. M_{yy}^T

f. M_{xy}^T

6.12. Repeat Problem 6.11 for a glass/epoxy structure.

6.13. Repeat Problem 6.11 for a Kevlar/epoxy structure.

6.14. A $[0/\pm 30/90]_s$ graphite/epoxy laminate is cured at 175°C and then cooled to room temperatures (20°C). Review Table 6.13, then use hand calculation to determine the following:

   a. Midplane strains and curvatures induced during cooldown

   b. Strains (ε_{xx}, ε_{yy}, γ_{xy}) induced in ply 2 during cooldown

   c. Stresses (σ_{xx}, σ_{yy}, τ_{xy}) induced in ply 2 during cooldown

   d. Stresses (σ_{11}, σ_{22}, τ_{12}) induced in ply 2 during cooldown

6.15. A $[0/\pm 30/90]_s$ graphite/epoxy laminate is cured at 175°C and then cooled to room temperatures (20°C). Although the moisture content immediately after cure was 0%, the laminate is stored in a humid environment and over the course of several months the moisture content is increased to 0.5%. Review Table 6.13, then use hand calculation to determine the following:

   a. Midplane strains and curvatures induced after moisture content increased to 0.5%

   b. Strains (ε_{xx}, ε_{yy}, γ_{xy}) induced in ply 2 after moisture content increased to 0.5%

   c. Stresses (σ_{xx}, σ_{yy}, τ_{xy}) induced in ply 2 after moisture content increased to 0.5%

   d. Stresses (σ_{11}, σ_{22}, τ_{12}) induced in ply 2 after moisture content increased to 0.5%

6.16. A $[0/\pm30/90]_s$ graphite/epoxy laminate is subjected to the following loads:

$$N_{xx} = N_{yy} = N_{xy} = 10,000 \text{ N/m} \quad M_{xx} = M_{yy} = M_{xy} = 0$$

Review Table 6.13, then use hand calculation to determine the following:

   a. Midplane strains and curvatures

   b. Strains (ε_{xx}, ε_{yy}, γ_{xy}) induced in ply 2

TABLE 6.13

Some Properties of the $[0/\pm30/90]_s$ Graphite/Epoxy Laminate Considered in Problems 6.14 through 6.18

- The $[\overline{Q}]$ matrix for ply 2 is

$$\begin{bmatrix} 10.76E10 & 2.606E10 & 4.813E10 \\ 2.606E10 & 2.722E10 & 2.152E10 \\ 4.813E10 & 2.152E10 & 3.605E10 \end{bmatrix}$$

- The $[ABD]$ matrix is

$$\begin{bmatrix} 9.906E7 & 1.454E7 & 0 & 0 & 0 & 0 \\ 1.454E7 & 5.885E7 & 0 & 0 & 0 & 0 \\ 0 & 0 & 2.452E7 & 0 & 0 & 0 \\ 0 & 0 & 0 & 11.89 & 1.032 & 0.7521 \\ 0 & 0 & 0 & 1.032 & 1.628 & 0.3362 \\ 0 & 0 & 0 & 0.7521 & 0.3362 & 1.864 \end{bmatrix}$$

- The $[abd]$ matrix is

$$\begin{bmatrix} 1.048E-8 & -0.2588E-8 & 0 & 0 & 0 & 0 \\ -0.2588E-8 & 1.763E-8 & 0 & 0 & 0 & 0 \\ 0 & 0 & 4.078E-8 & 0 & 0 & 0 \\ 0 & 0 & 0 & 9.029E-2 & -5.160E-2 & -2.713E-2 \\ 0 & 0 & 0 & -5.160E-2 & 0.6674 & -9.958E-2 \\ 0 & 0 & 0 & -2.713E-2 & -9.958E-2 & 0.5655 \end{bmatrix}$$

- If the laminate were cured at 175°C and then cooled to room temperatures (20°C), the thermal resultants induced during cooldown would be:

$$N_{xx}^T = -8607 \text{ N/m} \quad N_{yy}^T = -21825 \text{ N/m} \quad N_{xy}^T = M_{xx}^T = M_{yy}^T = M_{xy}^T = 0$$

- If the moisture content of the laminate immediately after cure is 0%, but over the course of several months is increased to 0.5%, then the moisture resultants induced by this slow moisture adsorption is

$$N_{xx}^M = 6092 \text{ N/m} \quad N_{yy}^M = 6098 \text{ N/m} \quad N_{xy}^M = M_{xx}^M = M_{yy}^M = M_{xy}^M = 0$$

---

    c. Stresses $(\sigma_{xx}, \sigma_{yy}, \tau_{xy})$ induced in ply 2

    d. Stresses $(\sigma_{11}, \sigma_{22}, \tau_{12})$ induced in ply 2

6.17. A $[0/\pm 30/90]_s$ graphite/epoxy laminate is cured at 175°C and then cooled to room temperatures (20°C). The moisture content immediately after cure was 0%. However, the laminate is stored in a humid environment and over the course of several months the moisture content is increased to 0.5%. The laminate is then subjected to the following loads:

$$N_{xx} = N_{yy} = N_{xy} = 10,000 \text{ N/m} \qquad M_{xx} = M_{yy} = M_{xy} = 0$$

Review Table 6.13, then use hand calculation to determine the following:

a. Midplane strains and curvatures
b. Strains (ε_{xx}, ε_{yy}, γ_{xy}) induced in ply 2
c. Stresses (σ_{xx}, σ_{yy}, τ_{xy}) induced in ply 2
d. Stresses (σ_{11}, σ_{22}, τ_{12}) induced in ply 2

6.18. A 10 cm × 10 cm square $[0/\pm30/90]_s$ graphite/epoxy laminate is supported between three infinitely rigid walls and frictionless rollers, as shown in Figure 6.35. A load $N_{xx} = -7500$ N/m is applied to the plate. Review Table 6.13, then (ignore the possibility of buckling) use hand calculation to determine N_{xx}, N_{yy}, N_{xy}, ε_{xx}^o, ε_{yy}^o, and γ_{xy}^o.

6.19. A $[20/65/-25]_s$ graphite/epoxy laminate is cured at 175°C and then cooled to room temperatures (20°C). Moisture content remains at 0%. The following loads are then applied:

$$N_{xx} = 30 \text{ kN/m} \qquad N_{yy} = -7 \text{ kN/m} \quad N_{xy} = 0$$
$$M_{xx} = 10 \text{ N-m/m} \quad M_{yy} = M_{xy} = 0$$

Using properties listed in Table 3.1 and assuming ply thicknesses of 0.125 mm:

a. Determine all ply strains and stresses using program CLT
b. Prepare plots similar to Figure 6.13, showing the through-thickness variation of strains ε_{xx}, ε_{yy}, and γ_{xy}
c. Prepare plots similar to Figure 6.14, showing the through-thickness variation of strains ε_{11}, ε_{22}, and γ_{12}
d. Prepare plots similar to Figure 6.15, showing the through-thickness variation of stresses σ_{xx}, σ_{yy}, and τ_{xy}
e. Prepare plots similar to Figure 6.16, showing the through-thickness variation of stresses σ_{11}, σ_{22}, and τ_{12}

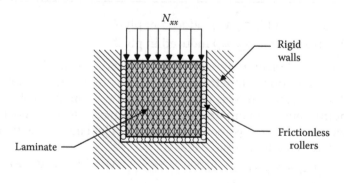

FIGURE 6.35
A $[0/\pm30/90]_s$ graphite/epoxy laminate supported between rigid walls and frictionless rollers.

6.20. Repeat Problem 6.19 for a glass/epoxy laminate.

6.21. Repeat Problem 6.19 for a Kevlar/epoxy laminate.

6.22. A $[20/65/-25]_s$ "hybrid" laminate is cured at 175°C and then cooled to room temperatures (20°C). Plies 1 and 6 are graphite/epoxy, plies 2 and 5 are glass/epoxy, and plies 3 and 4 are Kevlar/epoxy. The following loads are then applied:

$$N_{xx} = 30 \text{ kN/m} \quad N_{yy} = -7 \text{ kN/m} \quad N_{xy} = 0$$

$$M_{xx} = 10 \text{ N-m/m} \quad M_{yy} = M_{xy} = 0$$

Moisture content remains constant at 0%. Using properties listed in Table 3.1 and assuming the graphite/epoxy, glass/epoxy, and Kevlar/epoxy plies have thicknesses of 0.125 mm, 0.200 mm, and 0.15 mm, respectively:

a. Determine all ply strains and stresses using program *CLT*

b. Prepare plots similar to Figure 6.13, showing the through-thickness variation of strains ε_{xx}, ε_{yy}, and γ_{xy}

c. Prepare plots similar to Figure 6.14, showing the through-thickness variation of strains ε_{11}, ε_{22}, and γ_{12}

d. Prepare plots similar to Figure 6.15, showing the through-thickness variation of stresses σ_{xx}, σ_{yy}, and τ_{xy}

e. Prepare plots similar to Figure 6.16, showing the through-thickness variation of stresses σ_{11}, σ_{22}, and τ_{12}

6.23. Referring to Figure 6.21, the ANSYS analysis described in Example Problem 6.9, and the values for the in-plane strains and stresses at points B and C listed in Table 6.11, use program *UNIDIR* to confirm that the stresses and strains predicted at points B and C for each ply are consistent.

6.24. Referring to Figure 6.21, the ANSYS analysis described in Example Problem 6.10, and the values for the effective stresses at points A and C listed in Table 6.12, use program *CLT* to confirm that the strains predicted at points A and C are consistent with Equation 6.38.

6.25. Referring to Figure 6.21, the ANSYS analysis described in Example Problem 6.9, and the values for the in-plane strains and stresses at points A, B, and C listed in Table 6.11, determine the stresses relative to the 1–2 coordinate system induced in each ply.

a. at point A

b. at point B

c. at point C

References

1. Jones, R.M., *Mechanics of Composite Materials*, Hemisphere Publishing Corporation, New York, ISBN 0-89116-490-1, 1975.
2. Tsai, S.W. and Hahn, H.T., *Introduction to Composite Materials*, Technomic Publishing Co., Lancaster, PA ISBN0-87762-288-4, 1980.
3. Halpin, J.C., *Primer on Composite Materials: Analysis*, Technomic Publishing Co., Lancaster, PA, ISBN 87762-349-X, 1984.
4. Reddy, J.N., *Mechanics of Laminated Composite Plates and Shells*, 2nd edition, CRC Press, Boca Raton, FL, ISBN 0-8493-1592-1, 2004.
5. Ochoa, O.O. and Reddy, J.N., *Finite Element Analysis of Composite Laminates*, Kluwer Academic Publishers, Boston, MA, ISBN 0-79231-125-6, 1992.
6. Tenek, L.T. and Argyris, J., *Finite Element Analysis for Composite Structures*, Kluwer Academic Publishers, Boston, MA, ISBN 0-7923-4899-0, 1998.
7. Matthews, F.L., Davies, G.A.O., Hitchings, D., and Soutis, C., *Finite Element Modeling of Composite Materials and Structures*, CRC Press, Boca Raton, FL, ISBN 1-85573-422-2, 2000.
8. Barbero, E.J., *Finite Element Analysis of Composite Materials*, CRC Press Taylor & Francis Group, Boca Raton, FL, ISBN 978-1-4200-5433-0, 2008.
9. Pipes, R.B. and Daniel, I.M., Moiré analysis of the interlaminar shear edge effect in laminated composites, *Journal of Composite Materials*, 5, 255–259, 1971.
10. Oplinger, D.W., Parker, B.S., and Chiang, F-P, Edge-effect studies in fiber-reinforced laminates, *Experimental Mechanics*, 14(9), 347–354, 1974.
11. Czarnek, R., Post, D., and Herakovich, C.T., Edge effects in composites by moiré interferometry, *Experimental Techniques*, 7(1), 18–21, 1983.
12. Herakovich, C.T., Post, D., Buczek, M.B., and Czarnek, R., Free edge strain concentrations in real composite laminates: Experimental-theoretical correlation, *Journal of Applied Mechanics*, 52, 787–793, 1985.
13. Pipes, R.B. and Pagano, N.J., Interlaminar stresses in composite laminates under uniform axial extension, *Journal of Composite Materials*, 4, 538–548, 1970.
14. Rybicki, E.F., Approximate three-dimensional solutions for symmetric laminates under in-plane loading, *Journal of Composite Materials*, 5, 354–360, 1971.
15. Wang, A.S.D. and Crossman, F.W., Some new results on edge effects in symmetric composite laminates, *Journal of Composite Materials*, 11, 92–106, 1977.
16. Hsu, P.W. and Herakovich, C.T., Edge effects in angle-ply composite laminates, *Journal of Composite Materials*, 11, 422–428, 1977.
17. Herakovich, C.T., On the relationship between engineering properties and delamination of composite materials, *Journal of Composite Materials*, 15, 336–348, 1981.
18. Sancho, J. and Miravete, A., Design of composite structures including delamination studies, *Composite Structures*, 76, 283–290, 2006.
19. Puppo, A.H. and Evensen, H.A., Interlaminar shear in laminated composites under generalized plane stress, *Journal of Composite Materials*, 4, 204–220, 1970.
20. Pagano, N.J. and Pipes, R.B., The influence of stacking sequence on laminate strength, *Journal of Composite Materials*, January, 50–57, 1971.
21. Pagano, N.J., On the calculation of interlaminar normal stress in composite laminate, *Journal of Composite Materials*, January, 65–81, 1974.

22. Pagano, N.J., Free-edge stress fields in composite laminates, *Int Journal of Solids and Structures*, 14, 401–406, 1978.

23. Wang, S.S. and Choi, I., Boundary-layer effects in composite laminates: Part 1—Free-edge stress singularities, *Journal of Applied Mechanics*, 49, 541–548, 1982.

24. Wang, S.S. and Choi, I., Boundary-layer effects in composite laminates: Part 2—Free-edge stress solutions and basic characteristics, *Journal of Applied Mechanics*, 49, 549–560, 1982.

25. Zwiers, R.I., Ting, T.C.T., and Spilker, R.L., On the logarithmic singularity of free-edge stress in laminated composites under uniform extension, *Journal of Applied Mechanics*, 49, 561–569, 1982.

26. Ting, T.C.T., An excellent review and comparison of the Lekhnitskii and Stroh formalisms,in: *Anisotropic Elasticity Theory and Applications*, Oxford University Press, New York, ISBN 0-19-507447-5, 1996.

27. Kassapoglou, C. and Lagace, P.A., An efficient method for the calculation of interlaminar stresses in composite materials, *Journal of Applied Mechanics*, 53, 744–750, 1986.

28. Kassapoglou, C. and Lagace, P.A., Closed form solutions for the interlaminar stress field in angle-ply and cross-ply laminates, *Journal of Composite Materials*, 21, 292–308, 1987.

29. Becker, W. Closed-form solutions for the free-edge effect in cross-ply laminates, *Composite Structures*, 26, (1–2) 39–45, 1993.

30. Mittelstedt, C. and Becker, W., Free-edge effects in composite laminates, *Applied Mechanics Reviews*, 60, (5) 217–245, 2007.

7

Predicting Failure of a Multiangle Composite Laminate

7.1 Preliminary Discussion

Ideally the objective of this chapter would be to describe analytical tools that, when combined with appropriate experimental measurements, could be used to accurately predict failure of multiangle composite laminates under general thermomechanical loading conditions. Unfortunately, we will not be able to reach this objective. Although many theories have been proposed and progress has been made, predicting failure of multiangle composite laminates with a high degree of accuracy is still beyond the state of the art. In fact, an international survey has shown that predictions obtained using leading failure theories may differ by a factor of 2 or more, even when predicting failure of simple flat or cylindrical test specimens subjected to uniform loading and uniforms temperatures and moisture content [1,2].

The difficulty in predicting composite failure begins with a rather surprising fact: there is no widespread agreement on what constitutes "failure" of a composite laminate. That is, a composite that is considered to have failed by one engineer may not be considered to have failed by another. The difficulty in defining what composite "failure" means can be illustrated by considering the response of a quasi-isotropic $[0/45/90/-45]_s$ laminate to simple tensile loading. An idealized (but more-or-less representative) response is shown in Figure 7.1. It is assumed that the laminate is defect-free before loading, which implies that any preexisting thermal and/or moisture stresses are not high enough to have caused any ply failures or other defects. A uniaxial tensile load N_{xx} (or equivalently, a tensile effective stress $\bar{\sigma}_{xx} = N_{xx}/t$) is applied and steadily increased until final catastrophic fracture of the laminate occurs. As the laminate is symmetric, the tensile load does not cause bending and, ignoring free-edge stresses for the moment, it is assumed that the axial strain ε_{xx} is constant across the width and thickness of the specimen.

The initial slope of the $\bar{\sigma}_{xx}$ versus ε_{xx} curve represents the effective Young's modulus of the laminate $\bar{E}_{xx}$ and can be well predicted using CLT as described in Section 6.9. As $\bar{\sigma}_{xx}$ is increased the stresses and strains in individual plies are increased as well. Eventually stresses/strains reach a critical value and one or more plies yield and crack. For a $[0/45/90/-45]_s$ laminate subjected to tensile loading the first plies to yield or crack are usually the 90° plies. The difference between the load at which yielding is initiated and the load at which cracks begin to form within the 90° plies depends on the level of ductility exhibited by the matrix material. For most commercially important structural composites (e.g., graphite–epoxy), the difference is insignificant, so the effective stress level necessary to cause yielding or fracture of the 90° plies is essentially identical. The effective stress level that causes the 90° plies to fail is termed the *first-ply failure stress*. Multiple cracks form in the 90° plies at effective stress levels at and above the first-ply failure stress, as depicted in Figure 7.1. The cracks form perpendicular to the loading direction, lie within the y–z plane, and extend through the thickness of the 90° ply only (i.e., the cracks do not extend through the entire thickness of the laminate). Depending on the length of the specimen, tens or hundreds of cracks may form in the 90° plies. Note that if the laminate were a unidirectional $[90]_n$ laminate, then final fracture would occur at this point. As the laminate is instead a $[0/45/90/-45]_s$ laminate, and as no cracks have yet occurred within the ±45° or 0° plies, the laminate as a whole can support substantially higher load/stress levels.

Once the 90° plies have cracked/fractured, they can no longer contribute fully to the effective stiffness of the laminate. Hence, it would be expected that the slope of the $\overline{\sigma}_{xx}$ versus ε_{xx} curve (i.e., the effective Young's modulus, $\overline{E}_{xx}$) would decrease as cracks form in the 90° plies. Although a decrease

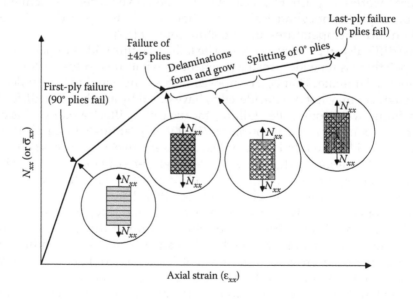

FIGURE 7.1
Idealized stress–strain plot for a $[0/45/90/-45]_s$ laminate, showing the evolution of internal damage.

does in fact occur, the decrease is often barely discernible. A pronounced decrease in slope has been shown in Figure 7.1 for illustrative purposes, but in practice the change in slope is usually far less than that implied by the figure. Failure of the 90° plies has little impact on the effective Young's modulus of the laminate because the stiffness of these plies in the x-direction, relative to the stiffness of the ±45° and 0° plies, is very low even before failure occurs.

As the effective stress level is further increased, cracks eventually begin to form within the ±45° plies. As depicted in Figure 7.1, extensive matrix cracking occurs within both the 90° and ±45° plies, before any significant defects occur within the 0° plies. Hence, the laminate as a whole can support still higher effective stress levels. An additional decrease in slope (i.e., a further decrease in apparent Young's modulus) occurs as cracks develop in the ±45° plies.

Thus far the cracks that have formed all lie within planes perpendicular to the $x–y$ plane. Note that, unlike isotropic materials, the orientation of the cracks is not governed by principal stresses or strains. Instead, the orientation of cracks is entirely dictated by the local fiber orientation.

Delamination failures begin to develop as the effective stress is increased further still. These new matrix cracks lie within planes that are *parallel* to the $x–y$ plane. That is, matrix cracks begin to form *between* plies. The initiation of delaminations near the edge of the specimen is due to the three-dimensional stresses induced near free edges, which were discussed in Section 6.13. However, as implied in Figure 7.1 delaminations have also been observed at interior regions, away from the edges of the specimen. It is speculated that at interior positions delaminations begin to form near the intersection of cracks in neighboring plies, for example near the intersection of cracks in neighboring 90° and +45° plies. As the applied stress is increased the delaminated regions grow in size and eventually coalesce, such that a delaminated region may extend across the entire width of the test specimen. Nevertheless, the 0° plies are still intact, so the specimen can support still higher loads.

As load is further increased matrix cracks finally begin to form within the 0° plies. These are the first defects within the 0° plies. The matrix cracks that form within the 0° plies are often referred to as "splitting" cracks, although these cracks lie within a plane perpendicular to the $x–y$ plane and are fundamentally no different that the matrix cracks that formed previously within the 90° and ±45° plies at lower load levels. Fiber–matrix debonding may also occur within the 0° plies. In this case, a crack forms around the periphery of a fiber, separating the fiber and matrix, so that load can no longer be transferred from the matrix to the fiber.

Finally, catastrophic *laminate* fracture is precipitated by failure of the fibers within the 0° plies. Note that these are the first extensive *fiber* failures that have occurred within the specimen. Although hundreds of matrix cracks may have formed within and between plies, very few (if any) fiber failures occur before final fracture of the 0° plies.

The effective stress level at which final laminate fracture occurs is called the *last-ply failure stress*. At final fracture, the composite laminate fractures into tens or hundreds of fragments, due to the extensive and preexisting matrix cracks and delaminations that previously occurred at lower stress levels as well as the large amount of energy release associated with fiber failures.

Four types of defects have been mentioned thus far in connection with the evolution of damage shown in Figure 7.1: matrix cracking, ply delaminations, fiber–matrix debonding, and fiber fractures. A fifth defect commonly encountered is fiber microbuckling. This may occur when a laminate is subjected to a loading that causes compressive stresses in one or more plies. The fibers within these plies may buckle, reducing the compressive stiffness exhibited by the ply or plies and ultimately leading to failure of the fibers due to the bending stresses induced. Fiber microbuckling is not normally observed during tensile tests of a $[0/45/90/-45]_s$ laminate.

Although the evolution of internal defects depicted in Figure 7.1 is for the specific case of a $[0/45/90/-45]_s$ laminate subjected to uniaxial tensile loading, this sequence of events is more-or-less representative of all multiangle laminates. A fundamental characteristic is that substantial internal damage occurs at load levels far below that required to cause final catastrophic fracture, where the term "damage" refers to the combination of literally hundreds of individual failure events including matrix cracks, ply delaminations, fiber–matrix debonding, fiber microbuckling, and fiber failures. Individually, none of these failure events have a measureable impact on the overall stiffness or strength of the laminate. It is only when the combined effects of all of these failures reaches a critical level (i.e., when damage reaches a critical level) that final catastrophic fracture occurs.

The difficulty in defining what constitutes "failure" of a composite should now be evident. What level of damage should be used to define failure? Should failure be defined on the basis of the initial formation of defects at the first-ply failure load (an extremely conservative definition), on the basis of the last-ply failure load (a very nonconservative definition), or on some criterion between these two limiting extremes? There is no single answer to this question, and in practice the definition of "failure" varies from case-to-case. Some of the factors to consider when defining "failure" are

- The intended service life of the structure. A structure intended for years of service will generally be designed to a much more conservative failure stress level compared with a structure intended for a single use.

- The cyclic nature of the applied loading. A structure subjected to cyclic fatigue loading will generally be designed to a more conservative failure stress level compared with a structure that will experience static or slowly varying loads during service.

- The consequences of structural failure. A structure whose fail ure will result in the loss of life and/or extensive property damage will generally be designed to a more conservative failure stress level compared with a structure whose failure is of lesser consequence.

7.2 Estimating Laminate Failure Strengths Using CLT

Predicting failure of polymeric composites with a high degree of accuracy has proven to be a formidable challenge, and methods of predicting fracture under general thermomechanical loading conditions remain an active area of research. Nevertheless, the practicing engineer requires some method of estimating the load-carrying capacity of a composite laminate for purposes of preliminary design. Toward that end, CLT can be used to estimate when failure of a multiangle laminate will occur. In essence, CLT is used to predict the ply stresses and/or strains that will be induced by a thermomechanical loading of interest. Once ply stresses/strains are known, a macromechanics-based failure criterion, for example one of the three criteria discussed in Section 5.5, is used to predict failure of individual plies.

Predicting first-ply failure loads using CLT will be discussed in the following section. As noted above, equating composite failure strength to first-ply failure is usually an overly conservative estimate, as (in general) a multiangle laminate can withstand a substantial increase in load beyond first-ply failure load. Last-ply failure loads can also be predicted by combining CLT with the so-called ply discount scheme, albeit with a substantial decrease in accuracy. Predicting last-ply failure loads will be discussed in a concluding section.

7.2.1 Using CLT to Predict First-Ply Failure

First-ply failure occurs when the combined effects of loading, temperature changes, and/or moisture changes cause nonlinear behavior (i.e., either yielding or cracking) to develop in any ply. The first ply failure stress for the specific case of a [0/45/90/−45]$_s$ laminate subjected to a uniaxial tensile loading was discussed in Section 7.1 and illustrated schematically in Figure 7.1. This same concept can be extended to laminates with different stacking sequences and/or to more general thermomechanical loading conditions. Although several different solution paths may be followed, one approach is illustrated in Figure 7.2. The flow diagram shown in this figure is based on the assumption that the laminate of interest is subjected to a uniform change in temperature and/or moisture content as well as some

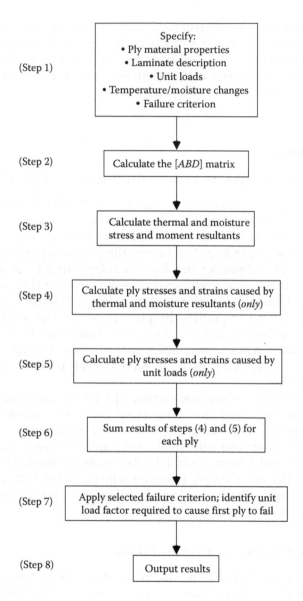

(Step 1) Specify:
• Ply material properties
• Laminate description
• Unit loads
• Temperature/moisture changes
• Failure criterion

(Step 2) Calculate the [ABD] matrix

(Step 3) Calculate thermal and moisture stress and moment resultants

(Step 4) Calculate ply stresses and strains caused by thermal and moisture resultants (*only*)

(Step 5) Calculate ply stresses and strains caused by unit loads (*only*)

(Step 6) Sum results of steps (4) and (5) for each ply

(Step 7) Apply selected failure criterion; identify unit load factor required to cause first ply to fail

(Step 8) Output results

FIGURE 7.2
A flow diagram illustrating calculation of the first-ply failure load using CLT.

combination of uniform external loads (N_{xx}, N_{yy}, N_{xy}, M_{xx}, M_{yy}, M_{xy}). It is further assumed that the relative magnitude of each load component is known. That is, if $N_{xx} = k_1 = 1$, then it is assumed that the magnitudes of the remaining loads are linearly related to $N_{yy} = k_2|N_{xx}|$, $N_{xy} = k_3|N_{xx}|$, $M_{xx} = k_4|N_{xx}|$, $M_{yy} = k_5|N_{xx}|$, $M_{xy} = k_6|N_{xx}|$. The relative magnitudes of each load component (i.e., constants k_i) are referred to as "unit loads" in Figure 7.2.

The calculation steps shown in Figure 7.2 can be summarized as follows:

Step 1: The first step is to define the problem. Ply properties, laminate stacking sequence, unit loads applied to the laminate, stress-free temperature (often assumed to be the cure temperature), service temperature, changes in moisture content, and the failure criterion to be employed must all be specified.

Step 2: The $[ABD]$ matrix for the laminate is determined. The reduced stiffness matrix, Q_{ij}, for each material used in the laminate is calculated (using Equations 5.11), and the transformed reduced stiffness matrix, $\overline{Q}_{ij}$, for each ply is determined (using Equations 5.31). This allows calculation of the A_{ij}, B_{ij}, and D_{ij} matrices, using Equations 6.27a, 6.27b, and 6.34, respectively, which are then assembled to form the $[ABD]$ matrix.

Step 3: Thermal and moisture stress and moment resultants corresponding to the specified ΔT and ΔM are calculated using Equations 5.25, 5.28, 6.41, and 6.42.

Step 4: Ply stresses and strains caused by the specified temperature and/or moisture changes (*only*) in the 1–2 coordinate system for each ply are calculated. In general, this involves calculation of the laminate midplane strains and curvatures using Equation 6.45, calculation of the ply strains and stresses in the $x–y$ coordinate system using Equations 6.12 and 5.30, respectively, and rotation of the ply strains and stresses to the 1–2 coordinate system using Equations 2.44 and 2.20, respectively.

Step 5: Ply stresses and strains caused by unit loads (*only*) in the 1–2 coordinate system for each ply are calculated. In general, this involves calculation of the laminate midplane strains and curvatures using Equation 6.45, calculation of the ply strains and stresses in the $x–y$ coordinate system using Equations 6.12 and 5.30, respectively, and rotation of the ply strains and stresses to the 1–2 coordinate system using Equations 2.44 and 2.20, respectively.

Step 6: The results of steps 4 and 5 are added together. This results in an expression for the stresses/strains in the 1–2 coordinate system caused by the combined effects of unit loads and temperature and/or moisture effects for each ply.

Step 7: The failure criterion selected for use is applied to each ply, and the unit load factor necessary to cause first-ply failure is determined.

Step 8: Results of the analysis are output.

The use of CLT to predict first-ply failure according to the flow diagram shown in Figure 7.2 is illustrated in Example Problems 7.1 and 7.2.

Example Problem 7.1

A [0/30/60]$_s$ graphite–epoxy specimen machined from a panel produced at a cure temperature of 175°C is tested at room temperature (20°C). A uniaxial tensile load N_{xx} is applied. Predict the first-ply failure load, assuming the laminate is stress-free at the cure temperature,[*] and that no change in moisture content occurs. Base the prediction on the Maximum Stress failure criterion, and use properties for graphite–epoxy listed in Table 3.1.

SOLUTION

Many of the calculations needed to solve this problem can be completed using program CLT. Referring to the flow diagram shown in Figure 7.2:

Step 1: It is completed by the problem statement itself:

- *Ply properties*: listed in Table 3.1
- *Laminate description*: [0/30/60]$_s$
- *Unit loads*: As $N_{xx} = k_1 = 1$, and it is assumed that $N_{yy} = k_2|N_{xx}|$, $N_{xy} = k_3|N_{xx}|$, $M_{xx} = k_4|N_{xx}|$, $M_{yy} = k_5|N_{xx}|$, and $M_{xy} = k_6|N_{xx}|$, then $k_1 = 1$, $k_2 = k_3 = k_4 = k_5 = k_6 = 0$
- *Temperature/moisture changes*: $\Delta T = 20°C - 175°C = -155°C$
$$\Delta M = 0\%$$
- *Failure criterion*: Maximum stress failure criterion

Step 2: Using ply properties for graphite–epoxy listed in Table 3.1 and program CLT, the [ABD] matrix for a [0/30/60]$_s$ laminate is

$$[ABD] = \begin{bmatrix} 76.44 \times 10^6 & 13.79 \times 10^6 & 17.41 \times 10^6 & 0 & 0 & 0 \\ 13.79 \times 10^6 & 36.23 \times 10^6 & 17.41 \times 10^6 & 0 & 0 & 0 \\ 17.41 \times 10^6 & 17.41 \times 10^6 & 21.27 \times 10^6 & 0 & 0 & 0 \\ 0 & 0 & 0 & 5.245 & 0.3461 & 0.4667 \\ 0 & 0 & 0 & 0.3461 & 0.6369 & 0.2588 \\ 0 & 0 & 0 & 0.4667 & 0.2588 & 0.6971 \end{bmatrix}$$

As an aside, the effective extensional stiffness of the laminate, based on the [ABD] matrix, is also returned by program *CLT* and equals $\overline{E}_{xx}^{ex} = 82.9\,\text{GPa}$.

Step 3: Using ply properties for graphite–epoxy listed in Table 3.1, the specified change in temperature ($\Delta T = -155°C$) and program CLT, the thermal stress and moment resultants are

[*] As discussed in Section 6.5, the stress-free temperature is likely to be lower than the cure temperature. For simplicity, it is assumed in this text that the final cure temperature defines the stress-free temperature.

$$\begin{Bmatrix} N_{xx}^T \\ N_{yy}^T \\ N_{xy}^T \\ M_{xx}^T \\ M_{yy}^T \\ M_{xy}^T \end{Bmatrix} = \begin{Bmatrix} -4803 \text{ N/m} \\ -18021 \text{ N/m} \\ 11447 \text{ N/m} \\ 0 \text{ N} - \text{m/m} \\ 0 \text{ N} - \text{m/m} \\ 0 \text{ N} - \text{m/m} \end{Bmatrix}$$

As there is no change in moisture content, moisture stress and moment resultants are zero for this problem.

Step 4: Program CLT returns the ply stresses caused by the thermal stress and moment resultants for each ply, relative to the 1–2 coordinate system. These are listed in Table 7.1.

Step 5: On the basis of the specified unit loads ($N_{xx} = k_1 = 1$, $k_2 = k_3 = k_4 = k_5 = k_6 = 0$), and insuring that temperature and moisture changes are zero, program CLT returns the ply stresses relative to the 1–2 coordinate system listed in Table 7.1.

Step 6: The sum of steps 4 and 5 gives the stresses induced in each ply by the specified temperature change and any value of N_{xx}. For example, stresses induced in the 0° plies (i.e., plies 1 and 6) are given by

$$\sigma_{11} = 2750 N_{xx} - 55.54 \times 10^6 \text{ (Pa)}$$

$$\sigma_{22} = 51.93 N_{xx} + 28.36 \times 10^6 \text{ (Pa)}$$

$$\tau_{12} = -174.8 N_{xx} + 22.83 \times 10^6 \text{ (Pa)}$$

The coefficients in these expressions are listed in Table 7.1. Analogous expressions for the stresses induced in the 30° and 60° plies can be constructed in the same way.

Step 7: The maximum stress failure criterion is to be applied in this problem. The criterion is applied to each ply in turn, using failure strengths listed in Table 3.1, thereby identifying the unit load factor necessary to cause failure of each ply.

TABLE 7.1

Numerical Results of the First-Ply Failure Analysis Described in Example Problem 7.1

Ply Fiber Angle (°)	Ply Stresses Caused by Temperature Change *Only*, 1–2 Coordinate System (MPa)			Ply Stresses Caused by Unit Loads *Only*, 1–2 Coordinate System (Pa)			Unit Load Scale Factor
	σ_{11}	σ_{22}	τ_{12}	σ_{11}	σ_{22}	τ_{12}	
0	−55.54	28.36	22.83	2750	51.93	−174.8	416.7×10^3
30	29.59	24.79	0	1112	120.6	−264.6	209.0×10^3
60	−55.54	28.36	−22.83	−209.8	176.0	−89.85	123.0×10^3

From Table 3.1, failure strengths are

$$\sigma_{11}^{fT} = 1500 \text{ MPa} \quad \sigma_{22}^{fT} = 50 \text{ MPa} \quad \tau_{12}^{f} = \pm 90 \text{ MPa}$$

The unit load factor necessary to cause failure of the 0° plies (i.e., plies 1 and 6) will be calculated to illustrate the process. Equating σ_{11} induced in the 0° plies by the temperature change and a unit load (determined in step 6), to the corresponding failure strength, we have

$$2750 N_{xx} - 55.54 \times 10^6 \text{ (Pa)} = 1500 \text{ MPa} \Rightarrow N_{xx}$$

$$= \frac{1500 \times 10^6 + 55.54 \times 10^6}{2750} = 565.7 \text{ kN/m}$$

Thus, in order for a fiber failure to occur in the 0° plies, the tensile unit load must be increased by a load factor of 565.7×10^3. Repeating this process for stresses σ_{22} and τ_{12} induced in the 0 plies, we find

$$51.93 N_{xx} + 28.36 \times 10^6 \text{(Pa)} = 50 \text{ MPa} \Rightarrow N_{xx} = \frac{50 \times 10^6 - 28.36 \times 10^6}{51.93}$$

$$= 416.7 \text{ kN/m}$$

$$-174.8 N_{xx} + 22.83 \times 10^6 \text{ (Pa)} = -90 \text{ Pa} \Rightarrow N_{xx} = \frac{-90 \times 10^6 - 22.83 \times 10^6}{-174.8}$$

$$= 645.5 \text{ kN/m}$$

As failure of the 0° plies would occur for the lowest value of N_{xx} that causes any stress component to reach a critical level, we conclude that the critical load factor for the 0° plies equals 416.7×10^3.

Repeating this process for the 30° and 60° plies results in critical unit load factors of 209.0×10^3 and 123.0×10^3, respectively. Hence, according to the maximum stress criterion, the first-ply failure load for the laminate is $N_{xx} = 123.0 \text{ kN/m}$, which corresponds to failure of the 60° plies (plies 3 and 4). Equivalently, the effective first-ply failure stress is

$$\bar{\sigma}_{xx} = \frac{N_{xx}}{t} = \frac{123.0 \text{ kN/m}}{6(0.125 \text{ mm})} = 164 \text{ MPa}$$

Example Problem 7.2

Repeat Example Problem 7.1, except use the Tsai–Wu criterion rather than the maximum stress criterion.

SOLUTION

We once again perform the analysis described by the flow diagram shown in Figure 7.2. The solution process is identical to that described in Example Problem 7.1, until we reach step 7. In this latter step we apply the Tsai–Wu failure criterion rather than the maximum stress criterion. For plane stress conditions, the Tsai–Wu criterion is given by Equation 5.62, repeated here for convenience:

$$X_1\sigma_{11}+X_2\sigma_{22}+X_{11}\sigma_{11}^2+X_{22}\sigma_{22}^2+X_{66}\tau_{12}^2+2X_{12}\sigma_{11}\sigma_{22}<1 \quad \text{(5.62) (repeated)}$$

Using Equations 5.56, 5.57, 5.59, and 5.61, and the properties listed for graphite–epoxy in Table 3.1, we find

$$X_1 = \frac{1}{\sigma_{11}^{fT}} - \frac{1}{\sigma_{11}^{fC}} = \frac{1}{1050\,\text{MPa}} - \frac{1}{690\,\text{MPa}} = \frac{-4.969e-10}{\text{Pa}}$$

$$X_{11} = \frac{1}{\sigma_{11}^{fT}\sigma_{11}^{fC}} = \frac{1}{(1050\,\text{MPa})(690\,\text{MPa})} = \frac{1.380e-18}{(\text{Pa})^2}$$

$$X_2 = \frac{1}{\sigma_{22}^{fT}} - \frac{1}{\sigma_{22}^{fC}} = \frac{1}{45\,\text{MPa}} - \frac{1}{120\,\text{MPa}} = \frac{1.852e-16}{\text{Pa}}$$

$$X_{22} = \frac{1}{\sigma_{22}^{fT}\sigma_{22}^{fC}} = \frac{1}{(45\,\text{MPa})(120\,\text{MPa})} = \frac{1.852e-16}{(\text{Pa})^2}$$

$$X_{66} = \left(\frac{1}{\tau_{12}^y}\right)^2 = \left(\frac{1}{40\,\text{MPa}}\right)^2 = \frac{6.250e-16}{(\text{Pa})^2}$$

$$X_{12} = \frac{-1}{2\sqrt{\sigma_{11}^{fT}\sigma_{11}^{fC}\sigma_{22}^{yT}\sigma_{22}^{yC}}} = \frac{-1}{2\sqrt{(1050\,\text{MPa})(690\,\text{MPa})(45\,\text{MPa})(120\,\text{MPa})}}$$

$$= \frac{-7.994e-18}{(\text{Pa})^2}$$

We now apply the Tsai–Wu criterion to the 0°, 30°, and 60° plies in turn. First consider the 0° plies (i.e., plies 1 and 6). From Example Problem 7.1, the stresses induced in these plies are given by

$$\sigma_{11} = 2750N_{xx} - 55.54 \times 10^6 \ (\text{Pa})$$

$$\sigma_{22} = 51.93N_{xx} + 28.36 \times 10^6 \ (\text{Pa})$$

$$\tau_{12} = -174.8N_{xx} + 22.83 \times 10^6 \ (\text{Pa})$$

Substituting these expressions and material constants into Equation 5.62, for the 0° plies the Tsai–Wu failure criterion becomes the following quadratic equation:

$$(8.667 \times 10^{-12})N_{xx}^2 - (1.730 \times 10^{-6})N_{xx} - 0.4353 = 0$$

The two roots to this equation are

$$N_{xx} = 345.1.\text{kN/m}, \quad -145.5.\text{kN/m}$$

As the problem statement specified that the applied loading is tensile, we select the positive root (if the applied load was compressive we would select the negative root). Thus, according to the Tsai–Wu criterion for a failure to occur in the 0° plies the tensile unit load would have to be increased by a load factor of 345.1×10^3.

Repeating this process for the 30° and 60° plies results in critical unit load factors of 153.7×10^3 and 81.3×10^3, respectively. Hence, according to the Tsai–Wu criterion the first-ply failure load is $N_{xx} = 81.3$ kN/m, which corresponds to failure of the 60° plies (plies 3 and 4). Equivalently, the effective first-ply failure stress is

$$\bar{\sigma}_{xx} = \frac{N_{xx}}{t} = \frac{81.3 \text{ kN/m}}{6(0.125 \text{ mm})} = 108 \text{ MPa}$$

Note that first ply failure is predicted to occur in the 60° plies by both the maximum stress and the Tsai–Wu failure criterions. However, the first-ply failure loads are quite different: according to the maximum stress criterion, first ply failure will occur at a load level of $N_{xx} = 123.0$ kN/m, whereas according to the Tsai–Wu criterion first ply failure will occur at a load level of $N_{xx} = 81.3$ kN/m, a difference of over 50%.

7.2.2 Predicting Last-Ply Failure

Last-ply failures predictions using CLT are based on the so-called "ply-discount" scheme. A flow diagram showing typical steps is presented in Figure 7.3. Steps 1 through 7 are virtually identical to those described in Section 7.2.1 for a first-ply failure prediction. Once step 7 is completed the unit loads necessary to cause one or more plies to fail has been calculated.

Assuming all plies in the laminate have not yet failed, then the next step is to discount (i.e., reduce) the elastic stiffness properties of the failed ply(ies). An unresolved question is: by what percentage should the properties of the ply be reduced? As the ply is presumed to have suffered a matrix failure, usually the matrix-dominated stiffnesses of a failed ply are reduced but fiber-dominated stiffnesses are not. If ply failure loads for a multiangle laminate

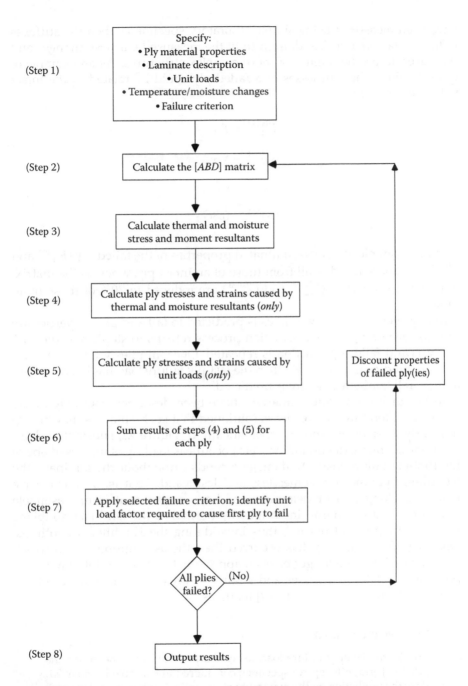

FIGURE 7.3
A flow diagram illustrating calculation of the last-ply failure load using CLT.

have been measured using simple laboratory specimens, then the stiffness reduction factors may be adjusted to fit the measured stiffness changes and then used to predict failure of more complex structures. As an example of typical values, the stiffnesses of a failed ply could be related to the intact values according to

$$E_{11}^{\text{failed}} = E_{11}$$

$$\nu_{12}^{\text{failed}} = \nu_{12}$$

$$E_{22}^{\text{failed}} = 0.3E_{22}$$

$$G_{12}^{\text{failed}} = 0.3G_{12}$$

In this example, the fiber-dominated properties of the failed ply (E_{11}^{failed} and ν_{12}^{failed}) are not reduced at all from those of an intact ply, whereas the matrix-dominated properties (E_{22}^{failed} and G_{12}^{failed}) are reduced to 30% of those of an intact ply.

In any event, once a ply or plies is predicted to fail and ply properties are reduced accordingly, the calculation process returns to step 2 and the CLT analysis and failure prediction is repeated. Once all plies within the laminate are predicted to have fail, then the last-ply failure load has been reached and overall laminate fracture is predicted.

Although last ply failure analyses have been described routinely in the composite literature, the reader is cautioned that such analyses may be significantly in error. As shown schematically in Figure 7.1, substantial damage (matrix cracks, delaminations, etc.) occurs as loading is increased above the first-ply failure load. As damage extends throughout the laminate, the Kirchhoff hypothesis becomes less and less valid. That is, once extensive matrix cracking and/or delaminations have occurred, it is unreasonable to assume that a straight line originally normal to the laminate midplane remains straight and normal, thus invalidating the Kirchhoff hypothesis. Once significant damage has occurred then the assumptions upon which CLT is based may no longer be even approximately true, and it may not be possible to relate ply stresses and strains to stress and moment resultants using the laminate [*ABD*] and [*abd*] matrices.

Example Problem 7.3

Predict the last-ply failure load using the ply discount scheme for the [0/30/60]$_s$ graphite–epoxy specimen considered in Example Problem 7.1. Base the prediction on the maximum stress failure criterion, and assume the matrix-dominated properties of a failed ply are reduced to 30% of their initial values, while fiber-dominated stiffnesses are not altered by matrix failures.

SOLUTION

As determined in Example Problem 7.1, the 60° plies (plies 3 and 4) are the first plies to fail, and the first-ply failure load is $N_{xx} = 122.9$ kN/m, or equivalently, the first-ply failure stress is $\bar{\sigma}_{xx} = 164$ MPa. The initial elastic modulus was also determined to be $\bar{E}_{xx}^{ex} = 82.9$ GPa, so at the first-ply failure load the axial strain is $\varepsilon_{xx} = 164$ MPa/82.9 GPa = 1978 µm/m.

As per the flow diagram shown in Figure 7.3, we now discount the properties of the failed 60° plies. That is, we assume that the stiffnesses of plies 3 and 4 are now:

$$E_{11}^{failed} = E_{11} = 170 \text{ GPa}$$

$$v_{12}^{failed} = v_{12} = 0.3$$

$$E_{22}^{failed} = 0.3E_{22} = 0.3(10 \text{ GPa}) = 3 \text{ GPa}$$

$$G_{12}^{failed} = 0.3G_{12} = 0.3(13 \text{ GPa}) = 3.9 \text{ GPa}$$

We repeat steps 2 through 7. The new [ABD] matrix, formed using the original stiffnesses for the 0° and 30° plies but with reduced stiffnesses for the 60° plies, is found to be

$$[ABD] = \begin{bmatrix} 73.54 \times 10^6 & 14.80 \times 10^6 & 16.87 \times 10^6 & 0 & 0 & 0 \\ 14.80 \times 10^6 & 34.12 \times 10^6 & 18.65 \times 10^6 & 0 & 0 & 0 \\ 16.87 \times 10^6 & 18.65 \times 10^6 & 20.54 \times 10^6 & 0 & 0 & 0 \\ 0 & 0 & 0 & 5.230 & 0.3514 & 0.4639 \\ 0 & 0 & 0 & 0.3514 & 0.6260 & 0.2652 \\ 0 & 0 & 0 & 0.4639 & 0.2652 & 0.6933 \end{bmatrix}$$

The new effective extensional modulus is $\bar{E}_{xx}^{ex} = 79.6$ GPa, a modest decrease from the original value $\bar{E}_{xx}^{ex} = 82.9$ GPa.

The process then proceeds as before: the new thermal stress and moment resultants are calculated (step 3), thermal stresses and strains for each ply are determined (step 4), ply stresses and strains caused by the unit load are determined and summed with the thermal stresses/strains (steps 5 and 6), and the failure criterion applied, in this case the maximum stress failure criterion (step 7). It is predicted that the 30° plies (plies 2 and 5) will fail at a load $N_{xx} = 199.0$ kN/m, or equivalently, at an effective stress $\bar{\sigma}_{xx} = 265$ MPa. These failures occur at an axial strain $\varepsilon_{xx} = 265$ MPa/79.6 GPa = 3329 µm m/m.

Finally, the stiffnesses properties of the 30° plies are reduced and the entire process repeated still again. It is found that the effective extensional modulus is reduced to $\bar{E}_{xx}^{ex} = 73.1$ GPa, and that the 0° plies will fail at a last-ply failure load $N_{xx} = 292.4$ kN/m, corresponding to an effective stress $\bar{\sigma}_{xx} = 390$ MPa and axial strain $\varepsilon_{xx} = 5334$ µm/m.

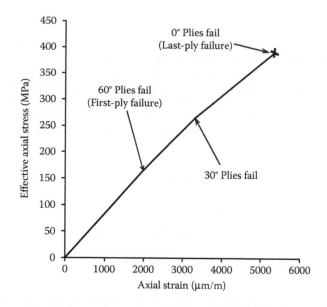

FIGURE 7.4
Predicted stress–strain curve for a $[0/30/60]_s$ graphite–epoxy laminate, based on the ply-discount scheme.

  As a uniaxial tensile load was considered in this problem, the stress–strain pairs predicted at each ply failure load can be plotted to produce a predicted stress–strain curve to failure for the $[0/30/60]_s$ graphite–epoxy laminate, as shown in Figure 7.4.

7.3 First-Ply Failure Envelopes

Methods used to determine the first-ply failure load for a laminate subjected to a specific single combination of stress and moment resultants were discussed in Section 7.2.1. Obviously, there are an infinite number of combinations of stress and moment resultants that will cause first-ply failure. Conceptually, those combinations of stress and moment resultants that (collectively) cause first-ply failure define a laminate *failure surface*. It is difficult to visualize such a surface, however, as there are a total of six stress and moment resultants involved and hence the failure surface involves six "dimensions." However, a *plane* that intersects the failure surface can be visualized by considering only two of the six stress resultants. For example, combinations of biaxial loads N_{xx} and N_{yy} that cause first-ply failure can be calculated using the approach described in Section 7.2.1, while assuming the remaining four resultants are zero ($N_{xy} = M_{xx} = M_{yy} = M_{xy} = 0$). A plot of the ($N_{xx}$, N_{yy}) combinations calculated under these assumptions is called a *first-ply failure envelope*. A typical first-ply failure envelope for a [0/45/90/−45]$_s$ graphite–epoxy laminate based on the maximum stress failure criterion is shown in Figure 7.5, where the effects of thermal and moisture stresses and strains have been ignored (i.e., this envelope was generated by assuming $\Delta T = \Delta M = 0$). As would be expected, the first plies predicted to fail depend on the combination of N_{xx} and N_{yy}. Also note that the envelope is not symmetric, even for a quasi-isotropic laminate subjected to in-plane loadings. The asymmetry occurs because composite strengths differ in tension and compression.

Distinctly different first-ply failure envelopes are usually predicted by differing failure criteria. For example, failure envelopes for a [0/45/90/−45]$_s$ graphite–epoxy laminate based on the maximum stress, Tsai–Hill, and Tsai–Wu failure criteria are superimposed in Figure 7.6 (again assuming $\Delta T = \Delta M = 0$).

The deleterious effects of thermal and moisture stresses are shown in Figure 7.7. Three first-ply failure envelopes are superimposed in this figure, all of which were generated based on the maximum stress failure criterion. In one case, thermal and moisture stresses and strains are ignored (i.e., this curve was generated by assuming $\Delta T = \Delta M = 0$, and is identical to Figure 7.5). In a second case the thermal stresses/strains caused during cooldown are included in the analysis but moisture effects are not (this curve was generated assuming $\Delta T = -155°C$, $\Delta M = 0$). This envelope is considerably smaller than the first and illustrates the deleterious effects of thermal stresses on first-ply failure loads, particularly in the first quadrant. Finally, the third envelope includes both thermal effects (based on $\Delta T = -155°C$) as well as a change in moisture content

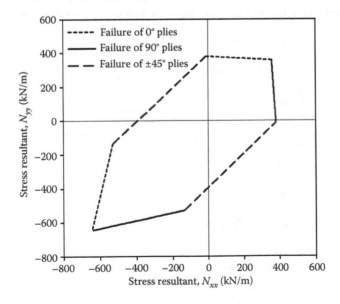

FIGURE 7.5
Biaxial first-ply failure envelopes for a [0/45/90/−45]$_s$ graphite–epoxy laminate based on the Maximum Stress failure criterion (environmental effects ignored).

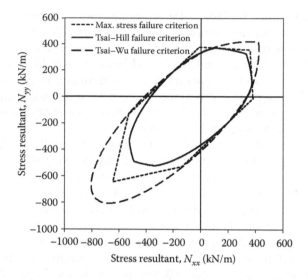

FIGURE 7.6
Biaxial first-ply failure envelopes for a $[0/45/90/-45]_s$ graphite–epoxy laminate based on the Maximum Stress, Tsai–Hill, and Tsai–Wu failure criteria (environment effects ignored).

of 1.0% (i.e., $\Delta M = 1.0\%$). As mentioned earlier, an increase in moisture content tends to relieve preexisting thermal stresses. This trend is evident in Figure 7.7, as first-ply failure envelope for the analysis that includes both thermal and moisture effects is larger than the analysis that includes only thermal effects.

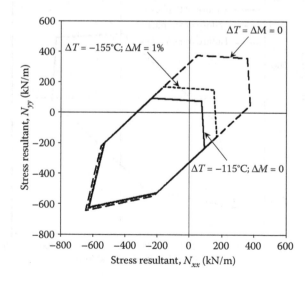

FIGURE 7.7
Biaxial first-ply failure envelopes for a $[0/45/90/-45]_s$ graphite–epoxy laminate for three different environmental conditions, based on the Maximum Stress failure criterion.

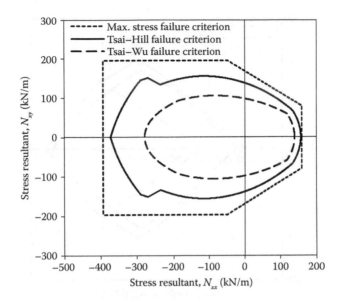

FIGURE 7.8
First-ply failure envelopes for a $[0/45/90/-45]_s$ graphite–epoxy laminate subjected to a combination of N_{xx} and N_{xy} ($\Delta T = -155°C$, $\Delta M = 1\%$).

It should be kept in mind that the first-ply failure load is a conservative design philosophy, and that the failure surfaces shown in Figures 7.4 through 7.7 do not represent load combinations that cause laminate *fracture*. Rather, these curves represent the loads that cause the first instance of ply failure on any kind, which in most cases is the onset of matrix cracking in one or more plies.

Similar two-dimensional failure envelopes can be developed based on any two of the six stress and moment resultants. For example, failure envelopes for a $[0/45/90/-45]_s$ graphite–epoxy laminate based on any combination of (N_{xx}, N_{xy}), (N_{xx}, M_{xx}), and (M_{xx}, M_{yy}) are shown in Figures 7.8, 7.9, and 7.10, respectively. Three envelopes are shown in each case, based on the maximum stress, Tsai–Hill, and Tsai–Wu failure criteria ($\Delta T = -155°C$ and $\Delta M = 1.0\%$ are also assumed in all cases).

7.4 Computer Programs LAMFAIL and PROGDAM

Computer programs LAMFAIL and PROGDAM are available to supplement the material discussed in this chapter. They can be downloaded at no cost from the following website:

http://depts.washington.edu/amtas/computer.html

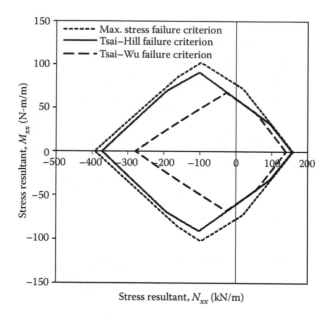

FIGURE 7.9
First-ply failure envelopes for a $[0/45/90/-45]_s$ graphite–epoxy laminate subjected to a combination of N_{xx} and M_{yy} ($\Delta T = -155°C$, $\Delta M = 1\%$).

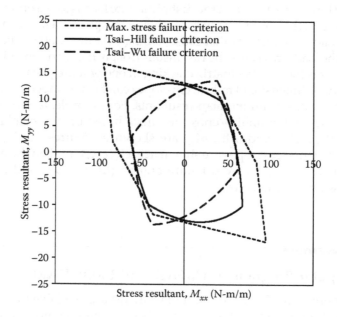

FIGURE 7.10
First-ply failure envelope for a $[0/45/90/-45]_s$ graphite–epoxy laminate based on combinations of M_{xx} and M_{yy} ($\Delta T = -155°C$, $\Delta M = 1\%$).

7.4.1 Program LAMFAIL

This program is used to perform a *laminate failure* analysis, and can be used to either:

- Perform a first-ply failure analysis for a specified combination of units loads, as described in Section 7.2.1 and illustrated in Example Problems 7.1 and 7.2; or
- Generate data needed to produce first-ply failure envelopes, as discussed in Section 7.3 and illustrated in Figures 7.5 through 7.10.

The program prompts the user to input all information necessary to perform the analysis. As previously discussed, an implicit assumption is that moduli and failure strengths input by the user correspond to the values exhibited by the composite at the temperature and moisture content of interest. Also, in the present context "failure" may represent *fracture* of the fibers (σ_{11}^{fT}, σ_{11}^{fC}) or *fracture* of the matrix (σ_{22}^{fT}, σ_{22}^{fC}, τ_{12}^{f}). The user specifies the number of plies within the laminate and material type and fiber angle for each ply. The user also selects from the maximum stress, Tsai–Hill, or Tsai–Wu failure criteria.[*]

If a first ply failure analysis for a particular combination of stress and moment resultants is to be performed, then the program prompts the user for unit loads. Alternatively, if the program is to be used to generate a failure envelope, then the particular pair of resultants to be considered during the analysis is specified by the user (i.e., any two of the six stress resultants N_{xx}, N_{yy}, N_{xy}, M_{xx}, M_{yy}, M_{xy}). Note that program LAMFAIL itself does not create a failure envelope. Rather, the program creates a file (named *Envelop.txt*) that contains the stress resultant pairs predicted to cause first-ply failure of the composite laminate specified by the user. A failure envelope may then be created using a second software package to import the data generated by program LAMFAIL. The first few lines of the file *Envelop.txt* created during a typical analysis are shown in Table 7.2. In this case the analysis was for a $[0/45/90/-45]_s$ graphite–epoxy laminate. The program had been directed to apply the maximum stress failure criterion, and to base the analysis on stress resultant pair N_{xx} and N_{yy}. The first few lines of the file represent header information. Notice that the file contains the ply number predicted to fail and the failure type (i.e., fiber, matrix, or shear), in addition to the combinations of N_{xx} and N_{yy} predicted to cause failure. Table 7.2 shows 12 such combinations of N_{xx} and N_{yy}. The complete file *Envelop.txt* contains a total of 2001 pairs. Thus, a first-ply failure envelope may be created by importing and plotting the N_{xx} and N_{yy} pairs.

---

[*] If the Tsai–Wu criterion is selected then the coupling strength term X_{12} is calculated according to Equation 5.61.

TABLE 7.2

First Few Lines of the File *Envelop.txt*, Created by Program LAMFAIL during an
Analysis of a [0/45/90/−45]$_s$ Graphite–Epoxy Laminate[a]

```
**PROGRAM LAMFAIL***
LAMINATE FAILURE PREDICTIONS BASED ON THE
MAXIMUM STRESS FAILURE CRITERION
STRESS-FREE TEMPERATURE = 175.0 DEGS
SERVICE TEMPERATURE = 20.0 DEGS
CHANGE IN TEMPERATURE = -155.0 DEGS
CHANGE IN MOISTURE CONTENT = 1.00%
```

Nxx	Nyy	FAILED PLY	FAILURE TYPE
.1590855E+06	.0000000E+00	3	MATRIX
.1590665E+06	.3187704E+03	3	MATRIX
.1590474E+06	.6387443E+03	3	MATRIX
.1590282E+06	.9599286E+03	3	MATRIX
.1590090E+06	.1282330E+04	3	MATRIX
.1589897E+06	.1605956E+04	3	MATRIX
.1589703E+06	.1930813E+04	3	MATRIX
.1589508E+06	.2256908E+04	3	MATRIX
.1589313E+06	.2584248E+04	3	MATRIX
.1589117E+06	.2912841E+04	3	MATRIX
.1588920E+06	.3242693E+04	3	MATRIX
.1588723E+06	.3573813E+04	3	MATRIX

[a] The maximum stress failure criterion had been selected for use, and stress resultants N_{xx} and N_{yy} were considered. The file also contains the number of the ply predicted to fail and the failure type (i.e., fiber, matrix, or shear) for each combination of N_{xx} and N_{yy} predicted to cause failure.

File *Envelop.txt* differs slightly if either the Tsai–Hill or Tsai–Wu criterion is selected for use. Specifically, if either of these criterion are selected then file *Envelop.txt* will not contain a column with the heading "failure type," as these criteria do not distinguish a particular failure mode in the same sense as does the maximum stress failure criterion. The first few lines of the file *Envelop.txt* created during an analysis for a [0/45/90/−45]$_s$ graphite–epoxy laminate in which the Tsai–Hill criterion was employed are shown in Table 7.3.

7.4.2 Program PROGDAM

This program is used to perform a progressive damage analysis of a composite laminate, based on the ply discount scheme described in Section 7.2.2. The program can generate data points that define a predicted stress–strain curve of a laminate, as described in Example Problem 7.3 and illustrated in Figure 7.4.

TABLE 7.3

First Few Lines of the File *Envelop.txt*, Created by Program LAMFAIL during an Analysis of a $[0/45/90/-45]_s$ Graphite–Epoxy Laminate[a]

```
**PROGRAM LAMFAIL***
LAMINATE FAILURE PREDICTIONS BASED ON THE
TSAI-HILL FAILURE CRITERION
STRESS-FREE TEMPERATURE = 175.0 DEGS
SERVICE TEMPERATURE = 20.0 DEGS
CHANGE IN TEMPERATURE = -155.0 DEGS
CHANGE IN MOISTURE CONTENT = 1.00%
```

		FAILED
Nxx	Nyy	PLY
.1565825E+06	.0000000E+00	3
.1565915E+06	.3138105E+03	3
.1566003E+06	.6289169E+03	3
.1566091E+06	.9453267E+03	3
.1566178E+06	.1263047E+04	3
.1566263E+06	.1582084E+04	3
.1566348E+06	.1902446E+04	3
.1566430E+06	.2224140E+04	3
.1566512E+06	.2547174E+04	3
.1566593E+06	.2871555E+04	6
.1566672E+06	.3197291E+04	3
.1566750E+06	.3524387E+04	3

[a] The Tsai–Hill failure criterion had been selected for use, and stress resultants N_{xx} and N_{yy} were considered. The file also contains the number of the ply predicted to fail for each combination of N_{xx} and N_{yy} predicted to cause failure.

The program prompts the user to input all information necessary to perform the analysis. As previously discussed, an implicit assumption is that moduli and failure strengths input by the user correspond to the values exhibited by the composite at the temperature and moisture content of interest. The user also selects from the maximum stress, Tsai–Hill, or Tsai–Wu failure criteria.[*]

The program creates a file (named *FAILURE.txt*) that contains the stress resultant that causes each successive ply failure, as well as the midplane strains and curvatures predicted to exist at each ply failure load. For example, the *FAILURE.txt* created during an analysis for a $[0/30/60]_s$ graphite–epoxy laminate and used to create Figure 7.4 is shown in Table 7.4. Note that the midplane strains and curvatures listed are those caused strictly by the mechanical load applied, and do not include components of the total midplane strains and curvatures due to thermal and moisture effects. However, the resultant associated with each ply failure does account for thermal and/or moisture

---

[*] If the Tsai–Wu criterion is selected then the coupling strength term X_{12} is calculated according to Equation 5.61.

TABLE 7.4

A Listing of the File *FAILURE.txt*, Created by Program PROGDAM during an Analysis of a [0/30/60]$_s$ Graphite–Epoxy Laminate[a]

```
DAMAGE PROGRESSION BASED ON THE MAXIMUM STRESS FAILURE CRITERION AND THE PLY
DISCOUNT SCHEME

***NOTE*** MIDPLANE STRAINS AND CURVATURES LISTED ARE THOSE CAUSED STRICTLY BY
THE MECHANICAL LOAD AND DO NOT INCLUDE COMPONENTS OF THE TOTAL STRAINS AND
CURVATURES DUE TO THERMAL OR MOISTURE EFFECTS

FAILURE                                                            FAILED
LOAD            MIDPLANE STRAIN AND CURVATURES
                                                                PLY  PLY
Nxx          EPSxx     EPSyy     GAMxy      Kxx      Kyy      Kxy   NO.  ANGLE

0.122994E+06 0.001978  0.000042 -0.001653 -0.000   -0.000   0.000   3   60.
0.122994E+06 0.001978  0.000042 -0.001653 -0.000   -0.000   0.000   4   60.

0.199028E+06 0.003335  0.000099 -0.002829 -0.000   -0.000   0.000   2   30.
0.199028E+06 0.003335  0.000099 -0.002829 -0.000   -0.000   0.000   5   30.

0.292442E+06 0.005332  0.000265 -0.005115 -0.000   -0.000   0.000   1   0.
0.292442E+06 0.005332  0.000265 -0.005115 -0.000   -0.000   0.000   6   0.
```

[a] Results were used to create Figure 7.4.

stresses. This reflects how strains are usually measured in practice. That is, strains measured in practice (using strain gages, for example) are usually measured relative to the service condition (at room temperature, say), rather than at the cure temperature of the composite.

HOMEWORK PROBLEMS

Notes:
Computer programs CLT, LAMFAIL, and PROGDAM are used in the following problems. These programs can be downloaded from the following website:

> http://depts.washington.edu/amtas/computer.html

7.1. Assume a new room-temperature cure graphite–epoxy prepreg system with properties listed in Table 3.1 has been developed. A $[0/\pm30]_s$ laminate is produced using this material, and is, therefore, initially stress-and-strain free at room temperature. The room-temperature laminate is then subjected to the biaxial tensile loads shown in Figure 7.11 (i.e., $N_{yy} = N_{xx}/2$ and $N_{xy} = M_{xx} = M_{yy} = M_{xy} = 0$). Use the following process to determine the tensile loads N_{xx} and N_{yy} necessary to cause first-ply failure, according to the maximum stress failure criteria:

   a. Use program CLT to determine the stresses (σ_{11}, σ_{22}, τ_{12}) induced in each ply by unit loads $N_{xx} = 1$, $N_{yy} = 1/2$ (i.e., use $k_2 = 0.5$).

   b. Determine which ply (or plies) is closest to the failure condition dictated by the maximum stress failure criterion, using the failure strengths listed for graphite/epoxy in Table 3.1.

   c. Calculate the increase in unit loads that will cause first-ply ply failure.

   d. Confirm the results of part (c) using program LAMFAIL.

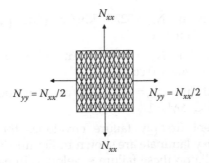

FIGURE 7.11
A $[0/\pm30]_s$ laminate loaded in biaxial tension, as described in Problem 7.1.

7.2. Repeat Problem 1 for a room-temperature cure glass/epoxy system.

7.3. Repeat Problem 1 for a room-temperature cure Kevlar/epoxy system.

7.4. Repeat Problem 1 for the following load condition:

$$N_{yy} = N_{xx}/2 \quad N_{xy} = 5N_{xx} \quad M_{xx} = M_{yy} = M_{xz} = 0$$

7.5. Repeat Problem 1, except assume the laminate is produced using a graphite–epoxy system cured at 175°C (350°F).

7.6. Repeat Problem 1, except assume the laminate is produced using a glass/epoxy system cured at 175°C (350°F).

7.7. Repeat Problem 1, except assume the laminate is produced using a Kevlar/epoxy system cured at 175°C (350°F).

7.8. Three different first-ply failure envelopes for a $[0/45/90/-45]_s$ graphite–epoxy laminate are shown in Figure 7.7. The three points listed below lie on these failure envelopes. In each case, use program CLT to determine which ply(ies) are predicted to fail, according to the maximum stress failure criterion:

  a. $N_{xx} = 327.3$ kN/m $N_{yy} = 0$ kN/m (this point lies on the $\Delta T = \Delta M = 0$ curve)

  b. $N_{xx} = 79.41$ kN/m $N_{yy} = 0$ kN/m (this point lies on the $\Delta T = -155°C, \Delta M = 0$ curve)

  c. $N_{xx} = 159.1$ kN/m $N_{yy} = 0$ kN/m (this point lies on the $\Delta T = -155°C, \Delta M = 1\%$ curve)

7.9. Three different first-ply failure envelopes for a $[0/45/90/-45]_s$, graphite–epoxy laminate are shown in Figure 7.7. The three points listed below lie on these failure envelopes. In each case, use program CLT to determine which ply(ies) are predicted to fail, according to the maximum stress failure criterion:

  a. $N_{xx} = 0$ kN/m $N_{yy} = 327.3$ kN/m (this point lies on the $\Delta T = \Delta M = 0$ curve)

  b. $N_{xx} = 0$ kN/m $N_{yy} = 79.41$ kN/m (this point lies on the $\Delta T = -155°C, \Delta M = 0$ curve)

  c. $N_{xx} = 0$ kN/m $N_{yy} = 159.1$ kN/m (this point lies on the $\Delta T = -155°C, \Delta M = 1\%$ curve)

7.10. Three different first-ply failure envelopes for a $[0/45/90/-45]_s$ graphite–epoxy laminate are shown in Figure 7.7. The three points listed below lie on these failure envelopes. In each case, use program *CLT* to determine which ply(ies) are predicted to fail, according to the maximum stress failure criterion:

a. $N_{xx} = -115.2$ kN/m $N_{yy} = 212.1$ kN/m (this point lies on the $\Delta T = \Delta M = 0$ curve)

b. $N_{xx} = -205.2$ kN/m $N_{yy} = 91.65$ kN/m (this point lies on the $\Delta T = -155°C$, $\Delta M = 0$ curve)

c. $N_{xx} = -172.7$ kN/m $N_{yy} = 154.6$ kN/m (this point lies on the $\Delta T = -155°C$, $\Delta M = 1\%$ curve)

7.11. Use program LAMFAIL to create first-ply envelopes for a $[0/\pm 45]_s$ graphite–epoxy laminate subjected to different combinations of N_{xx} and M_{yy}, based on the maximum stress, Tsai–Hill, and Tsai–Wu failure criteria. Assume the laminate was cured at 175°C, that it will be used at room temperatures (20°C) during service, and that it will not experience an increase in moisture content ($\Delta M = 0\%$). Use a plotting format similar to Figure 7.9.

7.12. A $[0/30/60]_s$ graphite–epoxy laminate is cured at 175°C and then cooled to room temperatures (20°C). Use program PROGDAM to generate predicted stress–strain curves for three different levels of moisture content:

a. $M = 0\%$

b. $M = 1\%$

c. $M = 2\%$.

Note that the stress–strain curve corresponding to part (a) appears in Figure 7.4. Explain why the stress–strain curves change with moisture content.

7.13. Repeat Problem 7.12, except use the Tsai–Hill failure criterion.

7.14. Repeat Problem 7.12, except use the Tsai–Wu failure criterion.

7.15. Repeat Problem 7.12, except assume a $[0/45/90/-45]_s$ laminate.

7.16. Repeat Problem 7.13, except assume a $[0/45/90/-45]_s$ laminate.

7.17. Repeat Problem 7.14, except assume a $[0/45/90/-45]_s$ laminate.

References

1. Soden, P.D., Hinton, M.J., and Kaddour, A.A., A comparison of the predictive capabilities of current failure theories for composite laminates, *Composite Science and Technologies*, 58, 1225–1254, 1998.
2. Hinton, M.J., Kaddour, A.A., and Soden, P.J., A Comparison of the predictive capabilities of current failure theories for composite laminates, judged against experimental evidence, *Composite Science and Technologies*, 62, 1725–1797, 2002.

8

Composite Beams

8.1 Preliminary Discussion

It is expected that readers of this chapter will have previously studied a subject known as "mechanics of materials" (also called "strength of materials"). One of the primary topics considered in mechanics of materials is the structural behavior of a *prismatic beam*. A prismatic beam is defined as a long, slender, and initially straight structural member whose cross-section does not vary along its length. Typical objectives are to determine the stresses, strains, and deflections induced in a prismatic beam by various types of external loads. Many excellent books are available to support such studies, a few of which are listed here as References 1–5. In virtually all of these references it is assumed that the beam is composed of an isotropic material, or at most is a beam in which different layers of isotropic materials are bonded or bolted together to form the overall beam.* In contrast, in this chapter we will consider the behavior of prismatic beams produced using laminated *anisotropic* composite materials.

The traditional pedagogical approach used in books devoted to mechanics of materials is to separate the types of loading commonly applied to prismatic beams/bars into three categories:

- Axial loads (i.e., loads acting parallel to the long axis of the beam, which tend to cause the beam to elongate or compress)
- Bending loads (i.e., loads acting transverse to the long axis of the beam, which tend to cause the initially straight beam to bend)
- Torques (i.e., moments acting about the long axis of the beam, which tend to cause the beam to twist about its long axis)

---

* A beam composed of two or more isotropic materials is sometimes called a "composite" beam. In such cases use of the term "composite" refers to the fact that the beam is composed of two or more *isotropic* materials. This is in contrast to the sort of composite beam discussed in this book, in which each individual composite ply is *anisotropic*.

Having made this classification, the stresses, strains, and deflections induced in a prismatic beam by each category of external load are considered in separate sections or chapters. It should be clear that this is an artificial distinction made purely for pedagogical purposes. In actual practice a prismatic beam or shaft may well be subjected to axial load(s), to bending load(s), and/or to torque(s) simultaneously. In such a case the stresses, strains, and deflections induced are due to the combined effects of all load components. Composite beams subjected to axial loads and/or bending loads will be discussed in this chapter. Beams subjected to torques will not be considered.

The beam analysis presented herein follows directly from CLT. A note of caution regarding free-edge stresses is, therefore, appropriate. As explained in Chapter 6, free-edge stresses occur because of a mismatch in the effective properties exhibited by adjacent plies in a multiangle composite laminate. Free-edge stresses cannot be predicted using CLT. Although free-edge stresses can be predicted using finite-element or finite-difference methods, free-edge stresses are rarely considered during preliminary structural design even when using these numerical techniques due to the high mesh densities and associated computational expense involved. Hence, in regions dominated by free-edge stresses the ply stresses and strains predicted on the basis of CLT and most finite-element and finite-difference analyses are invalid.

As a rule-of-thumb, the region over which significant free-edge stresses occur is approximately equal to the laminate thickness. Now, by definition a "beam" is a structure whose width is similar in magnitude to its thickness (i.e., the dimensions of the beam cross-section are much less than the beam length). CLT will be used in this chapter to predict the response of composite beams to axial and bending loads, and the effects of free-edge stresses will be ignored. The error in predicted response caused by neglecting free-edge stresses is difficult to estimate, in part because the magnitude of free-edge stresses depends on both stacking sequence and elastic properties of the material system used to fabricate the beam. In general, it is expected that free-edge stresses will have little impact on predicted beam stiffness, but may significantly influence beam failure loads.

8.2 Comparing Classical Lamination Theory to Isotropic Beam Theory

Before we begin our discussion of composite beams, it is instructive to compare CLT to results from isotropic beam theory. CLT is based on the behavior of a "thin plate." A plate with in-plane dimensions a and b and thickness t has been shown previously in Figure 6.1. The plate can be considered "thin" if the plate thickness is less than about one-tenth the in-plane dimensions, that is, if $t < a/10$ and $t < b/10$. As pointed out in Section 6.1, the description of a thin plate can be "converted" to the description of a beam if we let the x-direction define the beam axis and allow the in-plane width of the plate (dimension b) to approach the plate thickness: $b \approx t$. In this way we describe a beam with rectangular cross-section $b \times t$ and length a, as previously shown in Figure 6.3.

Recall that six types of loads are considered in CLT: three stress resultants (N_{xx}, N_{yy}, and N_{xy}) and three moment resultants (M_{xx}, M_{yy}, and M_{xy}). For present purposes, we will only consider two of these loads: stress resultant N_{xx} and moment resultant M_{xx}. That is, to compare CLT with isotropic beam theory we will specify that $N_{yy} = N_{xy} = M_{yy} = M_{xy} = 0$. A further distinction must also be made regarding the manner in which loads are described in CLT versus beam theory. In CLT, all stress and moment resultants are assumed to be uniformly distributed along the edge of a plate, and are described in units corresponding to this assumption. For example, stress resultant N_{xx} is described in units of N/m (or lbf/in.), and moment resultant M_{xx} is described in units of N-m/m (or lbf-in./in.). In contrast, in beam theory "point" or "concentrated" loads are usually specified. In beam theory, an axial load (corresponding to N_{xx}) is described in units of N (or lbf), whereas a bending moment (corresponding to M_{xx}) is described in units of N-m (or lbf-in.). In this book, the superscript "b" is used to differentiate between loads as defined in beam theory versus loads as defined in CLT. Thus, an axial load and bending moment applied to a beam will be denoted N_{xx}^b and M_{xx}^b, respectively. As the width of the beam is b, the two load definitions are related according to $N_{xx} = N_{xx}^b/b$ and $M_{xx} = M_{xx}^b/b$.

We are now ready to compare CLT with beam theory. From Section 6.6.2, the midplane strains and curvatures induced in a general multiangle composite laminate subjected to a combination of mechanical loads, a uniform change in temperature, and a uniform change in moisture content is given by Equation 6.45, repeated here for convenience:

$$
\begin{Bmatrix} \varepsilon_{xx}^o \\ \varepsilon_{yy}^o \\ \gamma_{xy}^o \\ \kappa_{xx} \\ \kappa_{yy} \\ \kappa_{xy} \end{Bmatrix} = \begin{bmatrix} a_{11} & a_{12} & a_{16} & b_{11} & b_{12} & b_{16} \\ a_{12} & a_{22} & a_{26} & b_{21} & b_{22} & b_{26} \\ a_{16} & a_{26} & a_{66} & b_{61} & b_{62} & b_{66} \\ b_{11} & b_{21} & b_{61} & d_{11} & d_{12} & d_{16} \\ b_{12} & b_{22} & b_{62} & d_{12} & d_{22} & d_{26} \\ b_{16} & b_{26} & b_{66} & d_{16} & d_{26} & d_{66} \end{bmatrix} \begin{Bmatrix} N_{xx} + N_{xx}^T + N_{xx}^M \\ N_{yy} + N_{yy}^T + N_{yy}^M \\ N_{xy} + N_{xy}^T + N_{xy}^M \\ M_{xx} + M_{xx}^T + M_{xx}^M \\ M_{yy} + M_{yy}^T + M_{yy}^M \\ M_{xy} + M_{xy}^T + M_{xy}^M \end{Bmatrix} \quad \text{(6.45) (repeated)}
$$

As our current objective is to compare CLT with isotropic beam theory, we will simplify our discussion by ignoring the effects of changes in temperature and moisture content (i.e., let $N_{ij}^T = M_{ij}^T = N_{ij}^M = M_{ij}^M = 0$). For an isotropic beam or plate substantial simplifications occur in the $[abd]$ matrix, as was already discussed in Section 6.7. Noting that we have already specified that $N_{yy} = N_{xy} = M_{yy} = M_{xy} = 0$, for an isotropic beam Equation 6.45 reduces to

$$
\begin{Bmatrix} \varepsilon_{xx}^o \\ \varepsilon_{yy}^o \\ \gamma_{xy}^o \\ \kappa_{xx} \\ \kappa_{yy} \\ \kappa_{xy} \end{Bmatrix} = \begin{bmatrix} a_{11} & a_{12} & 0 & 0 & 0 & 0 \\ a_{12} & a_{11} & 0 & 0 & 0 & 0 \\ 0 & 0 & 2(a_{11}-a_{12}) & 0 & 0 & 0 \\ 0 & 0 & 0 & d_{11} & d_{12} & 0 \\ 0 & 0 & 0 & d_{12} & d_{11} & 0 \\ 0 & 0 & 0 & 0 & 0 & 2(d_{11}-d_{12}) \end{bmatrix} \begin{Bmatrix} N_{xx}^b/b \\ 0 \\ 0 \\ M_{xx}^b/b \\ 0 \\ 0 \end{Bmatrix}
$$

where for an isotropic material the elements of the [*abd*] matrix reduce to (see Section 6.7)

$$
a_{11} = \frac{1}{Et} \quad a_{12} = \frac{-\nu}{Et}
$$

$$
d_{11} = \frac{12}{Et^3} \quad d_{12} = \frac{-12\nu}{Et^3}
$$

$$
d_{66} = 2(d_{11} - d_{12}) = \frac{24(1+\nu)}{Et^3}
$$

Let us first consider the midplane strains induced by N_{xx}^b alone (i.e., let $M_{xx}^b = 0$). From the above results, the axial midplane strain induced by N_{xx}^b (only) is given by

$$
\varepsilon_{xx}^o = (a_{11}) \frac{N_{xx}^b}{b} = \left(\frac{1}{Et} \right) \frac{N_{xx}^b}{b}
$$

As the cross-sectional area of the rectangular beam is $A = (t)(b)$, we can write this result as

$$
\varepsilon_{xx}^o = \frac{N_{xx}^b}{AE} = \frac{\sigma_{xx}}{E} \tag{8.1a}
$$

The quantity $\sigma_{xx} = N_{xx}^b/A$ is simply the uniaxial stress induced in a prismatic isotropic beam subjected to an axial load. Hence, CLT reduces to the uniaxial form of Hooke's law for an isotropic material. The product AE is known as the *axial rigidity* of an isotropic beam.

Similarly, the transverse midplane strain is given by

$$
\varepsilon_{yy}^o = (a_{12}) \frac{N_{xx}^b}{b} = \left(\frac{-\nu}{Et} \right) \frac{N_{xx}^b}{b} = \frac{-\nu N_{xx}^b}{AE} = \frac{-\nu\sigma_{xx}}{E} = -\nu\varepsilon_{xx}^o \tag{8.1b}
$$

Hence, for uniform axial loading, the transverse strain ε^o_{yy} is related to the axial strain ε^o_{xx} via Poisson's ratio, as expected for an isotropic beam subjected to a state of uniaxial stress. Finally, no midplane shear strain is predicted (also as expected for an isotropic beam):

$$\gamma^o_{xy} = 0 \tag{8.1c}$$

It is mentioned in passing that the stress tensor induced in the beam is uniaxial (i.e., $\sigma_{xx} \neq 0$, $\sigma_{yy} = \sigma_{zz} = \tau_{xy} = \tau_{xz} = \tau_{yz} = 0$). That is, the principal stresses are $\sigma_{p_1} = \sigma_{xx}$, $\sigma_{p_2} = \sigma_{p_3} = 0$, and the principal stress coordinate system is coincident with the x–y–z coordinate system. Similarly, as shear strain is zero the principal strain coordinate system is coincident with the x–y–z coordinate system and $\varepsilon_{p_1} = \varepsilon^o_{xx}$, $\varepsilon_{p_2} = \varepsilon^o_{yy}$. *The principal stress and principal strain coordinate systems are, therefore, coincident,* which is always the case for *isotropic* materials.

Now consider the midplane strains and curvatures induced by M^b_{xx} alone (i.e., let $N^b_{xx} = 0$). In this case, CLT predicts that all midplane strains are zero ($\varepsilon^o_{xx} = \varepsilon^o_{yy} = \gamma^o_{xy} = 0$). Therefore, the midplane represents the *neutral surface*, as expected for an isotropic beam with rectangular cross-section, subjected to pure bending. The midplane curvatures are

$$\kappa_{xx} = d_{11}\left(\frac{M^b_{xx}}{b}\right) \quad \kappa_{yy} = d_{12}\left(\frac{M^b_{xx}}{b}\right) \quad \kappa_{xy} = 0$$

The fact that the twist curvature κ_{xy} is zero implies that the x–z and y–z planes are the principal planes of curvature, as expected for a prismatic isotropic beam with symmetric rectangular cross-section. Substituting the reduced forms for d_{11} and d_{12} for an isotropic material, we find

$$\kappa_{xx} = \frac{12}{Et^3}\left(\frac{M^b_{xx}}{b}\right) \quad \kappa_{yy} = \frac{-12v}{Et^3}\left(\frac{M^b_{xx}}{b}\right)$$

The moment of inertia for a rectangular cross-section of width b and height t is $I = bt^3/12$. Therefore, the above expressions can be written as

$$\kappa_{xx} = \frac{M^b_{xx}}{EI} \tag{8.2a}$$

$$\kappa_{yy} = -\frac{vM^b_{xx}}{EI} \tag{8.2b}$$

Equations 8.2 are the well-known *moment-curvature equations* for isotropic beams. The product EI is known as the *flexural rigidity* of an isotropic beam. Together Equations 8.2a, 8.2b imply that $\kappa_{yy} = -\nu\kappa_{xx}$, which shows that anticlastic bending of an isotropic beam with symmetric rectangular cross-section is predicted by CLT, and further that κ_{yy} is related to κ_{xx} via Poisson's ratio, as expected.

The through-thickness variation in normal strains ε_{xx} and ε_{yy} is given by

$$\varepsilon_{xx} = \varepsilon^o_{xx} + z\kappa_{xx} = z\left(\frac{M^b_{xx}}{EI}\right)$$

$$\varepsilon_{yy} = \varepsilon^o_{yy} + z\kappa_{yy} = -z\left(\frac{\nu M^b_{xx}}{EI}\right)$$

Through-thickness variation in stresses can be calculated using Equation 5.30, which becomes (assuming isotropic properties and also that $\Delta T = \Delta M = 0$):

$$\begin{Bmatrix} \sigma_{xx} \\ \sigma_{yy} \\ \tau_{xy} \end{Bmatrix} = \begin{bmatrix} \dfrac{E}{(1-v^2)} & \dfrac{vE}{(1-v^2)} & 0 \\ \dfrac{vE}{(1-v^2)} & \dfrac{E}{(1-v^2)} & 0 \\ 0 & 0 & \dfrac{E}{2(1+v)} \end{bmatrix} \begin{Bmatrix} \varepsilon_{xx} \\ \varepsilon_{yy} \\ \gamma_{xy} \end{Bmatrix}$$

$$= \begin{bmatrix} \dfrac{E}{(1-v^2)} & \dfrac{vE}{(1-v^2)} & 0 \\ \dfrac{vE}{(1-v^2)} & \dfrac{E}{(1-v^2)} & 0 \\ 0 & 0 & \dfrac{E}{2(1+v)} \end{bmatrix} \begin{Bmatrix} \dfrac{zM^b_{xx}}{EI} \\ \dfrac{-zvM^b_{xx}}{EI} \\ 0 \end{Bmatrix}$$

On completing the matrix multiplication indicated above, we find

$$\begin{Bmatrix} \sigma_{xx} \\ \sigma_{yy} \\ \tau_{xy} \end{Bmatrix} = \begin{Bmatrix} \dfrac{M^b_{xx}z}{I} \\ 0 \\ 0 \end{Bmatrix}$$

Hence, the CLT analysis predicts that σ_{yy} and τ_{xy} are both zero, and that the axial stress σ_{xx} induced in an isotropic beam by M^b_{xx} (only) is given by the familiar *flexure formula*:

$$\sigma_{xx} = \frac{M_{xx}^b z}{I} \tag{8.3}$$

As before, as only an axial stress is induced, the stress tensor is everywhere uniaxial and principal stresses are $\sigma_{p_1} = \sigma_{xx}$, $\sigma_{p_2} = \sigma_{p_3} = 0$. The principal stress coordinate system is coincident with the x–y–z coordinate system. Similarly, as shear strain is zero the principal strain coordinate system is coincident with the x–y–z coordinate system and $\varepsilon_{p_1} = \varepsilon_{xx}^o$, $\varepsilon_{p_2} = \varepsilon_{yy}^o$. We again conclude that the principal stress and principal strain coordinate systems are coincident, which is always the case for isotropic materials.

On the basis of the above, we conclude that the analysis represented by CLT can be used to represent an isotropic beam with rectangular cross-section subjected to an axial load N_{xx}^b and bending moment M_{xx}^b, both of which are constant along the length of the beam. If $N_{xx}^b = 0$ then the loading condition represented by CLT corresponds to a state of pure bending. Now, there is an additional fundamental result from traditional beam theory that cannot be recovered by applying CLT to an isotropic beam. Specifically, traditional beam theory allows one to calculate the *shear stresses* induced in a beam. Recall that shear stresses τ_{xz} are given by the so-called shear formula [1–5]:

$$\tau_{xz} = \frac{VQ}{Ib}$$

where V is the shear force present at a specified beam cross-section, Q is the first moment of an area about the neutral axis (for a rectangular beam $Q = b[(t^2/4) - z^2]/2)$, I is the area moment of inertia of the entire cross-section about the neutral axis, and b is the width of the beam. The reason that CLT cannot be used to predict shear stress τ_{xz} is due to the fundamental assumption that the bending moment M_{xx}^b is *constant* along the length of the beam. That is, recall from earlier studies (see, e.g., References 1–5) that the shear force V is related to the bending moment M_{xx}^b according to

$$V = \frac{dM_{xx}^b}{dx}$$

As M_{xx}^b is assumed constant in a CLT analysis, $V = dM_{xx}^b/dx = 0$. Therefore, a CLT analysis implies that shear stress $\tau_{xz} = 0$, due to the assumed loading conditions.

The discussion presented in this section has been intended to show that CLT is entirely consistent with traditional isotropic beam theory.

The reader should carefully note, however, that in this discussion CLT has been *specialized to the case of isotropic beams*. The behavior of composite beams will be discussed in following sections. It will be seen that in some ways the behavior of composite and isotropic beams is similar, but in others they are quite different. In particular, for composite beams the flexure formula (Equation 8.3) is only valid if the beam is oriented in a specific way with respect to the applied loads. Also, for composite beams the principal stress and principal strain coordinate systems are not (in general) coincident.

8.3 Types of Composite Beams Considered

The types of composite beams considered in this chapter are summarized in Figure 8.1. An externally applied axial load N_{xx}^b, bending moment M_{xx}^b, and transverse distributed load $q(x)$ are also shown. Both the beam cross-section and applied loads are referenced to an x–y–z coordinate system, where the y- and z-axes are orthogonal to the long axis of the beam. Bending moment M_{xx}^b and transverse load $q(x)$ act within the x–z plane, and hence the x–z plane is called the *plane of loading*.

The type of "composite beam" implied during earlier sections shown in Figure 8.1a. In this case, the beam cross-section is rectangular, and all ply interfaces are *orthogonal* to the plane of loading. In contrast, a composite beam with rectangular cross-section but in a decidedly different orientation

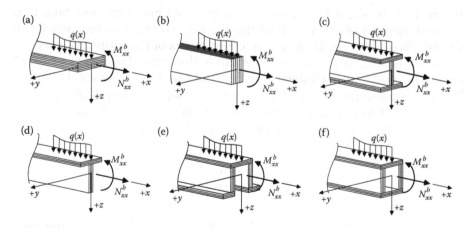

FIGURE 8.1
Composite beams with various cross-sections. (a) Rectangular beam, plies orthogonal to plane of loading; (b) rectangular beam, plies parallel to plane of loading; (c) I-beam; (d) T-beam; (e) hat-beam; (f) box-beam.

with respect to the plane of loading is shown in Figure 8.1b. In this case, the ply interfaces are *parallel* to the plane of loading. The effective axial rigidity exhibited by a composite beam is identical in either orientation. As will be seen, however, the effective flexural rigidity of a rectangular composite beam will differ substantially depending on whether the beam is orientated as shown in Figure 8.1a or b.

It is of course possible to manufacture and use composite beams with rectangular cross-sections. However, in practice it is far more common to use thin-walled composite beams, such as those shown in Figure 8.1c–f. Beams with thin-walled cross-sections are more commonly used because they provide far higher flexural rigidities per unit weight than a solid rectangular beam.

The method used to study thin-walled composite beams herein is similar to that described by Swanson [7]. The general approach is to approximate the cross-sections shown in Figure 8.1c–f as an assembly of flat rectangular laminates. The stresses and strains induced in each region of the cross-section will then be determined based on CLT. In all cases the plane of loading is defined as the x–z plane. Those regions of a beam cross-section that are parallel to the plane of loading will be called the "web laminate" (or "web laminates"), whereas those regions that are orthogonal to the plane of loading will be called the "flange laminate" (or "flange laminates"). Thus, for example, a composite *I*-beam has one web laminate and two flange laminates, whereas a composite box-beam has two flange laminates and two web laminates.

Several restrictions are placed on the types of composite beams considered. First and foremost, only beams in which all flange and web laminates are produced using symmetric stacking sequences will be considered. This restriction is imposed to avoid complications due to thermal and moisture effects. If a beam is produced using nonsymmetric web or flange laminates then *substantial* thermal and/or moisture moment resultants (M_{ij}^T and M_{ij}^M) are induced, which may lead to *substantial* warping of the beam, even before application of any external load. For this reason, composite beams used in practice are almost always produced using laminates based on a symmetric stacking sequence. As discussed in Section 6.7.1, for a symmetric stacking sequence $M_{ij}^T = M_{ij}^M = 0$, and hence this complication is eliminated.

Second, we require that all beam cross-sections possess both geometric *and* material symmetry about the plane of loading (the x–z plane). Having already stipulated that all laminates are symmetric, then the requirement of geometric and material symmetry about the x–z plane is achieved automatically for solid rectangular composite beams and in these cases requires no further discussion. However, symmetry about the x–z plane implies additional restrictions for thin-walled cross-sections involving multiple web laminates. For cross-sections with a single web laminate (an *I*-beam or *T*-beam), symmetry about the x–z plane is automatically satisfied if the web laminate is symmetric. However, for cross-sections with two web laminates (a hat- or box-beam), the two symmetric web laminate must represent a set of fiber angles that are symmetric *about the x–z plane,* as well as being symmetric about the local web laminate midplanes.

Note that symmetry about the *x–z* plane places no restrictions on the flange laminates. Thus, the top and bottom flanges in an *I-*, hat-, or box-beam are allowed to have different stacking sequences and thicknesses, although they must both be symmetric about their respective midplanes (the stacking sequences and widths of the two bottom flanges in a hat-beam must be identical, however). The beam cross-section may, therefore, be nonsymmetrical about the *x–y* plane.

In practice, it can be problematic to produce thin-walled composite beams that feature the symmetries just described. Difficulties may arise due to the manufacturing process used to produce the beam. Methods of producing a thin-walled composite beam using *unidirectional* prepreg tapes will be discussed to illustrate the practical difficulties encountered.

First, consider the processes that could be used to produce a composite *T*-beam. Two possibilities are shown in Figure 8.2. In Figure 8.2a the cross-section is formed by first curing the flange and web laminates separately. The two laminates are then bonded together in a second operation to form the desired *T*-cross-section, as indicated. An advantage to this approach is that the stacking sequences used in the web and flange laminates are completely independent. As any stacking sequence can be used in either laminate, it is easy to produce a *T*-beam with a symmetric stacking sequence in both web and flange laminates, and to insure symmetry about the *x–z* plane. However, a distinct disadvantage is that the web and flange laminates are joined solely by the adhesive bond—no continuous fibers cross the junction between web and flange. Hence, this approach is likely to result in a relatively low-strength *T*-beam. A more common method of producing a composite *T*-beam is illustrated in Figure 8.2b. In this case, the plies that

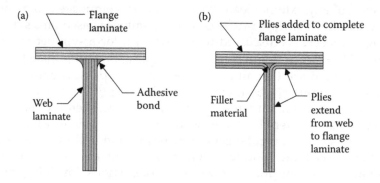

FIGURE 8.2
Methods to produce a composite *T*-beam using unidirectional prepreg tape. (a) Forming a *T*-cross section by bonding web and flange laminates; (b) forming a *T*-cross section by extending web plies into the flange.

exist within the web laminate are extended into (and become part of) the flange laminate. Additional plies, which span the width of the flange, are added to complete the flange laminate. During lay-up an internal v-shaped cavity is formed near the web-flange junction. This cavity is filled with some filler material (often preimpregnated unidirectional tow) prior to curing the *T*-beam. This second approach of producing a *T*-beam has the distinct advantage of providing continuous fibers across the junction between the web and flange. However, the stacking sequences used in the web and flange laminates are no longer independent. As the flange laminate is required to be symmetric (at least for the analysis presented in this section), the fiber angles used in the web and inner portions of the flange laminates must be repeated on the outer surface of the flange. Furthermore, if the web is produced using unidirectional prepreg tape, then it is *impossible* to produce a *T*-beam that has *both* a symmetric web and a symmetric flange laminate, unless only 0° or 90° plies are used in the web. This difficulty may not be immediately obvious, but is illustrated in Figure 8.3. The figure shows the formation of a symmetric two-ply web laminate, that is, a $[\theta°]_s$ web laminate (where $\theta \neq 0°$ or 90°). As indicated, extending the web plies into the flange results in a single flange ply in which the fiber angle is +θ-degrees on one side of the flange-web junction, but −θ-degrees on the other side. Hence, it is not possible to add additional plies across the entire width of the flange to produce both a symmetric laminate, unless web fiber angles are restricted to either θ = 0° or 90°. In some instances it may not be acceptable to restrict web angles to 0° or 90°, as this restriction results in a web laminate with relatively low shear stiffness. A similar difficulty is encountered when producing a beam with two web laminates such as a hat- or box-beam. A single ply that becomes part of both web laminates and both flange laminates in a box-beam is shown in Figure 8.4. In this case, it is not possible to produce web laminates that are symmetric about the *x–z* plane, as a web ply with a fiber angle of +θ° on one side of the *x–z* plane becomes a ply with a fiber angle of −θ-degrees on the other side of the *x–z* plane. As before, the only way to achieve symmetry about the *x–z* plane (if the beam is produced using unidirectional prepreg

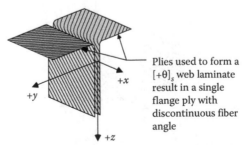

Plies used to form a $[+\theta]_s$ web laminate result in a single flange ply with discontinuous fiber angle

FIGURE 8.3
Illustration of why unidirectional prepregs extending from web to flange cannot produce symmetric web and flange laminates, unless, θ = 0° or 90°.

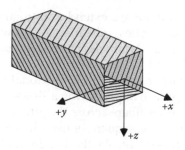

FIGURE 8.4
Illustration of why unidirectional prepregs cannot be used to produce web laminates that are symmetric about the x–z plane, unless $\theta = 0°$ or $90°$.

tape) is to use fiber angles of $\theta = 0°$ or $90°$ in the web, which may lead to web laminates with unacceptably low shear stiffness.

These difficulties in achieving symmetry may be avoided if the web is produced using a woven or braided prepreg fabrics rather than unidirectional prepreg, as shown in Figure 8.5. Recall from Section 1.4 that a single ply of a woven or braided fabric features two or three fiber directions, oriented symmetrically about the warp direction. As each individual woven or braided ply features a symmetric ±θ-degree fiber pattern, plies extending from the web into the flange (as in a *T*-beam) or completely around the circumference of the cross-section (as in a box-beam) retain the identical ±θ-degree

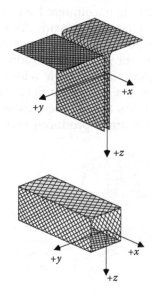

FIGURE 8.5
The use of woven or braided fabrics or prepreg can avoid the difficulties illustrated in Figures 8.3 and 8.4.

fiber pattern in both web and flange laminates. Note that use of a woven or braided ply to form the web laminate(s) does not preclude use of additional unidirectional plies in the flange laminate(s).

Finally, objectionable levels of thermal warping may still occur for some thin-walled composite beams, even though they satisfy all of the symmetry requirements described above. Thermal warping may occur because we have allowed the beam cross-section to be nonsymmetric about the x–y plane. Suppose, for example, that an I-beam is produced in which the symmetric top flange is produced using a very different stacking sequence than the symmetric bottom flange. This implies that the effective thermal expansion coefficient $\overline{\alpha}_{xx}$ of the top flange may be very different than that of the bottom flange. Consequently, if the beam experiences a uniform change in temperature then the top flange will tend to expand or contract at a different rate than the bottom flange, causing the beam to warp. The symmetric web laminate will tend to restrict such warping, but nevertheless this effect may be significant depending on details of a given beam design. This possibility will not be addressed in this chapter, although the theory presented can be easily modified to account for this effect.

8.4 Effective Axial Rigidity of Rectangular Composite Beams

A rectangular composite beam subjected to axial load N_{xx}^b is shown in Figure 8.6. The width and thickness of the beam cross-section are labeled b and t, respectively. We will now use CLT to predict the mechanical response of this beam to pure axial loading. As mentioned earlier, significant free-edge stresses may be present but will be neglected in the following discussion.

As the beam is assumed to be symmetric and the only applied load is N_{xx}^b, the midplane strains and curvatures induced in a symmetric

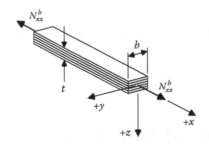

FIGURE 8.6
A composite beam with cross-section $b \times t$, subjected to axial load N_{xx}^b.

composite beam can be predicted using Equation 6.45, which becomes (for $N_{yy} = N_{xy} = M_{xx} = M_{yy} = M_{xy} = b_{ij} = M_{ij}^T = M_{ij}^M = 0$):

$$
\begin{Bmatrix} \varepsilon_{xx}^o \\ \varepsilon_{yy}^o \\ \gamma_{xy}^o \\ \kappa_{xx} \\ \kappa_{yy} \\ \kappa_{xy} \end{Bmatrix} = \begin{bmatrix} a_{11} & a_{12} & a_{16} & 0 & 0 & 0 \\ a_{12} & a_{22} & a_{26} & 0 & 0 & 0 \\ a_{16} & a_{26} & a_{66} & 0 & 0 & 0 \\ 0 & 0 & 0 & d_{11} & d_{12} & d_{16} \\ 0 & 0 & 0 & d_{12} & d_{22} & d_{26} \\ 0 & 0 & 0 & d_{16} & d_{26} & d_{66} \end{bmatrix} \begin{Bmatrix} (N_{xx}^b/b) + N_{xx}^T + N_{xx}^M \\ N_{yy}^T + N_{yy}^M \\ N_{xy}^T + N_{xy}^M \\ 0 \\ 0 \\ 0 \end{Bmatrix} \tag{8.4}
$$

Note that thermal and moisture stress resultants are (in general) present in the symmetric composite beam, even though the only externally applied load is N_{xx}^b. As discussed in earlier chapters, thermal and moisture resultants contribute substantially to the strains and stresses induced in each ply, and may lead to premature failure of a composite beam if not properly accounted for.

The preexisting ply stresses and strains are of course changed upon application of an external axial load N_{xx}^b. From Equation 8.4, it is easy to see that the incremental change in midplane strains caused by N_{xx}^b (only) are given by

$$
\varepsilon_{xx}^o = \frac{a_{11}}{b} N_{xx}^b \quad \varepsilon_{yy}^o = \frac{a_{12}}{b} N_{xx}^b \quad \gamma_{xy}^o = \frac{a_{16}}{b} N_{xx}^b \tag{8.5}
$$

The effective extensional Young's modulus, Poisson's ratio, and coefficient of mutual influence of the second kind for a composite laminate were all defined as follows in Section 6.9.1.1:

$$
\overline{E}_{xx}^{ex} = \frac{1}{t a_{11}} \quad \overline{\nu}_{xy}^{ex} = \frac{-a_{12}}{a_{11}} \quad \overline{\eta}_{xx,xy}^{ex} = \frac{a_{16}}{a_{11}}
$$

Substituting these effective properties into Equations 8.5, we find

$$
\varepsilon_{xx}^o = \frac{N_{xx}^b}{(t\,b)\overline{E}_{xx}^{ex}} = \frac{N_{xx}^b}{A\overline{E}_{xx}^{ex}} = \frac{\overline{\sigma}_{xx}}{\overline{E}_{xx}^{ex}} \tag{8.6a}
$$

$$
\varepsilon_{yy}^o = -\overline{\nu}_{xy}^{ex}\left(\frac{N_{xx}^b}{A\overline{E}_{xx}^{ex}}\right) = -\overline{\nu}_{xy}^{ex}\left(\frac{\overline{\sigma}_{xx}}{\overline{E}_{xx}^{ex}}\right) = -\overline{\nu}_{xy}^{ex}\varepsilon_{xx}^o \tag{8.6b}
$$

$$\gamma_{xy}^o = \overline{\eta}_{xx,xy}^{ex} \left(\frac{N_{xx}^b}{A\overline{E}_{xx}^{ex}} \right) = \overline{\eta}_{xx,xy}^{ex} \left(\frac{\overline{\sigma}_{xx}}{\overline{E}_{xx}^{ex}} \right) = \overline{\eta}_{xx,xy}^{ex} \varepsilon_{xx}^o \qquad (8.6c)$$

From Equation 8.6a we see that the effective axial rigidity of a rectangular composite beam can be written:

$$(A\overline{E}_{xx}^{ex}) = (t \ b)\overline{E}_{xx}^{ex} \qquad (8.7a)$$

where A is the cross-sectional area of the beam and $\overline{E}_{xx}^{ex}$ is the effective extensional Young's modulus of the laminate (defined in Section 6.9.1.1). Alternatively, the effective axial rigidity may be written as

$$(A\overline{E}_{xx}^{ex}) = \left(\frac{b}{a_{11}} \right) \qquad (8.7b)$$

Although the beam shown in Figure 8.6 is oriented such that ply interfaces are orthogonal to the x–z plane, the analysis is also applicable to a beam oriented such that ply interfaces are parallel to the x–z plane, as shown in Figure 8.1b. Hence, Equations 8.7a, 8.7b give the effective axial rigidity of a symmetric rectangular composite beam, regardless of ply orientation.

Note that Equation 8.6 are directly analogous to similar relations for an isotropic beam, as given by Equation 8.1. As only an axial load is applied the effective stress tensor is uniaxial (i.e., $\overline{\sigma}_{xx} \neq 0$, $\overline{\sigma}_{yy} = \overline{\sigma}_{zz} = \overline{\tau}_{xy} = \overline{\tau}_{xz} = \overline{\tau}_{yz} = 0$). That is to say, the *effective principal stresses* are $\overline{\sigma}_{p_1} = \overline{\sigma}_{xx}$, $\overline{\sigma}_{p_2} = \overline{\sigma}_{p_3} = 0$, and the effective principal stress coordinate system is coincident with the x–y–z coordinate system. However, for a composite beam an axial stress $\overline{\sigma}_{xx}$ will (in general) cause a *shear* strain γ_{xy}^o, due to the presence of a_{16} (or equivalently, due to the presence of $\overline{\eta}_{xx,xy}^{ex}$). Therefore, the principal strain coordinate system is not, in general, coincident with the effective principal stress coordinate system.

Recall from Section 6.7 that $a_{16} = 0$ for the following commonly used stacking sequences:

- Cross-ply
- Balanced
- Balanced angle-ply
- Quasi-isotropic

If an axially loaded symmetric composite beam is produced using any of these stacking sequences then shear strain γ_{xy}^o is zero and the effective principal stress and principal strain coordinate systems are coincident.

8.5 Effective Flexural Rigidities of Rectangular Composite Beams

An initially straight beam subjected to an external bending moment of equal magnitude at either end is said to be in a state of *pure bending*. The flexural rigidity of a beam is defined in pure bending. In this section, the effective flexural rigidities of composite beams with rectangular cross-sections will be determined. To do so we must consider two different cases. In the first case, we consider a beam oriented such that ply interfaces are orthogonal to the plane of loading, as previously shown in Figure 8.1a. In the second case, we consider a beam for which ply interfaces are parallel to the plane of loading, as was shown in Figure 8.1b.

8.5.1 Effective Flexural Rigidity of Rectangular Composite Beams with Ply Interfaces Orthogonal to the Plane of Loading

A composite beam oriented such that ply interfaces are orthogonal to the plane of loading is shown in Figure 8.7. The height and width of the beam cross-section are denoted as t and b, respectively. We will now use CLT to predict the mechanical response of this beam in pure bending. As mentioned earlier, significant free-edge stresses may be present (depending on the stacking sequence and material system used to produce the beam) but will be neglected in the following discussion.

As we have assumed that the beam is symmetric and subjected to M_{xx}^b only, the midplane strains and curvatures can be predicted using Equation 6.45,

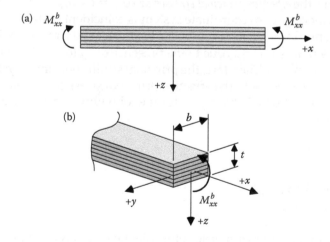

FIGURE 8.7
A composite beam with ply interfaces orthogonal to the x–z plane, subjected to pure bending. (a) A composite beam subjected to pure bending; (b) end view of beam.

which becomes (for $b_{ij} - N_{xx} = N_{yy} = N_{xy} - M_{yy} = M_{xy} = M_{ij}^T = M_{ij}^M = 0$, and $M_{xx} = M_{xx}^b/b$):

$$
\begin{Bmatrix}
\varepsilon_{xx}^o \\
\varepsilon_{yy}^o \\
\gamma_{xy}^o \\
\kappa_{xx} \\
\kappa_{yy} \\
\kappa_{xy}
\end{Bmatrix}
=
\begin{bmatrix}
a_{11} & a_{12} & a_{16} & 0 & 0 & 0 \\
a_{12} & a_{22} & a_{26} & 0 & 0 & 0 \\
a_{16} & a_{26} & a_{66} & 0 & 0 & 0 \\
0 & 0 & 0 & d_{11} & d_{12} & d_{16} \\
0 & 0 & 0 & d_{12} & d_{22} & d_{26} \\
0 & 0 & 0 & d_{16} & d_{26} & d_{66}
\end{bmatrix}
\begin{Bmatrix}
N_{xx}^T + N_{xx}^M \\
N_{yy}^T + N_{yy}^M \\
N_{xy}^T + N_{xy}^M \\
M_{xx}^b/b \\
0 \\
0
\end{Bmatrix}
\tag{8.8}
$$

Thermal and moisture stress resultants N_{ij}^T and N_{ij}^M are (in general) present in the symmetric composite beam. As discussed in earlier chapters, thermal and moisture stress resultants contribute substantially to the strains and stresses induced in each ply, and may lead to premature failure of a composite beam if not properly accounted for. The thermal and moisture stress resultants do not influence the initial *elastic* response of the beam, however. From Equation 8.8 it is easy to see that midplane strains are not changed upon application of M_{xx}^b, as $b_{ij} = 0$. Hence, the midplane of the beam cross-section represents the *neutral surface*, in the sense that application of M_{xx}^b does not contribute to (or alter) preexisting midplane strains. As noted in Section 8.2, the neutral surface for an *isotropic* beam with rectangular cross-section is also coincident with the midplane.

From Equation 8.8, the midplane curvatures are given by

$$
\kappa_{xx} = d_{11}\left(\frac{M_{xx}^b}{b}\right) \quad \kappa_{yy} = d_{12}\left(\frac{M_{xx}^b}{b}\right) \quad \kappa_{xy} = d_{16}\left(\frac{M_{xx}^b}{b}\right)
\tag{8.9}
$$

The fact that the twist curvature $\kappa_{xy} \neq 0$ (in general) shows that the x–z and y–z planes *are not the principal planes of curvature.* This is in direct contrast to an isotropic beam, as for a prismatic isotropic beam with symmetric rectangular cross-section the x–z and y–z planes represent the principal planes of curvature.

The effective flexural Young's modulus, Poisson's ratio, and coefficient of mutual influence of the second kind for a composite laminate were all defined as follows in Section 6.9.1:

$$
\overline{E}_{xx}^{fl} = \frac{12}{d_{11}t^3} \quad \overline{v}_{xy}^{fl} = \frac{-d_{12}}{d_{11}} \quad \overline{\eta}_{xx,xy}^{fl} = \frac{d_{16}}{d_{11}}
$$

Substituting these effective properties into Equations 8.9, we find

$$\kappa_{xx} = \frac{12}{\overline{E}_{xx}^{fl}t^3}\left(\frac{M_{xx}^b}{b}\right) \quad \kappa_{yy} = \frac{-12\overline{v}_{xy}^{fl}}{\overline{E}_{xx}^{fl}t^3}\left(\frac{M_{xx}^b}{b}\right) \quad \kappa_{xy} = \frac{12\overline{\eta}_{xx,xy}^{fl}}{\overline{E}_{xx}^{fl}t^3}\left(\frac{M_{xx}^b}{b}\right)$$

The moment of inertia for a rectangular cross-section of width b and height t is $I = bt^3/12$. Therefore, the above expressions can be written as

$$\kappa_{xx} = \frac{M_{xx}^b}{\overline{E}_{xx}^{fl}I} \tag{8.10a}$$

$$\kappa_{yy} = -\frac{\overline{v}_{xy}^{fl}M_{xx}^b}{\overline{E}_{xx}^{fl}I} \tag{8.10b}$$

$$\kappa_{xy} = \frac{\overline{\eta}_{xx,xy}^{fl}M_{xx}^b}{\overline{E}_{xx}^{fl}I} \tag{8.10c}$$

Equations 8.10 are *moment-curvature equations* for a composite beam with ply orientation orthogonal to the plane of loading, and should be compared with analogous results for isotropic beams (Equation 8.2). The product $\overline{E}_{xx}^{fl}I$ represents the *effective flexural rigidity* for a composite beam in this orientation. Thus, the effective flexural rigidity of a rectangular composite beam with ply orientation orthogonal to the plane of loading can be written as

$$(\overline{IE}_{xx}^{fl}) = \frac{bt^3\overline{E}_{xx}^{fl}}{12} = \left(\frac{b}{d_{11}}\right) \tag{8.11}$$

where $\overline{E}_{xx}^{fl}$ is the effective flexural Young's modulus of the laminate (defined in Section 6.9.1.2).

Equations 8.10a and 8.10b also imply $\kappa_{yy} = -\overline{v}_{xy}^{fl}\kappa_{xx}$, which shows that a composite beam with rectangular cross-section will exhibit anticlastic bending in the same manner as an isotropic beam. Equations 8.10a and 8.10c imply $\kappa_{xy} = \overline{\eta}_{xx,xy}^{fl}\kappa_{xx}$, which shows that a twist curvature is induced in a composite laminate, unless $\overline{\eta}_{xx,xy}^{fl} = 0$. A midplane twist curvature κ_{xy} does not occur for isotropic beams subjected to pure bending. Recalling that $\overline{\eta}_{xx,xy}^{fl} = d_{16}/d_{11}$, it is seen that $\overline{\eta}_{xx,xy}^{fl}$ is zero only for specially orthotropic laminates (i.e., if $d_{16} = 0$). This rarely occurs for composite beams used in practice. The only common stacking sequences that lead to a specially orthotropic beam are unidirectional $[0]_n$ or $[90]_n$ stacks, or symmetric cross-ply $[0/90]_{ns}$ stacks. The first two stacking sequences are essentially never used in practice because they possess significant strength and stiffness in one direction only. Cross-ply beams may be used on occasion, but suffer from very low shear stiffness. Beams produced using any other stacking sequence, for example a symmetric quasi-isotropic stacking sequence, will exhibit a twist curvature when subjected to a pure bending.

On the basis of the above definitions, we see that through-thickness variation in strains ε_{xx}, ε_{yy}, and γ_{xy} induced by pure bending are given by

$$\varepsilon_{xx}(z) = z\kappa_{xx} = z\left(\frac{M_{xx}^b}{\overline{E}_{xx}^{fl}I}\right) \tag{8.12a}$$

$$\varepsilon_{yy}(z) = z\kappa_{yy} = -z\left(\frac{\overline{v}_{xy}^{fl}M_{xx}^b}{\overline{E}_{xx}^{fl}I}\right) \tag{8.12b}$$

$$\gamma_{xy}(z) = z\kappa_{xy} = z\left(\frac{\overline{\eta}_{xx,xy}^{fl}M_{xx}^b}{\overline{E}_{xx}^{fl}I}\right) \tag{8.12c}$$

Strains are predicted to vary as linear functions of z, as expected from results in earlier chapters. Note, however, that ply stresses σ_{xx}, σ_{yy}, and τ_{xy} are *not* linear functions of z. It is not possible to develop a "flexure formula" for use with composite beams with ply orientation orthogonal to the plane of loading. Rather, stresses must be calculated using Hooke's law for an anisotropic material and based on the ply strain distributions given by Equations 8.12, as discussed in earlier chapters.

8.5.2 Effective Flexural Rigidity of Rectangular Composite Beams with Ply Interfaces Parallel to the Plane of Loading

We now consider a composite beam that is oriented such that the ply interfaces are parallel to the plane of loading. Such a beam in pure bending is shown in Figure 8.8. We will continue to label the beam cross-section using the same symbols as in earlier sections. Hence, the beam *depth* is denoted as b, whereas the beam *width* (which equals the thickness of the composite laminate) is denoted as t (compare Figures 8.7 and 8.8). As drawn in Figure 8.8, plies are numbered from left to right. That is, the outermost surface of ply 1 exists at $z' = -t/2$, whereas the outermost surface of ply n exists at $z' = +t/2$.

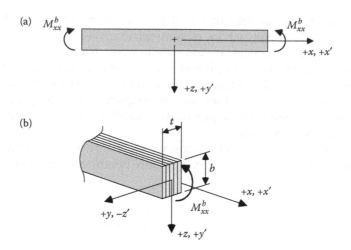

FIGURE 8.8
A composite beam with ply interfaces parallel to the x–z plane, subjected to pure bending.
(a) A composite beam subjected to pure bending; (b) end view of beam.

Ply fiber angles are referenced to the $+x'$-axis, and are measured positive *from* the $+x'$-axis *toward* the $+y'$-axis, in accordance with the right-hand rule.

Note that the bending moment is referenced to the x–y–z coordinate system, whereas the ply stacking sequence is referenced to a new x'–y'–z' coordinate system. As is apparent from Figure 8.8, the $+z$- and $+y'$-axes are coincident, as are the $+x$- and $+x'$-axes. The $+y$- and $+z'$-axes are parallel, but the $+z'$-direction is opposite to the $+y$-direction. The reader should carefully consider the two coordinate systems shown, *as they represent a change in nomenclature from our earlier discussion.* This is a subtle but potentially confusing change. For example, the bending moment M_{xx}^b shown in Figure 8.8 causes a very different state of stress in the composite beam than does the bending moment shown in Figure 8.7 (which is also denoted "M_{xx}^b"). This difference is of course due to the change in orientation of the composite beam.

Throughout this chapter we assume the stacking sequence is symmetric. Hence, both the beam cross-section as well as beam material properties are symmetric about the x–z plane. As the moment M_{xx}^b is constant along the length of the beam, all beam cross-sections must deform in an identical manner. Under these conditions the long axis of the beam *must* deform into a circular arc with a radius of curvature r_{xx}, as shown in Figure 8.9. Hence, *beam cross-sections that are initially plane and perpendicular to the axis of the beam must remain plane and perpendicular to the deformed axis of the beam following loading.* These are precisely the same conditions encountered in pure bending of isotropic prismatic beams (see article 102 of Reference 6). Hence, many conclusions based on isotropic beam theory are also applicable to the composite beam shown in Figures 8.8 and 8.9. In particular, the neutral axis must pass

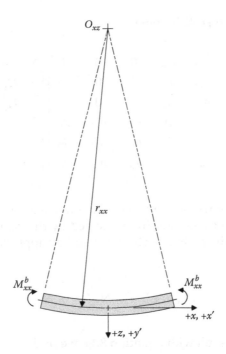

FIGURE 8.9
Deformations induced in a composite beam with ply interfaces parallel to the plane of loading, subjected to pure bending (deformations shown greatly exaggerated).

through the centroid of the beam cross-section, and axial normal strains are given by

$$\varepsilon_{xx} = z\kappa_{xx} \tag{8.13}$$

where curvature $\kappa_{xx} = 1/r_{xx}$. Hence, axial strain is a maximum at $z = \pm b/2$, and is zero at $z = 0$, the neutral axis. A subtle point is that in this case the curvature "κ_{xx}" represents bending in the x–z plane (the plane of loading), and *is not comparable* to the curvature "κ_{xx}" discussed previously. The curvature previously discussed occurs in the x'–z' plane, and would now be labeled $\kappa_{x'x'}$.

As we have assumed both a symmetric stacking sequence and symmetric rectangular cross-section, the axial strain induced at any position z will be uniform across the width of the beam. Hence, the axial strain given by Equation 8.13 represents the strain induced at the midplane of the laminate:

$$\varepsilon_{xx}^o = \varepsilon_{x'x'}^o = z\kappa_{xx} \tag{8.14}$$

We also note that for the loading condition shown in Figure 8.8:

$$N_{y'y'} = N_{x'y'} = M_{x'x'} = M_{y'y'} = M_{x'y'} = 0$$

Therefore, Equation 6.45 becomes

$$
\begin{Bmatrix}
\varepsilon^o_{x'x'} \\
\varepsilon^o_{y'y'} \\
\gamma^o_{x'y'} \\
\kappa_{x'x'} \\
\kappa_{y'y'} \\
\kappa_{x'y'}
\end{Bmatrix}
=
\begin{bmatrix}
a_{11} & a_{12} & a_{16} & 0 & 0 & 0 \\
a_{12} & a_{22} & a_{26} & 0 & 0 & 0 \\
a_{16} & a_{26} & a_{66} & 0 & 0 & 0 \\
0 & 0 & 0 & d_{11} & d_{12} & d_{16} \\
0 & 0 & 0 & d_{12} & d_{22} & d_{26} \\
0 & 0 & 0 & d_{16} & d_{26} & d_{66}
\end{bmatrix}
\begin{Bmatrix}
N_{x'x'} + N^T_{x'x'} + N^M_{x'x'} \\
N^T_{y'y'} + N^M_{y'y'} \\
N^T_{x'y'} + N^M_{x'y'} \\
0 \\
0 \\
0
\end{Bmatrix}
$$

Our current objective is to evaluate the effective flexural rigidity of the composite beam, so we ignore strains caused by thermal or moisture stress resultants. Hence, the midplane strain caused by application of stress resultant $N_{x'x'} = N_{xx}$ is

$$
\varepsilon^o_{x'x'} = \varepsilon^o_{xx} = a_{11}N_{x'x'} = a_{11}N_{xx}
$$

Combining this result with Equation 8.14, we find

$$
N_{xx} = z\left(\frac{\kappa_{xx}}{a_{11}}\right)
\tag{8.15}
$$

Now consider a free-body diagram showing the internal distribution of stress resultant N_{xx} at arbitrary cross-section A–A, as shown in Figure 8.10. Since the beam is in static equilibrium the externally applied moment M^b_{xx} must be balanced by the distribution of internal stress resultant N_{xx}. Recalling that the units of N_{xx} are force-per-length (e.g., N/m), the incremental axial force acting over an infinitesimal strip

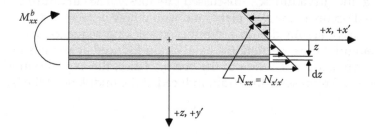

FIGURE 8.10
Free-body diagram used to relate external bending moment M^b_{xx} to internal stress resultant $N_{xx} = N_{x'x'}$.

of height dz, located at an arbitrary distance z from the neutral axis, is given by

$$dN_{xx}^b = N_{xx}dz$$

As this force acts at distance "z" from the neutral axis, the incremental internal moment associated with this force is

$$dM_{xx}^b = dN_{xx}^b z = N_{xx}zdz$$

The total moment is obtained by integrating over the height of the beam:

$$M_{xx}^b = \int_{-b/2}^{+b/2} N_{xx}zdz = \int_{-b/2}^{+b/2} \left(\frac{\kappa_{xx}}{a_{11}} \right) z^2dz$$

Neither κ_{xx} or a_{11} is function of z, so they can be removed from under the integral sign. Completing the integration indicated we find

$$M_{xx}^b = \frac{\kappa_{xx}b^3}{12a_{11}}$$

The effective extensional Young's modulus is given by Equation 6.63, repeated here for convenience:

$$\bar{E}_{xx}^{ex} = \frac{1}{ta_{11}} \qquad \text{(6.63) (repeated)}$$

Our expression for M_{xx}^b can, therefore, be written as

$$M_{xx}^b = \frac{\kappa_{xx}\bar{E}_{xx}^{ex}tb^3}{12}$$

Noting that the area moment of inertia about the y-axis for the rectangular cross-section is $I = tb^3/12$, we can rearrange this result to find:

$$\kappa_{xx} = \frac{M_{xx}^b}{\bar{E}_{xx}^{ex}I} \qquad \text{(8.16)}$$

Equation 8.16 is a *moment-curvature equation* for a composite beam with ply orientation parallel to the plane of loading. The product $\bar{E}_{xx}^{ex}I$ represents the *effective flexural rigidity* of a composite beam in this orientation. That is, the effective flexural rigidity of a rectangular composite beam with ply orientation parallel to the plane of loading can be written as

$$(\overline{IE^{ex}_{xx}}) = \frac{tb^3 \overline{E^{ex}_{xx}}}{12a_{11}} = \left(\frac{b^3}{12a_{11}}\right) \tag{8.17}$$

where $\overline{E^{ex}_{xx}}$ is the effective *extensional* modulus of the laminate (defined in Section 6.9.1.1). Equation 8.17 should be compared with our earlier expression for the effective flexural rigidity of a composite beam with plies orthogonal to the plane of loading (Equation 8.11). It is seen that the flexural rigidity of a composite beam with plies parallel to the plane of loading is dominated by extensional stiffnesses (i.e., $\overline{E^{ex}_{xx}}$ or a_{11}). In contrast, the flexural rigidity of a beam with plies orthogonal to the plane of loading is dominated by flexural stiffnesses (i.e., $\overline{E^{fl}_{xx}}$ or d_{11}).

Anticlastic bending will occur for the composite beam shown in Figure 8.10. Normal strains $\varepsilon_{yy} = \varepsilon_{z'z'}$ will vary linearly with distance from the neutral axis. Assuming a positive bending moment, then in regions below the neutral axis (i.e., for $z = y' > 0$) the transverse normal strains will be compressive, whereas in regions above the neutral axis (for $z = y' < 0$) these strains will be tensile. These transverse normal strains were neglected during our earlier development and hence are not included in CLT. These strains are not of immediate interest, so will not be further discussed here.

Next, consider the state of stress induced in the beam. Recall from Section 6.9.1 that the effective (or nominal) normal stress is given by $\overline{\sigma}_{x'x'} = \overline{\sigma}_{xx} = N_{xx}/t$. Substituting this definition into Equation 8.15 and rearranging, we have

$$\kappa_{xx} = \frac{\overline{\sigma}_{xx} t a_{11}}{z} = \frac{\overline{\sigma}_{xx}}{z\overline{E^{ex}_{xx}}}$$

Combining this result with Equation 8.16 and solving for effective stress, we have

$$\overline{\sigma}_{xx} = \frac{M^b_{xx} z}{I} \tag{8.18}$$

Equation 8.18 is analogous to the *flexure formula* for isotropic beams (Equation 8.3), and can be used to determine the effective stress at any position z. As would be expected, the effective stress $\overline{\sigma}_{xx}$ is at a maximum at the outer surfaces of the beam (at $z = \pm t/2$), and is zero at the neutral surface (at $z = 0$). Once the effective stress $\overline{\sigma}_{xx}$ has been calculated at a point of interest, then the stresses and strains present in individual plies can be determined using standard CLT analysis methods.

8.6 Effective Axial and Flexural Rigidities for Thin-Walled Composite Beams

In this section, the effective axial and flexural rigidities of thin-walled composite beams will be considered. The beam cross-section must conform to the symmetries described in Section 8.3. The process used to determine effective rigidities is similar for all cross-sections. A composite box-beam will be used to illustrate the process.

Referring to Figure 8.11, a box-beam is modeled using four symmetric rectangular laminates. The two web laminates are required to have identical stacking sequences, but the two symmetric flange laminates may differ. The thickness of the top and bottom flange laminates is labeled t_{tf} and t_{bf}, respectively. The width of both flanges is labeled b. The two identical web laminates have a thickness of t_w and height h. Thus, the overall width and depth of the beam cross-section are (b) and $(h + t_{tf} + t_{bf})$, respectively.

To define the axial rigidity, we must determine the axial strains ε_{xx} induced when the beam is subjected to an axial load N_{xx}^b (only). Furthermore, N_{xx}^b must be applied such that *uniform* axial strains are induced, that is, the beam must not bend. This is equivalent to saying that the line of action of N_{xx}^b must pass through the centroid of the beam cross-section. It is known *a priori* that the centroid lies along within the x–z plane, as we have required symmetry about this plane. However, we have not required symmetry about the x–y plane. Hence, the position of the centroid relative to either outer surface of the beam cross-section (i.e., coordinates z_t and z_b shown in Figure 8.11) is initially unknown. Note that as the z-axis is positive downwards, in all cases

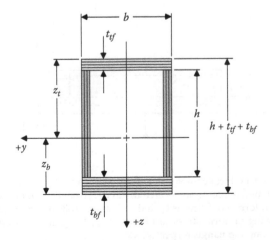

FIGURE 8.11
Cross-section of a composite box-beam. The positive x-axis is out of the plane of the figure.

$z_t < 0$ and $z_b > 0$. Our first objective is to determine either distance z_t or z_b. In the following derivation we will determine distance z_b. Note that once z_b is known z_t is also known, as $z_t = z_b - (h + t_{bf} + t_{tf})$.

Sketches of a composite box-beam subjected to axial load N_{xx}^b (only) are shown in Figure 8.12. In this figure, it is assumed that the line of action of N_{xx}^b *does* pass through the centroid. Consequently, a uniform axial strain ε_{xx} is induced at an arbitrary cross-section of the beam, as shown in Figure 8.12a. Since strain is uniform, the stress resultants induced in each segment of the beam can be written as

$$N_{xx}\big|_w = \frac{\varepsilon_{xx}}{a_{11}^w} = t_w \, \overline{E}_{xx}^{ex}\big|_w \, \varepsilon_{xx} \tag{8.19a}$$

$$N_{xx}\big|_{bf} = \frac{\varepsilon_{xx}}{a_{11}^{bf}} = t_{bf} \, \overline{E}_{xx}^{ex}\big|_{bf} \, \varepsilon_{xx} \tag{8.19b}$$

$$N_{xx}\big|_{tf} = \frac{\varepsilon_{xx}}{a_{11}^{tf}} = t_{tf} \, \overline{E}_{xx}^{ex}\big|_{tf} \, \varepsilon_{xx} \tag{8.19c}$$

Symbols "*w*," "*bf*," and "*tf*" have been used to denote variables associated with the web laminate, bottom flange laminate, and top flange laminate,

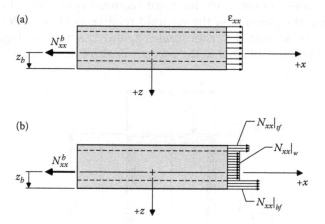

FIGURE 8.12
Side view of a composite box-beam subjected to axial load N_{xx}^b. The line-of-action of N_{xx}^b is assumed to pass through the centroid, located distance z_b from the bottom surface of the beam. Consequently, (a) uniform axial strains ε_{xx} induced at an arbitrary cross-section and (b) free-body diagram showing uniform stress resultants $N_{xx}\big|_w$, $N_{xx}\big|_{bf}$, and $N_{xx}\big|_{tf}$, induced within the web, bottom flange, and top flange, respectively.

respectively. For example, $\overline{E}_{xx}^{ex}\big|_{w}$ represents the effective extensional modulus of the web laminates.

The stress resultants given by Equations 8.19 are indicated in the free-body diagram shown in Figure 8.12b. To determine the location of the centroidal axis, we require that the moment associated with the internal stress resultants vanish. Summing moments about the y-axis and equating to zero $(\Sigma M_y = 0)$, we obtain

$$(2)\left[(N_{xx}\big|_w)(h)\right](z_b - t_{bf} - h/2) + \left[(N_{xx}\big|_{bf})(b)\right](z_b - t_{bf}/2)$$

$$+ \left[(N_{xx}\big|_{tf})(b)\right](z_b - t_{bf} - h - t_{tf}/2) = 0$$

Substituting Equation 8.19 and solving for distance z_b, we find

$$z_b = \frac{1}{2}\left[\frac{2A_w \overline{E}_{xx}^{ex}\big|_w (h + 2t_{bf}) + t_{bf} A_{bf} \overline{E}_{xx}^{ex}\big|_{bf} + A_{tf} \overline{E}_{xx}^{ex}\big|_{tf} (2t_b + 2h + t_{tf})}{2A_w \overline{E}_{xx}^{ex}\big|_w + A_{bf} \overline{E}_{xx}^{ex}\big|_{bf} + A_{tf} \overline{E}_{xx}^{ex}\big|_{tf}}\right] \tag{8.20a}$$

where

$A_w = t_w h$ = cross-sectional area of the web laminate
$A_{bf} = t_{bf}b$ = cross-sectional area of the bottom flange laminate
$A_{tf} = t_{tf}b$ = cross-sectional area of the top flange laminate

Noting that the effective extensional Young's moduli can be written $\overline{E}_{xx}^{ex}\big|_w = 1/t_w a_{11}^w$, $\overline{E}_{xx}^{ex}\big|_{bf} = 1/t_{bf} a_{11}^{bf}$, and $\overline{E}_{xx}^{ex}\big|_{tf} = 1/t_{tf} a_{11}^{tf}$, this result can also be expressed as

$$z_b = \frac{1}{2}\left\{\frac{2ha_{11}^{bf}a_{11}^{tf}(h + 2t_{bf}) + ba_{11}^w\left[t_{bf}a_{11}^{tf} + a_{11}^{bf}(2t_{bf} + 2h + t_{tf})\right]}{2ha_{11}^{bf}a_{11}^{tf} + ba_{11}^w(a_{11}^{bf} + a_{11}^{tf})}\right\} \tag{8.20b}$$

Equation 8.20 gives the distance from the centroid to the bottom surface of the composite box-beam. If the bottom and top flange laminates are identical (i.e., if $A_{bf} = A_{tf}$, $t_{bf} = t_{tf}$, and $\overline{E}_{xx}^{ex}\big|_{bf} = \overline{E}_{xx}^{ex}\big|_{tf}$), then Equation 8.20 reduces to $z_b = t_{bf} + h/2$, as would be expected.

Let us now determine the effective axial rigidity of the composite box beam. As a uniform axial strain (ε_{xx}) is induced, the stress resultants present in each segment of the beam to the axial strain are given by

$$N_{xx}\big|_w = \frac{\varepsilon_{xx}}{a_{11}^w} = t_w \, \overline{E}_{xx}^{ex}\Big|_w \varepsilon_{xx}$$

$$N_{xx}\big|_{bf} = \frac{\varepsilon_{xx}}{a_{11}^{bf}} = t_{bf} \, \overline{E}_{xx}^{ex}\Big|_{bf} \varepsilon_{xx} \qquad (8.21)$$

$$N_{xx}\big|_{tf} = \frac{\varepsilon_{xx}}{a_{11}^{tf}} = t_{tf} \, \overline{E}_{xx}^{ex}\Big|_{tf} \varepsilon_{xx}$$

Also, the total load applied to the beam as a whole (N_{xx}^b) must equal the sum of the forces acting in each beam segment:

$$N_{xx}^b = 2(N_{xx}\big|_w)(h) + (N_{xx}\big|_{bf})(b) + (N_{xx}\big|_{tf})(b)$$

Define the effective stress applied to the beam as the force applied to the beam divided by the area of the beam cross-section:

$$\overline{\sigma}_{beam} = \frac{N_{xx}^b}{A_{beam}} = \frac{2(N_{xx}\big|_w)(h) + (N_{xx}\big|_{bf})(b) + (N_{xx}\big|_{tf})(b)}{2A_w + A_{bf} + A_{tf}} \qquad (8.22)$$

Substituting Equation 8.21 into 8.22, we find

$$\overline{\sigma}_{beam} = \varepsilon_{xx} \left[\frac{2A_w \overline{E}_{xx}^{ex}\Big|_w + A_{bf} \overline{E}_{xx}^{ex}\Big|_{bf} + A_{tf} \overline{E}_{xx}^{ex}\Big|_{tf}}{2A_w + A_{bf} + A_{tf}} \right]$$

We can now define the effective Young's modulus of the beam as

$$\overline{E}_{xx}^{beam} = \frac{2A_w \overline{E}_{xx}^{ex}\Big|_w + A_{bf} \overline{E}_{xx}^{ex}\Big|_{bf} + A_{tf} \overline{E}_{xx}^{ex}\Big|_{tf}}{2A_w + A_{bf} + A_{tf}} \qquad (8.23)$$

The effective axial rigidity of the composite box beam is simply the effective beam Young's modulus multiplied by the cross-sectional area:

$$A\overline{E}_{xx}^{beam} = 2A_w \overline{E}_{xx}^{ex}\Big|_w + A_{bf} \overline{E}_{xx}^{ex}\Big|_{bf} + A_{tf} \overline{E}_{xx}^{ex}\Big|_{tf} \qquad (8.24a)$$

Alternatively, the effective axial rigidity of the box beam can be written as

$$\overline{AE}_{xx}^{\text{beam}} = \frac{2h}{a_{11}^w} + \frac{b}{a_{11}^{bf}} + \frac{b}{a_{11}^{tf}}$$

(8.24b)

If the bottom and top flange laminates are identical $\left(\text{i.e., if } A_{bf} = A_{tf}, \ t_{bf} = t_{tf},\right.$ and $\left.\overline{E}_{xx}^{ex}\big|_{bf} = \overline{E}_{xx}^{ex}\big|_{tf}\right)$, then Equation 8.24 reduce to

$$\overline{AE}_{\text{beam}} = 2\left(A_w\, \overline{E}_{xx}^{ex}\bigg|_{w} + A_{bf}\, \overline{E}_{xx}^{ex}\bigg|_{bf} \right) = 2\left(\frac{h}{a_{11}^w} + \frac{b}{a_{11}^{bf}} \right)$$

Next, consider the effective flexural rigidity of a composite box-beam. Recall that the flexural rigidity of a beam is defined under a condition of pure bending. A side view of a composite box-beam subjected to pure bending is shown in Figure 8.13a. As the beam cross-section is symmetric about the

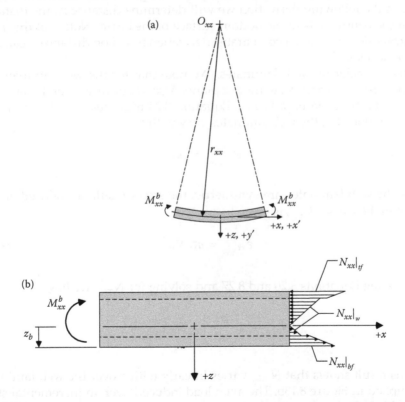

FIGURE 8.13
Side view of a composite box-beam subjected to pure bending. (a) Composite box-beam deformed into a circular are of radius r_{xx}; (b) distribution of internal stress resultants $N_{xx}\big|_w$, $N_{xx}\big|_{bf}$, and $N_{xx}\big|_{tf}$.

x–z plane, the long x-axis of the beam *must* deform into a circular arc with a radius of curvature r_{xx}, as shown. Hence, cross-sections of the box-beam that are initially plane and perpendicular to the axis of the beam must remain plane and perpendicular to the deformed axis of the beam following load-ing. These are precisely the same conditions encountered in pure bending of a rectangular composite beam with ply orientation parallel to the plane of loading (discussed Section 8.5.2), as well as in isotropic beams with sym-metric rectangular cross-sections. Axial normal strains vary linearly with z and are given by

$$\varepsilon_{xx} = z\kappa_{xx} \tag{8.25}$$

where curvature $\kappa_{xx} = 1/r_{xx}$. We assume that the axial strain induced at any position z is uniform across the width of the web and flange laminates; that is, we assume ε_{xx} is not a function of y.

Our first objective is to determine the location of the neutral axis. That is, we wish to determine either distance z_t or z_b, previously shown in Figure 8.11. In the following derivation we will determine distance z_b, the distance from the neutral axis to the bottom surface of the beam. Note that this is an arbitrary decision, and a comparable derivation based on distance z_t can eas-ily be developed.

First consider the web laminates. We note that for the web laminate the ply interfaces are parallel to the x–z plane. Also, we have assumed that strain ε_{xx} is not a function of y. Hence, Equation 8.25 indicates that the midplane strains induced in the web laminates vary with z:

$$\varepsilon^o_{xx}\big|_{\text{web}} = z\kappa_{xx} \tag{8.26}$$

As the web laminates are symmetric, the stress resultant induced at any point in the webs, $N_{xx}\big|_w$, is given by

$$\varepsilon^o_{xx}\big|_{web} = a^w_{11} N_{xx}\big|_w \tag{8.27}$$

Equating Equations 8.26 and 8.27 and solving for $N_{xx}\big|_w$ we have

$$N_{xx}\big|_w = z\left(\frac{\kappa_{xx}}{a^w_{11}}\right) = z t_w \overline{E}^{ex}_{xx}\big|_w \kappa_{xx}$$

This result shows that $N_{xx}\big|_w$ varies linearly with z over the web laminate, as implied in Figure 8.13b. The axial load induced over an incremental strip of width dz is

$$dN^w_{xx} = N_{xx}\big|_w dz$$

Hence, the total axial load supported the web, $N_{xx}^b\big|_w$, is given by

$$N_{xx}^b\big| = \int_{z_b-h-t_{bf}}^{z_b-t_{bf}} N_{xxw}\big|_w dz = \int_{z_b-h-t_{bf}}^{z_b-t_{bf}} \left(t_w \, \overline{E}_{xx}^{ex}\big|_w \kappa_{xx} \right) z\,dz$$

Evaluating this integral, we find

$$N_{xx}^b\big|_w = \frac{t_w \, \overline{E}_{xx}^{ex}\big|_w \kappa_{xx}}{2} \left[(z_b - t_{bf})^2 - (z_b - h - t_{bf})^2 \right] \tag{8.28}$$

Now consider the bottom flange laminate. Referring to Figure 8.11, note that the midplane of the bottom flange is located at $z = z_b - t_{bf}/2$. Substituting this value into Equation 8.25, the strain induced at the midplane of the bottom flange laminate is given by

$$\varepsilon_{xx}^o\big|_{bf} = \left(z_b - \frac{t_{bf}}{2} \right) \kappa_{xx} \tag{8.29}$$

As the bottom flange laminate is symmetric, the stress resultant induced in the bottom flange, $N_{xx}\big|_{bf}$, is

$$\varepsilon_{xx}^o\big|_{bf} = a_{11}^{bf} N_{xx}\big|_{bf} = \frac{N_{xx}\big|_{bf}}{t_{bf} \, \overline{E}_{xx}^{ex}\big|_{bf}} \tag{8.30}$$

Equating Equations 8.29 and 8.30 and solving for $N_{xx}\big|_{bf}$, we find

$$N_{xx}\big|_{bf} = \left(z_b - \frac{t_{bf}}{2} \right) t_{bf} \, \overline{E}_{xx}^{ex}\big|_{bf} \kappa_{xx} \tag{8.31}$$

This result gives the value of $N_{xx}\big|_{bf}$ induced at the midplane of the bottom laminate, shown in Figure 8.13b. The total axial force supported by the bottom flange, $N_{xx}^b\big|_{bf}$, is obtained by multiplying stress resultant $N_{xx}\big|_{bf}$ by the width of the bottom flange, b:

$$N_{xx}^b\big|_{bf} = b\,N_{xx}\big|_{bf} = \left(z_b - \frac{t_{bf}}{2} \right) A_{bf} \, \overline{E}_{xx}^{ex}\big|_{bf} \kappa_{xx} \tag{8.32}$$

Note that the cross-sectional area of the bottom flange, $A_{bf} = bt_{bf}$, has been used in Equation 8.32.

Finally, we must determine the total axial force supported by the top flange. From Figure 8.11 we see that the midplane of the top flange laminate exists at $z = (z_b - t_{bf} - h - t_{tf}/2)$. On the basis of this value, expressions analogous to

Equations 8.28 through 8.32 can be obtained for the top flange laminate. We find that the total axial load supported by the top flange is given by

$$N_{xx}^b\Big|_{tf} = \left(z_b - t_{bf} - h - \frac{t_{tf}}{2} \right) A_{tf}\, \overline{E}_{xx}^{ex}\Big|_{tf}\, \kappa_{xx} \tag{8.33}$$

All equations necessary to determine the position of the neutral axis have now been developed. The box beam is modeled as four rectangular laminates (two web laminates, the bottom flange laminate, and the top flange laminate) *and is in pure bending*. Hence, the axial loads present in all four laminates must sum to zero:

$$2N_{xx}^b\Big|_{w} + N_{xx}^b\Big|_{bf} + N_{xx}^b\Big|_{tf} = 0$$

Substituting Equations 8.28, 8.32, and 8.33 and solving for z_b, we obtain

$$z_b = \frac{1}{2}\left[\frac{ 2A_w\, \overline{E}_{xx}^{ex}\Big|_{w}\,(h + 2t_{bf}) + t_{bf}A_{bf}\, \overline{E}_{xx}^{ex}\Big|_{bf} + A_{tf}\, \overline{E}_{xx}^{ex}\Big|_{tf}\,(2t_b + 2h + t_{tf}) }{ 2A_w\, \overline{E}_{xx}^{ex}\Big|_{w} + A_{bf}\, \overline{E}_{xx}^{ex}\Big|_{bf} + A_{tf}\, \overline{E}_{xx}^{ex}\Big|_{tf} } \right] \tag{8.34}$$

Equation 8.34 is identical to Equation 8.20a. Hence, we conclude that in pure bending the neutral axis passes through the centroid of the beam cross-section. This same conclusion holds true for isotropic beams in pure bending.

Now that the position of the neutral axis has been identified the effective flexural rigidity of the composite box-beam can be determined. First consider the moment supported by a web laminate. The incremental moment associated with the force $dN_{xx}^b\big|_{w}$ located distance z from the neutral axis is given by

$$dM_{xx}^b\Big|_{w} = (d\,N_{xx}^b\big|_{w})z = \left(t_w\, \overline{E}_{xx}^{ex}\Big|_{w}\, \kappa_{xx} \right) z^2 dz$$

The total moment supported by a web laminate is then

$$M_{xx}^b\Big|_{w} = \int_{z_b - h - t_{bf}}^{z_b - t_{bf}} \left(t_w\, \overline{E}_{xx}^{ex}\Big|_{w}\, \kappa_{xx} \right) z^2 dz$$

Evaluating this integral, we find that the moment supported by a web laminate is given by

$$M_{xx}^b\Big|_{w} = \left(\frac{ t_w\, \overline{E}_{xx}^{ex}\Big|_{w}\, \kappa_{xx} }{3} \right) \left[(z_b - t_{bf})^3 - (z_b - h - t_{bf})^3 \right] \tag{8.35}$$

Now consider the moment supported by the bottom flange laminate. As the midplane of the bottom flange laminate is not located at the neutral axis of the overall beam cross-section, the flange laminate will experience both a midplane strain and a midplane curvature. The beam cross-section must remain plane and consequently the curvature experienced by the bottom flange is identical to that experienced by the beam as a whole: $\kappa_{xx}|_{bf} = \kappa_{xx}$. The moment resultant associated with this curvature is

$$M_{xx}|_{bf} = \kappa_{xx}/d_{11}^{bf} = t_{bf}^3 \, \overline{E}_{xx}^{fl}\Big|_{bf} \kappa_{xx}/12$$

This moment resultant contributes to the total moment supported by the bottom flange. In addition, the stress resultant induced $N_{xx}|_{bf}$ also contributes to the moment supported by the bottom flange and may be of equal or greater importance. The bottom flange laminate is located at $z = (z_b - t_{bf}/2)$. Substituting this coordinate into Equation 8.25, we find that the midplane strain induced in the bottom flange laminate is

$$\varepsilon_{xx}^o\big|_{bf} = (z_b - t_{bf}/2)\kappa_{xx}$$

This midplane strain is related to stress resultant $N_{xx}|_{bf}$ according to

$$\varepsilon_{xx}^o\big|_{bf} = a_{11}^{bf} N_{xx}\big|_{bf}$$

Equating and solving for $N_{xx}|_{bf}$, we find

$$N_{xx}\big|_{bf} = (z_b - t_{bf}/2)\left(\frac{\kappa_{xx}}{a_{11}^{bf}}\right) = (z_b - t_{bf}/2)\left(t_{bf}\, \overline{E}_{xx}^{ex}\Big|_{bf} \kappa_{xx}\right)$$

This stress resultant acts at a distance $(z_b - t_{bf}/2)$ from the neutral axis, and so the moment contributed by $N_{xx}|_{bf}$ is obtained by multiplying by this distance. Both $N_{xx}|_{bf}$ and $M_{xx}|_{bf}$ are resultants (i.e., loads per unit length) and must be multiplied by the width of the flange to determine the bending moment. On the basis of the above discussion, the total moment supported by the bottom flange laminate is

$$M_{xx}^b\big|_{bf} = \left[(z_b - t_{bf}/2)^2\left(bt_{bf}\, \overline{E}_{xx}^{ex}\Big|_{bf}\right) + \frac{bt_{bf}^3 \, \overline{E}_{xx}^{fl}\Big|_{bf}}{12}\right]\kappa_{xx}$$

This expression can be simplified by noting that the area of the bottom flange is $A_{bf} = bt_{bf}$, and further that the area moment of inertia of the bottom flange (taken about the local midplane of the bottom flange) is $I_{bf} = bt_{bf}^3/12$:

$$M_{xx}^b\big|_{bf} = \left[(z_b - t_{bf}/2)^2\left(A_{bf}\,\overline{E}_{xx}^{ex}\big|_{bf}\right) + I_{bf}\,\overline{E}_{xx}^{fl}\big|_{bf}\right]\kappa_{xx} \qquad (8.36)$$

In passing, it might be anticipated that Equation 8.36 could be further simplified by expressing the area moment of inertia of the bottom flange laminate about the neutral axis of the beam, through application of the parallel axis theorem. It turns out that this is not the case. That is, recall that through application of the parallel axis theorem the moment of inertia of the bottom flange laminate about the neutral axis of the beam cross-section is given by

$$I_{bf}\big|_{na} = (z_b - t_{bf}/2)^2 A_{bf} + I_{bf}$$

Comparing this expression with Equation 8.36, it is seen that the parallel axis theorem cannot be used to advantage for a composite box-beam because the effective *extensional* modulus of the bottom flange, $\overline{E}_{xx}^{ex}\big|_{bf}$, differs from the effective *flexural* modulus of the bottom flange, $\overline{E}_{xx}^{fl}\big|_{bf}$.

Finally, we consider the moment supported by the top flange. Following an equivalent process, we conclude that the moment support by the top flange laminate is given by

$$M_{xx}^b\big|_{tf} = \left[(z_b - t_{bf} - h - t_{tf}/2)^2\left(A_{tf}\,\overline{E}_{xx}^{ex}\big|_{tf}\right) + I_{tf}\,\overline{E}_{xx}^{fl}\big|_{tf}\right]\kappa_{xx} \qquad (8.37)$$

The total moment applied to the beam, M_o^b, must equal the sum of the moment components supported by the two web laminates, the bottom flange laminate, and the top flange laminate:

$$M_o^b = 2M_{xx}^b\big|_{web} + M_{xx}^b\big|_{bf} + M_{xx}^b\big|_{tf}$$

Also, by definition the effective flexural rigidity of the box beam is related to M_o^b according to

$$\overline{EI} = \frac{M_o^b}{\kappa_{xx}}$$

Substituting Equations 8.35 through 8.37, the effective flexural rigidity of a composite box-beam is given by

$$
\overline{EI} = \frac{2t_w \, \overline{E}_{xx}^{ex}\big|_w}{3} \left[(z_b - t_{bf})^3 - (z_b - h - t_{bf})^3 \right] + \left[(z_b - t_{bf}/2)^2 A_{bf} \, \overline{E}_{xx}^{ex}\big|_{bf} \right.
$$

$$
\left. + (z_b - t_{bf} - h - t_{tf}/2)^2 A_{tf} \, \overline{E}_{xx}^{ex}\big|_{tf} + I_{bf} \, \overline{E}_{xx}^{fl}\big|_{bf} + I_{tf} \, \overline{E}_{xx}^{fl}\big|_{tf} \right] \qquad (8.38)
$$

If a box-beam is considered for which the bottom and top flange laminates are identical $\left(\text{i.e., if } A_{bf} = A_{tf}, \ t_{bf} = t_{tf}, \text{ and } \overline{E}_{xx}^{ex}\big|_{bf} = \overline{E}_{xx}^{ex}\big|_{tf} \right)$, then Equation 8.38 reduces to

$$
\overline{EI} = 2I_w \, \overline{E}_{xx}^{ex}\big|_w + 2 \left[\frac{(t_{bf} + h)^2 A_{bf} \, \overline{E}_{xx}^{ex}\big|_{bf}}{4} + I_{bf} \, \overline{E}_{xx}^{fl}\big|_{bf} \right]
$$

where I_w = the area moment of inertia of a web laminate, taken about the neutral axis of the box-beam.

This concludes our analysis of a composite box-beam. To summarize, we have restricted our analysis to composite box beams that possess certain material and geometric symmetries. Specifically, both the flange and web laminates must be produced using a symmetric stacking sequence, and the beam cross-section must be symmetric about the plane of loading, the x–z plane. This implies that the two web laminates must be produced using an identical symmetric stacking sequence, such that the set of fiber angles represented by the two web laminates is symmetric about the x–z plane. The top and bottom flange laminates must also be symmetric, but they need not be identical. Therefore, the beam cross-section need not be symmetric about the y–z plane. Having defined a beam cross-section in this way, then the location of the centroid (which also defines the location of the neutral axis in pure bending) is given by Equations 8.20 or 8.34. The effective axial rigidity of the box beam is given by Equation 8.24, and the effective flexural rigidity is given by Equation 8.38. As will be discussed in following sections, once the effective axial and flexural rigidities of the composite box beam have been determined, then deflection of the beam caused by axial and/or transverse loads can be determined using the methods traditionally employed to study isotropic beams.

The general approach described above can also be applied to the other thin-walled cross-sections previously discussed in Section 8.3 and shown in Figure 8.1. A summary of results for these additional thin-walled cross-sections is provided in Tables 8.1 through 8.3.

Example Problem 8.1

Three box-beams are constructed using unidirectional graphite–epoxy prepreg tape with the material properties listed in Table 3.1. The first box beam is shown in Figure 8.14a. In this case, the web and flange laminates are produced using an identical eight-ply $[0/90]_{2s}$ stacking sequence. The second beam is shown in Figure 8.14b. In this case additional plies are added to the top flange. Hence, both web laminates and the bottom flange laminate are produced using an eight-ply $[0/90]_{2s}$ stacking sequence, but the top laminate is produced using a 20-ply $[(0/90)_{2s}/\pm 45]_s$ stacking sequence. Finally, the third beam is produced using $[0/90]_{2s}$ web laminates and $[(0/90)_{2s}/\pm 45]_s$ top and bottom flange laminates. In all three cases, the overall beam width and depth is 30 and 50 mm, respectively. Determine the position of the centroid and the effective axial and flexural rigidities for

a. The beam shown in Figure 8.14a
b. The beam shown in Figure 8.14b
c. The beam shown in Figure 8.14c

SOLUTION

a. As both webs and flanges are produced using an eight-ply stacking sequence, laminate thicknesses are

$$t_w = t_{bf} = t_{tf} = 8(0.125 \text{ mm}) = 1 \text{ mm}$$

The length of the bottom and top flange laminates equals the overall width of the beam (30 mm). The length of the web laminates are

$$h = 50 \text{ mm} - t_{bf} - t_{tf} = 50 \text{ mm} - 2(1 \text{ mm}) = 48 \text{ mm}$$

We can now calculate the area properties of the web and flange laminates:

$$A_w = (h)(t_w) = (48 \text{ mm})(1 \text{ mm}) = 48 \times 10^{-6} \text{m}^2$$

$$A_{bf} = A_{tf} = (b)(t_{bf}) = (30 \text{ mm})(1 \text{ mm}) = 30 \times 10^{-6} \text{m}^2$$

$$I_{bf} = I_{tf} = \frac{(b)(t_{bf})^3}{12} = \frac{(30 \text{ mm})(1 \text{ mm})^3}{12} = 2.5 \times 10^{-12} \text{m}^4$$

The effective moduli for the web and flange laminates are calculated using the process described in Section 6.9.1 and are found to be

$$\overline{E}_{xx}^{ex}\bigg|_w = \overline{E}_{xx}^{ex}\bigg|_{bf} = \overline{E}_{xx}^{ex}\bigg|_{tf} = 90.4 \text{ GPa}$$

$$\overline{E}_{xx}^{fl}\bigg|_{bf} = \overline{E}_{xx}^{fl}\bigg|_{tf} = 120 \text{ GPa}$$

TABLE 8.1

Centroid Location of Some Thin-Walled Composite Beams with Symmetric Web and Flange Laminates

Cross-Section	Location of Centroid, z_b
	$\dfrac{1}{2}\left[\dfrac{A_w \overline{E}_{xx}^{ex}\big|_w (h + 2t_{bf}) + t_{bf} A_{bf} \overline{E}_{xx}^{ex}\big|_{bf} + A_{tf} \overline{E}_{xx}^{ex}\big|_{tf} (2t_{bf} + 2h + t_{tf})}{A_w \overline{E}_{xx}^{ex}\big|_w + A_{bf} \overline{E}_{xx}^{ex}\big|_{bf} + A_{tf} \overline{E}_{xx}^{ex}\big|_{tf}}\right]$
	$\dfrac{1}{2}\left[\dfrac{A_w \overline{E}_{xx}^{ex}\big|_w h + A_{tf} \overline{E}_{xx}^{ex}\big|_{tf} (2h + t_{tf})}{A_w \overline{E}_{xx}^{ex}\big|_w + A_{tf} \overline{E}_{xx}^{ex}\big|_{tf}}\right]$
	$\dfrac{1}{2}\left[\dfrac{2A_w \overline{E}_{xx}^{ex}\big|_w (h + 2t_{bf}) + 2t_{bf} A_{bf} \overline{E}_{xx}^{ex}\big|_{bf} + A_{tf} \overline{E}_{xx}^{ex}\big|_{tf} (2t_{bf} + 2h + t_{tf})}{2A_w \overline{E}_{xx}^{ex}\big|_w + 2A_{bf} \overline{E}_{xx}^{ex}\big|_{bf} + A_{tf} \overline{E}_{xx}^{ex}\big|_{tf}}\right]$
	$\dfrac{1}{2}\left[\dfrac{2A_w \overline{E}_{xx}^{ex}\big|_w (h + 2t_{bf}) + t_{bf} A_{bf} \overline{E}_{xx}^{ex}\big|_{bf} + A_{tf} \overline{E}_{xx}^{ex}\big|_{tf} (2t_{bf} + 2h + t_{tf})}{2A_w \overline{E}_{xx}^{ex}\big|_w + A_{bf} \overline{E}_{xx}^{ex}\big|_{bf} + A_{tf} \overline{E}_{xx}^{ex}\big|_{tf}}\right]$

TABLE 8.2

Effective Axial Rigidities of Some Thin-Walled Composite Beams with Symmetric Web and Flange Laminates

Cross-Section	Effective Axial Rigidity, $\overline{AE}_{xx}\big|_{beam}$
	$A_w \overline{E}_{xx}^{ex}\big|_w + A_{bf} \overline{E}_{xx}^{ex}\big|_{bf} + A_{tf} \overline{E}_{xx}^{ex}\big|_{tf}$
	$A_w \overline{E}_{xx}^{ex}\big|_w + A_{tf} \overline{E}_{xx}^{ex}\big|_{tf}$
	$2A_w \overline{E}_{xx}^{ex}\big|_w + 2A_{bf} \overline{E}_{xx}^{ex}\big|_{bf} + A_{tf} \overline{E}_{xx}^{ex}\big|_{tf}$
	$2A_w \overline{E}_{xx}^{ex}\big|_w + A_{bf} \overline{E}_{xx}^{ex}\big|_{bf} + A_{tf} \overline{E}_{xx}^{ex}\big|_{tf}$

TABLE 8.3

Effective Flexural Rigidities of Some Thin-Walled Composite Beams with
Symmetric Web and Flange Laminates

Cross-Section	Effective Flexural Rigidity, $\overline{IE}_{xx}\big|_{\text{beam}}$
	$$\frac{t_w \left. \overline{E}_{xx}^{ex}\right|_w}{3}\left[(z_b - t_{bf})^3 - (z_b - h - t_{bf})^3\right]$$ $$+ \left[(z_b - t_{bf}/2)^2 A_{bf}\left.\overline{E}_{xx}^{ex}\right|_{bf} + (z_b - t_{bf} - h - t_{tf}/2)^2 A_{tf}\left.\overline{E}_{xx}^{ex}\right|_{tf}\right.$$ $$\left. + I_{bf}\left.\overline{E}_{xx}^{fl}\right|_{bf} + I_{tf}\left.\overline{E}_{xx}^{fl}\right|_{tf}\right]$$
	$$\frac{t_w h\left.\overline{E}_{xx}^{ex}\right|_w}{3}\left[3z_b^2 - 3z_b h + h^2\right]$$ $$+ \left[(z_b - h - t_{tf}/2)^2 A_{tf}\left.\overline{E}_{xx}^{ex}\right|_{tf} + I_{tf}\left.\overline{E}_{xx}^{fl}\right|_{tf}\right]$$
	$$\frac{2t_w\left.\overline{E}_{xx}^{ex}\right|_w}{3}\left[(z_b - t_{bf})^3 - (z_b - h - t_{bf})^3\right]$$ $$+ \left[2(z_b - t_{bf}/2)^2 A_{bf}\left.\overline{E}_{xx}^{ex}\right|_{bf} + (z_b - t_{bf} - h - t_{tf}/2)^2 A_{tf}\left.\overline{E}_{xx}^{ex}\right|_{tf}\right.$$ $$\left. + 2I_{bf}\left.\overline{E}_{xx}^{fl}\right|_{bf} + I_{tf}\left.\overline{E}_{xx}^{fl}\right|_{tf}\right]$$
	$$\frac{2t_w\left.\overline{E}_{xx}^{ex}\right|_w}{3}\left[(z_b - t_{bf})^3 - (z_b - h - t_{bf})^3\right]$$ $$+ \left[(z_b - t_{bf}/2)^2 A_{bf}\left.\overline{E}_{xx}^{ex}\right|_{bf} + (z_b - t_{bf} - h - t_{tf}/2)^2 A_{tf}\left.\overline{E}_{xx}^{ex}\right|_{tf}\right.$$ $$\left. + I_{bf}\left.\overline{E}_{xx}^{fl}\right|_{bf} + I_{tf}\left.\overline{E}_{xx}^{fl}\right|_{tf}\right]$$

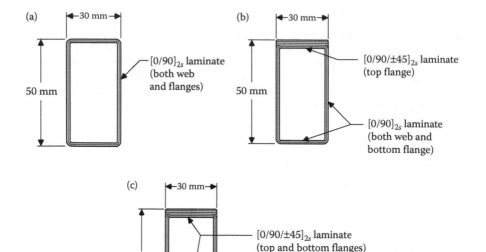

FIGURE 8.14
The composite box-beams considered in Example Problem 8.1 (not to scale). (a) Composite box-beam with identical stacking sequence in all web and flange laminates; (b) composite box-beam with additional plies added to the top flange laminate; (c) composite box-beam with additional plies added to both top and bottom flange laminates.

From Table 8.1, for a composite box beam the location of the centroid is given by

$$z_b = \frac{1}{2}\left[\frac{2A_w \overline{E}_{xx}^{ex}\Big|_w (h + 2t_{bf}) + t_{bf}A_{bf}\overline{E}_{xx}^{ex}\Big|_{bf} + A_{tf}\overline{E}_{xx}^{ex}\Big|_{tf}(2t_{bf} + 2h + t_{tf})}{2A_w \overline{E}_{xx}^{ex}\Big|_w + A_{bf}\overline{E}_{xx}^{ex}\Big|_{bf} + A_{tf}\overline{E}_{xx}^{ex}\Big|_{tf}}\right]$$

Substituting all known values and completing the indicated calculations, we find

$$z_b = 25\,\text{mm}$$

This result is as anticipated, as the bottom and top flange laminates are identical and the total depth of the beam is 50 mm.

From Table 8.2, for a composite box-beam the effective axial rigidity is

$$(\overline{AE}) = 2A_w \overline{E}_{xx}^{ex}\Big|_w + A_{bf}\overline{E}_{xx}^{ex}\Big|_{bf} + A_{tf}\overline{E}_{xx}^{ex}\Big|_{tf}$$

Substituting all known values and completing the indicated calculations, we find

$$(\overline{AE}) = 14.1 \times 10^6 \, \text{N}$$

From Table 8.3, for a composite box-beam the effective flexural rigidity is

$$\left(\overline{IE}\right) = \frac{2t_w \, \overline{E}^{ex}_{xx}\big|_w}{3} \left[(z_b - t_{bf})^3 - (z_b - h - t_{bf})^3\right] + \left[(z_b - t_{bf}/2)^2 A_{bf} \, \overline{E}^{bf}_{xx}\right.$$

$$+ (z_b - t_{bf} - h - t_{tf}/2)^2 A_{tf} \, \overline{E}^{ex}_{xx}\big|_{tf} + I_{bf} \, \overline{E}^{fl}_{xx}\big|_{bf} + I_{tf} \, \overline{E}^{fl}_{xx}\big|_{tf} \Big]$$

Substituting all known values and completing the indicated calculations, we find

$$\left(\overline{IE}\right) = 4.92 \times 10^3 \, \text{N-m}^2$$

b. The web and bottom flanges are produced using the same eight-ply stacking sequence considered in Part (a), but now the top flange is a 20-ply laminate. Adjusting our previous calculations, we find

$$t_w = t_{bf} = 8(0.125 \, \text{mm}) = 1 \, \text{mm}$$

$$t_{tf} = 20(0.125 \, \text{mm}) = 2.5 \, \text{mm}$$

$$h = 50 \, \text{mm} - t_{bf} - t_{tf} = 50 \, \text{mm} - 1 \, \text{mm} - 2.5 \, \text{mm} = 46.5 \, \text{mm}$$

$$A_w = (h)(t_w) = (46.5 \, \text{mm})(1 \, \text{mm}) = 46.5 \times 10^{-6} \, \text{m}^2$$

$$A_{bf} = (b)(t_{bf}) = (30 \, \text{mm})(1 \, \text{mm}) = 30 \times 10^{-6} \, \text{m}^2$$

$$A_{tf} = (b)(t_{tf}) = (30 \, \text{mm})(2.5 \, \text{mm}) = 75 \times 10^{-6} \, \text{m}^2$$

$$I_{bf} = \frac{(b)(t_{bf})^3}{12} = \frac{(30 \, \text{mm})(1 \, \text{mm})^3}{12} = 2.5 \times 10^{-12} \, \text{m}^4$$

$$I_{tf} = \frac{(b)(t_{tf})^3}{12} = \frac{(30 \, \text{mm})(2.5 \, \text{mm})^3}{12} = 39.1 \times 10^{-12} \, \text{m}^4$$

$$\overline{E}^{ex}_{xx}\big|_w = \overline{E}^{ex}_{xx}\big|_{bf} = 90.4 \, \text{GPa}$$

$$\left. \overline{E}_{xx}^{fl} \right|_{bf} = 120\,\text{GPa}$$

$$\left. \overline{E}_{xx}^{ex} \right|_{tf} = 83.3\,\text{GPa}$$

$$\left. \overline{E}_{xx}^{fl} \right|_{tf} = 94.0\,\text{GPa}$$

On the basis of the appropriate expression taken from Table 8.1, we find that the centroid is located at

$$z_b = 29.4\,\text{mm}$$

Note that the increased thickness of the top flange laminate has had the effect of moving the centroid 4.4 mm toward the top flange. The new axial and flexural rigidities are

$$(\overline{AE}) = 17.4 \times 10^6\,\text{N}$$

$$\left(\overline{IE} \right) = 6.35 \times 10^3\,\text{N-m}^2$$

The increased thickness of the top flange laminate has increased the effective axial and flexural rigidities by 23% and 29%, respectively.

c. For this cross-section, the web laminates are produced using the same eight-ply stacking sequence considered in part (a), but now both the bottom and top flanges consist of a 20-ply laminate. We find

$$t_w = 8(0.125\,\text{mm}) = 1\,\text{mm}$$

$$t_{tf} = t_{bf} = 20(0.125\,\text{mm}) = 2.5\,\text{mm}$$

$$h = 50\,\text{mm} - t_{bf} - t_{tf} = 50\,\text{mm} - 2.5\,\text{mm} - 2.5\,\text{mm} = 45\,\text{mm}$$

$$A_w = (h)(t_w) = (45\,\text{mm})(1\,\text{mm}) = 45 \times 10^{-6}\,\text{m}^2$$

$$A_{bf} = A_{tf} = (b)(t_{bf}) = (30\,\text{mm})(2.5\,\text{mm}) = 75 \times 10^{-6}\,\text{m}^2$$

$$I_{bf} = I_{tf} = \frac{(b)(t_{bf})^3}{12} = \frac{(30\,\text{mm})(2.5\,\text{mm})^3}{12} = 39.1 \times 10^{-12}\,\text{m}^4$$

$$\left. \overline{E}_{xx}^{ex} \right|_{w} = 90.4\,\text{GPa}$$

$$\left. \overline{E}_{xx}^{ex} \right|_{bf} = \left. \overline{E}_{xx}^{ex} \right|_{tf} = 83.3\,\text{GPa}$$

$$\left. \overline{E}_{xx}^{fl} \right|_{bf} = \left. \overline{E}_{xx}^{fl} \right|_{tf} = 94.0\,\text{GPa}$$

Employing the appropriate expressions from Tables 8.1 through 8.3, we find

$$z_b = 25.0 \text{ mm}$$

$$(\overline{AE}) = 20.6 \times 10^6 \text{ N}$$

$$\left(\overline{IE}\right) = 8.43 \times 10^3 \text{ N-m}^2$$

As the top and bottom flanges are identical, the centroid is centered in the overall beam cross-section, as was the case in part (a). There is now a further increase in effective rigidities. Compared with the values calculated in part (a) the increased thickness of the bottom and top flange laminates has increased the effective axial and flexural rigidities by 46% and 71%, respectively.

8.7 Statically Determinate and Indeterminate Axially Loaded Composite Beams

Deflections of composite beams subjected to axial loads (only) are considered in this section. Beam problems may be divided into two categories: those involving statically *determinate* beams and those involving statically *indeterminate* beams. For a statically determinate beam, all unknown reaction forces can be determined through application of the equations of equilibrium. In contrast, for a statically indeterminate beam the reaction forces are determined through the equations of equilibrium *and* a consideration of beam deflections.

A statically determinate composite beam is shown in Figure 8.15a. As indicated, the beam is clamped at one end and subjected to three known axial loads N_B^b, N_C^b, and N_D^b. In addition, the effective axial rigidity changes discretely at points B and C. In practice, a discrete change in effective axial rigidity is readily accomplished by adding or removing a ply (or plies) from the stacking sequence along the length of the beam. The internal force present in each beam segment can be determined through the use of the free-body diagrams shown in Figures 8.15b–d and the appropriate equation of equilibrium ($\Sigma F_x = 0$):

$$N_{xx}^b\big|_{AB} = N_B^b + N_C^b + N_D^b$$

$$N_{xx}^b\big|_{BC} = N_C^b + N_D^b$$

$$N_{xx}^b\big|_{CD} = N_D^b$$

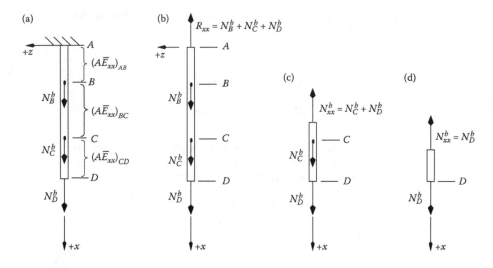

FIGURE 8.15
A statically determinate composite beam with changing effective axial rigidity subjected multiple axial loads. (a) Composite beam; (b) free-body diagram used to determine reaction force $R_{xx} = N_B^b + N_C^b + N_D^b$; (c) free-body diagram used to determine internal load in segment *BC* equals $N_{xx}^b = N_C^b + N_D^b$; (d) free-body diagram used to determine internal load in segment *CD* equals $N_{xx}^b = N_D^b$.

The change in length of a beam caused by axial loading is called the *elongation*, u_o. The elongation induced within each segment of the beam shown in Figure 8.15 is given by

$$u_o\big|_{AB} = \frac{N_{xx}^b\big|_{AB} L_{AB}}{(AE_{xx})_{AB}}$$

$$u_o\big|_{BC} = \frac{N_{xx}^b\big|_{BC} L_{BC}}{(AE_{xx})_{BC}}$$

$$u_o\big|_{CD} = \frac{N_{xx}^b\big|_{CD} L_{CD}}{(AE_{xx})_{CD}}$$

where L_{AB}, L_{BC}, and L_{CD} represent the length of beam segments *AB*, *BC*, and *CD*, respectively. The elongation of the bar as a whole is simply the sum of the elongation induced over each segment:

$$u_o = \sum_i \frac{N_{xx}^b\big|_i L_i}{(AE_{xx})_i} \tag{8.39}$$

A statically indeterminate composite beam is shown in Figure 8.16. In this case, the beam is clamped at *both* ends and subjected to two known axial

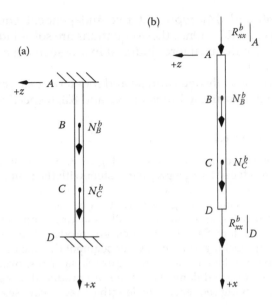

FIGURE 8.16
A statically indeterminate composite beam subjected to two known intermediate loads. (a) Original structure; (b) free-body diagram.

loads N_B^b and N_C^b. An unknown reaction force is induced at each end of the bar, $R_{xx}^b\big|_A$ and $R_{xx}^b\big|_D$. Note that both reaction forces have been assumed to act in the $+x$-direction in the Figure 8.16. Enforcing the appropriate equation of equilibrium ($\Sigma F_x = 0$) we find

$$R_{xx}^b\big|_A + N_B^b + N_C^b + R_{xx}^b\big|_D = 0 \qquad (8.40)$$

As $R_{xx}^b\big|_A$ and $R_{xx}^b\big|_D$ are unknown, a second independent equation is required to solve for the reaction forces. The remaining equations of equilibrium cannot be used to provide the necessary second independent equation, as only axial loads are present. The second independent equation is obtained by requiring that the beam satisfy known boundary conditions. That is, as the beam is clamped at both ends the total elongation of the beam must be zero. By requiring that the total elongation is zero we develop a second independent equation, known as an *equation of compatibility*. Treating reaction force $R_{xx}^b\big|_D$ as a *redundant force*, applying Equation 8.39, and equating the total elongation to zero, we find

$$u_o\big|_{AD} = \frac{\left(N_B^b + N_C^b + R_{xx}^b\big|_D\right)L_{AB}}{(AE_{xx})_{AB}} + \frac{\left(N_C^b + R_{xx}^b\big|_D\right)L_{BC}}{(AE_{xx})_{BC}} + \frac{\left(R_{xx}^b\big|_D\right)L_{CD}}{(AE_{xx})_{CD}} = 0 \qquad (8.41)$$

Equations 8.40 and 8.41 represent two independent equations in two unknowns, $R_{xx}^b|_A$ and $R_{xx}^b|_D$. Once these equations are solved to determine the reaction forces, then the axial force induced in any segment of the beam may also be determined.

The analysis of statically determinate and indeterminate composite beams will be illustrated in Example Problems 8.2 and 8.3, respectively.

Example Problem 8.2

Assume the composite beam shown in Figure 8.15 is a graphite–epoxy box-beam, produced using a prepreg system with the material properties listed in Table 3.1. Over segment *AC*, the beam has the cross-section previously shown in Figure 8.14c. That is, the two web laminates are produced using an eight-ply $[0/90]_{2s}$ stacking sequence, whereas the top and bottom flange laminates are produced using a 20-ply $[(0/90)_{2s}/\pm 45]_s$ stacking sequence. Over segment *CD*, the beam has the cross-section previously shown in Figure 8.14a. Thus, over this segment both the web and flange laminates are produced using an eight-ply $[0/90]_{2s}$ stacking sequence. The length of each beam segment is as follows:

$$L_{AB} = 250\,\text{mm}$$

$$L_{BC} = 500\,\text{mm}$$

$$L_{CD} = 250\,\text{mm}$$

Thus, the total length of the beam is 1.0 m. Finally, the applied loads are

$$N_B^b = 15\,\text{kN}$$

$$N_C^b = 10\,\text{kN}$$

$$N_D^b = 5\,\text{kN}$$

Determine the total beam elongation produced by this loading condition.

SOLUTION

The effective axial rigidities provided by the beam cross-sections involved were determined as a part of Example Problem 8.1 and found to be

$$\left.(\overline{AE_{xx}})\right|_{AB} = \left.(\overline{AE_{xx}})\right|_{BC} = 20.6 \times 10^6\,\text{N}$$

$$\left.(\overline{AE_{xx}})\right|_{CD} = 14.1 \times 10^6\,\text{N}$$

The elongation can be determined through application of Equation 8.39:

$$u_o = \sum_i \frac{N^b_{xx}\big|_i L_i}{(A\bar{E}_{xx})_i}$$

$$u_o = \frac{N^b_{xx}\big|_{AB} L_{AB}}{(A\bar{E}_{xx})_{AB}} + \frac{N^b_{xx}\big|_{BC} L_{BC}}{(A\bar{E}_{xx})_{BC}} + \frac{N^b_{xx}\big|_{CD} L_{CD}}{(A\bar{E}_{xx})_{CD}}$$

$$u_o = \frac{(30000N)(0.25\,\text{m})}{(20.6 \times 10^6\,\text{N})} + \frac{(15000N)(0.50\,\text{m})}{(20.6 \times 10^6\,\text{N})} + \frac{(5000N)(0.25\,\text{m})}{(14.1 \times 10^6\,\text{N})}$$

$$u_o = 0.817\,\text{mm}$$

Example Problem 8.3

Assume the composite beam shown in Figure 8.16 is a graphite–epoxy box-beam, produced using a prepreg system with the material properties listed in Table 3.1. Over segment *AC*, the beam has the cross-section previously shown in Figure 8.14c. That is, the two web laminates are produced using an eight-ply $[0/90]_{2s}$ stacking sequence, whereas the top and bottom flange laminates are produced using a 20-ply $[(0/90)_{2s}/\pm45]_s$ stacking sequence. Over segment *CD*, the beam has the cross-section previously shown in Figure 8.14a. Thus, over this segment both the web and flange laminates are produced using an eight-ply $[0/90]_{2s}$ stacking sequence. The length of each beam segment is as follows:

$$L_{AB} = 250\,\text{mm}$$

$$L_{BC} = 500\,\text{mm}$$

$$L_{CD} = 250\,\text{mm}$$

Thus, the total length of the beam is 1.0 m. Finally, loads $N^b_B = 15\,\text{kN}$ and $N^b_C = 10\,\text{kN}$.

Determine the reaction forces $R^b_{xx}\big|_A$ and $R^b_{xx}\big|_D$.

SOLUTION

The reaction forces are determined through solving Equations 8.40 and 8.41. The effective axial rigidities provided by the beam cross-sections involved were determined as a part of Example Problem 8.1 and found to be

$$(\overline{AE}_{xx})\big|_{AB} = (\overline{AE}_{xx})\big|_{BC} = 20.6 \times 10^6\,\text{N}$$

$$(\overline{AE}_{xx})\big|_{CD} = 14.1 \times 10^6\,\text{N}$$

Using the values provided, Equation 8.41 becomes:

$$\frac{\left(25\,\text{kN} + R_{xx}^b\big|_D\right)(0.25\,\text{m})}{(20.6 \times 10^6\,\text{N})} + \frac{\left(10\,\text{kN} + R_{xx}^b\big|_D\right)(0.5\,\text{m})}{(20.6 \times 10^6\,\text{N})} + \frac{\left(R_{xx}^b\big|_D\right)(0.25\,\text{m})}{(14.1 \times 10^6\,\text{N})} = 0$$

Solving for the redundant force:

$$R_{xx}^b\big|_D = -10.1\,\text{kN}$$

The fact that $R_{xx}^b\big|_D$ is negative implies that it acts in the direction opposite to that shown in Figure 8.16b.

The unknown reaction force $R_{xx}^b\big|_A$ can now be determined through application of the Equation 8.40:

$$R_{xx}^b\big|_A + N_B^b + N_C^b + R_{xx}^b\big|_D = R_{xx}^b\big|_A + 15\,\text{kN} + 10\,\text{kN} - 10.1\,\text{kN} = 0$$

$$R_{xx}^b\big|_A = -14.9\,\text{kN}$$

Reaction $R_{xx}^b\big|_A$ is also negative, implying that it acts the direction opposite to that shown in Figure 8.16b.

8.8 Statically Determinate and Indeterminate Transversely Loaded Composite Beams

Deflections associated with bending of composite beams are considered in this section. Three types of external loads that lead to beam bending are shown in Figure 8.17. A distributed load, $q(x)$, is a load distributed over a length of the beam and is specified in units of either Newtons/meter or pounds-force/in (N/m or lbf/in.). A point load, P, is a load acting transverse to the axis of the beam and applied at a specific cross-section. A point load is usually specified in units of either Newtons or pounds-force (N or lbf).

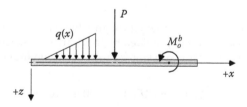

FIGURE 8.17
Illustration of external transverse loads that lead to beam bending.

Finally, a concentrated moment (also called a *couple*), M_o^b, is a moment applied at a particular cross-section and is specified in units of Newton-meters or pounds-force-in (N-m or lbf-in.). As in earlier sections a superscript "*b*" has been used to differentiate between a concentrated moment and a moment resultant, M_{xx}. In the following discussion the external beam loads shown in Figure 8.17 will be referred to as *transverse loads*.

In general, the *external* transverse forces applied to a beam induce *internal* forces at all cross-sections of a beam. To determine beam deflections, we must determine these internal forces. To do so the manner in which the beam is supported must first be defined. The three most common beam support conditions are illustrated in Figure 8.18. A *pinned support* is one that restrains the beam from deflection in the *x*- and *z*-directions, but does not prevent rotation. In contrast, a *roller support* restrains the beam from deflection in the *z*-direction, but does not restrain deflections in the *x*-direction nor does it restrict rotation. Finally, a *clamped* (or *fixed*) *support* does not allow the beam to rotate or deflect in the *x*- or *z*-directions.

Some combinations of beam support conditions occur so frequently that they are given special names. A *simply supported* beam is shown in Figure 8.19a. In this case, one end of the beam is restrained by a pin support, whereas the other is restrained by a roller support. A *cantilever beam* is shown in Figure 8.19b, and is defined as a beam clamped at one end and free at the other. A third common configuration is an *overhanging* beam, shown in Figure 8.19c. This condition is similar to a simply supported beam, except that the roller support is at an interior cross-section, such that one end of the beam is free.

Having specified the support conditions and transverse loading applied to a beam, the next step is to determine the *reaction forces*. In essence, reaction forces are external loads applied to the beam by the beam support(s). Determination of these forces begins by enforcing the six equations of equilibrium. As we have assumed that all loads act within the *x–z* plane

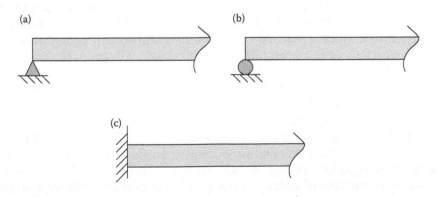

FIGURE 8.18
Common beam support conditions. (a) Pinned support; (b) roller support; (c) clamped support.

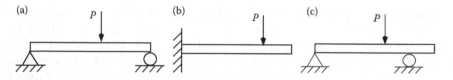

FIGURE 8.19
Common combinations of beam support conditions. (a) Simply supported beam; (b) cantilevered beam; (c) overhanging beam.

(the plane of loading), three equations of equilibrium are satisfied automatically: $\Sigma F_y = \Sigma M_x = \Sigma M_z = 0$. In this section, we do not consider any axial loads, therefore, $\Sigma F_x = 0$ is also satisfied automatically. Consequently, only two equations of equilibrium are available for use. Specifically, to remain in equilibrium it is required that all forces acting in the z-direction and all moments acting about the y-axis sum to zero:

$$\sum F_z = 0$$

$$\sum M_y = 0$$

Beam problems may now be divided into two main categories: statically *determinate* beams and statically *indeterminate* beams. For a statically determinate beam, the unknown reaction forces are determined solely through application of the equations of equilibrium. As we only have two remaining equilibrium equations, it is apparent that a statically determinate beam can have (at most) two unknown reaction forces. Free-body diagrams for simply supported, cantilevered, and overhanging beams are shown in Figure 8.20. All reaction forces are shown in an algebraically positive sense. In each case, only two unknown reaction forces are involved, and hence each of these beams is statically determinate. In contrast, if a beam problem involves three (or more) reaction forces then the beam is statically indeterminate and the reaction forces are determined through application of the equations of equilibrium *and* consideration of beam deflections. An example of a statically indeterminate beam is shown in Figure 8.21. In this case, the beam is clamped at one end, supported by a roller support at the other end, and subjected to a known transverse point load P. A total of three unknown reaction forces exist in this case ($R_z^{(A)}$, $R_z^{(B)}$, and M_o^b), as indicated in the accompanying free-body diagram. The three unknown reaction forces cannot be determined solely on the basis of equilibrium, as only two equations of equilibrium are available for use. The difference between the number of unknown reaction forces and the number of available equations of equilibrium is called the *degree of indeterminacy*: the beam shown in Figure 8.21 is statically indeterminate to the first degree. One way

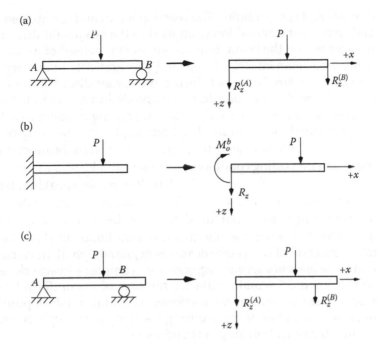

FIGURE 8.20
Free-body diagrams for common statically-determinate beams. (a) Simply supported beam; (b) cantilevered beam; (c) overhanging beam.

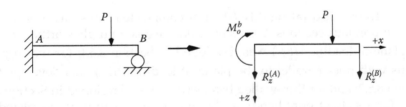

FIGURE 8.21
Free-body diagram for a beam statically indeterminate to the first degree.

to determine reaction force in this case is to enforce the requirement that the beam z-deflection at the roller support must equal zero, in addition to enforcing the two equations of equilibrium. This leads to third independent equation, a so-called *equation of compatibility*. Hence, together the two equations of equilibrium and the equation of compatibility represent three independent equations and can be used to solve for the three unknown reaction forces.

In any event, assume for the moment that all reaction forces have been determined for a beam of interest, either through application of the equations of equilibrium or through an appropriate combination of the equations of equilibrium and compatibility. The reaction forces and the applied transverse loads represent external forces applied to the beam, and these induce internal forces within the beam. Internal forces may consist of a shear force V_{xz}^b and bending moment M_{xx}^b. In general, V_{xz}^b and M_{xx}^b both vary along the length of the beam. Note that during the earlier discussion presented in Sections 8.5 and 8.6, we considered composite beams subjected to *pure bending*—in these earlier sections, the internal bending moment M_{xx}^b did not vary along the length of the beam. Furthermore, in pure bending no shear force exists: $V_{xz}^b = 0$. We now wish to determine transverse beam deflections under more general loading conditions, in which both V_{xz}^b and M_{xx}^b exist. This is often called a state of *non-uniform bending*. Rigorously speaking, both V_{xz}^b and M_{xx}^b contribute to transverse beam deflections. However, the deflections due to V_{xz}^b are less significant than the deflections due to M_{xx}^b and can usually be ignored. That is, we will assume that the contribution of shear force V_{xz}^b to the total deflection of composite beams is negligibly small. In essence, we assume that the *moment-curvature equations* for composite beams developed in Sections 8.5 and 8.6 remain valid for the case of nonuniform bending, even though these equations were developed assuming a state of pure bending. This same assumption is made during the study of isotropic beams. The moment-curvature equation for pure bending is

$$\kappa_{xx} = \frac{M_{xx}^b}{\overline{IE}_{xx}} \tag{8.42}$$

The effective flexural rigidity $(\overline{IE}_{xx})$ of composite beams has been developed in previous sections. For rectangular beams with plies orthogonal to the plane of loading $(\overline{IE}_{xx})$ was developed in Section 8.5.1; for rectangular beams with plies parallel to the plane of loading $(\overline{IE}_{xx})$ was developed in Section 8.5.2; and for thin-walled beams $(\overline{IE}_{xx})$ was developed in Section 8.6. Also, for a state of pure bending, the midplane curvature κ_{xx} is related to transverse beam deflections according to

$$\kappa_{xx} = -\frac{d^2w}{dx^2} \tag{8.43}$$

where $w = w(x)$ represents the z-deflection of the composite beam midplane.* Combining Equations 8.42 and 8.43, we find

$$\frac{d^2w}{dx^2} = -\frac{M_{xx}^b}{\overline{IE}_{xx}} \tag{8.44}$$

---

* For pure bending, Equation 8.43 is exact and is derived in any text devoted to mechanics of materials, including References 1–5, for example.

Most books devoted to mechanics of materials (References 1–5, for example) describe several techniques to determine transverse beam deflections. These include *direct integration* of the governing equation (i.e., direct integration of Equations 8.44), the *moment-area method*, the *method of superposition*, techniques based on the use of *discontinuity functions*, or the use of *Castigliono's Theorem* (which is a technique based on *energy methods*). Two techniques will be briefly discussed here: direct integration of the governing equations and the method of superposition.

Direct integration of the governing differential equation is perhaps the most straightforward approach. Once all external forces are known (including reaction forces) then it is a simple matter to express the internal bending moment as a function of x: $M_{xx}^b = M_{xx}^b(x)$. Hence, the beam deflection $w(x)$ is obtained by simply integrating Equation 8.44 twice. Constants of integration are determined by enforcing known beam support conditions (i.e., known boundary conditions).

The method of superposition is based on the observation that Equation 8.44 is a linear differential equation with constant coefficients. Consequently, the deflection of a beam subjected to several different load components acting simultaneously can be found by superimposing (i.e., adding together) the deflections caused by each load component acting separately. In practice then, tables of beam deflections caused by simple loading conditions and for commonly encountered beam support conditions are generated, and the deflections caused by a combination of two or more loads is obtained by adding together the deflection caused by each load component. Tables for deflections induced by common individual load components applied to cantilevered and simply supported beams are provided in Appendix B, Tables B.1 and B.2, respectively. Much more extensive tables, which provide solutions for other types of loading and support conditions, are available in engineering handbooks, for example, Reference 8.

Three example problems will now be presented to illustrate the process of determining composite beam deflections. The deflections of a statically determinate composite beam will be obtained by direct integration of Equation 8.44 in Example Problem 8.4. The method of superposition is then applied to a statically determinate composite beam subjected to a combination of loads in Example Problem 8.5. Finally, the method of superposition is used to determine the reaction forces and deflections for a statically indeterminate beam in Example Problem 8.6.

Example Problem 8.4

A cantilevered graphite–epoxy box-beam subjected to a linearly increasing distributed load $q(x) = (q_o/L)x$ is shown in Figure 8.22a:

   a. Obtain analytical expressions giving the reaction forces and transverse beam deflections.

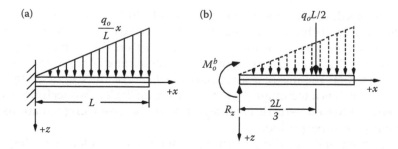

FIGURE 8.22
The cantilevered beam considered in Example Problem 8.4. (a) A cantilevered beam of length
L subjected to a linearly increasing distributed load $q(x) = (q_o/L)x$; (b) free-body diagram used
to determine the unknown reaction forces R_z and M_o^b (distributed load $q(x)$ replaced by statically
equivalent force $q_oL/2$, acting at $x = 2L/3$).

   b. Assume the beam has the box cross-section previously shown
      in Figure 8.14c and length $L = 1$ m. Obtain numerical values
      for reaction forces and plot transverse beam deflections if
      $q_o = 200$ N/m.

   SOLUTION

   a. A free-body diagram of the beam is shown in Figure 8.22b. As
      the beam is clamped two reaction forces may exist at the left-
      hand end: R_z and M_o^b. The first step is to determine the magni-
      tude and algebraic sign of these unknown reaction forces. The
      beam is statically determinate, as only two unknown reaction
      forces are present. The reaction forces are initially assumed to
      act in an algebraically positive sense, as shown in Figure 8.22b.
         To determine R_z and M_{xx}^b, we first replace the distributed
      load $q(x)$ with the statically equivalent force $q_oL/2$, located at
      $x = 2L/3$. This statically equivalent force is shown in Figure
      8.22b. By summing (a) forces in the z-direction, (b) moments
      about the y-axis, and (c) equating both to zero ($\Sigma F_z = \Sigma M_y = 0$),
      we obtain analytical expressions giving the reaction forces at
      the clamped end:

$$R_z = \frac{q_oL}{2} \qquad M_o^b = \frac{-q_oL^2}{3} \qquad \text{(a)}$$

      Note that if $q_o > 0$ then the bending moment M_o^b is algebra-
      ically negative and would act in the sense opposite to that
      shown in Figure 8.22b. The internal shear and bending moment
      induced at any cross-section located at arbitrary position x can

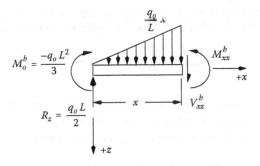

FIGURE 8.23
Free-body diagram used to determine internal shear and bending moment, V_{xz}^b and M_{xx}^b, respectively, acting at an arbitrary cross-section located at position x.

now be determined using the free-body diagram shown in Figure 8.23. Summing forces in the z-direction, we find

$$V_{xz}^b(x) = \frac{q_o}{2L}(L^2 - x^2)$$

A so-called *shear force diagram* is created by plotting $V_{xz}^b(x)$, as shown in Figure 8.24. Similarly, summing moments about an

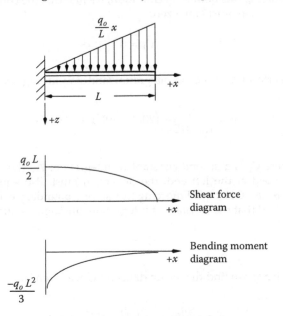

FIGURE 8.24
Shear and bending moment diagrams for the cantilevered beam considered in Example Problem 8.4.

axis passing through the left-hand side of the free-body diagram shown in Figure 8.23 and parallel to the y-axis we find

$$M_{xx}^b(x) = \frac{q_o}{6L}(-2L^3 + 3L^2x - x^3)$$

A so-called *bending moment diagram* is created by plotting $M_{xx}^b(x)$, as shown in Figure 8.24. Substituting the above expression for the bending moment $M_{xx}^b(x)$ into Equation 8.44, we find

$$\frac{d^2w}{dx^2} = -\frac{q_o}{6L(\overline{IE}_{xx})}(-2L^3 + 3L^2x - x^3)$$

Integrating once results in

$$\frac{dw}{dx} = -\frac{q_o}{6L(\overline{IE}_{xx})}\left(-2L^3x + \frac{3L^2x^2}{2} - \frac{x^4}{4}\right) + C_1$$

where C_1 is a constant of integration. As the beam is clamped at the left end, the slope must equal zero there: $dw/dx = 0$ at $x = 0$. Enforcing this boundary condition, we find that the constant of integration must equal zero:

$$C_1 = 0$$

Performing a second integration and simplifying, we obtain

$$w = \frac{q_o x^2}{120L(\overline{IE}_{xx})}\left(20L^3 - 10L^2x + x^3\right) + C_2$$

where C_2 is a second constant of integration. As the beam is clamped at the left end, the deflection must also equal zero there: $w = 0$ at $x = 0$. Enforcing this second boundary condition, we find that the constant of integration must again equal zero:

$$C_2 = 0$$

Hence, we find that beam deflections are given by

$$w = \frac{q_o x^2}{120L(\overline{IE}_{xx})}\left(20L^3 - 10L^2x + x^3\right) \qquad \text{(b)}$$

Note that this result is included in the list of solutions tabulated in Appendix B, Table B.1.

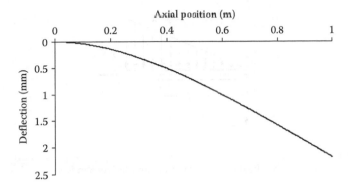

FIGURE 8.25
Deflections predicted for the graphite–epoxy composite box-beam considered in Example Problem 8.4.

b. The effective flexural rigidity of the box-beam shown in Figure 8.14c was determined as a part of Example Problem 8.1 and found to be

$$\left(\overline{IE_{xx}}\right) = 8.43 \times 10^3 \text{ N-m}^2$$

Substituting this and the other specified numerical values ($L = 1$ m, $q_o = 200$ N/m) into expression (b) above we find that transverse deflections are predicted to be (in millimeters):

$$w(x) = (0.1977x^2)\left(20L^3 - 10L^2x + x^3\right)$$

A plot of these beam deflections is shown in Figure 8.25. Note that both the deflection and slope are zero at the clamped end, as dictated by this boundary condition. A maximum deflection of 2.17 mm is predicted to occur at the free end.

Example Problem 8.5

Suppose the cantilevered graphite–epoxy beam considered in Example Problem 8.4 is subjected to both an external bending moment M_o^b applied at the free end and a linearly increasing distributed load $q(x) = (q_o/L)x$, as shown in Figure 8.26. Perform the following:

a. Use the method of superposition to obtain an analytical expression giving the predicted deflection $w(z)$.
b. Plot numerical values of the predicted deflections, assuming the following beam length and loading:

$$L = 1 \text{ m}$$

$$q_o = 200 \text{ N/m}$$

$$M_o^b = -40 \text{ N-m}$$

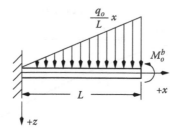

FIGURE 8.26
Cantilevered graphite–epoxy beam considered in Example Problem 8.5.

SOLUTION

a. Deflections induced by a concentrated moment M_o^b and distributed load $q(x)$ when acting separately are included in Table B.1. The deflections due to M_o^b acting alone are given by

$$w(x)\big|_{M_o^b} = \frac{M_o^b x^2}{2(\overline{IE}_{xx})}$$

The deflections due to the distributed load are given by

$$w(x)\big|_{q(x)} = \frac{q_o x^2}{120L(\overline{IE}_{xx})}(20L^3 - 10L^2 x + x^3)$$

Applying the principle of superposition, the beam deflections caused by both load components acting simultaneously is simply the sum of these two solutions:

$$w(x) = w(x)\big|_{M_o^b} + w(x)\big|_{q(x)} = \left(\frac{M_o^b x^2}{2(\overline{IE}_{xx})}\right) + \left(\frac{q_o x^2}{120L(\overline{IE}_{xx})}(20L^3 - 10L^2 x + x^3)\right)$$

b. Note that the specified numerical value of the concentrated bending moment is algebraically *negative*:

$$M_o^b = -40\,\text{N} - \text{m}$$

Hence, the concentrated moment is acting in a sense opposite to that shown in Figure 8.26. The distributed load $q(x)$ will tend to deflect the beam downwards (i.e., in the positive z-direction), whereas M_o^b will tend to deflect the beam upwards (in the negative z-direction).

Using the values specified for beam dimensions, material properties, and loads, deflections are given (in millimeters) by

$$w(x) = (0.1977x^2)(20 - 10x + x^3) - 2.372x^2$$

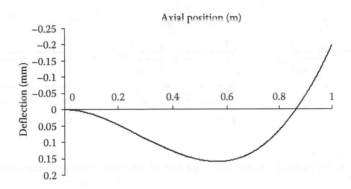

FIGURE 8.27
Deflections predicted for the beam considered in Example Problem 8.5.

which can be rearranged as

$$w(x) = 1.582x^2 - 1.977x^3 + 0.1977x^5$$

A plot of predicted deflections is shown in Figure 8.27. As before both the deflection and slope are zero at the left end, as required by the clamped boundary condition. A locally maximum positive (downwards) deflection of 0.160 mm occurs at an axial position of $x = 0.563$ m. The globally maximum deflection occurs at the free end, where a negative (upward) deflection of 0.198 mm occurs.

Example Problem 8.6

Consider a composite beam supported as previously shown in Figure 8.21.

- a. Obtain analytical expressions for the reaction forces.
- b. Assume the beam has the box cross-section previously shown in Figure 8.14c and length $L = 1$ m. Obtain numerical values for reaction forces and plot transverse beam deflections if $a = 3L/4 = 0.75$ m and $P = 1000$ N.

SOLUTION

- a. The beam shown in Figure 8.21 is statically indeterminate to the first degree, since there are three unknown reaction forces. Therefore, one of the unknown reaction forces must be selected to be a redundant force. Any one of the reaction forces ($R_z^{(A)}, R_z^{(B)}$, or M_o^b) can be treated as the redundant force. For present purposes $R_z^{(B)}$ will be treated as the redundant.
  This problem can be solved based on the method of superposition, as summarized in Figure 8.28. The first step is to determine the beam deflections that would occur if the redundant

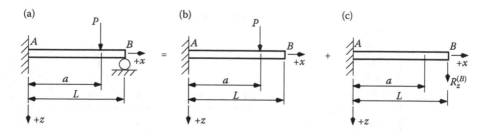

FIGURE 8.28
Summary of the method superposition applied to the indeterminate beam considered in Example Problem 8.4. (a) Original problem; (b) reduced problem; (c) redundant problem.

force was removed. Removal of redundant reaction force $R_z^{(B)}$ implies that the roller support at the right end is removed, as shown in Figure 8.28b. The process of obtaining beam deflections when the redundant force is removed is often called the *reduced* problem. From Table B.1, we find that beam deflections over the range $a \le x \le L$ associated with the reduced problem are given by

$$w(x)\big|_{\text{reduced}} = \frac{Pa^2}{6(\overline{IE}_{xx})}(3x - a)$$

The next step is to determine beam deflections that would be caused if only the redundant force was applied, as shown in Figure 8.28c. From Table B.1 we find that the beam deflections associated with the redundant force $R_z^{(B)}$ are given by

$$w(x)\big|_{R_z^{(B)}} = \frac{R_z^{(B)}x^2}{6(\overline{IE}_{xx})}(3L - x)$$

This expression is valid for any axial position $0 \le x \le L$.

We now superimpose the deflections associated with the reduced and redundant problems. In this case we have, for $a \le x \le L$:

$$w(x) = w(x)\big|_{\text{reduced}} + w(x)\big|_{R_z^{(B)}} = \frac{1}{6(\overline{IE}_{xx})}\left[Pa^2(3x - a) + R_z^{(B)}x^2(3L - x)\right] \quad \text{(a)}$$

We require that the total beam deflection equals zero at the right end (at $x = L$), as the beam is supported by a roller at that point. Hence,

$$w(x = L) = \frac{1}{6(\overline{IE}_{xx})}\left[Pa^2(3L - a) + R_z^{(B)}(L)^2(3L - L)\right] = 0$$

which reduces to

$$2R_z^{(B)}L^3 = Pa^2(a - 3L) \tag{b}$$

Expression (b) represents an equation of compatibility. Only one such equation is necessary in this problem, as the beam is indeterminate to the first degree.

Next, we enforce the equations of equilibrium, $\Sigma F_z = 0$, and $\Sigma M_y = 0$, resulting in (respectively)

$$R_z^{(A)} + R_z^{(B)} = -P \tag{c}$$

$$R_z^{(B)}L + M_o^b = -Pa \tag{d}$$

where expression (d) was obtained by summing moments about the left end of the beam (at $x = 0$).

Expressions (b), (c), and (d) represent three simultaneous equations in terms of the three unknown reaction forces. They can be rewritten in matrix form as

$$\begin{bmatrix} 0 & 2L^3 & 0 \\ 1 & 1 & 0 \\ 0 & L & 1 \end{bmatrix} \begin{Bmatrix} R_z^{(A)} \\ R_z^{(B)} \\ M_o^b \end{Bmatrix} = \begin{Bmatrix} Pa^2(a - 3L) \\ -P \\ -Pa \end{Bmatrix}$$

Solving this system of equations results in

$$R_z^{(A)} = -\frac{P}{2L^3}(a^3 + 2L^3 - 3a^2L) \tag{e}$$

$$R_z^{(B)} = -\frac{Pa^2}{2L^3}(3L - a) \tag{f}$$

$$M_o^b = -\frac{Pa}{2L^2}(a^2 + 2L^2 - 3aL) \tag{g}$$

b. On the basis of the specified numerical values, the reaction forces are

$$R_z^{(A)} = -\frac{(1000\ \text{N})}{2(1\ \text{m})^3}\left[(0.75\ \text{m})^3 + 2(1\ \text{m})^3 - 3(0.75\ \text{m})^2(1\ \text{m})\right] = -367\ \text{N}$$

$$R_z^{(B)} = -\frac{(1000\ \text{N})(0.75\ \text{m})^2}{2(1\ \text{m})^3}\left[3(1\ \text{m}) - (0.75\ \text{m})\right] = -633\ \text{N}$$

$$M_o^b = -\frac{(1000\ \text{N})(0.75\ \text{m})}{2(1\ \text{m})^2}\left[(0.75\ \text{m})^2 + 2(1\ \text{m})^2 - 3(0.75\ \text{m})(1\ \text{m})\right]$$

$$= -117\ \text{N-m}$$

Note that the calculated values for $R_z^{(A)}$, $R_z^{(B)}$, and M_o^b are all algebraically negative, which indicates that these reaction forces are acting in a sense opposite to that shown in Figure 8.21.

To plot beam deflections, note that expression (a), developed in the first part of this example problem, gives the deflection over the range $a \le x \le L$. Hence, we need an additional expression for beam deflections that is valid over $0 \le x \le a$. From Table B.1 we find that beam deflections over the range $0 \le x \le a$ associated with the reduced problem are given by

$$w(x)\big|_{\text{reduced}} = \frac{Px^2}{6(\overline{IE}_{xx})}(3a - x)$$

As before, from Table B.1 we find that the beam deflections associated with the redundant force $R_z^{(B)}$ are given by

$$w(x)\big|_{R_z^{(B)}} = \frac{R_z^{(B)}x^2}{6(\overline{IE}_{xx})}(3L - x)$$

This result is valid for any axial position $0 \le x \le L$. Superimposing these two results we obtain an expression for the total beam deflection, valid for $0 \le x \le a$:

$$w(x) = w(x)\big|_{\text{reduced}} + w(x)\big|_{R_z^{(B)}} = \frac{1}{6(\overline{IE}_{xx})}\left[Px^2(3a - x) + R_z^{(B)}x^2(3L - x)\right] \quad \text{(h)}$$

Recall from Example Problem 8.1 that the effective flexural rigidity for this beam is $\left(\overline{IE}_{xx}\right) = 8.43 \times 10^3\,\text{N-m}^2$. Substituting this value (and the other numerical values specific to this problem) into expressions (a) and (h), and plotting over the

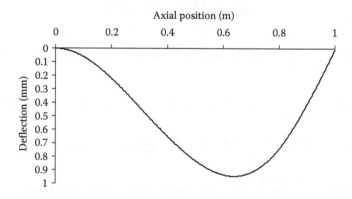

FIGURE 8.29
Deflections predicted for the beam considered in Example Problem 8.6.

appropriate ranges, results in Figure 8.29 As dictated by the specified boundary conditions, the beam deflection is zero at either end of the beam, and the slope of the beam is zero at the left end. A maximum positive (downward) deflection of 0.94 mm is predicted to occur at the axial location $x = 0.64$ m.

8.9 Computer Program BEAM

The computer program BEAM has been developed to supplement the material presented in Sections 8.4 through 8.6 of this chapter. This program can be downloaded at no cost from the following website:

http://depts.washington.edu/amtas/computer.html

Program BEAM can be used to calculate the centroidal location, effective axial rigidity, and effective flexural rigidity of a composite beam with either a rectangular cross section or with any of the thin-walled cross sections shown in Tables 8.1 through 8.3. The program prompts the user to input all information necessary to perform these calculations. Properties of up to five different materials may be defined. The user must input various numerical values using a consistent set of units. For example, the user must input elastic moduli for the composite material system(s) of interest. Using the properties listed in Table 3.1 and based on the SI system of units, the following numerical values would be input for graphite–epoxy:

$$E_{11} = 170 \times 10^9\,\text{Pa} \quad E_{22} = 10 \times 10^9\,\text{Pa} \quad \nu_{12} = 0.30 \quad G_{12} = 13 \times 10^9\,\text{Pa}$$

As 1 Pa = 1 N/m², all lengths must be input in meters. For example, ply thicknesses must be input in meters (not millimeters). A typical value would be $t_k = 0.000125$ m (corresponding to a ply thickness of 0.125 mm). Similarly, if an *I*-beam that involves a 50-mm-wide flange laminate was under consideration, then the width of the flange must be input as 0.050 m.

If the English system of units were used, then the following numerical values would be input for the same graphite–epoxy material system:

$$E_{11} = 25.0 \times 10^6\,\text{psi} \quad E_{22} = 1.5 \times 10^6\,\text{psi} \quad \nu_{12} = 0.30 \quad G_{12} = 1.9 \times 10^6\,\text{psi}$$

All lengths would be input in inches. A typical ply thickness would be $t_k = 0.005$ in, and for an *I*-beam a typical flange width would be 2.0 in.

Having determined the effective axial and flexural rigidities of a composite beam of interest, then beam deflections caused by axial loading or transverse loading can be calculated using the methods discussed in Sections 8.7 and 8.8, respectively.

HOMEWORK PROBLEMS

Note:

Computer programs CLT and BEAM are used in the following problems. These programs can be downloaded from the following website:

http://depts.washington.edu/amtas/computer.html

8.1.   A $[(0_2/\pm30/\pm45/\pm60)_3/90_2]_s$ composite beam with rectangular cross-section is shown in Figure 8.30. Assume this laminate is produced using a graphite/epoxy prepreg with properties listed in Table 3.1:

   a.   Use program CLT to determine numerical values of the [*abd*] matrix for this stacking sequence,

   b.   Use hand calculations to determine:

   i.   the effective axial rigidity of the beam,

   ii.   the effective flexural rigidity of the beam when ply interfaces are orthogonal to the plane of loading (Figure 8.30a), and

   iii.   the effective flexural rigidity of the beam when ply interfaces are parallel to the plane of loading (Figure 8.30b), and

   c.   Use program BEAM to determine the effective axial and flexural rigidities of the beam and compare these results with your hand calculations obtained in step (b).

8.2.   Repeat problem 8.1, except assume the beam is produced using a glass/epoxy prepreg with a thickness of 0.150 mm.

8.3.   Repeat problem 8.1, except assume the beam is produced using a Kevlar/epoxy prepreg with a thickness of 0.125 mm.

8.4.   Repeat problem 8.1, except assume that the 0°, +30°, and −30° plies are produced using a glass/epoxy prepreg with a thickness

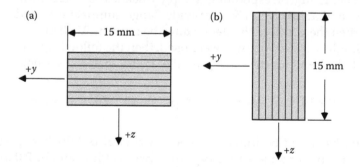

FIGURE 8.30

Composite beam with rectangular cross-section described in Problem 8.1 (not all ply interfaces shown). (a) Ply interfaces orthogonal to the plane of loading; (b) ply interfaces parallel to the plane of loading.

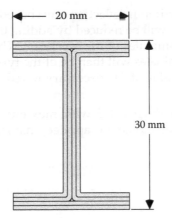

FIGURE 8.31
Composite *I*-beam described in Problem 8.5 (not all ply interfaces shown).

    of 0.150 mm, whereas the remaining plies are produced using a graphite/epoxy prepreg with a thickness of 0.125 mm.

8.5. A composite I-beam is shown in Figure 8.31. Assume the beam is produced using a graphite–epoxy prepreg with the properties listed in Table 3.1. Also, the stacking sequences of the top flange laminate, web laminate, and bottom flange laminate are: $[(0_2/90_2/0_2)_2/45_2/0/-45_2]_s$, $[0_2/90_2/0_2]_{2s}$, and $[(0_2/90_2/0_2)_2/45_2/0/-45_2]_s$, respectively. (Note that Figure 8.31 implies that the web plies extend into the top and bottom flange laminates.)

    a. Determine the centroid location, effective axial rigidity, and effective flexural rigidity of the beam using hand calculation and appropriate expressions from Tables 8.1 through 8.3. Determine the effective elastic moduli involved using program CLT.

    b. Use program BEAM to calculate the centroid location, effective axial rigidity, and effective flexural rigidity of the beam, and compare these results with your hand calculations obtained in step (a).

8.6. Repeat problem 8.5, except assume the beam is produced using a glass/epoxy prepreg with a thickness of 0.150 mm.

8.7. Repeat problem 8.5, except assume the beam is produced using a Kevlar/epoxy prepreg with a thickness of 0.125 mm.

8.8. Repeat problem 8.5, except assume that the stacking sequences of the top and bottom flange laminates are $[(0_2/90_2/0_2)_2/45_2/0/-45_2]_s$ and $[0_2/90_2/0_2]_{2s}$, respectively.

8.9. Refer to Example Problem 8.4, part (b). Suppose the maximum deflection of the cantilevered beam must be reduced by a factor of

2. That is, the maximum deflection must be reduced to 1.085 mm (or less). Deflections will be reduced by adding $0°$ plies to the top *and* bottom flange laminates of the box beam. The stacking sequence of the flange laminates will then be of the type $[(0/90)_{2s}/\pm ss45/0_n]_s$. Determine the value of "n" necessary to reduce deflections to the desired level.

8.10. Refer to Example Problem 8.5, what maximum deflection occurs if a positive bending moment is applied, that is, if $M_o^b = 40$ N-m?

References

1. Gere, J.M., and Timoshenko, S.P., 1997, *Mechanics of Materials*, 4th edition, PWS Publishing Co., Boston, MA, ISBN 0-534-93429-3.
2. Craig, R.R., 1996, *Mechanics of Materials*, John Wiley and Sons, New York, ISBN 0-471-50284-7.
3. Hibbeler, R.C., 2000, *Mechanics of Materials*, 4th edition, Prentice Hall, Upper Saddle River, NJ, ISBN 0-13-016467-4.
4. Bedford, A., and Liechti, K.M., 2000, *Mechanics of Materials*, Prentice Hall, Upper Saddle River, NJ, ISBN 0-201-89552-8.
5. Benham, P.P., Crawford, R.J., and Armstrong, C.G., 1996, *Mechanics of Engineering Materials*, 2nd edition, Longman Group Limited, Essex, ISBN 0-582-25164-8.
6. Timoshenko, S.P., and Goodier, J.N., 1970, *Theory of Elasticity*, 3rd edition, Section 124, McGraw-Hill Book Company, New York, ISBN 07-064720-8.
7. Swanson, S.R., 1997, *Introduction to Design and Analysis with Advanced Composite Materials*, Prentice-Hall Inc, Upper Saddle River, NJ, ISBN 0-02-418554-X.
8. Roark, R.J., and Young, W.C., 1989, *Formulas for Stress and Strain*, 6th edition, McGraw-Hill Book Company, New York, ISBN 0-07-072541-1.

9

Stress Concentrations Near an Elliptical Hole

9.1 Preliminary Discussion

Most structures contain one or more regions where a relatively rapid and localized change in geometry occurs. Familiar examples include bolt holes, notches, filets, or a sudden change in thickness. These geometrical features typically cause a localized increase in stress. The *stress concentration factor, K_t,* is used to quantify the increase in stress caused by these geometrical features.

The theory of anisotropic elasticity can be used to predict stress concentrations associated with through-thickness holes in many anisotropic panels. Lekhnitskii [1] and later Savin [2] developed solutions for rectangular, triangular, and elliptical holes in anisotropic panels that posses a certain class of anisotropic properties. Savin used a slightly different mathematical approach than Lekhnitskii, but the solutions are equally well known and are considered to be equivalent. These solutions have since been used to predict stress concentration factors near elliptical holes in composite panels [3,4].

One of the solutions developed by Savin will be described in this text. Namely, a solution for stresses induced in anisotropic panels with elliptical holes will be described. A detailed derivation of this solution is quite lengthy and is not provided here. Instead, the solution and calculation steps are summarized and illustrated via several numerical examples. The reader interested in a complete derivation of the solution described, and/or in the stress concentrations associated with rectangular or triangular holes in anisotropic plates, is referred to the original works by Lekhnitskii [1] and/or Savin [2].

Unfortunately, both the Lekhnitskii and Savin solutions can become indeterminate for some combinations of stacking sequences, hole dimensions, and loading conditions. To the authors' knowledge no closed-form solutions are available for these indeterminate cases. However, Tung [5] suggested an approximate approach that gives reasonably accurate solutions for indeterminate problems. The Tung approximation is described in this chapter and can be used when an indeterminate problem is encountered.

9.2 Summary of the Savin Solution for an Anisotropic Plate with Elliptical Hole

Consider the symmetric anisotropic composite panel with an elliptical hole, as shown in Figure 9.1. In this chapter, we will consider symmetric panels subjected to in-plane loads only, so bending is not considered and mid-plane curvatures will always equal zero: $M_{xx} = M_{yy} = M_{xy} = \kappa_{xx} = \kappa_{yy} = \kappa_{xy} = 0$. Also, as thermal and moisture resultants are self-equilibrating, they can be ignored for present purposes: let $N_{ij}^T = M_{ij}^T = N_{ij}^M = M_{ij}^M = 0$. Therefore, the midplane strains at any point (x, y) within in the panel are related to the stress resultants induced at that point and are given by Equations 6.45, which for the conditions assumed here become:

$$\begin{Bmatrix} \varepsilon_{xx}^o \\ \varepsilon_{yy}^o \\ \gamma_{xy}^o \end{Bmatrix} = \begin{bmatrix} a_{11} & a_{12} & a_{16} \\ a_{12} & a_{22} & a_{26} \\ a_{16} & a_{26} & a_{66} \end{bmatrix} \begin{Bmatrix} N_{xx} \\ N_{yy} \\ N_{xy} \end{Bmatrix} \qquad (9.1)$$

The major and minor axes of the hole (also called the major and minor diameters) are denoted $2a$ and $2b$ and are aligned with the x- and y-axes, respectively. The panel is considered to be "infinitely large." That is, the dimensions of the plate within the x–y plane are assumed to be large enough that the distribution of stress resultants near the hole is not influenced by the finite size of the panel. Note that as we have ignored thermal and moisture resultants, the midplane strains that appear in Equation 9.1 are solely due to mechanical loading.

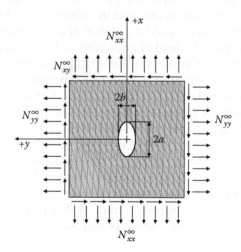

FIGURE 9.1
A large symmetric composite plate with elliptical hole subjected to remote uniform loads.

Stress resultants vary in regions near the hole, even though uniform loads N_{xx}^{∞}, N_{yy}^{∞}, and N_{xy}^{∞} are applied remotely. A superscript "∞" is used to distinguish between the uniform and remotely applied stress resultants and those that exist at an arbitrary point within the panel. That is,

$$N_{xx}(x,y)\big|_{x,y\to\infty} = N_{xx}^{\infty}$$

$$N_{yy}(x,y)\big|_{x,y\to\infty} = N_{yy}^{\infty}$$

$$N_{xy}(x,y)\big|_{x,y\to\infty} = N_{xy}^{\infty}$$

To be consistent with the Savin solution, it is convenient to express Equation 9.1 in terms of effective stresses rather than stress resultants. Toward that end, define the following terms:

$$\bar{a}_{ij} = t\, a_{ij} = \text{effective compliance matrix}$$

$$\bar{\sigma}_{ij} = \frac{N_{ij}}{t} = \text{effective stresses at any point}$$

$$\bar{\sigma}_{ij}^{\infty} = \frac{N_{ij}^{\infty}}{t} = \text{remotely applied effective stresses}$$

where
t = total laminate thickness

On the basis of these definitions, we can write Equation 9.1 as

$$\begin{Bmatrix} \varepsilon_{xx} \\ \varepsilon_{yy} \\ \gamma_{xy} \end{Bmatrix} = \begin{bmatrix} \bar{a}_{11} & \bar{a}_{12} & \bar{a}_{16} \\ \bar{a}_{12} & \bar{a}_{22} & \bar{a}_{26} \\ \bar{a}_{16} & \bar{a}_{26} & \bar{a}_{66} \end{bmatrix} \begin{Bmatrix} \bar{\sigma}_{xx} \\ \bar{\sigma}_{yy} \\ \bar{\tau}_{xy} \end{Bmatrix} \tag{9.2}$$

Now, the following fourth-order polynomial, called the "characteristic equation," is representative of any anisotropic panel:

$$\bar{a}_{11}s^4 - 2\bar{a}_{16}s^3 + (2\bar{a}_{12} + \bar{a}_{66})s^2 - 2\bar{a}_{26} + \bar{a}_{22} = 0 \tag{9.3}$$

As the characteristic equation is a fourth-order polynomial in s, it must have four roots. Muskhelishvili has shown [6] that these four roots cannot be real. That is, the four roots must either be complex or purely imaginary and can be written as

$$s_{1,3} = \alpha_1 \pm i\beta_1$$

$$s_{2,4} = \alpha_2 \pm i\beta_2 \tag{9.4}$$

where $i = \sqrt{-1}$, α_1, α_2, β_1, and β_2 are all real constants, and $\beta_{1,2} > 0$. Only roots s_1 and s_2 will be of further interest. These are called the *principal roots* and, as implied in Equation 9.4, these are the two roots with *positive* imaginary components:

$$s_1 = \alpha_1 + i\beta_1$$

$$s_2 = \alpha_2 + i\beta_2$$

$$(9.5)$$

Note that the value of the coefficients that appear in Equation 9.3, and, therefore, the principal roots s_1 and s_2, are dictated by elements of the effective compliance matrix (i.e., $\bar{a}_{11}$, $\bar{a}_{22}$, etc.). In turn, the compliance matrix reflects the ply properties and stacking sequence of the specific composite laminate under consideration. Ultimately then, roots s_1 and s_2 are governed solely by the ply properties and stacking sequence of the laminate and are not influenced by the hole dimensions or applied loading.

Referring again to Figure 9.1, the composite panel lies within the x–y plane, and the origin of the x–y coordinate system is at the center of the elliptical hole. We wish to calculate effective stresses induced within the panel at any point outside the elliptical hole. The x–y plane is called the "real plane." Points z_1 and z_2 are points within two complex planes that correspond to a point (x, y) in the real plane. The points z_1 and z_2 are defined by

$$z_1 = x + s_1 y = x + \alpha_1 y + i\beta_1 y$$

$$z_2 = x + s_2 y = x + \alpha_2 y + i\beta_2 y$$

$$(9.6)$$

Next, a mapping function is used to map points z_1 and z_2 to points ζ_1 and ζ_2 that exist within unit circles within "conformal" complex planes:

$$\zeta_1 = \frac{z_1 \pm \sqrt{z_1^2 - (a^2 + s_1^2 b^2)}}{a + is_1 b}$$

$$\zeta_2 = \frac{z_2 \pm \sqrt{z_2^2 - (a^2 + s_2^2 b^2)}}{a + is_2 b}$$

$$(9.7)$$

The algebraic signs that appear in Equations 9.7 are selected such that $|\zeta_1| \leq 1$ and $|\zeta_2| \leq 1$ at the current (x, y) position. The sign selected depends on the quadrant in the real plane, as defined by the current (x, y) point of interest. Note that ζ_1 and ζ_2 are functions of coordinates (x, y), the principal roots s_1 and s_2, and the dimensions a and b of the elliptical hole.

The following two functions $\phi_0(z_1)$ and $\psi_0(z_2)$ are written in terms of the variables defined above:

$$\phi_0(z_1) = \frac{-i\left[b\overline{\sigma}_{xx}^{\infty} + ias_2\overline{\sigma}_{yy}^{\infty} + (bs_2 + ia)\overline{\tau}_{xy}^{\infty}\right]}{2(s_1 - s_2)}\zeta_1$$

$$\psi_0(z_2) = \frac{i\left[b\overline{\sigma}_{xx}^{\infty} + ias_1\overline{\sigma}_{yy}^{\infty} + (bs_1 + ia)\overline{\tau}_{xy}^{\infty}\right]}{2(s_1 - s_2)}\zeta_2$$

(9.8)

Finally, the stresses at any point (x, y) within the panel, where the point defined in the real plane by coordinates (x, y) must be outside the elliptical hole, are given by

$$\overline{\sigma}_{xx} = \overline{\sigma}_{xx}^{\infty} + 2\Re\left[s_1^2\phi_0'(z_1) + s_2^2\psi_0'(z_2)\right]$$

$$\overline{\sigma}_{yy} = \overline{\sigma}_{yy}^{\infty} + 2\Re[\phi_0'(z_1) + \psi_0'(z_2)]$$

(9.9)

$$\overline{\tau}_{xy} = \overline{\tau}_{xy}^{\infty} - 2\Re[s_1\phi_0'(z_1) + s_2\psi_0'(z_2)]$$

The symbol "$\Re[]$" in Equations 9.9 denotes "the real part of" the expressions within the square brackets. The derivatives that appear in Equation 9.9 are given by

$$\phi_0'(z_1) = \frac{d\phi_0}{dz_1} = \frac{-i\left[b\overline{\sigma}_{xx}^{\infty} + ias_2\overline{\sigma}_{yy}^{\infty} + (bs_2 + ia)\overline{\tau}_{xy}^{\infty}\right]}{2(s_1 - s_2)}\left\{\frac{1}{a + is_1b}\left[1 \pm \frac{z_1}{\sqrt{z_1^2 - (a^2 + s_1^2b^2)}}\right]\right\}$$

$$\psi_0'(z_2) = \frac{d\psi_0}{dz_2} = \frac{i\left[b\overline{\sigma}_{xx}^{\infty} + ias_1\overline{\sigma}_{yy}^{\infty} + (bs_1 + ia)\overline{\tau}_{xy}^{\infty}\right]}{2(s_1 - s_2)}\left\{\frac{1}{a + is_2b}\left[1 \pm \frac{z_2}{\sqrt{z_2^2 - (a^2 + s_2^2b^2)}}\right]\right\}$$

The algebraic signs used in calculating $\phi_0'(z_1)$ and $\psi_0'(z_1)$ must be consistent with those used to calculate ζ_1 and ζ_2.

It is noted in passing that the Savin solution also allows calculation of the in-plane displacement fields $u(x, y)$ and $v(x,y)$ [2,7]. This aspect of the Savin solution will not be discussed here.

As mentioned earlier, the Savin solution becomes indeterminate for some problems. We are now in a position to explain where this indeterminacy arises. First, note that the terms $(a + is_1b)$ and $(a + is_2b)$ appear in the denominators of Equations 9.7. Therefore, if a problem is considered for which $(s_1 = ia/b)$ or $(s_2 = ia/b)$, then one or both of the denominator(s) of Equations 9.7 become zero and the problem is indeterminate. Second, the term $(s_1 - s_2)$ appears in the denominators of Equations 9.8. Hence, if the principal roots of the characteristic equation are repeated (i.e., if $s_1 = s_2$), then both denominators of Equations 9.8 become indeterminate and the Savin solution becomes indeterminate. One circumstance in which this latter circumstance occurs is for quasi-isotropic laminates, for which $s_1 = s_2 = i$.

Tung [5] suggested an approximate method of obtaining a solution for laminates with repeated principal roots. In these cases Tung suggested that the value of one of the principal roots be increased by one percent, and the value of the second principal roots be decreased by 1%. The Savin solution is then applied normally, using this artificial spread in the characteristic roots. The Tung approximation can be employed in those cases, where an indeterminate problem is considered.

For an elliptical hole the maximum stress will always occur *tangent* to the edge of the hole, and is called the effective circumferential stress, $\bar{\sigma}_{cc}$. Equations 9.9 will be used to calculate $\bar{\sigma}_{cc}$. A stress element at the edge of the hole and at an arbitrary angular position α is shown in Figure 9.2. In this figure the axes normal and tangential to the hole edge are labeled axes "n" and "c," respectively. As the hole is unloaded, the effective radial and shear stresses present at the edge must equal zero: $\bar{\sigma}_{nn} = \bar{\tau}_{nc} = 0$. The coordinate pairs (x, y) that define the hole are given by the equation of an ellipse:

$$\left(\frac{x}{a}\right)^2 + \left(\frac{y}{b}\right)^2 = 1$$

Rearranging, it is seen that at any point x on the edge of the hole the corresponding value of y is

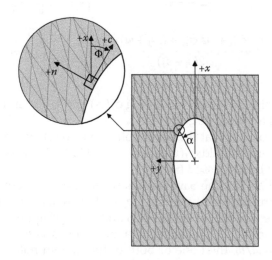

FIGURE 9.2
A stress element at an edge of an elliptical hole showing the angular position α, the normal and circumferential axes n and c, and the angle between the x- and c-axes (angle Φ).

$$y = \pm b \sqrt{1 - (x/a)^2}$$

where

$$-a \leq x \leq a \qquad (9.10)$$

The algebraic sign depends on quadrant. That is, referring to Figure 9.2, it is obvious that y is positive in the first and second quadrant of the x–y plane, and negative in the third and fourth quadrants. The slope of the c-axis is given by

$$\frac{\partial y}{\partial x} = \mp \frac{bx}{a^2 \sqrt{1 - (x/a)^2}}$$

Referring again to Figure 9.2, the angle Φ between the x- and the c-axes is calculated from the slope

$$\Phi = \tan^{-1}(\partial y / \partial x) = \tan^{-1} \left[\mp \frac{bx}{a^2 \sqrt{1 - (x/a)^2}} \right]$$

To calculate circumferential stress, Equations 9.9 are first used to calculate the effective stresses $\bar{\sigma}_{xx}$, $\bar{\sigma}_{yy}$, and $\bar{\tau}_{xy}$ at the edge of the elliptical hole, using (x, y) coordinates defined by Equation 9.10. The circumferential stress $\bar{\sigma}_{cc}$ is then calculated using the stress transformation equations, Equations 2.20:

$$\bar{\sigma}_{cc} = \bar{\sigma}_{xx} \cos^2 \Phi + \bar{\sigma}_{yy} \sin^2 \Phi + 2\bar{\tau}_{xy} \cos \Phi \sin \Phi \qquad (9.11)$$

As stress resultants N_{ij} may be calculated directly from the effective stresses, stress resultants everywhere along the edge of the hole are, therefore, obtained as well:

$$N_{nn} = 0$$

$$N_{cc} = t \, \bar{\sigma}_{cc}$$

$$N_{nc} = 0$$

However, due to free-edge effects (discussed in Section 6.13) individual *ply stresses* may not be accurately predicted near the edge of the hole based on these stress resultants, at least using classical lamination theory. Even though the effective stresses and stress resultants along the edge of the hole are accurately known, classical lamination theory cannot be used to calculate ply stresses at these locations. Determining ply stresses in these regions must be determined numerically using a finite-element or finite-difference analysis.

9.3 Circular Holes in Unidirectional Laminates

A graphite/epoxy laminate with a stacking sequence of $[\theta]_8$ and containing a 20-mm diameter circular hole (i.e., an ellipse with semi-axes $a = b = 10$ mm) is shown in Figure 9.3. The laminate is subjected to a remotely applied uniaxial effective stress $\bar{\sigma}_{xx}^{\infty} = N_{xx}^{\infty}/t$. The effective compliances and principal roots for $\theta = 0°, 30°, 45°, 60°$, and $90°$ (based on the properties listed for graphite-epoxy in Table 3.1) are listed in Table 9.1.

Normalized circumferential stresses $(\bar{\sigma}_{cc}/\bar{\sigma}_{xx}^{\infty})$ induced by this loading were calculated using Equations 9.9 and 9.11. Results are plotted as a function of angular position around the hole in Figure 9.4. For clarity, stresses have only been plotted over the range $0° \leq \alpha \leq 180°$. In general, distribution

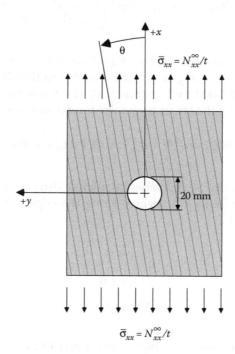

FIGURE 9.3
Unidirectional $[\theta]_8$ composite panel with 20-mm diameter circular hole subjected to uniaxial tensile loading.

TABLE 9.1

Effective Compliances and Principal Roots for Several Unidirectional Graphite–Epoxy Panels

		Stacking Sequence				
		$[0]_8$	$[30]_8$	$[45]_8$	$[60]_8$	$[90]_8$
Effective compliance terms $(m^2/N = 1/Pa)$	$\bar{a}_{11}$	0.58824e-11	0.23320e-10	0.44819e-10	0.70379e-10	0.10000e-9
	$\bar{a}_{12}$	−0.17647e-11	0.43269e-11	0.63575e-11	0.43269e-11	−0.17647e-11
	$\bar{a}_{16}$	0	−0.33720e-10	−0.47059e-10	−0.47788e-10	0
	$\bar{a}_{22}$	0.10000e-9	0.70379e-10	0.44819e-10	0.23320e-10	0.58824e-11
	$\bar{a}_{26}$	0	−0.47788e-10	−0.47059e-10	−0.33720e-10	0
	$\bar{a}_{66}$	0.76923e-10	0.10129e-9	0.10941e-9	0.10129e-9	0.76923e-10
Principal roots	s_1	0.0 + i(1.2477)	−0.21163+ i(1.0953)	−0.21774+ i(0.97601)	−0.17007+ i(0.88017)	0 + i(0.30261)
	s_2	0.0 + i(3.3046)	−1.2343+ i(0.94958)	−0.83227+ i(0.55442)	−0.50894+ i(0.39153)	0 + i(0.80147)

of stresses over the remaining portion of the hole edge, that is, over the range $180° \leq \alpha \leq 360°$, is either symmetric or antisymmetric to those shown.

Recall that the stress concentration factor for a circular hole in an infinite isotropic plate is 3.0. Also, for isotropic plates the maximum circumferential stress occurs at $\alpha = 90°$ (or equivalently, at $\alpha = -90°$). Inspection of Figure 9.4 reveals that substantially higher stress concentration factors can occur in unidirectional composite laminates. Furthermore, depending

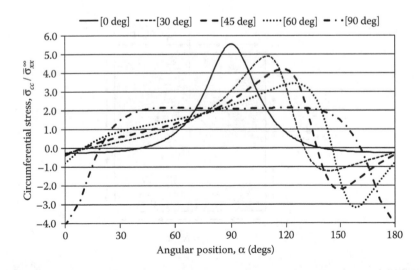

FIGURE 9.4

Normalized circumferential stresses along the edge of a circular hole in unidirectional laminates induced by remote effective stress $\bar{\sigma}_{xx}^{\infty}$.

on fiber orientation an effective stress $\bar{\sigma}_{xx}^{\infty}$ may cause significant tensile *and* compressive stress concentrations. Assuming $\bar{\sigma}_{xx}^{\infty}$ is tensile, for a $[0]_8$ laminate a maximum tensile stress concentration factor of about 5.5 occurs at an angular position of $\alpha = 90°$, and a modest compressive stress concentration factor of -0.24 occurs at $\alpha = 0°$, 180°. In contrast, for a $[45]_8$ laminate tensile and compressive stress concentration factors of 4.2, and -2.2, respectively, occur at angular positions of $\alpha = 118°$ and 150°, respectively. Similar trends are observed for the other laminates included in Figure 9.4. In fact, for a $[90]_8$ laminate a compressive stress concentration factor of -4.1 occurs at $\alpha = 0°$, 180°, which is almost twice as large as the maximum tensile stress concentration factor of 2.2, which occurs at $\alpha = 59°$, 121°. In other words, for this laminate the magnitude of the *compressive* stresses induced at the hole by an applied *tensile* load are almost twice as high as the tensile stresses induced at the hole.

Now consider a $[\theta]_8$ graphite/epoxy laminate with 20-mm diameter circular hole subjected to a remotely applied effective shear stress $\bar{\tau}_{xy}^{\infty} = N_{xy}^{\infty}/t$, as shown in Figure 9.5. Normalized circumferential stresses ($\bar{\sigma}_{cc}/\bar{\tau}_{xy}^{\infty}$) induced for $\theta = 0°$, 30°, 45°, 60°, and 90° are plotted as functions of angular position around the hole in Figure 9.6. In this case, stress concentrations range from ±6.4 (for $[0]_8$ and $[90]_8$ laminates) to as high as $+9.7$ for $[45]_8$ laminates.

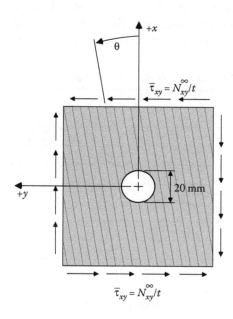

FIGURE 9.5
Unidirectional $[\theta]_8$ composite panel with 20-mm diameter circular hole subjected to pure shear loading.

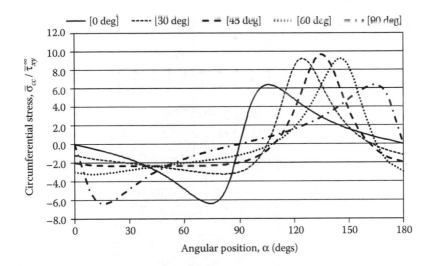

FIGURE 9.6
Normalized circumferential stresses along the edge of a circular hole in unidirectional laminates induced by remote effective stress $\bar{\tau}_{xy}^{\infty}$.

9.4 Elliptical Holes with an Aspect Ratio of Three in Unidirectional Laminates

A graphite/epoxy laminate with a stacking sequence of $[\theta]_8$ and containing an elliptical hole with major and minor diameters $2a = 30$ mm, $2b = 10$ mm (corresponding to an aspect ratio $a/b = 3$) is shown in Figure 9.7. The laminate is subjected to a remotely applied uniaxial effective stress $\bar{\sigma}_{xx}^{\infty} = N_{xx}^{\infty}/t$. The effective compliances and principal roots for this material system and for $\theta = 0°, 30°, 45°, 60°,$ and $90°$ have been previously listed in Table 9.1.

Normalized circumferential stresses induced by this loading are plotted as a function of angular position around the hole in Figure 9.8. As was the case for circular holes, depending on fiber orientation an effective stress $\bar{\sigma}_{xx}^{\infty}$ may cause significant tensile *and* compressive stress concentrations. For a $[0]_8$ laminate, the maximum tensile stress concentration factor is 2.52 and occurs at an angular position of $\alpha = 90°$, whereas a modest compressive stress concentration factor of -0.24 occurs at $\alpha = 0°, 180°$. In contrast, for the other laminates considered the maximum compressive stress concentration factors occur at angular positions ranging from $\alpha = 160°$ to $180°$, and reach a maximum of -4.1 for the $[90]_8$ laminate.

A $[\theta]_8$ graphite/epoxy laminate with elliptical hole subjected to a remotely applied effective shear stress $\bar{\tau}_{xy}^{\infty} = N_{xy}^{\infty}/t$, as shown in Figure 9.9. Normalized circumferential stresses $(\bar{\sigma}_{cc}/\bar{\tau}_{xy}^{\infty})$ induced for $\theta = 0°, 30°, 45°, 60°,$ and $90°$ are plotted as functions of angular position around the hole in Figure 9.10.

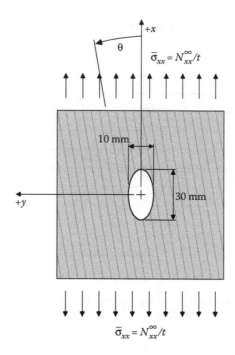

FIGURE 9.7
Unidirectional $[\theta]_8$ composite panel containing an elliptical hole with major and minor diameter $2a = 30$ mm, $2b = 10$ mm subjected to uniaxial tensile loading.

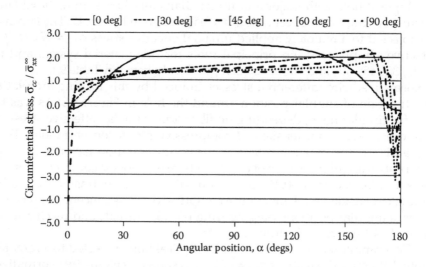

FIGURE 9.8
Normalized circumferential stresses along the edge of an elliptical hole in unidirectional laminates induced by remote effective stress $\bar{\sigma}_{xx}^{\infty}$.

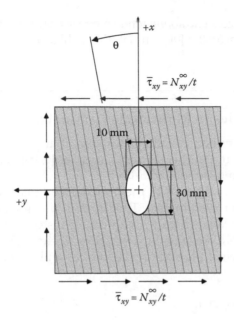

FIGURE 9.9
Unidirectional $[\theta]_8$ composite panel containing an elliptical hole with major and minor diameter $2a = 30$ mm, $2b = 10$ mm subjected to pure shear loading.

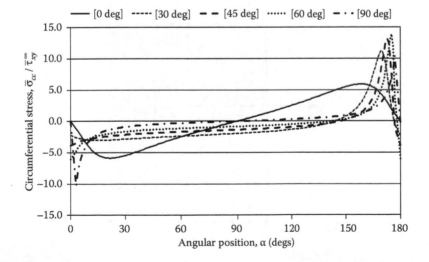

FIGURE 9.10
Normalized circumferential stresses along the edge of an elliptical hole in unidirectional laminates induced by remote effective stress $\bar{\tau}_{xy}^{\infty}$.

In this case, stress concentrations tend to occur near the ends of the elliptical hole, and range from 5.9 (for $[0]_8$ laminates) to as high as 13.9 (for $[60]_8$ laminates).

9.5 Circular Holes in Multiangle Laminates

Stress concentrations near circular holes in multiangle graphite–epoxy laminates will be explored in this section. Consider $[0/90]_{2s}$ cross-ply laminates containing a 20-mm diameter circular hole and subjected to uniaxial tensile loading. As shown in Figure 9.11, the uniaxial load may not be aligned with the principal axis of the laminate. That is, even though a cross-ply stacking sequence is used, relative to the *x–y* coordinate system the stacking sequence becomes $[\theta/(90 + \theta)]_{2s}$. Results for $\theta = 0°$, 15°, 30°, and 45° will be presented here. That is, four eight-ply stacking sequences will be considered:

- $[0/90]_{2s}$
- $[15/105]_{2s} = [60/-75]_{2s}$
- $[30/120]_{2s} = [30/-60]_{2s}$
- $[45/135]_{2s} = [45/-45]_{2s}$

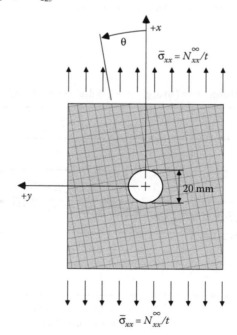

FIGURE 9.11
Cross-ply composite panel with 20-mm diameter circular hole subjected to uniaxial tensile loading.

TABLE 9.2

Effective Compliances and Principal Roots for Several Cross-Ply Graphite–Epoxy Panels

		Stacking Sequence			
		$[0/90]_{2s}$	$[15/-75]_{2s}$	$[30/-60]_{2s}$	$[45/-45]_{2s}$
Effective compliance terms ($m^2/N = 1/Pa$)	$\bar{a}_{11}$	0.11065e-10	0.14443e-10	0.21200e-10	0.24579e-10
	$\bar{a}_{12}$	−0.36882e-12	−0.37473e-11	−0.10504e-10	−0.13883e-10
	$\bar{a}_{16}$	0.0	−0.11704e-10	−0.11704e-10	0.0
	$\bar{a}_{22}$	0.11065e-10	0.14443e-10	0.21200e-10	0.24579e-10
	$\bar{a}_{26}$	0.0	0.11704e-10	0.11704e-10	0.0
	$\bar{a}_{66}$	0.76923e-10	0.63409e-10	0.36381e-10	0.22867e-10
Principal roots	s_1	$0.0 + i(0.38527)$	$0.059710+$ $i(0.508339)$	$0.39049+$ $i(0.70084)$	$0.74150+$ $i(0.67095)$
	s_2	$0.0 + i(2.5956)$	$-0.87003+$ $i(1.7494)$	$-0.94254+$ $i(0.81563)$	$-0.74150+$ $i(0.67095)$

The effective compliances and principal roots for these stacking sequences (based on the properties listed for graphite–epoxy in Table 3.1) are listed in Table 9.2.

Normalized circumferential stresses ($\bar{\sigma}_{cc}/\bar{\sigma}_{xx}^{\infty}$) induced by this loading are plotted as a function of angular position around the hole in Figure 9.12. The stress concentration factor depends on the angle between the principal material coordinate system and the direction of loading. A maximum stress concentration factor of 4.0 occurs for the $[0/90]_{2s}$ laminate.

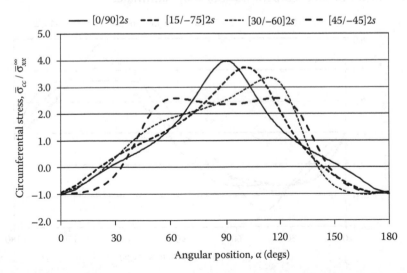

FIGURE 9.12

Normalized circumferential stresses along the edge of a circular hole in cross-ply laminates induced by remote effective stress $\bar{\sigma}_{xx}^{\infty}$.

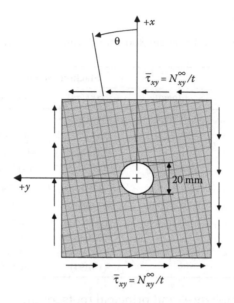

FIGURE 9.13
Cross-ply composite panel with 20-mm diameter circular hole subjected to pure shear loading.

A cross-ply laminate with 20-mm diameter circular hole subjected to a remotely applied effective shear stress $\overline{\tau}_{xy}^{\infty} = N_{xy}^{\infty}/t$ is shown in Figure 9.13. Normalized circumferential stresses $(\overline{\sigma}_{cc}/\overline{\tau}_{xy}^{\infty})$ induced for several cross-play laminates are shown in Figure 9.14. In this case, a maximum stress concentration factor of 5.8 occurs for the $[30/-60]_{2s}$ laminate.

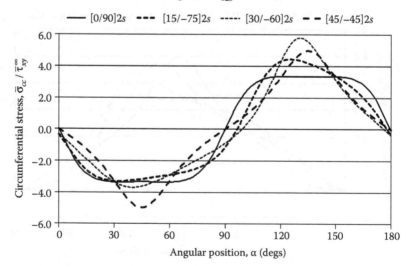

FIGURE 9.14
Normalized circumferential stresses along the edge of a circular hole in cross-ply laminates induced by remote effective stress $\overline{\tau}_{xy}^{\infty}$.

9.6 Computer Program HOLES

The computer program HOLES is available to supplement the material presented in this chapter. This program can be downloaded at no cost from the following website:

http://depts.washington.edu/amtas/computer.html

Program HOLES can be used to calculate the effective circumferential stresses induced at the edge of an elliptical opening in a symmetric laminate by any combination of remotely applied stress resultants N_{xx}^{∞}, N_{yy}^{∞}, and N_{xy}^{∞}. Note that the numerical results presented in this chapter are based on either uniaxial loading (N_{xx}^{∞} only) or pure shear loading (N_{xy}^{∞} only). Thus, all of the results presented here can be recreated using program HOLES, but the interested reader can also explore many other possible loading scenarios such as biaxial loading.

It is worth repeating that individual ply stresses at the hole edge cannot be accurately predicted, at least using classical lamination theory, due to free-edge effects. Thus, program HOLES provides an accurate prediction of the effective stresses and stress resultants along the edge of the hole, but determining ply stresses or strains in these regions must be determined numerically using a finite-element or finite-difference analysis.

When using program HOLES the user must define all variables using a consistent set of units. In particular, the user must input elastic moduli and hole dimensions consistently. For example, following numerical values would be input for graphite–epoxy and a 20-mm diameter circular hole:

$$E_{11} = 170 \times 10^9\,\text{Pa}, E_{22} = 10 \times 10^9\,\text{Pa}, v_{12} = 0.30, G_{12} = 13 \times 10^9\,\text{Pa}$$

$$a = 0.020\,\text{m} \quad b = 0.020\,\text{m}$$

HOMEWORK PROBLEMS

9.1. Prepare a plot similar to Figure 9.4 using properties for glass/epoxy (from Table 3.1) and program HOLES.

9.2. Prepare a plot similar to Figure 9.4 using properties for Kevlar/epoxy (from Table 3.1) and program HOLES.

9.3. Prepared a plot similar to Figure 9.4 for $[0]_8$, $[-30]_8$, $[-45]_8$, $[-60]_8$, and $[-90]_8$ laminates, using properties for graphite/epoxy (from Table 3.1) and program HOLES.

9.4. Prepared a plot similar to Figure 9.4 for $[0]_8$, $[-30]_8$, $[-45]_8$, $[-60]_8$, and $[-90]_8$ laminates, using properties for glass/epoxy (from Table 3.1) and program HOLES.

9.5. Prepared a plot similar to Figure 9.4 for $[0]_8$, $[-30]_8$, $[-45]_8$, $[-60]_8$, and $[-90]_8$ laminates, using properties for Kevlar/epoxy (from Table 3.1) and program HOLES.

9.6. Prepare a plot similar to Figure 9.8 using properties for glass/epoxy (from Table 3.1) and program HOLES.

9.7. Prepare a plot similar to Figure 9.8 using properties for Kevlar/epoxy (from Table 3.1) and program HOLES.

9.8. Prepared a plot similar to Figure 9.8 for $[0]_8$, $[-30]_8$, $[-45]_8$, $[-60]_8$, and $[-90]_8$ laminates, using properties for graphite/epoxy (from Table 3.1) and program HOLES.

9.9. Prepared a plot similar to Figure 9.8 for $[0]_8$, $[-30]_8$, $[-45]_8$, $[-60]_8$, and $[-90]_8$ laminates, using properties for glass/epoxy (from Table 3.1) and program HOLES.

9.10. Prepared a plot similar to Figure 9.8 for $[0]_8$, $[-30]_8$, $[-45]_8$, $[-60]_8$, and $[-90]_8$ laminates, using properties for Kevlar/epoxy (from Table 3.1) and program HOLES.

9.11. Prepared a plot similar to Figure 9.8 for $[0]_8$, $[-30]_8$, $[-45]_8$, $[-60]_8$, and $[-90]_8$ laminates, except consider an elliptical holes with $2a = 10$ and $2b = 30$ mm (i.e., with an aspect ratio of 1/3). Use properties for graphite/epoxy (from Table 3.1) and program HOLES.

9.12. Repeat Problem 9.11 for glass/epoxy laminates.

9.13. Repeat Problem 9.11 for Kevlar/epoxy laminates.

References

1. Lekhnitskii, S.G., *Anisotropic Plates*, translated by S.W. Tsai and T. Cheron, Gordan and Breach Pub. Co., ISBN 0-677-20670-4, 1968.
2. Savin, G.N., *Stress concentration Around Holes*, Pergamon Press, New York, 1961. See also NASA Technical Translation TT F-607, Nov 1970.
3. Greszczuk, L.B., Stress concentrations and failure criteria for orthotropic and anisotropic plates with circular openings, *Composites Materials: Testing and Design (Second Conference)*, ASTM STP 497, American Society for Testing and Materials, pp. 363–381, ISBN 978-0-8031-4606-8, 1972.
4. Tan, S.C., *Stress Concentrations in Laminated Composites*, Technomic Publishing Co., Lancaster, PA, ISBN 1-56676-0771-1, 1994.
5. Tung, T.K., On computation of stresses around holes in anisotropic plates, *Journal of Composite Materials*, 21(2), 100–104, 1987.
6. Muskhelishvili, N.I., *Some Basic Problems of the Mathematical Theory of Elasticity*, 4th edition, translated from the Russian by J.R.M. Radok, Noordhoff International Publishing, Leyden, The Netherlands, ISBN 9001-60701-2, 1977.
7. Chern, S.M. and Tuttle, M.E., On displacement fields in orthotropic laminates containing an elliptical hole, *Journal of Applied Mechanics*, 67, 527–539, 2000.

10

The Governing Equations of Thin-Plate Theory

10.1 Preliminary Discussion

Thin-plate theory, which describes the behavior of isotropic plates subject to in-plane and out-of-plane loads, was developed throughout the nineteenth and twentieth centuries and is now well established. Many textbooks devoted to isotropic plates theory have been published; two typical examples are References 1, 2. Although thin-plate theory is also applicable to anisotropic plates, relatively few texts devoted to this topic have appeared. Two of the best-known references are those by Whitney [3] and Turvey and Marshall [4].

A complete exposé of thin-plate theory as applied to composites is beyond the scope of this textbook. Rather, the objective of this and the following two chapters is to introduce the fundamental equations that govern the behavior of thin composite plates, and to present solutions to a few selected problems. It is hoped that this relatively brief treatment will form the basis for advanced study of this topic.

Solutions involving thin rectangular or elliptical plates with several different types of boundary conditions have been derived. In this chapter, discussion is limited to *rectangular* composite laminates with *symmetric* stacking sequences, subjected to simply supported boundary conditions along all four edges. The range of problems considered herein is, therefore, not as extensive as is presented elsewhere. The reader interested in a more detailed discussion of symmetric or nonsymmetric rectangular panels subject to alternate boundary conditions and/or a discussion of other panel shapes (e.g., elliptical composite panels) is referred to References 3, 4.

The rudiments of thin-plate theory have already been applied in Chapter 6. Specifically, during the development of classical lamination theory (CLT), we made several assumptions, two of which were

- All plies within a thin laminate are subjected to a state of plane stress ($\sigma_{zz} = \tau_{xz} = \tau_{yz} = 0$).

- The Kirchhoff hypothesis is valid: a straight line that is initially perpendicular to the midplane of a thin plate is assumed to remain straight and perpendicular to the midplane after deformation.

These assumptions are central to classical thin-plate theory, and we will continue to rely on them. However, in Chapter 6, we *also* assumed that the external loads applied to the laminate (i.e., stress and moment resultants N_{xx}, N_{yy}, N_{xy}, M_{xx}, M_{yy}, M_{xy}) were *constant* and *uniformly distributed* along the edge of the plate. Since the stress and moment resultants applied to the edge of the laminate were assumed to be uniform and constant, the resultants induced at all *interior regions* of the laminate were also constant and identically equal to the edge loads. Therefore, there was no need to distinguish between the resultants applied to the *edge* of a laminate and the resultants induced at any *interior* point.

In this chapter, we will relax this assumption and consider conditions in which *spatially varying* resultants are applied to the edge of the plate. Since the edge loads will now be allowed to vary, the stress and moment resultants induced at internal regions of the laminate will also vary and do not necessarily equal the stress and moment resultants applied along the edge of the laminate. We must, therefore, be careful to distinguish between the stress and moment resultants *applied to the edge* of the laminate and the resulting stress and moment versus resultants induced at *internal regions* of the laminate.

We will also include an additional type of loading in our analysis. Specifically, we will include the possibility that a distributed load acts *perpendicular* to the surface of the laminate. This transverse load can be visualized as a transverse pressure, and will be denoted $q(x, y)$.

A formal mathematical definition of the stress and moment resultants applied to the edge of the laminate will be given in Section 10.3. However, we will introduce some of the nomenclature used to describe external edge loads here. Examples of externally applied loads that vary over the edge of a laminate are shown schematically in Figures 10.1 and 10.2. As in earlier chapters, we assume the laminate is rectangular with plate edges parallel to the x- and y-axes. A sketch showing stress resultants N_{xx} acting on opposite edges of a thin laminate is presented in Figure 10.1a. Since N_{xx} is a load applied to the two plate edges that are parallel to the y-axis, N_{xx} cannot vary as a function of x. Therefore, along the plate edge, the externally applied resultant N_{xx} is *either a constant or, at most, a function of y only*

$$N_{xx} = N_{xx}(y).$$

There is no reason to expect that an identical distribution of loading is present on opposite sides of the laminate. Hence, we must distinguish between the resultant $N_{xx}(y)$ applied to the "negative" laminate edge (i.e., the edge whose outward normal "points" in the negative x-direction) and the resultant $N_{xx}(y)$ applied to the "positive" laminate edge. Therefore, the

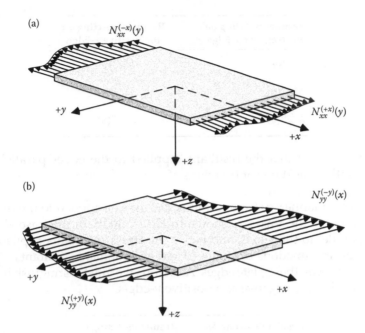

FIGURE 10.1
Illustration of varying stress resultants acting on opposite edges of a thin plate. (a) Variation of stress resultants $N_{xx}^{(+x)}(y)$ and $N_{xx}^{(-x)}(y)$; (b) variation of stress resultants $N_{yy}^{(+y)}(x)$ and $N_{yy}^{(-y)}(x)$.

load applied to the negative x-edge will be labeled $N_{xx}^{(-x)}(y)$, whereas the load applied to the positive x-edge will be labeled $N_{xx}^{(+x)}(y)$. Although not shown in Figure 10.1a, stress resultant N_{xy} and moment resultants M_{xx} and M_{xy} may also be applied to the two laminate edges parallel to the y-axis. These additional edge loads will be labeled:

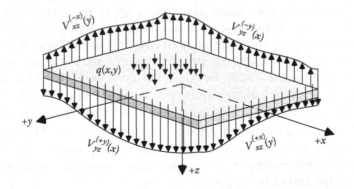

FIGURE 10.2
Illustration of varying transverse load $q(x, y)$ acting over the surface of a thin plate, perpendicular to the x–y plane, and the resulting out-of-plane shear stress resultants $V_{xz}^{(\pm x)}(y)$ and $V_{yz}^{(\pm y)}(x)$.

Resultants Acting on the Negative x-Edge	Resultants Acting on the Positive x-Edge
$N_{xy}^{(-x)}(y)$	$N_{xy}^{(+x)}(y)$
$M_{xx}^{(-x)}(y)$	$M_{xx}^{(+x)}(y)$
$M_{xy}^{(-x)}(y)$	$M_{xy}^{(+x)}(y)$

It is emphasized that the resultants applied to the edges parallel to the y-axis are either constants or functions of y only. *This will become an important point in the following discussions.*

In a similar manner, a stress resultant N_{yy} may be applied to the two plate edges parallel to the x-axis, as shown in Figure 10.1b. In this case, the plate edge is parallel to the x-axis, and hence the stress resultants applied along these edges are functions of x only. Stress and moment resultants N_{yx}, M_{yy}, and M_{yx} also act on these plate edges. As before, we must distinguish between loads applied to the negative and positive y-edge:

Resultants Acting on the Negative y-Edge	Resultants Acting on the Positive y-Edge
$N_{yy}^{(-y)}(x)$	$N_{yy}^{(+y)}(x)$
$N_{yx}^{(-y)}(x)$	$N_{yx}^{(+y)}(x)$
$M_{yy}^{(-y)}(x)$	$M_{yy}^{(+y)}(x)$
$M_{yx}^{(-y)}(x)$	$M_{yx}^{(+y)}(x)$

Since in our earlier analysis, we assumed that stress resultants were constant and uniformly distributed along the edge of the plate, there was no need to distinguish between shear stress resultants acting on adjacent faces. That is, an assumption made throughout Chapters 6 and 7 was $N_{xy} = N_{yx}$. We can no longer make this assumption. For example, the variation of the shear stress resultant over the finite length of the positive x-face of the laminate, $N_{xy}^{(+x)}(y)$, is independent of the variation of the shear stress resultant applied over the finite length of the positive y-face, $N_{yx}^{(+y)}(x)$. Similarly, the variation of the twisting moment applied to the positive x-face, $M_{xy}^{(+x)}(y)$, is now independent of the twisting moment applied to the positive y-face, $M_{yx}^{(+y)}(x)$. Exceptions to this statement occur at the four corners of the laminate. For example, at the corner $(x = a, y = b)$:

$$N_{xy}^{(+x)}(b) = N_{yx}^{(+y)}(a) \quad \text{and} \quad M_{xy}^{(+x)}(b) = M_{yx}^{(+y)}(a)$$

A distributed transverse force acting over the surface of the laminate, $q(x,y)$, is shown in Figure 10.2. The transverse force $q(x,y)$ has units of force per area, such as N/m^2 = Pascals or lbf/in^2 = psi. Since $q(x,y)$ is a distributed force acting in the z-direction, shear stress resultants $V_{xz}(y)$ and/or $V_{yz}(x)$ must be present along one or more edges of the laminate, so as to maintain static equilibrium (i.e., to maintain $\Sigma F_z = 0$). $V_{xz}^{(\pm)}(y)$ and $V_{yz}^{(\pm)}(x)$ are given by

$$V_{xz}^{(\pm)} = \int_{-t/2}^{t/2} \tau_{xz}^{(\pm)}dz \quad V_{yz}^{(\pm)} = \int_{-t/2}^{t/2} \tau_{yz}^{(\pm)}dz \tag{10.1}$$

The reader should object to the inclusion of $q(x,y)$, $V_{xz}^{(\pm)}(y)$, and $V_{yz}^{(\pm)}(x)$ in our analysis. After all, *they violate our assumption of plane stress.* That is, the presence of $q(x,y)$ implies that stress σ_{zz} will be induced in the plate, and furthermore Equations 10.1 imply that shear stresses τ_{xz} and τ_{yz} will also be induced. These objections are entirely valid. If a transverse load $q(x,y)$ is applied, then a 3D state of stress is, in fact, induced in the plate. However, *if the plate is thin,* then the magnitudes of the out-of-plane stresses are far lower than the magnitudes of the in-plane stresses: $\sigma_{zz}, \tau_{xyz}, \tau_{yz} \ll \sigma_{xx}, \sigma_{yy}, \tau_{xy}$. Hence, the stresses induced in the plate satisfy the plane stress assumption approximately. The inclusion of the out-of-plane transverse load $q(x,y)$ while still invoking the plane stress assumption is a fundamental discrepancy in thin-plate theory. In the discussion to follow, we will simply ignore this inconsistency.

To summarize, a total of five types of externally applied distributed loads on *each* of the four edges of a rectangular laminate are allowed in thin-plate theory: one normal stress resultant, one in-plane shear stress resultant, one out-of-plane shear stress resultant, one bending moment resultant, and one twisting moment resultant. All of these loads may vary along each laminate edge. In addition, a transverse loading $q(x,y)$ may be applied to the surface of the laminate. A laminate subjected to all external edge loads considered in thin-plate theory is shown in Figure 10.3.

Thus far, we have described the stress and moment resultants that may be specified along the edges of a rectangular laminate. Specified edge loading is a type of *boundary condition.* However, in many practical instances, the loads applied along the plate edges are unknown and hence cannot be used as specified boundary conditions. Rather, in many instances, midplane *displacements* along the plate edges are known. Specified midplane displacements represent a second type of boundary condition. The types of displacement boundary conditions have already been discussed in Chapter 8. For example, in beam theory, it is common to specify that one (or both) end(s) of a beam is (are) "clamped." If a beam end is clamped, then that end of the beam cannot move up/down, and the slope of the beam midplane must equal zero at the clamped end. Hence, when one specifies that a beam end is clamped, one has

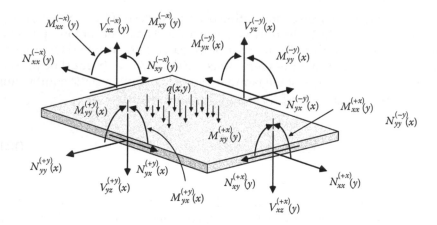

FIGURE 10.3
A thin plate subjected to stress resultants and transverse loading that vary along the length and width of the plate.

specified something about the displacement field in the beam, rather than something about the applied loading.

Similar displacement boundary conditions can also be defined for thin plates. Hence, to define a thin-plate problem of interest, we must specify the boundary conditions along each edge of the plate. The specified boundary conditions may involve (a) the loads applied to each edge of the plate, (b) the displacements imposed on each edge of the plate, or (c) some combination thereof. Boundary conditions that are consistent with the assumptions of thin-plate theory will be discussed in detail in Section 10.3.

The equations that govern the behavior of thin composite laminates will be developed in the following two sections. First, the *equations of equilibrium* will be developed in Section 10.2. *Boundary conditions* that are consistent with thin-plate theory will then be developed in Section 10.3. These equations and boundary conditions are quite general and govern the solution to any problem involving a thin plate, including composite laminates. Unfortunately, *it is often impossible to obtain exact solutions to the equations of equilibrium that also satisfy the prevailing boundary conditions.* Difficulties in obtaining exact solutions occur for either isotropic plates or anisotropic composite plates, and arise from two different sources. First, the boundary conditions encountered in a problem may preclude obtaining an exact solution. For example, an exact solution can be found for a flat rectangular isotropic plate subjected to transverse loading *if all four edges are simply supported,* but an exact solution cannot be found for an identical flat rectangular isotropic plate subjected to transverse loading *if all four edges are clamped.* Since boundary conditions are independent of material properties, the difficulties in obtaining exact solutions for certain boundary conditions are encountered for either isotropic or anisotropic composite plates.

The second source of difficulty has to do with the anisotropic nature of composites, and is not encountered during the study of isotropic plates. Specifically, exact solutions cannot be found for laminates that exhibit the following couplings terms:

- In-plane normal–shear coupling (i.e., laminates for which $A_{16}, A_{26} \neq 0$)
- Bending-twisting coupling (i.e., laminates for which $D_{16}, D_{26} \neq 0$)
- Coupling between in-plane loads and out-of-plane twisting (i.e., laminates for which $B_{16}, B_{26} \neq 0$)

All of these coupling terms exist for a composite laminate with an arbitrary stacking sequence, and hence an exact solution cannot, in general, be found for composite laminates with an arbitrary stacking sequence.

Still another complication associated with environmental factors exists for nonsymmetric laminates. Recall from Chapter 6 that nonsymmetric laminates will bend/warp if subjected to a change in temperature and/or moisture content, since $B_{ij}, M_{ij}^T, M_{ij}^M \neq 0$ for nonsymmetric laminates. For example, following cure at an elevated temperature a nonsymmetric laminate is likely to be warped at room temperature, and may warp further upon exposure to changes in humidity over time. It is for this reason that nonsymmetric laminates are rarely used in practice. An analysis of a nonsymmetric laminate must, therefore, account for the displacements due to externally applied loads as well as displacements (e.g., warping) due to changes in temperature or moisture content. Since the intention here is to provide a brief introduction to thin-plate theory as applied to composites, nonsymmetric laminates will not be considered.

In summary then, the equations of equilibrium and boundary conditions that will be developed in the following sections of this chapter are valid for any symmetric composite laminate. However, exact solutions to these governing equations can only be found for certain combinations of stacking sequences and boundary conditions. Some available exact solutions to the governing equations will be discussed in Chapter 10. Fortunately, techniques are also available that can be used to obtain *approximate* solutions for those laminates and boundary conditions that do not admit exact solutions. Although approximate, the accuracy of these methods is usually quite good and are suitable for most engineering applications. These approximate solution techniques will be discussed in Chapter 11.

10.2 Equations of Equilibrium for Symmetric Laminates

The *externally* applied edge loads shown in Figure 10.3 induce *internal* stress and moment resultants. In general, these internal resultants vary throughout the interior regions of the laminate. Ultimately, we wish to relate these varying internal stress and moment resultants to the externally applied loads that induce them. In order to do so, we must first develop the *equations of equilibrium* for a thin plate. In essence, the equations of equilibrium dictate how internal stress and moment resultants may vary over the length and width of the laminate such that static equilibrium is maintained.

The equations of equilibrium will be developed in Section 10.2.1 by applying the equations of statics. The equations developed in this manner are partial differential equations involving the internal stress and moment resultants, transverse loading $q(x, y)$, and the out-of-plane displacement field, $w(x, y)$. These equations will then be converted to a form more useful for composite studies in Section 10.2.2. This will involve expressing the internal stress and moment resultants in terms of elements of the $[ABD]$ matrix and midplane displacement fields, $u_o(x, y)$, $v_o(x, y)$, and $w(x, y)$.

10.2.1 Equations of Equilibrium Expressed in Terms of Internal Stress and Moment Resultants, Transverse Loading, and Out-of-Plane Displacements

Consider a thin composite laminate subjected to some combination of edge loads and/or edge displacements that induce stress and moment resultants at interior regions of the laminate. The laminate is assumed to be flat prior to the application of external edge loads and/or displacements. The laminate is assumed to be symmetric, and hence $B_{ij} = 0$. Since the laminate is symmetric, it will not warp if subjected to a change in temperature and/or moisture content ($M_{ij}^T = M_{ij}^M = 0$) prior to the application of external loads.

An infinitesimal element removed from an internal region of such a laminate is shown in Figure 10.4. For clarity, only the midsurface of the element is shown. The element thickness is actually equal to the total laminate thickness, t. The in-plane dimensions are dx and dy. The in-plane normal and shear stress resultants are shown in Figure 10.4a, while the moment resultants, out-of-plane shear resultants, and transverse loading are shown in Figure 10.4b (it should be understood that the resultants and transverse loads shown in Figure 10.4a and b are applied simultaneously to the element). Once again, it is important to realize that the resultants shown are induced at an *interior* region represented by the infinitesimal element, and do not necessarily equal the *external* loads applied along the edges of the laminate. A superscript (*) will be used to denote resultants present at *interior* regions of the laminate. Also, since the element is infinitesimal, we do not need to distinguish between shear resultants on adjacent faces, for example, $N_{yx}^* = N_{xy}^*$.

All resultants are assumed to vary slightly across the infinitesimal length and width of the element. For example, the stress resultant N_{xx}^* acting on the negative x-face of the element is assumed to vary slightly over distance dx, such that a stress resultant $N_{xx}^* + (\partial N_{xx}^*/\partial x)dx$ is present on the positive x-face of the element. The equations of equilibrium can be developed by applying the standard equations of statics. That is, by requiring that the sum of all

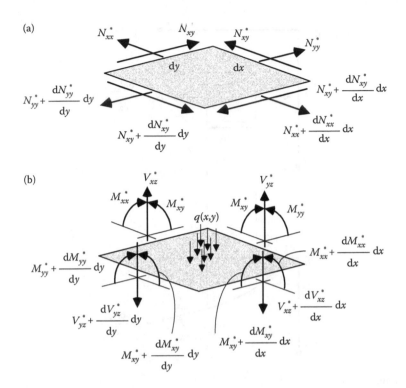

FIGURE 10.4

Internal stress resultants, moment resultants, and transverse loading acting at the midplane of an infinitesimal element. (a) In-plane normal and shear stress resultants; (b) moments resultants, out-of-plane shear stress resultants and traverse loading.

forces and moments acting on the infinitesimal element equate to zero: $\Sigma F_x = \Sigma F_y = \Sigma F_z = \Sigma M_x = \Sigma M_y = \Sigma M_z = 0.$

Let us first sum forces in the x-direction, based on the free-body diagram shown in Figure 10.5. Only those resultants that contribute toward forces in the x-direction are shown in this figure. The element has been deflected out of the original x–y plane by the transverse loading. A view of the x–y plane is shown in Figure 10.5a, while a view of the x–z plane is shown in Figure 10.5b. In the deflected condition, the left and right side of the element form angles α and $(\alpha + (\partial\alpha/\partial x)dx)$, respectively, with the original x–y plane. Consider the x-directed force associated with N_{xx}^*, acting on the left side of the element. Recall that the units of a stress resultant are (force/length), so to convert N_{xx}^* to a force we must multiply it by the distance over which it acts, namely by distance dy. Figure 10.5b shows that the line of action of N_{xx}^* is not precisely parallel to the x-axis, due to the out-of-plane deflection. The component of the force associated with N_{xx}^* that is parallel to the x-axis is, therefore, given by

$$-(N_{xx}^*)(dy)\cos\alpha$$

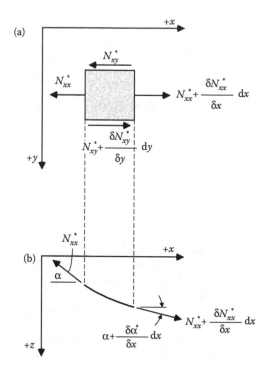

FIGURE 10.5
Free-body diagram of an infinitesimal element, showing only those resultants that contribute towards forces in the x-direction (deflection in x–z plane shown exaggerated for clarity). (a) x–y plane; (b) x–z plane.

We now assume that angle α is small (less than about 10 degrees or 0.1745 radians), such that the small-angle approximation can be applied: $\cos\alpha \approx 1$. With this assumption the component of the force associated with N_{xx}^* that is parallel to the x-axis reduces to

$$-(N_{xx}^*)(dy)$$

In effect, invoking the small-angle approximation implies that we will ignore the fact that N_{xx}^* is not precisely parallel to the x-axis during the summation of forces in the x-direction. In exactly the same way, the x-directed forces acting on the other three sides of the element are given by

- x-Directed force acting on the positive x-face: $(N_{xx}^* + (\partial N_{xx}^*/\partial x)dx)(dy)$
- x-Directed force acting on the negative y-face: $(N_{xy}^*)(dx)$
- x-Directed force acting on the positive y-face: $(N_{xy}^* + (\partial N_{xy}^*/\partial y)dy)(dx)$

Summing the force components acting on all four sides and equating to zero ($\Sigma F_x = 0$)

$$-(N_{xx}^*)dy + \left(N_{xx}^* + \frac{\partial N_{xx}^*}{\partial x} dx \right) dy - (N_{xy}^*)dx + \left(N_{xy}^* + \frac{\partial N_{xy}^*}{\partial y} dy \right) dx = 0$$

Simplifying, we obtain

$$\frac{\partial N_{xx}^*}{\partial x} + \frac{\partial N_{xy}^*}{\partial y} = 0 \tag{10.2a}$$

Equation 10.2a is the first equation of equilibrium. It shows that if N_{xx}^* does indeed vary in the x-direction at an interior point in the plate, then N_{xy}^* must also vary in the y-direction at a comparable rate, so that static equilibrium is maintained.

A second free-body diagram showing only those stress resultants whose line of action is parallel to the y-axis is shown Figure 10.6. As before, the x–y plane of the element is shown in Figure 10.6a while the y–z plane is shown in Figure 10.6b. Summing forces in the y-direction, invoking the small angle approximation, and equating to zero ($\Sigma F_y = 0$), we find

$$-\left[N_{yy}^* \right]dx + \left[N_{yy}^* + \frac{\partial N_{yy}^*}{\partial y} dy \right]dx - \left[N_{xy}^* \right]dy + \left[N_{xy}^* + \frac{\partial N_{xy}^*}{\partial x} dx \right]dy = 0$$

which simplifies to

$$\frac{\partial N_{yy}^*}{\partial y} + \frac{\partial N_{xy}^*}{\partial x} = 0 \tag{10.2b}$$

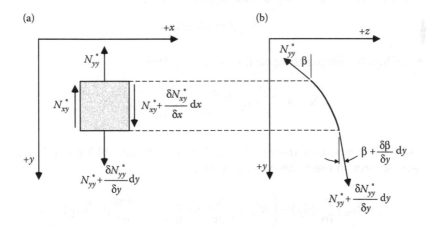

FIGURE 10.6
Free-body diagram of an infinitesimal element, showing only those resultants that contribute towards forces in the y-direction (deflection in y–z plane shown exaggerated for clarity). (a) x–y plane; (b) y–z plane.

Equation 10.2b is the second equation of equilibrium.

Let us now consider forces parallel to the z-axis. Due to out-of-plane deflection of the plate midsurface the stress resultants shown in Figures 10.5 and 10.6 have components in the z-direction. For example, from Figure 10.5b the component of the force associated with N_{xx}^* that is parallel to the z-axis is given by

$$-(N_{xx}^*)(dy)\sin\alpha$$

The small angle approximation (where α is expressed in radians) is

$$\sin\alpha \approx \alpha \approx \frac{\partial w}{\partial x}$$

Hence, the z-directed component of the force associated with N_{xx}^* is

$$-(N_{xx}^*)(dy)\frac{\partial w}{\partial x}$$

In exactly the same way, the z-directed forces acting on the other three sides of the element shown in Figure 10.5 are given by

- z-Directed force acting on the positive x-face:

$$(N_{xx}^* + (\partial N_{xx}^*/\partial x)dx)((\partial w/\partial x) + (\partial^2 w/\partial x^2)dx)(dy)$$

- z-Directed force acting on the negative y-face:

$$-(N_{xy}^*)(dx)(\partial w/\partial y)$$

- z-Directed force acting on the positive y-face:

$$\left(N_{xy}^* + \frac{\partial N_{xy}^*}{\partial y}dy\right)\left(\frac{\partial w}{\partial y} + \frac{\partial^2 w}{\partial x\partial y}dy\right)(dx)$$

Adding these results, the resultants shown in Figure 10.5 result in a component of force in the z-direction given by

$$-(N_{xx}^*)\frac{\partial w}{\partial x}(dy) + \left(N_{xx}^* + \frac{\partial N_{xx}^*}{\partial x}dx\right)\left(\frac{\partial w}{\partial x} + \frac{\partial^2 w}{\partial x^2}dx\right)(dy)$$

$$-(N_{xy}^*)\frac{\partial w}{\partial x}(dx) + \left(N_{xy}^* + \frac{\partial N_{xy}^*}{\partial y}dy\right)\left(\frac{\partial w}{\partial y} + \frac{\partial^2 w}{\partial x\partial y}dy\right)(dx)$$

When the algebra indicated is completed, two higher-order terms appear:

$$\left(\frac{\partial N_{xx}^*}{\partial x}\right)\left(\frac{\partial^2 w}{\partial x^2}\right)(dx)^2 dy \quad \text{and} \quad \left(\frac{\partial N_{xy}^*}{\partial y}\right)\left(\frac{\partial^2 w}{\partial x \partial y}\right)dx(dy)^2$$

Neglecting these higher-order terms, the above expression simplifies to

$$\left[N_{xx}^* \frac{\partial^2 w}{\partial x^2} + \frac{\partial N_{xx}^*}{\partial x}\frac{\partial w}{\partial x} + N_{xy}^* \frac{\partial^2 w}{\partial x \partial y} + \frac{\partial N_{xy}^*}{\partial y}\frac{\partial w}{\partial y}\right]dxdy \qquad \text{(a)}$$

Following an identical procedure, the stress resultants shown in Figure 10.6 represent a force in the z-direction given by

$$\left[N_{yy}^* \frac{\partial^2 w}{\partial y^2} + \frac{\partial N_{yy}^*}{\partial y}\frac{\partial w}{\partial y} + N_{xy}^* \frac{\partial^2 w}{\partial x \partial y} + \frac{\partial N_{xy}^*}{\partial x}\frac{\partial w}{\partial x}\right]dxdy \qquad \text{(b)}$$

The vertical shear resultants and transverse loading act directly in the z-direction, as previously shown in Figure 10.3b. Using a procedure similar to that described above, the sum of these forces in the z-direction is

$$-[V_{xz}^*]dy + \left[V_{xz}^* + \frac{\partial V_{xz}^*}{\partial x}dx\right]dy - [V_{yz}^*]dx + \left[V_{yz}^* + \frac{\partial V_{yz}^*}{\partial y}dy\right]dx + [q(x,y)]dxdy$$

which simplifies to

$$\left[\frac{\partial V_{xz}^*}{\partial x} + \frac{\partial V_{yz}^*}{\partial y} + q(x,y)\right]dxdy \qquad \text{(c)}$$

We are now ready to sum all forces in the z-direction. Adding expressions (a), (b), and (c), equating the resulting sum to zero ($\Sigma F_z = 0$), and rearranging, there results

$$N_{xx}^* \frac{\partial^2 w}{\partial x^2} + N_{yy}^* \frac{\partial^2 w}{\partial y^2} + 2N_{xy}^* \frac{\partial^2 w}{\partial x \partial y} + \left(\frac{\partial N_{xx}^*}{\partial x} + \frac{\partial N_{xy}^*}{\partial y}\right)\frac{\partial w}{\partial x} + \left(\frac{\partial N_{yy}^*}{\partial y} + \frac{\partial N_{xy}^*}{\partial x}\right)\frac{\partial w}{\partial y}$$

$$+ \frac{\partial V_{xz}^*}{\partial x} + \frac{\partial V_{yz}^*}{\partial y} + q(x,y) = 0$$

Notice that the terms within the two sets of parenthesis have been previously shown to equal zero, in accordance with Equations 10.2a and 10.2b. Hence, these terms may be dropped, and our third equation of equilibrium becomes

$$N_{xx}^* \frac{\partial^2 w}{\partial x^2} + N_{yy}^* \frac{\partial^2 w}{\partial y^2} + 2N_{xy}^* \frac{\partial^2 w}{\partial x \partial y} + \frac{\partial V_{xz}^*}{\partial x} + \frac{\partial V_{yz}^*}{\partial y} + q(x,y) = 0 \quad (10.2c)$$

Equations 10.2a, 10.2b, and 10.2c represent the requirement that all forces in the x-, y-, and z-directions, respectively, sum to zero.

Now consider moment equilibrium about the x-axis. A free-body diagram showing the resultants that contribute to the moment about the x-axis is shown in Figure 10.7a. Summing moments and equating to zero ($\Sigma M_x = 0$), we have

$$[M_{yy}^*]dx - \left[M_{yy}^* + \frac{\partial M_{yy}^*}{\partial y}dy \right]dx + [M_{xy}^*]dy - \left[M_{xy}^* + \frac{\partial M_{xy}^*}{\partial x}dx \right]dy$$

$$+ \left[V_{yz}^* + \frac{\partial V_{yz}^*}{\partial y}dy \right]dx(dy) + \left[V_{xz}^* + \frac{\partial V_{xz}^*}{\partial x}dx \right]dy\left(\frac{dy}{2} \right)$$

$$- [V_{xz}^*]dy\left(\frac{dy}{2} \right) + \left[q(x,y)dxdy \right]\left(\frac{dy}{2} \right) = 0$$

which simplifies to

$$-\frac{\partial M_{yy}^*}{\partial y}dxdy - \frac{\partial M_{xy}^*}{\partial x}dxdy + V_{yz}^*dxdy + \frac{\partial V_{yz}^*}{\partial y}dx(dy)^2$$

$$+ \frac{\partial V_{xz}^*}{\partial x}\frac{dx}{2}(dy)^2 + \frac{q(x,y)}{2}dx(dy)^2 = 0$$

The last three terms in this equation contain the higher-order factor $(dy)^2$. Hence, dropping the last three terms and simplifying, we obtain

$$V_{yz}^* = \frac{\partial M_{xy}^*}{\partial x} + \frac{\partial M_{yy}^*}{\partial y} \quad (10.3a)$$

A free-body diagram showing the stress and moment resultants that contribute to the moment about the y-axis is shown in Figure 10.7b. Summing moments ($\Sigma M_y = 0$), we have

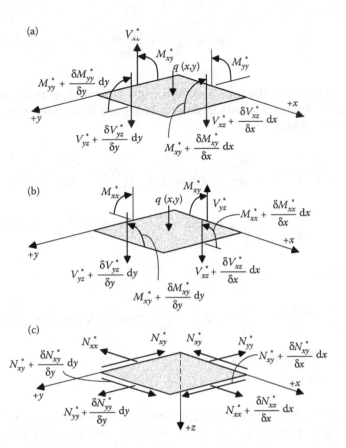

FIGURE 10.7
Free-body diagrams used to sum moments about the *x*-, *y*-, and *z*-axes. (a) Infinitesimal element used to sum moments about the *x*-axis; (b) infinitesimal element used to sum moments about the *y*-axis; (c) infinitesimal element used to sum moments about the *z*-axis.

$$[M_{xx}^*]dy - \left[M_{xx}^* + \frac{\partial M_{xx}^*}{\partial x}dx\right]dy + [M_{xy}^*]dx - \left[M_{xy}^* + \frac{\partial M_{xy}^*}{\partial y}dy\right]dx$$

$$+ \left[V_{xz}^* + \frac{\partial V_{xz}^*}{\partial x}dx\right]dy(dx) + \left[V_{yz}^* + \frac{\partial V_{yz}^*}{\partial y}dy\right]dx\left(\frac{dx}{2}\right) - [V_{yz}^*]dx\left(\frac{dx}{2}\right)$$

$$+ \left[q(x,y)dxdy\right]\left(\frac{dx}{2}\right) = 0$$

Upon completing the algebra indicated, a negligible higher-order term appears (in this case dx^2). Neglecting all terms that include this factor and simplifying, we obtain

$$V_{xz}^* = \frac{\partial M_{xx}^*}{\partial x} + \frac{\partial M_{xy}^*}{\partial y} \tag{10.3b}$$

Lastly, a free-body diagram shown the stress resultants that contribute to the moment about the z-axis is shown in Figure 10.7c. Summing moments ($\Sigma M_z = 0$), we obtain

$$(N_{xx}^* dy)\left(\frac{dy}{2}\right) - \left[\left(N_{xx}^* + \frac{\partial N_{xx}^*}{\partial x}dx\right)dy\right]\left(\frac{dy}{2}\right) + \left[\left(N_{xy}^* + \frac{\partial N_{xy}^*}{\partial x}dx\right)dy\right]dx$$

$$- \left[\left(N_{xy}^* + \frac{\partial N_{xy}^*}{\partial y}dy\right)dx\right]dy + \left[\left(N_{yy}^* + \frac{\partial N_{yy}^*}{\partial y}dy\right)dx\right]\left(\frac{dx}{2}\right) - (N_{yy}^* dx)\left(\frac{dx}{2}\right) = 0$$

Simplifying

$$-\frac{\partial N_{xx}^*}{\partial x}\frac{dxdy^2}{2} + \frac{\partial N_{xy}^*}{\partial x}dx^2dy - \frac{\partial N_{xy}^*}{\partial y}dxdy^2 + N_{xy}^*(dx^2dy - dxdy^2)$$

$$+ \frac{\partial N_{yy}^*}{\partial y}\frac{dx^2dy}{2} = 0$$

All terms in this equation contain a higher-order term (either dx^2 or dy^2). Hence, all terms on the left side of the equality are negligibly small, and the summation of moments about the z-axis provides no new information.

In summary, we have developed five useful equations through a summation of forces and moments. The first three, Equations 10.2a through 10.2c, were developed by requiring that the sum of forces in the x-, y-, and z-directions sum to zero, while the fourth and fifth equations, Equations 10.3a,b, were developed by requiring that the sum of moments about the x- and y-axes sum to zero. Note that Equations 10.3a,b relate the out-of-plane shear stress resultants, V_{xz}^* and V_{yz}^*, to the moment resultants M_{xx}^*, M_{yy}^*, and M_{xy}^*, and that the shear stress resultants also appear in Equation 10.2c. This means that *only three of the five equations we have developed are independent*. The fact that only three equations are independent is fundamentally due to the assumption of plane stress, since this assumption implies there are only three independent stress components. It is customary to substitute Equations 10.3a,b into Equation 10.2c, resulting in

$$N_{xx}^* \frac{\partial^2 w}{\partial x^2} + N_{yy}^* \frac{\partial^2 w}{\partial y^2} + 2N_{xy}^* \frac{\partial^2 w}{\partial x \partial y} + \frac{\partial^2 M_{xx}^*}{\partial x^2} + \frac{\partial^2 M_{yy}^*}{\partial y^2} + 2\frac{\partial^2 M_{xy}^*}{\partial x \partial y} + q(x,y) = 0$$

$$\tag{10.4}$$

Equations 10.2a, 10.2b, and 10.4 represent three independent equations of equilibrium for a thin plate subjected to a state of plane stress. If a thin plate is in static equilibrium then the distribution of internal stress and moment resultants induced at any interior region of a rectangular plate must satisfy these equations, which guarantee that $\Sigma F_i = 0$ and $\Sigma M_i = 0$. Since these equations will be referenced repeatedly in the following discussion, the three equations of equilibrium are collected here and renumbered for convenience:

$$\frac{\partial N_{xx}^*}{\partial x} + \frac{\partial N_{xy}^*}{\partial y} = 0 \tag{10.5a}$$

$$\frac{\partial N_{yy}^*}{\partial y} + \frac{\partial N_{xy}^*}{\partial x} = 0 \tag{10.5b}$$

$$N_{xx}^* \frac{\partial^2 w}{\partial x^2} + N_{yy}^* \frac{\partial^2 w}{\partial y^2} + 2N_{xy}^* \frac{\partial^2 w}{\partial x \partial y} + \frac{\partial^2 M_{xx}^*}{\partial x^2} + \frac{\partial^2 M_{yy}^*}{\partial y^2} + 2\frac{\partial^2 M_{xy}^*}{\partial x \partial y} + q(x,y) = 0$$

$$\tag{10.5c}$$

These results have been developed strictly on the basis of static equilibrium. Note that we have not specified what mechanism *caused* the internal resultants, and so Equations 10.5 are valid regardless of whether the internal resultants were caused by external edge loading, by enforced edge displacements, or by any combination thereof. Also, we have made no assumptions regarding material properties. Therefore Equations 10.5 are valid for any thin plate, including isotropic metallic plates or anisotropic composite plates. We have however made one important assumption: we have assumed that the angle α, defined by the slope of the deflected midplane and shown in Figures 10.5 and 10.6, is relatively small so that the small angle approximation can be invoked.

The third equation of equilibrium, Equation 10.5c, can be simplified for cases in which the loads applied to the plate result in a maximum out-of-plane deflection less than about half the laminate thickness: $w(x, y)|_{max} < t/2$. In such, a case the midplane slope (i.e., angle α) is exceedingly small and consequently the z-directed component of the in-plane stress resultants (N_{xx}^*, N_{yy}^*, and N_{xy}^*) can be neglected. If the in-plane resultants are ignored during the summation of forces in the z-direction (setting $(N_{xx}^* dy)\partial w/\partial x = 0$, for example) and the preceding derivation is repeated, the equations of equilibrium simplify as follows:

$$\frac{\partial N_{xx}^*}{\partial x} + \frac{\partial N_{xy}^*}{\partial y} = 0 \tag{10.6a}$$

$$\frac{\partial N_{yy}^*}{\partial y} + \frac{\partial N_{xy}^*}{\partial x} = 0 \tag{10.6b}$$

$$\frac{\partial^2 M_{xx}^*}{\partial x^2} + \frac{\partial^2 M_{yy}^*}{\partial y^2} + 2\frac{\partial^2 M_{xy}^*}{\partial x \partial y} + q(x,y) = 0 \qquad (10.6c)$$

Equations 10.6 are valid if the maximum out-of-plane displacement is less than about half the plate thickness: $w(x, y)|_{\max} < t/2$.

10.2.2 Equations of Equilibrium Expressed in Terms of the [*ABD*] Matrix, Transverse Loading, and Midplane Displacement Fields

In Chapter 6, we related stress and moment resultants to midplane strains and curvatures, temperature changes, and changes in moisture content. This relationship is summarized by Equation 6.44. We have since limited our discussion to symmetric laminates ($B_{ij} = M_{ij}^T = M_{ij}^M$), and are now using the superscript "*" to denote internal stress and moment resultants. Consequently, Equation 6.44 is modified slightly to become

$$\left\{\begin{matrix} N^* \\ M^* \end{matrix}\right\} = \begin{bmatrix} A & 0 \\ 0 & D \end{bmatrix}\left\{\begin{matrix} \varepsilon^o \\ \kappa \end{matrix}\right\} - \left\{\begin{matrix} N^T \\ 0 \end{matrix}\right\} - \left\{\begin{matrix} N^M \\ 0 \end{matrix}\right\} \qquad (10.7)$$

In this equation

- N^* and M^* represent internal stress and moment resultants associated with externally applied forces
- N^T represents thermal stress resultants associated with a uniform through-thickness change in temperature, ΔT
- N^M represents moisture stress resultants associated with a uniform through-thickness change in moisture content, ΔM

Recall that only *uniform* changes in temperature and/or moisture content are considered in this chapter. Consequently N^T and N^M should be viewed as constants, that is, they are not functions of x, y, or z.

Expanding the first of the six equations represented by Equation 10.7, we have

$$N_{xx}^* = A_{11}\varepsilon_{xx}^o + A_{12}\varepsilon_{yy}^o + A_{16}\gamma_{xy}^o - N_{xx}^T - N_{xx}^M$$

We will now express the midplane strains and curvatures that appear in this expression in terms of midplane displacement gradients. Before we do so, however, the discussion presented in Section 2.14 should be reiterated. In particular, a distinction between *finite* strains and *infinitesimal* strains was made at that point. Basically, strains can be considered to be infinitesimal when displacement gradients are small, so that the square of any displacement gradient can be neglected; $(\partial w/\partial x)^2 \approx 0$, for example. In this chapter, we have already assumed that the slope of the deflected plate midplane is small (i.e., $\partial w/\partial x$ and $\partial w/\partial y$ have already been assumed to be small), which allowed us to apply the small angle approximation. Consequently we will continue to treat strains as infinitesimal strains. As we will see, this assumption will ultimately lead to the conclusion that in-plane displacement fields $u_o(x, y)$ and $v_o(x, y)$ (as well as in-plane forces) are independent of the transverse load, $q(x, y)$. Rigorously speaking, this conclusion is incorrect. That is, if a thin plate is subjected to a transverse loading then in-plane displacement fields and/or in-plane forces will change, reflecting a dependence on transverse load. However, if displacement gradients are small, then the changes in-plane displacements or forces are also small and can usually be ignored. In effect, the assumption of infinitesimal strains has eliminated (in a mathematical sense) the coupling between transverse loads and in-plane displacements/forces. Occasionally, this leads to predictions that are nonintuitive. This will be pointed out at appropriate points in following sections (in particular, see Example Problem 11.2 in Section 11.4).

Although the interdependence between out-of-plane and in-plane displacements is ignored throughout most of this book, there is one important exception. This interdependence is included in Chapter 11, during formulation of an approximate analysis technique known as the *Ritz method*. The interdependence must be included there so as to obtain buckling predictions.

In any event, we now assume that strains are infinitesimal and can, therefore, be expressed in terms of midplane displacement fields in accordance with Equations 2.49:

$$\varepsilon_{xx}^o = \frac{\partial u_o}{\partial x}, \quad \varepsilon_{yy}^o = \frac{\partial v_o}{\partial y}, \quad \gamma_{xy}^o = \frac{\partial u_o}{\partial y} + \frac{\partial v_o}{\partial x}$$

$$\kappa_{xx} = -\frac{\partial^2 w}{\partial x^2}, \quad \kappa_{yy} = -\frac{\partial^2 w}{\partial y^2}, \quad \kappa_{xy} = -2\frac{\partial^2 w}{\partial x \partial y}$$

Therefore, the internal stress resultant N_{xx}^* can be written as

$$N_{xx}^* = A_{11}\frac{\partial u_o}{\partial x} + A_{12}\frac{\partial v_o}{\partial y} + A_{16}\left(\frac{\partial u_o}{\partial y} + \frac{\partial v_o}{\partial x}\right) - N_{xx}^T - N_{xx}^M$$

Given this result, the quantity $(\partial N_{xx}^*/\partial x)$ is easily obtained:

$$\frac{\partial N_{xx}^*}{\partial x} = A_{11}\frac{\partial^2 u_o}{\partial x^2} + A_{12}\frac{\partial^2 v_o}{\partial x \partial y} + A_{16}\left(\frac{\partial^2 u_o}{\partial x \partial y} + \frac{\partial^2 v_o}{\partial x^2}\right)$$

Note that this quantity appears as a term in the first equation of equilibrium, Equation 10.5a. Also note that neither the thermal nor moisture stress resultant is involved in this result, since as previously noted N^T and N^M are constants and, therefore, $\partial N^T_{xx}/\partial x = \partial N^M_{xx}/\partial x = 0$. Following an analogous procedure, the second term in Equation 10.5a is found to be

$$\frac{\partial N^*_{xy}}{\partial y} = A_{16}\frac{\partial^2 u_o}{\partial x \partial y} + A_{26}\frac{\partial^2 v_o}{\partial y^2} + A_{66}\left(\frac{\partial^2 u_o}{\partial y^2} + \frac{\partial^2 v_o}{\partial x \partial y}\right)$$

Adding these two results, Equation 10.5a can be written in terms of elements of the $[ABD]$ matrix and midplane displacements as follows:

$$A_{11}\frac{\partial^2 u_o}{\partial x^2} + (A_{12} + A_{66})\frac{\partial^2 v_o}{\partial x \partial y} + 2A_{16}\frac{\partial^2 u_o}{\partial x \partial y} + A_{16}\frac{\partial^2 v_o}{\partial x^2} + A_{26}\frac{\partial^2 v_o}{\partial y^2} + A_{66}\frac{\partial^2 u_o}{\partial y^2} = 0$$

$$(10.8a)$$

Following an identical procedure, the second and third equations of equilibrium (Equation 10.5b,c) can be written as

$$A_{16}\frac{\partial^2 u_o}{\partial x^2} + (A_{12} + A_{66})\frac{\partial^2 u_o}{\partial x \partial y} + 2A_{26}\frac{\partial^2 v_o}{\partial x \partial y} + A_{26}\frac{\partial^2 u_o}{\partial y^2} + A_{22}\frac{\partial^2 v_o}{\partial y^2} + A_{66}\frac{\partial^2 v_o}{\partial x^2} = 0$$

$$(10.8b)$$

$$\left[A_{11}\frac{\partial u_o}{\partial x} + A_{12}\frac{\partial v_o}{\partial y} + A_{16}\left(\frac{\partial u_o}{\partial y} + \frac{\partial v_o}{\partial x}\right) - N^T_{xx} - N^M_{xx}\right]\left(\frac{\partial^2 w}{\partial x^2}\right)$$

$$+ \left[A_{12}\frac{\partial u_o}{\partial x} + A_{22}\frac{\partial v_o}{\partial y} + A_{26}\left(\frac{\partial u_o}{\partial y} + \frac{\partial v_o}{\partial x}\right) - N^T_{yy} - N^M_{yy}\right]\left(\frac{\partial^2 w}{\partial y^2}\right)$$

$$+ 2\left[A_{16}\frac{\partial u_o}{\partial x} + A_{26}\frac{\partial v_o}{\partial y} + A_{66}\left(\frac{\partial u_o}{\partial y} + \frac{\partial v_o}{\partial x}\right) - N^T_{xy} - N^M_{xy}\right]\left(\frac{\partial^2 w}{\partial x \partial y}\right)$$

$$- D_{11}\frac{\partial^4 w}{\partial x^4} - 4D_{16}\frac{\partial^4 w}{\partial x^3 \partial y} - 4D_{26}\frac{\partial^4 w}{\partial x \partial y^3} - 2(D_{12} + 2D_{66})\frac{\partial^4 w}{\partial x^2 \partial y^2}$$

$$- D_{22}\frac{\partial^4 w}{\partial y^4} + q(x,y) = 0 \qquad (10.8c)$$

Equations 10.8 represent the equations of equilibrium for a symmetric laminate in terms of the $[ABD]$ matrix, transverse loading, and midplane displacement fields. These are valid for any thin symmetric laminate as long as out-of-plane displacements are not excessive. If out of plane displacements are very small (i.e., if $w(x, y)|_{\max} < t/2$) then the equations of equilibrium given by Equations 10.6 are applicable, which become

$$A_{11}\frac{\partial^2 u_o}{\partial x^2} + (A_{12} + A_{66})\frac{\partial^2 v_o}{\partial x \partial y} + 2A_{16}\frac{\partial^2 u_o}{\partial x \partial y} + A_{16}\frac{\partial^2 v_o}{\partial x^2} + A_{26}\frac{\partial^2 v_o}{\partial y^2} + A_{66}\frac{\partial^2 u_o}{\partial y^2} = 0$$

$$(10.9a)$$

$$A_{16}\frac{\partial^2 u_o}{\partial x^2} + (A_{12} + A_{66})\frac{\partial^2 u_o}{\partial x \partial y} + 2A_{26}\frac{\partial^2 v_o}{\partial x \partial y} + A_{26}\frac{\partial^2 u_o}{\partial y^2} + A_{22}\frac{\partial^2 v_o}{\partial y^2} + A_{66}\frac{\partial^2 v_o}{\partial x^2} = 0$$

$$(10.9b)$$

$$-D_{11}\frac{\partial^4 w}{\partial x^4} - 4D_{16}\frac{\partial^4 w}{\partial x^3 \partial y} - 4D_{26}\frac{\partial^4 w}{\partial x \partial y^3} - 2(D_{12} + 2D_{66})\frac{\partial^4 w}{\partial x^2 \partial y^2}$$

$$- D_{22}\frac{\partial^4 w}{\partial y^4} + q(x, y) = 0 \qquad\qquad (10.9c)$$

10.3 Boundary Conditions

A summary of the material presented in Sections 10.1 and 10.2 is as follows. We consider a thin rectangular plate of in-plane dimensions $a \times b$, whose edges are parallel to the x- and y-axes. The external edge loads considered in thin-plate theory were introduced in Section 10.1. A total of five types of externally applied distributed loads may be present on *each* of the four edges of a rectangular laminate: one normal stress resultant, one in-plane shear stress resultant, one out-of-plane shear stress resultant, one bending moment resultant, and one twisting moment resultant. Each of these loads is, at most, a function of the coordinate direction tangent to the plate edge. In addition, a transverse loading $q(x, y)$ may be applied to the surface of the laminate. Figure 10.3 provides a summary of all externally applied loads considered in thin-plate theory.

These externally applied edge loads induce a distribution of *internal* stress and moment resultants at all interior regions of the plate. Generally, internal stress and moment resultants are functions of both x and y and vary throughout the plate. The distribution of internal stress and moment resultants were investigated in Section 10.2.1 by considering free-body diagrams of an infinitesimal element removed from an interior point within the plate.

Requiring that the sum of all forces and moments equate to zero resulted in the equations of equilibrium, summarized as Equations 10.5. If the maximum out-of-plane displacement is less than half the laminate thickness, then a simplified version of the equations of equilibrium is applicable, summarized as Equations 10.6. A mathematically equivalent form for a symmetric composite laminate was obtained in Section 10.2.2, where the equations of equilibrium were written in terms of elements of the [ABD] matrix and midplane displacement fields, summarized as Equations 10.8 and 10.9.

In this section, we will formally define the *boundary conditions* of the plate. That is, we wish to precisely define what conditions exist along each of the four edges of the rectangular plate. Actually, we have already begun our discussion of boundary conditions, in the sense that the external stress and moment resultants that may be applied along the edges of the laminate were described in Section 10.1. However, in many instances the loads applied along the plate edges are unknown and hence cannot be used as specified boundary conditions. Rather, the midplane *displacements* along the edges are known, where displacements of the midplane in the x-, y-, and z-directions are denoted $u_o(x, y)$, $v_o(x, y)$, and $w(x,y)$, respectively, as in earlier chapters. Hence, two categories of boundary conditions can be defined. We can *either* specify components of the edge displacements, *or* we can specify components of the edge loads. Boundary conditions involving specified displacements are called *geometric* (or *kinematic*) boundary conditions, while boundary conditions involving specified edge loads are called *static* (or *natural*) boundary conditions.

10.3.1 Geometric (Kinematic) Boundary Conditions

Geometric boundary conditions are those that dictate some feature of midplane displacements along a plate edge. Each of the four edges of the plate is characterized by a normal and tangential direction. For example, for the positive x-edge shown in Figure 10.3 the x-direction is normal to the edge while the y-direction is tangent to the edge. Along this edge we may specify values for displacements in the x-, y-, and z-directions, as well as the slope of the laminate midplane measured in the normal direction. Since this edge is parallel to the y-axis, *the displacements imposed along this edge are either constant or, at most, functions of y; they are not functions of x.* Thus, we may specify values of $u_o^{(+x)}(y)$, $v_o^{(+x)}(y)$, $w^{(+x)}(y)$, and $(\partial w^{(+x)}(y))/\partial x$ along the edge $x = a$. As was the case for edge loads (discussed in Section 10.1), there is no reason to assume that the displacements on opposite edges are identical. Therefore the superscript (+x) has been used to indicate that these displacements are imposed along the positive x-edge of the laminate.

Note that by specifying a particular value of the midplane slope at the boundary, $(\partial w^{(+x)}(y)/\partial x)$, we have specified only one of several possible gradients in the three displacement fields. Initially one might suspect that, for reasons of mathematical symmetry perhaps, other displacement field gradients should also be specified. For example, perhaps $(\partial u_o^{(+x)}(y)/\partial y)$ or $(\partial v_o^{(+x)}(y)/\partial z)$

should be specified along $x = a$ as well. These other gradients need not be considered for two reasons. First, in thin-plate theory the boundary conditions are defined for the plate edge *at the midplane*. By definition the midplane has a thickness of zero. This fundamental definition precludes any midplane boundary conditions with a z-dependency; that is, the midplane displacement field gradients $(\partial u_o^{(+x)}(y))/\partial z$, $(\partial v_o^{(+x)}(y))/\partial z$, or $(\partial w^{(+x)}(y))/\partial z$ are undefined. This leaves us with possible gradients in the x- or y-directions. With the exception $(\partial w^{(+x)}(y))/\partial x$, a specified gradient in geometric boundary conditions along a plate edge in the x- or y-directions results in an overspecified problem. For example, having specified displacements in the x-direction along the edge $x = a$, namely $u_o^{(+x)}(y)$, the value of $(\partial u_o^{(+x)}(y))/\partial y$ is also specified as a consequence. A similar consideration of the other possible gradients results in a similar conclusion.

Recalling that displacements at any arbitrary *interior* point of the plate are denoted $u_o(x, y)$, $v_o(x, y)$, and $w(x,y)$, the geometric boundary conditions that may be specified along each of the four edges of the plate are

For Edge $x = a$	For Edge $x = 0$
$u_o(a,y) = u_o^{(+x)}(y)$	$u_o(0,y) = u_o^{(-x)}(y)$
$v_o(a,y) = v_o^{(+x)}(y)$	$v_o(0,y) = v_o^{(-x)}(y)$
$\dfrac{\partial w(a,y)}{\partial x} = \dfrac{\partial w^{(+x)}(y)}{\partial x}$	$\dfrac{\partial w(0,y)}{\partial x} = \dfrac{\partial w^{(-x)}(y)}{\partial x}$
$w(a,y) = w^{(+x)}(y)$	$w(0,y) = w^{(-x)}(y)$
For Edge $y = b$	**For Edge $y = 0$**
$u_o(x,b) = u_o^{(+y)}(x)$	$u_o(x,0) = u_o^{(-y)}(x)$
$v_o(x,b) = v_o^{(+y)}(x)$	$v_o(x,0) = v_o^{(-y)}(x)$
$\dfrac{\partial w(x,b)}{\partial y} = \dfrac{\partial w^{(+y)}(x)}{\partial y}$	$\dfrac{\partial w(x,0)}{\partial y} = \dfrac{\partial w^{(-y)}(x)}{\partial y}$
$w(x,b) = w^{(+y)}(x)$	$w(x,b) = w^{(-y)}(x)$

10.3.2 Static (Natural) Boundary Conditions

We will again use the positive x-edge to demonstrate possible static boundary conditions. Referring to Figure 10.3, note that five external stress and moment resultants may be specified along this edge: $N_{xx}^{(+x)}(y)$, $N_{xy}^{(+x)}(y)$, $V_{xz}^{(+x)}(y)$, $M_{xx}^{(+x)}(y)$, and $M_{xy}^{(+x)}(y)$. Values for $N_{xx}^{(+x)}(y)$, $N_{xy}^{(+x)}(y)$, and $M_{xx}^{(+x)}(y)$ can be specified independently, without consideration of any other stress or moment results. However, the remaining two resultants, $V_{xz}^{(+x)}(y)$ and $M_{xy}^{(+x)}(y)$, are not independent and must be considered together. The interdependence between $V_{xz}^{(+x)}(y)$ and $M_{xy}^{(+x)}(y)$ was first noted by Kirchhoff in 1850, and occurs because (as far as static equilibrium is concerned) a twisting moment $M_{xy}^{(+x)}(y)$ can be replaced by a statically equivalent couple involving two vertical shear resultants. This is illustrated in Figure 10.8. The twisting moments acting on two adjacent elements of length dy are shown in Figure 10.8a. The element on the right is subjected to a twisting moment $M_{xy}^{(+x)}(y)dy$, whereas the element on the left is subjected to a twisting moment $[M_{xy}^{(+x)}(y) + (dM_{xy}^{(+x)}(y)/dy)dy]dy$. These twisting moments may be replaced by an equivalent couple involving two pairs of vertical shear resultants, as shown in Figure 10.8b. The element on the right is subjected to two vertical *shear resultants* of magnitude $M_{xy}^{(+x)}(y)$, acting in the directions shown and separated by distance dy. Similarly, the element on the left is subjected to two vertical *shear resultants* of magnitude $[M_{xy}^{(+x)}(y) + (dM_{xy}^{(+x)}(y)/dy)]$, acting in the directions shown and separated by distance dy. Summing the force components present at the interface between the two infinitesimal elements, we obtain $\{[M_{xy}^{(+x)}(y) + (dM_{xy}^{(+x)}(y)/dy)] - M_{xy}^{(+x)}(y)\} = (dM_{xy}^{(+x)}(y)/dy)$. Hence, the *variation* of the twist moment $M_{xy}^{(+x)}(y)$ along the edge $x = a$ is statically equivalent to a vertical shear resultant of magnitude $(dM_{xy}^{(+x)}(y)/dy)$. Adding this statically equivalent vertical shear resultant component to the external

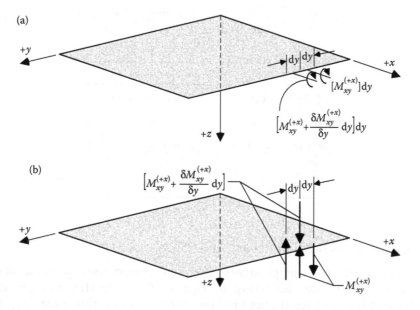

FIGURE 10.8

Two equivalent representations of the twisting moment resultant acting along the edge $x = a$. (a) Twisting moment resultants acting on two adjacent elements of length dy along the edge $x = a$; (b) two pairs of vertical shear resultants acting on adjacent elements along the edge $x = a$ that are statically equivalent to the twisting moments shown in (a).

vertical shear that is actually applied to the edge, $V_{xz}^{(+x)}(y)$, we define the "Kirchhoff" shear resultant acting along the edge $x = a$.

$$V_{xz}^{(+xK)}(y) = V_{xz}^{(+x)}(y) + \frac{dM_{xy}^{(+x)}(y)}{dy} \tag{10.10}$$

The superscript "+xK" is used to denote the Kirchhoff shear resultant (sometimes called the "effective" shear resultant) acting along the positive x-edge.

Now, the equations of equilibrium (developed in Section 10.2) must be satisfied at all points within the laminate, *including the laminate edge*. Therefore, the distribution of the external vertical shear resultant, $V_{xz}^{(+x)}(y)$, must satisfy Equation 10.3b. Along the edge $x = a$, this equation becomes

$$V_{xz}^{(+x)} = \frac{\partial M_{xx}^{(+x)}}{\partial x} + \frac{\partial M_{xy}^{(+x)}}{\partial y}$$

Recalling that *all boundary conditions along the positive x-edge are either constant or are functions of y only*, it must be that $(\partial M_{xx}^{(+x)}/\partial x) = 0$. Therefore, along the edge $x = a$.

$$V_{xz}^{(+x)} = \frac{\partial M_{xy}^{(+x)}}{\partial y} \tag{10.11}$$

Equation 10.11 shows that the variation of the externally applied shear $V_{xz}^{(+x)}(y)$ along the edge $x = a$ is intimately related to the variation of the externally applied twist moment $M_{xy}^{(+x)}(y)$ along this same edge, and that *the functional form of these two resultants cannot be specified independently*.* Suppose, for example, that the laminate is subjected to a vertical shear resultant that varies linearly over the edge $x = a$. That is, suppose $V_{xz}^{(+x)}(y) = Ay + B$, where A and B are constants. Under this circumstance, Equation 10.11 shows that a twist moment must *also* be present in order for static equilibrium to be maintained. Furthermore, the twist moment *must* vary according to $(dM_{xy}^{(+x)}(y)/dy) = Ay + B$, which implies that $M_{xy}^{(+x)}(y) = (A/2)y^2 + By + C$, where C is a constant of integration.

As a second case of interest, suppose that $V_{xy}^{(+x)}$ is constant, that is, assume $A = 0$ and hence that $V_{xz}^{(+x)}(y) = B$. In this case the rate of change in the twist moment *must* equal $(dM_{xy}^{(+x)}(y)/dy) = B$, otherwise Equation 10.11 is not satisfied and static equilibrium is not maintained. This shows that it is not

---

* It is noted that a similar interdependence occurs between the shear force and bending moment present in a prismatic beam. That is, from fundamental beam theory, the shear force present at any cross section within a prismatic beam is related to the bending moment according to $V = dM/dx$.

possible to apply constant vertical shear *only* along $x = a$; if a constant vertical shear is applied, then a corresponding twist moment that varies over the length of the plate edge must also be present. However, the converse is admissible. That is, if a constant twist moment $M_{xy}^{(+x)} = B$ is applied, then the vertical shear *must* equal zero: $(dM_{xy}^{(+x)}(y)/dy) = V_{xz}^{(+x)}(y) = 0$.

Substituting Equation 10.11 into Equation 10.10, we find that the effective Kirchhoff shear resultant *acting along the edge* $x = a$ can be written in two equivalent ways:

$$V_{xz}^{(+xK)}(y) = V_{xz}^{(+x)}(y) + \frac{dM_{xy}^{(+x)}(y)}{dy} \quad \text{(or)} \quad V_{xz}^{(+xK)}(y) = 2\frac{dM_{xy}^{(+x)}(y)}{dy} \qquad (10.12)$$

Note that the Kirchhoff shear resultant differs from the applied vertical shear only if there is variation in the twisting moment. That is, $V_{xz}^{(+xK)}(y)$ differs from $V_{xz}^{(+x)}(y)$ only if $(dM_{xy}^{(+x)}(y)/dx) \neq 0$.

Recalling that stress and moment resultants at any arbitrary *interior* point of the plate are denoted $N_{xx}^*(x,y)$, $N_{yy}^*(x,y)$, $N_{xy}^*(x,y)$, $N_{yx}^*(x,y)$, $M_{xx}^*(x,y)$, $M_{yy}^*(x,y)$, $M_{xy}^*(x,y)$, $M_{yx}^*(x,y)$, $V_{xz}^*(x,y)$, and $V_{yz}^*(x,y)$, the static boundary conditions that may be specified along the edges $x = a$ are as follows:

For edge $x = a$:

$$N_{xx}^*(a,y) = N_{xx}^{(+x)}(y)$$

$$N_{xy}^*(a,y) = N_{xy}^{(+x)}(y)$$

$$M_{xx}^*(a,y) = M_{xx}^{(+x)}(y)$$

$$\frac{\partial M_{xx}^*(a,y)}{\partial x} + 2\frac{\partial M_{xy}^*(a,y)}{\partial y} = V_{xz}^{(+x)}(y) + \frac{dM_{xy}^{(+x)}(y)}{dy} = 2\frac{dM_{xy}^{(+x)}(y)}{dy}$$

Static boundary conditions along the edge $x = a$ have been discussed above. Following an analogous procedure, static boundary conditions that may be present along the remaining three edges of the rectangular plate are as follows:

For edge $x = 0$:

$$N_{xx}^*(0,y) = N_{xx}^{(-x)}(y)$$

$$N_{xy}^*(0,y) = N_{xy}^{(-x)}(y)$$

$$M_{xx}^*(0,y) = M_{xx}^{(-x)}(y)$$

$$\frac{\partial M_{xx}^*(0,y)}{\partial x} + 2\frac{\partial M_{xy}^*(0,y)}{\partial y} = V_{xz}^{(-x)}(y) + \frac{dM_{xy}^{(-x)}(y)}{dy} = 2\frac{dM_{xy}^{(-x)}(y)}{dy}$$

For edge $y = b$:

$$N_{yy}^*(x,b) = N_{yy}^{(+y)}(x)$$

$$N_{yx}^*(x,b) = N_{yx}^{(+y)}(x)$$

$$M_{yy}^*(x,b) = M_{yy}^{(+y)}(x)$$

$$\frac{\partial M_{yy}^*(x,b)}{\partial y} + 2\frac{\partial M_{yx}^*(x,b)}{\partial x} = V_{yz}^{(+y)}(x) + \frac{dM_{yx}^{(+y)}(x)}{dx} = 2\frac{dM_{yx}^{(+y)}(x)}{dy}$$

For edge $y = 0$:

$$N_{yy}^*(x,0) = N_{yy}^{(-y)}(x)$$

$$N_{yx}^*(x,0) = N_{yx}^{(-y)}(x)$$

$$M_{yy}^*(x,0) = M_{yy}^{(-y)}(x)$$

$$\frac{\partial M_{yy}^*(x,0)}{\partial y} + 2\frac{\partial M_{yx}^*(x,0)}{\partial x} = V_{yz}^{(-y)}(x) + \frac{dM_{yx}^{(-y)}(x)}{dx} = 2\frac{dM_{yx}^{(-y)}(x)}{dy}$$

10.3.3 Combinations of Geometric and Static Boundary Conditions

Two fundamental categories of boundary conditions were defined in the preceding subsections: geometric boundary conditions and static boundary conditions. If either a geometric or static boundary condition is specified to equal zero, then that condition is called a *homogeneous* condition. In contrast, if a condition is required to take on a nonzero value, then the condition is called an *inhomogeneous* condition. Examples of homogenous and inhomogeneous geometric conditions along the edge $x = a$ are $u_o(a, y) = 0$ and $u_o(a, y) = 1$ mm, respectively. Examples of homogeneous and inhomogeneous static conditions along the edge $x = a$ are $N_{xx}(a, y) = 0$ and $N_{xx}(a, y) = 1000$ N/m, respectively.

Four potential conditions exist in each category for each of the four edges of a rectangular laminate. The possible geometric and static condition for each condition must be viewed as *complementary pairs*. That is, for each condition we may specify *either* a geometric requirement *or* a static requirement, *but we cannot specify both*. For example, suppose a problem is considered in which it is stipulated that displacements in the x-direction along the edge $x = a$ are zero. That is, we specify that $u_o(a, y) = 0$. Note that this requirement is an example of a homogeneous geometric boundary condition. In such a case, the geometric boundary condition is specified and is, therefore, known (i.e., $u_o(a, y) = 0$), whereas the corresponding static boundary condition is unknown. In fact, the question becomes, "what stress resultant $N_{xx}^{(+x)}(y)$ must exist along the edge $x = a$, so as to maintain the boundary condition $u_o(a, y) = 0$?"

Conversely, suppose in a different problem it is stipulated that the stress resultant acting normal to the edge $x = a$ must equal some constant value. For example, suppose we specify a boundary condition in which $N_{xx}^{(+x)}(y) = 1000\,\text{N/m}$. This is an example of an inhomogeneous static boundary condition, and in this case it is the geometric boundary condition that is unknown. In fact, the question becomes, "what displacements $u_o(a, y)$ exist along the edge $x = a$, given that the edge loading is known to be $N_{xx}^{(+x)}(y) = 1000\,\text{N/m}$?"

As stated above, there are four complementary pairs of boundary conditions for each edge of a rectangular plate. For each pair we may specify *either* a geometric requirement *or* a static requirement, *but we cannot specify both*. The conditions that must be specified for each edge of a rectangular plate are summarized below:

The following conditions can be specified for the edge $x = a$:

Geometric Condition		Static Condition
$u_o(a, y) = u_o^{(+x)}(y)$	(or)	$N_{xx}^*(a, y) = N_{xx}^{(+x)}(y)$
$v_o(a, y) = v_o^{(+x)}(y)$	(or)	$N_{xy}^*(a, y) = N_{xy}^{(+x)}(y)$
$\dfrac{\partial w(a, y)}{\partial x} = \dfrac{\partial w^{(+x)}(y)}{\partial x}$	(or)	$M_{xx}^*(a, y) = M_{xx}^{(+x)}(y)$
$w(a, y) = w^{(+x)}(y)$	(or)	$\dfrac{\partial M_{xx}^*(a, y)}{\partial x} + 2\dfrac{\partial M_{xy}^*(a, y)}{\partial y} = V_{xz}^{(+x)}(y) + \dfrac{dM_{xy}^{(+x)}(y)}{dy} = 2\dfrac{dM_{xy}^{(+x)}(y)}{dy}$

The following conditions can be specified for the edge $x = 0$:

Geometric Condition		Static Condition
$u_o(0, y) = u_o^{(-x)}(y)$	(or)	$N_{xx}^*(0, y) = N_{xx}^{(-x)}(y)$
$v_o(0, y) = v_o^{(-x)}(y)$	(or)	$N_{xy}^*(0, y) = N_{xy}^{(-x)}(y)$
$\dfrac{\partial w(0, y)}{\partial x} = \dfrac{\partial w^{(-x)}(y)}{\partial x}$	(or)	$M_{xx}^*(0, y) = M_{xx}^{(-x)}(y)$
$w(0, y) = w^{(-x)}(y)$	(or)	$\dfrac{\partial M_{xx}^*(0, y)}{\partial x} + 2\dfrac{\partial M_{xy}^*(0, y)}{\partial y} = V_{xz}^{(-x)}(y) + \dfrac{dM_{xy}^{(-x)}(y)}{dy} = 2\dfrac{dM_{xy}^{(-x)}(y)}{dy}$

The following conditions can be specified for the edge $y = b$:

Geometric Condition		Static Condition
$u_o(x,b) = u_o^{(+y)}(x)$	(or)	$N_{yx}^*(x,b) = N_{yx}^{(+y)}(x)$
$v_o(x,b) = v_o^{(+y)}(x)$	(or)	$N_{yy}^*(x,b) = N_{yy}^{(+y)}(x)$
$\dfrac{\partial w(x,b)}{\partial y} = \dfrac{\partial w^{(+y)}(x)}{\partial y}$	(or)	$M_{yy}^*(x,b) = M_{yy}^{(+y)}(x)$
$w(x,b) = w^{(+y)}(x)$	(or)	$\dfrac{\partial M_{yy}^*(x,b)}{\partial y} + 2\dfrac{\partial M_{yx}^*(x,b)}{\partial x} = V_{yz}^{(+y)}(x) + \dfrac{dM_{yx}^{(+y)}(x)}{dx} = 2\dfrac{dM_{yx}^{(+y)}(x)}{dy}$

The following conditions can be specified for the edge $y = 0$:

Geometric Condition		Static Condition
$u_o(x,0) = u_o^{(-y)}(x)$	(or)	$N_{yx}^*(x,0) = N_{yx}^{(-y)}(x)$
$v_o(x,0) = v_o^{(-y)}(x)$	(or)	$N_{yy}^*(x,0) = N_{yy}^{(-y)}(x)$
$\dfrac{\partial w(x,0)}{\partial y} = \dfrac{\partial w^{(-y)}(x)}{\partial y}$	(or)	$M_{yy}^*(x,0) = M_{yy}^{(-y)}(x)$
$w(x,0) = w^{(-y)}(x)$	(or)	$\dfrac{\partial M_{yy}^*(x,0)}{\partial y} + 2\dfrac{\partial M_{yx}^*(x,0)}{\partial x} = V_{yz}^{(-y)}(x) + \dfrac{dM_{yx}^{(-y)}(x)}{dx} = 2\dfrac{dM_{yx}^{(-y)}(x)}{dy}$

The number of different combinations that may be defined is enormous. Four conditions must be specified to define the boundary conditions along an edge. Since there are two possibilities in each case, a total of $2^4 = 16$ possible combinations exist along each edge. Since there are four edges, a total of $16^4 = 65,536$ different combinations of boundary conditions can be defined for a thin rectangular plate. However, some combinations are encountered so frequently that they have been given special names. Specifically, *free edges, uniformly loaded edges, simply supported edges,* and *clamped edges* are encountered very frequently in practice. These particular edge conditions will be illustrated below by considering the positive x-edge of the plate.

10.3.3.1 Free Edge

A *free edge* is defined as an edge that is entirely free of external loading. Hence, if the edge $x = a$ is a free edge, then the following four homogeneous static boundary conditions must be satisfied:

For $x = a$:

$$N_{xx}^*(a,y) = 0$$

$$N_{xy}^*(a,y) = 0$$

$$M_{xx}^*(a, y) = 0$$

$$\frac{\partial M_{xx}^*(a, y)}{\partial x} + 2\frac{\partial M_{xy}^*(a, y)}{\partial y} = 0$$

Analogous boundary conditions may be used to specify that any of the remaining three edges are free edges.

Each of the internal stress and moments resultants can be expressed in terms of the $[ABD]$ matrix and midplane displacement fields. Recalling that we have limited our discussion to symmetric laminates ($B_{ij} = M_{ij}^T = M_{ij}^M = 0$) the boundary conditions for a free edge may also be written as

For $x = a$:

$$A_{11}\frac{\partial u_o(a, y)}{\partial x} + A_{12}\frac{\partial v_o(a, y)}{\partial y} + A_{16}\left(\frac{\partial u_o(a, y)}{\partial y} + \frac{\partial v_o(a, y)}{\partial x}\right) - N_{xx}^T - N_{xx}^M = 0$$

$$A_{16}\frac{\partial u_o(a, y)}{\partial x} + A_{26}\frac{\partial v_o(a, y)}{\partial y} + A_{66}\left(\frac{\partial u_o(a, y)}{\partial y} + \frac{\partial v_o(a, y)}{\partial x}\right) - N_{xy}^T - N_{xy}^M = 0$$

$$-D_{11}\frac{\partial^2 w(a, y)}{\partial x^2} - D_{12}\frac{\partial^2 w(a, y)}{\partial y^2} - 2D_{16}\frac{\partial^2 w(a, y)}{\partial x \partial y} = 0$$

$$-D_{11}\frac{\partial^3 w(a, y)}{\partial x^3} - 4D_{16}\frac{\partial^3 w(a, y)}{\partial x^2 \partial y} - (D_{12} + 4D_{66})\frac{\partial^3 w(a, y)}{\partial x \partial y^2} - 2D_{26}\frac{\partial^3 w(a, y)}{\partial y^3} = 0$$

The results listed above are for the edge $x = a$. Comparable conditions are imposed if any of the other three edges of the plate are uniformly loaded.

10.3.3.2 Simply Supported Edges

A *simply supported* boundary condition is one in which the out-of-plane deflection and bending moment are zero along an edge. Hence, the $x = a$ edge of a rectangular plate is simply supported if $w(a, y) = 0$ and $M_{xx}(a, y) = 0$. Note that these are homogeneous geometric and static boundary conditions, respectively. The corresponding boundary condition in beam theory is a "pinned support," as discussed in Chapter 8.8. Recall that for the case of a pinned support as defined in beam theory only *two* requirements are involved ($w = M_x = 0$), whereas in thin-plate theory *four* conditions must be specified along each plate edge. Hence, for thin plates four distinct combinations of geometric and static boundary conditions may be classified as a "simple support." The possible simple-support boundary condition are often numbered S1 through S4, following a numbering scheme introduced by Almroth [5], and are defined along the edge $x = a$ as follows:

For $x = a$:

S1: $w(a,y) = 0 \quad M_{xx}^*(a,y) = 0, \quad u_o(a,y) = u_o^{(+x)}(y), \qquad v_o(a,y) = v_o^{(+x)}(y)$

S2: $w(a,y) = 0 \quad M_{xx}^*(a,y) = 0, \quad N_{xx}^*(a,y) = N_{xx}^{(+x)}(y), \quad v_o(a,y) = v_o^{(+x)}(y)$

S3: $w(a,y) = 0 \quad M_{xx}^*(a,y) = 0, \quad u_o(a,y) = u_o^{(+x)}(y), \qquad N_{xy}^*(a,y) = N_{xy}^{(+x)}(y)$

S4: $w(a,y) = 0 \quad M_{xx}^*(a,y) = 0, \quad N_{xx}^*(a,y) = N_{xx}^{(+x)}(y), \quad N_{xy}^*(a,y) = N_{xy}^{(+x)}(y)$

Note that a "simple support" is by definition a mixture of geometric and static boundary conditions. Also, depending on the values specified for $[u_o^{(+x)}(y)$ or $N_{xx}^{(+)}(y)]$ and $[v_o^{(+x)}(y)$ or $N_{xy}^{(+)}(y)]$, either homogeneous or inhomogeneous boundary conditions may be involved. The conditions involving stress and moment resultants can once again be expressed in terms of elements of the $[ABD]$ matrix and midplane displacement fields. For example, the four conditions that (collectively) define a simple support of type S4 for a symmetric laminate along the edge $x = a$ can be written as

S4 simple support, for $x = a$:

$$w(a, y) = 0$$

$$D_{11}\frac{\partial^2 w(a,y)}{\partial x^2} + D_{12}\frac{\partial^2 w(a,y)}{\partial y^2} - 2D_{16}\frac{\partial^2 w(a,y)}{\partial x \partial y} = 0$$

$$A_{11}\frac{\partial u_o(a,y)}{\partial x} + A_{12}\frac{\partial v_o(a,y)}{\partial y} + A_{16}\left(\frac{\partial u_o(a,y)}{\partial y} + \frac{\partial v_o(a,y)}{\partial x}\right) - N_{xx}^T - N_{xx}^M = N_{xx}^{(+x)}$$

$$A_{16}\frac{\partial u_o(a,y)}{\partial x} + A_{26}\frac{\partial v_o(a,y)}{\partial y} + A_{66}\left(\frac{\partial u_o(a,y)}{\partial y} + \frac{\partial v_o(a,y)}{\partial x}\right) - N_{xy}^T - N_{xy}^M = N_{xy}^{(+x)}$$

The other types of simple support can also be expressed in terms of midplane displacements following a comparable procedure. Analogous conditions may be imposed if any of the other three edges of a plate are simply supported.

10.3.3.3 Clamped Edges

A *clamped* edge (also called a *fixed* edge) is one in which the out-of-plane deflection and slope are zero along the edge. Hence, the $x = a$ edge of a rectangular plate is clamped if two homogenous geometric boundary conditions are satisfied: $w(a, y) = 0$ and $\partial w(a, y)/\partial x = 0$. Once again, the term "clamped end" is used during the study of beams, as discussed in Chapter 8.

However, during the study of beams only two geometric requirements are necessary to define a clamped boundary condition, whereas in the case of thin plates four conditions are required. Hence, four distinct combinations can be classified as a "clamped edge." The possible clamped boundary condition are often numbered C1 through C4 and are defined as follows:

For $x = a$:

$$\text{C1: } w(a,y) = 0, \quad \frac{\partial w(a,y)}{\partial x} = 0, \quad u_o(a,y) = u_o^{(+x)}(y), \quad v_o(a,y) = v_o^{(+x)}(y)$$

$$\text{C2: } w(a,y) = 0, \quad \frac{\partial w(a,y)}{\partial x} = 0, \quad N_{xx}^*(a,y) = N_{xx}^{(+)}(y), \quad v_o(a,y) = v_o^{(+x)}(y)$$

$$\text{C3: } w(a,y) = 0, \quad \frac{\partial w(a,y)}{\partial x} = 0, \quad u_o(a,y) = u_o^{(+x)}(y), \quad N_{xy}^*(a,y) = N_{xy}^{(+)}(y)$$

$$\text{C4: } w(a,y) = 0, \quad \frac{\partial w(a,y)}{\partial x} = 0, \quad N_{xx}^*(a,y) = N_{xx}^{(+)}(y), \quad N_{xy}^*(a,y) = N_{xy}^{(+)}(y)$$

A clamped boundary may or may not involve a mixture of geometric and static conditions, depending on the values specified for $[u_o^{(+x)}(y)$ or $N_{xx}^{(+)}(y)]$ and $[v_o^{(+x)}(y)$ or $N_{xy}^{(+)}(y)]$. Also, these conditions may be either homogeneous or inhomogeneous. In those cases in which the internal stress or moment resultants are required to take on specific values at the boundary (e.g., condition C4), the internal stress and moment resultant can be expressed in terms of elements of the [ABD] matrix and midplane displacement fields, as demonstrated above.

10.4 Representing Arbitrary Transverse Loads as a Fourier Series

In thin-plate theory, the distributed transverse loading is allowed to vary in any manner over the x–y plane, that is, $q = q(x, y)$. However, as will be seen it is particularly easy to obtain a solution if the transverse loading is distributed according to a double sinusoidal variation in x and y:

$$q(x,y) = q_o \sin\left(\frac{\pi x}{a}\right)\sin\left(\frac{\pi y}{b}\right) \tag{10.13}$$

where q_o is a constant and equals the magnitude of the distributed load at the center of the plate, that is, at $x = a/2$ and $y = b/2$. Solutions obtained for this particular distributed transverse loading will be described in Chapters 11 and 12.

Of course, a double sinusoidal variation is just one of an infinite number of possible transverse loads that may be encountered in practice. However, it can be shown that any *arbitrary* function $q(x, y)$ can be represented in terms of a double Fourier series:

$$q(x,y) = \sum_{m=1}^{\infty} \sum_{n=1}^{\infty} q_{mn} \sin\left(\frac{m\pi x}{a}\right) \sin\left(\frac{n\pi y}{b}\right) \tag{10.14}$$

Therefore, if a solution is obtained for the simple sinusoidal variation given by Equation 10.13, then a solution for an *arbitrary* transverse loading can be obtained by representing the arbitrary load as the sum of a series of double sinusoidal terms, in accordance with Equation 10.14. The constant coefficients q_{mn} that appear in Equation 10.14 are determined based on the functional form of $q(x, y)$, and correspond to a particular combination of m and n. To determine the values of q_{mn}, multiply Equation 10.14 by the factor $\sin(n'\,\pi y/b)dy$, and integrate from 0 to b:

$$\int_0^b q(x,y)\sin\left(\frac{n'\pi y}{b}\right)dy = \int_0^b \left(\sum_{m=1}^{\infty} \sum_{n=1}^{\infty} q_{mn} \sin\left(\frac{m\pi x}{a}\right)\sin\left(\frac{n\pi y}{b}\right) \right)\sin\left(\frac{n'\pi y}{b}\right)dy$$

$$\tag{10.15}$$

It can be shown that

$$\int_0^b \sin\left(\frac{n\pi y}{b}\right)\sin\left(\frac{n'\pi y}{b}\right)dy = 0, \quad \text{if } n \neq n'$$

$$\int_0^b \sin\left(\frac{n\pi y}{b}\right)\sin\left(\frac{n'\pi y}{b}\right)dy = \frac{b}{2}, \quad \text{if } n = n'$$

Consequently, upon performing the integration on the right side of Equation 10.15, only one term for the range $0 < n < \infty$ remains, specifically, the term for which $n = n'$. Hence, Equation 10.15 becomes

$$\int_0^b q(x,y)\sin\left(\frac{n'\pi y}{b}\right)dy = \frac{b}{2}\sum_{m=1}^{\infty} q_{mn'} \sin\left(\frac{m\pi x}{a}\right)$$

We now repeat this process for the variation in x. That is, multiply the preceding result by the factor $\sin(m'\,\pi x/a)dx$ and integrate from 0 to a:

$$\int\limits_0^a \int\limits_0^b q(x,y)\sin\left(\frac{m'\pi x}{a}\right)\sin\left(\frac{n'\pi y}{b}\right)dxdy$$

$$= \int\limits_0^a \left(\frac{b}{2}\sum_{m=1}^{\infty} q_{mn'}\sin\left(\frac{m\pi x}{a}\right)\right)\sin\left(\frac{m'\pi x}{a}\right)dx = \frac{ab}{4}q_{m'n'}$$

Solving this result for the coefficient $q_{m'n'}$, we find

$$q_{m'n'} = \frac{4}{ab}\int\limits_0^a \int\limits_0^b q(x,y)\sin\left(\frac{m'\pi x}{a}\right)\sin\left(\frac{n'\pi y}{b}\right)dxdy \qquad (10.16)$$

Equation 10.16 allows calculation of the constant coefficients, and hence a given distribution of transverse load $q(x, y)$ can be expressed as the double Fourier series shown in Equation 10.14.

Equation 10.14 involves a summation over an infinite number of terms (since $m, n \to \infty$). Of course, it is impossible to use an infinite number of terms; in practice a finite number of terms must be used. As the number of terms used is increased, the series representation given by Equation 10.14 resembles the actual load $q(x, y)$ more and more closely. That is, the series representation *converges* to the actual loading $q(x, y)$ as the number of terms is increased. In practice then, one should use the maximum number of terms that is reasonably possible, to insure a reasonable series representation of the applied load $q(x, y)$.

Three types of commonly encountered transverse loads will be used to illustrate the preceding discussion. First, consider the case of a constant uniform transverse load

$$q(x, y) = q_o$$

where q_o is the magnitude of the uniformly distributed load. Substituting $q(x, y) = q_o$ into Equation 10.16, we find the coefficients in the Fourier series expansion associated with specified value of m and n to be

$$q_{mn} = \frac{4q_o}{ab}\int\limits_0^a \int\limits_0^b \sin\left(\frac{m\pi x}{a}\right)\sin\left(\frac{n\pi y}{b}\right)dxdy = \frac{16q_o}{\pi^2 mn},$$

m, n = odd integers

$$q_{mn} = \frac{4q_o}{ab}\int\limits_0^a \int\limits_0^b \sin\left(\frac{m\pi x}{a}\right)\sin\left(\frac{n\pi y}{b}\right)dxdy = 0,$$

m, n = even integers

Hence, a constant transverse load $q(x, y) = q_o$ can be written as

$$q(x,y) = \frac{16q_o}{\pi^2} \sum_{m=1}^{\infty} \sum_{n=1}^{\infty} \frac{1}{mn} \sin\left(\frac{m\pi x}{a}\right)\sin\left(\frac{n\pi y}{b}\right),(m,n = 1,3,5,...) \qquad (10.17)$$

How well Equation 10.17 describes a constant transverse loading depends on the number of terms used. A normalized plot of the series representation given by Equation 10.17 along the plate centerline $y = b/2$ is presented in Figure 10.9. Curves are shown based on a 9-term expansion (i.e., m, $n = 1$, 3, 5), a 64-term expansion (m, $n = 1, 3, 5, ..., 15$), and a 169-term expansion (m, $n = 1, 3, 5, ..., 25$). The series clearly converges toward a constant loading as the number of terms used is increased, although even with 169 terms the series expansion represents a constant transverse loading in only an approximate sense.

As a second example, consider the case of a transverse force P, uniformly distributed over an internal rectangular region of dimensions $a_i \times b_i$, as shown in Figure 10.10. The center of the internal region is located at $x = \xi$ and $y = \eta$, as shown. It is emphasized that P is defined as a force, with units of Newtons or pounds-force, for example. Equation 10.16 becomes in this case

$$q_{mn} = \frac{4P}{aba_ib_i} \int_{\xi-a_i/2}^{\xi+a_i/2} \int_{\eta-b_i/2}^{\eta+b_i/2} \sin\left(\frac{m\pi x}{a}\right)\sin\left(\frac{n\pi y}{b}\right)dxdy$$

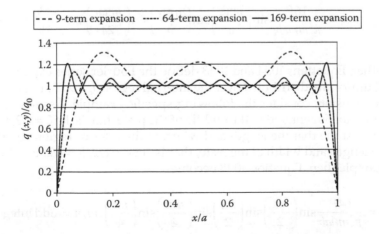

FIGURE 10.9
A normalized plot of Equation 10.17 along the plate centerline defined by $y = b/2$.

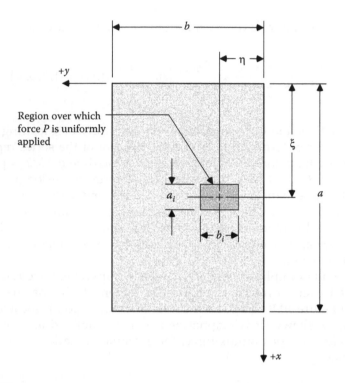

FIGURE 10.10
A rectangular plate with in-plane dimensions $a \times b$, subjected to a force P uniformly distributed over an interior region of dimensions $a_i \times b_i$. The interior region is centered at the point $x = \xi, y = \eta$.

Evaluating this integral we find

$$q_{mn} = \frac{16P}{\pi^2 mn a_i b_i} \sin\left(\frac{m\pi\xi}{a}\right) \sin\left(\frac{n\pi\eta}{b}\right) \sin\left(\frac{m\pi a_i}{2a}\right) \sin\left(\frac{n\pi b_i}{2b}\right) \qquad (10.18)$$

Together, Equations 10.14 and 10.18 define the Fourier series expansion of a force P, uniformly distributed over the internal region $a_i \times b_i$. To illustrate this series expansion, consider the following specific example. Assume that the interior region is centered in the middle of the plate, that is, let $\xi = a/2, \eta = b/2$. Further, assume that the length and width of the interior region equals one half the length and width of the plate, that is, let $a_i = a/2, b_i = b/2$. For this specific example then, Equation 10.18 becomes

$$q_{mn} = \frac{64P}{\pi^2 mnab} \sin\left(\frac{m\pi}{2}\right) \sin\left(\frac{n\pi}{2}\right) \sin\left(\frac{m\pi}{4}\right) \sin\left(\frac{n\pi}{4}\right), m, n = \text{odd integers}$$

$$q_{mn} = 0, m, n = \text{even integers}$$

Note that coefficients associated with even integers are zero only because we let $\xi = a/2$, $\eta = b/2$ in this example. This would not be true if the interior region were not centered on the plate. Substituting this result in Equation 10.14, we find

$$q(x,y) = \frac{64P}{\pi^2 ab} \sum_{m=1}^{\infty} \sum_{n=1}^{\infty} \left[\frac{1}{mn} \sin\left(\frac{m\pi}{2}\right) \sin\left(\frac{n\pi}{2}\right) \sin\left(\frac{m\pi}{4}\right) \sin\left(\frac{n\pi}{4}\right) \right] \sin\left(\frac{m\pi x}{a}\right)$$

$$\times \sin\left(\frac{n\pi y}{b}\right), m,n = \text{odd integers} \tag{10.19}$$

As before, how well Equation 10.19 describes a load P uniformly distributed over the interior region depends on the number of terms used. A normalized plot of the series representation given by Equation 10.19 along the plate centerline $y = b/2$ is presented in Figure 10.11. Curves are shown based on a 9-term expansion (i.e., m, $n = 1$, 3, 5), a 64-term expansion (m, $n = 1$, 3, 5, ..., 15), and a 169-term expansion (m, $n = 1$, 3, 5, ..., 25). The series clearly converges toward a constant loading over the central region of the plate, and approaches zero elsewhere.

As a third example, consider a *concentrated* transverse force P, located at $x = \xi$ and $y = \eta$. The Fourier coefficients for this case can be obtained from Equation 10.18 by allowing $a_i \to 0$, $b_i \to 0$, that is, the internal area over which load P acts is allowed to shrink to a point. It can be shown that in the limit Equation 10.18 becomes

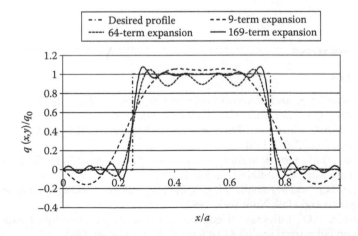

FIGURE 10.11
A normalized plot of Equation 10.19 along the plate centerline defined by $y = b/2$.

Structural Analysis of Polymeric Composite Materials

$$q_{mn} = \frac{4P}{ab} \sin\left(\frac{m\pi\xi}{a}\right) \sin\left(\frac{n\pi\eta}{b}\right) \tag{10.20}$$

Together, Equations 10.14 and 10.20 define the Fourier series expansion of a concentrated force P applied at the point $x = \xi$ and $y = \eta$. As a specific example, consider the case in which the concentrated load is applied at the center of the plate, that is, at $\xi = a/2$, $\eta = b/2$. In this case, Equation 10.20 becomes

$$q_{mn} = \begin{cases} \dfrac{4P}{ab}(-1)^{[(m+n)/2]-1}, & m,n = \text{odd integers} \\[2mm] 0, & m,n = \text{even integers} \end{cases}$$

Combining this result with Equation 10.14, we find that the Fourier series expansion for a concentrated load applied at the center of the plate is given by

$$q(x,y) = \frac{4P}{ab} \sum_{m=1}^{\infty} \sum_{n=1}^{\infty} \sin\left(\frac{m\pi x}{a}\right) \sin\left(\frac{n\pi y}{b}\right)(-1)^{[(m+n)/2]-1} \tag{10.21}$$

where $m,n = $ odd integers. As in the earlier examples, as the number of terms used in the series representation given by Equation 10.21 is increased the distribution of $q(x, y)$ resembles a concentrated force P more and more closely.

References

1. Timoshenko, S., and Woinowsky-Krieger, S., *Theory of Plates and Shells*, McGraw-Hill Book Co., New York, ISBN 0-07-0647798, 1987.
2. Ugural, A.C., *Stresses in Plates and Shells*, McGraw-Hill Book Co., New York, ISBN 0-07-065730-0, 1981.
3. Whitney, J.M., *Structural Analysis of Laminated Anisotropic Plates*, Technomic Pub. Co., Lancaster, PA, ISBN 87762-518-2, 1987.
4. Turvey, G.J., and Marshall, I.H., eds, *Buckling and Postbuckling of Composite Plates*, Chapman and Hall, New York, 1995.
5. Almroth, B.O., Influence of edge conditions on the stability of axially compressed cylindrical shells, *AIAA Journal*, 4(1), 134–140, 1966.

11

Some Exact Solutions for Specially
Orthotropic Laminates

11.1 Equations of Equilibrium for a Specially
Orthotropic Laminate

The equations of equilibrium for a symmetric composite laminate were presented in Sections 10.2. Boundary conditions consistent with thin-plate theory were then discussed in Section 10.3. Unfortunately, exact solutions to the equations of equilibrium that satisfy specified boundary conditions often cannot be found. Sometimes the difficulty in obtaining an exact solution is due to the boundary conditions involved; difficulties of this sort are encountered for both isotropic and anisotropic plates. In other cases, the difficulties are strictly due to the anisotropic nature of composites.

We have previously restricted our discussion to symmetric stacking sequences, and so $B_{ij} = M_{ij}^T = M_{ij}^M = 0$ in all cases considered. In addition, in this chapter, we will further restrict our discussion to so-called *specially orthotropic* laminates. A specially orthotropic laminate is one for which $A_{16} = A_{26} = D_{16} = D_{26} = N_{xy}^T = N_{xy}^M = 0$. It turns out that this is a very restrictive limitation. Indeed, a review of the stacking sequences presented in Section 6.7 will reveal that there are only three stacking sequences that eliminate these coupling stiffnesses and thermal/moisture resultants and can, therefore, be classified as "specially orthotropic." These are:

- Unidirectional $[0°]_n$ laminates (described in Section 6.7.1)
- Unidirectional $[90°]_n$ laminates (described in Section 6.7.1)
- Symmetric cross-ply laminates, for example, $[(0/90)_n]_s$ (described in Section 6.7.2)

It is mentioned in passing that the definition of a "specially orthotropic laminate" is sometimes relaxed from the definition just described. Specifically, the definition of a "specially orthotropic laminate" may be relaxed if the laminate is symmetric and is subjected to in-plane loading only. In problems of this type the coupling stiffnesses D_{16} and D_{26} play no role. Therefore, researchers studying the behavior of symmetric composites subjected to in-plane loads and displacements (only) often specify that a "specially ortho-tropic laminate" is one in which $A_{16} = A_{26} = N^T_{xy} = N^M_{xy} = 0$, but place no restriction on the values of D_{16} or D_{26}. For example, Lekhnitskii [1] and Savin [2] have obtained solutions for the in-plane stresses induced near elliptical, rectangular, or triangular holes in symmetric anisotropic composite plates subjected to in-plane loading (only). Since no out-of-plane bending is involved in these problems, the solutions only require that $A_{16} = A_{26} = N^T_{xy} = N^M_{xy} = 0$. Both symmetric balanced laminates and symmetric angle-ply laminates (see Section 6.7) are specially orthotropic under this relaxed definition. Nevertheless, in this chapter we consider situations in which the composite laminate is subjected to both in-plane and out-of-plane loading, and in these cases, the bending stiffnesses of the laminate plays an important role. Hence we must maintain the more restrictive definition and stipulate that a specially orthotropic laminate must have $A_{16} = A_{26} = D_{16} = D_{26} = N^T_{xy} = N^M_{xy} = 0$.

For a specially orthotropic laminate the equations of equilibrium for a symmetric laminate (i.e., Equations 10.8) are further simplified, and become:

$$A_{11}\frac{\partial^2 u_o}{\partial x^2} + (A_{12} + A_{66})\frac{\partial^2 v_o}{\partial x \partial y} + A_{66}\frac{\partial^2 u_o}{\partial y^2} = 0 \tag{11.1a}$$

$$(A_{12} + A_{66})\frac{\partial^2 u_o}{\partial x \partial y} + A_{22}\frac{\partial^2 v_o}{\partial y^2} + A_{66}\frac{\partial^2 v_o}{\partial x^2} = 0 \tag{11.1b}$$

$$\left[A_{11}\frac{\partial u_o}{\partial x} + A_{12}\frac{\partial v_o}{\partial y} - N^T_{xx} - N^M_{xx}\right]\left(\frac{\partial^2 w}{\partial x^2}\right)$$

$$+ \left[A_{12}\frac{\partial u_o}{\partial x} + A_{22}\frac{\partial v_o}{\partial y} - N^T_{yy} - N^M_{yy}\right]\left(\frac{\partial^2 w}{\partial y^2}\right) + 2\left[A_{66}\left(\frac{\partial u_o}{\partial y} + \frac{\partial v_o}{\partial x}\right)\right]\left(\frac{\partial^2 w}{\partial x \partial y}\right)$$

$$- D_{11}\frac{\partial^4 w}{\partial x^4} - 2(D_{12} + 2D_{66})\frac{\partial^4 w}{\partial x^2 \partial y^2} - D_{22}\frac{\partial^4 w}{\partial y^4} + q(x, y) = 0 \tag{11.1c}$$

In following sections, we will first obtain solutions for plate deflections under relatively simple loading conditions. It will then be seen that solutions for *simple* loading condition form the basis for predicting displacements caused by *complex* transverse loads.

There are, in essence, two techniques that can be used to obtain exact solutions for problems involving arbitrary transverse loads. Both methods place limitations on the type of boundary conditions that can be considered. The first approach was developed by Navier in about 1820, and is known as the Navier solution [3]. The Navier solution is applicable if all four edges of the specially orthotropic rectangular plate are simply supported. The Navier solution applied to the case of a simply supported specially orthotropic panel subjected to arbitrary transverse loading will be described in Section 11.6. The second exact solution technique was developed by Levy in 1899, and is known as the Levy solution [3]. In this case, two opposite edges of the plate must be simply supported, while the boundary conditions on the remaining two edges are arbitrary and may be clamped or free, for example. The Levy solution technique will not be discussed in this book, and the interested reader is referred to references [1,3,4,5] for a discussion of this method.

A final preliminary comment is that thin composite plates (or in fact any thin-walled structure) are prone to buckling if subjected to compressive in-plane loads and/or high shear loads. Buckling of specially orthotropic laminates is discussed in Sections 11.7 and 11.8, respectively.

11.2 In-Plane Displacement Fields in Specially Orthotropic Laminates

The overall goal in this chapter is to predict deflections induced in specially orthotropic composite panels that are simply supported along all four edges. As discussed in Section 10.3, four types of simple supports may be defined for a thin plate. For the edge $x = a$, these are:

S1: $w(a,y) = 0$

$$M_{xx}^*(a,y) = 0, u_o(a,y) = u_o^{(+x)}(y), v_o(a,y) = v_o^{(+x)}(y)$$

S2: $w(a,y) = 0$

$$M_{xx}^*(a,y) = 0, N_{xx}^*(a,y) = N_{xx}^{(+x)}(y), v_o(a,y) = v_o^{(+x)}(y)$$

S3: $w(a,y) = 0$

$$M_{xx}^*(a,y) = 0, u_o(a,y) = u_o^{(+x)}(y), N_{xy}^*(a,y) = N_{xy}^{(+x)}(y)$$

S4: $w(a,y) = 0$

$$M_{xx}^*(a,y) = 0, N_{xx}^*(a,y) = N_{xx}^{(+x)}(y), N_{xy}^*(a,y) = N_{xy}^{(+x)}(y)$$

Hence, the definition of a simply supported edge requires that either in-plane displacements, in-plane stress resultants, or some combination thereof must be specified along the edge, in addition to the requirement that out-of-plane displacements and bending moments vanish along the edge.

The following sequence of events is assumed to occur during fabrication and assembly of a simply supported composite plate. We assume the laminate is cured at an elevated temperature, and that the laminate is stress- and strain-free at the cure temperature.* Following cure, the laminate is cooled to room temperature, and, therefore, midplane displacements (as well as thermal stress resultants and associated ply strains and stresses) are induced during cooldown to room temperature. The laminate is then trimmed to the desired dimensions $a \times b$ and assembled in a surrounding structure that provides simple supports along all four edges. According to this scenario then, *midplane displacements have already been induced within the laminate prior to assembly in the simple supports.* Whether additional in-plane displacements subsequently occur, due to application of $q(x,y)$, a further change in temperature, and/or a change in moisture content, depends on the type of simple supports involved. That is, the development of additional in-plane displacements depends on whether simple supports of type $S1$, $S2$, $S3$, or $S4$ are imposed along each edge of the plate.

All problems considered here will be based on the following. First, we assume that opposite edges of the plate are subjected to the same type of simple support. For example, if the edge $x = 0$ is subjected to the type $S1$ simple support, then by assumption the edge $x = a$ is also subjected to type $S1$ supports. Secondly, we assume that stress resultants N_{xx}, N_{yy}, and/or N_{xy} applied to opposite edges of the plate (if any) are identical and uniformly distributed along the edge. This loading condition is precisely equivalent to that assumed during the development of CLT in Chapter 6. Therefore, we can develop general expressions giving the in-plane displacement fields caused by any combination of N_{ij}, ΔT, and/or ΔM using a standard CLT analysis. We will then specialize these general expressions to correspond to type $S1$, $S2$, $S3$, or $S4$ simple supports.

To begin this process we relate midplane strains and curvatures to stress resultants using Equation 6.45. For a symmetric specially orthotropic laminate subjected to uniform in-plane stress resultants, a uniform change in temperature, and/or a uniform change in moisture content, Equation 6.45 becomes

---

* As mentioned in Section 6.6.2, in practice the stress- and strain-free temperature is often 10–25°C below the final cure temperature. This complication has been ignored throughout this book, and it is assumed that the laminate is stress- and strain-free at the cure temperature.

$$
\begin{Bmatrix} \varepsilon^o_{xx} \\ \varepsilon^o_{yy} \\ \gamma^o_{xy} \\ \kappa_{xx} \\ \kappa_{yy} \\ \kappa_{xy} \end{Bmatrix} = \begin{bmatrix} a_{11} & a_{12} & 0 & 0 & 0 & 0 \\ a_{12} & a_{22} & 0 & 0 & 0 & 0 \\ 0 & 0 & a_{66} & 0 & 0 & 0 \\ 0 & 0 & 0 & d_{11} & d_{12} & 0 \\ 0 & 0 & 0 & d_{12} & d_{22} & 0 \\ 0 & 0 & 0 & 0 & 0 & d_{66} \end{bmatrix} \begin{Bmatrix} N_{xx} + N^T_{xx} + N^M_{xx} \\ N_{yy} + N^T_{yy} + N^M_{yy} \\ N_{xy} \\ 0 \\ 0 \\ 0 \end{Bmatrix}
$$

As in earlier chapters, we assume infinitesimal strain levels, and, therefore, midplane strains are related to midplane displacements according to Equation 6.10:

$$
\varepsilon^o_{xx} = \frac{\partial u_o}{\partial x}, \quad \varepsilon^o_{yy} = \frac{\partial v_o}{\partial y}, \quad \gamma^o_{xy} = \frac{\partial u_o}{\partial y} + \frac{\partial v_o}{\partial x}
$$

Consequently, midplane displacement fields are given by

$$
\frac{\partial u_o}{\partial x} = a_{11}\left(N_{xx} + N^T_{xx} + N^M_{xx}\right) + a_{12}\left(N_{yy} + N^T_{yy} + N^M_{yy}\right) \tag{11.2a}
$$

$$
\frac{\partial v_o}{\partial y} = a_{12}\left(N_{xx} + N^T_{xx} + N^M_{xx}\right) + a_{22}\left(N_{yy} + N^T_{yy} + N^M_{yy}\right) \tag{11.2b}
$$

$$
\frac{\partial u_o}{\partial y} + \frac{\partial v_o}{\partial x} = a_{66}(N_{xy}) \tag{11.2c}
$$

Integrating Equation 11.2a with respect to x, we find:

$$
u_o(x,y) = \left[a_{11}\left(N_{xx} + N^T_{xx} + N^M_{xx}\right) + a_{12}\left(N_{yy} + N^T_{yy} + N^M_{yy}\right)\right]x + f_1(y) + \lambda_1
$$

where $f_1(y)$ is an unknown function of y (only) and λ_1 is an unknown constant of integration. Similarly, integrating Equation 11.2b with respect to y we find:

$$
v_o(x,y) = \left[a_{12}(N_{xx} + N^T_{xx} + N^M_{xx}) + a_{22}(N_{yy} + N^T_{yy} + N^M_{yy})\right] y + f_2(x) + \lambda_2
$$

where $f_2(x)$ is an unknown function of x (only) and λ_2 is a second unknown constant. Without a loss in generality we assume that midplane displacements are zero at the origin (i.e., let $u_o = v_o = 0$ at $x = y = 0$), and consequently we conclude $\lambda_1 = \lambda_2 = 0$. Substituting the above expressions for $u_o(x, y)$ and $v_o(x, y)$ into Equation 11.2c, we find:

$$
\frac{\partial u_o}{\partial y} + \frac{\partial v_o}{\partial x} = \frac{\partial f_1}{\partial y} + \frac{\partial f_2}{\partial x} = a_{66} N_{xy}
$$

Since the terms that appear on the right side of the equality (i.e., a_{66} and N_{xy}) are known constants, it follows that f_1 and f_2 must be at most linear functions of y and x, respectively:

$$f_1(y) = \lambda_3 y$$

$$f_2(x) = \lambda_4 x$$

Hence, we can write:

$$\lambda_3 + \lambda_4 = a_{66} N_{xy}$$

Constants λ_3 and λ_4 can take on any value as long as they sum to the product $(a_{66} N_{xy})$. To determine particular values convenient for our use, we now require that the infinitesimal rotation vector in the x–y plane ω_{xy} (which represents rigid body motion of the plate) is zero. The infinitesimal rotation vector is given by [3]:

$$\omega_{xy} = \frac{1}{2}\left(\frac{\partial u_o}{\partial y} - \frac{\partial v_o}{\partial x} \right)$$

Requiring that $\omega_{xy} = 0$ leads to

$$\lambda_3 = \lambda_4 = \frac{1}{2} a_{66} N_{xy}$$

Combining the preceding results, we conclude that in-plane midplane displacement fields induced in a symmetric specially orthotropic composite panel by the combination of uniform in-plane stress resultants, a uniform change in temperature, and/or a uniform change in moisture contents are given by

$$u_o(x,y) = \left[a_{11}(N_{xx} + N_{xx}^T + N_{xx}^M) + a_{12}(N_{yy} + N_{yy}^T + N_{yy}^M) \right] x + \frac{1}{2}(a_{66} N_{xy})y$$

$$v_o(x,y) = \frac{1}{2}(a_{66} N_{xy})x + \left[a_{12}(N_{xx} + N_{xx}^T + N_{xx}^M) + a_{22}(N_{yy} + N_{yy}^T + N_{yy}^M) \right] y$$

These expressions can be simplified through the use of the effective thermal expansion coefficients of the laminate, defined by Equations 6.73a, and the effective moisture expansion coefficients, defined by Equations 6.76a. For a specially orthotropic laminate these become:

$$\bar{\alpha}_{xx} = \frac{1}{\Delta T}\left[a_{11} N_{xx}^T + a_{12} N_{yy}^T \right], \quad \bar{\alpha}_{yy} = \frac{1}{\Delta T}\left[a_{12} N_{xx}^T + a_{22} N_{yy}^T \right], \quad \bar{\alpha}_{xy} = 0$$

$$\bar{\beta}_{xx} = \frac{1}{\Delta M}\left[a_{11}N_{xx}^M + a_{12}N_{yy}^M\right], \bar{\beta}_{yy} = \frac{1}{\Delta M}\left[a_{12}N_{xx}^M + a_{22}N_{yy}^M\right], \bar{\beta}_{xy} = 0$$

Hence, we see that the midplane displacement fields can be written as

$$u_o(x,y) = \left[a_{11}N_{xx} + a_{12}N_{yy} + \Delta T\bar{\alpha}_{xx} + \Delta M\bar{\beta}_{xx}\right]x + \frac{1}{2}(a_{66}N_{xy})y \quad (11.3a)$$

$$v_o(x,y) = \frac{1}{2}(a_{66}N_{xy})x + \left[a_{12}N_{xx} + a_{22}N_{yy} + \Delta T\bar{\alpha}_{yy} + \Delta M\bar{\beta}_{yy}\right]y \quad (11.3b)$$

Note that the displacement fields are independent of the transverse load $q(x,y)$. As pointed out in Section 10.2, $u_o(x,y)$ and $v_o(x,y)$ are predicted to be independent of $q(x,y)$ because we have assumed displacement gradients are small, such that gradients squared can be ignored (e.g., $(\partial w/\partial x)^2 \approx 0$). Analyses that account for large displacement gradients are not considered in this book. If we had included large gradients in our analysis then expressions for $u_o(x,y)$ and $v_o(x,y)$ corresponding to Equations 11.3a,b would depend on transverse load $q(x,y)$.

Direct substitution of Equations 11.3a,b will reveal that these equations satisfy the equations of equilibrium, Equations 11.1a,b. In the following sections we will use these expressions to specify in-plane displacement fields associated with simple supports of type S1 through S4.

11.3 Specially Orthotropic Laminates Subject to Simple Supports of Type S1

In this section, we consider the specially orthotropic plate shown in Figure 11.1. The plate is assumed to be rectangular with thickness t and in-plane dimensions $a \times b$. All four edges of the plate are subject to simple supports of type S1, where it is assumed that the laminate was mounted in the structure that imposes type S1 supports following cooldown from the cure temperature to room temperature. After assembly, the plate is subjected to a uniform change in temperature and a transverse loading that varies over the x–y plane according to

$$q(x,y) = q_o \sin\left(\frac{\pi x}{a}\right)\sin\left(\frac{\pi y}{b}\right) \quad (11.4)$$

We will not consider a change in moisture content (i.e., let $\Delta M = 0$), although from earlier discussion it should be clear that a change in moisture content can be accounted for (in a mathematical sense) using the same techniques used to model uniform changes in temperature.

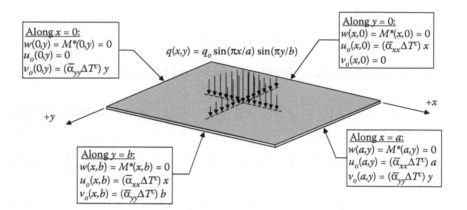

FIGURE 11.1
Thin rectangular plate of thickness t and in-plane dimensions $a \times b$, subjected to a transverse
load $q(x, y) = q_o \sin(\pi x/a)\sin(\pi y/b)$. All four edges of the plate are subject to simple supports of
type S1.

Let us define ΔT^c as the change in temperature from cure to room
temperature:

$$\Delta T^c = T_{RT} - T_C$$

where T_{RT} = room temperature and T_C = the cure temperature. For a symmetric specially orthotropic laminate the in-plane displacements caused by
ΔT^c can be calculated using Equations 11.3a,b with $N_{xx} = N_{yy} = N_{xy} = 0$:

$$u_o^c(x, y) = \left(\bar{\alpha}_{xx}\Delta T^c\right)x \tag{11.5a}$$

$$v_o^c(x, y) = \left(\bar{\alpha}_{yy}\Delta T^c\right)y \tag{11.5b}$$

These displacements are induced *before* assembly of the simply supported
plate. Note that these in-plane displacement fields satisfy the first two equations of equilibrium, Equation 11.1a,b. A type S1 simple support will simply
maintain these displacements during subsequent loading and/or temperature changes. Therefore, type S1 boundary conditions for all four edges of
the plate become:

For $x = 0$:   For $x = a$:

$w(0,y) = 0$ $w(a,y) = 0$

$$M_{xx}^*(0, y) = D_{11}\frac{\partial^2 w}{\partial x^2} + D_{12}\frac{\partial^2 w}{\partial y^2} = 0 \qquad M_{xx}^*(a, y) = D_{11}\frac{\partial^2 w}{\partial x^2} + D_{12}\frac{\partial^2 w}{\partial y^2} = 0$$

$$u_o(0,y) = u_o^{(-x)}(y) = 0 \qquad\qquad u_o(a,y) = u_o^{(+x)}(y) = \left(\bar{\alpha}_{xx}\Delta T^L\right)a$$

$$v_o(0,y) = v_o^{(-x)}(y) = \left(\bar{\alpha}_{yy}\Delta T^c\right)y \qquad\qquad v_o(0,y) = v_o^{(+x)}(y) = \left(\bar{\alpha}_{yy}\Delta T^c\right)y$$

For $y = 0$: \qquad\qquad\qquad\qquad For $y = b$:

$$w(x,0) = 0 \qquad\qquad\qquad\qquad\qquad w(x,b) = 0$$

$$M_{yy}^*(x,0) = D_{11}\frac{\partial^2 w}{\partial x^2} + D_{12}\frac{\partial^2 w}{\partial y^2} = 0 \qquad M_{yy}^*(x,b) = D_{11}\frac{\partial^2 w}{\partial x^2} + D_{12}\frac{\partial^2 w}{\partial y^2} = 0$$

$$u_o(x,0) = u_o^{(-y)}(x) = \left(\bar{\alpha}_{xx}\Delta T^c\right)x \qquad u_o(x,b) = u_o^{(+y)}(x) = \left(\bar{\alpha}_{xx}\Delta T^c\right)x$$

$$v_o(x,0) = v_o^{(-y)}(x) = 0 \qquad\qquad\qquad v_o(x,b) = v_o^{(+y)}(x) = \left(\bar{\alpha}_{yy}\Delta T^c\right)b$$

We wish to determine the out-of-plane displacement field $w(x,y)$ that satisfies these boundary conditions as well as the equations of equilibrium, when the plate is subjected to a transverse loading $q(x,y)$ and/or a further change in temperature. Guided by the functional form of the transverse pressure (i.e., Equation 11.4), we assume the out-of-plane displacement field is given by

$$w(x,y) = c\sin\left(\frac{\pi x}{a}\right)\sin\left(\frac{\pi y}{b}\right) \tag{11.6}$$

where c is an unknown constant. Substituting this assumed form into the boundary conditions will reveal that they are identically satisfied for any value of c. Hence the value of constant c must be determined by enforcing the third equation of equilibrium, Equation 11.1c. We perform the following operations:

a. Substitute Equations 11.4, 11.5a,b, and 11.6 into the third equation of equilibrium, Equation 11.1c.

b. Write the thermal stress resultants N_{xx}^T and N_{yy}^T in terms of effective thermal expansion coefficients (using Equations 6.73b).

c. Solve the resulting expression for constant c.

Following this process, we find:

$$c = \cfrac{q_0}{\pi^4\left[\begin{array}{c}\dfrac{D}{a^4} + \dfrac{2}{a^2 b^2}(D_{12} + 2D_{66}) + \dfrac{D_{22}}{b^4} + \dfrac{\left(\Delta T^c - \Delta T\right)}{\pi^2} \\[2mm] \times\left(\dfrac{1}{a^2}\left[A_{11}\bar{\alpha}_{xx} + A_{12}\bar{\alpha}_{yy}\right] + \dfrac{1}{b^2}\left[A_{12}\bar{\alpha}_{xx} + A_{22}\bar{\alpha}_{yy}\right]\right)\end{array}\right]} \tag{11.7a}$$

Notice that the temperature change as defined in earlier chapters (ΔT) appears in Equation 11.7a. That is,

$$\Delta T = (\text{current temperature}) - (\text{cure temperature})$$

Also note that if the temperature is not changed following assembly in the type S1 simple supports then $\Delta T = \Delta T^c$, and the effects of temperature cancel in Equation 11.7a.

Results from thin-plate theory are often expressed in terms of the so-called plate aspect ratio, $R = a/b$. Equation 11.7a can be rewritten using the aspect ratio as follows:

$$c = \cfrac{q_o R^4 b^4}{\pi^4 \left[\begin{array}{l} D_{11} + 2R^2(D_{12} + 2D_{66}) + R^4 D_{22} + \left((\Delta T^c - \Delta T)a^2/\pi^2\right) \\[2mm] \times \left\{ \bar{\alpha}_{xx}\left(A_{11} + R^2 A_{12}\right) + \bar{\alpha}_{yy}\left(A_{12} + R^2 A_{22}\right) \right\} \end{array} \right]} \tag{11.7b}$$

The predicted out-of-plane deflection is obtained by combining either Equation 11.7a or 11.7b with Equation 11.6. Using Equation 11.7b for example, we have:

$$w(x,y) = \cfrac{q_o R^4 b^4 \sin\left(\pi x/a\right)\sin\left(\pi y/b\right)}{\pi^4 \left[\begin{array}{l} D_{11} + 2R^2(D_{12} + 2D_{66}) + R^4 D_{22} + \left((\Delta T^c - \Delta T)a^2/\pi^2\right) \\[2mm] \times \left\{ \bar{\alpha}_{xx}\left(A_{11} + R^2 A_{12}\right) + \bar{\alpha}_{yy}\left(A_{12} + R^2 A_{22}\right) \right\} \end{array} \right]} \tag{11.8}$$

Equations 11.5a,b and 11.8 give the predicted displacement field induced in the plate and represent the solution to this problem.

To summarize, we have considered a symmetric specially orthotropic plate subjected to type S1 simple-supports. We have assumed that the plate is mounted within simple supports while at room temperature. The laminate has, therefore, likely experienced a change in temperature prior to assembly, since modern composites are typically cured at an elevated temperature. The change in temperature associated with cooldown to room temperature is represented by ΔT^c. After assembly the plate is subjected to a sinusoidally varying transverse loading and/or the temperature is changed away from room temperature. The resulting displacement fields are given by Equations 11.5a,b and 11.8. A typical application of this solution is discussed in Example Problem 11.1.

As a closing comment, it should be kept in mind that we have not yet considered the possibility of buckling. The solution presented here is not valid if the change in temperature is such that the resulting in-plane stress resultants are compressive and have a magnitude large enough to cause buckling. The phenomenon of buckling induced by a change in temperature is called "thermal buckling" and will be discussed in Section 11.8.

Example Problem 11.1

A $[(0_2/90)_2]_s$ graphite–epoxy laminate is cured at 175°C and then cooled to room temperature (20°C). After cooling the flat laminate is trimmed to in-plane dimensions of 300 × 150 mm and mounted in an assembly that provides type S1 simple supports along all four edges. The x-axis is defined parallel to the 300 mm edge (i.e., $a = 0.3$ m, $b = 0.15$ m). The laminate is then subjected to a transverse pressure given by $q(x,y) = 40 \sin (\pi x/a) \sin (\pi y/b)$(kPa) and a uniform temperature change. No change in moisture content occurs ($\Delta M = 0$). Plot the maximum out-of-plane displacement as a function of temperature, over the range –50°C < T < 20°C. Use the properties listed for graphite–epoxy in Table 3.1, and assume each ply has a thickness of 0.125 mm.

SOLUTION

The rectangular plate is a 12-ply laminate with total thickness $t = 12(0.125 \text{ mm}) = 1.5$ mm and aspect ratio $R = a/b = (0.3 \text{ m})/(0.15 \text{ m}) = 2.0$. Out-of-plane displacements are given by Equation 11.8, and hence elements of the $[ABD]$ matrix are required. Based on the properties listed in Table 3.1 for graphite–epoxy and the specified stacking sequence, the $[ABD]$ matrix is

$$[ABD] = \begin{bmatrix} 176 \times 10^6 & 4.52 \times 10^6 & 0 & 0 & 0 & 0 \\ 4.52 \times 10^6 & 95.6 \times 10^6 & 0 & 0 & 0 & 0 \\ 0 & 0 & 19.5 \times 10^6 & 0 & 0 & 0 \\ 0 & 0 & 0 & 40.1 & 0.848 & 0 \\ 0 & 0 & 0 & 0.848 & 10.8 & 0 \\ 0 & 0 & 0 & 0 & 0 & 3.66 \end{bmatrix}$$

where the units of A_{ij} are Pa–m and the units of D_{ij} are Pa–m^3.

We also require the effective thermal expansion coefficients. Based on the properties listed in Table 3.1 for graphite–epoxy and the specified stacking sequence these are

$$\bar{\alpha}_{xx} = 0.29 \text{ } \mu\text{m}/\text{m–°C}, \quad \bar{\alpha}_{yy} = 2.4 \text{ } \mu\text{m}/\text{m–°C}$$

Since the laminate is specially orthotropic the effective shear thermal expansion coefficient is zero ($\bar{\alpha}_{xy} = 0$). The cooldown from the cure temperature to room temperature is

$$\Delta T^c = (20°C) - (175°C) = -155°C$$

Following assembly, the temperature ranges from room temperature to as low as −50°C. Therefore:

$$-225°C < \Delta T < -155°C$$

Note that the temperatures to be considered are all at or below room temperature. Since the effective thermal expansion coefficients are algebraically positive, if the laminate were not constrained by the S1 simple supports it would tend to contract as temperature is lowered. Since it is in fact constrained by the simple supports, the in-plane stress resultants that develop as temperature is lowered tend to be tensile. Therefore, thermal buckling is not of concern.

Substituting all known values, Equation 11.8 becomes

$$w(x,y) = \frac{324}{\pi^4\left[278 + \left(88.6(\Delta T^c - \Delta T)/\pi^2\right)\right]}\sin\left(\frac{\pi x}{0.30}\right)\sin\left(\frac{\pi y}{0.15}\right)\text{(meters)}$$

This expression can be used to calculate the out-of-plane displacement induced at any point (x,y) over the surface of the plate. The maximum out-of-plane displacement occurs at the center of

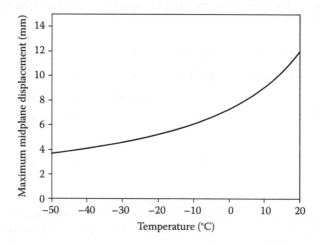

FIGURE 11.2
Maximum out-of-plane displacement for the graphite–epoxy plate considered in Example Problem 11.1 as a function of temperature.

the plate, that is, at $x = a/2 = 0.15$ m and $y = b/2 = 0.075$ m. Since $\Delta T^t = 20°C - 175°C = -155°C$, and temperatures ranging from $-50°C < T < 20°C$ are to be considered, the quantity $(\Delta T^t - \Delta T)$ ranges from $70°C \geq (\Delta T^t - \Delta T) \geq 0°C$ in this problem. A plot of maximum out-of-plane displacement as a function of temperature is shown in Figure 11.2. At room temperature (20°C) a maximum deflection of 12 mm is predicted. As would be expected, the plate becomes stiffer as the temperature is decreased, in the sense that out-of-plane displacements are decreased due to the in-plane tensile loads that develop as temperature is decreased. At the lowest temperature considered (–50°C) a maximum deflection of 3.7 mm is predicted.

11.4 Specially Orthotropic Laminates Subject to Simple Supports of Type S4

In this section, we consider the specially orthotropic plate shown in Figure 11.3. As in the preceding section, the plate is rectangular with thickness t and in-plane dimensions $a \times b$. However, we now assume that each edge of the plate is subject to simple supports of type S4, rather than type S1. Hence, displacement fields along the plate edges are not required to take on any specified value. Rather, we assume that a uniform normal stress resultant is applied to each edge, as shown in Figure 11.3. Note that shear loading is not considered: $N_{xy} = N_{yx} = 0$. The plate is also subjected to

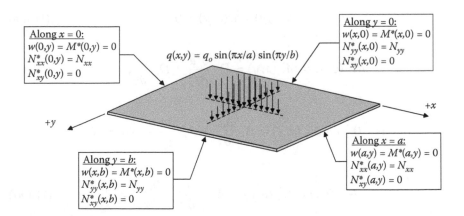

FIGURE 11.3
Thin rectangular plate of thickness t and in-plane dimensions $a \times b$, subjected to a transverse load $q(x, y) = q_0 \sin(\pi x/a)\sin(\pi y/b)$. All four edges of the plate are subject to simple supports of type S4 (compare with Figure 11.1).

a uniform change in temperature, ΔT, and a transverse loading that varies over the x–y plane according to

$$q(x,y) = q_o \sin\left(\frac{\pi x}{a}\right)\sin\left(\frac{\pi y}{b}\right) \tag{11.9}$$

We will not consider any change in moisture content, that is, $\Delta M = 0$. Also, we will not consider the possibility of buckling in this section, even though buckling is a possibility if either of the in-plane normal stress resultants is compressive. Buckling under type S4 simple supports will be discussed in Section 11.9.

The boundary conditions that define type S4 simple support were discussed in Section 10.3.3.2. Since the stress resultants applied along each edge are uniform and constant, we can write:

$$N_{xx}^*(0,y) = N_{xx}^*(a,y) = N_{xx}$$

$$N_{yy}^*(x,0) = N_{yy}^*(x,b) = N_{yy}$$

$$N_{xy}^*(0,y) = N_{xy}^*(a,y) = N_{yx}^*(x,0) = N_{yx}^*(x,b) = 0.$$

Since we have limited consideration to symmetric specially orthotropic laminates, $A_{16} = A_{26} = D_{16} = D_{26} = B_{ij} = N_{xy}^T = N_{xy}^M = M_{ij}^T = M_{ij}^M = 0$. Therefore, the boundary conditions can be written as

For $x = 0,a$:

$$w(0,y) = w(a,y) = 0 \tag{11.10a}$$

$$M_{xx}^*(0,y) = M_{xx}^*(a,y) = D_{11}\frac{\partial^2 w}{\partial x^2} + D_{12}\frac{\partial^2 w}{\partial y^2} = 0 \tag{11.10b}$$

$$N_{xx}^*(0,y) = N_{xx}^*(a,y) = A_{11}\frac{\partial u_o}{\partial x} + A_{12}\frac{\partial v_o}{\partial y} - N_{xx}^T = N_{xx} \tag{11.10c}$$

$$N_{xy}^*(0,y) = N_{xy}^*(a,y) = A_{66}\left(\frac{\partial u_o}{\partial y} + \frac{\partial v_o}{\partial x}\right) = 0 \tag{11.10d}$$

For $y = 0,b$:

$$w(x,0) = w(x,b) = 0 \tag{11.11a}$$

$$M_{yy}^*(x,0) = M_{yy}^*(x,b) = D_{12}\frac{\partial^2 w}{\partial x^2} + D_{22}\frac{\partial^2 w}{\partial y^2} = 0 \qquad (11.11b)$$

$$N_{yy}^*(x,0) = N_{yy}^*(x,b) = A_{12}\frac{\partial u_o}{\partial x} + A_{22}\frac{\partial v_o}{\partial y} - N_{yy}^T = N_{yy} \qquad (11.11c)$$

$$N_{yx}^*(x,0) = N_{yx}^*(x,b) = A_{66}\left(\frac{\partial u_o}{\partial y} + \frac{\partial v_o}{\partial x}\right) = 0 \qquad (11.11d)$$

Using Equations 11.3a,b and for the assumed conditions (i.e., $N_{xy} = \Delta M = 0$) in-plane displacements fields are

$$u_o(x,y) = \left[a_{11}N_{xx} + a_{12}N_{yy} + \Delta T\bar{\alpha}_{xx}\right]x \qquad (11.12a)$$

$$v_o(x,y) = \left[a_{12}N_{xx} + a_{22}N_{yy} + \Delta T\bar{\alpha}_{yy}\right]y \qquad (11.12b)$$

Let us confirm that these equations satisfy the appropriate boundary conditions. Substituting Equations 11.12a,b into boundary condition 11.10c and rearranging, we find that the following expression must be satisfied:

$$N_{xx}(A_{11}a_{11} + A_{12}a_{12}) + N_{yy}(A_{11}a_{12} + A_{12}a_{22}) + \Delta T(A_{11}\bar{\alpha}_{xx} + A_{12}\bar{\alpha}_{yy}) - N_{xx}^T = N_{xx}$$

Since the laminate is specially orthotropic,

$$[A] = \begin{bmatrix} A_{11} & A_{12} & 0 \\ A_{12} & A_{22} & 0 \\ 0 & 0 & A_{66} \end{bmatrix}$$

$$[a] = [A]^{-1} = \begin{bmatrix} \dfrac{A_{22}}{(A_{11}A_{22} - A_{12}^2)} & \dfrac{-A_{12}}{(A_{11}A_{22} - A_{12}^2)} & 0 \\ \dfrac{-A_{12}}{(A_{11}A_{22} - A_{12}^2)} & \dfrac{A_{11}}{(A_{11}A_{22} - A_{12}^2)} & 0 \\ 0 & 0 & \dfrac{1}{A_{66}} \end{bmatrix}$$

Therefore, by direct substitution we find:

$$(A_{11}a_{11} + A_{12}a_{12}) = 1$$

$$(A_{11}a_{12} + A_{12}a_{22}) = 0$$

Also, from Equation 6.73b, we find (for $A_{16} = 0$)

$$N_{xx}^T = \Delta T(A_{11}\overline{\alpha}_{xx} + A_{12}\overline{\alpha}_{yy})$$

Hence, the boundary condition represented by Equation 11.10c is satisfied identically by the in-plane displacement fields. A similar process can be used to confirm that the boundary condition given by Equation 11.11c is also satisfied. Both Equations 11.10d and 11.11d are satisfied as well, since $\partial u_o/\partial y = \partial v_o/\partial x = 0$.

Now consider out-of-plane displacements $w(x,y)$. Guided by the functional form of the transverse pressure (i.e., Equation 11.9), we once again assume the out-of-plane displacement field is given by

$$w(x,y) = c\sin\left(\frac{\pi x}{a}\right)\sin\left(\frac{\pi y}{b}\right) \tag{11.13}$$

where c is an unknown constant. Substituting this assumed form into boundary conditions Equation 11.10a,b and 11.11a,b will reveal that they are identically satisfied for any value of c. Hence the value of constant c must be determined by enforcing the third equation of equilibrium. Substituting Equations 11.12a,b, and 11.13 into the third equation of equilibrium, Equation 11.1c, and solving for constant c, we find:

$$c = \frac{q_o}{\pi^4\left[\begin{array}{c}\left(1/a^4\right)D_{11} + \left(2/a^2b^2\right)(D_{12} + 2D_{66}) + \left(1/b^4\right)D_{22} \\ + \left(1/\pi^2a^2\right)N_{xx} + \left(1/\pi^2b^2\right)N_{yy}\end{array}\right]} \tag{11.14a}$$

Using the definition of the plate aspect ratio, $R = a/b$, this result can also be written as

$$c = \frac{q_o R^4 b^4}{\pi^4\left[D_{11} + 2R^2(D_{12} + 2D_{66}) + R^4 D_{22} + (a^2/\pi^2)\left(N_{xx} + N_{yy}R^2\right)\right]} \tag{11.14b}$$

The predicted out-of-plane deflection is obtained by combining either Equation 11.14a or 11.14b with Equation 11.13. Using Equation 11.14b, for example, we have:

$$w(x,y) = \frac{q_o R^4 b^4 \sin(\pi x/a)\sin(\pi y/b)}{\pi^4\left[D_{11} + 2R^2(D_{12} + 2D_{66}) + R^4 D_{22} + (a^2/\pi^2)\left(N_{xx} + N_{yy}R^2\right)\right]} \tag{11.15}$$

Equations 11.12a,b and 11.15 give the predicted displacement fields induced in the plate and represent the solution to this problem.

To summarize, we have found the displacement fields induced in a symmetric specially orthotropic type S4 simply supported plate subjected to a sinusoidally varying transverse load, a uniform change in temperature ΔT, and uniform stress resultants N_{xx} and N_{yy}. A typical application of this solution is discussed in Example Problem 11.2. It should be kept in mind that we have not considered the possibility of buckling in this section. The solution we have obtained is not valid if N_{xx} and/or N_{yy} are compressive and have magnitudes large enough to cause buckling. Buckling under type S4 simple supports will be discussed in Section 11.9.

Example Problem 11.2

A $[(0_2/90)_2]_s$ graphite–epoxy laminate is cured at 175°C and then cooled to room temperature (20°C). After cooling the flat laminate is trimmed to in-plane dimensions of 300×150 mm and mounted in an assembly that provides type S4 simple supports along all four edges. The x-axis is defined parallel to the 300 mm edge (i.e., $a = 0.3$ m, $b = 0.15$ m). The laminate is then subjected to a uniform in-plane tensile loading (i.e., $N_{xx} = N_{yy}$) and transverse pressure given by $q(x,y) = 40 \sin (\pi x/a) \sin (\pi y/b)$ (kPa). Temperature remains constant and no change in moisture content occurs ($\Delta M = 0$).

    a. Plot the out-of-plane displacements induced along the centerline defined by $y = 0.075$ m, if in-plane loads $N_{xx} = N_{yy} = 50$ kN/m are applied.

    b. Plot the maximum out-of-plane displacement as a function of in-plane loads, over the range $0 < (N_{xx} = N_{yy}) < 70$ kN/m.

    c. Compare the maximum out-of-plane displacement caused by the specified transverse load at room temperature if the plate is subject to

       i. Type S1 simple supports (as discussed in Section 11.3), and

      ii. Type S4 simple supports, with $N_{xx} = N_{yy} = 0$

Use the properties listed for graphite–epoxy in Table 3.1, and assume each ply has a thickness of 0.125 mm.

SOLUTION

A $[(0_2/90)_2]_s$ graphite–epoxy laminate was also considered in Example Problem 11.1, and numerical values for the $[ABD]$ matrix are listed there. As before, the 12-ply laminate has a total thickness $t = 1.5$ mm and aspect ratio $R = a/b = 2.0$. Using these laminate stiffnesses, dimensions, and the specified transverse loading, Equations 11.15 becomes

$$w(x,y) = \left[\frac{324}{\pi^4 \left(278 + (0.090/\pi^2)\left(N_{xx} + 4N_{yy}\right)\right)} \right] \sin\left(\frac{\pi x}{0.3}\right) \sin\left(\frac{\pi y}{0.15}\right) \text{(meters)}$$

This expression can be used to calculate the out-of-plane displacement induced at any point (x,y) over the surface of the plate.

  a. Using $N_{xx} = N_{yy} = 50$ kN/m and $y = 0.075$ m, out-of-plane displacements are

$$w(x,y) = [1.30 \times 10^{-3}]\sin\left(\frac{\pi x}{0.3}\right)(\text{meters})$$

A plot of these displacements over the length of the plate $(0 < x < 0.3$ m is shown in Figure 11.4a. Displacements are zero at the edges defined by $x = 0, 0.3$ m, as dictated by the specified boundary conditions. As would be expected due to symmetry, out-of-plane displacement is maximum at the center of the plate, and equals 1.3 mm for the loading considered.

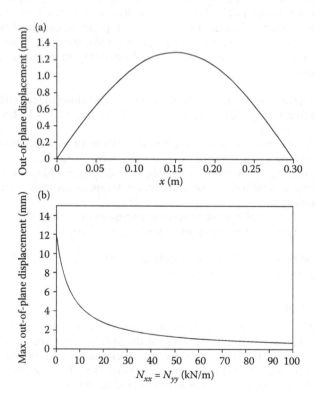

FIGURE 11.4
Out-of-plane displacements induced in the graphite–epoxy plate considered in Example Problem 11.2. (a) Out-of-plane displacements induced along centerline $y = 0.075$ m when $N_{xx} = N_{yy} = 50$ kN/m. (b) Maximum out-of-plane displacements as a function of in-plane loads $N_{xx} = N_{yy}$.

b. The maximum out-of-plane displacement occurs at the center of the plate, that is, at $x = a/2 = 0.15$ m and $y = b/2 = 0.075$ m, and at this point the maximum out-of-plane displacement is given by

$$w\Big|_{max} = \left[\frac{324}{\pi^4 \left(278 + (0.090/\pi^2)\left(N_{xx} + 4N_{yy}\right)\right)} \right] \text{(meters)}$$

A plot of maximum out-of-plane displacement as a function of in-plane tensile loads is shown in Figure 11.4b. As would be expected intuitively, the plate is stiffened by the application of in-plane loading. That is, the maximum out-of-plane displacement is decreased as in-plane tensile loads are increased. A maximum deflection of 12 mm occurs when $N_{xx} = N_{yy} = 0$, whereas the maximum deflection is reduced to 0.96 mm when $N_{xx} = N_{yy} = 70$ kN/m.

c. This same panel and transverse loading was considered in Example Problem 11.1, except type S1 simple supports were assumed. Thus, in Example Problem 11.1 in-plane displacements were fixed and were not allowed to change when the transverse load was applied. In contrast, type S4 simple supports are assumed in this problem; in-plane stress resultants are specified rather than in-plane displacements.

Referring to the results presented in these two Example Problems, we find that identical deflections are predicted, despite the differences in boundary conditions. That is, a maximum deflection of 12 mm is predicted at room temperature for type S1 condition, and an identical 12 mm deflection is predicted if $N_{xx} = N_{yy} = 0$ for type S4 conditions. This result may seem non-intuitive and (rigorously speaking) is incorrect. That is, for type S4 boundary conditions a transverse loading will cause a change in in-plane displacements. Therefore one might anticipate that the out-of-plane displacement for type S4 conditions would be increased, relative to type S1 conditions. However, the relative increase is very small if displacement gradients are small. Thus, the relative increase in out-of-plane displacements for type S4 conditions is not predicted *because we have based our analysis on infinitesimal strains*. The consequences of the infinitesimal strain assumption were alluded to in Section 10.2.2. It was noted there that this assumption ultimately leads to the conclusion that in-plane displacement fields $u_o(x,y)$ and $v_o(x,y)$ are independent of the transverse load, $q(x,y)$. The comparison between the results of Example Problems 11.1 and 11.2 presented here is an illustration of this independence.

Of course, results for the two different boundary conditions are identical because we have considered the case in which $N_{xx} = N_{yy} = 0$. If N_{xx} and/or $N_{yy} \neq 0$, then the transverse displacements for a plate supported by type S4 simple supports is quite different from that of a plate supported by type S1 supports.

11.5 Specially Orthotropic Laminates with Two Simply Supported Edges of Type S1 and Two Edges of Type S2

In this section, we consider the specially orthotropic plate shown in Figure 11.5. As in the preceding section, the plate is rectangular with thickness t and in-plane dimensions $a \times b$. We now assume that the two edges $x = 0,a$ are subjected to simple supports of type S2. That is, we specify that these two edges are subjected to known stress resultants N_{xx} and known displacements v_o. In contrast, the two edges $y = 0,b$ are subject to simple supports of type S1, which means that we require these edges to maintain known in-plane displacements u_o and v_o. The plate is also subjected to a uniform change in temperature, ΔT, and a transverse loading that varies over the x–y plane according to

$$q(x,y) = q_o \sin\left(\frac{\pi x}{a}\right)\sin\left(\frac{\pi y}{b}\right) \tag{11.16}$$

We will not consider a change in moisture content, that is, $\Delta M = 0$. Also, we will not consider the possibility of buckling in this section, even though buckling is a possibility if N_{xx} *is* compressive or if the change in temperature ΔT tends to cause the laminate to expand.

As in earlier sections, we assume the laminate was cured at an elevated temperature and cooled to room temperature prior to assembly in a surrounding structure that provides simple supports. Consider the midplane

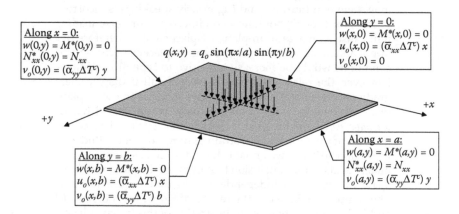

FIGURE 11.5

Thin rectangular plate of thickness t and in-plane dimensions $a \times b$, subjected to a transverse load $q(x,y) = q_o \sin(\pi x/a)\sin(\pi y/b)$. Edges $x = 0,a$ are subject to simple supports of type S2, whereas edges $y = 0,b$ are subject to simple support of type S1 (compare with Figures 11.1 and 11.3).

strains $v_o(x,y)$ induced during cooldown. Using Equation 11.3b, this can be written (with $N_{xx} = N_{yy} = N_{xy} = \Delta M = 0$):

$$v_o(x,y) = (\Delta T^c \bar{\alpha}_{yy})y \tag{11.17}$$

We can use this expression to specify known values of v_o along all four edges of the plate. Since a known stress resultant N_{xx} is applied to the two edges $x = 0$ and $x = a$, all quantities necessary to define the type S2 simple support along these two edges are known. To define the type S1 simple supports along the edges $y = 0,b$ we must specify known values of u_o. Towards that end, equate Equations 11.3b and 11.17:

$$\frac{1}{2}(a_{66}N_{xy})x + \left[a_{12}N_{xx} + a_{22}N_{yy} + \Delta T \bar{\alpha}_{yy}\right]y = (\Delta T^c \bar{\alpha}_{yy})y$$

This relation must be satisfied along the edges $y = 0,b$. Substituting $y = 0$ it is seen that N_{xy} must vanish ($N_{xy} = 0$). Substituting $y = b$, and solving for N_{yy}, we obtain:

$$N_{yy} = \frac{\bar{\alpha}_{yy}(\Delta T^c - \Delta T) - a_{12}N_{xx}}{a_{22}} \tag{11.18}$$

Equation 11.18 gives the stress resultant N_{yy} that must be provided by the simple supports along the $y = 0,b$ so as to maintain the stipulated value of v_o represented by Equation 11.17. Note that if temperature does not change from room temperature (i.e., if $\Delta T = \Delta T^c$), and if no stress resultant N_{xx} is applied (i.e., if $N_{xx} = 0$), then $N_{yy} = 0$, as would be expected.

We can now calculate the displacement u_o implied by these conditions, using Equation 11.3b:

$$u_o(x,y) = \left[a_{11}N_{xx} + a_{12}\left(\frac{\bar{\alpha}_{yy}(\Delta T^c - \Delta T) - a_{12}N_{xx}}{a_{22}}\right) + \Delta T \bar{\alpha}_{xx}\right]x$$

This expression can be written as

$$u_o(x,y) = \left[N_{xx}\left\{a_{11} - \frac{a_{12}^2}{a_{22}}\right\} + \frac{a_{12}}{a_{22}}\bar{\alpha}_{yy}(\Delta T^c - \Delta T) + \bar{\alpha}_{xx}\Delta T\right]x \tag{11.19}$$

Equations 11.17 and 11.19 can be used to specify the displacement boundary conditions along all four edge of the plate. A summary of all boundary conditions associated with the problem considered in this section is

For $x = 0, a$:

$$w(0,y) = w(a,y) = 0 \tag{11.20a}$$

$$M_{xx}^*(0,y) = M_{xx}^*(a,y) = D_{11}\frac{\partial^2 w}{\partial x^2} + D_{12}\frac{\partial^2 w}{\partial y^2} = 0 \tag{11.20b}$$

$$N_{xx}^*(0,y) = N_{xx}^*(a,y) = A_{11}\frac{\partial u_o}{\partial x} + A_{12}\frac{\partial v_o}{\partial y} - N_{xx}^T = N_{xx} \tag{11.20c}$$

$$v_o(0,y) = v_o(a,y) = (\Delta T^c \overline{\alpha}_{yy})y \tag{11.20d}$$

For $y = 0,b$:

$$w(x,0) = w(x,b) = 0 \tag{11.21a}$$

$$M_{yy}^*(x,0) = M_{yy}^*(x,b) = D_{12}\frac{\partial^2 w}{\partial x^2} + D_{22}\frac{\partial^2 w}{\partial y^2} = 0 \tag{11.21b}$$

$$u_o(x,0) = u_o(x,b) = \left[N_{xx}\left\{a_{11} - \frac{a_{12}^2}{a_{22}}\right\} + \frac{a_{12}}{a_{22}}\overline{\alpha}_{yy}(\Delta T^c - \Delta T) + \overline{\alpha}_{xx}\Delta T \right]x \tag{11.21c}$$

$$v_o(x,0) = 0, \quad v_o(x,b) = (\Delta T^c \overline{\alpha}_{yy})b \tag{11.21d}$$

Now consider out-of-plane displacements $w(x,y)$. Guided by the functional form of the transverse pressure (i.e., Equation 11.16), we once again assume the out-of-plane displacement field is given by

$$w(x,y) = c\sin\left(\frac{\pi x}{a}\right)\sin\left(\frac{\pi y}{b}\right) \tag{11.22}$$

where c is an unknown constant. Substituting this assumed form into boundary conditions Equation 11.20a,b and 11.21a,b will reveal that they are identically satisfied for any value of c. Hence the value of constant c must be determined by enforcing the third equation of equilibrium. Substituting Equations 11.17, 11.19, and 11.22 into the third equation of equilibrium, Equation 11.1c, and solving for constant c, we find:

$$c = \cfrac{q_0}{\pi^4 \left[\begin{array}{l}[(D_{11}/a^4) + (2(D_{12} + 2D_{66})/a^2b^2) + (D_{22}/b^4) + (N_{xx}/\pi^2)] \\ \times ((1/a^2) - (A_{12}/A_{11}b^2)) + (\overline{\alpha}_{yy}(A_{11}A_{22} - A_{12}^2)/\pi^2b^2A_{11})(\Delta T^c - \Delta T)\end{array}\right]}$$

(11.23)

Using the definition of the plate aspect ratio, $R = a/b$, this result can also be written as

$$c = \cfrac{q_0 R^4 b^4}{\pi^4 \left[\begin{array}{l}D_{11} + 2R^2(D_{12} + 2D_{66}) + R^4 D_{22} + (R^4 b^2/\pi^2) \\ \left\{N_{xx}((1/R^2) + (A_{12}/A_{11})) + (\overline{\alpha}_{yy}(A_{11}A_{22} - A_{12}^2)/A_{11})(\Delta T^c - \Delta T)\right\}\end{array}\right]}$$

(11.24)

The predicted out-of-plane deflection is obtained by combining either Equation 11.23 or 11.24 with Equation 11.22. Using Equation 11.24, for example, we have:

$$w(x,y) = \cfrac{q_0 R^4 b^4 \sin(\pi x/a)\sin(\pi y/b)}{\pi^4 \left[\begin{array}{l}D_{11} + 2R^2(D_{12} + 2D_{66}) + R^4 D_{22} + (R^4 b^2/\pi^2) \\ \left\{N_{xx}((1/R^2) + (A_{12}/A_{11})) + (\overline{\alpha}_{yy}(A_{11}A_{22} - A_{12}^2)/A_{11})(\Delta T^c - \Delta T)\right\}\end{array}\right]}$$

(11.25)

Equations 11.17, 11.19, and 11.25 give the predicted displacement fields induced in the plate and represent the solution to this problem.

To summarize, we have found the displacement fields induced in a symmetric specially orthotropic plate subjected to a sinusoidally varying transverse load and a uniform change in temperature ΔT. The two edges $x = 0, a$ are subject to simple supports of type S2, whereas the two edges $y = 0, b$ are subject to simple supports of type S1. A typical application of this solution is discussed in Example Problem 11.3. It should be kept in mind that the possibility of buckling has not been considered. Buckling is a possibility if N_{xx} is compressive or if the change in temperature ΔT tends to cause the laminate to expand.

Example Problem 11.3

A $[(0_2/90)_2]_s$ graphite–epoxy laminate is cured at 175°C and then cooled to room temperature (20°C). After cooling, the flat laminate is trimmed to in-plane dimensions of 300 × 150 mm and mounted in an assembly that provides type S2 simple supports along the two edges $x = 0$ and $x = a$, and type S1 simple supports along the two edges $y = 0$ and $y = b$. The x-axis is defined parallel to the 300 mm edge (i.e., $a = 0.3$ m, $b = 0.15$ m). The laminate is then subjected to a uniform in-plane tensile loading N_{xx} along $x = 0,a$ and a transverse pressure given by $q(x,y) = 40 \sin (\pi x/a) \sin (\pi y/b)$ (kPa).

a. Plot the maximum out-of-plane displacement as a function of tensile load over the range $0 < N_{xx} < 70$ kN/m, assuming temperature remains constant at room temperature.
b. Plot the maximum out-of-plane displacement as a function of temperature, over the range $-50°C < T < 20°C$, assuming a constant in-plane tensile load.

$$N_{xx} = 50 \text{ kN/m}$$

Use the properties listed for graphite–epoxy in Table 3.1, and assume each ply has a thickness of 0.125 mm.

SOLUTION

A $[(0_2/90)_2]_s$ graphite–epoxy laminate was also considered in Example Problem 11.1, and numerical values for the $[ABD]$ matrix and effective thermal expansion coefficients are listed there. As before, the 12-ply laminate has a total thickness $t = 1.5$ mm and aspect ratio $R = a/b = 2.0$. Using these laminate stiffnesses, dimensions, and the specified transverse loading, Equations 11.25 becomes

$$w(x,y) = \left[\frac{324}{\pi^4 \left(278 + (0.099/\pi^2)N_{xx} + (82.5/\pi^2)(\Delta T^c - \Delta T) \right)} \right]$$
$$+ \sin\left(\frac{\pi x}{0.3} \right) \sin\left(\frac{\pi y}{0.15} \right) \text{(meters)}$$

This expression can be used to calculate the out-of-plane displacement induced at any point (x,y) over the surface of the plate.

a. The maximum out-of-plane displacement occurs at the center of the plate, that is, at $x = a/2 = 0.15$ m and $y = b/2 = 0.075$. Since the plate remains at room temperature $\Delta T = \Delta T^c$. Under these conditions the maximum out-of-plane displacement is given by

$$w|_{max} = \frac{324}{\pi^4 \left(278 + (0.099/\pi^2)N_{xx} \right)} \text{(meters)}$$

A plot of maximum out-of-plane displacement as a function of N_{xx} is shown in Figure 11.6a. As would be expected intuitively, the plate is stiffened as N_{xx} is increased. A maximum

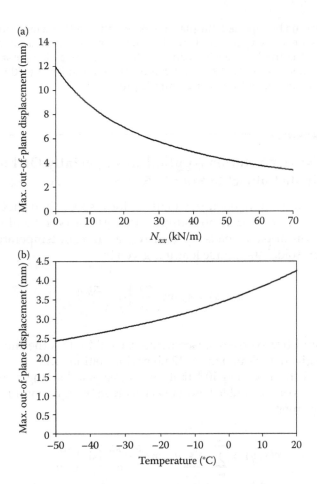

FIGURE 11.6
Out-of-plane displacements induced in the graphite–epoxy plate considered in Example Problem 11.3. (a) Maximum out-of-plane displacements at room temperature as a function of in-plane load N_{xx}; (b) maximum out-of-plane displacements as a function of temperature ($N_{xx} = 50$ kN/m).

deflection of 12 mm occurs when $N_{xx} = 0$, whereas the maximum deflection is reduced to 3.4 mm when $N_{xx} = 70$ kN/m.

b. As before, the maximum out-of-plane displacement occurs at $x = a/2 = 0.15$ m and $y = b/2 = 0.075$. Since a constant in-plane tensile load $N_{xx} = 50$ kN/m is applied, the maximum out-of-plane displacement is given by

$$w\big|_{max} = \frac{324}{\pi^4 \left[781 + (82.5/\pi^2)(\Delta T^c - \Delta T) \right]} \text{(meters)}$$

A plot of maximum out-of-plane displacement as a function of temperature is shown in Figure 11.6b. At room temperature (20°C), a maximum deflection of 4.3 mm is predicted. As would be expected, the plate becomes stiffer as the temperature is decreased. Out-of-plane displacements are decreased due to the in-plane tensile load N_{yy} that develops as temperature is decreased. At the lowest temperature considered (–50°C) a maximum deflection of 2.4 mm is predicted.

11.6 The Navier Solution Applied to a Specially Orthotropic Laminate Subject to Simple Supports of Type S4

In Section 11.4, we developed the solution for a specially orthotropic laminate subjected to homogeneous simple supports of type S4 along all four edges, uniform in-plane loads N_{xx} and N_{yy}, a uniform temperature change ΔT, and a sinusoidal transverse loading given by

$$q(x,y) = q_o \sin\left(\frac{\pi x}{a}\right)\sin\left(\frac{\pi y}{b}\right)$$

In-plane and transverse displacements caused by this thermomechanical loading are given by Equations 11.12a,b and Equation 11.15.

Now, recall from Section 10.4 that *any* transverse loading may be represented in terms of the double Fourier series given by Equation 10.14, repeated here for convenience:

$$q(x,y) = \sum_{m=1}^{\infty} \sum_{n=1}^{\infty} q_{mn} \sin\left(\frac{m\pi x}{a}\right)\sin\left(\frac{n\pi y}{b}\right)$$

Thus, any transverse loading can be viewed as the sum of a large number of sinusoidal load components. The displacements caused by an arbitrary transverse loading applied to a plate with simple supports of type S4 can, therefore, be obtained using the same approach as that used in Section 11.4. The only differences are that the transverse load is given by Equation 10.14 and the transverse deflections $w(x,y)$ are assumed to be of the form:

$$w(x,y) = \sum_{m=1}^{\infty} \sum_{n=1}^{\infty} c_{mn} \sin\left(\frac{m\pi x}{a}\right)\sin\left(\frac{n\pi y}{b}\right) \tag{11.26}$$

where c_{mn} are unknown constants to be determined using the equations of equilibrium. The assumed form and solution for in-plane displacements $u_o(x,y)$ and $v_o(x,y)$ remain unchanged and are given by Equations 11.12a,b. Substituting Equations 11.26 and Equation 10.14 into the third equation of equilibrium, Equation 11.1c, and equating coefficients (following the same procedure as used in Section 11.4), we obtain:

$$c_{mn} = \cfrac{q_{mn}a^4b^4}{\pi^4 \begin{bmatrix} D_{11}(mb)^4 + 2(D_{12} + 2D_{66})(mnab)^2 + D_{22}(na)^4 \\[2mm] + \left(\dfrac{ab}{\pi}\right)^2 \left\{N_{xx}(mb)^2 + N_{yy}(na)^2\right\} \end{bmatrix}} \qquad (11.27a)$$

Using the definition of the plate aspect ratio, $R = a/b$, this result can also be written as

$$c_{mn} = \cfrac{q_{mn}R^4b^4}{\pi^4 \begin{bmatrix} D_{11}m^4 + 2(D_{12} + 2D_{66})(mnR)^2 + D_{22}(nR)^4 \\[2mm] + (a^2/\pi^2)\left\{N_{xx}m^2 + N_{yy}(nR)^2\right\} \end{bmatrix}} \qquad (11.27b)$$

Substituting this result into Equation 11.26 completes the solution to the problem. To summarize, the midplane displacements induced in a symmetric specially orthotropic laminate subjected to an arbitrary transverse loading given by Equation 10.14, constant and uniform in-plane loads N_{xx} and N_{yy}, a uniform temperature change ΔT, and homogeneous simple supports of type S4 along all four edges are given by

$$u_o(x,y) = \left[\left(\frac{A_{22}N_{xx} - A_{12}N_{yy}}{A_{11}A_{22} - A_{12}^2}\right) + \bar{\alpha}_{xx}\Delta T\right]x \qquad (11.28a)$$

$$v_o(x,y) = \left[\left(\frac{A_{11}N_{yy} - A_{12}N_{xx}}{A_{11}A_{22} - A_{12}^2}\right) + \bar{\alpha}_{yy}\Delta T\right]y \qquad (11.28b)$$

$$w(x,y) = \frac{R^4b^4}{\pi^4}\sum_{m=1}^{\infty}\sum_{n=1}^{\infty} \cfrac{q_{mn}\sin(m\pi x/a)\sin(n\pi y/b)}{\begin{bmatrix} D_{11}m^4 + 2(D_{12} + 2D_{66})(mnR)^2 + D_{22}(nR)^4 \\[2mm] + (a^2/\pi^2)\left\{N_{xx}m^2 + N_{yy}(nR)^2\right\} \end{bmatrix}} \qquad (11.28c)$$

The method of using a double Fourier series expansion to represent the transverse loading and out-of-plane displacements was first suggested by Navier in about 1820, and is known as the Navier solution. An application of this technique is illustrated in Example Problem 11.4. The solution presented here is for a plate subjected to type S4 simple supports, but the same approach may be used to study simply supported plates with type S1 supports (as in Section 11.3), or plates with mixed simple supports (as in Section 11.5).

As previously mentioned, a second technique known as the Levy solution [1,3,4,5] can also be used to obtain exact solutions for symmetric specially orthotropic laminates. While the Navier solution requires that all four edges be simply supported, the Levy solution only requires that two opposite edges of the plate are simply supported; the boundary conditions on the remaining two edges are arbitrary and may be clamped or free, for example. The Levy solution technique will not be discussed in this book, and the interested reader is referred to the references cited for a discussion of this approach.

Example Problem 11.4

A $[(0_2/90)_2]_s$ graphite–epoxy laminate is cured at 175°C and then cooled to room temperature (20°C). After cooling, the flat laminate is trimmed to in-plane dimensions of 300×150 mm and mounted in an assembly that provides type S4 simple supports along all four edges. The *x*-axis is defined parallel to the 300 mm edge (i.e., $a = 0.3$ m, $b = 0.15$ m). The laminate is then subjected to a uniform in-plane tensile loading $N_{xx} = N_{yy} = 50$ kN/m and uniform transverse loading $q(x,y) = 100$ kPa. Temperature remains constant and no change in moisture content occurs ($\Delta M = 0$). Plot the out-of-plane displacements induced along the centerline defined by $y = 0.075$ m. Use the properties listed for graphite–epoxy in Table 3.1, and assume each ply has a thickness of 0.125 mm.

SOLUTION

A $[(0_2/90)_2]_s$ graphite–epoxy laminate was also considered in Example Problem 11.1, and numerical values for the $[ABD]$ matrix are listed there. As before, the 12-ply laminate has a total thickness $t = 1.5$ mm and aspect ratio $R = a/b = 2.0$.

The double Fourier series expansion of a uniform transverse loading was discussed in Section 10.4. The coefficients in the Fourier series expansion were found to be

$$q_{mn} = \frac{16q_o}{\pi^2 mn}, m, n = \text{odd integers}$$

Combining these coefficients with Equation 11.28c allows prediction of out-of-plane displacements. A plot of these displacements along the centerline of the plate defined by $y = b/2 = 0.075$ m is shown in Figure 11.7. Curves are shown based on a 1-term expansion (i.e., m, $n = 1$), a 4-term expansion (m, $n = 1, 3$), and a 9-term expansion (m, $n = 1, 3, 5$). As would be expected due to symmetry, out-of-plane displacement is maximum at the center of the plate. The solution converges rapidly. The maximum displacement predicted on the basis of a 1-, 4-, and 9-term expansion equals 5.27, 4.71, and 4.78 mm³, respectively. If 100 terms were used ($m, n = 1,3, \ldots, 19$) the maximum predicted displacement is 4.77 mm.

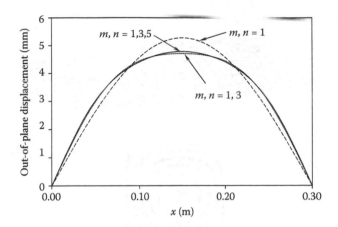

FIGURE 11.7
Out-of-plane displacements induced in the graphite–epoxy plate considered in Example Problem 11.4, along centerline $y = 0.075$ m.

11.7 Buckling of Rectangular Specially Orthotropic Laminates Subject to Simple Supports of Type S4

This section is devoted to the phenomenon known as plate "buckling." A brief discussion of what the term "buckling" means in the context of a thin plate is in order. Consider an initially flat symmetric laminate subjected to constant and uniform boundary edge loads $N_{xx}^{(-x)} = N_{xx}^{(+x)} = N_{xx}$ and $N_{yy}^{(-y)} = N_{yy}^{(+y)} = N_{yy}$ (shear resultants are assumed zero: $N_{xy} = N_{yx} = 0$). As we have seen in preceding sections, coupling stiffnesses $B_{ij} = 0$ for all symmetric laminates. Hence, according to our earlier analyses we would not expect these edge loads to cause out-of-plane displacements, since in-plane loads and out-of-plane displacements are (apparently) uncoupled for a symmetric laminate.

This statement is always true if both N_{xx} and N_{yy} are tensile; a flat symmetric laminate subjected to tensile edge loads N_{xx} and/or N_{yy} will remain flat regardless of the magnitude of load (unless, of course, the loads are high enough to cause fracture). However, suppose that either N_{xx} or N_{yy} (or both) are compressive. If the magnitude of the compressive load(s) is (are) relatively low, then the initially flat laminate remains flat, that is, only in-plane displacements $u_o(x,y)$ and $v_o(x,y)$ are induced and out-of-plane displacements remain zero: $w(x,y) = 0$. However, as the compressive load(s) is (are) increased the laminate may exhibit a sudden out-of-plane displacement, $w(x,y) \neq 0$. This is the phenomenon is known as "buckling."

The load level at which out-of-plane displacements initially occur is called the *critical buckling load*. In general, a thin plate does not collapse at the critical buckling load, and can support a further increase in load. However,

(a)

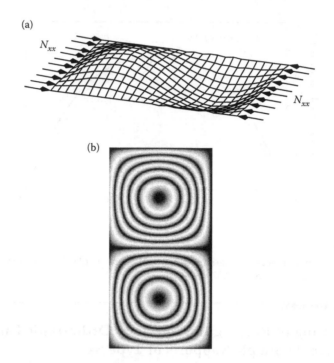

(b)

FIGURE 11.8
A typical mode [2,1] buckling mode, caused in a rectangular plate by a uniaxial compressive loading N_{xx}. (a) 3-D view. (b) 2-D view.

out-of-plane displacements increase rapidly as in-plane loads are increased beyond the initial buckling load, signaling imminent structural failure.

The out-of-plane displacements that occur at the onset of buckling exhibit a characteristic pattern. This characteristic pattern is called the *bucking mode*. A typical buckling mode is illustrated in Figure 11.8. A three-dimensional view of the buckling mode is shown in Figure 11.8a, while a 2-D view (which can be thought of as a "topographical map" of out-of-plane displacements) is shown in Figure 11.8b. In the discussion to follow predicted buckling modes will be illustrated in a form similar to Figure 11.8b.

The buckling mode exhibited by a given laminate depends on the stiffness of the laminate, plate aspect ratio, applied loading, and the boundary conditions. A particular buckling mode is described in terms of the number of points of relative maximum out-of-plane displacements in the x- and y-directions. For the buckling mode shown in Figure 11.8, there are two points of relative maximum displacement in the x-direction, whereas there is only one point of relative maximum displacement in the y-direction. Hence, the pattern shown in Figure 11.8 is called "mode [2,1]."

In this section, we consider buckling of an initially flat rectangular specially orthotropic laminate subjected to simple supports of type S4 and *uniform* edge loads $N_{xx}^{(-x)} = N_{xx}^{(+x)} = N_{xx}$ and $N_{yy}^{(-y)} = N_{yy}^{(+y)} = N_{yy}$, at least one of which is compressive. If the laminate was cured at an elevated temperature and then cooled to room temperature, then pre-existing thermal stress resultants associated with cooldown, N_{xx}^T and N_{yy}^T, are present. It is assumed that no transverse loading is applied and that no change moisture content occurs: $q(x,y) = \Delta M = 0$.

Since the applied edge loads are constant and uniform, *prior to buckling* the internal stress resultants at all points within the laminate are also constant and uniform, and are equal to the external edge loads. Hence, prior to buckling the internal stress resultants at all points within the laminate are uniform and given by

$$N_{xx}^*(x,y) = A_{11}\frac{\partial u_o}{\partial x} + A_{12}\frac{\partial v_o}{\partial y} - N_{xx}^T = N_{xx}^{(-x)} = N_{xx}^{(+x)} = N_{xx} \quad (11.29a)$$

$$N_{yy}^*(x,y) = A_{12}\frac{\partial u_o}{\partial x} + A_{22}\frac{\partial v_o}{\partial y} - N_{yy}^T = N_{yy}^{(-y)} = N_{yy}^{(+y)} = N_{yy} \quad (11.29b)$$

$$N_{xy}^*(x,y) = A_{66}\left(\frac{\partial u_o}{\partial y} + \frac{\partial v_o}{\partial x}\right) = 0 \quad (11.29c)$$

It is emphasized that Equations 11.29 are valid *prior to buckling*. After buckling has occurred and significant out-of-plane displacements have developed the internal stress and moment resultants may no longer be constant and uniform, and may not equal the applied edge loads.

The remaining type S4 simple support boundary conditions are:

Along $x = 0$:

$$w(0,y) = 0 \quad (11.30a)$$

$$M_{xx}^*(0,y) = D_{11}\frac{\partial^2 w(0,y)}{\partial x^2} + D_{12}\frac{\partial^2 w(0,y)}{\partial y^2} = 0 \quad (11.30b)$$

Along $x = a$:

$$w(a,y) = 0 \quad (11.31a)$$

$$M_{xx}^*(a,y) = D_{11}\frac{\partial^2 w(a,y)}{\partial x^2} + D_{12}\frac{\partial^2 w(a,y)}{\partial y^2} = 0 \quad (11.31b)$$

Along $y = 0$:

$$w(x,0) = 0 \quad (11.32a)$$

$$M_{yy}^*(x,0) = D_{12}\frac{\partial^2 w(x,0)}{\partial x^2} + D_{22}\frac{\partial^2 w(x,0)}{\partial y^2} = 0 \qquad (11.32b)$$

Along $y = b$:

$$w(x,b) = 0 \qquad (11.33a)$$

$$M_{yy}^*(x,b) = D_{12}\frac{\partial^2 w(x,b)}{\partial x^2} + D_{22}\frac{\partial^2 w(x,b)}{\partial y^2} = 0 \qquad (11.33b)$$

The equation of equilibrium governing out-of-plane displacements, Equation 11.1c, is (with $q(x,y) = \Delta M = 0$):

$$\left[A_{11}\frac{\partial u_o}{\partial x} + A_{12}\frac{\partial v_o}{\partial y} - N_{xx}^T\right]\left(\frac{\partial^2 w}{\partial x^2}\right) + \left[A_{12}\frac{\partial u_o}{\partial x} + A_{22}\frac{\partial v_o}{\partial y} - N_{yy}^T\right]\left(\frac{\partial^2 w}{\partial y^2}\right)$$

$$+ 2\left[A_{66}\left(\frac{\partial u_o}{\partial y} + \frac{\partial v_o}{\partial x}\right)\right]\left(\frac{\partial^2 w}{\partial x \partial y}\right) - D_{11}\frac{\partial^4 w}{\partial x^4} - 2(D_{12} + 2D_{66})\frac{\partial^4 w}{\partial x^2 \partial y^2}$$

$$-D_{22}\frac{\partial^4 w}{\partial y^4} = 0 \qquad (11.34)$$

Equations 11.29 are valid *up to the onset* of buckling, that is, we assume that internal stress resultants are constant and uniform as the buckling phenomenon is approached. Hence, substituting Equation 11.29 into Equation 11.34, we have:

$$D_{11}\frac{\partial^4 w}{\partial x^4} + 2(D_{12} + 2D_{66})\frac{\partial^4 w}{\partial x^2 \partial y^2} + D_{22}\frac{\partial^4 w}{\partial y^4} = N_{xx}\left(\frac{\partial^2 w}{\partial x^2}\right) + N_{yy}\left(\frac{\partial^2 w}{\partial y^2}\right) \qquad (11.35)$$

Notice that thermal stress and moment resultants do not appear in Equation 11.35. This reveals that the buckling response of symmetric specially orthotropic laminates is independent of temperature, *when subject to simple supports of type S4*. As will be discussed in Section 11.8, if the laminate were subject to boundary conditions that constrain in-plane displacements, for example simple supports of type S1, then buckling may be caused by a temperature change.

We now assume out-of-plane displacements are given by

$$w(x,y) = c\sin\left(\frac{m\pi x}{a}\right)\sin\left(\frac{n\pi y}{b}\right) \qquad (11.36)$$

where m and n are positive integers and c is an unknown constant representing the maximum out-of-plane deflection. It is easy to show that Equation 11.36 satisfies the boundary conditions Equations 11.30 through 11.33, so the next step is to evaluate the conditions under which the equation of equilibrium governing out-of-plane deflections is satisfied. Substituting Equation 11.36 into Equation 11.35 and simplifying, we find that in order for equilibrium to be maintained it is required that:

$$D_{11}\left(\frac{m\pi}{a}\right)^4 + 2\left(D_{12} + 2D_{66}\right)\left(\frac{m\pi}{a}\right)^2\left(\frac{n\pi}{b}\right)^2 + D_{22}\left(\frac{n\pi}{b}\right)^4 = -N_{xx}\left(\frac{m\pi}{a}\right)^2$$

$$- N_{yy}\left(\frac{n\pi}{b}\right)^2 \tag{11.37}$$

The unknown constant c has canceled out and does not appear in the requirement for equilibrium, Equation 11.37. Referring to Equation 11.36, note that integers m and n dictate the spatial variation of the out-of-plane displacement field, that is, m and n define how $w(x,y)$ varies with x and y, and consequently define the buckling mode. Constant c represents the magnitude of $w(x,y)$. Hence, while we are able to determine m and n and thus predict the buckling mode on the basis of Equation 11.37, we cannot predict the *magnitude* of out-of-plane displacements.

Although Equation 11.37 is valid for any combination of N_{xx} and N_{yy}, it is convenient to rearrange Equation 11.37 for three different loading conditions. As explained above, N_{xx} and/or N_{yy} must be compressive to cause buckling. Assume for the moment that a constant transverse tension $N_{yy} \geq 0$ is applied, in which N_{xx} case must be compressive to cause buckling. Solving Equation 11.37 for N_{xx}, we find:

$$N_{xx} = \frac{-\pi^2}{m^2 a^2 b^4}\left[D_{11}\left(mb\right)^4 + 2\left(D_{12} + 2D_{66}\right)\left(mnab\right)^2 + D_{22}\left(na\right)^4 + N_{yy}\left(\frac{na^2b}{\pi}\right)^2\right]$$

$$\tag{11.38}$$

Using the definition of the plate aspect ratio, $R = a/b$, this result can also be written as

$$N_{xx} = \frac{-\pi^2}{\left(ma\right)^2}\left[D_{11}m^4 + 2\left(D_{12} + 2D_{66}\right)\left(mnR\right)^2 + D_{22}\left(nR\right)^4 + N_{yy}\left(\frac{naR}{\pi}\right)^2\right]$$

$$\tag{11.39}$$

Hence, given some value for $N_{yy} \geq 0$ and assumed integer values for m and n, either Equation 11.38 or 11.39 can be used to calculate a corresponding value for N_{xx}. Since m and n can be any combination of positive integers, there are an infinite number of values for N_{xx} that satisfy Equation 11.38 or 11.39. The critical buckling load, denoted N_{xx}^c, corresponds to the particular combination of m and n that leads to the value of N_{xx} with lowest magnitude. The combination of m and n that correspond to this lowest load define the predicted critical buckling mode.

The following observations are based on inspection of Equation 11.38 or 11.39. Since we have assumed for the moment that $N_{yy} \geq 0$, all variables that appear on the right side of the equality sign are algebraically positive. Hence, N_{xx}^c must be algebraically negative, that is, the critical buckling load N_{xx}^c is predicted to be compressive, as would be expected. Second, the minimum magnitude of N_{xx}^c will always correspond to $n = 1$, since $N_{yy} \geq 0$. Finally, note that if the constant transverse load N_{yy} is increased then the magnitude of N_{xx}^c is increased. That is, a transverse tension will cause an increase in the critical buckling load.

As a second loading condition of interest, let us now assume that a constant $N_{xx} \geq 0$ is applied, and that a compressive load in the y-direction causes buckling. Solving Equation 11.37 for N_{yy}, we find:

$$N_{yy} = \frac{-\pi^2}{n^2 a^4 b^2}\left[D_{11}(mb)^4 + 2(D_{12} + 2D_{66})(mnab)^2 + D_{22}(na)^4 + N_{xx}\left(\frac{mab^2}{\pi}\right)^2 \right]$$

(11.40)

Using the plate aspect ratio, $R = a/b$, this result can be written as

$$N_{yy} = \frac{-\pi^2}{(nbR^2)^2}\left[D_{11}m^4 + 2(D_{12} + 2D_{66})(mnR)^2 + D_{22}(nR)^4 + N_{xx}\left(\frac{ma}{\pi}\right)^2 \right]$$

(11.41)

As before, m and n can be any combination of positive integers, and, therefore, there are an infinite number of values for N_{yy} that satisfy Equation 11.40 or 11.41. The critical buckling load N_{yy}^c corresponds to the particular combination of m and n that leads to the value of N_{yy} with lowest magnitude, and the combination of m and n that correspond to this load define the predicted buckling mode. Since it has been assumed that $N_{xx} \geq 0$, $m = 1$ in all cases, and if the constant tensile load N_{xx} is increased the magnitude of the critical buckling load N_{yy}^c will increase.

Finally, consider a third loading condition in which buckling is caused by the simultaneous increase in both N_{xx} and N_{yy}. Further, assume the two loads are linearly related. That is, assume $N_{yy} = kN_{xx}$, where k is a known constant. Substituting this relation into Equation 11.37 and solving for N_{xx}, we find:

$$N_{xx} = -\frac{\pi^2}{(ab)^2} \frac{\left[D_{11}(mb)^4 + 2(D_{12} + 2D_{66})(mnab)^2 + D_{22}(na)^4 \right]}{\left[(mb)^2 + (kna)^2 \right]}$$ (11.42)

As before, m and n can be any combination of positive integers, and, therefore, there are an infinite number of values for N_{xx} that satisfy Equation 11.42. The critical buckling condition, defined by the two simultaneous loads N_{xx}^c and $N_{yy}^c = kN_{xx}^c$, corresponds to the particular combination of m and n that leads to the value of N_{xx} (and N_{yy}) with lowest magnitude. The combination of m and n that correspond to this load condition defines the predicted buckling mode. Note that if $k < 0$, then N_{xx} and N_{yy} are of opposite algebraic signs. This implies that there are two distinct buckling load conditions, one in which ($N_{xx} < 0$, $N_{yy} > 0$), and a second in which ($N_{xx} > 0$, $N_{yy} < 0$).

Numerical examples illustrating buckling predictions for symmetric specially orthotropic simply supported laminates will be discussed in Example Problems 11.5 and 11.6. Two important concluding comments are made in passing. First, from the preceding discussion the reader may have inferred that buckling is only caused by N_{xx}, by N_{yy}, or by some combination thereof. This is not the case. Other forms of loading may cause buckling, for example a shear load N_{xy}. In this chapter only buckling caused by N_{xx} and/or N_{yy} is considered. The reader interested in buckling caused by other types of loading is referred to [4,5].

Second, recall that the analysis presented above does not allow prediction of the magnitude of out-of-plane displacements induced at buckling. Buckling of thin plates is rarely catastrophic, and in general finite out-of-plane displacements occur at and above the critical buckling load. That is, a thin plate does not collapse once the buckling load is reached and can support a further increase in load, albeit at a reduced level of stiffness. Methods to predict the magnitude of out-of-plane displacement for load levels at or above the initial buckling load are known as "post-buckling" analyses, and are not discussed in this book.

Example Problem 11.5

A $[(0_2/90)_2]_s$ graphite–epoxy laminate is cured at 175°C and then cooled to room temperature (20°C). After cooling the flat laminate is trimmed to in-plane dimensions of 300×150 mm and mounted in an assembly that provides type S4 simple supports along all four edges. The x-axis is defined parallel to the 300 mm edge (i.e., $a = 0.3$ m, $b = 0.15$ m).

    a. Predict the critical buckling load N_{xx}^c and mode for this laminate, if $0 \le N_{yy} \le 400$ kN/m.

    b. Predict the critical buckling load N_{yy}^c and mode for this laminate, if $0 \le N_{xx} \le 400$ kN/m.

Use the properties listed for graphite–epoxy in Table 3.1, and assume each ply has a thickness of 0.125 mm.

SOLUTION

A $[(0_2/90)_2]_s$ graphite–epoxy laminate was also considered in Example Problem 11.1, and numerical values for the $[ABD]$ matrix are listed there. As before, the 12-ply laminate has a total thickness $t = 1.5$ mm and aspect ratio $R = a/b = 2.0$.

a. It is noted that $n = 1$, since $N_{yy} \geq 0$. Equation 11.39 becomes in this case:

$$N_{xx} = \frac{-\pi^2}{0.09m^2}\left[40.1m^4 + 65.34m^2 + 172.8 + N_{yy}\left(\frac{0.6}{\pi}\right)^2\right]$$

A plot of the critical buckling load for $0 \leq N_{yy} \leq 400$ kN/m is presented in Figure 11.9a. As expected, N_{xx}^c is increased as N_{yy} is increased. A change in buckling mode also occurs as N_{yy} is increased. The plate buckles in mode [2,1] over the range $0 \leq N_{yy} < 35$ kN/m, in mode [3,1] over the range 35 kN/m $\leq N_{yy} < 150$ kN/m, and in mode [4,1] over the range 150 kN/m $\leq N_{yy} < 400$ kN/m. These buckling modes are illustrated in Figures 11.9b,c,d, respectively.

b. It is noted that $m = 1$, since $N_{xx} \geq 0$. Equation 11.41 becomes in this case:

$$N_{yy} = \frac{-\pi^2}{0.36n^2}\left[40.1 + 65.34n^2 + 172.8(n)^4 + N_{xx}\frac{0.09}{\pi^2}\right]$$

A plot of the critical buckling load for $0 \leq N_{xx} \leq 400$ kN/m is presented in Figure 11.10a. As expected, N_{yy}^c is increased as N_{xx} is increased. A change in buckling mode also occurs as N_{xx} is increased. The plate buckles in mode [1,1] over the range $0 \leq N_{xx} < 72$ kN/m, and in mode [1,2] over the range 72 kN/m $\leq N_{xx} < 400$ kN/m. These buckling modes are illustrated in Figures 11.10b,c, respectively.

Note that the magnitudes of N_{yy}^c calculated in part (b) are far lower than those calculated for N_{xx}^c in part (a). This pronounced difference is largely due to the stacking sequence involved. For the $[(0_2/90)_2]_s$ laminate under consideration, 8 of 12 plies are $0°$ plies (i.e., plies with fibers parallel to the x-axis). Hence, the resistance to buckling due to a compressive load N_{xx} is far higher than resistance to buckling due to a compressive N_{yy}.

Example Problem 11.6

A structure is being designed that will involve a $[(0_2/90)_2]_s$ graphite–epoxy laminate with a width (in the y-direction) of 150 mm. The length of the panel (in the x-direction) has not yet been established, and could be anywhere from 150 to 750 mm². During service the panel will be subjected to a compressive load N_{xx} (only), and simple supports of type S4 along all four edges. Buckling is, therefore, of concern. Predict the buckling load and mode for the panel, for any panel length ranging from 150 to 750 mm².

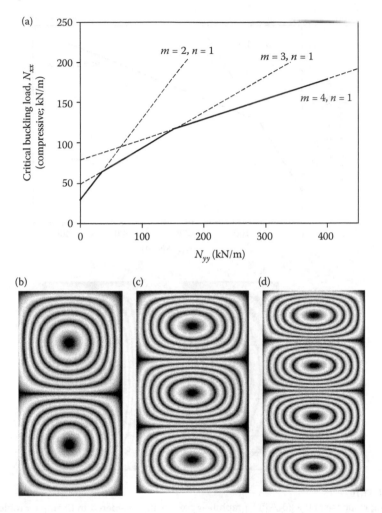

FIGURE 11.9

Buckling response of the $[(0_2/90)_2]_s$ graphite–epoxy plate considered in Example Problem 11.5. (a) Critical buckling load N_{xx}^c over the range $0 \le N_{yy} \le 400$ kN/m; (b) buckling mode for $0 \le N_{yy} < 35$ kN/m; (c) buckling mode for 35 kN/m $< N_{yy} < 150$ kN/m; (d) buckling mode for 150 kN/m $< N_{yy} < 400$ kN/m.

SOLUTION

A $[(0_2/90)_2]_s$ graphite–epoxy laminate was also considered in Example Problem 11.1, and numerical values for the $[ABD]$ matrix are listed there. As before, the 12-ply laminate has a total thickness $t = 1.5$ mm. Buckling loads and modes will be predicted using Equation 11.39. According to the problem statement, $b = 0.15$ m, and 0.15 m $< a < 0.75$ m. The plate aspect ratio, therefore, varies over $1 \le R \le 5$. Since transverse loading is zero ($N_{yy} = 0$), $n = 1$ and Equation 11.39 becomes

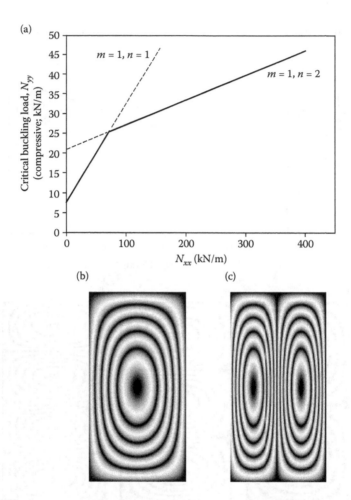

FIGURE 11.10
Buckling response of the $[(0_2/90)_2]_s$ graphite–epoxy plate considered in Example Problem 11.5. (a) Critical buckling load N_{yy}^c over the range $0 \leq N_{xx} \leq 400$ kN/m; (b) buckling mode for $0 \leq N_{xx} < 72$ kN/m; (c) buckling mode for 72 kN/m $< N_{xx} < 400$ kN/m.

$$N_{xx} = \frac{-\pi^2}{(ma)^2}\left[40.1m^4 + 16.34(mR)^2 + 10.8R^4\right]$$

A plot of the predicted critical buckling load over the specified range in aspect ratio is presented in Figure 11.11. The buckling mode is predicted to increase as the aspect ratio increases: mode [1,1] is predicted over the range $1 < R < 1.96$, mode [2,1] is predicted over the range $1.96 < R < 3.40$, mode [3,1] is predicted for $3.40 < R < 4.81$, and mode [4,1] is predicted for $4.81 < R < 5.00$. Still higher buckling modes would occur at higher aspect ratios.

The predicted buckling load generally decreases with aspect ratio, although a local maximum in the buckling load occurs at each

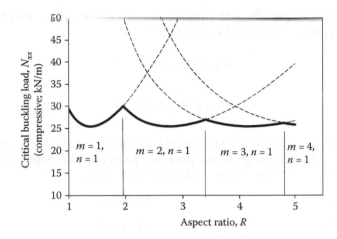

FIGURE 11.11
Predicted buckling loads and modes as a function of aspect ratio for the $[(0_2/90)_2]_s$ graphite–epoxy panel considered in Example Problem 11.6.

aspect ratio corresponding to a change in mode shape. At an aspect ratio $R = 1$ (i.e., for a square plate) buckling is predicted to occur at $N_{xx}^c = 29.5 \, \text{kN/m}$.

11.8 Thermal Buckling of Rectangular Specially Orthotropic Laminates Subject to Simple Supports of Type S1

Buckling caused by direct application of uniform external edge loads N_{xx} and/or N_{yy} was considered in the Section 11.7. The analysis was conducted for a specially orthotropic laminate subjected to inhomogeneous simple supports of type S4. The initially flat symmetric laminate was subjected to uniform boundary edge loads $N_{xx}^{(-x)} = N_{xx}^{(+x)}$ and $N_{yy}^{(-y)} = N_{yy}^{(+y)}$. Since we specified in-plane edge loads, we did not specify in-plane edge displacements. In essence, edge displacements were allowed to vary with changes in the specified edge loads.

We now wish to consider buckling caused by a different mechanism. Namely, we wish to consider buckling caused by a change in temperature. In this case, we will specify inhomogeneous simple supports of type S1. That is, we will specify in-plane edge displacements, but will not specify in-plane edge loads. These boundary conditions were also considered in Section 11.3. As was discussed there, we assume that pre-existing midplane displacements (as well as ply strains) are induced *prior to assembly of the laminate in the simply supported configuration*. That is, midplane displacements are induced during cooling from an elevated cure temperature to room temperature.

For a symmetric specially orthotropic laminate the midplane displacements caused by cooling are given by Equation 11.5:

$$u_o^c(x,y) = \left(\overline{\alpha}_{xx}\Delta T^c\right)x$$

$$v_o^c(x,y) = \left(\overline{\alpha}_{yy}\Delta T^c\right)y$$

where $\overline{\alpha}_{xx}$ and $\overline{\alpha}_{yy}$ are the effective thermal expansion coefficients of the laminate and ΔT^c is the change in temperature from cure to room temperature. The boundary conditions along all four edges are therefore:

For $x = 0$:

$$w(0,y) = 0$$

$$M_{xx}^*(0,y) = D_{11}\frac{\partial^2 w}{\partial x^2} + D_{12}\frac{\partial^2 w}{\partial y^2} = 0$$

$$u_o(0,y) = u_o^{(-x)}(y) = 0$$

$$v_o(0,y) = v_o^{(-x)}(y) = \left(\overline{\alpha}_{yy}\Delta T^c\right)y$$

For $y = 0$:

$$w(x,0) = 0$$

$$M_{yy}^*(x,0) = D_{11}\frac{\partial^2 w}{\partial x^2} + D_{12}\frac{\partial^2 w}{\partial y^2} = 0$$

$$u_o(x,0) = u_o^{(-y)}(x) = \left(\overline{\alpha}_{xx}\Delta T^c\right)x$$

$$v_o(x,0) = v_o^{(-y)}(x) = 0$$

For $x = a$:

$$w(a,y) = 0$$

$$M_{xx}^*(a,y) = D_{11}\frac{\partial^2 w}{\partial x^2} + D_{12}\frac{\partial^2 w}{\partial y^2} = 0$$

$$u_o(a,y) = u_o^{(+x)}(y) = \left(\overline{\alpha}_{xx}\Delta T^c\right)a$$

$$v_o(0,y) = v_o^{(+x)}(y) = \left(\overline{\alpha}_{yy}\Delta T^c\right)y$$

For $y = b$:

$$w(x,b) = 0$$

$$M_{yy}^*(x,b) = D_{11}\frac{\partial^2 w}{\partial x^2} + D_{12}\frac{\partial^2 w}{\partial y^2} = 0$$

$$u_o(x,b) = u_o^{(+y)}(x) = \left(\overline{\alpha}_{xx}\Delta T^c\right)x$$

$$v_o(x,b) = v_o^{(+y)}(x) = \left(\overline{\alpha}_{yy}\Delta T^c\right)b$$

After assembly, we assume the laminate is subjected to a uniform change in temperature. The change in temperature is referenced to the strain-free temperature (assumed to be the cure temperature) and is represented by ΔT = (current temperature) − (cure temperature). If the laminate was not simply supported and instead was free to expand or contract then a uniform change in temperature would simply cause a change in midplane displacements and no external edge loads would result. Since the laminate is instead subject to simple supports of type S1, no changes in midplane displacements are allowed to occur and external edge loads develop as temperature changes. Depending on the magnitude of the temperature change, these thermally induced edge loads may cause the laminate to buckle.

Although the edge loads are thermally induced, they are external mechanical loads nevertheless and consequently the analysis presented in Section 11.7 is still applicable. The onset of thermal buckling occurs if Equation 11.35 is satisfied, repeated here for convenience:

$$D_{11}\left(\frac{m\pi}{a}\right)^4 + 2(D_{12} + 2D_{66})\left(\frac{m\pi}{a}\right)^2\left(\frac{n\pi}{b}\right)^2 + D_{22}\left(\frac{n\pi}{b}\right)^4 = N_{xx}\left(\frac{m\pi}{a}\right)^2 N_{yy}\left(\frac{n\pi}{b}\right)^2$$

In the present case, loads N_{xx} and N_{yy} are caused by a deviation from room temperature, and are given by

$$N_{xx} = (\Delta T^c - \Delta T)(A_{11}\bar{\alpha}_{xx} + A_{12}\bar{\alpha}_{yy}) \tag{11.43a}$$

$$N_{yy} = (\Delta T^c - \Delta T)(A_{12}\bar{\alpha}_{xx} + A_{22}\bar{\alpha}_{yy}) \tag{11.43b}$$

Substituting Equations 11.43a,b into Equation 11.35, solving for the temperature difference, and using the definition of the plate aspect ratio, $R = a/b$, we obtain:

$$(\Delta T^c - \Delta T) = \frac{-\pi^2}{b^2 R^4}\left[\frac{D_{11}m^4 + 2(D_{12} + 2D_{66})(mnR)^2 + D_{22}(nR)^4}{n^2(A_{12}\bar{\alpha}_{xx} + A_{22}\bar{\alpha}_{yy}) + \left(\frac{m}{R}\right)^2(A_{11}\bar{\alpha}_{xx} + A_{12}\bar{\alpha}_{yy})}\right] \tag{11.44a}$$

It is convenient to express the temperature differences involved as

$$\Delta T^c - \Delta T = T_{RT} - T$$

where
T_{RT} = room temperature (i.e., the temperature at which the laminate was mounted in the simple support fixture)
T = current temperature

With this change in notation, we have:

$$T = \frac{\pi^2}{b^2 R^4}\left[\frac{D_{11}m^4 + 2(D_{12} + 2D_{66})(mnR)^2 + D_{22}(nR)^4}{n^2(A_{12}\bar{\alpha}_{xx} + A_{22}\bar{\alpha}_{yy}) + \left(\frac{m}{R}\right)^2(A_{11}\bar{\alpha}_{xx} + A_{12}\bar{\alpha}_{yy})}\right] + T_{RT} \tag{11.44b}$$

Equations 11.44a and b are entirely equivalent, and both are based on the assumption that the simply supported laminate is assembled at room temperature, T_{RT}. In practice, it is conceptually simplest to use Equation 11.44b. By specifying a particular laminate and room temperature, all variables on the right side of the equality sign in Equation 11.44b are known, except for variable m and n. Since m and n can be any combination of positive integers, there are an infinite number of values of temperature T that will satisfy Equation 11.44b. The temperature at which thermal buckling will occur, denoted T_{bk}, corresponds to the particular combination of m and n that leads to the value of T of lowest magnitude. The combination of m and n that correspond to this lowest temperature define the predicted critical thermal buckling mode.

Now, based on the results of Example Problem 11.6, one might anticipate that the thermal buckling mode exhibited would vary as a function of aspect ratio. That is, based on earlier analyses one would expect that the values of m and n that correspond to the critical thermal buckling load would vary with aspect ratio. Equations 11.44a,b allow for such dependence. However, if physically reasonable material properties are used during the evaluation of Equations 11.44a,b, then numerical experiments show that $m = n = 1$ in all cases. Thus, a thermal buckling mode [1,1] is always predicted when realistic material properties are used (at least for the simply supported boundary conditions considered in this section), regardless of the aspect ratio or stacking sequence considered.

Assuming that this observation always holds true and that $m = n = 1$, Equation 11.44b reduces to

$$T_{bk} = \frac{\pi^2}{b^2 R^4}\left[\frac{D_{11} + 2(D_{12} + 2D_{66})R^2 + D_{22}R^4}{(A_{12}\overline{\alpha}_{xx} + A_{22}\overline{\alpha}_{yy}) + \frac{1}{R^2}(A_{11}\overline{\alpha}_{xx} + A_{12}\overline{\alpha}_{yy})}\right] + T_{RT}$$

This latter result shows that the thermal buckling temperature decreases with an increase in aspect ratio. In the limit (i.e., as $R \to \infty$) the thermal buckling temperature becomes

$$T_{bk} = \frac{\pi^2 D_{22}}{b^2(A_{12}\overline{\alpha}_{xx} + A_{22}\overline{\alpha}_{yy})} + T_{RT}$$

Example Problem 11.7

Two $[(0_2/90)_2]_s$ graphite–epoxy laminates are cured at 175°C and then cooled to room temperature (20°C). After cooling one laminate is trimmed to in-plane dimensions of 300×150 mm², whereas the second is trimmed to in-plane dimensions of 3000×150 mm². Both laminates are then mounted in assemblies that provide type S1 simple supports along all four edges, and subjected to a uniform increase in temperature.

Determine the temperature at which each plate will buckle. Use the properties listed for graphite–epoxy in Table 3.1, and assume each ply has a thickness of 0.125 mm.

SOLUTION

A $[(0_2/90)_2]_s$ graphite–epoxy laminate was also considered in Example Problem 11.1, and numerical values for the $[ABD]$ matrix and effective thermal expansion coefficients are listed there. The aspect ratios involved in this problem are:

$$R = 300 \text{ mm}/150 \text{ mm} = 2$$

and

$$R = 3000 \text{ mm}/150 \text{ mm} = 20$$

The temperature at which thermal buckling is predicted to occur is obtained through application of Equation 11.44b. It is found that both laminates are predicted to buckle in mode [1,1]. For the laminate with aspect $R = 2$ thermal buckling is predicted to occur when temperature is raised to 51°C, whereas for the laminate with aspect ratio $R = 20$ thermal buckling is predicted at a temperature of 40°C.

11.9 Computer Program *SPORTHO*

The computer program *SPORTHO* is provided with this book to supplement the material presented in this chapter. This program can also be downloaded at no cost from the following website:

http://depts.washington.edu/amtas/computer.html

Program *SPORTHO* is applicable to symmetric specially orthotropic laminates. The program can be used to calculate the transverse deflections according to the analyses presented in Sections 11.3 through 11.6, to calculate buckling loads according to the analysis presented in Section 11.7, or to calculate the temperature at which thermal buckling will occur, according to the analysis presented in Section 11.8. The user is prompted to input all information necessary to perform these calculations. Properties of up to five different materials may be defined. Numerical values must be defined using a consistent set of units in all cases. For example, the user must input elastic moduli for the composite material system(s) of interest. Using the properties listed in Table 3.1 and based on the SI system of units, the following numerical values would be input for graphite–epoxy:

$$E_{11} = 170 \times 10^9 \text{Pa}, \quad E_{22} = 10 \times 10^9 \text{Pa}, \quad v_{12} = 0.30, \quad G_{12} = 13 \times 10^9 \text{Pa}$$

Structural Analysis of Polymeric Composite Materials

Since 1 Pa = 1 N/m², all other lengths must be input in meters. For example, ply thicknesses must be input in meters (not millimeters). A typical value would be $t_k = 0.000125$ m (corresponding to a ply thickness of 0.125 mm). In-plane plate dimensions must also be input in meters.

If the English system of units were used, then the following numerical values would be input for the same graphite–epoxy material system:

$$E_{11} = 25.0 \times 10^6\,\text{psi}, \quad E_{22} = 1.5 \times 10^6\,\text{psi}, \quad v_{12} = 0.30, \quad G_{12} = 1.9 \times 10^6\,\text{psi}$$

In this case, all lengths would be input in inches.

References

1. Lekhnitskii, S.G., 1968. *Anisotropic Plates*, translated by S.W. Tsai and T. Cheron, Gordan and Breach Pub. Co., ISBN 0-677-20670-4.
2. Savin, G.N., 1961. *Stress concentration Around Holes*, Pergamon Press, New York.
3. Timoshenko, S., and Woinowsky-Krieger, S., 1987. *Theory of Plates and Shells*, McGraw-Hill Book Co., New York, ISBN 0-07-0647798.
4. Whitney, J.M., 1987. *Structural Analysis of Laminated Anisotropic Plates*, Technomic Pub. Co., Lancaster, PA, ISBN 87762-518-2.
5. Turvey, G.J., and Marshall, I.H., eds, 1995. *Buckling and Postbuckling of Composite Plates*, Chapman & Hall, New York.

12

Some Approximate Solutions for Symmetric Laminates

12.1 Preliminary Discussion

The equations of equilibrium for a thin symmetric composite laminate were derived in Section 10.2 based on a summation of forces and moments. Boundary conditions consistent with thin-plate theory were then discussed in Section 10.3. It turns out that exact solutions to these equations and boundary conditions can only be obtained if $A_{16} = A_{26} = D_{16} = D_{26} = N_{xy}^T = N_{xy}^M = 0$, that is, exact solutions are only available for specially orthotropic laminates. A few exact solutions for simply supported and symmetric specially orthotropic laminates were presented in Chapter 11. Unfortunately, many stacking sequences widely used in practice are not specially orthotropic. For example, symmetric quasi-isotropic laminates are not specially orthotropic, since D_{16}, $D_{26} \neq 0$ for this stacking sequence. Hence, the solutions presented in Chapter 11 are not rigorously valid for many laminates encountered in practice. Fortunately, approximate numerical solutions are available that are suitable for use with any laminate, including symmetric quasi-isotropic laminates. A brief introduction to these approximate solutions techniques will be presented in this chapter. Readers interested in a more detailed discussion are referred to References 1, 2.

It is appropriate to clarify the distinction between "exact" and "approximate" solutions in the present context. In particular, confusion may arise in that "exact" solutions often involve a series with an infinite number of terms (as in the Navier solutions presented in Section 11.6, for example). A numerical evaluation of these solutions will contain some level of error, since a finite number of terms must obviously be used during any practical application. However, the source of error in these cases arises from an inability to describe the applied loading in an exact analytical sense, rather than the solution itself. Further, in a practical application the number of terms used can be increased to obtain numerical results that satisfy all boundary conditions and the equations of equilibrium to any desired number of significant figures.

Structural Analysis of Polymeric Composite Materials

The distinction between an "exact" and "approximate" solution does not refer to the need to use a finite number of terms during a numerical evaluation. Rather, the distinction refers to the rigor with which the *boundary conditions* are satisfied. Recall from Section 10.3 that it is necessary to specify four boundary conditions along each edge of a thin plate. Further recall that any boundary condition can be classified as either a geometric boundary condition or a static (also called a "natural") boundary condition. Geometric boundary conditions are those that dictate some feature of midplane displacements along a plate edge, whereas static boundary conditions are those that dictate some external load applied along a plate edge. Many common edge conditions consist of a combination of geometric and static boundary conditions. For example, a simply supported edge is one in which out-of-plane deflection is zero along the edge (a geometric condition) and the bending moment is zero along the edge (a static condition); the remaining two conditions may be either geometric or static. Consequently there are four "types" of simply supported edges (see Section 10.3.3.2).

Now, during derivation of an "exact" solution all four boundary conditions along each edge of the plate are accounted for during derivation of the solution, and are satisfied exactly. For example, in the solutions for simply supported specially orthotropic plates discussed in Chapter 11 all geometric and static boundary conditions are satisfied exactly by the solutions presented. In contrast, in an "approximate" solution *only the geometric boundary conditions are specified and enforced directly*. Static boundary conditions are not enforced directly. Therefore, solutions obtained using approximate analysis techniques may, or may not, satisfy the static boundary conditions.

The approximate solutions to be discussed in this chapter are based on the *principle of minimum potential energy*. Solution techniques that follow from this principle are known as *energy methods*. Consider a solid body subjected to some external loading. Briefly stated, the principle of minimum potential energy states that, for all possible displacement fields that satisfy given boundary conditions, the displacement field that actually occurs is one in which the potential energy of the solid body assumes a minimum value. This statement is "exact." In fact, it is possible to re-derive the equations of equilibrium and associated boundary conditions (and hence all of the results presented in Chapters 10 and 11) based on this principle.

Energy methods will not be used to re-derive the equations of equilibrium and associated boundary conditions in this textbook; the reader interested in this derivation is referred to the text by Whitney [1]. Rather, the reason energy methods are of interest here is that they form the basis for approximate numerical solution techniques that are based on the principle of minimum potential energy. The two most common approximate numerical techniques are the *Ritz method* and the *Galerkin method*. The Ritz method will be developed and applied to several problems in this Chapter. The Galerkin method will not be discussed, and the reader interested in this method is referred to References 1, 2.

The essential elements of the Ritz method are as follows. The potential energy of a solid body is denoted Π. The mathematical form of Π depends on details of the problem under consideration and will be discussed in later sections. At this point suffice it to say that the potential energy Π of a solid body can be calculated if (a) the elastic properties of the body are known, (b) the external forces applied to the body are known and (c) the resulting displacement fields induced in the body are known. During a typical structural analysis the elastic properties and external forces are known, *but the displacement fields are not* (in fact, the objective during a structural analysis is often to *determine* the displacement fields induced by some specified loading). During application of the Ritz method it is assumed, in effect, that the *functional form* of the displacement fields is known. For general thin-plate problems, the midplane displacement fields are assumed to be of the form

$$u_o(x,y) = \sum_{m=1}^{M_1} \sum_{n=1}^{N_1} a_{mn} U_{mn}(x,y) \tag{12.1a}$$

$$v_o(x,y) = \sum_{m=1}^{M_2} \sum_{n=1}^{N_2} b_{mn} V_{mn}(x,y) \tag{12.1b}$$

$$w(x,y) = \sum_{m=1}^{M_3} \sum_{n=1}^{N_3} c_{mn} W_{mn}(x,y) \tag{12.1c}$$

where a_{mn}, b_{mn}, and c_{mn} are unknown constants, and U_{mn}, V_{mn}, and W_{mn} are known functions that vary over x and y. In general, the number of terms used to describe each displacement field may differ (e.g., M_1 does not necessarily equal M_2 or M_3). Also, the number of terms used to describe the variation in x and y may differ (e.g., M_1 does not necessarily equal N_1).

Having assumed the functional form for the displacement fields as represented by Equations 12.1, then the potential energy Π of an elastic plate subjected to specified external loading can be calculated. Note that the values of constants a_{mn}, b_{mn}, and c_{mn} in Equations 12.1 effectively define the *magnitudes* of the displacement fields. Hence, as constants a_{mn}, b_{mn}, and/or c_{mn} are increased or decreased (and assuming that elastic properties and external forces remain constant), the potential energy of the body is increased or decreased accordingly. The principle of minimum potential energy states that the displacement field actually adopted by a solid body is one in which the potential energy is a minimum. This condition is, therefore, defined by the following criteria:

$$\frac{\partial \Pi}{\partial a_{mn}} = 0 \qquad \begin{cases} m = 1, \ 2,..., M_1 \\ n = 1, \ 2,..., N_1 \end{cases} \qquad (12.2a)$$

$$\frac{\partial \Pi}{\partial b_{mn}} = 0 \qquad \begin{cases} m = 1, \ 2,..., M_2 \\ n = 1, \ 2,..., N_2 \end{cases} \qquad (12.2b)$$

$$\frac{\partial \Pi}{\partial c_{mn}} = 0 \qquad \begin{cases} m = 1, \ 2,..., M_3 \\ n = 1, \ 2,..., N_3 \end{cases} \qquad (12.2c)$$

Equations 12.2 lead to $(M_1 \times N_1) + (M_2 \times N_2) + (M_3 \times N_3)$ equations that must be satisfied simultaneously. Hence, by solving these equations for constants a_{mn}, b_{mn}, and c_{mn}, the magnitudes of the displacement fields given by Equations 12.1 that correspond to the minimum potential energy are known and the problem is solved. The validity of the solution obtained hinges on whether Equations 12.1 adequately represent the displacement fields actually induced in the structure.

It is seen, therefore, that solutions obtained using the Ritz method are based on the functions U_{mn}, V_{mn}, and W_{mn} that appear in Equations 12.1. These functions are more or less arbitrarily selected, but must possess two important characteristics: (a) they must be continuous and differentiable to at least the second order, and (b) they must satisfy the *geometric* boundary conditions. This latter characteristic is the source of the approximate nature of the Ritz analysis. That is, the mathematical forms of functions $U_{mn}(x,y)$, $V_{mn}(x,y)$, and $W_{mn}(x,y)$ are selected to satisfy the prevailing *geometric* boundary conditions, but the *static* boundary conditions are not considered during this selection. Hence, these functions may, or may not, satisfy the prevailing *static* boundary conditions. If the functions $U_{mn}(x,y)$, $V_{mn}(x,y)$, and $W_{mn}(x,y)$ do not satisfy *both* geometric and static boundary conditions then the solution obtained using the Ritz approach will be "approximate." On the other hand, if the forms selected for U_{mn}, V_{mn}, and W_{mn} do happen to satisfy both geometric and static boundary conditions, then the solution obtained using the Ritz method is "exact."

The challenge of identifying functional forms for U_{mn}, V_{mn}, and W_{mn} that satisfy both geometric and static boundary conditions will be illustrated for a simply supported plate. Functions U_{mn}, V_{mn}, and W_{mn} are usually selected to be separable functions of x and y. That is, function W_{mn} (for example), is usually selected to be of the form

$$W_{mn} = X_m(x)Y_n(y)$$

For simply supported plates, it is common to assume that X_m and Y_n are sinusoidal functions of x and y:

$$X_m(x) = \sin\left(\frac{m\pi x}{a}\right) \quad Y_n(y) = \sin\left(\frac{n\pi y}{b}\right)$$

where a and b are the length and width of a rectangular composite plate, respectively. Thus, based on this selection Equation 12.1c becomes

$$w(x,y) = \sum_{m=1}^{M_3}\sum_{n=1}^{N_3} c_{mn} \sin\left(\frac{m\pi x}{a}\right)\sin\left(\frac{n\pi y}{b}\right) \tag{12.3}$$

These functional forms for X_m and Y_n are appropriate for simply supported plates because they satisfy the *geometric* boundary conditions, regardless of the magnitudes of c_{mn}. That is, Equation 12.3 gives $w(x,y) = 0$ for $x = 0$, a and $y = 0$, b for all m, n. They do not necessarily satisfy the *static* boundary conditions for a simply supported plate, however. Recall from Section 10.3 that the bending moment must vanish along a simply supported edge, resulting in the following static boundary condition along the two edges $x = 0$, a (a comparable condition must be satisfied along edges $y = 0$, b):

$$D_{11}\frac{\partial^2 w}{\partial x^2} + D_{12}\frac{\partial^2 w}{\partial y^2} - 2D_{16}\frac{\partial^2 w}{\partial x \partial y} = M_{xx}^* = 0$$

Upon substituting the assumed displacement field, Equation 12.3, into this static boundary condition, the first two terms lead to

$$D_{11}\frac{\partial^2 w}{\partial x^2} = -D_{11}\left[\sum_{m=1}^{M_3}\sum_{n=1}^{N_3} c_{mn}\left(\frac{m\pi}{a}\right)^2 \sin\left(\frac{m\pi x}{a}\right)\sin\left(\frac{n\pi y}{b}\right)\right] = 0, \quad \text{for } x = 0, a$$

$$D_{12}\frac{\partial^2 w}{\partial x^2} = -D_{12}\left[\sum_{m=1}^{M_3}\sum_{n=1}^{N_3} c_{mn}\left(\frac{n\pi}{b}\right)^2 \sin\left(\frac{m\pi x}{a}\right)\sin\left(\frac{n\pi y}{b}\right)\right] = 0, \quad \text{for } x = 0, a$$

However, the third term leads to

$$2D_{16}\frac{\partial^2 w}{\partial x^2} = 2D_{16}\left[\sum_{m=1}^{M_3}\sum_{n=1}^{N_3} c_{mn}\left(\frac{m\pi}{a}\right)\left(\frac{n\pi}{b}\right)\cos\left(\frac{m\pi x}{a}\right)\cos\left(\frac{n\pi y}{b}\right)\right]$$

Evaluating this third term along $x = 0, a$:

$$2D_{16}\frac{\partial^2 w}{\partial x^2} = \frac{2\pi^2 D_{16}}{ab}\sum_{m=1}^{M_3}\sum_{n=1}^{N_3}(mn)\,c_{mn}\cos\left(\frac{n\pi y}{b}\right),\quad \text{for } x = 0, a$$

Hence, the static boundary condition associated with simply supported edges $x = 0, a$ is satisfied only if

$$\sum_{m=1}^{M_3}\sum_{n=1}^{N_3}(mn)c_{mn}\cos\left(\frac{n\pi y}{b}\right) = 0,\quad \text{for } 0 \le y \le b \qquad (12.4)$$

If only a single term is used to describe the displacement field (i.e., if $M_3 = N_3 = 1$), then the static boundary condition is clearly *not* satisfied even approximately, since

$$c_{11}\cos\left(\frac{\pi y}{b}\right) \ne 0,\quad \text{for all } 0 \le y \le b$$

However, if the number of terms used is increased (i.e., as M_3 and/or N_3 are increased), then Equation 12.4 may be satisfied more and more exactly, through proper selection of the values of constants c_{mn}. For example, if four terms are used (i.e., if $M_3 = N_3 = 2$), then the static boundary condition along edges $x = 0, a$ is satisfied if

$$c_{11}\cos\left(\frac{\pi y}{b}\right) + 2c_{12}\cos\left(\frac{2\pi y}{b}\right) + 2c_{21}\cos\left(\frac{\pi y}{b}\right) + 4c_{22}\cos\left(\frac{2\pi y}{b}\right) = 0,$$

for all $0 \le y \le b$

It is now possible to satisfy the static boundary condition along edges $x = 0$, a exactly, by setting $c_{21} = -c_{11}/2$ and $c_{22} = -c_{12}/2$. Of course, selecting constants that satisfy these requirements may not lead to the displacement field that represents the state of minimum potential energy, since constants $c_{11}, c_{12}, c_{21},$ and c_{22} must also satisfy the static boundary conditions along the edges $y = 0$, b. Still, it is apparent that by increasing the number of terms used to describe the displacement field it is possible to satisfy the static boundary conditions along all four edges more and more exactly.

In general, then the validity of a solution based on the Ritz approach is increased as the number of terms is increased. Assuming functions X_m and Y_n are selected to satisfy the geometric boundary conditions, then a Ritz analysis will *converge* toward the exact solution as M_3 and/or N_3 are increased. It is also appropriate to note that the difficulty in obtaining an exact solution is due to the D_{16} term. If $D_{16} = 0$ (i.e., if the laminate were specially orthotropic) then the geometric and static boundary conditions associated with the simply supported edges $x = 0, a$ would be satisfied exactly by the assumed sinusoidal form for X_m and Y_n.

A relatively general description of how the Ritz approach is applied has been presented in the preceding paragraphs. We will now specialize this approach for application to the specific problems discussed in this chapter. First, as in the previous chapters, we limit our discussion to rectangular symmetric laminates, simply supported along all four edges. Second, for all problems considered herein the in-plane displacement fields $u_o(x,y)$ and $v_o(x,y)$ that exist prior to application of $q(x,y)$ can be deduced based on a CLT analysis (as discussed in the next section), *and are known a priori.* Hence, for present purposes there is no need to express $u_o(x,y)$ or $v_o(x,y)$ in terms of the double series as listed as Equations 12.1a,b, nor to determine the magnitudes of in-plane displacement fields using Equations 12.2a,b. For the problems considered herein only $w(x,y)$ is unknown. The out-of-plane displacement field will be expressed using the double series listed as Equation 12.1c, and the magnitude of out-of-plane displacements will be determined through application of Equation 12.2c. Functions X_m and Y_n will be assumed to be sinusoidal functions, so the assumed form for $w(x,y)$ is given by Equation 12.3. To simplify nomenclature, let $M_3 \to M$ and $N_3 \to N$, so $w(x,y)$ will be written as

$$w(x,y) = \sum_{m=1}^{M}\sum_{n=1}^{N} c_{mn} \sin\left(\frac{m\pi x}{a}\right)\sin\left(\frac{n\pi y}{b}\right) \qquad (12.5)$$

Although not discussed in this chapter, solutions are also available based on alternate (nonsinusoidal) functional forms for X_m and Y_n. Alternate forms include polynomials in x and y:

$$X_m(x) = (x^2 - ax)^2 x^{m-1}$$

$$Y_n(y) = (y^2 - ay)^2 y^{n-1}$$

forms involving other trigonometric functions:

$$X_m(x) = \left[1 - \cos\left(\frac{2m\pi x}{a}\right)\right]$$

$$Y_n(y) = \left[1 - \cos\left(\frac{2n\pi y}{b}\right)\right]$$

or forms involving so-called beam functions:

$$X_m(x) = \gamma_m \cos\left(\frac{\lambda_m x}{a}\right) - \gamma_m \cosh\left(\frac{\lambda_m x}{a}\right) + \sin\left(\frac{\lambda_m x}{a}\right) - \sinh\left(\frac{\lambda_m x}{a}\right)$$

$$Y_n(x) = \gamma_n \cos\left(\frac{\lambda_n x}{b}\right) - \gamma_n \cosh\left(\frac{\lambda_n x}{b}\right) + \sin\left(\frac{\lambda_n x}{b}\right) - \sinh\left(\frac{\lambda_n x}{b}\right)$$

These alternate forms may be used to model other boundary conditions such as a clamped edge. The reader interested application of these alternate forms is referred to References 1, 2.

12.2 In-Plane Displacement Fields

In this section, we will derive expressions that give the midplane displacement fields induced in a symmetric laminate for specific loading conditions. The derivation is based on the following sequence of events. First, assume the laminate is cured at an elevated temperature, and that the laminate is stress- and strain-free at the cure temperature.* Following cure, the laminate is cooled to room temperature. The laminate is then trimmed to the desired final rectangular dimensions $a \times b$. Finally, the laminate is subjected to constant and uniformly distributed stress resultants (N_{xx}^*, N_{yy}^*, and/or N_{xy}^*) along all four edges, and/or a *further* change in temperature. Although we will not consider the possibility of a uniform change in moisture content directly, from the previous discussion, it should be clear that changes in moisture content could be accounted for (in a mathematical sense) using the same techniques used to model uniform changes in temperature.

The sequence of events described above are precisely those assumed during the development of CLT in Chapter 6. Since the externally applied stress resultants are uniformly distributed along each edge, the stress resultants induced at all interior points are equal to the edge loads:

$$N_{xx}^*(0,y) = N_{xx}^*(a,y) = N_{xx}$$

$$N_{yy}^*(x,0) = N_{yy}^*(x,b) = N_{yy}$$

$$N_{xy}^*(0,y) = N_{xy}^*(a,y) = N_{yx}^*(x,0) = N_{yx}^*(x,b) = N_{xy}$$

---

* As mentioned in Section 6.6.2, in practice the stress- and strain-free temperature is often 10–25°C below the final cure temperature. This complication has been ignored throughout this text, and it is assumed that the laminate is stress- and strain-free at the cure temperature.

For a symmetric laminate subjected to uniform in-plane stress resultants and a change in temperature, Equation 6.45 becomes

$$
\begin{Bmatrix}
\varepsilon_{xx}^o \\
\varepsilon_{yy}^o \\
\gamma_{xy}^o \\
\kappa_{xx} \\
\kappa_{yy} \\
\kappa_{xy}
\end{Bmatrix}
=
\begin{bmatrix}
a_{11} & a_{12} & a_{16} & 0 & 0 & 0 \\
a_{12} & a_{22} & a_{26} & 0 & 0 & 0 \\
a_{16} & a_{26} & a_{66} & 0 & 0 & 0 \\
0 & 0 & 0 & d_{11} & d_{12} & d_{16} \\
0 & 0 & 0 & d_{12} & d_{22} & d_{26} \\
0 & 0 & 0 & d_{16} & d_{26} & d_{66}
\end{bmatrix}
\begin{Bmatrix}
N_{xx} + N_{xx}^T \\
N_{yy} + N_{yy}^T \\
N_{xy} + N_{xy}^T \\
0 \\
0 \\
0
\end{Bmatrix}
$$

As in earlier chapters, we assume infinitesimal strain levels, and, therefore, midplane strains are related to midplane displacements according to Equation 6.10:

$$
\varepsilon_{xx}^o = \frac{\partial u_o}{\partial x} \quad \varepsilon_{yy}^o = \frac{\partial v_o}{\partial y} \quad \gamma_{xy}^o = \frac{\partial u_o}{\partial y} + \frac{\partial v_o}{\partial x}
$$

Consequently, midplane displacement fields are related to the stress and thermal resultants as follows:

$$
\frac{\partial u_o}{\partial x} = a_{11}(N_{xx} + N_{xx}^T) + a_{12}(N_{yy} + N_{yy}^T) + a_{16}(N_{xy} + N_{xy}^T) \tag{12.6a}
$$

$$
\frac{\partial v_o}{\partial y} = a_{12}(N_{xx} + N_{xx}^T) + a_{22}(N_{yy} + N_{yy}^T) + a_{26}(N_{xy} + N_{xy}^T) \tag{12.6b}
$$

$$
\frac{\partial u_o}{\partial y} + \frac{\partial v_o}{\partial x} = a_{16}(N_{xx} + N_{xx}^T) + a_{26}(N_{yy} + N_{yy}^T) + a_{66}(N_{xy} + N_{xy}^T) \tag{12.6c}
$$

Integrating Equation 12.6a with respect to x, we find

$$
u_o(x,y) = \left[a_{11}(N_{xx} + N_{xx}^T) + a_{12}(N_{yy} + N_{yy}^T) + a_{16}(N_{xt} + N_{xy}^T) \right] x + f_1(y) + \lambda_1
$$

where $f_1(y)$ is an unknown function of y (only) and λ_1 is an unknown constant of integration. Similarly, integrating Equation 12.6b with respect to y, we find

$$
v_o(x,y) = \left[a_{12}(N_{xx} + N_{xx}^T) + a_{22}(N_{yy} + N_{yy}^T) + a_{26}(N_{xy} + N_{xy}^T) \right] y + f_2(x) + \lambda_2
$$

where $f_2(x)$ is an unknown function of x (only) and λ_2 is a second unknown constant. Without a loss in generality we assume that midplane displacements are zero at the origin (i.e., at $x = y = 0$), and consequently $\lambda_1 = \lambda_2 = 0$. Substituting these expressions for $u_o(x,y)$ and $v_o(x,y)$ into Equation 12.6c, we find

$$\frac{\partial u_o}{\partial y} + \frac{\partial v_o}{\partial x} = \frac{\partial f_1}{\partial y} + \frac{\partial f_2}{\partial x} = a_{16}(N_{xx} + N_{xx}^T) + a_{26}(N_{yy} + N_{yy}^T) + a_{66}(N_{xy} + N_{xy}^T)$$

Since all terms on the right side of the equality are known constants, it follows that f_1 and f_2 must be at most linear functions of y and x, respectively:

$$f_1(y) = \lambda_3 y$$

$$f_2(x) = \lambda_4 x$$

Hence, we can write

$$\lambda_3 + \lambda_4 = a_{16}(N_{xx} + N_{xx}^T) + a_{26}(N_{yy} + N_{yy}^T) + a_{66}(N_{xy} + N_{xy}^T)$$

Since all quantities that appear in this relation are constants, λ_3 and λ_4 can take on any value as long as they sum to the expression on the right side of the equality. To determine particular values that satisfy this expression, we now require that the infinitesimal rotation vector in the x–y plane ω_{xy} (which represents rigid body motion of the plate) is zero. The infinitesimal rotation vector is given by [3,4]

$$\omega_{xy} = \frac{1}{2}\left(\frac{\partial u_o}{\partial y} - \frac{\partial v_o}{\partial x}\right)$$

Requiring that $\omega_{xy} = 0$ leads to

$$\lambda_3 = \lambda_4 = \frac{1}{2}\left[a_{16}(N_{xx} + N_{xx}^T) + a_{26}(N_{yy} + N_{yy}^T) + a_{66}(N_{xy} + N_{xy}^T)\right]$$

Combining the preceding results, we conclude that the in-plane midplane displacements induced in a symmetric composite panel by the combination of uniform in-plane stress resultants and a change in temperature are given by

$$u_o(x,y) = \left[a_{11}(N_{xx} + N_{xx}^T) + a_{12}(N_{yy} + N_{yy}^T) + a_{16}(N_{xy} + N_{xy}^T)\right]x$$

$$+ \frac{1}{2}\left[a_{16}(N_{xx} + N_{xx}^T) + a_{26}(N_{yy} + N_{yy}^T) + a_{66}(N_{xy} + N_{xy}^T)\right]y$$

$$v_o(x,y) = \frac{1}{2}\Big[a_{16}(N_{xx} + N_{xx}^T) + a_{26}(N_{yy} + N_{yy}^T) + a_{66}(N_{xy} + N_{xy}^T)\Big]x$$

$$+ \Big[a_{12}(N_{xx} + N_{xx}^T) + a_{22}(N_{yy} + N_{yy}^T) + a_{26}(N_{xy} + N_{xy}^T)\Big]y$$

These expressions can be simplified through the use of the effective thermal expansion coefficients of the laminate, defined by Equations 6.73a and repeated here for convenience:

$$\bar{\alpha}_{xx} = \frac{1}{\Delta T}\Big[a_{11}N_{xx}^T + a_{12}N_{yy}^T + a_{16}N_{xy}^T\Big]$$

$$\bar{\alpha}_{yy} = \frac{1}{\Delta T}\Big[a_{12}N_{xx}^T + a_{22}N_{yy}^T + a_{26}N_{xy}^T\Big] \qquad (6.73a)$$

$$\bar{\alpha}_{xy} = \frac{1}{\Delta T}\Big[a_{16}N_{xx}^T + a_{26}N_{yy}^T + a_{66}N_{xy}^T\Big]$$

Hence, we see that the midplane displacement fields can be written as

$$u_o(x,y) = \Big[a_{11}N_{xx} + a_{12}N_{yy} + a_{16}N_{xy} + \Delta T\bar{\alpha}_{xx}\Big]x$$

$$+ \frac{1}{2}\Big[a_{16}N_{xx} + a_{26}N_{yy} + a_{66}N_{xy} + \Delta T\bar{\alpha}_{xy}\Big]y$$

$$\qquad (12.7)$$

$$v_o(x,y) = \frac{1}{2}\Big[a_{16}N_{xx} + a_{26}N_{yy} + a_{66}N_{xy} + \Delta T\bar{\alpha}_{xy}\Big]x$$

$$+ \Big[a_{12}N_{xx} + a_{22}N_{yy} + a_{26}N_{xy} + \Delta T\bar{\alpha}_{yy}\Big]y$$

Note that Equations 12.7 are valid only if both stress resultants and temperature changes are uniform.

In the following sections, we will use these in-plane displacement fields to obtain solutions based on the Ritz method for simply supported composite plates. Recall from Section 10.3 that four distinct combinations of geometric and static boundary conditions, numbered S1 through S4, can be defined as a "simple-support." The distinction between the different types of simple supports has to do with the boundary condition assumed for the in-plane displacement fields. For example, to define a simple support of type S1 one specifies known values of in-plane displacements, whereas to define a simple support of type S4 one specifies known values of in-plane stress resultants. For the problems considered here, we are able to calculate the midplane displacement field induced by a specified combination of edge loads (or vice versa). This is possible because we have assumed all stress resultants applied at the edge of the plate are *constant and uniform*. Of course, if the stress resultants were not uniform but rather varied along the plate edges then it would be more difficult (and in most cases impossible) to determine associated in-plane displacement fields.

Since we have limited discussion to cases in which stress resultants applied to the edges are constant and uniform, the midplane displacement fields given by Equation 12.7 can be used to obtain solutions for any of the four types of simple support boundary conditions, S1 through S4. Only one type of simple support will be considered in this chapter, since this discussion is intended to be a brief introduction to the Ritz method. Specifically, type S4 simple supports will be assumed throughout the remainder of this chapter. Thus, solutions will be obtained using the Ritz methods for problems in which uniform in-plane stress resultants N_{xx}, N_{yy}, and/or N_{xy} are applied to the edge of the plate. Although not discussed herein, similar analyses can be performed assuming type S1, S2, or S3 simple supports, or combinations thereof.

12.3 Potential Energy in a Thin Composite Plate

In this section, we will develop the equations necessary to calculate the potential energy of a thin elastic plate subjected to a combination of loads and uniform temperature changes. The possibility of a uniform change in moisture content will not be considered, although from the previous discussion, it should be clear that changes in moisture content can be accounted for (in a mathematical sense) using the same techniques used to model uniform changes in temperature. Type S4 boundary conditions are assumed for all four edges of the plate.

Two energy terms will be encountered in the following discussion. First, we will consider the *work* done when a transverse load $q(x,y)$ is applied to a thin plate. This energy term is denoted W. Recall that the fundamental definition of "work" is force multiplied by the distance through which it travels. Also recall that the transverse load applied to a plate $q(x,y)$ has been defined using units of force/area. The product $(q)(dx)(dy)$ represents the transverse force provided by $q(x,y)$ acting over an infinitesimal element of area $(dx)(dy)$. The distance through which this force travels equals the out-of-plane deflection of the plate at that point, $w(x,y)$. Therefore, the *total work* done by the transverse load acting over the entire surface of the plate is given by

$$W = \int\int \big[q(x,y)\big]\big[w(x,y)\big]dx\,dy \qquad (12.8)$$

Second, we will consider the *strain energy* within the plate in the deformed condition. This energy term is denoted U. A general expression giving the strain energy within a linear elastic solid body subjected to an arbitrary state of stress is

$$U = \frac{1}{2}\iiint\left(\sigma_{xx}\varepsilon_{xx} + \sigma_{yy}\varepsilon_{yy} + \sigma_{zz}\varepsilon_{zz} + \tau_{yz}\gamma_{yz} + \tau_{xz}\gamma_{xz} + \tau_{xy}\gamma_{xy}\right)dx\,dy\,dz$$

Note that evaluation of strain energy involves integration over the entire volume of the body. In our case, we wish to calculate the strain energy within a thin symmetric composite laminate subjected to a state of plane stress. We will, therefore, specialize the above general expression of strain energy for present purposes. First, since plane stress is assumed ($\sigma_{zz} = \tau_{yz} = \tau_{xz} = 0$), we can immediately discard terms involving these stress components:

$$U = \frac{1}{2}\iiint\left(\sigma_{xx}\varepsilon_{xx} + \sigma_{yy}\varepsilon_{yy} + \tau_{xy}\gamma_{xy}\right)dx\,dy\,dz \tag{12.9}$$

Next, the stresses in ply k of the laminate are given by Equations 5.30, which become (for $\Delta M = 0$):

$$\begin{Bmatrix} \sigma_{xx} \\ \sigma_{yy} \\ \tau_{xy} \end{Bmatrix}_k = \begin{bmatrix} \overline{Q}_{11} & \overline{Q}_{12} & \overline{Q}_{16} \\ \overline{Q}_{12} & \overline{Q}_{22} & \overline{Q}_{26} \\ \overline{Q}_{16} & \overline{Q}_{26} & \overline{Q}_{66} \end{bmatrix}_k \begin{Bmatrix} \varepsilon_{xx} - \Delta T\alpha_{xx} \\ \varepsilon_{yy} - \Delta T\alpha_{yy} \\ \gamma_{xy} - \Delta T\alpha_{xy} \end{Bmatrix}_k$$

Substituting these expressions for ply stresses into Equation 12.9 and rearranging, we find

$$\begin{aligned} U = \frac{1}{2}\iiint\Big[& \overline{Q}_{11}^{(k)}\varepsilon_{xx}^2 + 2\overline{Q}_{12}^{(k)}\varepsilon_{xx}\varepsilon_{yy} + 2\overline{Q}_{16}^{(k)}\varepsilon_{xx}\gamma_{xy} + 2\overline{Q}_{26}^{(k)}\varepsilon_{yy}\gamma_{xy} + \overline{Q}_{22}^{(k)}\varepsilon_{yy}^2 + \overline{Q}_{66}^{(k)}\gamma_{xy}^2 \\ & - \Delta T\left(\alpha_{xx}^{(k)}\overline{Q}_{11}^{(k)} + \alpha_{yy}^{(k)}\overline{Q}_{12}^{(k)} + \alpha_{xy}^{(k)}\overline{Q}_{16}^{(k)}\right)\varepsilon_{xx} - \Delta T\left(\alpha_{xx}^{(k)}\overline{Q}_{12}^{(k)} + \alpha_{yy}^{(k)}\overline{Q}_{22}^{(k)} + \alpha_{xy}^{(k)}\overline{Q}_{26}^{(k)}\right)\varepsilon_{yy} \\ & - \Delta T\left(\alpha_{xx}^{(k)}\overline{Q}_{16}^{(k)} + \alpha_{yy}^{(k)}\overline{Q}_{26}^{(k)} + \alpha_{xy}^{(k)}\overline{Q}_{66}^{(k)}\right)\gamma_{xy}\Big]dx\,dy\,dz \end{aligned} \tag{12.10}$$

We now invoke the Kirchhoff hypothesis, which allows us to relate ply strains at any through-thickness position z to midplane strains and curvatures, in accordance with Equations 6.12, repeated here for convenience:

$$\varepsilon_{xx} = \varepsilon_{xx}^o + z\kappa_{xx}$$

$$\varepsilon_{yy} = \varepsilon_{yy}^o + z\kappa_{yy} \tag{6.12}$$

$$\gamma_{xy} = \gamma_{xy}^o + z\kappa_{xy}$$

Substitution of Equations 6.12 into Equation 12.10 results in

$$
U = \frac{1}{2}\iiint \left\{ \overline{Q}_{11}\left[\left(\varepsilon_{xx}^o \right)^2 + 2z\varepsilon_{xx}^o \kappa_{xx} + z^2\kappa_{xx}^2 \right] \right.
$$

$$
+ 2\overline{Q}_{12}\left[\varepsilon_{xx}^o \varepsilon_{yy}^o + z\varepsilon_{xx}^o \kappa_{yy} + z\varepsilon_{yy}^o \kappa_{xx} + z^2\kappa_{xx}\kappa_{yy} \right]
$$

$$
+ 2\overline{Q}_{16}\left[\varepsilon_{xx}^o \gamma_{xy}^o + z\varepsilon_{xx}^o \kappa_{xy} + z\gamma_{xy}^o \kappa_{xx} + z^2\kappa_{xx}\kappa_{xy} \right]
$$

$$
+ 2\overline{Q}_{26}\left[\varepsilon_{yy}^o \gamma_{xy}^o + z\varepsilon_{yy}^o \kappa_{xy} + z\gamma_{xy}^o \kappa_{yy} + z^2\kappa_{yy}\kappa_{xy} \right]
$$

$$
+ \overline{Q}_{22}\left[\left(\varepsilon_{yy}^o \right)^2 + 2z\varepsilon_{yy}^o \kappa_{yy} + z^2\kappa_{yy}^2 \right]
$$

$$
+ \overline{Q}_{66}\left[\left(\gamma_{xy}^o \right)^2 + 2z\gamma_{xy}^o \kappa_{xy} + z^2\kappa_{xy}^2 \right]
$$

$$
- \Delta T\left[\alpha_{xx}\overline{Q}_{11} + \alpha_{yy}\overline{Q}_{12} + \alpha_{xy}\overline{Q}_{16} \right]\left(\varepsilon_{xx}^o + z\kappa_{xx} \right)
$$

$$
- \Delta T\left[\alpha_{xx}\overline{Q}_{12} + \alpha_{yy}\overline{Q}_{22} + \alpha_{xy}\overline{Q}_{26} \right]\left(\varepsilon_{yy}^o + z\kappa_{yy} \right)
$$

$$
\left. - \Delta T\left[\alpha_{xx}\overline{Q}_{16} + \alpha_{yy}\overline{Q}_{26} + \alpha_{xy}\overline{Q}_{66} \right]\left(\gamma_{xy}^o + z\kappa_{xy} \right) \right\} dx\,dy\,dz
$$

(12.11)

Next, integrate Equation 12.11 over the thickness of the laminate, that is, over the range $-t/2 \le z \le t/2$. During this process, a number of integrals will be encountered that were previously evaluated in Chapter 6. A few specific examples are

$$
\int_{-t/2}^{t/2} \overline{Q}_{11}\,dz, \text{ which after integration becomes } A_{11}
$$

$$
\int_{-t/2}^{t/2} \overline{Q}_{11}\,zdz, \text{ which after integration becomes } B_{11}
$$

$$
\int_{-t/2}^{t/2} \overline{Q}_{11}\,z^2dz, \text{ which after integration becomes } D_{11}
$$

$$
\int_{-t/2}^{t/2} \Delta T\left[\alpha_{xx}\overline{Q}_{11} + \alpha_{yy}\overline{Q}_{12} + \alpha_{xy}\overline{Q}_{16} \right]dz, \text{ which after integration becomes } N_{xx}^T
$$

$$
\int_{-t/2}^{t/2} \Delta T\left[\alpha_{xx}\overline{Q}_{11} + \alpha_{yy}\overline{Q}_{12} + \alpha_{xy}\overline{Q}_{16} \right]dz, \text{ which after integration becomes } M_{xx}^T
$$

Hence, after integration, Equation 12.11 can be written as

$$
U = \frac{1}{2} \iint \Big\{ A_{11}\left(\varepsilon_{xx}^o\right)^2 + 2A_{12}\varepsilon_{xx}^o\varepsilon_{yy}^o + A_{22}\left(\varepsilon_{yy}^o\right)^2 + 2\left(A_{16}\varepsilon_{xx}^o + A_{26}\varepsilon_{yy}^o\right)\gamma_{xy}^o + A_{66}\gamma_{xy}^2
$$
$$
+ 2B_{11}\varepsilon_{xx}^o\kappa_{xx} + 2B_{12}\left(\varepsilon_{xx}^o\kappa_{yy} + \varepsilon_{yy}^o\kappa_{xx}\right) + 2B_{16}\left(\varepsilon_{xx}^o\kappa_{xy} + \gamma_{xy}^o\kappa_{xx}\right)
$$
$$
+ 2B_{26}\left(\varepsilon_{yy}^o\kappa_{xy} + \gamma_{xy}^o\kappa_{yy}\right) + 2B_{22}\varepsilon_{yy}^o\kappa_{yy} + 2B_{66}\gamma_{xy}^o\kappa_{xy} + D_{11}\kappa_{xx}^2
$$
$$
+ 2D_{12}\kappa_{xx}\kappa_{yy} + 2\left(D_{16}\kappa_{xx} + D_{26}\kappa_{yy}\right)\kappa_{xy} + D_{22}\kappa_{yy}^2 + D_{66}\kappa_{xy}^2
$$
$$
- N_{xx}^T\varepsilon_{xx}^o - N_{yy}^T\varepsilon_{yy}^o - N_{xy}^T\gamma_{xy}^o - M_{xx}^T\kappa_{xx} - M_{yy}^T\kappa_{yy} - M_{xy}^T\kappa_{xy} \Big\} dx\,dy \qquad (12.12)
$$

Since we have limited discussion to symmetric laminates, $B_{ij} = M_{xx}^T = M_{yy}^T = M_{xy}^T = 0$ in all cases considered. Equation 12.12, therefore, simplifies to

$$
U = \frac{1}{2} \iint \Big\{ A_{11}\left(\varepsilon_{xx}^o\right)^2 + 2A_{12}\varepsilon_{xx}^o\varepsilon_{yy}^o + A_{22}\left(\varepsilon_{yy}^o\right)^2 + 2\left(A_{16}\varepsilon_{xx}^o + A_{26}\varepsilon_{yy}^o\right)\gamma_{xy}^o + A_{66}\gamma_{xy}^2
$$
$$
+ D_{11}\kappa_{xx}^2 + 2D_{12}\kappa_{xx}\kappa_{yy} + 2\left(D_{16}\kappa_{xx} + D_{26}\kappa_{yy}\right)\kappa_{xy} + D_{22}\kappa_{yy}^2 + D_{66}\kappa_{xy}^2
$$
$$
- N_{xx}^T\varepsilon_{xx}^o - N_{yy}^T\varepsilon_{yy}^o - N_{xy}^T\gamma_{xy}^o \Big\} dx\,dy \qquad (12.13)
$$

We have now developed expressions for the two energy terms necessary for our purposes: the work W done by a transverse load applied to a laminate, Equation 12.8, and the strain energy U within a deformed laminate, Equation 12.13. We wish to form an appropriate combination of these terms so as to represent the total potential energy of a symmetric composite laminate. An itemized conceptual description of how U and W are combined is presented below. Mathematical implementation of these concepts for the particular class of problems considered in this chapter is then discussed in separate subsections. The reader is urged to carefully consider the follow conceptual description *before* considering the mathematical formulation that follows.

Step 1: We assume that in-plane stress resultants N_{ij} are applied to the laminate *first*, before the application of any other load(s) that cause bending. Our first step is, therefore, to calculate the strain energy within a laminate subjected to N_{ij} *only*, even if other loads are involved in the problem under consideration. We will label this component of strain energy U_I. Calculation of U_I is straightforward, since we have limited our discussion to symmetric laminates. That is, for a symmetric laminate stress resultants N_{ij} are solely responsible for the development of midplane strains, and do not cause curvatures to develop. If we had included nonsymmetric laminates in our analysis then $B_{ij} \neq 0$. If this were the case, then N_{ij} would also contribute to curvatures, and further both M_{ij} and $q(x,y)$ would contribute to midplane strains. These coupling effects would greatly complicate our analysis. Since symmetry *has* been assumed, the laminate *remains flat* following application of N_{ij}. Therefore, the only loads that contribute to midplane strains are stress resultants N_{ij}. The strain energy associated with N_{ij} *only* (i.e., the strain energy component U_I) can, therefore, be obtained from Equation 12.13 simply by setting $\kappa_{xx} = \kappa_{yy} = \kappa_{xy} = 0$. Ultimately our expression for U_I will involve in-plane displacements $u(x,y)$ and $v(x,y)$, but will *not* involve out-of-plane displacements $w(x,y)$.

Step 2: Next, we calculate strain energy associated with midplane curvatures *only*. That is, we ignore any preexisting midplane strains and base our calculation of strain energy based solely on curvatures κ_{xx}, κ_{yy}, and κ_{xy}. We will label this component of the strain energy U_{II}. Since the laminate is symmetric, midplane curvatures are caused by the combined effects of the transverse load $q(x,y)$ and/or bending moment resultants M_{ij}, but are independent of stress resultants N_{ij}. Strain energy component U_{II} can be obtained from Equation 12.13 simply by setting $\varepsilon_{xx}^o = \varepsilon_{yy}^o = \gamma_{xy}^o = 0$. Ultimately our expression for U_{II} will involve out-of-plane displacements $w(x,y)$, but will *not* involve in-plane displacements $u(x,y)$ or $v(x,y)$.

Step 3: As described above, in step 1, we calculate the strain energy U_I associated with stress resultants N_{ij} *prior to the application of any load(s) that cause the laminate to bend*. Of course, when transverse loads and/or bending moments are subsequently applied, the laminate will bend. Now, once bending has occurred then calculation of strain energy as performed in step 1 is incomplete. That is, as the laminate begins to bend, the in-plane strains that exist *prior to bending* will change, resulting in a change in the strain energy associated with stress resultants. Therefore, in step 3 we calculate the change in strain energy caused by bending and associated with in-plane stress resultants. We will label this component of the strain energy U_{III}. We assume that the stress resultants N_{ij} that exist prior to bending *remain constant during bending*. That is, we assume that only in-plane strains change as bending occurs. The change in in-plane strains will be related to out-of-plane displacements, $w(x,y)$. Hence, our expression for U_{III} will involve $w(x,y)$, but will *not* involve in-plane displacements $u(x,y)$ or $v(x,y)$.

Step 4: In step 4, we calculate the work W done by the transverse load $q(x,y)$, in accordance with Equation 12.6. The mathematical form of our expression for W will obviously depend on the nature of the transverse load. Thus, for example, the mathematical form of W for a uniform transverse load, $q(x,y) = q_o$, will differ from the mathematical form of W if the transverse load varies sinusoidally: $q(x,y) = q_o \sin(\pi x/a) \sin(\pi y/b)$.

Step 5: Finally, in step 5, we form the desired expression for the total potential energy Π in the composite plate. The total potential energy is given by $\Pi = U_I + U_{II} + U_{III} - W$.

12.3.1 Evaluation of Strain Energy Component U_I

Strain energy component U_I is obtained from Equation 12.11 by setting $\kappa_{xx} = \kappa_{yy} = \kappa_{xy} = 0$:

$$U_I = \frac{1}{2}\iint \Big\{ A_{11}\big(\varepsilon_{xx}^o\big)^2 + 2A_{12}\varepsilon_{xx}^o\varepsilon_{yy}^o + A_{22}\big(\varepsilon_{yy}^o\big)^2 + 2\big(A_{16}\varepsilon_{xx}^o + A_{26}\varepsilon_{yy}^o\big)\gamma_{xy}^o + A_{66}\gamma_{xy}^2$$

$$- N_{xx}^T\varepsilon_{xx}^o - N_{yy}^T\varepsilon_{yy}^o - N_{xy}^T\gamma_{xy}^o \Big\}\,dxdy$$

As in earlier chapters, we assume infinitesimal strain levels, and, therefore, midplane strains are related to midplane displacements according to Equation 6.10:

$$\varepsilon_{xx}^o = \frac{\partial u_o}{\partial x} \quad \varepsilon_{yy}^o = \frac{\partial v_o}{\partial y} \quad \gamma_{xy}^o = \frac{\partial u_o}{\partial y} + \frac{\partial v_o}{\partial x}$$

With this substitution, we have

$$U_I = \frac{1}{2}\iint \Bigg\{ A_{11}\left(\frac{\partial u_o}{\partial x}\right)^2 + 2A_{12}\left(\frac{\partial u_o}{\partial x}\right)\left(\frac{\partial v_o}{\partial y}\right) + A_{22}\left(\frac{\partial v_o}{\partial y}\right)^2$$

$$+ 2\left(A_{16}\frac{\partial u_o}{\partial x} + A_{26}\frac{\partial v_o}{\partial y}\right)\left(\frac{\partial u_o}{\partial y} + \frac{\partial v_o}{\partial x}\right) + A_{66}\left(\frac{\partial u_o}{\partial y} + \frac{\partial v_o}{\partial x}\right)^2$$

$$- N_{xx}^T\left(\frac{\partial u_o}{\partial x}\right) - N_{yy}^T\left(\frac{\partial v_o}{\partial y}\right) - N_{xy}^T\left(\frac{\partial u_o}{\partial y} + \frac{\partial v_o}{\partial x}\right)\Bigg\}\,dx\,dy \tag{12.14}$$

Equation 12.14 gives the strain energy component U_I for any in-plane displacement fields $u_o(x,y)$ and $v_o(x,y)$. We will now integrate this expression using the displacement fields induced by uniform stress resultants and a uniform change in temperature, as given by Equation 12.7. To avoid a very lengthy expression, we make the following change in notation:

$$C_1 = \big[a_{11}N_{xx} + a_{12}N_{yy} + a_{16}N_{xy} + \Delta T\bar{\alpha}_{xx} \big]$$

$$C_2 = \big[a_{12}N_{xx} + a_{22}N_{yy} + a_{26}N_{xy} + \Delta T\bar{\alpha}_{yy} \big] \tag{12.15}$$

$$C_3 = \frac{1}{2}\big[a_{16}N_{xx} + a_{26}N_{yy} + a_{66}N_{xy} + \Delta T\bar{\alpha}_{xy} \big]$$

With this change in notation, Equation 12.7 can be written as

$$u_o(x,y) = C_1 x + C_3 y$$

$$v_o(x,y) = C_3 x + C_2 y$$

Substituting these expressions into Equation 12.14, we have

$$U_I = \frac{1}{2}\int_0^b\int_0^a \Big\{ A_{11}(C_1)^2 + 2A_{12}(C_1)(C_2) + A_{22}(C_2)^2$$

$$+ 2(A_{16}C_1 + A_{26}C_2)(2C_3) + A_{66}(2C_3)^2$$

$$- N_{xx}^T(C_1) - N_{yy}^T(C_2) - N_{xy}^T(2C_3) \Big\} dx\,dy$$

Evaluating this definite integral, we find

$$U_I = \Big[A_{11}(C_1)^2 + 2A_{12}(C_1)(C_2) + A_{22}(C_2)^2$$

$$+ 4\big(A_{16}C_1C_3 + A_{26}C_2C_3 + A_{66}C_3^2\big) - N_{xx}^T C_1 - N_{yy}^T C_2 - 2N_{xy}^T C_3 \Big](ab)$$

Substituting Equations 6.73b, which give the thermal stress resultants in terms of elements of the [A] matrix, and Equations 12.5, we find that U_I can be written as

$$U_I = \big[C_1 N_{xx} + C_2 N_{yy} + 2C_3 N_{xy} \big](ab) \tag{12.16}$$

12.3.2 Evaluation of Strain Energy Component U_{II}

As previously discussed, for symmetric laminates strain energy component U_{II} can be obtained from Equation 12.13 by setting $\varepsilon_{xx}^o = \varepsilon_{yy}^o = \gamma_{xy}^o = 0$:

$$U_{II} = \frac{1}{2}\int\int \Big\{ D_{11}\kappa_{xx}^2 + 2D_{12}\kappa_{xx}\kappa_{yy} + 2\big(D_{16}\kappa_{xx} + D_{26}\kappa_{yy}\big)\kappa_{xy} + D_{22}\kappa_{yy}^2 + D_{66}\kappa_{xy}^2 \Big\} dx\,dy$$

Midplane curvatures are related to out-of-plane displacements according to Equation 6.10:

$$\kappa_{xx} = -\frac{\partial^2 w}{\partial x^2} \quad \kappa_{yy} = -\frac{\partial^2 w}{\partial y^2} \quad \kappa_{xy} = -2\frac{\partial^2 w}{\partial x\,\partial y}$$

With this substitution, we have

$$U_{II} = \frac{1}{2}\int\int\left\{D_{11}\left(\frac{\partial^2 w}{\partial x^2}\right)^2 + 2D_{12}\left(\frac{\partial^2 w}{\partial x^2}\right)\left(\frac{\partial^2 w}{\partial y^2}\right) + 4\left[D_{16}\left(\frac{\partial^2 w}{\partial x^2}\right) + D_{26}\left(\frac{\partial^2 w}{\partial y^2}\right)\right]\frac{\partial^2 w}{\partial x \partial y}\right.$$

$$\left. + D_{22}\left(\frac{\partial^2 w}{\partial y^2}\right)^2 + 4D_{66}\left(\frac{\partial^2 w}{\partial x \partial y}\right)^2\right\}dx\,dy \qquad (12.17)$$

We will now integrate Equation 12.17 for the class of problems considered in this chapter. In all problems, we consider simply supported plates and assume out-of-plane displacements are given by Equation 12.6, repeated here for convenience:

$$w(x,y) = \sum_{m=1}^{M}\sum_{n=1}^{N} c_{mn} \sin\left(\frac{m\pi x}{a}\right)\sin\left(\frac{n\pi y}{b}\right)$$

Since Equation 12.6 will be used in all problems considered, we will integrate Equation 12.17 based on this displacement field. The following derivatives appear in Equation 12.17:

$$\frac{\partial^2 w}{\partial x^2} = -\sum_{m=1}^{M}\sum_{n=1}^{M} c_{mn}\left(\frac{m\pi}{a}\right)^2 \sin\left(\frac{m\pi x}{a}\right)\sin\left(\frac{n\pi y}{b}\right)$$

$$\frac{\partial^2 w}{\partial y^2} = -\sum_{m=1}^{M}\sum_{n=1}^{N} c_{mn}\left(\frac{n\pi}{b}\right)^2 \sin\left(\frac{m\pi x}{a}\right)\sin\left(\frac{n\pi y}{b}\right)$$

$$\frac{\partial^2 w}{\partial x \partial y} = \sum_{m=1}^{M}\sum_{n=1}^{N} c_{mn}\left(\frac{m\pi}{a}\right)\left(\frac{n\pi}{b}\right)\cos\left(\frac{m\pi x}{a}\right)\cos\left(\frac{n\pi y}{b}\right)$$

Consider the first term under the integral sign in Equation 12.17. Upon substituting the expression for $(\partial^2 w/\partial x^2)$ listed above, this term becomes

$$\frac{1}{2}\int_0^b\int_0^a D_{11}\left(\frac{\partial^2 w}{\partial x^2}\right)^2 dx\,dy = \frac{1}{2}\int_0^b\int_0^a D_{11}\left[-\sum_{m=1}^{M}\sum_{n=1}^{N} c_{mn}\left(\frac{m\pi}{a}\right)^2 \sin\left(\frac{m\pi x}{a}\right)\sin\left(\frac{n\pi y}{b}\right)\right]$$

$$\left[-\sum_{i=1}^{M}\sum_{j=1}^{N} c_{ij}\left(\frac{i\pi}{a}\right)^2 \sin\left(\frac{i\pi x}{a}\right)\sin\left(\frac{j\pi y}{b}\right)\right]dx\,dy$$

The evaluation of this integral is greatly simplified by noting the following identities:

$$\int_0^a \sin\left(\frac{m\pi x}{a}\right)\sin\left(\frac{i\pi x}{a}\right)dx = 0, \quad \text{if } m \neq i$$

$$\int_0^b \sin\left(\frac{n\pi y}{b}\right)\sin\left(\frac{i\pi y}{b}\right)dy = 0, \quad \text{if } n \neq j$$

Hence, we retain only those terms for which $m = i$ and $n = j$, so the integration becomes:

$$\frac{1}{2}\int_0^b\int_0^a D_{11}\left(\frac{\partial^2 w}{\partial x^2}\right)^2 dx\,dy = \frac{1}{2}\int_0^b\int_0^a D_{11}\sum_{m=1}^{M}\sum_{n=1}^{M}\left[c_{mn}^2\left(\frac{m\pi}{a}\right)^4\sin^2\left(\frac{m\pi x}{a}\right)\right.$$

$$\left.\sin^2\left(\frac{n\pi y}{b}\right)\right]dx\,dy$$

After integration and evaluation, this term becomes

$$\frac{1}{2}\int_0^b\int_0^a D_{11}\left(\frac{\partial^2 w}{\partial x^2}\right)^2 dx\,dy = \frac{\pi^4 b}{8a^3}D_{11}\sum_{m=1}^{M}\sum_{n=1}^{N}m^4 c_{mn}^2$$

The following terms also appear in Equation 12.17 and are evaluated in a similar manner:

$$\frac{1}{2}\int_0^b\int_0^a 2D_{12}\left(\frac{\partial^2 w}{\partial x^2}\right)\left(\frac{\partial^2 w}{\partial y^2}\right)dx\,dy = \frac{\pi^4}{4ab}D_{12}\sum_{m=1}^{M}\sum_{n=1}^{N}m^2 n^2 c_{mn}^2$$

$$\frac{1}{2}\int_0^b\int_0^a D_{22}\left(\frac{\partial^2 w}{\partial y^2}\right)^2 dx\,dy = \frac{\pi^4 a}{8b^3}D_{22}\sum_{m=1}^{M}\sum_{n=1}^{N}n^4 c_{mn}^2$$

$$\frac{1}{2}\int_0^b\int_0^a 4D_{66}\left(\frac{\partial^2 w}{\partial x\partial y}\right)^2 dx\,dy = \frac{\pi^4}{2ab}D_{66}\sum_{m=1}^{M}\sum_{n=1}^{N}m^2 n^2 c_{mn}^2$$

The remaining terms in Equation 12.17 involve D_{16} and D_{26}. Upon substituting the appropriate derivatives, the first of these becomes

$$\frac{1}{2}\int_0^b\int_0^a 4D_{16}\left(\frac{\partial^2 w}{\partial x^2}\right)\left(\frac{\partial^2 w}{\partial x \partial y}\right)dx\,dy$$

$$= \frac{1}{2}\int_0^b\int_0^a 4D_{16}\left[-\sum_{m=1}^M\sum_{n=1}^N C_{mn}\left(\frac{m\pi}{a}\right)^2\sin\left(\frac{m\pi x}{a}\right)\sin\left(\frac{n\pi y}{b}\right)\right]$$

$$\left[\sum_{i=1}^M\sum_{j=1}^N C_{ij}\left(\frac{i\pi}{a}\right)\left(\frac{j\pi}{b}\right)\cos\left(\frac{i\pi x}{a}\right)\cos\left(\frac{j\pi y}{b}\right)\right]dx\,dy$$

This integration does not simplify as readily as those considered above. We make use of the following identities:

$$\int_0^a \sin\left(\frac{m\pi x}{a}\right)\cos\left(\frac{i\pi x}{a}\right)dx = \begin{cases} 0, & \text{if } (m+i) \text{ is even} \\[2mm] \dfrac{2\,ma}{\pi\left[m^2 - i^2\right]}, & \text{if } (m+i) \text{ is odd} \end{cases}$$

$$\int_0^b \sin\left(\frac{n\pi x}{b}\right)\cos\left(\frac{j\pi x}{b}\right)dy = \begin{cases} 0, & \text{if } (n+j) \text{ is even} \\[2mm] \dfrac{2\,nb}{\pi\left[n^2 - j^2\right]}, & \text{if } (n+j) \text{ is odd} \end{cases}$$

On the basis of these identities, we find after integration and evaluation

$$\frac{1}{2}\int_0^b\int_0^a 4D_{16}\left(\frac{\partial^2 w}{\partial x^2}\right)\left(\frac{\partial^2 w}{\partial x \partial y}\right)dxdy$$

$$= \frac{-2\pi^2}{a^2}D_{16}\sum_{m=1}^M\sum_{n=1}^N\sum_{i=1}^M\sum_{j=1}^N C_{mn}C_{ij}\left\{\frac{m^3 nij}{[m^2 - i^2][n^2 - j^2]}\right\}(MI)(NJ)$$

where

$$(MI) = [(-1)^m(-1)^i - 1]$$

$$(NJ) = [(-1)^n(-1)^j - 1]$$

Notice that

$$(MI) = \begin{cases} 0, & \text{if } (m + i) \text{ is even} \\ \\ -2, & \text{if } (m + i) \text{ is odd} \end{cases}$$

$$(NJ) = \begin{cases} 0, & \text{if } (n + j) \text{ is even} \\ \\ -2, & \text{if } (n + j) \text{ is odd} \end{cases}$$

The final term that appears in Equation 12.17 is evaluated in a similar manner, and becomes

$$\frac{1}{2} \int_0^b \int_0^a 4D_{26} \left(\frac{\partial^2 w}{\partial y^2} \right) \left(\frac{\partial^2 w}{\partial x \partial y} \right) dx dy$$

$$= \frac{-2\pi^2}{b^2} D_{26} \sum_{m=1}^M \sum_{n=1}^N \sum_{i=1}^M \sum_{j=1}^N c_{mn} c_{ij} \left\{ \frac{mn^3 ij}{[m^2 - i^2][n^2 - j^2]} \right\} (MI)(NJ)$$

We have now integrated all terms that appear in Equation 12.17. Combining these results and rearranging, the integrated form of Equation 12.17 can be written as

$$U_{II} = \sum_{m=1}^M \sum_{n=1}^N \left\{ \frac{\pi^4}{8} c_{mn}^2 \left[\frac{bm^4}{a^3} D_{11} + \frac{2m^2 n^2}{ab} (D_{12} + 2D_{66}) + \frac{an^4}{b^3} D_{22} \right] \right.$$

$$\left. -2\pi^2 mn c_{mn} \sum_{i=1}^M \sum_{j=1}^N \left[c_{ij} \left(\frac{ij}{[m^2 - i^2][n^2 - j^2]} \right) \left(\frac{m^2}{a^2} D_{16} + \frac{n^2}{b^2} D_{26} \right) (MI)(NJ) \right] \right\}$$

$$(12.18)$$

12.3.3 Evaluation of Strain Energy Component U_{III}

Strain energy component U_{III} represents the *change* in strain energy associated with stress resultants N_{ij} caused by bending. We assume that the stress resultants N_{ij} that exist prior to bending *remain constant*; only in-plane strains are changed during bending. We must, therefore, evaluate the change in in-plane strains caused by the development of out-of-plane displacements, $w(x,y)$. The change in midplane strain ε_{xx}^o may be determined by means of Figure 12.1. The figure shows an element of length dx, which represents an infinitesimal element of the midplane *that has already been deformed* by stress

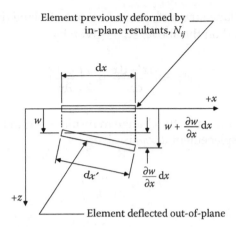

FIGURE 12.1
Sketch used to determine change in in-plane strain caused by out-of-plane deflections.

resultants N_{ij} during step 1, as previously discussed. The length dx is further increased to length dx' if the plate is deflected out-of-plane. From Figure 12.1 we see that

$$dx' = \left[dx^2 + \left(\frac{\partial w}{\partial x}\right)^2 dx^2 \right]^{1/2} = dx\left[1 + \left(\frac{\partial w}{\partial x}\right)^2\right]^{1/2}$$

The quantity within the square bracket and raised to the 1/2 power can be expanded in terms of a binomial power series expansion. A general statement of this series expansion is given by [5]

$$(1 + \xi)^{1/2} = 1 + \frac{1}{2}\xi - \left(\frac{1}{2}\right)\left(\frac{1}{4}\right)\xi^2 + \left(\frac{1}{2}\right)\left(\frac{1}{4}\right)\left(\frac{3}{6}\right)\xi^3 + \cdots$$

We adopt this general formula for our use by retaining only the first two terms and letting $\xi = (\partial w/\partial x)^2$. Hence

$$\left[1 + \left(\frac{\partial w}{\partial x}\right)^2\right]^{1/2} \approx 1 + \frac{1}{2}\left(\frac{\partial w}{\partial x}\right)^2$$

With this approximation, the new length of the element is given by

$$dx' = dx\left[1 + \frac{1}{2}\left(\frac{\partial w}{\partial x}\right)^2\right]$$

The change in-plane strain ε_{xx}^o caused by out-of-plane displacement $w(x,y)$ is labeled $\varepsilon_{xx}^{o,b}$. Based on the above, $\varepsilon_{xx}^{o,b}$ is given by

$$\varepsilon_{xx}^{o,b} = \frac{dx' - dx}{dx} = \frac{1}{2}\left(\frac{\partial w}{\partial x}\right)^2 \tag{12.19a}$$

Using a similar approach, the change in in-plane strains ε_{yy}^o and γ_{xy}^o caused by out-of-plane displacement $w(x,y)$ is given by

$$\varepsilon_{yy}^{o,b} = \frac{1}{2}\left(\frac{\partial w}{\partial y}\right)^2 \tag{12.19b}$$

$$\gamma_{xy}^{o,b} = \left(\frac{\partial w}{\partial x}\right)\left(\frac{\partial w}{\partial y}\right) \tag{12.19c}$$

We can now calculate the strain energy associated with the change in in-plane strains caused by bending. We assume in-plane stress resultants N_{ij} *remain constant* as bending develops. Consider, for example, the force represented by stress resultant N_{xx} acting over an infinitesimal element with length and width dx and dy. The x-directed force is $(N_{xx}dy)$, and the distance through which this force moves as bending develops equals $\varepsilon_{xx}^{o,b}dx$. The incremental strain energy associated with this force and caused by bending is, therefore, $dU_{III} = (N_{xx}dy)(\varepsilon_{xx}^{o,b}dx)$. Analogous expressions hold for stress resultants N_{yy} and N_{xy}. Hence, strain energy U_{III} is given by

$$U_{III} = \iint\left[N_{xx}\varepsilon_{xx}^{o,b} + N_{yy}\varepsilon_{yy}^{o,b} + N_{xy}\gamma_{xy}^{o,b}\right]dxdy \tag{12.20}$$

Equation 12.20 represents a general expression for U_{III}. Recall that during application of the Ritz method we are able to specify geometric boundary conditions directly, but are not able to specify static boundary conditions, at least directly. Therefore, in the present form, Equation 12.20 is inconvenient for use with the Ritz method. That is, we wish to express stress resultants N_{xx}, N_{yy}, and N_{xy} in terms of displacement fields, which will ultimately allow us to specify geometric boundary conditions that represent known values of N_{xx}, N_{yy}, and N_{xy}. From Equation 6.44, we can write (for $B_{ij} = \Delta M = 0$)

$$N_{xx} = A_{11}\varepsilon_{xx}^o + A_{12}\varepsilon_{yy}^o + A_{16}\gamma_{xy}^o - N_{xx}^T = A_{11}\left(\frac{\partial u_o}{\partial x}\right) + A_{12}\left(\frac{\partial v_o}{\partial y}\right)$$

$$+ A_{16}\left(\frac{\partial u_o}{\partial y} + \frac{\partial v_o}{\partial x}\right) - N_{xx}^T$$

Similarly,

$$N_{yy} = A_{12}\left(\frac{\partial u_o}{\partial x}\right) + A_{22}\left(\frac{\partial v_o}{\partial y}\right) + A_{26}\left(\frac{\partial u_o}{\partial y} + \frac{\partial v_o}{\partial x}\right) - N_{yy}^T$$

$$N_{xy} = A_{16}\left(\frac{\partial u_o}{\partial x}\right) + A_{26}\left(\frac{\partial v_o}{\partial y}\right) + A_{66}\left(\frac{\partial u_o}{\partial y} + \frac{\partial v_o}{\partial x}\right) - N_{xy}^T$$

Substituting these expressions as well as Equations 12.19a,b,c into Equation 12.20, we obtain

$$
U_{III} = \frac{1}{2}\int\int\left[\left\{A_{11}\left(\frac{\partial u_o}{\partial x}\right) + A_{12}\left(\frac{\partial v_o}{\partial y}\right) + A_{16}\left(\frac{\partial u_o}{\partial y} + \frac{\partial v_o}{\partial x}\right) - N_{xx}^T\right\}\left(\frac{\partial w}{\partial x}\right)^2\right.
$$
$$
+\left\{A_{12}\left(\frac{\partial u_o}{\partial x}\right) + A_{22}\left(\frac{\partial v_o}{\partial y}\right) + A_{26}\left(\frac{\partial u_o}{\partial y} + \frac{\partial v_o}{\partial x}\right) - N_{yy}^T\right\}\left(\frac{\partial w}{\partial y}\right)^2
$$
$$
\left. +2\left\{A_{16}\left(\frac{\partial u_o}{\partial x}\right) + A_{26}\left(\frac{\partial v_o}{\partial y}\right) + A_{66}\left(\frac{\partial u_o}{\partial y} + \frac{\partial v_o}{\partial x}\right) - N_{xy}^T\right\}\left(\frac{\partial w}{\partial x}\right)\left(\frac{\partial w}{\partial y}\right)\right]dx\,dy
$$

$$(12.21)$$

We will now integrate Equation 12.21 for the class of problems considered in this chapter. We assume the plate is simply supported and that out-of-plane displacements are given by Equation 12.6. We also assume the laminate is symmetric and subjected to uniform stress resultants N_{ij} and/or a temperature change ΔT. For these conditions the in-plane displacement fields are given by Equation 12.7. Substituting Equation 12.6 and Equation 12.7 into Equation 12.21 (and utilizing the simplifying change in notation introduced as Equation 12.15), we have

$$
U_{III} = \frac{1}{2}\int_0^b\int_0^a\left[\left\{A_{11}C_1 + A_{12}C_2 + 2A_{16}C_3 - N_{xx}^T\right\}\right.
$$
$$
\left(\sum_{m=1}^{M}\sum_{n=1}^{N}c_{mn}\left(\frac{m\pi}{a}\right)\cos\left(\frac{m\pi x}{a}\right)\sin\left(\frac{n\pi y}{b}\right)\right)^2
$$
$$
+\left\{A_{12}C_1 + A_{22}C_2 + 2A_{26}C_3 - N_{yy}^T\right\}\left(\sum_{m=1}^{M}\sum_{n=1}^{N}c_{mn}\left(\frac{n\pi}{b}\right)\sin\left(\frac{m\pi x}{a}\right)\cos\left(\frac{n\pi y}{b}\right)\right)^2
$$

$$+ 2\left\{A_{16}C_1 + A_{26}C_2 + 2A_{66}C_3 - N_{xy}^T\right\}\left(\sum_{m=1}^{M}\sum_{n=1}^{N} c_{mn}\left(\frac{m\pi}{a}\right)\cos\left(\frac{m\pi x}{a}\right)\sin\left(\frac{n\pi y}{b}\right)\right)$$

$$\left(\sum_{i=1}^{M}\sum_{j=1}^{N} c_{ij}\left(\frac{j\pi}{b}\right)\sin\left(\frac{i\pi x}{a}\right)\cos\left(\frac{j\pi y}{b}\right)\right)\right]dxdy$$

Integration of this expression is simplified through the use of the trigono-metric identities listed in preceding section. We obtain

$$U_{III} = \sum_{m=1}^{M}\sum_{n=1}^{N}\left\{\frac{\pi^2}{8}\,c_{mn}^2\left[\begin{array}{c}\frac{m^2 b}{a}\left(A_{11}C_1 + A_{12}C_2 + 2A_{16}C_3 - N_{xx}^T\right)\\[2mm]+\frac{n^2 a}{b}\left(A_{12}C_1 + A_{22}C_2 + 2A_{26}C_3 - N_{yy}^T\right)\end{array}\right]\right.$$

$$+ 2m^2 n c_{mn}\left[A_{16}C_1 + A_{26}C_2 + 2A_{66}C_3 - N_{xy}^T\right]$$

$$\left.\sum_{i=1}^{M}\sum_{j=1}^{N}\left[c_{ij}\left(\frac{j}{(m^2 - i^2)(n^2 - j^2)}\right)(MI)(NJ)\right]\right\}$$

This result can be further simplified by substituting Equations 6.73b, which give the thermal stress resultants in terms of elements of the [A] matrix, and Equations 12.5. We finally find

$$U_{III} = \sum_{m=1}^{M}\sum_{n=1}^{N}\left\{\frac{\pi^2}{8}\,c_{mn}^2\left[\frac{m^2 b}{a}N_{xx} + \frac{n^2 a}{b}N_{yy}\right]\right.$$

$$\left.+ 2m^2 n c_{mn} N_{xy}\sum_{i=1}^{M}\sum_{j=1}^{N}\left[c_{ij}\left(\frac{j}{(m^2 - i^2)(n^2 - j^2)}\right)(MI)(NJ)\right]\right\} \quad (12.22)$$

As an aside, it is interesting to note that the change in in-plane strains given by Equations 12.18a,b,c are similar to the nonlinear terms that appear in Green's strain tensor, mentioned in Section 2.14. Recall that Green's strain tensor represents a definition of "finite" strains, which must be accounted for during analyses involving large displacement gradients. For example, finite strain ε_{xx} is defined as

$$\varepsilon_{xx} = \frac{\partial u}{\partial x} + \frac{1}{2}\left[\left(\frac{\partial u}{\partial x}\right)^2 + \left(\frac{\partial v}{\partial x}\right)^2 + \left(\frac{\partial w}{\partial x}\right)^2\right]$$

Since we have incorporated the term $\varepsilon_{xx}^{o,b} = 1/2(\partial w/\partial x)^2$ during our calculation of U_{III}, it would be natural to conclude that our analysis is valid for finite strain levels. This conclusion would be incorrect. Our analysis is based on infinitesimal strains, despite the inclusion of Equations 12.18 during calculation of U_{III}. To perform an analysis based on energy methods that accounts for finite strain levels we would need to develop new expressions comparable to Equations 12.16, 12.18, and 12.20 (i.e., our current expressions for U_I, U_{II}, and U_{III}, respectively), based on the Green strain tensor. In turn, this would require a new derivation of results from Chapter 6; that is, an analysis based on finite strains would require a re-derivation of CLT. Such an analysis is beyond the scope of this textbook and will not be discussed.

12.3.4 Evaluation of Work Done by Transverse Loads

The work done by the transverse load $q(x,y)$ is denoted as W and is calculated in accordance with Equation 12.8. Only simply supported plates are considered in this chapter, and $w(x,y)$ is assumed to be given by Equation 12.6 in all cases. Therefore, the work done by transverse loads is given by

$$W = \iint \{q(x,y)\} \left\{ \sum_{m=1}^{M} \sum_{n=1}^{N} c_{mn} \sin\left(\frac{m\pi x}{a}\right) \sin\left(\frac{n\pi x}{b}\right) \right\} dx\, dy \qquad (12.23)$$

Integration of Equation 12.23 depends on the functional form of transverse load $q(x,y)$. Problems involving various types of transverse loads will be considered in following sections. Equation 12.23 will be integrated as needed during the discussion to follow.

12.4 Symmetric Composite Laminates Subject to Simple Supports of Type S4

In this section, the transverse deflections of simply supported symmetric composite panels will be predicted on the basis of a Ritz analysis. The general approach is to first obtain an expression for the total potential energy Π of the plate, given by

$$\Pi = U_I + U_{II} + U_{III} - W$$

In general, Π is a function of the elastic properties of the plate, plate dimensions, and midplane displacement fields $u_o(x,y)$, $v_o(x,y)$, and $w(x,y)$. For the problems considered herein in-plane displacement fields $u_o(x,y)$ and $v_o(x,y)$

are known while the out-of-plane displacement field $w(x,y)$ is unknown. The out-of-plane displacement field is assumed to be of the form

$$w(x,y) = \sum_{m=1}^{M} \sum_{n=1}^{N} c_{mn} \sin\left(\frac{m\pi x}{a}\right) \sin\left(\frac{n\pi y}{b}\right)$$

The magnitude of out-of-plane deflections are obtained by applying the principle of minimum potential energy, which requires

$$\frac{\partial \Pi}{\partial c_{mn}} = 0 \qquad \begin{cases} m = 1,\ 2,...,M \\ n = 1,\ 2,...,N \end{cases}$$

This process leads to $(M \times N)$ equations that must be satisfied simultaneously. Hence, by solving these equations for constants c_{mn}, the magnitude of out-of-plane displacements that corresponds to the state of minimum potential energy is determined and the problem is solved.

The equations for strain energy components U_I, U_{II}, and U_{III} are identical for all problems considered herein, and were developed in Section 12.3. The work done by the transverse load, W, depends on the nature of the applied load. Solutions for a few common transverse loads are presented in the following subsections.

12.4.1 Deflections due to a Uniform Transverse Load

Consider a composite plate subjected to a constant and uniform transverse load, $q(x,y) = q_0$. In this case, Equation 12.23 becomes

$$W = q_0 \int_0^b \int_0^a \left\{ \sum_{m=1}^{M} \sum_{n=1}^{N} c_{mn} \sin\left(\frac{m\pi x}{a}\right) \sin\left(\frac{n\pi x}{b}\right) \right\} dx\,dy$$

During integration of this expression, we note the following:

$$\int_0^a \sin\left(\frac{m\pi x}{a}\right) dx = \begin{cases} 0, & \text{if } m \text{ is even} \\[2mm] \dfrac{2a}{m\pi}, & \text{if } m \text{ is odd} \end{cases}$$

$$\int_0^b \sin\left(\frac{n\pi x}{b}\right) dy = \begin{cases} 0, & \text{if } n \text{ is even} \\[2mm] \dfrac{2b}{n\pi}, & \text{if } n \text{ is odd} \end{cases}$$

Hence, the work done by a uniform transverse load can be written as

$$W = \sum_{m=1}^{M} \sum_{n=1}^{N} \left\{ \frac{abq_o c_{mn}}{\pi^2 mn} \left[(-1)^m - 1 \right] \left[(-1)^n - 1 \right] \right\} \tag{12.24}$$

The total potential energy can now be obtained by combining Equations 12.16, 12.18, 12.22, and 12.24:

$$\Pi = \left\{ \left[C_1 N_{xx} + C_2 N_{yy} + 2C_3 N_{xy} \right] (ab) \right\}$$

$$+ \left\{ \begin{array}{l} \displaystyle\sum_{m=1}^{M} \sum_{n=1}^{N} \left\{ \frac{\pi^4}{8} c_{mn}^2 \left[\frac{bm^4}{a^3} D_{11} + \frac{2m^2 n^2}{ab} (D_{12} + 2D_{66}) + \frac{an^4}{b^3} D_{22} \right] \right. \\[2.5em] \displaystyle\left. -2\pi^2 mn c_{mn} \sum_{i=1}^{M} \sum_{j=1}^{N} \left[c_{ij} \left(\frac{ij}{[m^2 - i^2][n^2 - j^2]} \right) \left(\frac{m^2}{a^2} D_{16} + \frac{n^2}{b^2} D_{26} \right) (MI)(NJ) \right] \right\} \end{array} \right\}$$

$$+ \left\{ \begin{array}{l} \displaystyle\sum_{m=1}^{M} \sum_{n=1}^{N} \left\{ \frac{\pi^2}{8} c_{mn}^2 \left[\frac{m^2 b}{a} N_{xx} + \frac{n^2 a}{b} N_{yy} \right] \right. \\[2.5em] \displaystyle\left. + 2m^2 n c_{mn} N_{xy} \sum_{i=1}^{M} \sum_{j=1}^{N} \left[c_{ij} \left(\frac{j}{(m^2 - i^2)(n^2 - j^2)} \right) (MI)(NJ) \right] \right\} \end{array} \right\}$$

$$- \left\{ \sum_{m=1}^{M} \sum_{n=1}^{N} \left\{ \frac{abq_o c_{mn}}{\pi^2 mn} \left[(-1)^m - 1 \right] \left[(-1)^n - 1 \right] \right\} \right\} \tag{12.25}$$

The four individual energy components U_I, U_{II}, U_{III}, and W are shown within the large braces in Equation 12.25. This expression is unwieldy, so a change in notation will be made to facilitate inspection of the mathematical structure of Π. Toward that end, define the following constants:

$$F_1 = \frac{\pi^4 b}{8a^3} D_{11} \qquad F_2 = \frac{2\pi^4}{8ab}(D_{12} + 2D_{66}) \qquad F_3 = \frac{\pi^4 a}{8b^3} D_{22}$$

$$F_4 = \frac{\pi^2 b}{8a} N_{xx} \qquad F_5 = \frac{\pi^2 a}{8b} N_{yy} \qquad F_6 = \frac{-2\pi^2}{a^2} D_{16}$$

$$F_7 = \frac{-2\pi^2}{b^2} D_{26} \qquad F_8 = 2N_{xy} \qquad F_9 = \frac{-abq_o}{\pi^2}$$

Note that these terms are all known constants for a given laminate and loading condition. On the basis of these definitions, Equation 12.25 can be rearranged as follows:

$$
\Pi = U_I + \sum_{m=1}^{M} \sum_{n=1}^{N} \begin{bmatrix} c_{mn}^2 \left\{ F_1 m^4 + F_2 m^2 n^2 + F_3 n^4 + F_4 m^2 + F_5 n^2 \right\} \\[2mm] + c_{mn} \left\{ \dfrac{F_9}{mn} \left[(-1)^m - 1\right]\left[(-1)^n - 1\right] \right\} \\[2mm] + c_{mn} \left\{ \sum_{i=1}^{M} \sum_{j=1}^{N} \begin{bmatrix} \dfrac{c_{ij}}{(m^2 - i^2)(n^2 - j^2)} \\[2mm] \left\{ F_6(m^3 nij) + F_7(mn^3 ij) + F_8(m^2 nj) \right\}(MI)(NJ) \end{bmatrix} \right\} \end{bmatrix}
$$

$$(12.26)$$

To further explore the Ritz method, we must now expand the expression for Π, based on some specified values of M and N. In general, the accuracy of the Ritz approach is improved as M and N are increased. Although not required, it is usual practice to let $M = N$, which means that the number of terms with x- and y- dependency in Equation 12.6 is identical. Often, 100 terms or more ($M = N = 10$, or more) are necessary to obtain a reasonable convergence of the Ritz solution. Writing down the expanded form of Π based on values of M and N as high as 10 is obviously untenable. For purposes of illustration, we will expand Π using $M = N = 2$, which will allow us to explore the essential elements of the Ritz analysis.

Hence, expanding our Equation 12.26 based on $M = N = 2$, we find

$$
\Pi = U_I + c_{11}^2 \left(F_1 + F_2 + F_3 + F_4 + F_5 \right)
$$

$$
+ c_{12}^2 (F_1 + 4F_2 + 16F_3 + F_4 + 4F_5)
$$

$$
+ c_{21}^2 (16F_1 + 4F_2 + F_3 + 4F_4 + F_5)
$$

$$
+ 4c_{22}^2 (4F_1 + 4F_2 + 4F_3 + F_4 + F_5)
$$

$$
+ \frac{40}{9} c_{11} c_{22} \left(2F_6 + 2F_7 + F_8 \right)
$$

$$
- \frac{40}{9} c_{12} c_{21} (2F_6 + 2F_7 + F_8)
$$

$$
- 4c_{11} F_9
$$

$$(12.27)$$

Two important features of Equation 12.27 should be noted. First, the expression for Π is a second-order polynomial in terms of the unknown coefficients c_{mn}. Second, the term involving F_9 is a linear function of c_{mn}. Referring to the definitions of F_1 through F_9 listed above, it is seen that F_9 is the only term related to the transverse load q_o. Hence, that portion of the total expression for Π which is related to the transverse loading is a *linear* function of the unknown coefficients c_{mn}. These two characteristics always hold for Π, regardless of the values of M and N or the nature of the transverse loading. That is, for Π is always a second-order polynomial in c_{mn}, and terms involving the transverse load are always linear functions of c_{mn}.

Proceeding with the Ritz analysis, we now apply the principle of minimum potential energy. That is, we wish to identify the particular values of c_{mn} dictated by

$$\frac{\partial \Pi}{\partial c_{mn}} = 0$$

Since $M = N = 2$ in this example, we must take four partial derivatives of Π, each of which will represent an independent equation that is then equated to zero. For example, taking the derivative of Π with respect to c_{11} and equating to zero, we have

$$\frac{\partial \Pi}{\partial c_{11}} = 2(F_1 + F_2 + F_3 + F_4 + F_5)c_{11} + \frac{40}{9}(2F_6 + 2F_7 + F_8)c_{22} - 4F_9 = 0$$

Three additional equations are also formed ($\partial \Pi/\partial c_{12} = \partial \Pi/\partial c_{21} = \partial \Pi/\partial c_{22} = 0$). The four equations can be represented using matrix notation as shown in Figure 12.2a. These equations may be easily solved for coefficients c_{mn} by multiplying both sides of the equation by the inverse of the [4×4] array, as shown in Figure 12.2b. Once coefficients c_{mn} have been determined the out-of-plane deflections can be calculated using Equation 12.6 and the problem is solved.

Referring to the definitions of F_1 through F_9 listed above, it is seen that normal stress resultants N_{xx} and N_{yy} appear only in terms F_4 and F_5, respectively, while the shear stress resultant N_{xy} appears only in term F_8. Inspection of the [4×4] array shown on the left side of the equality in Figure 12.2a reveals that F_4 and F_5 appear only along the main diagonal of the matrix, whereas F_8 appears only in off-diagonal positions. This pattern always occurs, regardless of the value of the value of M and N or the nature of the transverse loading—normal stress resultants appear only along the main diagonal while the shear stress resultant appears only in off-diagonal positions.

In summary, Figure 12.2 represents the solution for a simply supported symmetric composite panel subjected to a uniform transverse pressure q_o, uniform in-plane stress resultants N_{xx}, N_{yy}, and N_{xy}, and a uniform change in temperature, ΔT, based on $M = N = 2$. Depending on details of a specific

(a)

$$
\begin{bmatrix}
2(F_1+F_2+F_3+F_4+F_5) & 0 & 0 & \frac{40}{9}(2F_6+2F_7+F_8) \\
0 & 2(F_1+4F_2+16F_3+F_4+4F_5) & -\frac{40}{9}(2F_6+2F_7+F_8) & 0 \\
0 & -\frac{40}{9}(2F_6+2F_7+F_8) & 2(16F_1+4F_2+F_3+4F_4+F_5) & 0 \\
\frac{40}{9}(2F_6+2F_7+F_8) & 0 & 0 & 8(4F_1+4F_2+4F_3+F_4+F_5)
\end{bmatrix}
\begin{bmatrix} c_{11} \\ c_{12} \\ c_{21} \\ c_{22} \end{bmatrix}
=
\begin{bmatrix} 4F_9 \\ 0 \\ 0 \\ 0 \end{bmatrix}
$$

(b)

$$
\begin{bmatrix} c_{11} \\ c_{12} \\ c_{21} \\ c_{22} \end{bmatrix}
=
\begin{bmatrix}
2(F_1+F_2+F_3+F_4+F_5) & 0 & 0 & \frac{40}{9}(2F_6+2F_7+F_8) \\
0 & 2(F_1+4F_2+16F_3+F_4+4F_5) & -\frac{40}{9}(2F_6+2F_7+F_8) & 0 \\
0 & -\frac{40}{9}(2F_6+2F_7+F_8) & 2(16F_1+4F_2+F_3+4F_4+F_5) & 0 \\
\frac{40}{9}(2F_6+2F_7+F_8) & 0 & 0 & 8(4F_1+4F_2+4F_3+F_4+F_5)
\end{bmatrix}^{-1}
\begin{bmatrix} 4F_9 \\ 0 \\ 0 \\ 0 \end{bmatrix}
$$

FIGURE 12.2
Summary of the solution obtained for a simply-supported laminate subjected to a uniform transverse pressure using a Ritz analysis and $M = N = 2$. (a) The set of simultaneous equations obtained by enforcing $\partial\Pi/\partial c_{mn} = 0$, for $M = N = 2$; (b) solving the set of simultaneous equations shown in part (a).

problem, the number of terms required to insure convergence may be substantially higher than only four terms; it is not uncommon to require 100 terms or more ($M = N = 10$ or more). Obviously, the solution process presented above is rarely (if ever) performed by hand calculation. Rather, a computer-based routine is typically used to expand Equation 12.26 (or equivalently, Equation 12.25) for specified values of M and N, to perform the required partial differentiation, to determine the inverse of the resulting $[M \times N]$ matrix, and to complete the final matrix multiplication that gives the coefficients c_{mn}. Typical solutions obtained on the basis of the Ritz approach are illustrated in the following three example problems.

Example Problem 12.1

A $[(\pm45/0)_2]_s$ graphite–epoxy laminate is cured at 175°C and then cooled to room temperature (20°C). After cooling, the flat laminate is trimmed to in-plane dimensions of 300×150 mm and mounted in an assembly that provides type S4 simple supports along all four edges. The x-axis is defined parallel to the 300 mm edge (i.e., $a = 0.3$ m, $b = 0.15$ m). The laminate is then subjected to a uniform transverse load $q(x,y) = 30$ kPa. No in-plane loads are applied (i.e., $N_{xx} = N_{yy} = N_{xy} = 0$). Determine the maximum out-of-plane displacement based on a Ritz analysis and plot the out-of-plane displacement field. Use the properties listed for graphite–epoxy in Table 3.1, and assume each ply has a thickness of 0.125 mm.

SOLUTION

Based on the properties listed in Table 3.1 for graphite–epoxy, the $[ABD]$ matrix for a $[(\pm45/0)_2]_s$ laminate is

$$[ABD] = \begin{bmatrix} 145.2 \times 10^6 & 35.3 \times 10^6 & 0 & 0 & 0 & 0 \\ 35.3 \times 10^6 & 64.8 \times 10^6 & 0 & 0 & 0 & 0 \\ 0 & 0 & 50.2 \times 10^6 & 0 & 0 & 0 \\ 0 & 0 & 0 & 22.3 & 7.97 & 2.20 \\ 0 & 0 & 0 & 7.97 & 14.3 & 2.20 \\ 0 & 0 & 0 & 2.20 & 2.20 & 10.8 \end{bmatrix}$$

where the units of A_{ij} are Pa – m and the units of D_{ij} are Pa – m^3. Notice that neither D_{16} nor D_{26} equal zero, and hence the laminate is generally orthotropic. The 12-ply laminate has a total thickness $t = 1.5$ mm and aspect ratio $R = a/b = 2.0$.

The computer program *SYMM* (described in Section 12.6) can be used to perform the required Ritz analysis. Several analyses were performed using increasing values of M (and N), to evaluate whether the solution has converged to a reasonably constant value. Solutions were obtained using values of M (and N) ranging from 1 through 10 (i.e., analyses were performed in which the number of terms used to describe the displacement field ranged from 1 through 100). Maximum predicted displacement is plotted as a function of M and N in Figure 12.3. As indicated, the maximum displacement converges to a value of 8.03 mm when $M = N = 10$.

A contour plot of out-of-plane displacements predicted using $M = N = 10$ is shown in Figure 12.4. As would be expected, the maximum displacement occurs at the center of the plate (i.e., at $x = 150$ mm, $y = 75$ mm). Careful examination of these contours will

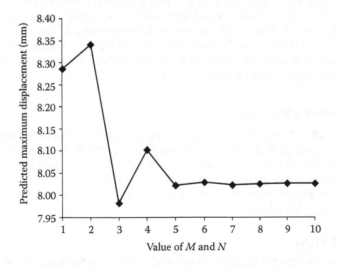

FIGURE 12.3
Convergence of predicted plate deflections based on a Ritz analysis as M and N are increased from 1 to 10 (i.e., as the number of terms is increased from 1 to 100).

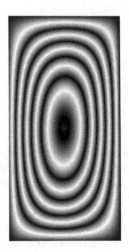

FIGURE 12.4
A contour plot of out-of-plane displacements for the $[(\pm45/0)_2]_s$ laminate considered in Example Problem 12.1.

reveal that the contours are very slightly distorted. This distortion (which is barely discernible in Figure 12.4) occurs because the plate is generally orthotropic. That is, for a $[(\pm45/0)_2]_s$ laminate D_{16}, $D_{26} \neq 0$. However, for this problem the magnitudes of D_{16} and D_{26} (relative to D_{11} and D_{22}) are very small. Specifically, for the laminate considered in this problem $D_{16}/D_{11} = D_{26} D_{11} = 0.0986$ and $D_{16}/D_{22} = D_{26} D_{22} = 0.153$. Consequently, distortion of out-of-plane displacements is very slight. The out-of-plane displacement induced by a uniform transverse load applied to a laminate with relatively higher values of D_{16} and D_{26} is considered in Example Problem 12.3. As will be seen, the distortion of displacement contours is much more pronounced in that case, due to the relatively higher values of D_{16} and D_{26}.

Example Problem 12.2

The $[(\pm45/0)_2]_s$ graphite–epoxy laminate described in Example Problem 12.1 is again subjected to type S4 simple supports along all four edges and a uniform transverse load $q(x,y) = 30$ kPa. However, the plate is now also subjected to uniform in-plane stress resultants $N_{xx} = N_{yy}$. Use a Ritz analysis to determine the maximum out-of-plane displacement for $0 < N_{xx} = N_{yy} < 100$ kN/m.

SOLUTION

As was the case for Example Problem 12.1, solutions for this problem can be obtained using program *SYMM* (described in Section 12.6). Multiple solutions were obtained using the specified range in N_{xx} and N_{yy}, and 100 terms were used in the displacement field in all cases.

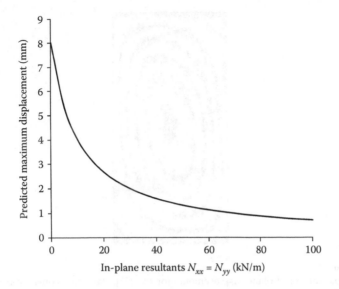

FIGURE 12.5
Predicted maximum deflections of the plate considered in Example Problem 12.2 ($M = N = 10$ in all cases).

Results are summarized in Figure 12.5. The maximum displacement occurs at the center of the plate (i.e., at $x = 150$ mm, $y = 75$ mm). As would be expected, the in-plane tensile stress resultants tend to reduce out-of-plane displacement. For $N_{xx} = N_{yy} = 0$ a maximum out-of-plane displacement of 8.03 mm is predicted. In contrast, if tensile stress resultants $N_{xx} = N_{yy} = 100$ kN/m are applied the maximum out-of-plane displacement is reduced to 0.71 mm.

Example Problem 12.3

A $[25°]_{12}$ graphite–epoxy laminate is trimmed to in-plane dimensions of 300×150 mm and mounted in an assembly that provides type S4 simple supports along all four edges. The laminate is then subjected to a uniform transverse load $q(x,y) = 30$ kPa. No in-plane loads are applied (i.e., $N_{xx} = N_{yy} = N_{xy} = 0$). Determine the maximum out-of-plane displacement based on a Ritz analysis and plot the out-of-plane displacement field. Use the properties listed for graphite–epoxy in Table 3.1, and assume each ply has a thickness of 0.125 mm.

SOLUTION

Note that the plate has an aspect ratio $R = 150/300 = 2.0$, as was the case for the laminates considered in Example Problems 12.1 and 12.2. A rather unusual fiber angle of 25° has been selected for consideration in this problem because it results in high relative values of D_{16} and D_{26},

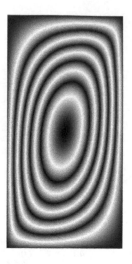

FIGURE 12.6
A contour plot of out-of-plane displacements for the $[25]_{12}$ laminate considered in Example Problem 12.3.

resulting in an interesting distortion of the predicted out-of-plane displacement field. Specifically, for this laminate:

$$D_{16}/D_{11} = 0.370 \qquad D_{16}/D_{22} = 2.21$$

$$D_{26}/D_{11} = 0.0.126 \qquad D_{26}/D_{22} = 0.755$$

These relative values of D_{16} and D_{26} are quite high, at least as compared to those exhibited by the $[(\pm45/0)_2]_s$ laminate considered in Example Problem 12.1.

A solution for this problem was obtained using program *SYMM*, using $M = N = 10$. A maximum displacement of 16.1 mm is predicted to occur at the center of the plate. A contour plot of out-of-plane displacements is shown in Figure 12.6. Distortion of the displacement field due to the generally orthotropic nature of the $[25°]_{12}$ panel is obvious, especially when compared to the very slightly distorted pattern for a $[(\pm45/0)_2]_s$ laminate, previously shown in Figure 12.4.

12.4.2 Deflections due to a Sinusoidal Transverse Load

Consider a simply supported symmetric composite plate subjected to a transverse load that varies sinusoidally over the surface of the plate:

$$q(x,y) = q_o \sin\left(\frac{\pi x}{a}\right)\sin\left(\frac{\pi y}{b}\right)$$

In this case, the work done by the transverse load, calculated using Equation 12.23, is given by

$$W = q_o \int_0^b \int_0^a \left\{ \sin\left(\frac{\pi x}{a}\right) \sin\left(\frac{\pi y}{b}\right) \right\} \left\{ \sum_{m=1}^M \sum_{n=1}^N c_{mn} \sin\left(\frac{m\pi x}{a}\right) \sin\left(\frac{n\pi x}{b}\right) \right\} dx\, dy$$

During integration of this expression, we note the following:

$$\int_0^a \sin\left(\frac{\pi x}{a}\right) \sin\left(\frac{m\pi x}{a}\right) dx = \begin{cases} 0, & \text{if } m \neq 1 \\[2mm] \dfrac{a}{2}, & \text{if } m = 1 \end{cases}$$

$$\int_0^b \sin\left(\frac{\pi y}{b}\right) \sin\left(\frac{n\pi x}{b}\right) dy = \begin{cases} 0, & \text{if } n \neq 1 \\[2mm] \dfrac{b}{2}, & \text{if } n = 1 \end{cases}$$

Hence, the work done by a sinusoidal transverse load is simply

$$W = \frac{q_o abc_{11}}{4}$$

The total potential energy is $\Pi = U_I + U_{II} + U_{III} + W$. Our earlier expressions for U_I, U_{II}, and U_{III} are not altered by the change in transverse load and are given by Equations 12.16, 12.18, and 12.22, respectively. Hence, the total potential energy is

$$\Pi = \left\{ \left[C_1 N_{xx} + C_2 N_{yy} + 2C_3 N_{xy} \right](ab) \right\}$$

$$+ \left\{ \begin{aligned} &\sum_{m=1}^M \sum_{n=1}^N \left\{ \frac{\pi^4}{8} c_{mn}^2 \left[\frac{bm^4}{a^3} D_{11} + \frac{2m^2 n^2}{ab}(D_{12} + 2D_{66}) + \frac{an^4}{b^3} D_{22} \right] \right. \\ &\left. - 2\pi^2\, mn c_{mn} \sum_{i=1}^M \sum_{j=1}^N \left[c_{ij} \left(\frac{ij}{[m^2 - i^2][n^2 - j^2]} \right) \left(\frac{m^2}{a^2} D_{16} + \frac{n^2}{b^2} D_{26} \right)(MI)(NJ) \right] \right\} \end{aligned} \right\}$$

$$+\left\{\begin{array}{l}\displaystyle\sum_{m=1}^{M}\sum_{n=1}^{N}\left\{\frac{\pi^2}{8}\,c_{mn}^2\left[\frac{m^2b}{a}N_{xx}+\frac{n^2a}{b}N_{xx}\right]\right.\\[2ex]\left.+\,2m^2nc_{mn}N_{xy}\sum_{i=1}^{M}\sum_{j=1}^{N}\left[c_{ij}\left(\frac{j}{(m^2-i^2)(n^2-j^2)}\right)(MI)(NJ)\right]\right\}\end{array}\right\}-\left\{\frac{q_0abc}{4}11\right\}$$

(12.28)

As before, the next step is to apply the principle of minimum potential energy:

$$\frac{\partial\Pi}{\partial c_{mn}}=0 \qquad \begin{cases}m=1,\,2,...,M\\ n=1,\,2,...,N\end{cases}$$

This process leads to $(M\times N)$ equations (similar in form to those shown in Figure 12.2) that must be satisfied simultaneously. Hence, by solving these equations for constants c_{mn} the out-of-plane displacements caused by a transverse load that varies sinusoidally over the surface of the plate can be calculated using Equation 12.6, and the problem is solved.

12.4.3 Deflections due to a Transverse Load Distributed over an Interior Region

A composite panel subjected to a transverse force P uniformly distributed over an internal rectangular region of dimensions $a_i\times b_i$ was previously shown in Figure 10.10. The center of the interior region is located at $x=\xi$ and $y=\eta$. As discussed in Section 10.4, this loading can be expressed in terms of a Fourier series expansion:

$$q(x,y)=\frac{16P}{\pi^2 a_ib_i}\sum_{m=1}^{\infty}\sum_{n=1}^{\infty}\left\{\left[\frac{1}{mn}\sin\left(\frac{m\pi\xi}{a}\right)\sin\left(\frac{n\pi\eta}{b}\right)\sin\left(\frac{m\pi a_i}{2a}\right)\sin\left(\frac{n\pi b_i}{2b}\right)\right]\sin\left(\frac{m\pi x}{a}\right)\sin\left(\frac{n\pi y}{b}\right)\right\}$$

Substituting the above into the expression representing the work done by transverse loads W (Equation 12.23), and integrating, it will be found

$$W=\frac{4abP}{\pi^2 a_ib_i}\sum_{m=1}^{M}\sum_{n=1}^{N}\left\{\frac{c_{mn}}{mn}\sin\left(\frac{m\pi\xi}{a}\right)\sin\left(\frac{n\pi\eta}{b}\right)\sin\left(\frac{m\pi a_i}{2a}\right)\sin\left(\frac{n\pi b_i}{2b}\right)\right\}$$ (12.29a)

The total potential energy is $\Pi = U_I + U_{II} + U_{III} + W$. Our earlier expressions for U_I, U_{II}, and U_{III} are not altered by the change in transverse load and are given by Equations 12.16, 12.18, and 12.22, respectively. Hence, the total potential energy is

$$\Pi = \left\{\left[C_1 N_{xx} + C_2 N_{yy} + 2C_3 N_{xy}\right](ab)\right\}$$

$$+ \left\{
\begin{array}{l}
\displaystyle\sum_{m=1}^{M}\sum_{n=1}^{N}\left\{\frac{\pi^4}{8}\, c_{mn}^2 \left[\frac{bm^4}{a^3}D_{11} + \frac{2m^2 n^2}{ab}(D_{12} + 2D_{66}) + \frac{an^4}{b^3}D_{22}\right]\right. \\[3ex]
\displaystyle\left. - 2\pi^2 mnc_{mn}\sum_{i=1}^{M}\sum_{j=1}^{N}\left[c_{ij}\left(\frac{ij}{[m^2 - i^2][n^2 - j^2]}\right)\left(\frac{m^2}{a^2}D_{16} + \frac{n^2}{b^2}D_{26}\right)(MI)(NJ)\right]\right\}
\end{array}
\right\}$$

$$+ \left\{
\begin{array}{l}
\displaystyle\sum_{m=1}^{M}\sum_{n=1}^{N}\left\{\frac{\pi^2}{8}\, c_{mn}^2\left[\frac{m^2 b}{a}N_{xx} + \frac{n^2 a}{b}N_{yy}\right]\right. \\[3ex]
\displaystyle\left. + 2m^2 nc_{mn}N_{xy}\sum_{i=1}^{M}\sum_{j=1}^{N}\left[c_{ij}\left(\frac{j}{(m^2 - i^2)(n^2 - j^2)}\right)(MI)(NJ)\right]\right\}
\end{array}
\right\}$$

$$- \left\{\frac{4abP}{\pi^2 a_i b_i}\sum_{m=1}^{M}\sum_{n=1}^{N}\left\{\frac{c_{mn}}{mn}\sin\left(\frac{m\pi\xi}{a}\right)\sin\left(\frac{n\pi\eta}{b}\right)\sin\left(\frac{m\pi a_i}{2a}\right)\sin\left(\frac{n\pi b_i}{2b}\right)\right\}\right\}$$

(12.29b)

As before, the next step is to apply the principle of minimum potential energy:

$$\frac{\partial\Pi}{\partial c_{mn}} = 0 \qquad \begin{cases} m = 1,\ 2, ..., M \\ n = 1,\ 2, ..., N \end{cases}$$

This process leads to $(M \times N)$ equations (similar in form to those shown in Figure 12.2) that must be satisfied simultaneously. Hence, by solving these equations for constants c_{mn} the out-of-plane displacements caused by a transverse force P uniformly distributed over an interior rectangular region of dimensions $a_i \times b_i$ can be calculated using Equation 12.6, and the problem is solved.

12.4.4 Deflections due to a Transverse Point Load

The work done by a concentrated point load P applied at $x = \xi$ and $y = \eta$ can be obtained by allowing $a_i \to 0$, $b_i \to 0$ in Equation 12.29. In the limit, we obtain

$$W = P \sum_{m=1}^{M} \sum_{n=1}^{N} \left[c_{mn} \sin\left(\frac{m\pi\xi}{a}\right) \sin\left(\frac{n\pi\eta}{b}\right) \right]$$

The total potential energy is

$$\Pi = \left\{ \left[C_1 N_{xx} + C_2 N_{yy} + 2C_3 N_{xy} \right] (ab) \right\}$$

$$+ \left\{ \begin{array}{l} \displaystyle\sum_{m=1}^{M} \sum_{n=1}^{N} \left\{ \frac{\pi^4}{8} c_{mn}^2 \left[\frac{bm^4}{a^3} D_{11} + \frac{2m^2 n^2}{ab} (D_{12} + 2D_{66}) + \frac{an^4}{b^3} D_{22} \right] \right. \\ \\ \left. - 2\pi^2 mn c_{mn} \displaystyle\sum_{i=1}^{M} \sum_{j=1}^{N} \left[c_{ij} \left(\frac{ij}{[m^2 - i^2][n^2 - j^2]} \right) \left(\frac{m^2}{a^2} D_{16} + \frac{n^2}{b^2} D_{26} \right) (MI)(NJ) \right] \right\} \end{array} \right\}$$

$$+ \left\{ \begin{array}{l} \displaystyle\sum_{m=1}^{M} \sum_{n=1}^{N} \left\{ \frac{\pi^2}{8} c_{mn}^2 \left[\frac{m^2 b}{a} N_{xx} + \frac{n^2 a}{b} N_{yy} \right] \right. \\ \\ \left. + 2m^2 n c_{mn} N_{xy} \displaystyle\sum_{i=1}^{M} \sum_{j=1}^{N} \left[c_{ij} \left(\frac{j}{(m^2 - i^2)(n^2 - j^2)} \right) (MI)(NJ) \right] \right\} \end{array} \right\}$$

$$- \left\{ P \sum_{m=1}^{M} \sum_{n=1}^{N} \left[c_{mn} \sin\left(\frac{m\pi\xi}{a}\right) \sin\left(\frac{n\pi\eta}{b}\right) \right] \right\} \qquad (12.30)$$

We apply the principle of minimum potential energy:

$$\frac{\partial \Pi}{\partial c_{mn}} = 0 \qquad \begin{cases} m = 1,\ 2,...,M \\ n = 1,\ 2,...,N \end{cases}$$

This process leads to $(M \times N)$ equations (similar in form to those shown in Figure 12.2) that must be satisfied simultaneously. Hence, by solving these equations for constants c_{mn} the out-of-plane displacements caused by a concentrated load P applied at $x = \xi$ and $y = \eta$ can be calculated using Equation 12.6, and the problem is solved.

12.5 Buckling of Symmetric Composite Plates Subject to Simple Supports of Type S4

In this section, buckling of type S4 simply supported symmetric composite panels will be considered. A brief explanation of what is meant by "buckling" was presented in Section 11.4. In essence, the term "buckling" refers to the fact that (under the proper circumstances) in-plane stress resultants N_{xx}, N_{yy}, and/or N_{xy} can cause out-of-plane displacements $w(x,y)$. A coupling between in-plane loads and out-of-plane displacement is of course expected for non-symmetric laminates, since $B_{ij} \neq 0$ in this case. We have limited discussion to symmetric laminates, however, so we are considered with a coupling between in-plane loads and out-of-plane displacement that is not predicted by the CLT analysis developed in Chapter 6.

The buckling phenomenon can be explained on the basis of the principle of minimum potential energy as follows. Based on our earlier discussion, the total potential energy Π of an initially flat plate subjected to in-plane loading is given by

$$\Pi = U_I + U_{II} + U_{III}$$

Note that the work term W no longer appears in our expression for Π, since we assume that no transverse loads are applied. General expressions for strain energy components U_I, U_{II}, U_{III} for a composite panel have been presented as Equations 12.14, 12.17, and 12.21, respectively. Inspection of these equations reveals that U_I is a function of in-plane displacements $u_o(x,y)$ and $v_o(x,y)$, that U_{II} is a function of out-of-plane displacements $w(x,y)$, and that U_{III} is a function of u_o, v_o, and w. If an initially flat plate is subjected to relatively low levels of in-plane loads, then only in-plane displacements occur—that is, the plate remains flat, because at low load levels the flat configuration corresponds to the state of minimum potential energy. In this case, U_I is the only nonzero strain energy component: $U_{II} = U_{III} = 0$, since $w(x,y) = 0$. However, as the magnitude of in-plane loading is increased then the configuration that corresponds to the state of minimum potential energy may no longer be flat but rather bent—the configuration that corresponds to the state of minimum potential energy involves out-of-plane displacement $w(x,y)$ and the plate "buckles." Once buckling occurs then U_I, U_{II}, $U_{III} \neq 0$.

In Sections 12.3 and 12.4, we developed expressions for the potential energy of a simply supported symmetric composite panel subjected to both in-plane and transverse loads. These earlier results can be adopted for present purposes by simply discarding those terms involving the transverse load. Hence, from Equation 12.26, we can immediately write

$$\Pi = U_I + \sum_{m=1}^{M} \sum_{n=1}^{N} \left[c_{mn}^2 \left\{ F_1 m^4 + F_2 m^2 n^2 + F_3 n^4 + F_4 m^2 + F_5 n^2 \right\} \right.$$

$$+ c_{mn} \left\{ \sum_{i=1}^{M} \sum_{j=1}^{N} \left[\begin{array}{c} \dfrac{c_{ij}}{(m^2 - i^2)(n^2 - j^2)} \\ \left\{ F_6 \left(m^3 n i j \right) + F_7 \left(m n^3 i j \right) + F_8 (m^2 n j \right\} (MI)(NJ) \end{array} \right] \right\} \right] \qquad (12.31)$$

The constants F_1 through F_8 were defined in Section 12.4, in conjunction with Equation 12.26. Note that one constant, namely F_9, no longer appears. This term has been dropped because it represents a transverse load that has now been assumed to be zero.

We can predict the onset of buckling by applying the principle of minimum potential energy:

$$\frac{\partial \Pi}{\partial c_{mn}} = 0 \qquad \begin{cases} m = 1, \ 2, ..., M \\ n = 1, \ 2, ..., N \end{cases}$$

This process leads to $(M \times N)$ equations that must be satisfied simultaneously. For purposes of illustration, consider the set of equations that are obtained using $M = N = 2$. In this case, we obtain the same set of equations as those previously illustrated in Figure 12.2a, except that now $F_9 = 0$. Referring to Figure 12.2a, we see that in the present case, all terms on the right side of the equality are zero.

We must now specify the loading condition of interest. That is, we must specify the loading conditions under which buckling is to be predicted. There are many possibilities, even though we have limited discussion in this chapter to type S4 simple supports. For example, we may wish to determine the value of N_{xx} that will cause buckling, given some constant values for N_{yy} and N_{xy}. Alternatively, we may wish to determine the value of N_{xy} that will cause buckling, given some constant values for N_{xx} and N_{yy}. A third possibility is a loading situation in which all three resultants are increased proportionately (i.e., $N_{yy} = k_1 N_{xx}$ and $N_{xy} = k_2 N_{xx}$, where k_1 and k_2 are known constants). In this case, we are interested in buckling caused by a simultaneous increase in N_{xx}, N_{yy}, and N_{xy}.

For illustrative purposes let us assume we are interested in buckling caused by uniaxial loading: that is, for the loading condition $N_{xx} \neq 0$, $N_{yy} = N_{xy} = 0$. In this case, we have

$$F_5 = \frac{\pi^2 a}{8b} N_{yy} = 0 \qquad F_8 = 2N_{xy} = 0$$

Also, $F_9 = 0$, since no transverse load is applied. For this situation, the set of equations shown in Figure 12.2a reduce to

$$
\begin{bmatrix}
2(F_1+F_2+F_3+F_4) & 0 & 0 & \dfrac{40}{9}(2F_6+2F_7) \\[2mm]
0 & 2(F_1+4F_2+16F_3+F_4) & -\dfrac{40}{9}(2F_6+2F_7) & 0 \\[2mm]
0 & -\dfrac{40}{9}(2F_6+2F_7) & 2(16F_1+4F_2+F_3+4F_4) & 0 \\[2mm]
\dfrac{40}{9}(2F_6+2F_7) & 0 & 0 & 8(4F_1+4F_2+4F_3+F_4)
\end{bmatrix}
\begin{Bmatrix} c_{11} \\ c_{12} \\ c_{21} \\ c_{22} \end{Bmatrix}
=
\begin{Bmatrix} 0 \\ 0 \\ 0 \\ 0 \end{Bmatrix}
$$

The unknown buckling load is contained in the term $F_4 = \pi^2 b N_{xx}/8a$. We can, therefore, rearrange the above expression by bringing all terms involving F_4 to the right side of the equality

$$
\begin{bmatrix}
2(F_1+F_2+F_3) & 0 & 0 & \dfrac{40}{9}(2F_6+2F_7) \\[2mm]
0 & 2(F_1+4F_2+16F_3) & -\dfrac{40}{9}(2F_6+2F_7) & 0 \\[2mm]
0 & -\dfrac{40}{9}(2F_6+2F_7) & 2(16F_1+4F_2+F_3) & 0 \\[2mm]
\dfrac{40}{9}(2F_6+2F_7) & 0 & 0 & 8(4F_1+4F_2+4F_3)
\end{bmatrix}
\begin{Bmatrix} c_{11} \\ c_{12} \\ c_{21} \\ c_{22} \end{Bmatrix}
$$

$$
= F_4
\begin{bmatrix}
-2 & 0 & 0 & 0 \\
0 & -2 & 0 & 0 \\
0 & 0 & -8 & 0 \\
0 & 0 & 0 & -8
\end{bmatrix}
\begin{Bmatrix} c_{11} \\ c_{12} \\ c_{21} \\ c_{22} \end{Bmatrix}
= N_{xx}
\begin{bmatrix}
\dfrac{-\pi^2 b}{4a} & 0 & 0 & 0 \\[2mm]
0 & \dfrac{-\pi^2 b}{4a} & 0 & 0 \\[2mm]
0 & 0 & \dfrac{-\pi^2 b}{a} & 0 \\[2mm]
0 & 0 & 0 & \dfrac{-\pi^2 b}{a}
\end{bmatrix}
\begin{Bmatrix} c_{11} \\ c_{12} \\ c_{21} \\ c_{22} \end{Bmatrix}
$$

$$\tag{12.32}$$

Equation 12.32 is in the form of a so-called generalized eigenvalue problem. In general, there are $(M \times N)$ eigenvalues that satisfy a generalized eigenvalue problem, and for each eigenvalue there exists a corresponding eigenvector. For this particular example the eigenvalues corresponds to N_{xx}, and the eigenvector corresponds to the column matrix containing coefficients c_{ij}. Since we have assumed $M = N = 2$, there are four values of N_{xx} that will satisfy Equation 12.32. The eigenvalue with lowest magnitude represents the critical buckling load, N_{xx}^c, and the coefficients c_{ij} that correspond to this eigenvalue represent the critical buckling mode, since $w(x,y) = \sum_{m=1}^{M} \sum_{n=1}^{N} c_{mn} \sin(m\pi x/a) \sin(n\pi y/b)$. As in all analyses based on the Ritz method, several predictions of buckling load should be obtained using increased values of M and N, to insure convergence of the predicted buckling load and mode.

Equation 12.32 represents the generalized eigenvalue problem for the case $N_{xx} \neq 0$, $N_{yy} = N_{xy} = 0$, and $M = N = 2$. As a second example, consider buckling caused by a pure shear load. That is, assume $N_{xy} \neq 0$, $N_{xx} = N_{yy} = 0$. We now have $F_4 = F_5 = F_9 = 0$. Following a process identical to that described above, we arrive at the following generalized eigenvalue problem:

$$\begin{bmatrix} 2(F_1 + F_2 + F_3) & 0 & 0 & \frac{40}{9}(2F_6 + 2F_7) \\ 0 & 2(F_1 + 4F_2 + 16F_3) & -\frac{40}{9}(2F_6 + 2F_7) & 0 \\ 0 & -\frac{40}{9}(2F_6 + 2F_7 + F_8) & 2(16F_1 + 4F_2 + F_3) & 0 \\ \frac{40}{9}(2F_6 + 2F_7) & 0 & 0 & 8(4F_1 + 4F_2 + 4F_3) \end{bmatrix} \begin{Bmatrix} c_{11} \\ c_{12} \\ c_{21} \\ c_{22} \end{Bmatrix}$$

$$= N_{xy} \begin{Bmatrix} \begin{bmatrix} 0 & 0 & 0 & -\frac{80}{9} \\ 0 & 0 & \frac{80}{9} & 0 \\ 0 & \frac{80}{9} & 0 & 0 \\ -\frac{80}{9} & 0 & 0 & 0 \end{bmatrix} \begin{Bmatrix} c_{11} \\ c_{12} \\ c_{21} \\ c_{22} \end{Bmatrix} \end{Bmatrix} \tag{12.33}$$

Once again, there will be four eigenvalue/eigenvector pairs that satisfy Equation 12.33, since we have assumed $M = N = 2$. The eigenvalue with lowest magnitude represents the critical buckling shear load, N_{xy}^c, and the corresponding eigenvector represents the critical buckling mode.

12.6 Computer Program *SYMM*

The solutions described in this chapter are implemented in a computer program called *SYMM*. This program can be downloaded at no cost from the following website:

http://depts.washington.edu/amtas/computer.html

Program *SYMM* is based on the Ritz method, and is applicable to any symmetric composite laminate. The program prompts the user to input all information necessary to perform these calculations. Properties of up to five different materials may be defined. The user must input various numerical values using a consistent set of units. For example, the user must input elastic moduli for the composite material system(s) of interest. Using the properties listed in Table 3.1 and based on the SI system of units, the following numerical values would be input for graphite–epoxy:

$$E_{11} = 170 \times 10^9 \, \text{Pa} \quad E_{22} = 10 \times 10^9 \, \text{Pa} \quad v_{12} = 0.30 \quad G_{12} = 13 \times 10^9 \, \text{Pa}$$

Since $1 \, \text{Pa} = 1 \, \text{N/m}^2$, all lengths must be input in meters. For example, ply thicknesses must be input in meters (not millimeters). A typical value would be $t_k = 0.000125 \, \text{m}$ (corresponding to a ply thickness of 0.125 mm). Similarly, if an analysis of a plate with a length and width of 500 cm × 300 cm were being performed, then the length and width of the plate must be input as 5.00 and 3.00 m, respectively.

If the English system of units were used, then the following numerical values would be input for the same graphite–epoxy material system:

$$E_{11} = 25.0 \times 10^6 \, \text{psi} \quad E_{22} = 1.5 \times 10^6 \, \text{psi} \quad v_{12} = 0.30 \quad G_{12} = 1.9 \times 10^6 \, \text{psi}$$

All lengths would be input in inches. A typical ply thickness might be $t_k = 0.005 \, \text{in.}$, and the length and width of a plate might be 36 in. and 20 in., for example.

References

1. Whitney, J.M., 1987. *Structural Analysis of Laminated Anisotropic Plates*, Technomic Pub. Co., Lancaster, PA, ISBN 87762-518-2.
2. Turvey, G.J. and Marshall, I.H., eds., 1995. *Buckling and Postbuckling of Composite Plates*, Chapman and Hall, New York.
3. Fung, Y.C., 1969. *A First Course in Continuum Mechanics*, Prentice-Hall, Englewood Cliffs, NJ.

4. Timoshenko, S. and Woinowsky-Krieger, 1987. *Theory of Plates and Shells*, McGraw-Hill Book Co., New York, ISBN 0-07-0647798.
5. Korn, G.A., and Korn, T.M, eds., 1968. *Mathematical Handbook for Scientists and Engineers*, 2nd edition, McGraw-Hill Book Co., New York, Table E-6.

Appendix A: Experimental Methods Used to Measure In-Plane Elastic Properties

A brief discussion of tensile tests used to measure in-plane properties E_{11}, E_{22}, v_{12}, and G_{12} is provided in this Appendix. The reader interested in a more detailed discussion of composite experimental methods should consult References 1–3, or review the many test standards published by ASTM International [4].

The most widely used tensile test specimen geometries used with composites are based on ASTM Standard D3039 "Test Method for Tensile Properties of Polymer Matrix Composite Materials" [5]. Typical specimens that conform to this standard are shown in Figures A.1 through A.4. As indicated, tensile specimens are flat and straight-sided with rectangular cross-section. Very often adhesively bonded end tabs are used. The $[0]_n$ specimen, the "off-axis" $[\theta]_n$ specimen, and the $[90]_n$ specimen (Figures A.1, A.2, and A.3, respectively) are often equipped with adhesively bonded end tabs. Since the use of bonded

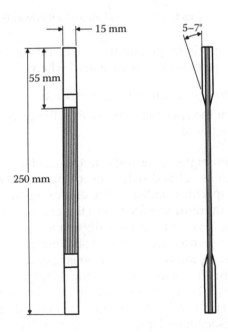

FIGURE A.1
A $[0]_n$ tensile specimen, showing typical dimensions and adhesively bonded end tabs.

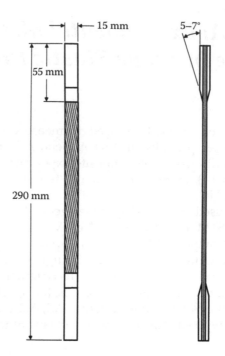

FIGURE A.2

A $[\theta]_n$ tensile specimen, showing typical dimensions and adhesively bonded end tabs.

end-tabs increases specimen preparation time and cost, they are only used when necessary. In general, end tabs are used when:

- The *fracture* stress or strain is to be measured
- It is found that the specimen fails within the grip region if bonded end tabs are not used

End tabs used with high-strength specimens (such as the $[0]_n$ or $[\theta]_n$ specimens shown in Figures A.1 and A.2, respectively) must be beveled at a shallow angle to avoid specimen failure at the end of the tab. The shallow bevel provides for a smooth transfer of load from the grip region to the gage region of the specimen. The bevel angle is typically about 5° to 7°, although angles as high as 30° are used on occasion. Use of a shallow bevel angle becomes less important when testing low-strength specimens, and in fact end tabs with a bevel angle of 90° are often used when testing $[90]_n$ specimens (Figure A.3).

As a general rule all tensile test specimens are long and narrow. That is, they have a high aspect ratio. Aspect ratio is defined as specimen length divided by width (specimen length is defined as the tab-to-tab distance). High-strength composite specimens must have an exceptionally high aspect ratio. Recommended aspect ratios of the $[0]_n$ and $[\theta]_n$ specimens shown in Figures A.1 and A.2 are 9.3 and 12.0, respectively. In contrast,

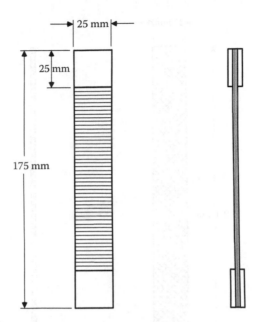

FIGURE A.3
A $[90]_n$ tensile specimen, showing typical dimensions and adhesively bonded end tabs.

the recommended aspect ratio of the relatively low-strength $[90]_n$ specimen (Figure A.3) is 5.0.

Bonded end tabs are often not necessary when testing symmetric and balanced laminates, such as the $[\pm45]_{ns}$ specimen shown in Figure A.4. In these cases so-called "friction" tabs may be used. A friction tab is held in place by the pressure of the grips. Often an abrasive paper (such as emery cloth) is placed between the tab and surface of the specimen to enhance friction.

Bonded end tabs are commonly made using a cross-ply $[0/90]_{ns}$ E-glass/polymer composite, although end tabs made using steel or aluminum have also been successfully used.

Strains induced during a tensile test are measured using either bonded resistance strain gages or extensometers. Properties E_{11} and v_{12} are measured using a $[0]_n$ specimen while E_{22} is measured using a $[90]_n$ specimen. It is possible to measure v_{21} as well, using a $[90]_n$ specimen. However, for most advanced unidirectional composites the numerical value of v_{21} (the so-called "minor" Poisson ratio) is at least an order of magnitude smaller than the "major" Poisson ratio v_{12}. For example, for graphite/epoxy the major Poisson ratio $v_{12} \approx 0.3$ whereas the minor Poisson ratio $v_{12} \approx 0.01$. Measurement of v_{21} requires measurement of a very small transverse strain, often leading to a high percent measurement error. Further, v_{21} can be calculated using the inverse relation:

$$v_{21} = v_{12} \frac{E_{22}}{E_{11}}$$

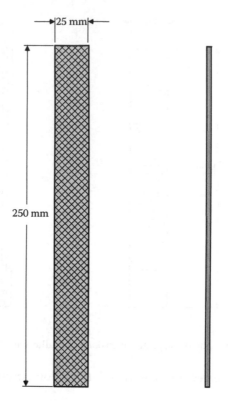

FIGURE A.4
A $[\pm 45]_{ns}$ tensile specimen, showing typical dimensions.

Owing to these factors, ν_{21} is rarely measured in practice.

Several techniques by which the shear modulus G_{12} can be measured using a tensile specimen are available. One approach is based on an off-axis $[\theta]_n$ specimen. Referring to Equation 5.38c, the axial modulus E_{xx} of such a specimen is given by

$$E_{xx} = \frac{1}{(\cos^4(\theta)/E_{11}) + ((1/G_{12}) - (2\nu_{12}/E_{11}))\cos^2(\theta)\sin^2(\theta) + (\sin^4(\theta)/E_{22})}$$

Solving this equation for G_{12}, we obtain:

$$G_{12} = \frac{E_{xx}E_{11}E_{22}\cos^2(\theta)\sin^2(\theta)}{E_{11}E_{22} - E_{xx}[E_{22}\cos^4(\theta) + 2\nu_{12}E_{22}\cos^2(\theta)\sin^2(\theta) - E_{11}\sin^4(\theta)]} \quad \text{(A.1)}$$

Thus, if E_{xx} is measured for a $[\theta]_n$ specimen (where θ is known), and assuming that E_{11}, v_{12}, and E_{22} have also been measured and are known, then the shear modulus G_{12} can be calculated using Equation A.1. If $\theta = 45°$ (the most common case) then Equation A.1 reduces to

$$G_{12} = \frac{E_{xx}E_{11}E_{22}}{4E_{11}E_{22} - E_{xx}\left[E_{22} + 2v_{12}E_{22} - E_{11}\right]} \tag{A.2}$$

An alternate method is based on the use of a $[\pm 45]_{ns}$ specimen. In this case, the axial and transverse strains (ε_{xx} and ε_{yy}) induced by a uniaxial tensile stress (σ_{xx}) are measured, usually using biaxial strains gages. For this stacking sequence, it can be shown [6] that the shear stress (τ_{12}) and shear strain (γ_{12}) induced in the $+45°$ plies are given by

$$\tau_{12}\big|_{+45°} = -\sigma_{xx}/2$$

$$\gamma_{12}\big|_{+45°} = -(\varepsilon_{xx} - \varepsilon_{yy})$$

In contrast, the shear stress and strain induced in the $-45°$ plies are given by

$$\tau_{12}\big|_{-45°} = \sigma_{xx}/2$$

$$\gamma_{12}\big|_{-45°} = (\varepsilon_{xx} - \varepsilon_{yy})$$

Hence, for the $[\pm 45]_{ns}$ stacking sequence, the shear modulus is given by

$$G_{12} = \frac{\tau_{12}}{\gamma_{12}} = \frac{\sigma_{xx}}{2(\varepsilon_{xx} - \varepsilon_{yy})}$$

References

1. Whitney, J.M., Daniel, I.M., and Pipes, R.B., 1984. *Experimental Mechanics of Fiber Reinforced Composite Materials*, 2nd Edition, SEM Monograph 4, Society for Experimental Mechanics, Bethel, CT, ISBN 0-912053-01-1.
2. Carlsson, L.A., and Pipes, R.B., 1997. *Experimental Characterization of Advanced Composite Materials*, 2nd Edition, Technomic Pub. Co., Lancaster, PA, ISBN 1-56676-433-5.
3. Jenkins, C.H., ed., 1998. *Manual on Experimental Methods of Mechanical Testing of Composites*, 2nd Edition, Society for Experimental Mechanics, Bethel, CT, ISBN 0-88173-284-2.

4. The main ASTM International website is www.astm.org/

5. ASTM Standard D3039/D3039M-08, *Standard Test Method for Tensile Properties of Polymer Matrix Composite Materials*, ASTM International, West Conshohocken, PA, 2006, DOI: 10.1520/D3039_D3039M-08, www.astm.org

6. Rosen, B.W., 1972. A simple procedure for experimental determination of the longitudinal shear modulus of unidirectional composites, *Journal of Composite Materials*, 6, 552.

Appendix B: Tables of Beam Deflections and Slopes

TABLE B.1

Deflection and Slopes of Cantilever Beams

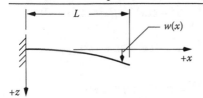

$w(x)$ = beam deflection (positive downwards)

$w'(x) = dw/dx$ = slope of the deflection curve

$\overline{IE}_{xx}$ = effective flexural rigidity

$$w(x) = \frac{Px^2}{6\overline{IE}_{xx}}(3L - x)$$

$$w'(x) = \frac{Px}{2\overline{IE}_{xx}}(2L - x)$$

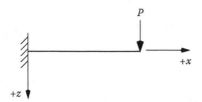

for $0 \leq x \leq a$:

$$w(x) = \frac{Px^2}{6\overline{IE}_{xx}}(3a - x) \qquad w'(x) = \frac{Px}{2\overline{IE}_{xx}}(2a - x)$$

for $a \leq x \leq L$:

$$w(x) = \frac{Pa^2}{6\overline{IE}_{xx}}(3x - a) \qquad w'(x) = \frac{Pa^2}{2\overline{IE}_{xx}}$$

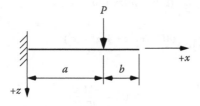

$$w(x) = \frac{qx^2}{24\overline{IE}_{xx}}(6L^2 - 4Lx + x^2)$$

$$w'(x) = \frac{qx}{6\overline{IE}_{xx}}(3L^2 - 3Lx + x^2)$$

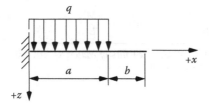

for $0 \leq x \leq a$:

$$w(x) = \frac{qx^2}{24\overline{IE}_{xx}}(6a^2 - 4ax + x^2)$$

$$w'(x) = \frac{qx}{6\overline{IE}_{xx}}(3a^2 - 3ax + x^2)$$

for $a \leq x \leq L$:

$$w(x) = \frac{qa^3}{24\overline{IE}_{xx}}(4x - a) \qquad w'(x) = \frac{qa^3}{6\overline{IE}_{xx}}$$

continued

TABLE B.1 (continued)

Deflection and Slopes of Cantilever Beams

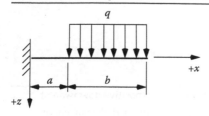

for $0 \leq x \leq a$:

$$w(x) = \frac{qbx^2}{12\overline{IE}_{xx}}(3L + 3a - 2x) \quad w'(x) = \frac{qbx}{2\overline{IE}_{xx}}(L + a - x)$$

for $a \leq x \leq L$:

$$w(x) = \frac{q}{24\overline{IE}_{xx}}(x^4 - 4Lx^3 + 6L^2x^2 - 4a^3x + a^4)$$

$$w'(x) = \frac{q}{6\overline{IE}_{xx}}(x^3 - 3Lx^2 + 3L^2x - a^3)$$

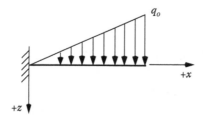

$$w(x) = \frac{q_o x^2}{120L\overline{IE}_{xx}}(20L^3 - 10L^2x + x^3)$$

$$w'(x) = \frac{q_o x}{24L\overline{IE}_{xx}}(8L^3 - 6L^2x + x^3)$$

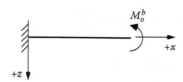

$$w(x) = \frac{q_o x^2}{120L\overline{IE}_{xx}}(10L^3 - 10L^2x + 5Lx^2 - x^3)$$

$$w'(x) = \frac{q_o x}{24L\overline{IE}_{xx}}(4L^3 - 6L^2x + 4Lx^2 - x^3)$$

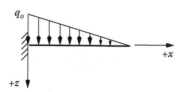

$$w(x) = \frac{-M_o^b x^2}{2\overline{IE}_{xx}} \qquad w'(x) = \frac{-M_o^b x}{\overline{IE}_{xx}}$$

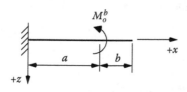

for $0 \leq x \leq a$:

$$w(x) = \frac{-M_o^b x^2}{2\overline{IE}_{xx}}$$

$$w'(x) = \frac{-M_o^b x}{\overline{IE}_{xx}}$$

for $a \leq x \leq L$:

$$w(x) = \frac{-M_o^b a}{2\overline{IE}_{xx}}(2x - a)$$

$$w'(x) = \frac{-M_o^b a}{\overline{IE}_{xx}}$$

Source: Adapted from Gere, J.M., and Timoshenko, S.P., 1997, *Mechanics of Materials*, 4th edition, PWS Publishing Co., Boston, MA, ISBN 0-534-93429-3.

TABLE B.2

Deflections and Slopes of Simply Supported Beams

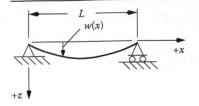

$w(x)$ = beam deflection (positive downwards)

$w'(x) = dw/dx$ = slope of the deflection curve

$\overline{IE}_{xx}$ = effective flexural rigidity

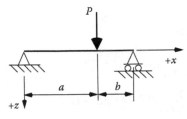

for $0 \leq x \leq a$:

$$w(x) = \frac{Pbx}{6L\overline{IE}_{xx}}(L^2 - b^2 - x^2)$$

$$w'(x) = \frac{Pb}{6L\overline{IE}_{xx}}(L^2 - b^2 - 3x^2)$$

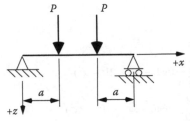

for $0 \leq x \leq a$:

$$w(x) = \frac{Px}{6\overline{IE}_{xx}}(3aL - 3a^2 - x^2)$$

$$w'(x) = \frac{P}{2\overline{IE}_{xx}}(aL - a^2 - x^2)$$

for $a \leq x \leq L - a$

$$w(x) = \frac{Pa}{6\overline{IE}_{xx}}(3Lx - 3x^2 - a^2) \quad w'(x) = \frac{Pa}{2\overline{IE}_{xx}}(L - 2x)$$

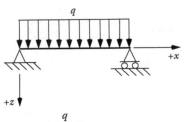

$$w(x) = \frac{qx}{24\overline{IE}_{xx}}(L^3 - 2Lx^2 + x^3)$$

$$w'(x) = \frac{q}{24\overline{IE}_{xx}}(L^3 - 6Lx^2 - 4x^3)$$

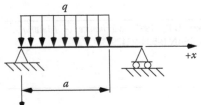

for $0 \leq x \leq a$:

$$w(x) = \frac{qx}{24L\overline{IE}_{xx}}(a^4 - 4a^3L + 4a^2L^2$$
$$+ 2a^2x^2 - 4aLx^2 + Lx^3)$$

$$w'(x) = \frac{q}{24L\overline{IE}_{xx}}(a^4 - 4a^3L + 4a^2L^2$$
$$+ 6a^2x^2 - 12aLx^2 + 4Lx^3)$$

for $a \leq x \leq L$:

$$w(x) = \frac{qa^2}{24L\overline{IE}_{xx}}(-a^2L + 4L^2x + a^2x - 6Lx^2 + 2x^3)$$

$$w'(x) = \frac{qa^2}{24L\overline{IE}_{xx}}(4L^2 + a^2 - 12Lx + 6x^2)$$

continued

TABLE B.2   (continued)

Deflections and Slopes of Simply Supported Beams

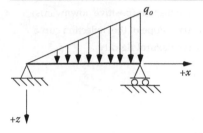

$$w(x) = \frac{q_o x}{360 L \overline{IE}_{xx}}(7L^4 - 10L^2 x^2 + 3x^4)$$

$$w'(x) = \frac{q_o}{360 L \overline{IE}_{xx}}(7L^4 - 30L^2 x^2 + 15x^4)$$

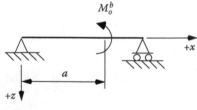

for $0 \le x \le L/2$:

$$w(x) = \frac{q_o x}{960 L \overline{IE}_{xx}}(5L^2 - 4x^2)^2$$

$$w'(x) = \frac{q_o}{192 L \overline{IE}_{xx}}(5L^2 - 4x^2)(L^2 - 4x^2)$$

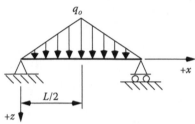

for $0 \le x \le a$:

$$w(x) = \frac{M_o^b x}{6 L \overline{IE}_{xx}}(6aL - 3a^2 - 2L^2 - x^2)$$

$$w'(x) = \frac{M_o^b}{6 L \overline{IE}_{xx}}(6aL - 3a^2 - 2L^2 - 3x^2)$$

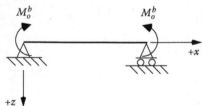

$$w(x) = \frac{M_o^b x}{2 \overline{IE}_{xx}}(L - x)$$

$$w'(x) = \frac{M_o^b}{2 \overline{IE}_{xx}}(L - 2x)$$

Source: Adapted from Gere, J.M., and Timoshenko, S.P., 1997, *Mechanics of Materials*, 4th edition, PWS Publishing Co., Boston, MA, ISBN 0-534-93429-3.

Reference

1. Gere, J.M., and Timoshenko, S.P., 1997, *Mechanics of Materials*, 4th edition, PWS Publishing Co., Boston, MA, ISBN 0-534-93429-3.

Index

9 780367 380588